DJAIRO G. de FIGUEIREDO –
Selected Papers

Djairo G. de Figueiredo

Photo by Christina Figueiredo Prudencio

DJAIRO G. de FIGUEIREDO – Selected Papers

Edited by

David G. Costa

impa
instituto nacional de
matemática
pura e aplicada

SBM

Springer

Djairo G. de Figueiredo
IMECC-UNICAMP
Cidade Universitaria Instituto de Matematica
Estatistica e Computação Cientifica
Campinas
São Paulo
Brazil

Editor
David G. Costa
Department of Mathematical Sciences
University of Nevada Las Vegas
Las Vegas, NV
USA

ISBN 978-3-319-02855-2 ISBN 978-3-319-02856-9 (eBook)
DOI 10.1007/978-3-319-02856-9
Springer Cham Heidelberg New York Dordrecht London

Library of Congress Control Number: 2012954381

Math. Subj. Classification (2010): 35J20 35J50, 35J60, 35J65, 45G99, 46E35, 58E99

Printed on acid-free paper

Springer is part of Springer Science+Business Media (www.springer.com)

Selected Works of Outstanding Brazilian Mathematicians

Mathematics in Brazil is a very young endeavor. Regular research activities and graduate studies go back to the 1950s, when the biannual Brazilian Mathematical Colloquium was launched and Brazil joined the International Mathematical Union (IMU). The Instituto de Matemática Pura e Aplicada (IMPA) was founded in 1952. Creation of the SBM, Sociedade Brasileira de Matemática, would follow, in 1969.

Half a century later, Brazil has reached Group IV in the IMU's ranking and is responsible for about 1.8 % of the world's mathematical production (according to MathSciNet). The Brazilian mathematical community is growing steadily, and has earned international reputation in several fields of research. While many great challenges remain ahead, one can certainly be proud of what has been achieved in such a short time.

This progress was largely due to a number of leading mathematicians who, in addition to their own fundamental contributions to Mathematics, created genuine "schools" around them that raised entire research fields in Brazil to the best international standards. Through this collection, the SBM and IMPA aim to highlight their scientific trajectories and to pay tribute to their pioneering efforts.

Editors
Hilário Alencar
César Camacho
André Nachbin
Marcelo Viana

Preface

Djairo Guedes de Figueiredo has had a major influence in the development of the area of Analysis, particularly in its applications to problems in Partial Differential Equations (PDEs) for equations and systems of elliptic type by means of variational and topological techniques. As a mathematician living in his native country of Brazil, his guidance of students, his publication of 29 books and monographs, and collaboration with many colleagues (myself included) has been instrumental in creating a whole generation of researchers and various seminal groups in PDEs throughout the country. In addition, his respected status in the international mathematical community has contributed to the establishment of several collaborative programs of Brazil with other countries, especially Belgium and Italy.

The present volume contains a selection of papers of Djairo (as he is commonly addressed) spanning the areas of geometry of Banach spaces, monotone operators, and boundary value problems for scalar equations and systems of elliptic PDEs. In addition, a complete list of his research output of over 100 papers, including 40 collaborators, and the names and professional affiliations of his 26 (and counting) Ph.D. students is provided. Finally, a biographical account of Djairo's professional life (by Djairo, himself) is included in this volume of his selected works.

More often than not, especially through meetings in honor of Djairo, his students and colleagues have expressed their admiration and thanks for his guidance, inspiration and friendship.

I feel honored and privileged to have known, collaborated with, and enjoyed Djairo's friendship for the past nearly four decades.

Las Vegas *David G. Costa*

Acknowledgments

The Brazilian Mathematical Society (SBM), the Institute for Pure and Applied Mathematics (IMPA) and Springer-Verlag express their gratitude to the following copyright holders for granting permisson to reprint the publications below. Also find below a list of publications from Springer-Verlag whose reprints will be included in this volume.

American Mathematical Society

De Figueiredo, D.G.; Karlovitz, L.A., On the Radial Projection in Normed Spaces, Bull. Amer. Math. Soc. **73** (1967), 364–368.

Browder, F.E.; De Figueiredo, D.G.; Gupta, C.P., Maximal Monotone Operators and Nonlinear Integral Equations of Hammerstein Type, Bull. Amer. Math. Soc. **76** (1970), 700–705.

De Figueiredo, D.G.; Gupta, C.P., On the Variational Method for the Existence of Solutions of Nonlinear Equations of Hammerstein Type, Proc. Amer. Math. Soc. **40** (1973), 470–476.

De Figueiredo, D.G., Positive Solutions for Some Classes of Semilinear Elliptic Problems, Nonlinear Functional Analysis and its Applications, Proc. Sympos. Pure Math. **45**, Part 1, Amer. Math. Soc. (1986), 371–379.

De Figueiredo, D.G.; Felmer, P., On Superquadratic Elliptic Systems, Trans. Amer. Math. Soc. **343** (1994), 99–116.

De Figueiredo, D.G.; Ding, Y.H, Strongly Indefinite Functionals and Multiple Solutions of Elliptic Systems, Trans. Amer. Math. Soc. **355** (2003), 2973–2989.

Annali della Scuola Normale Superiore

De Figueiredo, D.G.; Felmer, P., A Liouville-Type Theorem for Elliptic Systems, Ann. Scuola Norm. Sup. Pisa **XXI** (1994), 387–397.

American Institute of Mathematical Sciences

De Figueiredo, D.G.; Ding, Y., Solutions of a Nonlinear Schrodinger Equation, Discrete Cont. Dyn. Syst. **8** (2002), 563–584.

De Gruyter

De Figueiredo, D.G.; Ruf, B., On a Superlinear Sturm-Liouville Equation and a Related Bouncing Problem, J. Reine Angew. Math. **421** (1991), 1–22.

Elsevier

De Figueiredo, D.G.; Karlovitz, L.A., The Extension of Contractions and the Intersection of Balls in Banach Spaces, J. Funct. Anal. **11** (1972), 168–178.

De Figueiredo, D.G.; Gossez, J.-P., Nonlinear Perturbations of a Linear Elliptic Problem near its First Eigenvalue, J. Differential Equations **30** (1978), 1–19.

De Figueiredo, D.G.; Ni, W.-M.; Perturbations of Second Order Linear Elliptic Problems by Nolinearities without Landesman-Lazer Condition, Nonlinear Anal. **3** (1979), 629–634.

De Figueiredo, D,G,, Semilinear Equations at Resonance: Higher Eigenvalues and Unbounded Nonlinearities, Recent Advances in Differential Equations, Ed. Roberto Conti, Academic Press (1981), 89–99.

De Figueiredo, D.G.; Lions P.-L.; Nussbaum, R.D., A Priori Estimates and Existence of Positive Solutions of Semilinear Elliptic Equations, J. Math. Pures Appl. **61** (1982), 41–63.

De Figueiredo, D.G., On the Superlinear Ambrosetti-Prodi Problem, Nonlineal Anal. **8** (1984), 655–665.

Costa, D.G.; De Figueiredo, D.G., Radial Solutions of a Dirichlet Problem in a Ball, J. Differential Equations **60** (1985), 80–89.

Costa, D.G.; De Figueiredo, D.G.; Gonçalves, J.V.A., On the Uniqueness of Solution for a Class of Superlinear Elliptic Problems, J. Math. Anal. Appl. **123** (1987), 170–180.
Costa, D.G.; De Figueiredo, D.G.; Srikanth, P.N., The Exact Number of Solutions for a Class of Ordinary Differential Equations through Morse Index Computation, J. Differential Equations **96** (1992), 185–199.
De Figueiredo, D.G.;Yang, J., Decay, Symmetry and Existence of Solutions of Semilinear Elliptic Systems, Nonlinear Anal. **33** (1998), 211–234.
De Figueiredo, D.G.; Gossez, J.-P.; Ubilla, P., Local Superlinearity and Sublinearity for Indefinite Semilinear Elliptic Problems, J. Funct. Anal. **199** (2003), 452–467.
De Figueiredo, D.G.; Do Ó, J.M.; Ruf, B., An Orlicz-Space Approach to Superlinear Elliptic Systems, J. Funct. Anal. **224** (2005), 471–496.
De Figueiredo, D.G.; Sirakov, B., On the Ambrosetti-Prodi Problem for Non-Variational Elliptic Systems, J. Differential Equations **240** (2007), 357–374.
De Figueiredo, D.G.; Lopes, O., Solitary Waves for some Nonlinear Schrodinger Systems, Ann. Inst. H. Poincare, Anal. Non Lineaire **25** (2008), 149–161.
De Figueiredo, D.G.; Gossez, J.-P.; Ubilla, P., Local 'Superlinearity' and 'Sublinearity' for the p-Laplacian, J. Funct. Anal. **257** (2009), 721–752.

Indiana University

De Figueiredo, D.G.; Lions, P.-L., On Pairs of Positive Solutions for a Class of Semilinear Elliptic Problems, Indiana Univ. Math. J. **34** (1985), 591–606.

Rendiconti dell'Istituto di Matematica dell'Università di Trieste

De Figueiredo, D.G.; Mitidieri, E., Maximum Principles for Linear Elliptic Systems, Univ. Trieste, Vol. **XXII** (1990), 36–66.

Rocky Mountain Mathematics Consortium

De Figueiredo, D.G., On Superlinear Elliptic Problems with Nonlinearities Interacting only with Higher Eigenvalues, Rocky Mountain J. Math. **18** (1988), 287–303.

Royal Netherlands Academy of Arts and Sciences

Browder, F.E.; De Figueiredo, D.G., J-Monotone Nonlinear Operators in Banach Spaces, Koninkl. Nederl. Akad. Wet. **28** (1966), 412–420.

Society for Industrial and Applied Mathematics

De Figueiredo, D.G.; Mitidieri, E., A Maximum Principle for an Elliptic System and Applications to Semilinear Problems, SIAM J. Math. Anal. **17** (1986), 836–849.

Springer-Verlag

De Figueiredo, D.G., The Dirichlet Problem for Nonlinear Elliptic Equations: A Hilbert Space Approach, Lecture Notes in Mathematics, Vol. **446** (1975), 144–165.

De Figueiredo, D.G.; Gossez, J.-P., Nonresonance Below the First Eigenvalue for a Semilinear Elliptic Problem, Math. Ann. **281** (1988), 589–610.

De Figueiredo, D.G., Monotonicity and Symmetry of Solutions of Elliptic Systems in General Domains, NoDEA Nonlinear Differential Equations and Appl. **1** (1994), 119–123.

De Figueiredo, D.G.; Miyagaki, O.H.; Ruf, B., Elliptic Equations in R2 with Nonlinearities in the Critical Growth Range, Calc. Var. and Partial Differential Equations **3** (1995), 139–153.

Bartsch, T.; De Figueiredo, D.G., Infinitely Many Solutiond of Nonlinear Elliptic Systems, Prog. Nonlinear Differential Equations Appl. **35** (The Herbert Amann Anniversary Volume) (1999), 51–68.

Boccardo, L.; De Figueiredo, D.G., Some Remarks on a System of Quasilinear Elliptic Equations, NoDEA Nonlinear Differential Equations and Appl. **9** (2002), 309–323.

De Figueiredo, D.G.; Sirakov, B., Liouville Type Theorems, Monotonicity Results and A Priori Bounds for Positive Solutions of Elliptic Systems, Math. Ann. **133** (2005), 231–260.

De Figueiredo, D.G.; Do Ó, J.M.; Ruf, B., Non-variational Elliptic Systems in Dimension Two: A Priori Bounds and Existence of Positive Solutions, J. Fixed Point Theory Appl. **4** (2008), 77–96.

De Figueiredo, D.G.; Peral, I; Rossi, J.D., The Critical Hyperbola for a Hamiltonian Elliptic System with Weights, Ann. Mat. Pura Appl **187** (2008), 531–545.

Taylor & Francis

De Figueiredo, D.G.; Berestycki, H., Double Resonance in Semilinear Elliptic Problems, Comm. Partial Differential Equations **6** (1981), 91–120.

De Figueiredo, D.G.; Solimini, S., A Variational Approach to Superlinear Elliptic Problems, Comm. Partial Differential Equations **9** (1984), 699–717.

De Figueiredo, D.G.; Gossez, J.-P., Strict Monotonicity of Eigenvalues and Unique Continuation, Comm. Partial Differential Equations **17** (1992), no. 1–2, 339–346.

Clément, Ph.; De Figueiredo, D.G.; Mitidieri, E., Positive Solutions of Semilinear Elliptic Systems, Partial Differential Equations **17** (1992), 923–940.

De Figueiredo, D.G., Yang, J., A Priori Bounds for Positive Solutions of a Non-Variational Elliptic System, Comm. Partial Differential Equations **26** (2001), no. 11–12, 2305–2321.

Contents

A Short Summary of my Scientific Life xix
D. G. de Figueiredo

J-Monotone Nonlinear Operators in Banach Spaces 1
F. E. Browder and D. G. De Figueiredo

On the Radial Projection in Normed Spaces 11
D. G. De Figueiredo and L. A. Karlovitz

Maximal Monotone Operators and Nonlinear Integral Equations of Hammerstein Type 17
Felix E. Browder, Djairo G. de Figueiredo and Chaitan P. Gupta

The Extension of Contractions and the Intersection of Balls in Banach Spaces 23
D. G. de Figueiredo and L. A. Karlovitz

On the Variational Method for the Existence of Solutions of Nonlinear Equations of Hammerstein Type 35
Djairo G. de Figueiredo and Chaitan P. Gupta

The Dirichlet Problem for Nonlinear Elliptic Equations: A Hilbert Space Approach 43
Djairo Guedes de Figueiredo

Nonlinear Perturbations of a Linear Elliptic Problem Near Its First Eigenvalue 65
Djairo G. de Figueiredo and Jean-Pierre Gossez

Perturbations of Second Order Linear Elliptic Problems by Nonlinearities Without Landesman–Lazer Condition 85
Djairo Guedes de Figueiredo and Wei-Ming Ni

Double Resonance in Semilinear Elliptic Problems 91
Henri Berestycki and Djairo Guedes de Figueiredo

Semilinear Elliptic Equations at Resonance: Higher Eigenvalues and Unbounded Nonlinearities 121
Djairo G. de Figueiredo

A Priori Estimates and Existence of Positive Solutions of Semilinear Elliptic Equations 133
D. G. De Figueiredo, P. -L. Lions and R. D. Nussbaum

On The Superlinear Ambrosetti–Prodi Problem 157
Djairo G. de Figueiredo

A Variational Approach to Superlinear Elliptic Problems 169
Djairo G. de Figueiredo and Sergio Solimini

Radial Solutions for a Dirichlet Problem in a Ball 189
D. G. Costa and D. G. De Figueiredo

On Pairs of Positive Solutions for a Class of Semilinear Elliptic Problems 199
Djairo G. De Figueiredo and Pierre-Louis Lions

Positive Solutions for Some Classes of Semilinear Elliptic Problems 215
Djairo G. de Figueiredo

A Maximum Principle for an Elliptic System and Applications to Semilinear Problems 225
Djairo G. de Figueiredo and Enzo Mitidieri

On the Uniqueness of Solution for a Class of Semilinear Elliptic Problems 239
D. G. Costa, D. G. De Figueiredo and J. V. A. Goncalves

On Superlinear Elliptic Problems with Nonlinearities Interacting Only with Higher Eigenvalues 251
Djairo G. De Figueiredo

Nonresonance Below the First Eigenvalue for a Semilinear Elliptic Problem 269
Djairo G. De Figueiredo and Jean-Pierre Gossez

Maximum Principles for Linear Elliptic Systems 291
Djairo G. De Figueiredo and Enzo Mitidieri

On a Superlinear Sturm-Liouville Equation and a Related Bouncing Problem 323
D. G. Figueiredo and B. Ruf

The Exact Number of Solutions for a Class of Ordinary Differential Equations Through Morse Index Computation 345
D. G. Costa, D. G. De Figueiredo and P. N. Srikanth

Strict Monotonicity of Eigenvalues and Unique Continuation 361
Djairo G. de Figueiredo and Jean-Pierre Gossez

Positive Solutions of Semilinear Elliptic Systems 369
Ph. Clément, D. G. de Figueiredo and E. Mitidieri

On Superquadratic Elliptic Systems 387
Djairo G. de Figueiredo and Patricio L. Felmer

A Liouville-Type Theorem for Elliptic Systems 405
D. G. de Figueiredo and P. L. Felmer

Monotonicity and Symmetry of Solutions of Elliptic Systems in General Domains 417
Djairo G. De Figueiredo

Elliptic Equations in $\mathbf{R}^2$ with Nonlinearities in the Critical Growth Range 423
D. G. de Figueiredo, O. H. Miyagaki and B. Ruf

Decay, Symmetry and Existence of Solutions of Semilinear Elliptic Systems 439
Djairo G. de Figueiredo and Yang Jianfu

Infinitely Many Solutions of Nonlinear Elliptic Systems 463
Thomas Bartsch and Djairo G. de Figueiredo

A Priori Bounds for Positive Solutions of a Non-variational Elliptic System 483
Djairo Guedes de Figueiredo and Jianfu Yang

Some Remarks on a System of Quasilinear Elliptic Equations 501
Lucio Boccardo and Djairo Guedes De Figueiredo

Solutions of a Nonlinear Schrödinger Equation 517
D. G. de Figueiredo and Y. H. Ding

Strongly Indefinite Functionals and Multiple Solutions of Elliptic Systems 539
D. G. De Figueiredo and Y. H. Ding

Local Superlinearity and Sublinearity for Indefinite Semilinear Elliptic Problems 557
Djairo G. De Figueiredo, Jean-Pierre Gossez and Pedro Ubilla

An Orlicz-Space Approach to Superlinear Elliptic Systems 573
Djairo G. de Figueiredo, João Marcos do 'O and Bernhard Ruf

Liouville Type Theorems, Monotonicity Results and a Priori Bounds for Positive Solutions of Elliptic Systems 599
Djairo G. de Figueiredo and Boyan Sirakov

On the Ambrosetti–Prodi Problem for Non-variational Elliptic Systems 629
Djairo G. de Figueiredo and Boyan Sirakov

Solitary Waves for Some Nonlinear Schrödinger Systems 647
Djairo G. de Figueiredo and Orlando Lopes

Non-variational Elliptic Systems in Dimension Two: A Priori Bounds and Existence of Positive Solutions 661
Djairo G. de Figueiredo, João Marcos do Ó and Bernhard Ruf

The Critical Hyperbola for a Hamiltonian Elliptic System with Weights 681
Djairo G. de Figueiredo, Ireneo Peral and Julio D. Rossi

Local 'Superlinearity' and 'Sublinearity' for the*p*-Laplacian 697
Djairo G. de Figueiredo, Jean-Pierre Gossez and Pedro Ubilla

List of Publications of Djairo G. de Figueiredo 729

List of Ph.D. Students of Djairo G. de Figueiredo 737

A Short Summary of my Scientific Life

Djairo Guedes de Figueiredo

Born in 1934 in the small town of Limoeiro do Norte in the interior of the state of Ceará, Brazil, I completed all my primary schooling and middle school studies in the Ginásio Diocesano Padre Anchieta. In those days, a small town did not have schools offering higher education. So, in 1949, I went to Fortaleza, the state capital of Ceará, to conclude the last 3 years of High School at the Colégio Castelo Branco. There I had the fortune of having not only an excellent Professor of Mathematics, Francisco Mena Barreto, but also a mentor whose motivation and ingenuity in teaching ignited my initial delight in Mathematics. It was natural to consider, after graduating from High School, a career in Engineering. However, the difficulties were insurmountable if I remained in Ceará, so my father, Miguel Angelo Figueiredo, under a considerable financial sacrifice made it possible that I move to Rio de Janeiro, the capital of Brazil at the time, to work towards fulfilling my dream. Upon my arrival, in February 1952, I enrolled in a competitive examination in order to gain entrance to the National School of Engineering at the University of Brazil. Owing to the solid education I received in Ceará, I was classified in seventh place in a group of over 1,000 candidates.

At this school, once again, I had the fortuitous opportunity to have as a Professor of Rational Mechanics, the famous Brazilian mathematician Maurício Peixoto who wholeheartedly was devoted to Mathematics. As an extra curricular activity, he organized a Saturday Seminar where students were stimulated to further their mathematical studies in addition to the regular program of study of the School of Engineering and of course there could be no better suited person for this endeavor. I think it is fair to say that these seminars turned out to be pivotal in my decision to embrace a mathematical career. The thought-provoking environment in these seminars led me to consider the possibility of graduate studies abroad. Once again under Peixoto's guidance, I was able to choose a distinguished university in the United States and received a full scholarship from the recently created program of the Brazilian National Research Council (CNPq).

D. G. Costa (ed.), *Djairo G. de Figueiredo - Selected Papers*,
DOI: 10.1007/978-3-319-02856-9,

Djairo Guedes de Figueiredo

I arrived in September 1957 to the Institute of Mathematical Sciences of New York University, (today known as the Courant Institute of Mathematical Sciences, CIMS) where I was soon immersed in an invigorating mathematical atmosphere. Although my initial plans were to study Probability and Statistics, I did not resist the charm of the "Wizards of Waverly Place" (the CIMS functioned at a building on a street with that name). Of course, the keenness and fervor of mathematicians such as K.O. Friedrichs, Fritz John, P.D. Lax, Lipman Bers, J. Keller, and Louis Nirenberg had a major impact on my decision to work in the field of Partial Differential Equations. I was extremely fortunate to have my doctoral work supervised by Louis Nirenberg and whose influence marked my research in the forthcoming years.

During those years in New York I discovered many things, but by far the most important was that my happiness would not be complete without Maruja, my beloved wife, who has been my lifelong companion and whose enthusiastic support has been a source of inspiration.

Upon the completion of my doctoral work, I returned to Brazil in 1961 at the invitation of the enthusiastic and idealist Darcy Ribeiro, President of the University of Brasília, to establish together with other mathematicians a Department of Mathematics in this nascent university located at the newly founded capital of Brazil. In the midst of the expansion of the university, erupted the 1964 Revolution, which brought about an exodus of intellectuals from Brazil.

In 1965, I was awarded a Guggenheim Fellowship and I returned to the United States to spend a year at the Mathematics Research Center of the University of Wisconsin in Madison. Subsequently, I spent a semester at the University of Chicago and the following year at the University of Maryland at College Park. In the ensuing year, I obtained a position at the University of Illinois in Chicago. I remained there until 1971, when once again I was invited to return to the University of Brasília to help, together with other colleagues, revitalize the Mathematics Department.

Calling to mind those years from 1965 to 1973, it is clear that the close research collaboration with Felix E. Browder and C.P. Gupta on the newly developed theory of Monotone Operators was very advantageous for my career. Also, worthy of note was my collaboration during this period with Les Karlowitz on the Geometry of Banach Spaces. Once relocated to Brazil, my research shifted toward the theory of Nonlinear Elliptic Equations and Systems focusing on studying existence, uniqueness, multiplicity and symmetric properties of solutions, and also their behavior at the boundary and infinity. In this branch of study, I had over 30 collaborators from Brazil, Argentina, Chile, the United States, China, India, Belgium, France, Germany, Holland, Italy, Portugal, and Spain.

I must end by paying tribute to all my collaborators and friends with whom throughout the years I have had the privilege to work and to share the excitement of pushing the frontiers of knowledge a little further.

KONINKL. NEDERL. AKADEMIE VAN WETENSCHAPPEN – AMSTERDAM
Reprinted from Proceedings, Series A, 69, No. 4 and Indag. Math., 28, No. 4, 1966

MATHEMATICS

J-MONOTONE NONLINEAR OPERATORS IN BANACH SPACES

BY

F. E. BROWDER AND D. G. DE FIGUEIREDO [1])

(Communicated by Prof. H. FREUDENTHAL at the meeting of January 29, 1966)

Introduction

Let X be a real Banach space, X^* its conjugate space, and let (w, u) denote the pairing between w in X^* and u in X.

The recent literature has seen a great development of the theory of monotone nonlinear operators T from X to X^*, where T is said to be monotone if for every pair u and v of elements of X:

$$(Tu-Tv, u-v) \geqslant 0.$$

(Cf. especially BROWDER [2] and MINTY [13]. An extensive survey of the literature and of the applications of the theory of monotone operators to boundary value problems for nonlinear partial differential equations is given in BROWDER [7]. Further references to related Russian papers are given in KACHUROVSKI [11].)

It is our object in the present paper to carry over some of the basic results of the theory of continuous (or demicontinuous) monotone operators from X to X^* to a somewhat different but related class of nonlinear operators T from X to X, the J-monotone operators defined in BROWDER [8].

Definition 1: *Let X be a Banach space, X^* its adjoint space, $\mu(r)$ a continuous strictly increasing real-valued function on R^1 with $\mu(0)=0$. If J is a mapping of X into X^*, then J is said to be a duality mapping of X into X^* with gauge function μ if for each u in X, the following two conditions hold:*

$$(Ju, u) = \|u\| \cdot \mu(\|u\|),$$

and

$$\|Ju\| = \mu(\|u\|).$$

When $X=H$, a Hilbert space, then X^* can be identified with H by the inner product and the simplest duality mapping is the identity mapping I. If X is the Banach space l^p, $1<p<+\infty$, then $X^*=l^{p/(p-1)}$ and the simplest duality mapping is the *canonical* mapping

$$J_0(\{x_k\}) = \{|x_k|^{p-2}x_k\}.$$

[1]) The work of the first-named author was partially supported by N.S.F. Grant GP-3552, that of the second-named author by AFOSR Grant, AF-AFOSR-736-65.

Originally published in: Koninkl. Nederl. Akad. Wet. 28 (1966), 4 12-420,

DOI: 10.1007/978-3-319-02856-9_1

In both these examples (as noted in BROWDER [8]), the duality mappings we have singled out have the property of being continuous both in the pair of weak topologies of X and X^* and the pair of strong topologies of X and X^*.

If X^* is strictly convex, the duality mapping J is uniquely determined by its gauge function μ and exactly one duality mapping J exists for each gauge function. Moreover, J is continuous from the strong topology of X to the weak* topology of X^*. If in addition X^* is uniformly convex, J is strongly continuous, i.e. continuous from the strong topology of X to the strong topology of X^*. However, the uniform convexity of X and X^* does not imply the *weak continuity* of J, (i.e. its continuity from the weak topology of X to the weak* topology of X^*) for any duality mapping J of X into X^*. As we show in Section 2 below, for any L^p space on the interval [0,1] with $1<p<+\infty$ and $p\neq 2$, there are no weakly continuous duality mappings. (This conclusion was already established by an example for $p=4$ in BROWDER [8]. Let us also note here the analytical results concerning duality mappings involved in the Beurling-Livingston theorem (BEURLING-LIVINGSTON [1], BROWDER [5], [6]. In particular, every duality mapping is a monotone map from X to X^*.)

Definition 2: *Let X be a Banach space, X^* its conjugate space, J a mapping of X into X^* (and in particular, a duality mapping). Then if T is a mapping of X into X, T is said to be J-monotone if for all u and v of X:*

$$(J(u-v), Tu-Tv)\geqslant 0.$$

J-monotone mappings were introduced in connection with the study of nonexpansive mappings in Banach spaces in BROWDER [8]. If S is a mapping of X into X, S is said to be nonexpansive if for all u and v of X,

$$\|Su-Sv\|\leqslant\|u-v\|.$$

If S is nonexpansive and J is any duality mapping of X into X^*, it was observed in [8] that the mapping $T=I-S$ of X into X is J-monotone.

If H is a Hilbert space, the class of I-monotone maps of H into H coincides with the class of monotone mappings as previously defined. Results for J-monotone operators with J linear were given in BROWDER [3], [4] and KACHUROVSKI [11]. If on the other hand, T is linear, the concept of J-monotone mappings T with J a duality map corresponds to the definition that $(-T)$ be dissipative in the generalized sense of LUMER and PHILLIPS [12]. [2])

[2]) *Added in proof:* A somewhat different application of a related generalization of monotonicity to nonlinear equations of evolution in Hilbert space has been given by PHILIP HARTMAN in a paper which came to our attention after the manuscript of the present note had been transmitted (P. HARTMAN, Liapounoff functions and nonlinear functional equations, Annali Di. Roma, 69–70, (1965)).

Definition 3: *The Banach space X is said to have property $(\pi)_1$ if there exists a directed set $\{F_\alpha\}$ of finite dimensional subspaces of X ordered by inclusion and a corresponding set of projection mappings P_α of X with range F_α such that the union of the F_α is dense in X, while for each α, $\|P_\alpha\|=1$.*

Hilbert spaces obviously have property $(\pi)_1$, and more generally so do the l^p spaces for $1<p<+\infty$. In the latter case, the subspaces F_α are taken as the subspaces G_k of all the sequences whose components vanish after the *k-th*. The corresponding projection mappings P_α are the mappings which wipe out all components after the *k-th*. More generally still, we show in Section 2 by a construction which we owe to Richard Beals that every σ-finite measure space, $L^p(m)$ has the property $(\pi)_1$ for $1\leqslant p<\infty$. We remark finally, that in separable spaces X, property $(\pi)_1$ is a consequence of the existence of a monotone Schauder basis in the sense of DAY [9] (p. 67). Property $(\pi)_1$ was applied to the study of weakly continuous mappings from X to X in DE FIGUEIREDO [10].

Our principal results are contained in the following three theorems:

Theorem 1: *Let X be a reflexive Banach space with a strictly convex conjugate space X^* with X satisfying property $(\pi)_1$. Let J be a duality mapping of X into X^* which is both weakly and strongly continuous. Let T be a mapping of X into X such that T is continuous from the strong topology of X to the weak topology of X and satisfies both of the following conditions:*

(*a*) *T is J-monotone, i.e. for all u and v of X,*

$$(J(u-v), Tu-Tv)\geqslant 0.$$

(*b*) *T is J-coercive, i.e. there exists a real-valued function $c(r)$ on R^1 with $c(r)\to+\infty$ as $r\to+\infty$, such that for all u in X*

$$(Ju, Tu)\geqslant c(\|u\|)\|Ju\|.$$

Then T maps X onto X, i.e. the equation $Tu=f$ has a solution u in X for every f in X.

As a specialization of Theorem 1, we have the following:

Theorem 2: *Let T be a mapping of the space l^p into itself $(1<p<+\infty)$, and let J_0 be the canonical duality map of l^p into $l^{p/(p-1)}$. Suppose that T is continuous from the strong to the weak topology of l^p, and that T_0 is both J_0-monotone and J_0-coercive.*

Then T maps l^p onto l^p.

Theorem 2 obviously follows from Theorem 1 together with the facts already noted that l^p has property $(\pi)_1$ and that the duality map J_0 of l^p into $l^{p/(p-1)}$ is both weakly and strongly continuous.

We shall obtain Theorem 1 as a consequence of the following more precise result:

Theorem 3: *Let X be a reflexive Banach space with a strictly convex conjugate space X^* such that X has property $(\pi)_1$ and such that there exists a duality map J of X into X^* which is both strongly and weakly continuous. Let T be a J-monotone mapping of X, into X, with T continuous from the strong topology of X to the weak topology of X. For a given $r>0$, let B_r be the closed ball of radius r about the origin in X, and suppose that for $\|u\|=r$,*

$$(Ju, Tu) \geq 0.$$

Then there exists a point u_0 in B_r such that $Tu_0=0$.

Proof of Theorem 1 from Theorem 3: Let w be a fixed element of X, and consider the mapping T_w of X into X given by $T_w u = Tu - w$. Since T is J-monotone, so is T_w. For $\|u\|=r$, for any $r>0$, we have

$$(Ju, T_w u) = (Ju, Tu) - (Ju, w) \geq c(r)\|Ju\| - \|w\|\,\|Ju\| \geq 0$$

if $c(r) \geq \|w\|$. Since $c(r) \to +\infty$ as $r \to +\infty$, there exists $R(\|w\|)$ such that $c(r) \geq \|w\|$ for $r \geq R(\|w\|)$. For such a choice of r, it follows from Theorem 3 that T_w has a zero u_0 in B_r, i.e. $Tu_0 = w$. q.e.d.

We carry through the proof of Theorem 3 based upon four Lemmas in Section 1. In Section 2, we verify the remarks already made about the L^p spaces.

Section 1:

Lemma 1: *Let X be a Banach space having property $(\pi)_1$ with X^* strictly convex. Let $\{F_\alpha\}$ and $\{P_\alpha\}$ be families of spaces and projections for which the conditions of property $(\pi)_1$ hold, and let $P_\alpha{}^*$ be the adjoint map of P_α, where $P_\alpha{}^*$ is an idempotent selfmapping of X^*. Let J be any duality mapping of X into X^*.*

Then for any u in F_α,

$$P_\alpha{}^* Ju = Ju.$$

Proof of Lemma 1: We remark first that if X^* is strictly convex and J is a duality mapping of X into X^* with gauge function μ, then for any element v of X^* and an element u of X such that

$$(v, u) = \|u\| \cdot \mu(\|u\|), \text{ and } \|v\| \leq \mu(\|u\|),$$

we must have $v = Ju$. Indeed, it follows from the first equality that $\|u\|\mu(\|u\|) \leq \|v\| \cdot \|u\|$. If $u=0$, the second inequality implies that $v=0=Ju$. If $u \neq 0$, we may divide the derived inequality by $\|u\|$, and obtain

$$\|v\| \geq \mu(\|u\|),$$

and hence $\|v\| = \mu(\|u\|)$. Finally, the set

$$\{v \mid v \in X^*, (v, u) = \|u\|\mu(\|u\|), \|v\| \leq \mu(\|u\|)\}$$

is a convex subset of a sphere in X^* for each fixed u in X. It contains the point Ju and by the strict convexity of X^* it must consist of a single point. Hence $v=Ju$.

To prove Lemma 1, therefore, it suffices to show for each u in F_α that P_α^*Ju satisfies the two conditions characterizing Ju. However,

$$(P_\alpha^*Ju, u)=(Ju, P_\alpha u)=(Ju, u)=\|u\|\mu(\|u\|),$$

while

$$\|P_\alpha^*Ju\|\leqslant\|P_\alpha^*\|\cdot\|Ju\|=\|Ju\|=\mu(\|u\|)$$

since $\|P_\alpha^*\|=\|P_\alpha\|=1$. Hence $P_\alpha^*Ju=Ju$ for $u\in F_\alpha$. q.e.d.

Lemma 2: *Let F be a finite dimensional Banach space, J a duality mapping of F into F^*. Let T be a continuous mapping of F into F, and suppose that for a given $r>0$, with B_r the closed ball of radius r about the origin in F, for all u in F with $\|u\|=r$,*

$$(Ju, Tu)\geqslant 0.$$

Then there exists a point u_0 in B_r such that $Tu_0=0$.

Proof of Lemma 2: Let $U=I-T$, with I the identity map of F. It suffices to show that U has a fixed point in B_r. Let V be the map of B_r into B_r given by

$$Vu=\begin{cases} Uu, \text{ if } Uu \text{ lies in } B_r \\ r\dfrac{Uu}{\|Uu\|}, \text{ if } Uu \text{ lies outside } B_r. \end{cases}$$

V is a continuous map of B_r into B_r and has a fixed point u_0 in B_r by the Brouwer fixed point theorem. If $\|u_0\|<r$, the $Vu_0=u_0$ lies inside the interior of B_r and $Uu_0=Vu_0=u_0$. If $\|u_0\|=r$, then $\|Uu_0\|\geqslant r$, and we have

$$Uu_0=\lambda u_0,\ \lambda=(\|Uu_0\|/r)\geqslant 1.$$

If $\lambda>1$, we have a contradiction since for $\|u_0\|=r$

$$(Ju_0, Uu_0)=(Ju_0, u_0)-(Ju_0, Tu_0)\leqslant(Ju_0, u_0)$$

while

$$(Ju_0, Uu_0)=\lambda(Ju_0, u_0)>(Ju_0, u_0).$$

Hence $\lambda=1$, and $Uu_0=u_0$. q.e.d.

Lemma 3: *Under the hypotheses of Theorem 3, let $\{F_\alpha\}$ and $\{P_\alpha\}$ be families of finite-dimensional subspaces of X and the corresponding projection mappings satisfying the conditions for property $(\pi)_1$ for X. For a given α, let T_α be the mapping of F_α into F_α given by $T_\alpha u=P_\alpha Tu$.*

Then T_α has a zero in $B_r\cap F_\alpha$.

Proof of Lemma 3: Since P_α maps X onto F_α, P_α^* maps X^* onto F_α^*, i.e. the restriction of elements of X^* to their action on F_α maps $P_\alpha^*(X^*)$ isometrically onto F_α^*. If we identify these two spaces, we remark that the mapping J_α of F_α into F_α^* given by $J_\alpha u = P_\alpha^* Ju$ is a duality mapping of F_α into F_α^*. Indeed by Lemma 1, $J_\alpha u = Ju$ for each u in F_α, so that if $\mu(r)$ is the gauge function of J, $\|J_\alpha u\| = \|Ju\| = \mu(\|u\|)$ and $(J_\alpha u, u) = (Ju, u)$ for each u in F_α. On the other hand,

$$(J_\alpha u, T_\alpha u) = (Ju, P_\alpha Tu) = (P_\alpha^* Ju, Tu) = (Ju, Tu) \geq 0$$

for each u in F_α with $\|u\| = r$. Applying Lemma 2, we see that T_α has a zero in $B_r \cap F_\alpha$. q.e.d.

Lemma 4: *Let J be a duality map of X into X^*, T a mapping of X into X which is continuous from line segments in X to the weak topology of X. Suppose that X is reflexive and X^* strictly convex. Let u_0 and w_0 be elements of X such that for all u in X,*

$$(J(u-u_0), Tu-w_0) \geq 0.$$

Then $Tu_0 = w_0$.

Proof of Lemma 4: For an arbitrary v in X and any $t > 0$, let $u_t = u_0 + tv$. We have

$$0 \leq (J(u_t - u_0), Tu_t - w_0) = (J(tv), Tu_t - w_0).$$

By the definition of duality mapping and the strict convexity of X^*, for each $t > 0$, there exists a real number $\zeta_v(t) > 0$ such that $J(tv) = \zeta_v(t) Jv$. Hence

$$\zeta_v(t)(Jv, Tu_t - w_0) \geq 0.$$

Cancelling the positive multiplier $\zeta_v(t)$, we obtain

$$(Jv, Tu_t - w_0) \geq 0.$$

Letting $t \to 0+$, it follows from the weak continuity of T on segments in X that

$$(Jv, Tu_0 - w_0) \geq 0.$$

Replacing v by $(-v)$ and using the fact that $J(-v) = -Jv$, we see that

$$(Jv, Tu_0 - w_0) \leq 0,$$

and hence that

$$(Jv, Tu_0 - w_0) = 0.$$

Finally by the generalized Beurling-Livingston theorem (Browder [5]), J maps X onto X^* since X is reflexive and X^* is strictly convex. Hence $Tu_0 = w_0$. q.e.d.

Proof of Theorem 3 completed: Let $\{F_\alpha\}$ and $\{P_\alpha\}$ be the spaces and projections corresponding to property $(\pi)_1$ for $\cdot X$. Let T_α, for each α, be the mapping of F_α into F_α given by $T_\alpha u = P_\alpha T u$. By Lemma 3, the equation $T_\alpha u_\alpha = 0$ has a solution u_α in $B_r \cap F_\alpha$.

For each γ in the directed set of indices, let

$$V_\gamma = \bigcup_{F_\gamma \subset F_\alpha} \{u_\alpha\}.$$

By the weak compactness of the closed ball B_r in the reflexive Banach space X, there exists a point u_0 of B_r which belongs to the intersection of the weak closures of the sets V_γ. We shall show that $Tu_0 = 0$.

Let v be any element of $\bigcup_\gamma F_\gamma$, and choose γ such that v lies in F_γ. For any u_α in the corresponding V_γ, we have

$$(J(v-u_\alpha), Tv - Tu_\alpha) \geqslant 0$$

by the J-monotonicity of T. Since both v and u_α lie in F_α, however, we have

$$(J(v-u_\alpha), Tu_\alpha) = (P_\alpha{}^* J(v \quad u_\alpha), Tu_\alpha) = (J(v-u_\alpha), P_\alpha T u_\alpha) - 0$$

since $P_\alpha T u_\alpha = 0$, (while by Lemma 1, $P_\alpha{}^* J(v-u_\alpha) = J(v-u_\alpha)$). Thus

$$(J(v-u_\alpha), Tv) \geqslant 0.$$

The function h from B_r to the reals given by $h(w) = (J(v-w), Tv)$ is continuous in the weak topology on B_r since J is weakly continuous. Since h is non-negative on V_γ, it is non-negative on the weak closure of V_γ and in particular $h(u_0) \geqslant 0$, i.e.

$$(J(v-u_0), Tv) \geqslant 0$$

for all v in $\bigcup_\gamma F_\gamma$.

Let v be any element of X. Since $\bigcup_\gamma F_\gamma$ is dense in X, we may find a sequence $\{v_k\}$ in this dense subset converging strongly to v. Since T is continuous from the strong topology to the weak topology of X, Tv_k converges weakly to Tv. Since J is strongly continuous, $J(v_k - u_0)$ converges strongly to $J(v-u_0)$. Hence $(J(v_k - u_0), Tv_k) \to (J(v-u_0), Tv)$. Since for each k,

$$(J(v_k - u_0), Tv_k) \geqslant 0,$$

it follows that

$$(J(v-u_0), Tv) \geqslant 0$$

for all v in X. Finally, applying Lemma 4, we see that $Tu_0 = 0$. q.e.d.

Section 2:

Proposition 2.1: *Let X be the Banach space $L^p([0,1])$ for $1 < p < +\infty$ with $p \neq 2$. Then there exists no weakly continuous duality mapping J of X into X^*.*

Proof of Proposition 2.1: Since X^* is strictly convex, if J is any duality mapping of X into X^* for $X = L^p([0,1])$, there exists a continuous function ζ on $(0, \infty)$ such that for $u \neq 0$,

$$Ju = \zeta(\|u\|)J_0 u$$

where J_0 is the canonical duality mapping

$$(J_0 u)(x) = |u(x)|^{p-2}u(x).$$

We shall construct a sequence $\{u_k\}$ with $\|u_k\|_{L^p} = 1$ such that $u_k \to 0$ weakly in L^p while $J_0 u_k$ converges weakly to a non-zero limit. It follows from the form of J that Ju_k also converges weakly to a non-zero limit and hence that J is not weakly continuous.

To construct the sequence $\{u_k\}$, we choose a function u_0 in $L^p([0,1])$ such that

$$\int_0^1 u_0(x)\, dx = 0$$

while

$$\int_0^1 |u_0(x)|^{p-2}u_0(x)\, dx = c_0 \neq 0.$$

We normalize u_0 by the condition that

$$\int_0^1 |u_0(x)|^p\, dx = 1,$$

and extend its definition to $[0, \infty)$ by making it periodic of period 1.

For $k \geqslant 1$, let

$$u_k(x) = u_0(kx),\ x \in [0,1].$$

Then

$$\|u_k\|^p_{L^p([0,1])} = \int_0^1 |u_0(kx)|^p\, dx = (1/k) \int_0^k |u_0(y)|^p dy = k/k = 1.$$

To show that u_k converges weakly to 0 in $L^p([0,1])$, it suffices for each c in $[0,1]$ to show that $\int_0^c u_k(x)\, dx \to 0$. However,

$$\int_0^c u_k(x)\, dx = \int_0^c u_0(kx)\, dx = 1/k \int_0^{kc} u_0(y)\, dy = 1/k \int_0^{kc-[kc]} u_0(y)\, dy = 0(1/k) \to 0.$$

On the other hand

$$\int_0^1 (J_0 u_k)(x)\, dx = \int_0^1 |u_k(x)|^{p-2}u_k(x)\, dx = 1/k\int_0^k |u_0(y)|^{p-2}u_0(y)dy = c_0 \neq 0.$$

Hence $J_0 u_k$ does not converge weakly to zero, and, as follows from a similar computation as done for u_k, converges weakly to $c_0 \neq 0$. q.e.d.

Proposition 2.2: *For any p with $1 < p < \infty$, $L^p(m)$ satisfies property $(\pi)_1$ for a σ-finite measure m.*

Proof: Let $\{S_1, \ldots, S_r\}$ be any finite family of disjoint sets of finite measure, $\{f_1, \ldots, f_r\}$ their characteristic functions. We define a projection P on the space spanned by $\{f_1, \ldots, f_r\}$ by setting

$$Pf = \sum_{j=1}^{r} (1/m(S_j)) \int_{S_j} f(x)\, dx.$$

It follows from the Holder inequality that $\|P\|_{L^p(m)} = 1$. q.e.d.

Department of Mathematics
University of Chicago

REFERENCES

1. Beurling, A. and A. E. Livingston, A theorem on duality mappings in Banach spaces, Ark. Mat., **4**, 405–411 (1961).
2. Browder, F. E., Nonlinear elliptic boundary value problems, Bull. Amer. Math. Soc., **69**, 862–874 (1963).
3. ———, Nonlinear elliptic problems, II, Bull. Amer. Math. Soc., **70**, 299–302 (1964).
4. ———, Nonlinear elliptic boundary value problems, II, Trans. Amer. Math. Soc., **117**, 530–550 (1965).
5. ———, On a theorem of Beurling and Livingston, Canad. Math. Jour. **17**, 367–372 (1965).
6. ———, Multivalued monotone nonlinear mappings and duality mappings in Banach spaces, Trans. Amer. Math. Soc., **118**, 338–351 (1965).
7. ———, Existence and uniqueness theorems for solutions of nonlinear boundary value problems, Proc. Symposia on Applied Mathematics, Vol. 17, American Math. Soc., 24–49 (1965).
8. ———, Fixed point theorems for nonlinear semicontractive mappings in Banach spaces, Archives Rat. Mech. and Anal. **21**, 259–270 (1966).
9. Day, M., Normed Linear spaces, Ergebnisse der Math., Heft 21, 1962.
10. Figueiredo, D. G. de, A fixed point theorem for weakly continuous mappings of Banach spaces, Technical Report, U.S. Army Research Center, Madison, Wisc. (1966).
11. Kachurovski, R. I., Monotone nonlinear operators in Banach spaces, Dokladi Akad. Nauk SSSR, **163**, 559–562 (1965).
12. Lumer, G. and R. S. Phillips, Dissipative operators in Banach spaces, Pacific Jour. Math., **11** 679–698 (1961).
13. Minty, G. J., On a "monotonicity" method for the solution of nonlinear equations in Banach spaces, Proc. Nat. Acad. Sci. U.S.A., **50**, 1038–1041 (1963).

ON THE RADIAL PROJECTION IN NORMED SPACES

BY D. G. DeFIGUEIREDO AND L. A. KARLOVITZ[1]

Communicated by F. E. Browder, December 8, 1966

1. Let X be a real normed space with norm $\| \ \|$, T the *radial projection mapping* defined by

$$Tx = x, \quad \text{if } \|x\| \leqq 1, \quad \text{and} \quad Tx = x/\|x\|, \quad \text{if } \|x\| \geqq 1.$$

Our concern is with the Lipschitz constant of T; i.e. with the constant K such that $\|Tx-Ty\| \leqq K\|x-y\|$ for all x, y in X. In particular, we wish to determine under what conditions on the space X the mapping T will be *nonexpansive*, i.e. $K=1$.

T is a special case of a proximity mapping defined by a convex set in a normed vector space, i.e. a mapping which assigns to each point of X, the nearest point of the convex set C. There has been a good deal of interest in recent years in proximity maps, nonexpansive mappings, and their interrelations (Moreau, Browder, Petryshyn, Kirk, De Prima, Lions and Stampacchia, and others). It is easy to see that if X is an inner product space (and in particular, a Hilbert space), then T and every proximity map is nonexpansive. More precisely, Kirk and Smiley [1] proved that X is an inner product space if and only if for all nonzero x and y in X

$$\|x/\|x\| - y/\|y\|\,\| \leqq 2/(\|x\| + \|y\|)\|x - y\|.$$

For an arbitrary normed space X, Dunkl and Williams [2] have proved that for all nonzero x, y in X

$$\|x/\|x\| - y/\|y\|\,\| \leqq 4/(\|x\| + \|y\|)\|x - y\|.$$

From this it can be seen that $K \leqq 2$. This bound is the best possible because it is easily seen that for l_1, $K=2$.

It is of great interest in nonlinear functional analysis to know if there is a normed space which is not an inner product space and for which $K=1$. We prove the following theorem which shows that such spaces exist only for the trivial case of dimension two.

THEOREM. *If X has dimension not less than three, then X is an inner product space if and only if T is nonexpansive.*

If X has dimension two then the nonexpansiveness of T does not imply that X is an inner product space.

---

[1] The first author was supported by N.S.F. Grant GP-4921, and the second by N.S.F. Grant GP-3666.

Originally published in: Bull. Amer. Math. Soc. 73 (1967), 364-368,

DOI: 10.1007/978-3-319-02856-9_2

It is also reasonable to ask about the relation of K to other geometric entities of the space. A discussion of this question will appear in another paper.

The theorem is split into the following three propositions which are proved in the sequel.

PROPOSITION 1. *T is nonexpansive if and only if X has the following property*

(P) *For every ordered pair of nonzero vectors (x, y) such that $\|x\| = \|y\|$ and $\|x-\lambda_0 y\| \leqq \|x-\lambda y\|$ for all λ, it follows that $|\lambda_0| \leqq 1$.*

REMARK. If X has a strictly convex norm and if it has property (P), then $|\lambda_0| < 1$ if x and y are linearly independent. The proof of this fact is similar to the second part of the proof of Proposition 1.

PROPOSITION 2. *If X has dimension not less than three, then X has property* (P) *if and only if X is an inner product space.*

PROPOSITION 3. *X has property* (P) *if and only if X has the following property*

(J) *For every ordered pair of nonzero vectors (x, y) such that $\|x\| \leqq \|x-\lambda y\|$ for all λ, it follows that $\|y\| \leqq \|y-\lambda x\|$ for all λ.*

REMARK. Property (J) expresses that orthogonality in the sense of James [3] is symmetric. Birkhoff [4] discusses property (J) and gives an example of a 2-dimensional space which has this property but which is not an inner product space.

2. Corresponding to each ordered pair of nonzero vectors (x, y) we define the real-valued function

$$\phi(\lambda) = \phi(\lambda; x, y) = \|x - \lambda y\|, \qquad -\infty < \lambda < \infty.$$

It follows that ϕ is a convex function, it assumes its minimum value for λ in a finite interval $[\lambda_1, \lambda_2]$, and it is strictly decreasing (increasing) for $\lambda < \lambda_1$ $(\lambda > \lambda_2)$. If X has a strictly convex norm, then ϕ is a strictly convex function and $\lambda_1 = \lambda_2$. In terms of the function ϕ, we may restate property (P) as follows.

(P) For every ordered pair of nonzero vectors (x, y), such that $\|x\| = \|y\|$, the function ϕ is strictly decreasing (increasing) for $\lambda \leqq -1$ $(\lambda \geqq 1)$.

PROOF OF PROPOSITION 1. We first assume that X has property (P) and prove that T is nonexpansive. It suffices to prove the following inequalities

$$\|x - y/\|y\|\,\| \leqq \|x - y\|, \qquad \text{for } \|x\| \leqq 1 < \|y\|. \tag{1}$$

$$\|x/\|x\| - y/\|y\|\,\| \leqq \|x - y\|, \qquad \text{for } 1 \leqq \|x\| \leqq \|y\|. \tag{2}$$

The inequality (1) is a direct consequence of property (P) applied to the pair $(x, \|x\|/\|y\|y)$. If $1 \leqq \|x\| \leqq \|y\|$, an application of property (P) to the pair $(x/\|x\|, y/\|y\|)$ yields

$$\|x/\|x\| - y/\|y\|\,\| \leqq \|x/\|x\| - (\|y\|/\|x\|)y/\|y\|\,\| = (1/\|x\|)\|x - y\|,$$

which implies the inequality (2).

Next we assume that T is nonexpansive and prove that X has property (P). It is sufficient to consider an ordered pair of vectors (x, y) such that $\|x\| = \|y\| = 1$ because $|a|\phi(\lambda; x, y) = \phi(\lambda; ax, ay)$. For such a pair, the nonexpansiveness of T implies

$$\|x - y\| \leqq \|x - \lambda y\| \quad \text{and} \quad \|x + y\| \leqq \|x + \lambda y\| \qquad \text{for all } \lambda \geqq 1.$$

Since $\phi(\lambda; x, y)$ is convex, this shows that ϕ is increasing (decreasing) for $\lambda \geqq 1$ $(\lambda \leqq -1)$. To finish the proof we need to show that ϕ is *strictly* increasing (*strictly* decreasing) for $\lambda \geqq 1$ $(\lambda \leqq -1)$. Suppose, on the contrary, that ϕ is not strictly increasing for $\lambda \geqq 1$. Then there exists a constant $d > 1$ so that $\phi(\lambda; x, y) = \phi(1, x, y)$ for $1 \leqq \lambda \leqq d$. Let $z = \alpha x + (1-\alpha)y$, $0 < \alpha < 1$. Then

$$\begin{aligned}\phi(1; x, z/\|z\|) &= \|x - z/\|z\|\,\| \\ &= (1 - \alpha/\|z\|)\|x - (1 - \alpha)/(\|z\| - \alpha)y\| \\ &= (1 - \alpha/\|z\|)\phi((1 - \alpha)/(\|z\| - \alpha); x, y),\end{aligned} \tag{3}$$

and

$$\begin{aligned}\phi(1 + \delta; x, z/\|z\|) &= \|x - (1 + \delta)z/\|z\|\,\| \\ &= (1 - \alpha(1 + \delta)/\|z\|) \\ &\quad \cdot\|x - (1 + \delta)(1 - \alpha)/(\|z\| - (1 + \delta)\alpha)y\| \\ &= (1 - \alpha(1 + \delta)/\|z\|) \\ &\quad \cdot\phi((1 + \delta)(1 - \alpha)/(\|z\| - (1 + \delta)\alpha); x, y).\end{aligned}$$

The limit, as $\alpha \to 0$, of $(1-\alpha)/(\|z\| - \alpha) = 1$. Hence we may choose $\alpha > 0$ and $\delta > 0$ so that

$$1 \leqq (1 - \alpha)/(\|z\| - \alpha) \leqq d, \tag{4}$$

and

$$1 \leqq (1 + \delta)(1 - \alpha)/(\|z\| - (1 + \delta)\alpha) \leqq d.$$

Then it follows, from (3), (4) and the assumption that $\phi(\lambda; x, y)$ is constant on the interval $[1, d]$, that

$$\phi(1; x, z/\|z\|) = (1 - \alpha/\|z\|)\phi(1; x, y), \tag{5}$$

and

$$\phi(1 + \delta; x, z/\|z\|) = (1 - \alpha(1 + \delta)/\|z\|)\phi(1; x, y).$$

Since $\alpha, \delta > 0$, (5) implies that $\phi(1; x, z/\|z\|) > \phi(1+\delta; x, z/\|z\|)$. This contradicts property (P) for the pair $(x, z/\|z\|)$, and hence $\phi(\lambda; x, y)$ is strictly increasing for $\lambda \geqq 1$. Similarly we can show that $\phi(\lambda; x, y)$ is strictly decreasing for $\lambda \leqq -1$, and the proof is finished.

3. **Proof of Proposition 2.** We may assume that X has dimension three. If X is an inner product space, then it is clear that X has property (P). The proof of the converse is based on a characterization of inner product spaces given by Kakutani [5]. Accordingly it is sufficient to show that for every two dimensional linear subspace Y of X there exists a linear projection P of X onto Y which has norm one. To define such a projection, we choose $x_0 \in X$, $x_0 \notin Y$. There exists $y_0 \in Y$ so that $\|x_0 - y_0\| \leqq \|x_0 - y\|$ for all $y \in Y$. Let $x_1 = x_0 - y_0$. Every $x \in X$ may be uniquely decomposed in the form $x = \alpha x_1 + y$, $y \in Y$. A linear projection P, of X onto Y, is then defined by $Px = P(\alpha x_1 + y) = y$. It remains to show that P has norm one. We consider an arbitrary $x \in X$ with decomposition $x = \alpha x_1 + y$. If $y = 0$, $\|Px\| = 0 \leqq \|x\|$, hence we assume $y \neq 0$. From the construction of x_1 and the decomposition of x, there follows: $\|x - y\| = \|\alpha x_1\| \leqq \|\alpha x_1 - \lambda y\|$ for all λ. Letting $z = (\|x\|/\|y\|)y$, it follows from this inequality that: $\|x - (\|y\|/\|x\|)z\| \leqq \|x - \lambda z\|$ for all λ. Since $\|x\| = \|z\|$ and X has property (P), this implies $\|y\|/\|x\| \leqq 1$. Hence $\|Px\| \leqq \|x\|$, and the proof is finished.

4. **Proof of Proposition 3.** We first assume that X has property (P) and prove that X has property (J). Suppose, on the contrary, that there exist nonzero vectors x and y such that $\|x\| \leqq \|x - \lambda y\|$ for all λ and such that $\|y + \lambda_1 x\| < \|y\|$ for some $\lambda_1 \neq 0$. Let $m = \|y + \lambda_1 x\|/\|y\| < 1$. The vectors $y + \lambda_1 x$ and my have equal norms. Hence property (P) implies that $\phi(1; y + \lambda_1 x, my) < \phi(1/m; y + \lambda_1 x, my)$. This inequality implies that $\|x - ((m-1)/\lambda_1)y\| < \|x\|$, which is a contradiction.

To show that if X has property (J) then X has property (P), we consider an arbitrary pair of nonzero vectors (x, y) such that $\|x\| = \|y\|$. Let $[\lambda_1, \lambda_2]$ be the finite interval on which $\phi(\lambda; x, y)$ achieves its minimum. We wish to show that $-1 \leqq \lambda_1 \leqq \lambda_2 \leqq 1$. If $\lambda_1 = \lambda_2 = 0$, we are finished. If not, let $\lambda_0 \neq 0$ so that $\lambda_1 \leqq \lambda_0 \leqq \lambda_2$. Let $x_1 = x - \lambda_0 y$. Then $\|x_1\| \leqq \|x_1 - \lambda y\|$ for all λ; and it follows from property (J) that

$\|y\| \leq \|y-\lambda x_1\|$ for all λ. If in particular we let $\lambda = -1/\lambda_0$, we may conclude that:

$$\|y\| \leq \|y - \lambda x_1\| = \|y + (1/\lambda_0)(x - \lambda_0 y)\| = |1/\lambda_0| \|x\|.$$

Hence $|\lambda_0| \leq 1$, and the proof is finished.

We wish to thank Professor W. V. Petryshyn for calling our attention to this problem.

References

1. W. A. Kirk and M. F. Smiley, *Another characterization of inner product*, Amer. Math. Monthly **71** (1964), 890–891.

2. C. F. Dunkl and K. S. Williams, *A simple norm inequality*, Amer. Math. Monthly **71** (1964), 53–54.

3. R. C. James, *Orthogonality and linear functionals in normed linear spaces*, Trans. Amer. Math. Soc. **61** (1947), 265–292.

4. G. Birkhoff, *Orthogonality in linear metric spaces*, Duke J. Math. **1** (1935), 169–172.

5. S. Kakutani, *Some characterizations of Euclidean space*, Japan. J. Math. **16** (1939), 93–97.

University of Maryland

MAXIMAL MONOTONE OPERATORS AND NONLINEAR INTEGRAL EQUATIONS OF HAMMERSTEIN TYPE

BY FELIX E. BROWDER, DJAIRO G. DE FIGUEIREDO AND CHAITAN P. GUPTA

Communicated by December 29, 1969

A nonlinear integral equation of Hammerstein type is one of the form

$$u(x) + \int_G K(x, y)f(y, u(y))dy = w(x) \tag{1}$$

where G is a measure space with a σ-finite measure dy, the given function $w(x)$ and the unknown function $u(x)$ are defined on G. In operator-theoretic terms, the problem of determining the solutions of equation (1), with u, w lying in a given Banach space of functions on G, can be put in the form of a nonlinear functional equation

$$u + AN(u) = w \tag{2}$$

with the linear and nonlinear mappings A and N respectively given by

$$Av(x) = \int_G K(x, y)v(y)dy, \quad Nu(x) = f(x, u(x)). \tag{3}$$

In the present note, we apply the theory of maximal monotone operators in Banach spaces to establish general results on the existence of solutions of equation (2) for the reflexive Banach space X. Our results generalize the results of Browder-Gupta [8], Amann [1],

---

AMS Subject Classifications. Primary 4530, 4770, 4780.

Key Words and Phrases. Monotone operators, maximal monotone operators, pseudomonotone operators, hemicontinuous mappings, coercive mappings, integral equations of Hammerstein type, completely continuous mappings.

Originally published in: Bull. Amer. Math. Soc. 76 (1970), 700-705,

DOI: 10.1007/978-3-319-02856-9_3

Kolomý [11], Brezis [2], Kolodner [10], Dolph-Minty [9] and Vaĭnberg [14], [15].

We employ the following definitions: If X is a real Banach space, X^* its conjugate space, we let (w, u) denote the duality pairing between the element w in X^* and the element u in X. A subset G of the Cartesian product $X \times X^*$ is said to be *monotone* if $(v_1 - v_2, u_1 - u_2) \geqq 0$ for $[u_1, v_1] \in G$ and $[u_2, v_2] \in G$. A monotone set G is said to be *maximal monotone* if it is not properly contained in any other monotone set. Let T be a mapping of X into the set 2^{X^*} of all subsets of X^*.

The effective domain $D(T)$ of T is the subset of X given by $D(T) = \{u \mid u \in X,\ T(u) \neq \emptyset\}$. The range $R(T)$ of T is the subset of X^* given by $R(T) = \cup\{T(u) \mid u \in X\}$. The graph $G(T)$ of T is the subset of the Cartesian product $X \times X^*$ given by $G(T) = \{[u, v] \mid u \in X, v \in T(u)\}$.

T is said to be *monotone* if the graph $G(T)$ of T is a monotone set. T is said to be *maximal monotone* if the graph $G(T)$ of T is a maximal monotone set. The graph of the inverse mapping T^{-1} of X^* into 2^X is the subset of the Cartesian product $X^* \times X$ given by $G(T^{-1}) = \{[v, u] \mid [u, v] \in G(T)\}$.

It is immediate that a mapping T of a reflexive Banach space X into 2^{X^*} is maximal monotone if and only if the inverse mapping T^{-1} of X^* into 2^X is maximal monotone. T is said to be *pseudomonotone* if for any sequence $\{u_j\}$ in X which converges weakly to an element u_0 in X with

$$\limsup (T(u_j), u_j - u_0) \leqq 0$$

we have

$$\liminf (T(u_j), u_j - v) \geqq (Tu_0, u_0 - v) \quad \text{for all } v \in X.$$

Let N be a mapping of X into X^*. N is said to be *hemicontinuous* if it is continuous from each line segment in X to the weak topology in X^*. N is said to be *coercive* if $(N(v), v)/\|v\| \to \infty$ as $\|v\| \to \infty$.

THEOREM 1. *Let X be a reflexive Banach space, X^* its conjugate space. Let A be a maximal monotone mapping of X^* into 2^X and N a coercive hemicontinuous mapping of X into X^* which is monotone or bounded pseudomonotone.*

Then the range $R(I+AN)$ of the mapping $I+AN$ is all of X.

PROOF. For given w in X, we seek v in X with $N(v)$ in the domain $D(A)$ of A and

$$w \in v + AN(v). \tag{4}$$

The relation (4) is equivalent to

$$0 \in N(v) - A^{-1}(w - v). \tag{5}$$

Let T_w be the mapping of X into 2^{X^*} given by $T_w(v) = -A^{-1}(w-v)$ for v in X. Since $G(T_w) = -G(A^{-1}) + [w, 0]$, it follows that T_w is a maximal monotone mapping of X into 2^{X^*}. By hypothesis, we know in addition that N is a coercive hemicontinuous mapping of X into X^* with N monotone or bounded pseudomonotone. Hence it follows from results of Browder (Theorem 1, [5]) and Rockafellar [12], (see also Brezis-Crandall-Pazy [4]) that the range $R(T_w+N)$ is the whole of X^*.

Thus there exists a v_0 in X such that $0 \in T_w(v_0) + N(v_0) = N(v_0) - A^{-1}(w - v_0)$. Such an element v_0 is a solution of the relation (5) and hence of the relation (4). q.e.d.

COROLLARY 1. *Let X be a reflexive Banach space, A a hemicontinuous monotone mapping of X^* into X, N a coercive hemicontinuous monotone mapping of X into X^*.*

Then for every w in X, the equation $v + AN(v) = w$ has a solution v in X.

It suffices to note that the mapping A described in Corollary 1 is a maximal monotone mapping from X^* to X.

COROLLARY 2. *Let X be a reflexive Banach space and A a densely defined closed linear monotone mapping of X^* into X. Suppose that A^* is also a monotone mapping from X^* to X (this will be the case if in particular A is bounded).*

Let N be a coercive hemicontinuous monotone mapping of X into X^*.

Then for each w in X, there exists a v in X with $N(v)$ lying in $D(A)$ such that $v + AN(v) = w$.

By a theorem of Brezis [3], each closed densely defined monotone linear mapping A of X^* into X with A^* monotone is maximal monotone.

THEOREM 2. *Let X be a reflexive Banach space. Let A be a bounded linear monotone mapping of X^* into X and N be a hemicontinuous monotone mapping of X into X^*. Suppose that there is a nondecreasing nonnegative function ϕ defined on the set R^+ of nonnegative real numbers such that if for any u in X there is a v in X satisfying*

$$v + AN(v) = u$$

then $\|v\| \leq \phi(\|u\|)$.

Then the range $R(I+AN)$ of the mapping $I+AN$ is the whole of X.

PROOF. Let w be an arbitrary element of X. As in the proof of Theorem 1, it suffices to show that $0\in R(T_w+N)$ where T_w is the mapping of X into 2^{X^*} defined by

$$T_w(v) = -A^{-1}(w-v)$$

for v in X. Since A is a bounded linear monotone mapping of X^* into X, A is a maximal monotone mapping. Hence, as in the proof of Theorem 1, the mapping T_w of X into 2^{X^*} is maximal monotone. We first observe that the mapping T_w+N of X into 2^{X^*} is maximal monotone by the results of Browder [5] and Rockafellar [13].

Now for v in X and u in $(T_w+N)(v)$ we have that $u\in -A^{-1}(w-v)+N(v)$ i.e. $N(v)-u\in A^{-1}(w-v)$. This implies that $AN(v)-A(u)=w-v$ i.e. $v+AN(v)=w+A(u)$.

It now follows from our hypothesis that

$$\|v\| \leqq \phi(\|w+A(u)\|) \leqq \phi(\|w\|+\|A\|\,\|u\|).$$

Since ϕ is a nonnegative, nondecreasing function, it follows from the above inequality that the mapping $(T_w+N)^{-1}$ of X^* into 2^X is bounded. It then follows from the results of Browder [5] and Rockafellar [12] that the range $R(T_w+N)$ is whole of X^*. In particular, $0\in R(T_w+N)$. So there is v in X such that $0\in -A^{-1}(w-v)+N(v)$. This gives that $v+ANv=w$ and hence the theorem. q.e.d.

THEOREM 3. *Let X be a reflexive Banach space, A a bounded pseudo-monotone mapping of X^* into X, N a monotone hemicontinuous mapping of X into X^*. Suppose $k(r)$ and $c(r)$ are real valued functions of r in R^+ such that $k(r)+c(r)\to\infty$ as $r\to\infty$, that the following two inequalities are satisfied:*

(1) $(w, Aw) \geqq k(\|w\|)\|w\|, \qquad (w\in X^*)$

(2) $(w, u) \geqq c(\|w\|)\|w\|, \qquad (w\in X^*)$ *for all* $u\in N^{-1}(w)$.

Then $R(I+AN)=X$.

PROOF. $(I+AN)N^{-1}=N^{-1}+A$. Hence it suffices to prove that $R(N^{-1}+A)=X$. Since N is monotone and hemicontinuous, N^{-1} is maximal monotone from X^* to 2^X. Since A is bounded and pseudo-monotone, $A+N^{-1}$ will be onto X if $A+N^{-1}$ is coercive (Browder [7]). For w in X^* and u in $N^{-1}(w)$ we have

(*) $$(w, u+A(w)) \geqq [c(\|w\|)+k(\|w\|)]\cdot\|w\|.$$

This shows that $A+N^{-1}$ is coercive. Hence the theorem. q.e.d.

REMARK. We note that the inequality (*) will hold if condition (1) of Theorem 3 is replaced by the following:

$$(1') \qquad \|A(w)\| \leqq k_1(\|w\|), \qquad (w \in X^*)$$

where $k_1(r)$ is a real valued function of r in R^+ and $c(r)-k_1(r)\to\infty$ as $r\to\infty$.

COROLLARY 3. *Let X be a reflexive Banach space. A a bounded hemicontinuous monotone mapping of X^* into X, C a completely continuous mapping of X^* into X and N a hemicontinuous monotone mapping of X onto X^*. Suppose* (1) *and* (2) *hold with A replaced by $A+C$.*
Then $R(I+(A+C)N)=X$.

Corollary 3 follows from Theorem 3 immediately since $A+C$ is a bounded pseudomonotone mapping of X^* into X (Browder [**7**]).

We remark that the result of Corollary 3 is analogous to the results of Vaĭnberg for the equation (2) when A is a quasi-negative linear mapping of X^* into X (Vaĭnberg [**15**], [**14**]).

BIBLIOGRAPHY

1. H. Amann, *Ein existenz und eindeutigkeitssatz für die Hammersteinsche gleichung in Banachraumen*, Math. Z. **111** (1969), 175–190.

2. H. Brezis, *Equations et inéquations non-linéaires dans les espaces vectoriels en dualité*, Ann. Inst. Fourier (Grenoble) **18** (1968), 115–176.

3. ———, *On some degenerate nonlinear parabolic equations*, Proc. Sympos. Pure Math., vol. 18, Amer. Math. Soc., Providence, R.I., 1970, pp. 28–38.

4. H. Brezis, M. G. Crandall and A. Pazy, *Perturbations of nonlinear maximal monotone sets in Banach space*, Comm. Pure. Appl. Math. **23** (1970), 123–144.

5. F. E. Browder, *Nonlinear maximal monotone operators in Banach space*, Math. Ann. **175** (1968), 89–113. MR **36** #6989.

6. ——— *Nonlinear monotone and accretive operators in Banach spaces*, Proc. Nat. Acad. Sci. U.S.A. **61** (1968), 388–393.

7. ———, *Nonlinear operators and nonlinear equations of evolution in Banach spaces*, Proc. Sympos. Pure Math., vol. 18, Amer. Math Soc., Providence, R.I.

8. F. E. Browder and C. P. Gupta, *Monotone operators and nonlinear integral equations of Hammerstein type*, Bull. Amer. Math. Soc. **75** (1969), 1347–1353.

9. C. L. Dolph and G. J. Minty, *On nonlinear integral equations of the Hammerstein type*, Proc. Advanced Seminar Math. Res. Center, Nonlinear Integral Equations (Madison, Wis., 1963), Univ. of Wisconsin Press, Madison, Wis., 1964, pp. 99–154. MR **28** #4322.

10. I. I. Kolodner, *Equations of Hammerstein type in Hilbert spaces*, J. Math. Mech. **13** (1964), 701–750. MR **30** #1415.

11. J. Kolomý, *The solvability of nonlinear integral equations*, Comment. Math. Univ. Carolinae **8** (1967), 273–289. MR **35** #5878.

12. R. T. Rockafellar, *Local boundedness of nonlinear maximal monotone operators*, Michigan Math. J. **16** (1969), 397–407.

13. ———, *On the maximality of sums of nonlinear monotone operators*, Trans. Amer. Math. Soc. **149** (1970), (to appear).

14. M. Vaĭnberg, *Variational methods for the study of nonlinear operators*, GITTL Moscow, 1956, English transl., Holden-Day, San Francisco, Calif., 1964. MR **19**, 567; MR **31** #638.

15. ———, *Nonlinear equations with potential and monotonic operators*, Dokl. Akad. Nauk SSSR **183** (1968), 747–749 = Soviet Math. Dokl. **9** (1968), 1427–1430. MR **39** #832.

University of Chicago, Chicago, Illinois 60637

University of Illinois, Chicago, Illinois 60680

Northern Illinois University, DeKalb, Illinois 60115

JOURNAL OF FUNCTIONAL ANALYSIS 11, 168–178 (1972)

The Extension of Contractions and the Intersection of Balls in Banach Spaces*

D. G. deFigueiredo

Department of Mathematics, University of Illinois at Chicago Circle, Chicago 60680

AND

L. A. Karlovitz[†]

Institute for Fluid Dynamics and Applied Mathematics, University of Maryland, College Park, Maryland 20742

Communicated by Alberto P. Calderon

Received January 15, 1971

Let X be a normed space. A mapping $T : D \to X$ is called a contraction if $\| Tx - Ty \| \leqslant \| x - y \|$ for all x and y in D. The space X has the contraction extension property if every contraction T defined on a subset D of X can be extended to a contraction $\hat{T}$ defined on all of X.

A characterization of classes of spaces having this property is given.

Introduction

Let X be a normed linear space. A mapping T defined on a subset D of X with values in X is said to be a *contraction* (also called a *nonexpansive mapping*) if $\| Tx - Ty \| \leqslant \| x - y \|$ for all $x, y \in D$. It is interesting, and in some applications important, to seek a contraction $\hat{T}$ defined on all of X which extends T, i.e., $\hat{T}x = Tx$ for all $x \in D$. We say that *the space X has the contraction extension property*, abbreviated *property* (E), if for each subset $D \subset X$ and each contraction $T: D \to X$ there exists a contraction $\hat{T}: X \to X$ which extends T. It is our purpose here to characterize classes of spaces having property (E). In this we will make use of certain related combinatorial–geometric

* Presented at the Conference on Mapping Techniques and Problems, University of Houston, November 6 and 7, 1970 in honor of Professor David Bourgin.

† Partially supported by the National Science Foundation under Grant NSF GP 20555.

Originally published in: J. Funct. Anal. 11 (1972), 168-178,

Reprint by Springer-Verlag Berlin Heidelberg 2013 with kind permission,
DOI: 10.1007/978-3-319-02856-9_4

properties of X. We employ the usual notation for balls: $B(y, r) = \{x: x \in X, \| x - y \| \leqslant r\}$; and we refer to $B(0, 1)$ as the *unit ball.*

We say that the *space X has the Kirszbraun property for the cardinal* γ, abbreviated *property* (K, γ), if for each set of points $x_i, y_i \in X$ and reals $r_i > 0$, $i \in I$, such that I is an index set with cardinality γ, such that $\| y_i - y_j \| \leqslant \| x_i - x_j \|$, for all $i, j \in I$, and such that $\bigcap_{i \in I} B(x_i, r_i)$ is nonempty, it follows that $\bigcap_{i \in I} B(y_i, r_i)$ is also nonempty. We say that the *space X has the Kirszbraun property*, abbreviated *property (K)*, if X has property (K, γ) for all γ. It is wellknown and readily verified that:

(1) X has property (E) if and only if it has property (K).

Kirszbraun [7] showed that every finite dimensional Hilbert space has property (K), and this result was extended by Valentine [12] to all Hilbert spaces. We note parenthetically that, in analogy to (1), X has property (K, γ) if and only if X has the following restricted contraction extension property: For each subset $D \subset X$ with cardinality $D \leqslant \gamma$, each point $x_0 \in X$, and each contraction $T: D \to X$ there exists a contraction $\tilde{T}: D \cup \{x_0\} \to X$ which extends T.

We say that the *space X has the* γ, 2 *intersection property for the cardinal* $\gamma \geqslant 2$, abbreviated *intersection property* $(\gamma, 2)$, if for each set of balls $B(x_i, r_i)$, $i \in I$, such that I is an index set of cardinality γ and such that $B(x_i, r_i) \cap B(x_j, r_j)$ is nonempty for all pairs $i, j \in I$, it follows that the intersection $\bigcap_{i \in I} B(x_i, r_i)$ is nonempty. We say that X has the *finite-two intersection property* if X has the intersection property $(\gamma, 2)$ for all finite γ. We say that X has the *binary intersetion property* if X has the intersection property $(\gamma, 2)$ for all γ. The following relation between the properties follows immediately from the definitions.

(2) If X has the intersection property $(\gamma, 2)$ then X has property (K, γ).

Combining (2) with the Kirszbraun–Valentine results we find,

(3) If X has the intersection property $(\gamma, 2)$ or if X is a Hilbert space then X has property (K, γ).

The following theorem states a partial converse for (3).

THEOREM 1. *Let X be a real Banach space. Suppose that the unit ball of X contains at least one extreme point. Then the following two properties are equivalent:*

(4) *X has property* $(K, 3)$,

(5) *X is either a Hilbert space or X has the intersection property* (3, 2).

Moreover, the following three properties are equivalent:

(6) *X has property* $(K, 4)$,

(7) *X has property* (K, n) *for all finite n*,

(8) *X is either a Hilbert space or X has the finite-two intersection property.*

Remark 1. It is not known whether the hypothesis about the existence of an extreme point is necessary for the validity of the theorem.

By virtue of work of Aronszajn and Panitchpakdi [1], Goodner [4], Kelley [6], Lindenstrauss [8], and Nachbin [9] on the intersection property $(\gamma, 2)$ and the aforementioned work of Kirszbraun and Valentine, we can readily derive the following theorem, characterizing spaces having property (E), as a corollary of Theorem 1. Results from Lindenstrauss [8] are also used in the proof of Theorem 1.

THEOREM 2. *Let X be a real Banach space. Suppose that X is a conjugate space. (For example, X may be a reflexive space.) Then the following properties are equivalent:*

(9) *X has property* (E),

(10) *X has property* $(K, 4)$,

(11) *X is either a Hilbert space or a* P_1 *space*, i.e., *a space isometric to a* $C(\Omega)$, *where* Ω *is an extremally disconnected compact Hausdorff space.*

Remark 2. The equivalence between (9) and (11) was shown by Grünbaum [5] for the case $\dim X = 2$, extended by Schönbeck [10] to the case $\dim X < \infty$, and extended hereby to the case that X is a conjugate space. It is worth mentioning that we know of no example of a Banach space which satisfies (9) without also satisfying (11).

A Banach space X is said to have *property* $(\prod_1)$ if there exists a set of finite dimensional linear subspaces M_α, $\alpha \in A$, directed by inclusion, such that closure $\bigcup_{\alpha\in A} M_\alpha = X$ and such that each M_α is the range of a linear projection L_α of norm 1, i.e., $L_\alpha: X \to M_\alpha$, $L_\alpha^2 = L_\alpha$, $L_\alpha x = x$ for all $x \in M_\alpha$, and $\|L_\alpha\| = 1$. A Banach space is said to be an $\mathcal{N}_1$ *space* if there exists a set of finite dimensional linear subspaces N_α, $\alpha \in A$, directed by inclusion, such that each N_α is a P_1 space and such that closure $\bigcup_{\alpha\in A} N_\alpha = X$. Every P_1 space is an $\mathcal{N}_1$ space, Lindenstrauss [8].

Theorem 3. *Let X be a real Banach space which has property* (Π_1). *If in addition X has property* (E), *then X is either a Hilbert or an* $\mathcal{N}_1$ *space.*

The proofs of all three theorems appear in Section 1 below. The proof of Theorem 3 depends on a result of a paper by the present authors, which is noted in the following remark.

Remark 3. Alongside the problem of determining whether or not a space X has the contraction extension property, it is equally important and interesting to determine whether or not it has the extension property for contractions defined in *closed convex subsets* of X. In [3] the present authors considered various aspects of the latter problem. However, it is far from being well understood.

1. *Proofs.* The proof of Theorem 1 will depend on the following sequence of lemmas. Throughout, X is a real normed space of dimension not less than two. We make use of the quantity $r(x_1, x_2, x_3)$ defined for all $x_1, x_2, x_3 \in X$ by

$$(12) \quad r(x_1, x_2, x_3) = \inf_{z \in X} \max_{1 \leqslant i \leqslant 3} \| z - x_i \|.$$

We say that the *number* $r(x_1, x_2, x_3)$ *is achieved for* (x_1, x_2, x_3) if (12) is achieved as a minimum; i.e., if there exists $z = z(x_1, x_2, x_3)$ such that $\max\{\| z - x_i \| : i = 1, 2, 3\} = r(x_1, x_2, x_3)$.

Lemma 1. *Let X be a real normed linear space with property* $(K, 3)$. *For each* λ, $0 \leqslant \lambda \leqslant 1$, *let*

$$(13) \quad \Delta(\lambda) = \{(x_1, x_2, x_3) : x_i \in X, \| x_1 - x_2 \| = \| x_1 - x_3 \| = 1, \\ \| x_2 - x_3 \| = \lambda\}.$$

For each λ there exists a number $m(\lambda)$ such that

$$(14) \quad 1/2 \leqslant m(\lambda) = r(x_1, x_2, x_3) \quad \textit{for all} \quad (x_1, x_2, x_3) \in \Delta(\lambda).$$

Moreover if the number $m(\lambda)$ is achieved for some $(x_1, x_2, x_3) \in \Delta(\lambda)$ then it is achieved for all $(x_1, x_2, x_3) \in \Delta(\lambda)$.

Proof. It follows immediately from the triangle inequality that $r(x_1, x_2, x_3) \geqslant 1/2$ for each $(x_1, x_2, x_3) \in \Delta(\lambda)$. If

$$r = r(x_1, x_2, x_3) < r(y_1, y_2, y_3) \quad \text{for} \quad (x_1, x_2, x_3), \ (y_1, y_2, y_3) \in \Delta(\lambda)$$

for some λ, then, by virtue of the definition (12), for $\epsilon > 0$ sufficiently small, we have

$$\bigcap_{i=1}^{3} B(x_i, r + \epsilon) \neq \phi \quad \text{and} \quad \bigcap_{i=1}^{3} B(y_i, r + \epsilon) = \phi.$$

Since $\| x_i, - x_j \| = \| y_i - y_j \|$, $i, j = 1, 2, 3$, this contradicts the assumption that X has property $(K, 3)$. This proves (14). The last part of the lemma is proved in a similar way. This finishes the proof of Lemma 1.

By virtue of Lemma 1 we may define for each space X with property $(K, 3)$ and with $\dim X \geqslant 2$ the set

$$\Lambda(X) = \{\lambda: 0 \leqslant \lambda \leqslant 1,\ m(\lambda) = 1/2 \text{ and it is achieved for each } (x_1, x_2, x_3) \in \Delta(\lambda)\}.$$

Note, for example, that $0 \in \Lambda(X)$ for all such X. For another example if X has the intersection property $(3, 2)$ then $\Lambda(X) = [0, 1]$. The following two lemmas characterize $\Lambda(X)$ in terms of certain geometric properties of X. We will make use of the following

Notation. For $x, y \in X$ we *denote the line segment*

$$\{tx + (1 - t)y: 0 \leqslant t \leqslant 1\} \quad \textit{by} \quad [x, y].$$

We *denote by* S *the boundary of the unit ball* of X; i.e., $S = \{x: x \in X, \| x \| = 1\}$.

LEMMA 2. *Let X be a real normed linear space with property* $(K, 3)$ *and with* $\dim X \geqslant 2$. *Then* $\lambda \in \Lambda(X)$ *if and only if there exist* $x, y \in X$ *such that*

(15) $[x, y] \subset S$ *and* $\| x - y \| = 2\lambda$,

i.e., *if and only if there exists a line segment of length* 2λ *in the boundary of the unit ball. Consequently, if* $\lambda \in \Lambda(X)$ *and* $0 \leqslant \lambda' \leqslant \lambda$, *then* $\lambda' \in \Lambda(X)$.

Proof. Assume first that there exist x, y satisfying (15). Let $x_1 = 0$, $x_2 = (3x + y)/4$, $x_3 = (x + 3y)/4$, and $z = (x + y)/4$. It follows from (15) that $(x_1, x_2, x_3) \in \Delta(\lambda)$ and that $\| z - x_i \| = 1/2$, $i = 1, 2, 3$. Hence $r(x_1, x_2, x_3) = 1/2$ and it is achieved. It follows from Lemma 1 that $\lambda \in \Lambda(X)$.

Next assume that $\lambda \in \Lambda(X)$. We first show that there exists a line segment of length λ in the boundary of the unit ball. To this end, choose $(x_1, x_2, x_3) \in \Delta(\lambda)$ such that $x_1 = 0$. Since $\lambda \in \Lambda(X)$, there exists $z \in X$ such that

$$\| 2z \| = \| x_2 \| = \| x_3 \| = \| 2x_2 - 2z \| = \| 2x_3 - 2z \| = 1.$$

Since, moreover, $x_2 = (1/2)(2z) + (1/2)(2x_2 - 2z)$ and $x_3 = (1/2)(2z) + (1/2)(2x_3 - 2z)$ it follows from the convexity of the unit ball that the line segments $[2z, 2x_2 - 2z]$ and $[2z, 2x_3 - 2z]$ lie in S. The lengths of the segments, $\| 2x_2 - 4z \|$ and $\| 2x_3 - 4z \|$, satisfy

$$2\lambda = \| 2x_2 - 2x_3 \| \leqslant \| 2x_2 - 4z \| + \| 2x_3 - 4z \|.$$

Hence, at least one of the segments, say $[2z, 2x_2 - 2z]$, has length at least λ and lies in S. Consequently, we can choose $y_2, y_3 \in [2z, 2x_2 - 2z]$ such that $(0, y_2, y_3) \in \Delta(\lambda)$ and $[y_2, y_3] \subset S$. Since $\lambda \in \Lambda(X)$, there exists v such that

$$\| 2v \| = \| y_2 \| = \| y_3 \| = \| 2y_2 - 2v \| = \| 2y_3 - 2v \| = 1.$$

Let $w = 2v$, $x = 2y_2 - 2v$, and $y = 2y_3 - 2v$. Then,

(16) $\|(1/2)w + (1/4)x + (1/4)y\| = \|(y_2 + y_3)/2\| = 1,$

because $[y_1, y_2] \subset S$. Since $\| w \| = \| x \| = \| y \| = 1$ it follows from (16) and the convexity of the unit ball that the convex hull of w, x and y lies in S. In particular;

$$[x, y] \subset S.$$

Moreover, the length of $[x, y]$ is given by

$$\| x - y \| = \| 2y_2 - 2y_3 \| = 2\lambda.$$

This proves that if $\lambda \in \Lambda(X)$, then there exists x, y satisfying (15).

The last part of the lemma follows immediately from (15). This finishes the proof of Lemma 2.

Lemma 3. *Let X be a real normed linear space with property $(K, 3)$ and with* dim $X \geqslant 2$. *Suppose that the unit ball of X is not strictly convex and that it contains at least one extreme point e. Let M be a 2-dimensional linear subspace of X such that $e \in M$. Then M is a P_1 space. Moreover, $\Lambda(X) = [0, 1]$.*

Proof. Let

(17) $\lambda^* = \sup \Lambda(X).$

Since the unit ball of X is not strictly convex, it follows from Lemma 2 that $\lambda^* > 0$. Let e be an extreme point of the unit ball of X, and let M be a 2-dimensional linear subspace of X such that $e \in M$. Let f be a linear functional defined on M such that $\|f\| = 1$ and $f(e) = 0$. Let λ_n be a sequence such that $\lambda_n \in \Lambda(X)$ and $\lambda_n \to \lambda^*$. Choose sequences x_n and $y_n \in M$ such that $\| x_n \| = \| y_n \| = 1$, such that $\| x_n - e \| = \| y_n - e \| = \lambda_n$, and such that $f(x_n) < 0 < f(y_n)$. It follows that there exists $x^*, y^* \in M$ such that

$$(18) \quad x_n \to x^* \text{ and } y_n \to y^* \text{ as } n \to \infty.$$

Moreover, using the fact that $0 < \lambda^* \leqslant 1$, we also have

$$(19) \quad \| x^* - e \| = \| y^* - e \| = \lambda^* \text{ and } f(x^*) < 0 < f(y^*).$$

Since $(0, x_n, e) \in \Delta(\lambda_n)$ and $\lambda_n \in \Lambda(X)$ it follows that there exists z_n such that

$$\max\{\| z_n \|, \| x_n - z_n \|, \| e - z_n \|\} = 1/2, \qquad n = 1, 2, \ldots .$$

By the triangle inequality this is equivalent to

$$(20) \quad \| z_n \| = \| x_n - z_n \| = \| e - z_n \| = 1/2, \; n = 1, 2, \ldots .$$

Since e is an extreme point, $\| z_n \| = \| e - z_n \| = 1/2$ implies

$$(21) \quad z_n = e/2, \; n = 1, 2, \ldots .$$

Hence, by (20), $\| 2x_n - e \| = 1$. Since, moreover,

$$x_n = (1/2)(2x_n - e) + (1/2)e \qquad \text{and} \qquad \| x_n \| = \| e \| = 1,$$

it follows from the convexity of the unit ball that $[2x_n - e, e] \subset S$, for all $n \geqslant 1$. By the same argument, $[2y_n - e, e] \subset S$, for all $n \geqslant 1$. In the limit, letting $u = 2x^* - e$ and $v = 2y^* - e$, we find

$$(22) \quad [u, e], [v, e] \subset S \text{ and } \| u - e \| = \| v - e \| = 2\lambda^*.$$

We next prove that $u = -v$ and $\lambda^* = 1$. To this end we consider the quadrilateral with vertices 0, u, v, and e. Since $\| u \| = \| v \| = \| e \| = 1$ and, by (19), $f(u) = f(2x^*) < 0 = f(e) < f(2y^*) = f(v)$, we can choose numbers α and β such that

$$(23) \quad \alpha u + (1 - \alpha)v = \beta e, \text{ and } 0 < \alpha < 1, \; -1 \leqslant \beta \leqslant 1.$$

Hence

$$(24) \quad -u = ((1 - \alpha)/\alpha)v + (-\beta/\alpha)e \text{ and } -v = (\alpha/(1 - \alpha))u + (-\beta/(1 - \alpha))e.$$

If $\beta \leqslant 0$ it follows from (24) and $[u, e], [v, e] \subset S$ that

$$(1-\alpha)/\alpha - \beta/\alpha = 1 = \alpha/(1-\alpha) - \beta/(1-\alpha),$$

and consequently $\alpha = 1/2$ and $\beta = 0$. Hence in (23) β may be restricted by $0 \leqslant \beta \leqslant 1$. This means, in particular, that the quadrilateral with vertices 0, u, e, and v is convex. By the triangle inequality, any convex quadrilateral lying in a 2-dimensional normed linear space has the property that twice the sum of the lengths of its diagonals is greater than or equal to the sum of the lengths of its sides. Applying this to our case, we find

$$2\| u - v \| + 2\| e \| \geqslant \| u \| + \| v \| + \| u - e \| + \| v - e \|.$$

Equivalently,

(25) $\| u - v \| \geqslant 2\lambda^*.$

We note that $\| u - v \|/2 = \| x^* - y^* \| \in \Lambda(X)$ because

$$(0, x^*, y^*) \in \Delta(\| x^* - y^* \|)$$

and, by virtue of (18), (20) and (21), $r(0, x^*, y^*) = 1/2$ and it is achieved. Hence by Lemma 1, $\| x^* - y^* \| \in \Lambda(X)$; and, by (17), $\| x^* - y^* \| \leqslant \lambda^*$. Combining this with (25), we find

(26) $\| u - v \| = 2\lambda^*.$

Applying the triangle inequality to 0, u, βe and to u, βe, e and keeping in mind that $0 \leqslant \beta \leqslant 1$, we find

(27) $\| u - \beta e \| + \beta \geqslant 1$ and $1 - \beta + \| u - \beta e \| \geqslant 2\lambda^*.$

Addition of the two inequalities in (27) yields $\| u - \beta e \| \geqslant \lambda^*$. By the same argument: $\| v - \beta e \| \geqslant \lambda^*$. Since $\| u - v \| = \| u - \beta e \| + \| v - \beta e \|$ we can combine the latter two inequalities with (26) to conclude

(28) $\| u - \beta e \| = \| v - \beta e \| = \lambda^*.$

Substituting $\| u - \beta e \| = \lambda^*$ in (27) we find

(29) $\beta = 1 - \lambda^*.$

Let $x_1 = (u - \beta e)/2\lambda^*$, $x_2 = (v - \beta e)/2\lambda^*$, and $x_3 = (e - \beta e)/2\lambda^*$. Then, by (22) and (26), $\| x_i - x_j \| = 1, i \neq j$, i.e., $(x_1, x_2, x_3) \in \Delta(1)$, and $\| x_i - 0 \| = 1/2$, $i = 1, 2, 3$, i.e., $r(x_1, x_2, x_3) = 1/2$ and it is achieved. Hence, by Lemma 1, $1 \in \Lambda(X)$; and, by (17), $\lambda^* = 1$.

Substitution of $\lambda^* = 1$ in (29) yields $\beta = 0$. Substituting $\beta = 0$ in (23) and using $\| u \| = \| v \| = 1$ we finally find $u = -v$.

Since $u = -v$ we conclude from (22) that $[u, e]$, $[-u, e] \subset S$. Consequently $\| \gamma u + \delta e \| = | \gamma | + | \delta |$, for all real γ and δ; and hence M is a P_1 space. Since $\lambda^* = 1 \in \Lambda(X)$ it follows from Lemma 2 that $\Lambda(X) = [0, 1]$. This finishes the proof of Lemma 3.

LEMMA 4. (*Contained in Edelstein and Thompson* [2, *Theorem* 2.1] *and in Schönbeck* [11, *Theorem* 1]). *Let X be a Banach space with property* $(K, 3)$. *Suppose that the unit ball of X is strictly convex. Then X is a Hilbert space.*

Proof of Theorem 1. If X has dimension one, there is nothing to prove. Hence we assume throughout that $\dim X \geqslant 2$.

(5) $\Rightarrow$ (4). If X is a Hilbert space, this follows from the aforementioned result of Valentine; and if X has the intersection property (3, 2), this follows from (3).

(4) $\Rightarrow$ (5): If the unit ball of X is strictly convex, this follows from Lemma 4. If the unit ball of X is not strictly convex, we propose to show that X has the intersection property (3, 2). Thus we need to show that if $B(y_i, r_i)$, $i = 1, 2, 3$ are balls such that

$$B(y_i, r_i) \cap B(y_j, r_j) \neq \phi, \qquad i, j = 1, 2, 3,$$

then $\bigcap_{i=1}^{3} B(y_i, r_i) \neq \phi$. By a result of Lindenstrauss [8, Theorem 4.3], we may assume $r_1 = r_2 = r_3 = r$. Since the unit ball of X has an extreme point we deduce from Lemma 3 that $1 \in \Lambda(X)$, and hence there exists $w_1, w_2, w_3, v \in X$ such that $\| w_i - w_j \| = 1$, $i \neq j$, and $\| w_i - v \| = 1/2$, $i = 1, 2, 3$. Let $x_i = (2r) w_i$, $i = 1, 2, 3$ and $w = (2r)v$. Then $\| y_i - y_j \| \leqslant 2r = \| x_i - x_j \|$, $i \neq j$, and $\| x_i - w \| = r$, $i = 1, 2, 3$. Since X has property $(K, 3)$ this implies that there exists $y \in X$ such that $y \in \bigcap_{i=1}^{3} B(y_i, r)$, as desired.

(8) $\Rightarrow$ (7): The proof of this is completely analogous to the proof of (5) $\Rightarrow$ (4).

(7) $\Rightarrow$ (6): This is obvious.

(6) $\Rightarrow$ (8): If the unit ball of X is strictly convex, this follows from Lemma 4. If the unit ball of X is not strictly convex, we propose to show that X has the finite-two intersection property. By a result of Lindenstrauss [8, Theorem 4.1] this is equivalent to showing that X has the intersection property (4,2). Thus we need to show that if

$B(y_i, r_i)$, $i = 1, 2, 3, 4$ are balls such that $B(y_i, r_i) \cap B(y_j, r_j) \neq \phi$, $i, j = 1,..., 4$, then $\bigcap_{i=1}^{4} B(y_i, r_i) \neq \phi$. For the same reason as in the proof of (4) $\Rightarrow$ (5), we may assume that $r_1 = r_2 = r_3 = r_4 = r$. Since the unit ball of X has an extreme point we can apply Lemma 3 to deduce that X has a 2-dimensional subspace M spanned by u and v such that $\| su + tv \| = | s | + | t |$ for real s and t. Let $x_1 = ru$, $x_2 = -ru$, $x_3 = rv$, and $x_4 = -rv$. Then $\| y_i - y_j \| \leqslant 2r = \| x_i - x_j \|$, $i \neq j$ and $\| x_i - 0 \| = \| x_i \| = r$, $i = 1,..., 4$. Since X has property $(K, 4)$ this implies that there exists $y \in X$ such that $y \in \bigcap_{i=1}^{4} B(y_i, r)$, as desired. This finishes the proof of Theorem 1.

Proof of Theorem 2. (9) $\Rightarrow$ (10): By (1) property (E) is equivalent to property (K); and, by definition, property (K) implies property $(K, 4)$.

(10) $\Rightarrow$ (11): Since X is a conjugate space its unit ball contains at least one extreme point, and Theorem 1 is applicable. Hence X having property $(K, 4)$ is equivalent to X either being a Hilbert space or having the finite-two intersection property. If the former holds, we are finished. If the latter holds, we can combine the weak-star compactness of the unit ball with the finite-two intersection property in an obvious way to deduce that X has the binary intersection property. By the work on the intersection property $(\gamma, 2)$, by the authors noted in the introduction, this is equivalent to X being a P_1 space. Alternatively, it was shown by Lindenstrauss [8, Corollary 3 of Theorem 6.1] that if X is a conjugate space and has intersection property (4, 2) then X is a P_1 space.

(11) $\Rightarrow$ (9): As noted in the proof of (10) $\Rightarrow$ (11), X being a P_1 space is equivalent to X having the binary intersection property. Hence, using (3), (11) implies that X has the property (K, γ) for all γ. By definition this is equivalent to X having property (K) which, by (1), is equivalent to property (E). This finishes the proof of Theorem 2.

Proof of Theorem 3. Let M_α, $\alpha \in A$, be a set of linear subspaces of X satisfying the definition of property $(\prod_1)$. Since X has property (E) and since each M_α is finite dimensional and the range of a linear projection of norm one, the hypotheses of a theorem of the present authors [3, Theorem 1] are satisfied for each $\alpha \in A$. We can conclude that either X is a Hilbert space or M_α is a P_1 space for each α. By definition, the latter alternative means that X is an N_1 space. This finishes the proof of Theorem 3.

References

1. N. Aronszajn and P. Panitchpakdi, Extensions of uniformely continuous transformations and hyperconvex metrix spaces, *Pacific J. Math.* **6** (1956), 405–439.
2. M. Edelstein and A. C. Thompson, Contractions, isometries and some properties of inner product spaces, *Nederl. Akad. Wetensch. Proc. Ser. A* **70** (1967), 326–332.
3. D. G. deFifueiredo and L. A. Karlovitz, "On the extension of contractions on normed spaces," Proceedings Symposium Pure Mathematics, Vol. 18, Part 1, pp. 95–104, American Mathematical Society, Providence, RI, 1970.
4. D. G. Goodner, Projections in normed spaces, *Trans. Amer. Math. Soc.* **69** (1950), 89–108.
5. B. Grünbaum, On a theorem of Kirszbraun, *Bull. Res. Counc. Israel* **17F** (1958), 129–132.
6. J. L. Kelley, Banach spaces with the extension property, *Trans. Amer. Math. Soc.* **72** (1952), 323–326.
7. M. D. Kirszbraun, Über die zusammenziehende und Lipschitzsche Transformationen, *Fund. Math.* **22** (1934), 77–108.
8. J. Lindenstrauss, "Extension of compact operators," *Mem. Amer. Math. Soc.* **48** (1964).
9. L. Nachbin, A theorem of the Hahn–Banach type for linear transformations, *Trans. Amer. Math. Soc.* **68** (1950), 28–46.
10. S. O. Schönbeck, On the extension of Lipschitz maps, *Ark Mat.* **7** (1967), 201–209.
11. S. O. Schönbeck, Extension of nonlinear contractions, *Bull. Amer. Math. Soc.* **72** (1966), 99–101.
12. F. A. Valentine, A Lipschitz condition preserving extension for a vector function, *Amer. J. Math.* **67** (1945), 83–93.

PROCEEDINGS OF THE
AMERICAN MATHEMATICAL SOCIETY
Volume 40, Number 2, October 1973

ON THE VARIATIONAL METHOD FOR THE EXISTENCE OF SOLUTIONS OF NONLINEAR EQUATIONS OF HAMMERSTEIN TYPE

DJAIRO G. DEFIGUEIREDO AND CHAITAN P. GUPTA[1]

ABSTRACT. Let X be a real Banach space and X^* its conjugate Banach space. Let A be an unbounded monotone linear mapping from X to X^* and N a potential mapping from X^* to X. In this paper we establish the existence of a solution of the equation $u+ANu=v$ for a given v in X^* using variational method. Our method consists in using a splitting of A via an auxiliary Hilbert space and solving an equivalent equation in this auxiliary Hilbert space. In §2, we prove the same result in the case when X is a Hilbert space using the natural splitting of A in terms of its square root. We do this to compare and contrast the proofs in the two cases.

Introduction. Let X be a real Banach space and X^* denote its conjugate Banach space. Let A be an unbounded monotone linear mapping from X to X^* and N a nonlinear mapping from X^* to X satisfying no monotone hypothesis. In this paper we study the solvability of the equation

$$u + ANu = v \tag{1}$$

for a given element v in X^*. Equations of the form given in (1) are known as equations of Hammerstein type. The most interesting example of (1) is when A is a linear integral mapping and N is a so-called Nemitsky mapping. The case when A is an unbounded linear mapping corresponds to the case of linear integral mappings having Carleman kernels. We may mention that our study of equation (1) when A is an unbounded linear mapping makes it possible to apply our results not only to nonlinear integral equations of Hammerstein type but also to certain boundary value problems in nonlinear differential equations.

In recent years there has been extensive work on these equations under various monotonicity assumptions on N. A recent paper of Browder [1] contains a fairly complete bibliography on this subject. The authors in their previous works [3], [4] had considered the case when the linear

Received by the editors April 7, 1971.
AMS (MOS) subject classifications (1970). Primary 47H15, 49G99, 45G05, 45G99.
Key words and phrases. Variational method, integral equations of Hammerstein type, weak *-gradient, Carleman kernels.

[1] Part of the research of this author supported by NSF grant #GP28735.

Originally published in: Proc. Amer. Math. Soc. 40 (1973), 470-476,

DOI: 10.1007/978-3-319-02856-9_5

mapping A is unbounded and N satisfied some sort of monotonicity conditions. In case N does not satisfy any monotonicity condition it seems appropriate to use variational methods. However, one is bound to have several difficulties stemming from the unbounded nature of the linear mapping A. In a Hilbert space these difficulties could be handled because the unbounded monotone linear mapping A has a natural splitting in terms of its square root, while no such thing had been known in the case of Banach spaces. It is for these reasons (it seems) that Vainberg and Lavrentiev [**8**] in their joint work impose several extraneous conditions on the Banach spaces and on the linear mapping A, for then they are able to reduce equation (1) to an equivalent equation in a Hilbert space and use the known splitting there. In this paper, we have been able to study (1) without any such restrictions by using a *new* splitting of A and our results generalize their results.

A subset D of $X \times X^*$ is said to be *monotone* if $(x-y, x^*-y^*) \geqq 0$ for all pairs $[x, x^*]$, $[y, y^*]$ in D. Here (x, x^*) denotes the duality pairing between the elements x of X and x^* of X^* and in the following we make the convention that $(x^*, x) = (x, x^*)$. A monotone set is said to be *maximal monotone* if it is not a proper subset of any other monotone set. A mapping $T: D(T) \to X^*$, where $D(T) \subset X$, is *monotone* if its graph $G(T) = \{[x, Tx]: x \in D(T)\}$ is a monotone set, and T is *maximal monotone* if its graph is a maximal monotone set. A linear mapping $A: D(A) \to X^*$, $D(A) \subset X$, is *symmetric* if $(Ax, y) = (Ay, x)$ for every x, y in $D(A)$. A mapping $N: X^* \to X$ is called a *weak*-gradient* if there exists a functional $f: X^* \to \boldsymbol{R}$ (where $\boldsymbol{R}$ denotes the set of real numbers) such that

$$(Nx^*, y^*) = \lim_{t \to 0} \frac{f(x^* + ty^*) - f(x^*)}{t}$$

for every x^*, y^* in X^*. In particular, if X is reflexive the notion of weak*-gradient coincides with that of weak-gradient. We remark that if X is not reflexive and a given $f: X^* \to \boldsymbol{R}$ is weakly-Gateaux differentiable then its weak-gradient is a mapping from X^* to X^{**} and so f does not give rise, in general, to a mapping of the type considered here.

Our main result is as follows:

THEOREM 1. *Let X be a real Banach space such that the unit ball in its conjugate X^* is weak*-sequentially-compact. (In particular, this is the case if X is either reflexive or separable.) Let $A: D(A) \to X^*$, $D(A) \subset X$, be a linear maximal monotone symmetric densely defined mapping from X to X^*. Let $N: X^* \to X$ be the weak*-gradient of a functional $f: X^* \to \boldsymbol{R}$ which is assumed to be lower-weak*-semicontinuous and such that*

(2) $$f(x^*) \geqq \omega(\|x^*\|) \quad \text{for all } x^* \text{ in } X^*$$

where the numerical function $\omega(r)$ is bounded below for $r \geqq 0$ and $\omega(r) \to +\infty$ as $r \to +\infty$. Then, there exists $x^* \in X^*$ such that $x^* + ANx^* = y^*$ for each given y^* in X^*.

REMARKS. (i) The bound (2) on f required in Theorem 1 is satisfied if

$$f(x^*) \geqq a_1 \|x^*\|^2 + a_2 \|x^*\|^\theta + a_3 \tag{3}$$

where $a_1 > 0$, a_2 and a_3 are any given real numbers, $0 < \theta < 2$. The estimate (3) arises in several situations for the potential of Nemitsky mappings, see **[9]**.

(ii) Compare the above theorem with Theorem 4 of **[8]** where even under more stringent conditions only a sort of generalized solution is obtained.

(iii) Without loss of generality we may assume that $y^* = 0$. For if this is not the case we replace N by the mapping N_1 defined by $N_1(x^*) = N(x^* + y^*)$ and it is easy to verify that N_1 satisfies the conditions of Theorem 1, above.

In the proof of Theorem 1, we use a splitting of the unbounded linear mapping A, which is given in Theorem 2, below. To obtain this splitting we follow Browder-Gupta **[2]** who considered the case of bounded A. We may also mention that Theorem 1, above, extends a previous result of Gupta **[5]** for the case of A bounded.

THEOREM 2. *Let $A: D(A) \to X^*$ be a linear monotone symmetric densely defined mapping from X to X^*. Then there exists a Hilbert space H, a linear mapping $S: D(A) \to H$ such that $A = S^*S$, where $S^*: D(S^*) \to X^*$, $D(S^*) \subset H$, is the adjoint of S.*

The proofs of above theorems are given in §§1 and 2. In §3, we consider the Hilbert space case, where unlike the Banach space case we succeed in reducing the variational problem to the one of minimizing a functional on a complete set. In this manner we obtain some results which are comparable with results announced by Lavrentiev in a Doklady note **[7]**.

1. **Proof of Theorem 2.** Consider in $D(A)$ the following bilinear form, $b[x, y] = (Ax, y)$, which in view of the monotonicity of A is such that $b[x, x] \geqq 0$. Consequently one has Cauchy-Schwarz inequality

$$|b[x, y]| \leqq (b[x, x])^{1/2} (b[y, y])^{1/2}. \tag{4}$$

Next we define the set $\mathcal{N} = \{x \in D(A) : b[x, x] = 0\}$, which in view of (4) is also given by

$$\mathcal{N} = \{x \in D(A) : b[x, y] = 0 \text{ for all } y \text{ in } D(A)\}.$$

Thus $\mathcal{N}$ is a closed linear subspace of $D(A)$. So the quotient space $H_0 = D(A)/\mathcal{N}$ is a prehilbertian space with our inner product $[\ ,\ ]$ defined as follows. Let $\pi: D(A) \to H_0$ be the canonical mapping, and let $\bar{x}, \bar{y} \in H_0$, Then $[\bar{x}, \bar{y}] = (Ax, y)$ where $x \in \pi^{-1}(\bar{x})$ and $y \in \pi^{-1}(\bar{y})$.

Now let us denote by H the completion of H_0 with respect to this inner product. Let us finally denote by S the composition of π with the embedding mapping of H_0 into H and let us also use $[\ ,\ ]$ for the inner product in H.

So the mapping $S: D(A) \to H$ is densely defined and has a dense range, by construction. It then follows that the adjoint mapping $S^*: D(S^*) \to X^*$ is injective. Now, we claim that $R(S) \subset D(S^*)$, which implies, in particular, that S^* is densely defined. To prove the claimed inclusion let $Sx \in R(S)$, $x \in D(A)$; the linear functional $l: D(A) \to \boldsymbol{R}$ defined by $l(y) = [Sx, Sy]$ is bounded because

$$|[Sx, Sy]| = |(Ax, y)| \leqq \|Ax\| \cdot \|y\|.$$

So l can be extended to a bounded linear functional on X, which proves that $S(x) \in D(S^*)$ and $[Sx, Sy] = (S^*Sx, y)$ for all y in $D(A)$. On the other hand, $[Sx, Sy] = (Ax, y)$, by definition. It then follows from the density of $D(A)$ in X that $S^*S = A$, as we wanted to prove. Q.E.D.

2. **Proof of Theorem 1.** Let us denote the norm in H by $|\cdot|$, and let us define a functional $\Phi: D(\Phi) \to \boldsymbol{R}$, $D(\Phi) \subset H$, by

$$\Phi(u) = \tfrac{1}{2}|u|^2 + f(S^*u) \tag{5}$$

where S^* is the adjoint of the mapping S defined in Theorem 2. So $D(\Phi) = D(S^*)$. In view of the hypothesis on f we obtain $\Phi(u) \geqq \frac{1}{2}|u|^2 + \omega(\|S^*u\|)$. Now let $M = \inf\{\omega(r): r \geqq 0\}$. Thus $\Phi(u) \geqq \frac{1}{2}|u|^2 + M$ for all $u \in D(\Phi)$ and the functional Φ is bounded from below on $D(\Phi)$. Let

$$d = \inf\{\Phi(u): u \in D(\Phi)\}$$

and let $\{u_n\}$ be a minimizing sequence for Φ. It is clear that, for all n sufficiently large,

$$d + 1 \geqq \Phi(u_n) \geqq \tfrac{1}{2}|u_n|^2 + \omega(\|S^*u_n\|) > M. \tag{6}$$

It then follows from (6) that, the sequences $\{u_n\}$ and $\{S^*u_n\}$ are bounded. In view of the weak compactness of the ball in H and the weak*-sequential compactness of the ball in X^* we may assume (by going to a subsequence, if necessary) that there exist elements u_0 in H and y^* in X^* such that $u_n \rightharpoonup u_0$ in H and $S^*u_n \rightharpoonup^* y^*$ in X^*, where $\rightharpoonup$ denotes the weak convergence and $\rightharpoonup^*$ weak* convergence. Now we claim that $u_0 \in D(S^*)$. Indeed, all we have to do is to prove that the linear functional $l: D(A) \to \boldsymbol{R}$ defined by $l(x)=[u_0, Sx]$ is bounded. And this is so because

$$(7)\qquad \begin{aligned} &[u_0, Sx] = \lim[u_n, Sx] = \lim(S^*u_n, x) = (y^*, x), \quad \text{and} \\ &|[u_0, Sx]| \leqq \|y^*\| \cdot \|x\|; \end{aligned}$$

from (7) it follows that $S^*u_0=y^*$. Next we prove that $\Phi(u)$ assumes its infimum at u_0, i.e., $\Phi(u_0)=d$. Indeed, since the norm in H is weakly-lower-semicontinuous and f is weak*-lower-semicontinuous we obtain that

$$\tfrac{1}{2}|u_0|^2 + f(S^*u_0) \leqq \liminf\{\tfrac{1}{2}|u_n|^2 + f(S^*u_n)\} = \lim \Phi(u_n) = d.$$

Now, since Φ has a minimum at u_0, it follows that for all $u \in D(\Phi)$, we have

$$\lim_{t\to 0} \frac{\Phi(u_0 + tu) - \Phi(u_0)}{t} = 0$$

when such a limit exists. It is, however, easy to see that this limit exists for all $u \in D(\Phi)$, and we obtain

$$(8)\qquad [u_0, u] + (N(S^*u_0), S^*u) = 0.$$

We, now, claim that $N(S^*u_0) \in D(A)$ and, moreover, $ANS^*u_0= -S^*u_0$. This we can do if we prove that

$$(9)\qquad (Ax + S^*u_0, x - NS^*u_0) \geqq 0$$

for all $x \in D(A)$, in view of the maximal monotonicity of A. Using (8) first with $u=u_0$ and then with $\iota=Sx$ we obtain

$$-(N(S^*u_0), S^*u_0) = [u_0, u_0], \qquad -(N(S^*u_0), S^*Sx) = [u_0, Sx].$$

This reduces (9) to the inequality

$$(10)\qquad (Ax, x) + (S^*u_0, x) + [u_0, u_0] + [u_0, Sx] \geqq 0.$$

Now let $\{x_n\} \subset D(A)$ be such that $Sx_n \to u_0$ in H. It is then easy to see that (10) is obtained from the inequality $(Ax+Ax_n, x+x_n)= |Sx+Sx_n|^2 \geqq 0$ by taking the limit as $n\to\infty$.

Finally, from the relation $S^*u_0+ANS^*u_0=0$ it follows that $x^*=S^*u_0$ is a solution of $x^*+ANx^*=0$. Q.E.D.

3. **Equations in a Hilbert space**. We begin with some remarks on self-adjoint (not necessarily bounded) linear mappings in Hilbert spaces. If $A: D(A)\to H$, where $D(A)$ is a dense subspace of the Hilbert space H, is selfadjoint and monotone, then it follows that A is closed, densely defined and maximal monotone. So A is m-accretive in the sense of Kato [**6**]. Thus A has a square root $A^{1/2}$, which is also selfadjoint and m-accretive, see [**6**].

The next result is an immediate consequence of Theorem 1. However we include another proof in the hope that the ideas expounded there may be useful in other situations.

THEOREM 3. *Let $A:D(A)\to H$ be a selfadjoint monotone linear mapping in a real Hilbert space H. Let $f:H\to \mathbf{R}$ be a weakly Gateaux differentiable functional which is also weakly lower semicontinuous and is such that $f(x)\geqq a_1\|x\|^2+a_2\|x\|^\theta+a_3$ where $a_1>0$, $a_2\leqq 0$, $a_3\leqq 0$, $0<\theta<2$. Let N denote the weak-gradient of f.*

Then the equation $x+ANx=y$, for a given y in H, has at least one solution.

PROOF. We can see as in the Remark (iii) after Theorem 1 that there is no loss of generality in assuming that $y=0$. Define in $D(A^{1/2})$ the following inner-product $[x,y]=(x,y)+(A^{1/2}x, A^{1/2}y)$, which gives the norm $|x|=(\|x\|^2+\|A^{1/2}x\|^2)^{1/2}$.

The linear subspace $D(A^{1/2})$ with this new norm is complete and consequently is a Hilbert space which we denote by H'. In fact, if $\{u_n\}$ is a Cauchy sequence in H' then $\{u_n\}$ and $\{A^{1/2}u_n\}$ are both Cauchy sequences in H. So there are u, v in H such that $u_n\to u$ and $A^{1/2}u_n\to v$ in H. Since $A^{1/2}$ is closed we see that $u\in D(A^{1/2})$ and $v=A^{1/2}u$. So $u\in H'$ and clearly $u_n\to u$ in H'.

Now, consider the functional $\Phi:H'\to\mathbf{R}$ defined by $\Phi(u)=\frac{1}{2}\|u\|^2+f(A^{1/2}u)$. We shall prove, now, that Φ is weakly-lower-semicontinuous in H'. To do this we first prove that if $u_n\rightharpoonup u$ in H', then $u_n\rightharpoonup u$ in H and $A^{1/2}u_n\rightharpoonup A^{1/2}u$ in H. In fact,

$$\begin{aligned}[u_n, v] &= (u_n, v) + (A^{1/2}u_n, A^{1/2}v)\\ &= (u_n, v + Av) \quad \text{for all } v\in D(A^{1/2}).\end{aligned}\tag{11}$$

Since $D(A)\subset D(A^{1/2})$ and $R(I+A)=H$ by virtue of m-accretiveness of A it follows from (11) that $u_n\rightharpoonup u$ in H. Similarly, using the expression

$$[u_n, A^{1/2}v] = (u_n, A^{1/2}v) + (A^{1/2}u_n, Av) = (A^{1/2}u_n, v + Av)$$

we prove that $A^{1/2}u_n\rightharpoonup A^{1/2}u$ in H. Thus we have

$$\|u\| \leqq \liminf \|u_n\|, \qquad f(A^{1/2}u) \leqq \liminf f(A^{1/2}u_n),$$

whenever $u_n\rightharpoonup u$ in H'. These two inequalities then imply that $\Phi(u)\leqq \liminf \Phi(u_n)$ whenever $u_n\rightharpoonup u$ in H'. Thus Φ is weakly-lower-semicontinuous in H'.

Moreover, from the assumptions on f we have

$$\begin{aligned}\Phi(u) &\geqq \tfrac{1}{2}\|u\|^2 + a_1\|A^{1/2}u\|^2 + a_2\|A^{1/2}u\|^\theta + a_3\\ &\geqq \min(\tfrac{1}{2}, a_1)\,|u|^2 + a_2\|A^{1/2}u\|^\theta + a_3,\end{aligned}$$

which implies that $\Phi(u)\to+\infty$ as $|u|\to+\infty$. Thus there exists a $u_0\in H'$ where Φ assumes a minimum. At this point the gradient of Φ vanishes and this gives

$$(u_0, v) + (NA^{1/2}u_0, A^{1/2}v) = 0.$$

Since this holds for all $v\in D(A^{1/2})$ we obtain that $NA^{1/2}u_0\in D(A^{1/2})$ and $A^{1/2}NA^{1/2}u_0=-u_0$. Finally, setting $x_0=A^{1/2}u_0$, we obtain $x_0+ANx_0=0$. Q.E.D.

THEOREM 4. *Let $A: D(A)\to H$ be a selfadjoint monotone linear mapping in a real Hilbert space H. Let $f: H\to \boldsymbol{R}$ be a weakly Gateaux differentiable functional which is also weakly-lower-semicontinuous. Let N denote its weak-gradient. Assume that there is an $m>0$ such that $(Nu, u)\geqq m\|u\|^2$ for all u in H. Then the equation $u+ANu=0$ has at least one solution in H.*

The proof of this theorem follows the same lines as the proof of Theorem 3.

BIBLIOGRAPHY

1. F. E. Browder, *Nonlinear functional analysis and nonlinear integral equations of Hammerstein and Urysohn type*, Contributions to Nonlinear Functional Analysis, Edited by E. H. Zarantonello, Academic Press, New York, 1971.

2. F. E. Browder and C. P. Gupta, *Monotone operators and nonlinear integral equations of Hammerstein type*, Bull. Amer. Math. Soc. **75** (1969), 1347–1353. MR **40** #3381.

3. D. G. de Figueiredo and C. P. Gupta, *Nonlinear integral equations of Hammerstein type involving unbounded monotone linear mappings*, J. Math. Anal. Appl. **39** (1972), 37–48.

4. ———, *Nonlinear integral equations of a Hammerstein type with indefinite linear kernel in a Hilbert space*, Indag. Math. **34** (1972), 335–344.

5. C. P. Gupta, *On existence of solutions of nonlinear integral equations of Hammerstein type in a Banach space*, J. Math. Anal. Appl. **32** (1970), 617–620. MR **42** #3630.

6. T. Kato, *Perturbation theory for linear operators*, Die Grundlehren der math. Wissenschaften, Band 132, Springer-Verlag, New York, 1966. MR **34** #3324.

7. I. M. Lavrent'ev, *Solvability of nonlinear equations*, Dokl. Akad. Nauk SSSR **175** (1967), 1219–1221=Soviet Math. Dokl. **8** (1967), 993–996. MR **36** #6990.

8. M. M. Vaĭnberg and I. M. Lavrent'ev, *Equations with monotonic and potential operators in Banach spaces*, Dokl. Akad. Nauk SSSR **187** (1969), 711–714=Soviet. Math. Dokl. **10** (1969), 907–910. MR **42** #5127.

9. M. M. Vaĭnberg, *Variational methods for the study of nonlinear operators*, GITTL, Moscow, 1956; English transl., Holden-Day, San Francisco, Calif., 1964. MR **19**, 567; **31** #638.

DEPARTMENT OF MATHEMATICS, UNIVERSITY OF ILLINOIS, CHICAGO, ILLINOIS 60680 (Current address of D. G. deFigueiredo)

DEPARTMENT OF MATHEMATICS, UNIVERSITY OF BRASILIA, BRASILIA, BRAZIL

DEPARTMENT OF MATHEMATICS, NORTHERN ILLINOIS UNIVERSITY, DEKALB, ILLINOIS 60115 (Current address of C. P. Gupta)

which implies that [illegible]. Thus there exists [illegible] at this point. The gradient of [illegible] and this gives

[illegible]

Since this holds for [illegible] we obtain that [illegible] and [illegible]. Finally, setting [illegible] we obtain [illegible].

Q.E.D.

Theorem 4. [illegible]

The proof of this Theorem follows the same lines as the proof of Theorem [illegible].

References

1. [illegible] Contributions to Nonlinear Functional Analysis [illegible] Academic Press, New York, 1971.
2. [illegible]
3. [illegible]
4. [illegible]
5. [illegible]
6. [illegible]
7. [illegible]
8. [illegible]
9. [illegible]

[illegible]

[illegible]

[illegible]

THE DIRICHLET PROBLEM FOR NONLINEAR ELLIPTIC EQUATIONS: A HILBERT SPACE APPROACH

by

DJAIRO GUEDES DE FIGUEIREDO*

Department of Mathematics
Universidade de Brasília
Brasília, D. F., Brasil

1. INTRODUCTION

Let Ω be a bounded domain in R^N, and

(1) $$Lu = \sum_{\substack{|\alpha|\le m \\ |\beta|\le m}} (-1)^{|\beta|} D^\beta (a_{\alpha\beta}(x) D^\alpha u)$$

be a uniformly strongly elliptic operator acting on functions defined in Ω. Besides the existence theory of generalized solutions for the Dirichlet problem associated with L, we shall also use the regularity theory of solutions. For that matter, we will assume some regularity conditions on the coefficients of L and on the boundary $\partial\Omega$ of Ω. Let $F(x,\eta)$ be a real-valued function defined on $\Omega \times R^M$, where M is the number of all derivatives D^α, for $0 \le |\alpha| \le 2m - 1$.

---

* The material presented here corresponds roughly to the lectures given by the author at Tulane, in a program sponsored by the Ford Foundation.

Originally published in: Lecture Notes in Mathematics, Vol. 446 (1975), 144-165,

Reprint by Springer-Verlag Berlin Heidelberg 2013,
DOI: 10.1007/978-3-319-02856-9_6

In this paper we study a Hilbert space theory for the Dirichlet problem for the equation

$$(2) \qquad Lu = F(x, Du(x)), \quad \text{in } \Omega,$$

where $Du(x)$ stands for the M-vector whose components are the derivatives $D^{\alpha}u(x)$ of u, for $0 \le |\alpha| \le 2m-1$. As usual the Dirichlet boundary conditions are $\frac{\partial^j u}{\partial \nu^j} = 0$ on the boundary $\partial\Omega$, for all $0 \le j \le m-1$, where $\partial/\partial\nu$ stands for the outward normal derivative at a point on $\partial\Omega$. We will consider the generalized Dirichlet problem

$$(3) \quad a[u,\phi] = (F(\cdot,Du), \phi), \quad \text{for all } \phi \in C_o^{\infty}(\Omega),$$

where $a[\cdot,\cdot]$ is the Dirichlet form associated to L, and $(\cdot,\cdot)$ stands for the L_2-inner product. A solution of (3) will be a function in some appropriate Sobolev space, namely $H^{2m}(\Omega) \cap H_o^m(\Omega)$.

In Section 2 we collect some known results on the theory of the Dirichlet problem for linear elliptic equations, stating them in the form that they are needed here. The short Section 3 contains a simple proof that the Niemytski mapping $Nu = F(x, Du)$ is continuous from H^{2m-1} to L^2.

In Section 4 we consider the Dirichlet problem when the nonlinearity $F(x, Du)$ grows at most linearly in u, and in a certain sense is bounded away from the eigenvalues of L. For that reason, we say that, in this case, there is no resonance. The main theorem of Section 4 has been proved by the author in [2], and generalizes results of Landesman-Lazer [8], Adams [1] and Schultz [12]. The proof of Theorem 2 presented here is simpler than the one given in [2], and follows a suggestion of F. E. Browder.

In Section 5 we consider nonlinearities $F(x, Du)$ that grow like powers k $(0 \le k < 1)$ of u and its derivatives of order $\le 2m - 1$. In this case we do not assume that $F(x, Du)$ stays away from the eigenvalues of L. For that reason we say that there is resonance. The result presented in this section generalizes a result of [2], and uses the regularity theory, as suggested by H. Brezis and L. Nirenberg. Theorem 3 generalizes also previous results of Nečas [10], Fučik-Kučera-Nečas [5], Hess [6], Landesman-Lazer [8], Nirenberg [11], Williams [14] and the author [3].

2. LINEAR ELLIPTIC EQUATIONS

Let us consider a linear differential operator

$$Lu = \sum_{|\alpha|,|\beta|\le m} (-1)^{|\beta|} D^{\beta}(a_{\alpha\beta}(x) D^{\alpha} u) \tag{2.1}$$

acting on real-valued functions $u(x)$ defined on a bounded domain Ω in R^N, whose boundary $\partial\Omega$ is assumed to be of class C^{2m}. No regularity of the boundary would be necessary if we were to use only the existence of H_o^m-solutions for the Dirichlet problem, as we did in [2]. The regularity of $\partial\Omega$ will be required to improve the class of differentiability of these solutions through the so-called regularity theory. The following assumptions will be made on L throughout this paper.

(L - 1) *The coefficients* $a_{\alpha\beta}$ *are real-valued functions defined in* $\bar{\Omega}$ *and* $a_{\alpha\beta} \in C^{|\beta|}(\bar{\Omega})$, *for all*

$0 \leq |\alpha|, \quad |\beta| \leq m.$

(L - 2) L *is uniformly strongly elliptic, i.e., there exists a constant* $c > 0$ *such that*

$$(2.2) \qquad \sum_{|\alpha|=|\beta|=m} a_{\alpha\beta}(x)\xi^{\alpha}\xi^{\beta} \geq c|\xi|^{2m}$$

for all $\xi \in R^N$ *and all* $x \in \overline{\Omega}$.

The Dirichlet from associated with L is given by

$$(2.3) \quad a[\psi,\phi] = \sum_{|\alpha|,|\beta|\leq m} \int_{\Omega} a_{\alpha\beta}(x) D^{\alpha}\psi(x) D^{\beta}\phi(x)\,dx,$$

which, in view of (L-1), is a bounded bilinear form in the space $C_o^{\infty}(\Omega)$ of the infinitely differentiable functions with compact support in Ω, endowed with the m-norm. If k is a non-negative integer and ϕ is a continuously differentiable function up to order k, the k-norm is defined by

$$(2.4) \qquad \|\phi\|_k = \left[\sum_{|\alpha|\leq k} \int_{\Omega} |D^{\alpha}\phi|^2\right]^{1/2}.$$

Let us denote by H_o^m the space obtained by completion of $C_o^{\infty}(\Omega)$ with respect to the m-norm. H_o^m is a Hilbert space, $H_o^m \subset L^2$, and $H_o^o = L^2$. The Dirichlet form (2.3) can be extended by H_o^m.

Using (L - 1) and (L - 2) we can prove Gårding's inequality: there exist real constants $c_o > 0$ and k_o such that, for all $\phi \in H_o^m$, one has

$$(2.5) \qquad a[\phi,\phi] \geq c_o\|\phi\|_m^2 - k\|\phi\|_o^2.$$

We remark that the whole power of (L - 1) is not used in the derivation of Gårding's inequality. Namely (2.5) holds under the following weaker assumptions on the coefficients: (i) $a_{\alpha\beta} \in L^{\infty}(\Omega)$, for $0 \leq |\alpha|, |\beta| \leq m$, (ii) $a_{\alpha\beta}$, for $|\alpha| = |\beta| = m$, are uniformly continuous in Ω. See Friedman [4] for the proof of Gårding's inequality and other results stated in this section.

The classical Dirichlet problem is the following. Given a real-valued function $f: \Omega \to \mathbb{R}$, to find $u \in C^{2m}(\Omega) \cap C^{m-1}(\overline{\Omega})$ such that

$$Lu = f, \quad \text{in } \Omega$$

$$\frac{\partial^j u}{\partial \nu^j} = 0, \quad \text{on } \partial\Omega, \quad \text{for } j = 0,1,\ldots,m-1.$$

It is known that this problem does not always have a solution. Indeed there are continuous functions f such that $\Delta u = f$, in Ω, and $u = 0$, on $\partial\Omega$, has no solution $u \in C^2(\Omega) \cap C^0(\overline{\Omega})$. However, the existence of solution is guaranteed whenever f is Hölder-continuous. It is possible to deal directly with the classical Dirichlet problem, using Schauder estimates and working in spaces of Hölder continuous functions. It became clear in the 50's that a better approach to treat the Dirichlet problem is through a Hilbert space formulation of the problem. In this way, a generalized Dirichlet problem is set and solved using essentially two tools, Gårding's inequality and the Lax-Milgram lemma. The solution so obtained is in the Sobolev space H_0^m. Then comes the much harder part, called the regularity theory, which consists in proving that this solution is actually in a Sobolev space H^k, for some $k > m$. Once this done, one can obtain information about continuity and differentiability of this solution, via the Sobolev imbedding theorem.

The generalized Dirichlet problem is the following. Given a real valued function $f \in L^2(\Omega)$, to find $u \in H_o^m$ such that

$$a[u,\phi] = (f,\phi), \quad \text{for all} \quad \phi \in H_o^m, \tag{2.6}$$

where $(\, , \,)$ stands for the L^2-inner product.

The following result is a complete answer to the existence and uniqueness questions for the generalized Dirichlet problem.

PROPOSITION 1. (Fredholm Alternative) *Assume* (L-1) *and* (L-2) *above. Then, one and only one of the following possibilities holds true:*

(i) There exists a unique $u \in H_o^m$, *for each given* $f \in L^2$, *such that* (2.6) *holds. In particular* $a[u,\phi] = 0$, *for all* $\phi \in H_o^m$, *implies* $u = 0$.

(ii) There exists $0 \neq u \in H_o^m$ *such that*

$$a[u,\phi] = 0, \quad \text{for all} \quad \phi \in H_o^m. \tag{2.7}$$

Furthermore, when (i) holds, the solution operator $A: L^2 \to H_o^m$, *defined by* $Af = u$, *is linear and bounded. In the case when (ii) holds, the set of solutions of* (2.7) *is a finite dimensional subspace of* H_o^m, *whose dimension is the same as the dimension of*

$$N^* = \{v \in H_o^m : a[\phi,v] = 0, \quad \text{for all} \quad \phi \in H_o^m\}.$$

Moreover, still in case (ii) equation (2.6) *has a solution if and only if* $(f,v) = 0$, *for all* $v \in N^*$.

The eigenvalue problem associated with the generalized

Dirichlet problem consists in the following. To find λ, such that for some $0 \neq u \in H_o^m$ one has

(2.8) $\quad a[u,\phi] = \lambda(u,\phi), \quad$ for all $\phi \in H_o^m$.

λ is an eigenvalue, u is an eigenfunction corresponding to the eigenvalue λ, and the set of all eigenfunctions corresponding to λ form the eigenspace corresponding to the eigenvalue λ.

The differential operator L is said to be symmetric if $a_{\alpha\beta} = a_{\alpha\beta}$, for all $0 \leq |\alpha|, \; |\beta| \leq m$. In this case $a[\phi,\psi] = a[\psi,\phi]$. Now we state the main facts about the eigenvalue problem for a symmetric operator L.

PROPOSITION 2. *Assume* (L-1) *and* (L-2) *above, and that* L *is symmetric. Then:*

(a) The eigenvalues of (2.8) *are all real, they form a sequence* (λ_n) *such that* $\lambda_n \to +\infty$, *and the corresponding eigenspaces are finite dimensional.*

(b) If part (i) of the alternative holds, then the operator A *thought as an operator from* L^2 *to* L^2 *is symmetric, bounded and its norm is given by*

(2.9) $\quad \|A\|_{L^2\to L^2} = \max\{|\lambda|^{-1} : \lambda$ *an eigenvalue of* (2.8)$\}$.

Now we state the regularity result needed for the present paper. Let us first introduce the Sobolev spaces H^k, where k is some fixed non-negative integer. Let $\hat{C}^k(\Omega)$ denote the space of all functions $u: \Omega \to \mathbb{R}$, which are of class C^k and $\|u\|_k < \infty$, where this norm is defined in (2.4). Now define H^k as the completion of $\hat{C}^k(\Omega)$ with respect to this norm. When $\partial\Omega$ is of class C^k, this definition is equivalent to taking the completion, in the norm (2.4), of the space $C^k(\overline{\Omega})$ of all functions $u: \overline{\Omega} \to \mathbb{R}$, which are continuously differentiable up to order k in $\overline{\Omega}$.

PROPOSITION 3. *Assume* (L-1) *and* (L-2) *above, and that* $\partial\Omega$ *is of class* C^{2m}. *Then, if* $u \in H_o^m$ *is a solution of* (2.6), *it follows that* $u \in H^{2m}$ *and*

$$\|u\|_{2m} \leq c_1(\|f\|_o + \|u\|_o). \tag{2.10}$$

Thus, when part (i) of the Alternative holds the solution operator A maps L^2 onto $H_o^m \cap H^{2m}$. Moreover, A viewed as an operator from L^2 to H^{2m} is bounded. Indeed, it follows from (2.10) that

$$\|Af\|_{2m} \leq c_1(\|f\|_o + \|Af\|_o),$$

and, from Proposition 1, $\|Af\|_o \leq \|Af\|_m \leq \text{const}\|f\|_o$. Consequently $\|Af\|_{2m} \leq \text{const}\|f\|_o$ for all $f \in L^2$.

3. NIEMYTSKI MAPPINGS

We shall consider perturbations of the linear elliptic equations studied in Section 2 by a nonlinear term of the form $F(x, Du)$ where Du stands for the M-vector whose components are $D^\alpha u$, for all $0 \leq |\alpha| \leq 2m - 1$. The following basic assumption is made on F:

(F-1) *The function* $F: \Omega \times R^M \to R$ *satisfies Carathéodory's conditions, i.e., (i) for each fixed* $\eta \in R^M$, *the function* $x \mapsto F(x,\eta)$ *is measurable in* Ω, *(ii) for fixed* $x \in \Omega$ *(a.e.), the function* $\eta \mapsto F(x,\eta)$ *is continuous in* R^M.

The above assumption (F-1) implies that if the mapping $H: \Omega \to R^M$ is measurable, then the function $x \mapsto F(x, H(x))$ is also measurable, see Krasnoselski [7] or Vainberg [13]. We shall need the following result, whose proof can also be seen in the afore-mentioned books. However, we will include here a proof, which seems to be of some interest.

PROPOSITION 4. *Suppose that a function* $F: \Omega \times R^M \to R$ *satisfies Carathéodory's conditions* (F-1), *and moreover, that there exist constants* $b_\alpha \geq 0$ *and a function* $b \in L^2$ *such that*

$$(3.1) \qquad |F(x,\eta)| \leq \sum_{|\alpha| \leq 2m-1} b_\alpha |\eta_\alpha| + b(x)$$

for all $\eta \in R^M$ *and* $x \in \Omega$ *(a.e.). Then the Niemytski mapping* N *given by* $(Nu)(x) = F(x, Du(x))$, *where* $Du(x)$ *is the* M-*vector whose components are* $D^\alpha u(x)$, *is bounded and continuous from* H^{2m-1} *to* L^2.

PROOF. It follows from Minkowski inequality that

$$\|Nu\|_o \leq \sum_{|\alpha| \leq 2m-1} b_\alpha \|D^\alpha u\|_o + \|b\|_o$$

which implies, via the Cauchy-Schwarz inequality, that

$$(3.2) \qquad \|Nu\|_o \leq B\|u\|_{2m-1} + \|b\|_o ,$$

where B is a constant depending only on the b_α's. Thus (3.2) implies that N is a mapping from H^{2m-1} into L^2, and also that such a mapping is bounded. To prove the continuity of N we use the following characterization of L^1-convergence: "Let $f_n: \Omega \to R$ be a sequence of measurable functions which converges almost everywhere to a function f. Then f_n converges to f in L^p, $p \geq 1$, if and only if, given any subsequence $(f_{n'})$ of (f_n), there exists a further subsequence $(f_{n''})$ of $(f_{n'})$ and an L^p-function g (depending on this last subsequence) such that $|f_{n''}(x)| \leq g(x)$, for all n'' and $x \in \Omega$ (a.e.)." Now let $u_n \in H^{2m-1}$ be a sequence converging to u in the H^{2m-1}-norm, and we claim that $F_n(x) = F(x, Du_n(x))$ converges in L^2 to $F(x) = F(x, Du(x))$. To apply the result stated above, we take a subsequence $(F_{n'})$ of (F_n). Then, using the above result, we take a subsequence $(u_{n''})$ of $(u_{n'})$ such that $|D^\alpha u_{n''}(x)| \leq g(x)$, for $0 \leq |\alpha| \leq 2m - 1$, where $g \in L^2$. Using now inequality (3.1) we get

$$|F_{n''}(x)| = |F(x, Du_{n''}(x))| \leq \left(\sum_{|\alpha| \leq 2m-1} b_\alpha\right) g(x) + b(x),$$

where the last term of the inequality is an L^2-function. So we apply the result above once more to get the convergence of F_n to F in the L^2-sense.

We recall that a mapping $N: X \to Y$ from a Banach space X into another Banach space Y is <u>compact</u> if it is continuous and takes bounded sets of X into relatively compact sets of Y.

COROLLARY. *Same hypothesis as in Proposition* 4. *Then the Niemytski mapping* N *is a compact mapping from* H^{2m} *to* L^2.

PROOF. This is a direct consequence of Proposition 4 and the fact that H^{2m} is compactly imbedded in H^{2m-1}.

The generalized Dirichlet problem for

$$Lu = F(x, Du) \tag{3.3}$$

where F satisfies (F-1) and an inequality of type (3.1) above, consists in finding $u \in H^{2m} \cap H_o^m$ such that

$$a[u,\phi] = (Nu,\phi), \quad \text{for all } \phi \in H_o^m. \tag{3.4}$$

4. THE DIRICHLET PROBLEM WHEN THERE IS NO RESONANCE

First we consider the generalized Dirichlet problem (3.4) in the case when there is no resonance. This terminology is justified by the fact that in the differential equation (3.3) the perturbation $F(x, Du)$ "stays away" from the eigenvalues of L, in a sense that will be made precise below.

THEOREM 1. *Suppose that the differential operator* L *is symmetric and satisfies assumptions* (L-1) *and* (L-2) *above. Let* $\lambda_1 < \lambda_2$ *be two consecutive eigenvalues of the eigenvalue problem* (2.8). *Let* λ *and* γ *be such that* $\lambda_1 < \lambda < \lambda_2$ *and* $0 \le \lambda < \min\{\lambda-\lambda_1, \lambda_2-\lambda\}$. *Assume that the function* F *satisfies* (F-1) *and the following assumption:*

(F-2) *There exists constants* $0 \le \kappa < 1$, $c_\alpha \ge 0$, *for* $|\alpha| \le 2m-1$, *and an* L^2*-function* $b(x)$ *such that*

$$|F(x,\eta) - \lambda\eta_o| \le \gamma|\eta_o| + \sum_{|\alpha|\le 2m-1} c_\alpha|\eta_\alpha|^\kappa + b(x), \tag{4.1}$$

for all $\eta \in R^M$ *and* $x \in \Omega$ *(a.e.). Here* η_o *stands for the first component of the vector* η*, which corresponds to* u *in the vector* Du.

Then there exists $u \in H^{2m} \cap H_o^m$ *such that*

(4.2) $\quad a[u,\phi] = (Nu,\phi)$, *for all* $\phi \in H_o^m$.

PROOF. First of all we observe that in view of the fact that Ω is bounded, it follows from (4.1) that the function $F(x,\eta)$ satisfies an inequality like (3.1). So Proposition 4 and its corollary apply. Now since (4.2) is equivalent to

(4.3) $\quad a[u,\phi] - \lambda(u,\phi) = (Nu - \lambda u,\phi)$, for all $\phi \in H_o^m$,

the present theorem will follow as a consequence of Theorem 2, below, and the following remarks:

(i) The problem

(4.4) $\quad a[u,\phi] - \lambda(u,\phi) = (f,\phi)$, for all $\phi \in H_o^m$

is uniquely solvable, for each $f \in L^2$; see Propositions 1 and 2, Section 2.

(ii) The mapping $A: L^2 \to H^{2m}$ defined by $Af = u$, where u is the solution of (4.4), is linear and continuous, and, moreover, its range $\mathcal{R}(A) \subset H_o^m$, and its norm as mapping from L^2 to L^2 can be estimated as follows

$$\|A\|_{L^2 \to L^2} \leq \max\left\{\frac{1}{\lambda-\lambda_1}, \frac{1}{\lambda_2-\lambda}\right\} < \frac{1}{\gamma}.$$

The next result will be stated separately because it holds without the symmetry of L.

THEOREM 2. *Suppose that the differential operator* L *satisfies assumptions* (L-1) *and* (L-2), *and that part (i) of the Alternative in Proposition* 1 *holds. Assume that the function* F *satisfies* (F-1) *and*

(F-3) *There exist constants* $0 \le c' < \|A\|^{-1}_{L^2 \to L^2}$, $0 \le \kappa < 1$, $c_\alpha \ge 0$, *for* $|\alpha| \le 2m - 1$, *and an* L^2-*function* $b(x)$, *such that*

$$|F(x,\eta)| \le c'|\eta_0| + \sum_{|\alpha| \le 2m-1} c_\alpha |\eta_\alpha|^\kappa + b(x) \tag{4.5}$$

for all $\eta \in R^M$ *and* $x \in \Omega$ *(a.e.).*

Then there exists $u \in H^m_o \cap H^{2m}$ *such that*

$$a[u,\phi] = (Nu,\phi), \quad \textit{for all} \quad \phi \in H^m_o. \tag{4.6}$$

PROOF. In view of Proposition 1, the problem of solving (4.6) is equivalent to finding $u \in H^{2m}$ such that

$$u = ANu. \tag{4.7}$$

Observe that (4.7) says that $u \in \mathcal{R}(A)$, and so any solution of (4.7) is also in H^m_o. Now Propositions 3 and 4 imply that $AN: H^{2m} \to H^{2m}$ is compact. So the theorem will be proved if we show that AN has a fixed point. First we observe that (4.7) has a solution whenever the equation

$$v = NAv \tag{4.8}$$

has a solution $v \in L^2$. Indeed, if v is a solution of (4.8) then $u = Av$ is in H^{2m} and u is a solution of (4.7). To prove that (4.8) has a solution we use Schauder fixed point theorem. So we show that there exists a ball of radius R about the origin in L^2 which is mapped into itself by NA. Using (4.5) and Hölder's inequality we get

$$(4.9) \qquad \|Nu\|_o \le c' \|u\|_o + c'' \|u\|_{2m}^{\kappa} + \|b\|_o ,$$

for all $u \in H^{2m}$, where c'' is a constant depending on the c_α's and on κ. On the other hand it follows from (2.10) that

$$(4.10) \quad \|Av\|_{2m} \le c_1(\|v\|_o + \|Av\|_o) \le c_1(1 + \|A\|_{L^2 \to L^2})\|v\|_o$$

for all $v \in L^2$. Now using (4.9) we can obtain the following estimate:

$$(4.11) \quad \|NAv\|_o \le c' \|Av\|_o + c'' \|Av\|_{2m}^{\kappa} + \|b\|_o$$

for all $v \in L^2$. Using (4.10) in (4.11) we conclude that there are constants c_2 and c_3 such that

$$\|NAv\|_o \le c' \|A\|_{L^2 \to L^2} \|v\|_o + c_2 \|v\|_o^{\kappa} + c_3$$

for all $v \in L^2$. Since $c' \|A\|_{L^2 \to L^2} < 1$ and $0 \le \kappa < 1$, we see that, for R sufficiently large

$$\|v\|_o \le R \text{ implies } \|NAv\|_o \le R.$$

Thus the proof of Theorem 2 is complete.

5. THE DIRICHLET PROBLEM WHEN THERE IS RESONANCE

Now we consider the case when the perturbation $F(x, Du)$ does not stay properly between the eigenvalues of L. In fact it could touch at most one eigenvalue.

More precisely, we assume

(F-4) *Let* λ_1 *be an eigenvalue of the eigenvalue problem* (2.8), *where* L *is assumed to be symmetric. There exist constants* $c_\alpha \geq 0$, $0 \leq \kappa < 1$ *and an* L^2-*function* $d(x)$ *such that*

$$(5.1) \quad |F(x,\eta) - \lambda_1\eta_o| \leq \sum_{|\alpha|\leq 2m-1} c_\alpha|\eta_\alpha|^\kappa + d(x)$$

for all $\eta \in R^M$ *and* $x \in \Omega$ *(a.e.).*

Compare this condition with (F-2): Observe that if λ_1 and λ_2 are consecutive eigenvalues of the eigenvalue problem (2.8), and we take λ, such that $\lambda_1 < \lambda < (\lambda_1 + \lambda_2)/2$, then inequality (5.1) implies

$$(5.2) \quad |F(x,\eta) - \lambda\eta_o| \leq (\lambda-\lambda_1)|\eta_o| + \sum_{|\alpha|\leq 2m-1} c_\alpha|\eta_\alpha|^\kappa + d(x).$$

So, it is $\lambda - \lambda_1$ that corresponds to the constant γ of (4.1), and, unlike γ, it cannot be taken strictly smaller than $\min\{\lambda - \lambda_1, \lambda_2 - \lambda\}$, no matter how we choose λ.

THEOREM 5-3. *Suppose that the operator* L *is symmetric and satisfies conditions* (L-1) *and* (L-2). *Assume that* F *satisfies* (F-1), (F-4) *and the two additional conditions below:*

(F-5) *For each* $\eta \in S_1 = \{\eta \in R^M : \sum_{|\alpha|\leq 2m-1} |\eta_\alpha|^2 = 1\}$,

and $x \in \Omega$ *(a.e.), the limit below exists*

$$(5.3)\qquad \lim_{n\to+\infty} \frac{F\left(x,\, r_n\eta^{(n)}\right) - \lambda_1 r_n \eta_o^{(n)}}{r_n^{\kappa}} = h(x,\, \eta),$$

for all sequences (r_n), $r_n \in R$ *and* $(\eta^{(n)})$, $\eta^{(n)} \in S_1$, *such that* $r_n \to +\infty$ *and* $\eta^{(n)} \to \eta \in S_1$.

(F-6) *For all* $v \in W = \left\{v \in H_o^m : a[v,\phi] = \lambda_1(v,\phi),\right.$ *for all* $\left.\phi \in H_o^m\right\}$, *with* $\|v\|_{2m} = 1$, *one has*

$$(5.4)\qquad \int_\Omega h\left(x,\, \frac{Dv(x)}{|Dv(x)|}\right)|Dv(x)|^{\kappa} v(x)\,dx > 0,$$

where $h(x,\eta)$ *is a real measurable function defined in* $\Omega \times S_1$, *and* $|h(x,\eta)| \le g(x)$, *for* $x \in \Omega$ *(a.e.) and* $\eta \in S_1$, *with a* $g \in L^{2/(1-\kappa)}(\Omega)$.

Then there exists $u \in H_o^m \cap H^{2m}$ *such that*

$$(5.5)\qquad a[u,\phi] = (Nu,\phi), \quad \textit{for all } \phi \in H_o^m.$$

REMARKS. 1) If F depends only on x and u, conditions (F-4), (F-5) and (F-6) take the following simpler forms.

$(F\text{-}4)_o$ *There exist constants* $c > 0$, $0 \le \kappa < 1$ *and an* L^2*-function* $d(x)$ *such that*

$$|F(x,s) - \lambda_1 s| \le c|s|^{\kappa} + d(x)$$

for all $s \in R$ *and* $x \in \Omega$ *(a.e.)*.

$(F\text{-}5)_o$ *There exist functions* $h_+ \in L^{2/(1-\kappa)}$ *and* $h_- \in L^{2/(1-\kappa)}$ *such that*

$$\lim_{s\to\pm\infty} \frac{F(x,s) - \lambda_1 s}{|s|^{\kappa}} = h_{\pm}(x)$$

for $x \in \Omega$ *(a.e.).*

$(F\text{-}6)_0$ *For all* $v \in W = \{v \in H_0^m : a[v,\phi] = \lambda_1(v,\phi),$ *for all* $\phi \in H_0^m\}$, *with* $\|v\|_{2m} = 1$ *one has*

$$\int_{v>0} h_+|v|^{1+\kappa} - \int_{v<0} h_-|v|^{1+\kappa} > 0.$$

These are the conditions we have given in [2]. We remark that assumption (F-4) of [2] is superfluous.

2) If F depends only on x and u, and $\kappa = 0$, conditions $(F\text{-}4)_0$, $(F\text{-}5)_0$ and $(F\text{-}6)_0$ coincide with the ones given by Nirenberg [11], although he considers more general linear operators L.

3) If F depends only on u and $\kappa = 0$, conditions $(F\text{-}4)_0$, $(F\text{-}5)_0$ and $(F\text{-}6)_0$ coincide with the ones given by Landesman-Lazer [9], for second order symmetric operators L. The condition $\kappa = 0$ implies that the function $F(u)$ is bounded and continuous for $-\infty < u < \infty$. This condition has been assumed also in the papers of Hess [6], Nečas [10] and Williams [14].

4) Nonlinearities F depending on derivatives of u have been considered before in [11], Fučik-Kučera-Nečas [5], and the author [3]. The conditions in [11] seem hard to check.

PROOF. Consider the approximate equations, for each positive integer n,

(5.6) $a[u,\phi] - \frac{1}{n}(u,\phi) = (Nu,\phi)$, for all $\phi \in H_0^m$.

Let λ_2 be first eigenvalue of (2.8) larger than λ_1. So $\lambda_1 - \frac{1}{n}$ and $\lambda_2 - \frac{1}{n}$ are consecutive eigenvalues of the eigenvalue problem

$$a[u,\phi] - \frac{1}{n}(u,\phi) = \mu(u,\phi), \quad \text{for all} \quad \phi \in H_o^m.$$

For n sufficiently large, $\lambda_1 - \frac{1}{n} < \lambda_1 < \lambda_2 - \frac{1}{n}$. So we see, using (F-4), that for equation (5.6) condition (F-2) holds with $\lambda = \lambda_1$ and $\gamma = 0$. Thus Theorem 1 applies to (5.6) and we obtain $u_n \in H_o^m \cap H^{2m}$ such that

$$(5.7) \quad a[u_n, \phi] - \frac{1}{n}(u_n, \phi) = (Nu_n, \phi), \quad \text{for all} \quad \phi \in H_o^m.$$

Now we claim that $||u_n||_{2m}$ is bounded. Assuming for the moment that this has been established, the proof of the theorem is completed as follows. Selecting subsequences, if necessary, we can assume that $u_n \rightharpoonup u$ in H^{2m}, $u_n \to u$ in H_o^m, $Nu_n \rightharpoonup g$ in L^2. So passing to the limit in (5.7) we get

$$(5.8) \qquad a[u,\phi] = (g,\phi), \quad \text{for all} \quad \phi \in H_o^m.$$

From (5.7) and (5.8) we get

$$(5.9) \quad a[u_n - u, \phi] = (Nu_n - g + \frac{1}{n}u_n, \phi), \quad \text{for all} \quad \phi \in H_o^m.$$

Using (2.10) we obtain from (5.9)

$$||u_n - u||_{2m} \leq c_1(\,||Nu_n - g + \frac{1}{n}u_n||_o + ||u_n - u||_o)$$

which implies that $u_n \to u$ in H^{2m}. So $Nu_n \to Nu$ in L^2, and then $g = Nu$. Thus (5.8) gives

$$a[u,\phi] = (Nu,\phi), \quad \text{for all } \phi \in H_o^m,$$

and then the theorem would be proved. So all we have to do is to prove that $(\|u_n\|_{2m})$ is a bounded sequence. Let us assume by contradiction that $\|u_n\|_{2m} \to \infty$. Let $v_n = u_n / \|u_n\|_{2m}$. Taking subsequences, if necessary, we may assume that there exist $v \in H_o^m \cap H^{2m}$ such that

$$v_n \rightharpoonup v \text{ in } H^{2m}, \quad v_n \to v \text{ in } H_o^m,$$

$$D^\alpha v_n \to D^\alpha v \text{ in } \Omega \text{ (a.e.)}$$

for all $0 \le |\alpha| \le 2m - 1$. From (5.7) we have

$$(5.10) \quad a[v_n, \phi] - \lambda_1 (v_n, \phi) - \frac{1}{n}(v_n, \phi) = \left(\frac{Nu_n - \lambda_1 u_n}{\|u_n\|_{2m}}, \phi \right),$$

for all $\phi \in H_o^m$. On the other hand, (F-4) and Hölder's inequality implies that there is a constant c_4, such that

$$(5.11) \quad \|Nu_n - \lambda_1 u_n\|_o \le c_4 \|u\|_{2m-1}^{\kappa} + \|d\|_o .$$

Now passing to the limit in (5.10) and using (5.11) we have

$$(5.12) \quad a[v,\phi] - \lambda_1 (v,\phi) = 0, \quad \text{for all } \phi \in H_o^m,$$

i.e. $v \in W$. Now we proceed to prove that $v \neq 0$, which is a consequence of the fact that $v_n \to v$ in H^{2m}. And this last fact is readily proved in the following way. It follows from (5.10) and (5.12) that

$$a[v_n - v, \phi] = \left(\frac{1}{n} v_n + \frac{Nu_n - \lambda_1 u_n}{\|u_n\|_{2m}} + \lambda_1 (v_n - v), \phi \right)$$

which implies, in view of (2.10), that

$$\|v_n - v\|_{2m} \le c_1 \left(\left\| \frac{1}{n} v_n + \frac{Nu_n - \lambda_1 u_n}{\|u_n\|_{2m}} + \lambda_1 (v_n - v) \right\|_o + \|v_n - v\|_o \right).$$

Using (5.11) and the fact that $v_n \to v$ in L^2, we then obtain that $v_n \to v$ in H^{2m}. Now taking $\phi = v$ in (5.10) and using the symmetry of L we obtain

$$-\frac{1}{n}(v_n , v) = \left(\frac{Nu_n - \lambda_1 u_n}{\|u_n\|_{2m}} , v \right).$$

Thus, for large n, we have

$$(Nu_n - \lambda_1 u_n , v) < 0$$

or, equivalently,

$$\int_\Omega \frac{F(x, Du_n(x)) - \lambda_1 u_n(x)}{\|u_n\|_{2m}^{\kappa}} \, v(x) ds < 0. \tag{5.13}$$

Let $\Omega_o = \{x \in \Omega : v(x) \neq 0\}$. So, for each $x \in \Omega_o$,

$$|Dv(x)| = \left[\sum_{|\alpha| \le 2m-1} |D^\alpha v(x)|^2 \right]^{1/2} > 0.$$

Thus, for each fixed $x \in \Omega_o$ (a.e.), there is an n_o (depending on x) such that, for $n \ge n_o$:

$$Du_n(x) = \|u_n\|_{2m} Dv_n(x) = \|u_n\|_{2m} |Dv_n(x)| \frac{Dv_n(x)}{|Dv_n(x)|} .$$

Now since $\|u_n\|_{2m} |Dv_n(x)| \to +\infty$, for $x \in \Omega_o$ (a.e.),

and $\dfrac{Dv_n(x)}{|Dv_n(x)|} \longrightarrow \dfrac{Dv(x)}{|Dv(x)|}$, for each fixed $x \in \Omega_o$

(a.e.), we use (F-5) to conclude that

$$(5.14) \qquad \frac{F(x, Du_n(x)) - \lambda_1 u_n(x)}{\|u_n\|_{2m}^{\kappa}} \longrightarrow |Dv(x)|^{\kappa} h\left(x, \frac{Dv(x)}{|Dv(x)|}\right) .$$

On the other hand, using (F-4), we have

$$(5.15) \qquad \left| \frac{F(x, Du_n(x)) - \lambda_1 u_n(x)}{\|u_n\|_{2m}^{\kappa}} \right| \le \frac{1}{\|u_n\|_{2m}^{\kappa}} \sum_{|\alpha| \le 2m-1} c_\alpha |Du_n(x)|^{\kappa} + d(x) .$$

Now using Hölder's inequality, we see that the right side of (5.15) is an L^2-function. So we can apply the Lebesgue dominated convergence theorem to (5.13) to get, in view of (5.14),

$$\int_\Omega h\left(x, \frac{Dv(x)}{|Dv(x)|}\right) |Dv(x)|^{\kappa} v(x)\,dx \le 0 ,$$

and this is a contradiction to assumption (F-6).

REFERENCES

1. ADAMS, R.A., *A quasi-linear elliptic boundary value problem*, Canad. J. Math. 18 (1966), 1105-1112.

2. DE FIGUEIREDO, D. G., *Some remarks on the Dirichlet Problem for semilinear elliptic equations*, Anais Acad. Brasil. Ci., to appear.

3. DE FIGUEIREDO, D. G., *On the range of nonlinear operators with linear asymptotes which are not invertible*, Comm. Math. Univ. Carolinae, to appear.

4. FRIEDMAN, A., *Partial Differential Equations*, Holt, Rinehart and Winston, New York, 1969.

5. FUČIK, KUČERA, NEČAS, *Ranges of nonlinear asymptotically linear operators*, to appear.

6. HESS, P., *On a theorem by Landesman and Lazer*, Indiana U. Math. J. 23 (1974), 827-829.

7. KRASNOSELSKII, M. A., *Topological Methods in the Theory of Nonlinear Integral Equations*, Pergamon Press, New York, 1964.

8. LANDESMAN, E. M., and LAZER, A. C., *Linear eigenvalues and a nonlinear boundary value problem*, Pacific J. Math. 33 (1970), 311-328.

9. LANDESMAN, E. M., and LAZER, A. C., *Nonlinear perturbations of linear elliptic boundary value problems at resonance*, J. Math. Mech. 19 (1970), 609-623.

10. NEČAS, J., *On the range of nonlinear operators with linear asymptotes which are not invertible*, Comm. Math. Univ. Carolinae 14 (1973), 63-72.

11. NIRENBERG, L., *An application of generalized degree to a class of nonlinear problems*, Proc. Symp. Fctl. Anal. Liège, Sept, 1971.

12. SHULTZ, M. H., *Quasi-linear elliptic boundary value problems*, J. Math. Mech. 18 (1968), 21-25.

13. VAINBERG, M., *Variational Methods for the Study of Nonlinear Operators*, Holden Day, San Francisco, 1964.

14. WILLIAMS, S. A., *A sharp sufficient condition for solution of a nonlinear elliptic boundary value problem*, J. Diff. Eq. 8 (1970), 580-586.

JOURNAL OF DIFFERENTIAL EQUATIONS 30, 1–19 (1978)

Nonlinear Perturbations of a Linear Elliptic Problem Near Its First Eigenvalue

DJAIRO G. DE FIGUEIREDO

Departamento de Matematica, Universidade de Brasilia, Instituto de Ciencias Exatas, 70,000 Brasilia, D. Federal, Brazil

AND

JEAN-PIERRE GOSSEZ

Department of Mathematics, University of Brussels, Brussels, Belgium

Received January 20, 1977; revised November 1, 1977

1. INTRODUCTION

In this paper we investigate the existence of solutions for the Dirichlet problem

$$Lu = f(x, u), \qquad \text{in } \Omega \tag{1}$$

$$\partial^j u/\partial\nu^j = 0 \qquad (j = 0, 1,..., m - 1), \qquad \text{on } \partial\Omega, \tag{2}$$

where Ω is a bounded domain in $\mathbb{R}^N$, $\partial/\partial\nu$ is the derivative in the direction of the outward normal to the boundary $\partial\Omega$ of Ω, and L is a symmetric uniformly strongly elliptic operator of order $2m$, $m \geqslant 1$. The main point here is a successful treatment of cases when there is a fast growth of f as $u \to \pm\infty$, as well as a certain interaction of f with the spectrum of L, giving rise to resonance. In two previous papers, [1, 2], the first named author has considered problems of the type studied here, but he has essentially been able to treat the case when f grows linearly as $u \to \pm\infty$. When $m = 1$, results similar to the ones presented here have been obtained by Kazdan and Warner [3], using completely different techniques. We emphasize that the method used in [3] cannot be extended to treat higher order equations, since it is based on maximum principles only available for second-order elliptic equations. We establish the existence of generalized solutions only; very little is known about regularity for higher order equations, cf. [16].

A very crucial role will be played here by the limits $k_-(x)$ and $k_+(x)$ defined by

$$k_\pm(x) = \limsup_{s\to\pm\infty} \frac{f(x, s)}{s}, \tag{3}$$

0022-0396/78/0301-0001$02.00/0

Originally published in: J. Differential Equations 30 (1978), 1-19,

DOI: 10.1007/978-3-319-02856-9_7

which, in the situations considered below, will take values in $\mathbb{R} \cup \{-\infty\}$. If both k_- and k_+ are strictly less than the first eigenvalue λ_1 of L we have nonresonant problems, and the novelty seen in the present work is the possibility of treating cases when either or both k_- and k_+ are equal to $-\infty$ on sets of positive measure. We remark that, since f is not assumed to be decreasing, a direct use of the monotone operator theory in the sense of Browder–Minty is precluded. See Theorem 1 below. We remark also that the results of the second named author [4, 5] on elliptic problems in Orlicz–Sobolev spaces do not apply here; indeed the main point in those results was the introduction of rapidly (or slowly) increasing nonlinearities in the upper order terms.

If either k_- or k_+ equals λ_1 on some set of positive measure, resonance arises. In Section 4 we treat such problems even when the other k_- or k_+ equals $-\infty$ on sets of positive measure. Consequently our Theorem 4 extends the results of [2]. An anouncement of the results presented here appeared in [6]. After the completion of this paper we have been informed that a result related to our Theorem 4 has been obtained by Brezis–Nirenberg [16] using different techniques.

2. Some Preliminaries

Let Ω be a bounded open subset of $\mathbb{R}^N$, and

$$Lu = \sum_{|\alpha|,|\beta| \leqslant m} (-1)^{|\beta|} D^\beta(a_{\alpha\beta}(x)\, D^\alpha u)$$

be a differential operator acting on real-valued functions $u(x)$ defined in Ω, satisfying the following conditions.

(L-1) L is uniformly elliptic, that is, there is a constant $c_0 > 0$ such that

$$\sum_{|\alpha|=|\beta|=m} a_{\alpha\beta}(x)\, \xi^\alpha \xi^\beta \geqslant c_0 \,|\, \xi \,|^{2m}$$

for all $x \in \Omega$, and all $\xi \in \mathbb{R}^N$ ($|\,\xi\,|^2 = \xi_1^2 + \cdots + \xi_N^2$).

(L-2) The coefficients $a_{\alpha\beta}$ are real-valued functions in $L^\infty(\Omega)$ and the higher order coefficients $a_{\alpha\beta}$, $|\,\alpha\,| = |\,\beta\,| = m$, are uniformly continuous functions in $\bar{\Omega}$.

(L-3) L is symmetric, that is, $(L\phi, \psi)_0 = (\phi, L\psi)_0$, for all $\phi, \psi \in C_0^\infty(\Omega)$. $(\;,\;)_0$ denotes L^2-inner product. Under assumptions (L-1) and (L-2) we have Gårding's inequality: there are real constants $c > 0$ and c' such that:

$$a[\phi, \phi] \geqslant c \,\|\, \phi \,\|_m^2 - c' \,\|\, \phi \,\|_0^2, \qquad \phi \in C_0^\infty(\Omega),$$

where $a[\cdot, \cdot]$ is the so-called Dirichlet bilinear form associated to L

$$a[\phi, \psi] = \int_\Omega \sum_{|\alpha|, |\beta| \leqslant m} a_{\alpha\beta}(x)\, D^\alpha\phi(x)\, D^\beta\psi(x)\, dx,$$

and $\|\cdot\|_m$ is the Sobolev norm

$$\|\phi\|_m^2 = \int_\Omega \sum_{|\alpha| \leqslant m} |D^\alpha\phi(x)|^2\, dx.$$

Let us denote by $H_0^m(\Omega)$, or simply H_0^m, the completion of $C_0^\infty(\Omega)$ with respect to the $\|\cdot\|_m$-norm. In particular, $H_0^0(\Omega) = L^2(\Omega)$. It is easy to see that H_0^m is a Hilbert space under the following inner product

$$(u, v)_m = \int_\Omega \sum_{|\alpha| \leqslant m} D^\alpha u(x)\, D^\alpha v(x)\, dx.$$

The generalized Dirichlet problem for L is defined as follows. "Given $g \in L^2(\Omega)$, find $u \in H_0^m(\Omega)$ such that $a[u, \phi] = (g, \phi)_0$, for all $\phi \in C_0^\infty(\Omega)$." We see that such a u is a solution of $Lu = g$ in the distribution sense.

A real number λ is an eigenvalue for L if there is a nonzero distribution solution, $u \in H_0^m$, of $Lu = \lambda u$. From now on we also assume (L-3). As a consequence of Gårding's inequality, Lax–Milgran lemma, the Riesz–Schauder theory of linear compact operators, and the theory of symmetric compact operators in Hilbert spaces, we have the following conclusions. L has a sequence of (real) eigenvalues $\lambda_1 < \lambda_2 < \cdots$, with $\lambda_n \to \pm\infty$, whose corresponding eigenspaces are finite-dimensional subspaces of H_0^m. For each $g \in L^2$, the equation $Lu = g$ has one and only one distribution solution $u \in H_0^m$, provided 0 is not an eigenvalue of L. Moreover we have the following inequality

$$a[u, u] \geqslant \lambda_1 \|u\|_0^2, \qquad u \in H_0^m. \tag{4}$$

Details and proofs of most of the forementioned results can be seen in Friedman [7].

The nonlinearity is a function $f\colon \Omega \times \mathbb{R} \to \mathbb{R}$ which is always assumed to satisfy

(f-1) Carathéodory's conditions, that is, for each fixed $s \in \mathbb{R}$, the function $x \to f(x, s)$ is measurable, and for almost all $x \in \Omega$ fixed, the function $s \to f(x, s)$ is continuous.

(f-2) For each $r > 0$, there is a function $\alpha_r \in L^2(\Omega)$ such that $|f(x, s)| \leqslant \alpha_r(x)$, for all $x \in \Omega$ and all $|s| \leqslant r$.

The point of assumption (f-1) is to obtain $f(x, u(x))$ as a measurable function of x, provided u is. Assumption (f-2) is to prevent certain wild behaviors of f

with respect to x. For example, we do not want functions like $f(x, s) = sx^{-1}$, when $\Omega = (0, 1)$. Observe that (f-2) is not a restriction whatsoever to the growth of f as $|s| \to \infty$; it is trivially satisfied when f does not depend on x: $f(x, s) = f(s)$.

As we remarked in the Introduction an important rôle is played by the positioning of the two limits defined in (3) with respect to the eigenvalues of L.

Nonresonant problems are those where k_- and k_+ are away of the eigenvalues. In [1] it is considered the case when

$$|f(x, s) - \lambda s| \leqslant \gamma |s| + b(x) \tag{$*$}$$

where $b(x) \in L^2(\Omega)$, $\lambda_n < \lambda < \lambda_{n+1}$, $0 < \gamma < \min(\lambda - \lambda_n, \lambda_{n+1} - \lambda)$, and λ_n, λ_{n+1} are two consecutive eigenvalues of L. The above inequality implies, the following two inequalities:

$$\lambda_n < \eta_n \leqslant k_-, \; k_+ \leqslant \eta_{n+1} < \lambda_{n+1}, \tag{$**$}$$

or

$$\lambda_n < \eta_n \leqslant l_-, \; l_+ \leqslant \eta_{n+1} < \lambda_{n+1} \tag{$***$}$$

where $l_\pm$ is defined as $k_\pm$, with lim sup replaced by lim inf. In this paper the nonresonant problem corresponds to the following assumption:

(f-3) $\qquad k_-, k_+ \leqslant \eta_1 < \lambda_1, \qquad$ uniformly in x,

where λ_1 is the first eigenvalue of L. This assumption does not imply that f has a linear growth as $s \to \pm\infty$. By (f-3) we mean more precisely that there is an η, with $\eta_1 < \eta < \lambda_1$, an $s_0 > 0$ and $0 \leqslant \beta(x) \in L^2(\Omega)$ such that

$$\frac{f(x, s)}{s} \leqslant \eta + \frac{\beta(x)}{|s|}, \qquad x \in \Omega, \; |s| \geqslant s_0, \tag{5}$$

which is equivalent to the two relations below

$$f(x, s) \leqslant \eta s + \beta(x), \qquad x \in \Omega, \quad s \geqslant s_0, \tag{6}$$

and

$$f(x, s) \geqslant \eta s - \beta(x), \qquad x \in \Omega, \quad s \leqslant -s_0. \tag{7}$$

So the growth of f as $s \to +\infty$ is bounded above by a linear function, but it is unrestricted from below. A similar statement can be made for the growth of f as $s \to -\infty$.

In the next section we shall prove the following theorem.

THEOREM 1. *Assume* (L-1), (L-2), (L-3), (f-1), (f-2), *and* (f-3). *Then there exists* $u \in H_0^m$, *with* $f(x, u(x)) \in L^1(\Omega)$ *and* $f(x, u(x))\, u(x) \in L^1(\Omega)$ *which is a distribution solution of* $Lu = f(x, u)$.

Remark. Assumptions (f-2) and (f-3) do not imply that the Niemytskii operator $F: u \to f(x, u)$ maps $L^2(\Omega)$ into itself. This causes a problem when, following Dolph [8] and [1], one tries to use the theory of (nonlinear) compact operators to solve a Hammerstein equation of the form $u = AFu$, which is equivalent to the Dirichlet problem for $Lu = f(x, u)$, with A essentially equal to L^{-1}. To overcome this difficulty we will approximate F by Niemytskii maps which are well defined in L^2.

Resonance arises when either k_+, k_-, l_+, or l_- are equal to eigenvalues on sets of positive measure, or when a crossing of eigenvalues occurs, such as $k_- < \lambda_n < k_+$. These problems are much more difficult, and they are not yet completely understood. The case $l_- = k_- = l_+ = k_+ = \lambda_n$ has been treated by several authors, [9, 10, 11, 1], and others. Again this implies a linear growth of f as $s \to \pm\infty$. For a recent survey on problems of this sort we mention Fucik [12]. The crossing of the first eigenvalue has been studied by Ambrosetti and Prodi [13]. In the present paper the resonant problem considered corresponds to the assumption

$$\text{(f-4)} \qquad k_-, k_+ \leqslant \lambda_1, \qquad \text{uniformly in } x,$$

with either k_- or k_+, or both, being equal to λ_1 on sets of positive measure. The possibility of also having k_+ or k_- equal to $-\infty$ somewhere is not excluded.

3. Proof of Theorem 1

Observe initially that we can without loss of generality assume that $\lambda_1 > 0$. Indeed, equation $Lu = f(x, u)$ is equivalent to $L_1 u = f_1(x, u)$, where $L_1 u = Lu + \lambda u$ and $f_1(x, u) = f(x, u) + \lambda u$. If λ is sufficiently large the first eigenvalue $\lambda_1 + \lambda$ of L_1 is positive, and the function f_1 satisfies the hypotheses (f-1), (f-2), and (f-3) with respect to L_1. The number η_1 of condition (f-3) will accordingly also be assumed >0.

We shall use the following lemma which is a special case of Theorem 1.

Lemma 1. *Assume* (L-1), (L-2), (L-3), (f-1), *and* (f-3). *Moreover suppose that f has a linear growth, that is, there is a constant $a > 0$ and an $L^2(\Omega)$-function $b(x)$ such that $|f(x, s)| \leqslant a|s| + b(x)$, for all $x \in \Omega$ and all $s \in \mathbb{R}$. Then there is a distribution solution $u \in H_0^m$ of $Lu = f(x, u)$.*

This lemma has been proved in [2] when $\beta(x) \equiv 0$ using Leray–Schauder fixed point theorem; the proof given there carries over immediately to the present situation. To prove the general case we approximate f by functions f_n which have a linear growth. For each positive integer n define

$$f_n(x, s) = \begin{cases} f(x, s), & \text{if } |s| \leqslant n \\ f(x, n), & \text{if } s \geqslant n \\ f(x, -n), & \text{if } s \leqslant -n. \end{cases} \tag{8}$$

It follows readily that

f_n satisfies Carathéodory's conditions, (8i)

$f_n(x, s)/s \leqslant \eta + \beta(x)/|s|$, for all $x \in \Omega$, and all $|s| \geqslant s_0$, (8ii)

$f_n(x, s)$ has a linear growth, or even better it is bounded by an L^2-function. For, in view of (f-2), one has $|f_n(x, s)| \leqslant \alpha_n(s)$, for all $x \in \Omega$ and all $s \in \mathbb{R}$. (8iii)

So Lemma 1 implies the existence of a distribution solution $u_n \in H_0{}^m$ of $Lu_n = f_n(x, u_n)$. It is also obtained that $f_n(x, u_n(x)) \in L^2(\Omega)$.

LEMMA 2. *The following estimates hold true*

$$\| u_n \|_0 \leqslant K, \tag{9}$$

$$0 \leqslant \int_\Omega f_n(x, u_n)u_n \leqslant K', \tag{10}$$

$$\| u_n \|_m \leqslant K'', \tag{11}$$

where K, K', and K" are constants independent of n.

Proof. Using (4) we have

$$\lambda_1 \| u_n \|_0^2 \leqslant a[u_n, u_n] = \int_\Omega f_n(x, u_n)u_n \tag{12}$$

and using (5) and (f-2)

$$\int_{|u_n|<s_0} f_n(x, u_n)u_n + \int_{|u_n|\geqslant s_0} f_n(x, u_n)u_n$$
$$\leqslant s_0 \| \alpha_{s_0} \|_0 \, |\Omega|^{1/2} + \eta \| u_n \|_0^2 + \| \beta \|_0 \| u_n \|_0, \tag{13}$$

where $|\Omega|$ denotes the Lebesgue measure of Ω. Since $\lambda_1 > \eta$, inequalities (12) and (13) readily imply the estimates (9) and (10). Finally using Gårding's inequality and estimates (9) and (10) we obtain (11). Q.E.D.

Estimate (11) allows us to assert, passing to subsequences if necessary, that there exists $u \in H_0{}^m$ such that $u_n \to u$, weakly in $H_0{}^m$, that $u_n \to u$, in L^2 and that $u_n \to u$, a.e. in Ω. Thus $a[u_n, \phi] \to a[u, \phi]$ for all $\phi \in C_0{}^\infty(\Omega)$. So if we can prove that $f(x, u(x)) \in L^1(\Omega)$ and that $f_n(x, u_n(x)) \to f(x, u(x))$ in L^1, the existence of a distribution solution of $Lu = f(x, u)$ will be established. For that matter we first observe that $f_n(x, u_n(x)) \to f(x, u(x))$, a.e. in Ω. Indeed, let $x \in \Omega$, be fixed, and take n_0 such that $|u_n(x)| < |u(x)| + 1$ for all $n \geqslant n_0$; then for $n \geqslant \max\{n_0, |u(x)| + 1\}$ we have $f_n(x, u_n(x)) = f(x, u(x))$, in view of (8). Thus we have just to prove Lemma 3 below and then the result follows from Vitali's theorem, cf. Dunford–Schwartz [14, p. 150]. The proof of Lemma 3 uses an idea of Strauss [15] and has been simplified after a suggestion of A. Damlamian.

LEMMA 3. *The functions* $g_n(x) = f_n(x, u_n(x))$ *are equiabsolutely integrable, that is, given* $\epsilon > 0$ *there is* $\delta > 0$ *such that for every* $\Omega_0 \subset \Omega$, *with Lebesgue measure* $|\Omega_0| < \delta$, *one has* $\int_{\Omega_0} |g_n| < \epsilon$, *for all* n.

Before proving the above lemma, let us introduce a partition of Ω made up of the following sets

$$A_n = \{x \in \Omega\colon |u_n(x)| < s_0\}$$
$$B_n = \{x \in \Omega\colon |u_n(x)| \geqslant s_0\,;\ u_n(x)\, g_n(x) \geqslant 0\}$$
$$C_n = \{x \in \Omega\colon |u_n(x)| \geqslant s_0\,;\ u_n(x)\, g_n(x) < 0\}.$$

and let us establish the estimates

$$\left| \int_{A_n} g_n u_n \right| \leqslant K_1, \tag{14}$$

$$0 \leqslant \int_{B_n} g_n u_n \leqslant K_2, \tag{15}$$

$$0 \leqslant \int_{C_n} -g_n u_n \leqslant K_3, \tag{16}$$

where K_1, K_2, and K_3 are constants independent of n. Estimate (14) is a direct consequence of (f-2), and, in fact, K_1 can be taken as $s_0 \| \alpha_{s_0} \|_0 \, | \Omega |^{1/2}$, cf. (13). Estimate (15) follows directly from (8ii), and, in fact, K_2 can be taken as $\eta K^2 + K \| \beta \|_0$, where K is the constant in (9). Finally, (16) follows from (10), (14), and (15), and K_3 can be taken as $\eta K^2 + s_0 \| \alpha_{s_0} \|_0 \, | \Omega |^{1/2} + K \| \beta \|_0 + K'$, where K' is the constant in (10).

Proof of Lemma 3. For the given $\epsilon > 0$, choose $r \geqslant s_0$ such that

$$(1/r)[\eta K^2 + s_0 \| \alpha_{s_0} \|_0 \, | \Omega |^{1/2} + K \| \beta \|_0 + K'] < \epsilon,$$

and let

$$A_n' = \{x \in \Omega : |u_n(x)| < r\}$$
$$B_n' = \{x \in \Omega : |u_n(x)| \geqslant r, \qquad u_n(x)\, g_n(x) \geqslant 0\}$$
$$C_n' = \{x \in \Omega : |u_n(x)| \geqslant r, \qquad u_n(x)\, g_n(x) < 0\}.$$

We break Ω_0 as the union $(\Omega_0 \cap A_n') \cup (\Omega_0 \cap B_n') \cup (\Omega_0 \cap C_n')$, and estimate the integral of $|g_n|$ over each one of these sets; it suffices to consider $n \geqslant r$. First, from (f-2) and (8) we have:

$$\int_{\Omega_0 \cap A_n'} |g_n| = \int_{\Omega_0 \cap A_n'} |f(x, u_n)| \leqslant \| \alpha_r \|_{L^1(\Omega_0)} .$$

Next, using (15)

$$\int_{\Omega_0 \cap B_n'} |g_n| \leqslant \frac{1}{r} \int_{\Omega_0 \cap B_n'} |u_n| \, |g_n| = \frac{1}{r} \int_{\Omega_0 \cap B_n'} u_n g_n \leqslant \frac{1}{r} \int_{B_n} u_n g_n$$
$$\leqslant \frac{1}{r} K_2 < \epsilon,$$

and finally using (16)

$$\int_{\Omega_0 \cap C_n'} |g_n| \leqslant \frac{1}{r} \int_{\Omega_0 \cap C_n'} |u_n| \, |g_n| = \frac{1}{r} \int_{\Omega_0 \cap C_n'} -u_n g_n \leqslant \frac{1}{r} \int_{C_n} -u_n g_n$$
$$\leqslant \frac{1}{r} K_3 < \epsilon.$$

Now the proof will be complete by choosing $\delta > 0$ in such a way that $\| \alpha_r \|_{L^1(\Omega_0)} < \epsilon$ and $\int_{\Omega_0} |g_n| < \epsilon$, $n < r$, for all $|\Omega_0| < \delta$. Q.E.D.

The proof of Theorem 1 will be completed by proving the final lemma.

LEMMA 4. *The function $f(x, u(x))\, u(x)$ is in $L^1(\Omega)$.*

Proof. Let us denote by h^+ and h^-, respectively, the positive and negative parts of a function $h(x)$, $x \in \Omega$. Since $f(x, u_n)u_n \to f(x, u)u$, a.e. in Ω, it follows that $[f(x, u_n)u_n]^+ \to [f(x, u)u]^+$, a.e. in Ω, and similarly for the negative parts. Now

$$\int_{\Omega} [f_n(x, u_n)u_n]^+ \leqslant \int_{A_n} |f_n(x, u_n)u_n| + \int_{B_n} f_n(x, u_n)u_n ,$$

which is uniformly bounded in view of estimates (14) and (15). Thus, by Fatou's lemma

$$\int_{\Omega} [f(x, u)u]^+ < \infty.$$

On the other hand

$$\int_{\Omega} [f_n(x, u_n)u_n]^- \leqslant \int_{A_n} |f_n(x, u_n)u_n| + \int_{C_n} -f_n(x, u_n)u_n ,$$

which is again uniformly bounded in view of estimates (14) and (16). Using Fatou's lemma once more we get

$$\int_{\Omega} [f(x, u)u]^- < \infty,$$

and Lemma 4 is proved. Q.E.D.

4. Problems wiht Resonance

As already mentioned, resonance in this paper corresponds to the following situation

$$\limsup_{|s|\to\infty} \frac{f(x, s)}{s} \leqslant \lambda_1, \text{ uniformly in } x, \tag{f-4}$$

with equality holding on subsets of Ω with positive measure. By (f-4) we mean more precisely that, given $\epsilon > 0$, there is $s_0 > 0$ and $0 \leqslant \beta(x) \in L^2(\Omega)$ such that

$$\frac{f(x, s)}{s} \leqslant \lambda_1 + \epsilon + \frac{\beta(x)}{|s|}, \; x \in \Omega, \; |s| \geqslant s_0. \tag{17}$$

The idea for treating the Dirichlet problem (1)–(2) for functions satisfying (17) is to approximate Eq. (1) by nonresonant equations of the type $Lu + (1/n)u = f(x, u)$, whereupon Theorem 1 applies. This is a kind of argument originally used by Hess [11]. We start with the following auxiliary result.

Theorem 2. *Assume* (L-1), (L-2), (L-3), (f-1), (f-2), *and* (f-4). *Then the approximating equations*

$$Lu_n + \frac{1}{n} u_n = f(x, u_n) \tag{18}$$

have distribution solutions $u_n \in H_0^m$, *with* $f(x, u_n) \in L^1$ *and* $f(x, u_n)u_n \in L^1$. *Moreover, either*

(i) $\| u_n \|_m$ *is bounded and then* $Lu = f(x, u)$ *has a distribution solution* $u \in H_0^m$ *with* $f(x, u) \in L^1$ *and* $f(x, u)u \in L^1$.

or

(ii) *there is a subsequence of* $\| u_n \|_m$, *which we denote also by* $\| u_n \|_m$, *going to* $+\infty$. *In this case, for a further subsequence,* $v_n = u_n/\| u_n \|_m$ *converges in* H_0^m *to a nonzero* λ_1*-eigenfunction* v *of* L, *and the following inequality holds*

$$\int_\Omega [f(x, u_n) - \lambda_1 u_n] v_n > 0 \tag{19}$$

for n *sufficiently large.*

Proof. As in the previous section we may assume $\lambda_1 > 0$. The existence of solutions u_n of the approximating equations is a direct consequence of Theorem 1. Suppose now that $\| u_n \|_m$ is bounded. Then, going to subsequences if necessary, we conclude that there is $u \in H_0^m$ such that $u_n \to u$, weakly in H_0^m, strongly in L^2, and a.e. in Ω. So $a[u_n, \phi] + 1/n(u_n, \phi)_0 \to a[u, \phi]$ for all $\phi \in C_0^\infty(\Omega)$. In order to prove that u is a distribution solution of $Lu = f(x, u)$

we will show that $f(x, u) \in L^1$ and that $f(x, u_n) \to f(x, u)$ in L^1. Since $f(x, u_n(x)) \to f(x, u(x))$ a.e. in Ω, it suffices to prove that the functions $f(x, u_n(x))$ are equi-absolutely integrable and then use Vitali's theorem. The procedure to be used here is completely analogous to the one in the previous section, and so we omit it. The proof that $f(x, u)u \in L^1$ follows the same lines as the proof of Lemma 4. Finally the proof of the statements on part (ii) follows in a similar way that of Theorem 2 of [2], and again we omit it. Q.E.D.

Now we proceed to give some applications of Theorem 2.

THEOREM 3. *Assume* (L-1), (L-2), (L-3), (f-1), (f-2), *and* (f-4). *Suppose also that for all nonzero eigenfunctions v of L corresponding to the eigenvalue λ_1, we have*

$$\int_{v>0} k_+ v^2 + \int_{v<0} k_- v^2 < \lambda_1 \| v \|_0^2, \tag{20}$$

where k_+ and k_- are defined in (3). *Then there exists a distribution solution $u \in H_0^m$ of $Lu = f(x, u)$, with $f(x, u) \in L^1$ and $f(x, u)u \in L^1$.* (Note that the integrals on the left side of (20) are well defined in $[-\infty, +\infty[$.)

Proof. If alternative (i) of Theorem 2 holds then the present theorem is proved. Alternatively if (ii) of Theorem 2 holds, we proceed as follows to get a contradiction. Inequality (19) gives

$$\lambda_1 \| v_n \|_0^2 < \int_\Omega f(x, u_n) \| u_n \|_m^{-1} v_n$$

for large n. So

$$\lambda_1 \| v \|_0^2 \leqslant \limsup \int_\Omega f(x, u_n) \| u_n \|_m^{-1} v_n \tag{21}$$

and let us estimate the right side of (21). Introduce the following partition of Ω:

$$\begin{aligned} R_n &= \{x \in \Omega: | u_n(x)| \leqslant s_0\} \\ S_n &= \{x \in \Omega: v(x) > 0, | u_n(x)| > s_0\} \\ Z_n &= \{x \in \Omega: v(x) < 0, | u_n(x)| > s_0\} \\ W_n &= \{x \in \Omega: v(x) = 0, | u_n(x)| > s_0\}. \end{aligned}$$

Using (f-2) we have

$$\left| \int_{R_n} f(x, u_n) \| u_n \|_m^{-1} v_n \right| \leqslant \| \alpha_{s_0} \|_0 \| u_n \|_m^{-1} \| v_n \|_0 \to 0.$$

To estimate the integral over S_n, let us denote by χ_n the characteristic function of the set S_n. Observe that, since $u_n(x) = v_n(x) \| u_n \|_m$ and $v_n(x) \to v(x)$, then $u_n(x) \to +\infty$ for $x \in \Omega$ such that $v(x) > 0$. So $\chi_n(x) \to 1$ and $\limsup f(x, u_n(x))\, u_n(x)^{-1} = k_+(x)$, for all x in the set $\{v > 0\}$. Applying these observations and Fatou's lemma to the identity

$$\int_{S_n} f(x, u_n) \| u_n \|_m^{-1} v_n = \int_{v>0} \chi_n \left\{ \frac{f(x, u_n)}{u_n} - (\lambda_1 + \epsilon) - \frac{\beta(x)}{| u_n |} \right\} v_n^2$$
$$+ \int_{v>0} \left(\lambda_1 + \epsilon + \frac{\beta(x)}{| u_n |} \right) \chi_n v_n^2,$$

we get

$$\limsup \int_{S_n} f(x, u_n) \| u_n \|_m^{-1} v_n \leqslant \int_{v>0} k_+ v .$$

In a similar way we prove

$$\limsup \int_{Z_n} f(x, u_n) \| u_n \|_m^{-1} v_n \leqslant \int_{v<0} k_- v^2.$$

The integral over W_n is estimated as follows:

$$\int_{W_n} f(x, u_n) \| u_n \|_m^{-1} v_n \leqslant \int_{W_n} \left(\lambda_1 + \epsilon + \frac{\beta(x)}{| u_n |} \right) v_n^2$$
$$\leqslant (\lambda_1 + \epsilon) \int_{v=0} v_n^2 + \| u_n \|_m^{-1} \int_\Omega \beta \, | v_n | \to 0.$$

Consequently (21) gives

$$\lambda_1 \| v \|_0^2 \leqslant \int_{v>0} k_+ v^2 + \int_{v<0} k_- v^2,$$

yielding in this manner a contradiction to our assumption (20). Q.E.D.

COROLLARY 1. *Assume* (L-1), (L-2), (L-3), (f-1), (f-2), (f-4), *and that the solutions of* $Lu = \lambda_1 u$ *have the unique continuation property. That is, if* $u \in H_0^m$ *is a distribution solution of* $Lu = \lambda_1 u$, *which vanishes on a subset of* Ω *with positive measure, then* $u \equiv 0$. *Suppose that there exists a subset* Ω' *of* Ω *with positive measure, such that*

$$k_+(x),\, k_-(x) < \lambda_1 , \qquad x \in \Omega'.$$

Then there exists a distribution solution $u \in H_0^m$ *of* $Lu = f(x, u)$, *with* $f(x, u) \in L^1$ *and* $f(x, u)u \in L^1$.

Proof. It suffices to show that under the above hypotheses condition (20) of Theorem 3 holds true. Suppose by contradiction that this is not the case, i.e.,

$$\int_{v>0} (k_+ - \lambda_1)\, v^2 + \int_{v<0} (k_- - \lambda_1)\, v^2 \geqslant 0,$$

for some nonzero λ_1-eigenfunction v. Using the fact that (f-4) implies k_+, $k_- \leqslant \lambda_1$, in Ω, we conclude that $k_+ = \lambda_1$ in $v > 0$ and $k_- = \lambda_1$ in $v < 0$. This together with the fact that the set $\{v = 0\}$ has measure zero, gives a contradiction.

Q.E.D.

COROLLARY 2. *Assume* (L-1), (L-2), (L-3), (f-1), (f-2), (f-4), *and that there exists a subset* Ω' *of* Ω, *with positive measure, where* $k_+ < \lambda_1$. *Suppose also that* (20) *holds (only) for the nonzero* λ_1*-eigenfunctions* v *which are* $\leqslant 0$ *on* Ω'. *Then there exists a distribution solution* $u \in H_0^m$ *of* $Lu = f(x, u)$, *with* $f(x, u)$ *and* $f(x, u)u$ *in* L^1.

Proof. It suffices to show that under the present hypotheses condition (20) of Theorem 3 holds true. Proceeding as in the proof of the previous corollary, we conclude that $k_+ = \lambda_1$ on $v > 0$ and $k_- = \lambda_1$ on $v < 0$. In view of our hypotheses it follows that $v(x) \leqslant 0$ on Ω', and this leads to a contradiction with our assumption.

Q.E.D.

Of course an analogous result holds assuming the following hypothesis on k_-, instead of k_+: $k_- < \lambda_1$ on some subset Ω' of Ω, with positive measure.

Some Examples of Application of Theorem 3 *and Its Corollaries*

(1) Suppose $k_+ < \lambda_1$ on the whole of Ω; as a special case one could have $k_+ \equiv -\infty$. Then, by Corollary 2, a solution exists provided

$$\int_\Omega k_- v^2 < \lambda_1 \| v \|_0^2 \tag{22}$$

for all nonzero λ_1-eigenfunctions $v \leqslant 0$.

(2) In many applications the function $f(x, u)$ appears in the form $k(x, u)u + h(x)$, where $h(x) \in L^2(\Omega)$ and k satisfies Carathéodory's conditions and a hypothesis like (f-2). So a nontrivial example of application of Corollary 1, when $\lambda_1 = 0$, would be in the case that $k(x, u) = -(e^u + 1)\chi_{\Omega'}$, where $\chi_{\Omega'}$ is the characteristic function of a subset Ω' of Ω with $0 < |\Omega'| < |\Omega|$. Indeed, $k_+(x) = -\infty$ on Ω', $k_+(x) = 0$ on $\Omega \backslash \Omega'$, $k_-(x) = -1$ on Ω', and $k_-(x) = 0$ on $\Omega \backslash \Omega'$.

Our next application is to equations of the form

$$Lu - \lambda_1 u = g(x, u) + h(x), \tag{23}$$

where $h \in L^2(\Omega)$ and g satisfies the following hypotheses:

(g-1) Carathéodory's conditions.

(g-2) For each $r > 0$ there exists $\alpha_r(x) \in L^2(\Omega)$ such that

$$|g(x, s)| \leqslant \alpha_r(x), \qquad x \in \Omega, \quad |s| \leqslant r.$$

(g-3) $\limsup_{|s|\to\infty} g(x, s)/s \leqslant 0$, uniformly in $x \in \Omega$.

(g-4) There exist two L^2-functions $a(x) \geqslant 0$ and $b(x) \leqslant 0$ such that

$$g(x, s) \leqslant a(x), \qquad x \in \Omega, \quad s \geqslant 0 \tag{24}$$

$$g(x, s) \geqslant b(x), \qquad x \in \Omega, \quad s \leqslant 0. \tag{25}$$

Remark. Condition (g-4) is a stronger restriction on the one-sided growth of g than (g-3) is. Indeed, compare (24) and (25) with the following equivalent statement of (g-3): given $\epsilon > 0$ there is $s_0 > 0$ and $0 \leqslant \beta(x) \in L^2(\Omega)$ such that

$$g(x, s) \leqslant \epsilon s + \beta(x), \qquad x \in \Omega, \quad s \geqslant s_0$$
$$g(x, s) \geqslant \epsilon s - \beta(x), \qquad x \in \Omega, \quad s \leqslant -s_0 .$$

Observe however that g is unrestricted from below as $s \to +\infty$. And similarly g is unrestricted from above as $s \to -\infty$. Finally note that (g-2) implies that condition (g-4) is equivalent to the requirement that (24) and (25) hold only for $|s| \geqslant$ some s_0.

THEOREM 4. *Assume* (L-1), (L-2), (L-3), (g-1), (g-2), *and* (g-4). *Then, for a given* $h \in L^2$, *there exists a distribution solution* $u \in H_0^m$ *of* (23) *with* $g(x, u)$ *and* $g(x, u)u$ *in* L^1, *provided*

$$\int_\Omega h(x)\, v(x) < -\int_{v>0} G_+(x)\, v(x) - \int_{v<0} G_-(x)\, v(x) \tag{26}$$

for all nonzero λ_1*-eigenfunctions* v *of* L, *where*

$$G_+(x) = \limsup_{s\to+\infty} g(x, s)$$

and

$$G_-(x) = \liminf_{s\to-\infty} g(x, s).$$

Remarks. (a) Note that the integrals on the right side of (26) are well defined, with values in $[-\infty, \infty[$. Note also that if $G_-(x) \geqslant g(x, s) \geqslant G_+(x)$, for all $x \in \Omega$, $s \in \mathbb{R}$, then condition (26) with $<$ replaced by $\leqslant$ is essentially necessary for the existence of a solution $u \in H_0^m$ of (23) with $g(x, u)$ in L^1. Indeed, assuming some regularity on L and Ω (or that $2m > n$, and Ω satisfies the cone condition) so that any λ_1-eigenfunction v of L is continuous on $\bar{\Omega}$, one deduces from (23) that

$$\int_\Omega h(x)\, v(x)\, dx = -\int_\Omega g(x, u)\, v(x)\, dx$$

and the conclusion follows readily.

(b) Theorems 3 and 4 are not deductible one from the other. We next give an example where Theorem 3 applies but Theorem 4 does not. Let the function f in (1) be given by:

$$\begin{aligned} f(x, s) &= c_+(x)\, s + d_+(x)\, g_+(s), \qquad s \geqslant 0 \\ &= c_-(x)\, s + d_-(x)\, g_-(s), \qquad s \leqslant 0 \end{aligned}$$

where $0 \geqslant c_\pm \in L^2(\Omega)$, $d_\pm \in L^\infty(\Omega)$, $g_\pm$ are continuous with $g_\pm(0) = 0$, $g_+(s)s^{-1} \to 0$ as $s \to +\infty$ and $g_-(s)s^{-1} \to 0$ as $s \to -\infty$. Assume also that L in (1) is a second-order uniformly strongly elliptic operator with $\lambda_1 = 0$; this implies that the λ_1-eigenfunctions are of the same sign in Ω. Then, assuming that $c_\pm \neq 0$ on sets of positive measure, we see that Theorem 3 applies to guarantee that there exists a distribution solution $u \in H_0^m$ of $Lu = f(x, u)$. We also see that Theorem 4 does not apply since condition (g-4) is not verified in this case. On the other hand, the next example exhibits a situation where Theorem 4 applies but Theorem 3 does not. Let the function f in (1) be given by

$$\begin{aligned} f(x, s) &= \quad a_+(x) + b_+(x)\, g_+(s), \qquad s \geqslant 0 \\ &= -a_-(x) - b_-(x)\, g_-(s), \qquad s \leqslant 0 \end{aligned}$$

where $0 \geqslant a_\pm \in L^2(\Omega)$, $b_\pm \in L^2(\Omega)$, $g_\pm$ are continuous, with $g_\pm(0) = 0$, $g_\pm(s) \to 0$ as $s \to \pm\infty$. Assume also that L in (1) is a second-order uniformly strongly elliptic operator with $\lambda_1 = 0$. Then, assuming that $a_\pm(x) \neq 0$ on sets of positive measure, we see that (26) holds, and consequently Theorem 4 applies. We also see that $k_\pm(x) \equiv 0$ and then Theorem 3 does not apply.

Proof. Let $f(x, u) = \lambda_1 u + g(x, u) + h(x)$. Such an f satisfies all the hypotheses of Theorem 2. So if alternative (i) of Theorem 2 holds, then (23) has a distribution solution $u \in H_0^m$ with $g(x, u)$ and $g(x, u)u$ are in $L^1(\Omega)$. To complete the proof let us show that (26) implies that alternative (ii) of Theorem 2 cannot hold. Accordingly suppose by contradiction that (19) holds, which implies

$$\int_\Omega h(x)\, v_n > \int_\Omega -g(x, u_n)\, v_n\,. \tag{27}$$

The left side of (27) converges to $\int_\Omega h(x)v$; to evaluate the lim inf of the right side of (27) let us break the integral over Ω as a sum of six integrals whose lim inf's will be calculated separately.

$$\text{(i)} \quad \int_{v>0,\, u_n>0} = \int_{v>0} \chi_n[-g(x, u_n)\, v_n + a(x)\, v_n] - \int_{v>0} \chi_n a(x)\, v_n$$

where χ_n is the characteristic function of the set $\{u_n \geqslant 0\}$. Since $u_n =$

$v_n \| u_n \|_m \to +\infty$ in $\{v > 0\}$, it follows that $\chi_n \to 1$ a.e. in $\{v > 0\}$, and by Fatou's lemma

$$\begin{aligned}
&\liminf \int_{v>0} \chi_n[-g(x, u_n)\, v_n + a(x)\, v_n] \\
&\quad \geqslant \int_{v>0} \liminf \chi_n[-g(x, u_n)\, v_n + a(x)\, v_n] \\
&\quad \geqslant -\int_{v>0} \limsup \chi_n g(x, u_n)\, v_n + \int_{v>0} \liminf \chi_n a(x)\, v_n \\
&\quad = -\int_{v>0} G_+(x)\, v(x) + \int_{v>0} a(x)\, v.
\end{aligned}$$

Next we observe that

$$\lim \int_{v>0} \chi_n a(x)\, v_n = \int_{v>0} a(x)\, v$$

and consequently

$$\liminf \int_{v>0, u_n \geqslant 0} \geqslant -\int_{v>0} G_+(x)\, v(x).$$

(ii) $$\int_{v<0, u_n<0} = \int_{v<0} \varphi_n[-g(x, u_n)\, v_n + b(x)\, v_n] - \int_{v<0} \varphi_n b(x)\, v_n,$$

where φ_n is the characteristic function of $\{u_n < 0\}$. Since $u_n = v_n \| u_n \|_m \to -\infty$ in $\{v < 0\}$, it follows that $\varphi_n \to 1$ a.e. in $\{v < 0\}$. By arguments similar to the ones in (i) above we conclude that

$$\liminf \int_{v<0, u_n<0} \geqslant -\int_{v<0} G_-(x)\, v(x).$$

(iii) $$\int_{v>0, u_n<0} = \int_{v>0} -\varphi_n g(x, u_n)\, v_n \geqslant \int_{v>0} -\varphi_n b(x)\, v_n,$$

where the last integral converges to zero in view of the observation that $\varphi_n \to 0$ a.e. in $\{v > 0\}$. So

$$\liminf \int_{v>0, u_n<0} \geqslant 0.$$

(iv) $$\int_{v<0, u_n \geqslant 0} = \int_{v<0} -\chi_n g(x, u_n)\, v_n \geqslant \int_{v<0} -\chi_n a(x)\, v_n,$$

and by an argument similar to the one in (iii) we get

$$\liminf \int_{v<0, u_n \geqslant 0} \geqslant 0.$$

(v) $$\int_{v=0, u_n \geqslant 0} = \int_{v=0} -\chi_n g(x, u_n)\, v_n \geqslant \int_{v=0} -\chi_n a(x)\, v_n,$$

where the last integral converges to zero since $v_n \to v$ in $L^2(\Omega)$ and thus $v_n \to 0$ in $L^2(\{v = 0\})$. So

$$\liminf \int_{v=0,\, u_n \geq 0} \geq 0.$$

(vi) $$\int_{v=0,\, u_n<0} = \int_{v=0} - \varphi_n g(x, u_n)\, v_n \geq \int_{v=0} - \varphi_n b(x)\, v_n$$

and we proceed as in (v). So

$$\liminf \int_{v=0,\, u_n<0} \geq 0.$$

Putting together the estimates in (i),..., (vi) above, we get from (27)

$$\int_\Omega h(x)\, v(x) \geq -\int_{v>0} G_+(x)\, v(x) - \int_{v<0} G_-(x)\, v(x),$$

which contradicts (26). The proof of Theorem 4 is complete. Q.E.D.

COROLLARY 3. *Assume* (L-1),..., (g-4) *as in Theorem* 4. *Then, for any given* $h \in L^2$ *orthogonal to the* λ_1*-eigenspace of* L, *there exists a distribution solution* $u \in H_0^m$ *of* (23), *with* $g(x, u)$ *and* $g(x, u)u$ *in* L^1, *provided*

$$G_-(x) > 0 > G_+(x), \qquad x \in \Omega.$$

Proof. If (26) does not hold, then

$$0 \geq -\int_{v>0} G_+ v - \int_{v<0} G_- v$$

for some nonzero λ_1-eigenfunction v of L. But both $-G_+ v$ and $-G_- v$ in the above integrals are ≥ 0. Consequently $G_+ v = 0$ on $\{v > 0\}$ and $G_- v = 0$ on $\{v < 0\}$, and thus $v = 0$ a.e. in Ω, which is a contradiction. Q.E.D.

COROLLARY 4. *Assume* (L-1),..., (g-4) *as in Theorem* 4. *Suppose that* $G_+(x) = -\infty$ *on a subset* Ω' *of* Ω, *with positive measure. Then, for a given* $h \in L^2$, (23) *has a distribution solution* $u \in H_0^m$ *with* $g(x, u)$ *and* $g(x, u)u$ *in* L^1 *provided*

$$\int_\Omega hv < -\int_{v>0} G_+ v - \int_{v<0} G_- v \tag{28}$$

for all nonzero λ_1*-eigenfunctions* $v \leq 0$ *in* Ω'.

Proof. We claim that (26) holds (for *all* nonzero λ_1-eigenfunctions). Suppose by contradiction that there is $v \neq 0$ such that

$$\int_\Omega hv \geqslant -\int_{v>0} G_+ v - \int_{v<0} G_- v. \tag{29}$$

Since $G_+ = -\infty$ in Ω', (29) implies that $v \leqslant 0$ in Ω'. Thus (29) would hold for a nonzero λ_1-eigenfunction $v \leqslant 0$ in Ω', which contradicts (28). Q.E.D.

COROLLARY 5. *Assume* (L-1),..., (g-4) *as in Theorem* 4. *Suppose that* $G_+ = -\infty$ *in the whole of* Ω. *Then, for a given* $h \in L^2$, *a distribution solution* $u \in H_0^m$ *of* (23), *with* $g(x, u)$ *and* $g(x, u)u$ *in* L^1, *exists provided*

$$\int_\Omega hv < -\int_\Omega G_- v \tag{30}$$

for all nonzero λ_1*-eigenfunction* $v \leqslant 0$ *in* Ω.

Proof. Special case of Corollary 3. Q.E.D.

Of course results analogous to Corollaries 4 and 5 hold for $G_-(x) = +\infty$ (instead of $G_+(x) = -\infty$) on some $\Omega' \subset \Omega$.

COROLLARY 6. *Assume* (L-1),..., (g-4) *as in Theorem* 4. *Suppose also that* $G_+ = -\infty$ *and* $G_- = +\infty$ *in the whole of* Ω. *Then a distribution solution* $u \in H_0^m$ *of* (23), *with* $g(x, u)$ *and* $g(x, u)u$ *in* L^1, *exists for all* $h \in L^2$.

Proof. Direct consequence of Corollary 5. Q.E.D.

The conclusion of Corollary 6 still holds if $G_+ = -\infty$ and $G_- = +\infty$ on $\Omega' \subset \Omega$ with $|\Omega'| > 0$, provided that the solutions of $Lu = \lambda_1 u$ enjoy the unique continuation property. This follows from (26).

COROLLARY 7 (Landesman–Lazer [9]). *Suppose* (L-1), (L-2), (L-3), *and* (g-1). *Assume also that there is* $c(x) \in L^2(\Omega)$ *such that* $|g(x, s)| \leqslant c(x)$, *for all* $s \in \mathbb{R}$, $x \in \Omega$. *Then, for a given* $h \in L^2$, *Eq.* (23) *has a distribution solution* $u \in H_0^m$, *with* $g(x, u)$ *and* $g(x, u)u$ *in* L^1, *provided*

$$\int_\Omega hv < -\int_{v>0} G_+ v - \int_{v<0} G_- v \tag{31}$$

for all nonzero λ_1*-eigenfunctions.*

Proof. It suffices to observe that, under the present hypotheses, conditions (g-2), (g-3), and (g-4) follow immediately. Q.E.D.

COROLLARY 8. (*A nonresonant problem, a reformulation of Theorem* 1). *Assume* (L-1), (L-2), (L-3), (g-1), *and* (g-2). *Suppose also that there are* $\epsilon > 0$ *and* $s_0 > 0$ *and* $0 \leqslant \beta(x) \in L^2(\Omega)$ *such that*

$$g(x, s)/s \leqslant -\epsilon + \beta(x)/|s|, \qquad x \in \Omega, \quad |s| > s_0. \tag{32}$$

Then (23) *has a distribution solution* $u \in H_0^m$, *with* $g(x, u)$ *and* $g(x, u)u$ *in* $L^1(\Omega)$, *for each* $h \in L^2(\Omega)$ *given.*

Proof. (32) implies (g-4). Moreover $G_+ \equiv -\infty$, $G_- \equiv +\infty$. So the result follows from Corollary 6. Q.E.D.

Some Examples of Application of Theorem 4 *and Its Corollaries*

(I) Let $g(x, s) = (-s)^{1/2}$, for $s \leqslant 0$ and $g(x, s) = -se^s$, for $s > 0$. Then for each given $h \in L^2$, Eq. (23) has a distribution solution $u \in H_0^m$, with $g(x, u)$ and $g(x, u)u$ in $L^1(\Omega)$. To see that, just observe that $G_+ \equiv -\infty$ and $G_- \equiv +\infty$. All the other conditions of Corollary 6 are readily verified. Let us remark that this is a resonant problem since $\lim_{s \to -\infty} g(x, s)/s = 0$.

(II) Let $g(x, s) = -g(x)e^s$, where $g(x) \in L^2(\Omega)$, $g(x) \geqslant 0$, and the set $\Omega' = \{x \in \Omega: g(x) > 0\}$ has positive measure. Then, for a given $h \in L^2(\Omega)$, there exists a distribution solution $u \in H_0^m$ of $Lu - \lambda_1 u = -g(x)e^u + h(x)$, with $g(x, u)$ and $g(x, u)u$ in L^1, provided $\int_\Omega hv < 0$ for all eigenfunctions $v \leqslant 0$ in Ω'. To prove this we use Corollary 4. Since $G_+ = -\infty$ in Ω', $G_+ = 0$ in $\Omega \backslash \Omega'$, and $G_- \equiv 0$, (30) reduces to $\int_\Omega hv < 0$. Again observe that this is a resonant problem since $\lim_{s \to -\infty} g(x, s)/s = 0$, for $x \in \Omega$, and $\lim_{s \to +\infty} g(x, s)/s = -\infty$, for $x \in \Omega'$. This example has been studied in [3] for second-order operators.

(III) Let $g(x, s) = g(x)e^{-s}$, where $g(x) \in L^2$, $g(x) \geqslant 0$, and let $\Omega' = \{x \in \Omega: g(x) > 0\}$. Then, for a given $h \in L^2(\Omega)$, there exists a distribution solution $u \in H_0^m$ of $Lu - \lambda_1 u = g(x)e^{-u} + h(x)$, with ge^{-u} and $ge^{-u}u$ in L^1, provided $\int_\Omega hv < 0$ for all nonzero λ_1-eigenfunction $v \geqslant 0$ in Ω'. Indeed, $G_+(x) \equiv 0$, $G_-(x) = +\infty$ on Ω' and 0 on $\Omega \backslash \Omega'$, and it suffices to apply Corollary 4 (with the roles of G_+ and G_- interchanged). Again this is a resonant problem.

REFERENCES

1. D. G. DE FIGUEIREDO, "The Dirichlet Problem for Nonlinear Elliptic Equations: A Hilbert Space Approach," Springer Lecture Notes, Vol. 445, pp. 144–165, Springer-Verlag, New York, 1974.
2. D. G. DE FIGUEIREDO, Semilinear elliptic problems with nonlinearities near the first eigenvalue, *in* "Dynamical Systems" (Proc. Univ. Florida International Symposium) (A. R. Bednarek and L. Cesari, Eds.), pp. 59–70, Academic Press, New York, 1976.
3. J. KAZDAN AND F. W. WARNER, Remarks on some quasilinear elliptic equations, *Comm. Pure Appl. Math.* **28** (1975), 567–597.
4. J. P. GOSSEZ, Nonlinear elliptic boundary value problems for equations with rapidly or slowly increasing coefficients, *Trans. Amer. Math. Soc.* **190** (1974), 163–205.

5. J. P. Gossez, Surjectivity results for pseudo-monotone mappings in complementary systems, *J. Math. Anal. Appl.* **53** (1976), 484–494.
6. D. G. de Figueiredo and J. P. Gossez, Perturbation non linéaire d'un problème elliptique linéaire près de sa première valuer propre, *C. R. Acad. Sci. Paris* **284** (1977), 163–166.
7. A. Friedman, "Partial Differential Equations," Holt, Rinehart & Winston (1969).
8. C. L. Dolph, Nonlinear integral equations of the Hammerstein type, *Trans. Amer. Math. Soc.* **66** (1949), 289–307.
9. E. M. Landesman and A. C. Lazer, Nonlinear perturbations of linear elliptic boundary value problems at resonance, *J. Math. Mech.* **19** (1970), 609–623.
10. L. Nirenberg, An application of generalized degree to a class of nonlinear problems, *Troisième Colloq. Anal. Fonct. Liege* (Sept. 1970), 57–74.
11. Peter Hess, On a theorem by Landesman and Lazer, *Indiana Univ. Math. J.* **23** (1974), 827–829.
12. S. Fucik, Open problems in the solvability of nonlinear equations, to appear.
13. A. Ambrosetti and G. Prodi, On the inversion of some differentiable mappings with singularities between Banach spaces, *Ann. Mat.* **93** (1972).
14. D. Schwartz, "Linear Operators," Part 1: "General Theory," Interscience, New York, 1958.
15. W. A. Strauss, On weak solutions of semi-linear hyperbolic equations, *An. Acad. Brasil Ci.* **42** (1970), 645–651.
16. H. Brezis and L. Nirenberg, Characterizations of the ranges of some nonlinear operators and applications to boundary value problems, *Ann. Scuola Norm. Sup. Pisa*, to appear.

Nonlinear Analysis, Theory, Methods & Applications. Vol. 3, No. 5, pp. 629–634 0362-546X/79/0901–0629 $02.00/0
© Pergamon Press Ltd. 1979. Printed in Great Britain

PERTURBATIONS OF SECOND ORDER LINEAR ELLIPTIC PROBLEMS BY NONLINEARITIES WITHOUT LANDESMAN–LAZER CONDITION

DJAIRO GUEDES DE FIGUEIREDO
Universidade de Brasilia, Departmento de Matemática, Brasilia, D.F., Brazil
and
WEI-MING NI
Courant Institute of Mathematical Sciences

(*Received* 30 *August* 1978)

Key Words: Nonlinear elliptic problems, maximum principle, *a priori* bounds, Leray–Schauder degree theory.

1.

LET $\mathscr{L}$ be a second order symmetric uniformly elliptic operator with smooth coefficients acting on real valued functions defined in a bounded smooth domain Ω in $\mathbf{R}^N$. Let us denote by λ_1 the first eigenvalue of the eigenvalue problem $\mathscr{L}u = \lambda u$ in Ω and $u = 0$ on $\partial\Omega$. It is well known [1] that λ_1 is a simple eigenvalue and that there is a corresponding smooth eigenfunction $\phi > 0$ in Ω. It follows from the Hopf–Giraud maximum principle [2, p. 151] that the outward normal derivative $\partial\phi/\partial v$ is <0 on the boundary. In this paper we consider the Dirichlet problem

$$\mathscr{L}u - \lambda_1 u + g(u) = h \qquad \text{in } \Omega, \qquad u = 0 \text{ on } \partial\Omega, \tag{1}$$

where $g:\mathbf{R} \to \mathbf{R}$ is a bounded continuous function. An intensive research in this problem started after the 1970 paper [3] of Landesman and Lazer. Their result is essentially as follows. Problem (1) has a solution $u \in W_0^{2,1}(\Omega) \cap W^{2,2}(\Omega)$ for every $h \in L^2(\Omega)$ provided

$$g_- \int_\Omega \phi < \int_\Omega h\phi < g_+ \int_\Omega \phi \tag{2}$$

where $g_\pm = \lim_{u \to \pm\infty} g(u)$. ($W^{m,p}(\Omega)$ is the Sobolev space of L^p-functions whose derivatives in the distribution sense are in $L^p(\Omega)$, see, for example, Adams [4]). In fact they do not restrict to the first eigenvalue, but they suppose that $g_- \leqslant g(u) \leqslant g_+$, for all $u \in \mathbf{R}$. The hypothesis on the simplicity of the eigenvalue, the condition $g_- \leqslant g(u) \leqslant g_+$, the order of the elliptic operator and even the condition on the boundedness of g have all been relaxed by several authors in recent years. However in all cases a condition like (2) has been assumed. Recently Fučik and Krbec [5] and Hess [6] have been able to prove results on the existence of solutions of (1) for h orthogonal to ϕ in L^2 in the case that $g_+ = g_- = 0$. But in every case there are hypotheses on the speed that g goes to 0 at ∞. Namely they assume that g is *odd* and

$$\liminf_{t \to +\infty} \left[t \min_{a \leqslant \tau \leqslant t} g(\tau) \right] > 0$$

Originally published in: Nonlinear Anal. - TMA 3 (1979), 629-634,

DOI: 10.1007/978-3-319-02856-9_8

for some suitable constant $a > 0$. In contrast we make no assumption on the behaviour of g at ∞, except in Section 3 where we show how our main theorem implies a result of Hess. In the next section we prove a result (Theorem 2) which implies the following theorem.

THEOREM 1. Let $\mathscr{L}$ be as above and $g:\mathbf{R} \to \mathbf{R}$ be a bounded continuous function such that

$$g(t)t \leqslant 0, \qquad t \in \mathbf{R}. \tag{3}$$

Then, for every L^∞-function h orthogonal to ϕ in L^2, equation (1) has a solution $u \in W_0^{1,2} \cap W^{2,p}$ for every $p \geqslant 2$. In particular $u \in C^{1,\lambda}(\bar{\Omega})$ for $0 < \lambda < 1$ and $u = 0$ on $\partial\Omega$ in the classical sense. ($C^{1,\lambda}(\bar{\Omega})$ is the space of the C^1-functions in $\bar{\Omega}$ whose first derivatives are Hölder–continuous of exponent λ).

Remark. Once the solution is proved to be in $W^{2,p}(\Omega)$ for every $p \geqslant 2$, the second part of the conclusion in the above theorem follows from the Sobolev imbedding theorem (4, p. 98]. One should not expect a better degree of regularity for the solution on the mere assumption of continuity of g and the essential boundedness of h. It is well known that the solution of the linear Dirichlet problem $\Delta u = \rho$ in Ω, $u = 0$ on $\partial\Omega$, for a continuous function ρ, is not necessarily in $C^2(\Omega)$.

2.

Instead of proving Theorem 1 directly, we prefer to state and prove a more general result for the operator A defined below. The advantage of this procedure is to isolate the properties of $\mathscr{L}$ which are really essential. Let $A:D(A) \subset L^2(\Omega) \to L^2(\Omega)$ be a linear operator with closed range $R(A)$ and compact resolvent $(A - \lambda)^{-1}:L^2(\Omega) \to L^1(\Omega)$ for all λ in the resolvent set $\rho(A)$, which is supposed to be not empty. Moreover the nullspace $N(A)$ of A is generated by a function $\phi \in C^1(\bar{\Omega})$ such that $\phi > 0$ in Ω, $\phi = 0$ on $\partial\Omega$ and $\partial\phi/\partial\nu < 0$ on $\partial\Omega$, where $\partial/\partial\nu$ is the outward normal derivative. We also suppose that $L^2 = R(A) \oplus N(A)$, and so A^{-1} is a well-defined operator from $R(A)$ onto $D(A) \cap R(A)$. Finally we assume that

$$A^{-1}(R(A) \cap L^p) \subset W^{2,p} \cap W_0^{1,2}, \qquad p \geqslant 2, \tag{4}$$

and

$$\|A^{-1}f\|_{W^{2,p}} \leqslant c_p\|f\|_{L^p}, \qquad f \in R(A) \cap L^p. \tag{5}$$

The operator $\mathscr{L} - \lambda_1$ satisfies all the above properties. (4) and (5) are just the statements of the regularity theory of the Dirichlet problem, see, for example, [7].

Let us denote by P and Q the orthogonal projections on L^2 into $N(A)$ and $R(A)$ respectively.

Let $g:\mathbf{R} \to \mathbf{R}$ be a continuous and bounded function, and let $b > 0$ be such that $|g(t)| \leqslant b$, for all $t \in \mathbf{R}$. Then the Niemytskii mapping $u \mapsto g(u)$ is bounded and continuous in every L^p, $1 \leqslant p \leqslant \infty$, and

$$\|g(u)\|_{L^p} \leqslant b|\Omega|^{1/p}, \qquad u \in L^p. \tag{6}$$

Now let us state our main result.

THEOREM 2. Let A be the linear operator defined above and $g: \mathbf{R} \to \mathbf{R}$ a bounded continuous function satisfying the following condition
(*) for any given $R > 0$ there is $t_0 > 0$ such that

$$\int g(u)\phi \cdot \int g(v)\phi < 0 \tag{7}$$

for all $u, v \in L^2(\Omega)$ with $u \geqslant t_0\phi$, $v \leqslant -t_0\phi$, $\|Qu\|_{L^2} \leqslant R$ and $\|Qv\|_{L^2} \leqslant R$. Then, for every L^∞-function h orthogonal in L^2 to ϕ, the equation

$$Au + g(u) = h \tag{8}$$

has a solution $u \in W_0^{1,2} \cap W^{2,p}$, for all $p \geqslant 2$.

Theorem 2 $\Rightarrow$ *Theorem* 1. It suffices to prove that (3) implies (*). In fact, we may suppose that there is $t_1 > 0$ and $t_2 < 0$ such that $g(t_1) < 0$ and $g(t_2) > 0$, for otherwise $g(t) \equiv 0$ for $t \leqslant 0$ or $g(t) \equiv 0$ for $t \geqslant 0$. And in this case we may construct solutions of the problem (1) in the following way: let $\omega = A^{-1}h$, which by the Sobolev imbedding theorem [4, p. 98] belongs to $C^{1,\lambda}(\bar{\Omega})$. Then there is a $t > 0$ such that $u_+ \equiv t\phi + \omega \geqslant 0$ and $u_- \equiv -t\phi + \omega \leqslant 0$. If $g(t) \equiv 0$ for $t \geqslant 0$, u_+ is a solution of (1), and similarly the other case. Alternatively if there are the numbers t_1 and t_2 as specified above take t_0 such that $t_0 \max[\phi(x): x \in \Omega] > t_1$ and $-t_0 \max[\phi(x): x \in \Omega] < t_2$. Then clearly

$$\int g(u)\phi < 0 \quad \text{and} \quad \int g(v)\phi > 0$$

for $u \geqslant t_0\phi$ and $v \leqslant -t_0\phi$. Q.E.D.

Remark. Under the hypothesis of Theorem 1, except (3) being replaced by

$$g(t)t \geqslant 0, \qquad t \in \mathbf{R}, \tag{9}$$

the same conclusion of that theorem holds. Such a result has been observed by Fučik and Krbec [5] for odd functions g. We remark however that in this case (even without the oddness of g) the result follows readily from our Theorem 2. Moreover this is not very exciting, for in this case the method of monotone convergence [8, 9] applies, since we can show the existence of a subsolution u_- and a supersolution u_+ such that $u_- \leqslant u_+$; these are precisely the functions that appear in the above proof of Theorem 1.

Example. The existence of solution for the Dirichlet problem for

$$-\Delta u - \lambda_1 u - u\,e^{-u^2} = h, \qquad h \perp \phi,$$

follows from our Theorem 1. This case is not covered by the results of Fučik and Krbec or Hess. It seems also that there are no subsolution u^- and supersolution u^+, with $u^- \leqslant u^+$; then the monotone convergence method would not apply.

Our proof of Theorem 2 makes use of the rather surprising fact that the solutions of (8) have an *a priori* bound. In general *a priori* bounds are obtained when one knows the behaviour of g at ∞.

LEMMA. Under the hypotheses of Theorem 2, there exists a constant c, depending only on A, $\|h\|_{L^\infty}$ and the bound b on g given in (6) such that

$$\|u\|_{L^2} \leqslant c$$

for all eventual solutions $u \in L^2(\Omega)$ of (8).

Proof. Let $u = t\phi + \omega$ be an eventual solution of (8), where $\omega \in R(A)$ and $t \in \mathbf{R}$. Since $g(u) \in L^\infty$ it follows from (4) that $\omega \in W^{2,p} \cap W_0^{1,2}$ for all p, and from (5)

$$\|\omega\|_{W^{2,p}} \leqslant c_p'(b + \|h\|_{L^\infty}).$$

By the Sobolev imbedding theorem $\omega \in C^{1,\lambda}(\bar{\Omega})$ and

$$\|\omega\|_{C^1} \leqslant c \Rightarrow \|\omega\|_{L^2} \leqslant c' \tag{10}$$

where c and c' are constants independent of the particular solution u. Finally we see that t has to be bounded also, for otherwise for large $t > 0$ we would have in view of (10) that $t\phi + \omega \geqslant t_0\phi$. Applying condition (*) we would get $\int g(u)\phi \neq 0$, which is impossible since from (8) we see that this integral must be 0. In a similar way we prove that $t < 0$ and large in absolute value is not possible. So $\|t\phi\|_{L^2}$ is bounded for every solution $u = t\phi + \omega$ of (8). This together with (10) proves the lemma. Q.E.D.

Proof of Theorem 2. Let $\lambda \in \rho(A)$. Then equation (8) is equivalent to

$$u = (A + \lambda I)^{-1}(\lambda u - g(u) + h) \equiv Tu, \tag{11}$$

where $T: L^2 \to L^2$ is compact. Let us define the operator

$$T_k = \frac{1}{k+1} Tu - \frac{k}{k+1} T(-u), \qquad 0 \leqslant k \leqslant 1 \tag{12}$$

which is also compact and $T_0 = T$ and T_1 is an odd operator. It is known [10] that the Leray–Schauder degree

$$\deg(I - T_1, B_R, 0) \neq 0$$

for any ball $B_R = \{u \in L^2 : \|u\|_{L^2} \leqslant R\}$. If we show that there is an R such that

$$u - T_k u \neq 0, \qquad \|u\|_{L^2} = R, \qquad 0 \leqslant k \leqslant 1 \tag{13}$$

then $\deg(T, B_R, 0) \neq 0$. Thus (11) would have a solution in L^2 and from (4) our result would follow. So let us prove that the eventual solutions of

$$u - T_k u = 0 \tag{14}$$

have an *a priori* bound. Equation (14) is equivalent to

$$Au + \frac{1}{k+1} g(u) - \frac{k}{k+1} g(-u) = \frac{1-k}{1+k} h.$$

Observe that, for every $0 \leqslant k \leqslant 1$,

$$g_k(t) = \frac{1}{k+1} g(t) - \frac{k}{k+1} g(-t), \qquad t \in \mathbf{R}$$

is a bounded continuous function which satisfies (*). Also $|g_k(t)| \leqslant b$, for all $t \in \mathbf{R}$ and $0 \leqslant k \leqslant 1$, and $\|(1-k)/(1+k)h\|_{L^\infty} \leqslant \|h\|_{L^\infty}$. Then apply the Lemma to conclude that the eventual solutions of (14) have an *a priori* bound. The proof of Theorem 2 is thus complete. Q.E.D.

3.

In this section we show that a result similar to one of Hess [11] can be obtained as a corollary of our Theorem 2.

COROLLARY. Let A be as in Section 2 and let $g:\mathbf{R} \to \mathbf{R}$ be a bounded continuous function such that

$$\limsup_{t \to +\infty} tg(t) = -\alpha < 0 \qquad \text{and} \qquad \limsup_{t \to -\infty} tg(t) = -\beta < 0. \tag{15}$$

Then equation (8) has a solution $u \in W_0^{1,2} \cap W^{2,p}$ for all $p \geqslant 2$.

Proof. We first observe that it follows from the hypotheses that

$$\liminf_{t \to +\infty} g(t) \leqslant 0 \qquad \text{and} \qquad \limsup_{t \to -\infty} g(t) \geqslant 0.$$

We then assume that these limits are actually both equal to zero, for otherwise the existence of solution for equation (8) would follow from known results [3]. In order to apply Theorem 2 it suffices to prove that condition (*) holds for functions u and v with the additional restriction that $Qu, Qv \in L^\infty$ and $\|Qu\|_{L^\infty}, \|Qv\|_{L^\infty} \leqslant k$ for some constant k; check the proof of the lemma where (*) is used. We now prove that there is $t_0 > 0$ such that

$$\int_\Omega g(u)\phi < 0 \tag{16}$$

for $u \geqslant t_0\phi$, $\|Qu\|_{L^\infty} \leqslant k$. Let $K > 0$ be such that $tg(t) \leqslant -\alpha/2$ and $|g(t)| < \alpha/(4k)$ for $t > K$. Let $t > 0$ and define

$$\Omega_t = \{x \in \Omega : \phi(x) \leqslant (k + K)t^{-1}\}.$$

Observing that, for $x \in \Omega\backslash\Omega_t$, $u(x) \geqslant K$ we can estimate the integral in (16) as follows:

$$\int_\Omega g(u)t\phi = \int_{\Omega\backslash\Omega_t} g(u)u - \int_{\Omega\backslash\Omega_t} g(u)Qu + \int_{\Omega_t} g(u)t\phi$$

$$\leqslant -\frac{\alpha}{2}|\Omega\backslash\Omega_t| + \frac{\alpha}{4}|\Omega\backslash\Omega_t| + b(k+K)|\Omega_t|$$

which is negative for large positive t. Here $|\Omega\backslash\Omega_t|$ denotes the Lebesgue measure of $\Omega\backslash\Omega_t$. In an analogous way we show that $\int_\Omega g(v)\phi > 0$ for $v \leqslant -t_0\phi$, $\|Qv\|_{L^\infty} \leqslant k$ and large t_0. Q.E.D.

Remark. Observe that (15) implies that $tg(t) < 0$ for $|t| > d$ for some $d > 0$. It is clear that Theorem 1 does not hold with condition (3) replaced by $tg(t) \leqslant 0$ for $|t| \geqslant d$, for some $d > 0$. Indeed, consider the example

$$g(t) = \begin{cases} 1 & t \leqslant 1 \\ 0 & t \geqslant 2 \\ \text{linear} & 1 \leqslant t \leqslant 2 \end{cases}$$

where the linear piece is taken so that g is continuous. The Dirichlet problem for $-\Delta u - \lambda_1 u + g(u) = h$, where h is orthogonal in L^2 to ϕ, has no solution, since the necessary condition $\int g(u)\phi = 0$ cannot ever be satisfied. Yet this condition is sufficient for the existence of periodic solutions for ordinary differential equations [12], which in the partial differential equation case would correspond to Neumann problem.

Acknowledgement—The authors wish to express their gratitude to Professor Louis Nirenberg for suggesting this problem and his help.

REFERENCES

1. KREIN M. G. & RUTMAN M. A., Linear operators leaving invariant a cone in a Banach space. *Am. math. Soc. Transl. Ser. 1* **10**, 199–325 (1962).
2. BERS L., JOHN F. & SCHECHTER M., *Partial Differential Equations*. Interscience, New York (1964).
3. LANDESMAN E. M. & LAZER A. C., Nonlinear perturbations of linear elliptic boundary value problems at resonance, *J. Math. Mech.* **19**, 609–623 (1970).
4. ADAMS R. A., *Sobolev Spaces*. Academic Press, New York (1975).
5. FUČIK S. & KRBEC M., Boundary value problem with bounded nonlinearity and general nullspace of linear part. *Math. Z.*, to appear.
6. HESS P., A remark on the preceding paper of Fučik and Krbec, *Math. Z.*, to appear.
7. FRIEDMAN A., *Partial Differential Equations*. Holt, Rinehart & Winston, New York (1969).
8. AMANN H., On the existence of positive solutions of nonlinear elliptic boundary value problems, *Indiana Univ. math. J.* **21**, 125–146 (1971).
9. SATTINGER D. H., *Topics in Stability and Bifurcation Theory*. Springer-Verlag, Berlin (1973).
10. NIRENBERG L., Topics in nonlinear functional analysis, Courant Inst. Math. Sci. Lecture Notes (1974).
11. HESS P., Nonlinear perturbations of linear elliptic and parabolic problems at resonance: existence of multiple Solutions, to appear.
12. LAZER A. C., On Schauder's fixed point theorem and forced second-order nonlinear oscilations, *Differential and Functional Equations*, (Edited by W. A. Harris, Jr and Y. Sibuya). Benjamin, New York (1967).

COMM. IN PARTIAL DIFFERENTIAL EQUATIONS, 6(1), 91-120 (1981)

DOUBLE RESONANCE IN SEMILINEAR ELLIPTIC PROBLEMS

Henri Berestycki
CNRS and Université Paris 6
4, Place Jussieu
75230 Paris Cedex 05
France

Djairo Guedes de Figueiredo
Departamento de Matemática
Universidade de Brasília
70.910 - Brasília-DF
Brazil

INTRODUCTION. In this paper we establish some existence results for problems of the type

$$Au + Bu = h, \tag{0.1}$$

where h is given in $L^2(\Omega)$; Ω is a bounded domain of R^N. A is a linear operator in $L^2(\Omega)$. One typical model of such an A is a self-adjoint elliptic operator with zero Dirichlet data on the boundary. However, our results apply to more general cases since, for instance, we do not require the self-adjointness of A. In the sequel, B designates a Nemytskii mapping defined for functions $u \in L^2(\Omega)$ by $(Bu)(x) = g(x, u(x))$, where $g:\bar{\Omega}\times R \to R$ is some given function with linear growth. A particular emphasis is given here to the cases where the behavior

Originally published in: Comm. Partial Differential Equations 6 (1981), 91-120,

Reprint by Springer-Verlag Berlin Heidelberg 2013
with kind permission, DOI: 10.1007/978-3-319-02856-9_9

of $g(x,u)$ as $u \to \pm\infty$ produces resonance at more than one eigenvalue, and in particular, at two consecutive eigenvalues, which we call "double resonance". Problems of this type have received much attention in recent years, after the 1970 paper of Landesman-Lazer [17]. In the case that A is self-adjoint and the function $g(x,u)$ is monotone increasing (or decreasing) in u, necessary conditions for the solvability of (0.1) can be obtained taking inner products of the equation with eigenfunctions of A. Then, in [17], on the basis of those conditions, some sufficient conditions for the solvability of (0.1) were established, in the case of 2nd order self-adjoint elliptic problems. These conditions, known in the literature as Landesman-Lazer conditions have proved to be very useful. In the work of Brézis-Nirenberg [5], equation (0.1) is analysed as a particular case of their theory on the characterization of the range of nonlinear operators. We refer the reader to the bibliography in [5] for further references on this type of problems, up to 1978. We would also like to mention the more recent papers of Ambrosetti-Amann-Mancini [3], Amann-Mancini [2], de Figueiredo [9], de Figueiredo-Ni [11], Fucik-Krbec [12], Gonçalves [13], Hess [14], [15], Mawhin [18]. In all these papers, conditions are set in such a way that resonance occurs at one eigenvalue; an exception is the paper of

Dancer [8], where some second order self-adjoint problems are considered. In order to illustrate some of our results let us be more specific about the assumptions on equation (0.1). We suppose that $A:D(A) \subset L^2(\Omega) \to L^2(\Omega)$ is a closed densely defined linear operator with a closed range $R(A)$ and a finite dimensional null-space $N(A) \neq \{0\}$, such that $R(A) = N(A)^{\perp}$. Then the restriction of A to $R(A)$ has a well-defined (singlevalued) inverse from $R(A)$ into itself. Let it be denoted by $A^{-1}:R(A) \to R(A)$. We suppose also that A^{-1} is compact. It is easy to show (see Section 1) that one of the following possibilities occur: i) either there is a maximal real number $\alpha > 0$ such that $-\alpha^{-1}\|Au\|^2 \leq (Au,u)$, for all $u \in D(A)$, ii) or there is no such a number, and that is the case when A is monotone. In the latter case we take $\alpha = +\infty$. It will be shown later that α (when $\alpha < \infty$) is an eigenvalue of the operator $-A$, provided A satisfies some further condition; this is the case in particular when $A = A^*$. The Nemytskii mapping B is defined by $(Bu)(x) = g(x, u(x))$, where $g:\bar{\Omega}\times R \to R$ is a function satisfying Carathéodory's conditions (a Carathéodory function, for short) and with a linear growth, i.e., there are a constant $\gamma > 0$ and an L^2-function $b(x)$ such that

$$(0.2) \qquad |g(x,u)| \leq \gamma|u| + b(x), \quad u \in R, \quad x \in \Omega.$$

It is implicitly understood that $x \in \Omega$ means $x \in \Omega$ a.e.. Inequality (0.2) above implies that B is a well defined mapping from $L^2(\Omega)$ into itself. And moreover it is continuous and bounded (i.e., takes bounded sets into bounded sets). For simplicity at the moment, we assume that the following limits exist

$$(0.3) \qquad k_{\pm}(x) = \lim_{u \to \pm\infty} \frac{g(x,u)}{u}.$$

In Section 2, we actually state more general results that do not require the existence of these limits. One of the main features of this paper is allowing resonance at two eigenvalues, namely here, 0 and α. So we will assume

$$(0.4) \qquad 0 \leq k_{\pm}(x) \leq \alpha, \quad x \in \Omega.$$

Let us remark that the case when $0 \leq k_{\pm}(x) < \beta < \alpha$, which corresponds to resonance at one eigenvalue, has been extensively discussed in [5] and [9]. We can now state one of our results, which will be proved in Section 2.

Theorem 1. Assume the above hypotheses on A and B. Suppose further that $\alpha < \infty$ and that $A^{-1}+\alpha^{-1}I:R(A)\to R(A)$ is strictly monotone (see definition in Section 1). If

$$(0.5) \qquad k_{+}(x)v^{+}(x)+k_{-}(x)v^{-}(x) \not\equiv 0, \quad \forall\, 0 \neq v \in N(A), \text{ and}$$

(0.6) $(\alpha-k_+(x))v^+(x)+(\alpha-k_-(x))v^-(x)\not\equiv 0, \quad \forall\ 0\neq v \in N(A+\alpha I)$

then equation (0.1) is solvable for all $h \in L^2(\Omega)$, i.e. $A+B$ is surjective. [Here $v^+(x) = \max(v(x), 0)$ is the positive part of v and v^- is its negative part. $\not\equiv 0$ is taken in a.e. sense].

Remark. Conditions (0.5) and (0.6) are equivalent to

(0.7) $$\int_{v>0} k_+v^2 + \int_{v<0} k_-v^2 > 0, \quad \forall\ 0\neq v \in N(A), \text{ and}$$

(0.8) $$\int_{v>0} (\alpha-k_+)v^2 + \int_{v<0} (\alpha-k_-)v^2 > 0, \quad \forall\ 0\neq v \in N(A+\alpha I).$$

This type of conditions appears in the work of J.P. Gossez and one of the authors [10].

The main model for an operator A satisfying the requirements in Theorem 1 is a uniformly strongly elliptic linear differential operator in Ω of order $2m$, with domain $D(A) = H^{2m}(\Omega) \cap H_0^m(\Omega)$, and which is self-adjoint. In particular, the theorem applies to the Dirichlet problem for $\Delta u + \lambda_k u + g(x,u) = h$, where λ_k is an eigenvalue of $-\Delta$. In this case, $\alpha = \lambda_{k+1} - \lambda_k$. Hypotheses (0.5) and (0.6) are verified for a wide class of nonlinearities g. In particular, they are implied by $0 < \varepsilon \leq k_+(x) \leq \alpha$ and $0 \leq k_-(x) \leq \alpha-\varepsilon < \alpha$, or $0 \leq k_+(x) \leq \alpha-\varepsilon < \alpha$ and $0 < \varepsilon \leq k_-(x) \leq \alpha$, in the case that the eigenfunctions in $N(A)$ and $N(A+\alpha I)$ change sign. In Section 3, we consider the special case when A is a self-adjoint second order elliptic operator and zero is the first eigenvalue of A. In this case the eigenfunctions in $N(A)$ have a definite sign. This situation arise, for example, in the Dirichlet problem for $\Delta u + \lambda_1 u + g(x,u) = h$. Here $\alpha = \lambda_2 - \lambda_1$. For these problems we treat the case when $k_+(x) \equiv 0$, replacing condition (0.5) by a Landesman-Lazer condition. We also allow a "crossing" of several eigenvalues. That is, the hypothesis on the behavior of $g(x,u)$ at $-\infty$ is described by the condition $0 \leq k_-(x) \leq \delta$, where δ may be any positive number, not necessarily $\leq \alpha$. This extends results of Aguinaldo-Schmitt [1] and Castro [7] on the ordinary differential equation case, and a previous result of Dancer [8]. The results in this paper are all proved by simple applications of the Leray-Schauder degree theory. We remark that the Landesman-Lazer condition and conditions like (0.5) and (0.6) yield a priori estimates on the solutions of (0.1).

This paper is organized as follows. In Section 1 we prove some useful abstract results for the class of linear operators A considered here. In Section 2 we prove Theorem 1 stated above and a more general form of it. These theorems model more efficiently the case when 0 and α are higher order eigenvalues, or better, the case when the eigenfunctions in $N(A)$ and $N(A+\alpha I)$ change sign. For that reason, in Section 3, we take up the case when the eigenfunctions in $N(A)$ have a definite sign.

1. SOME PRELIMINARY ABSTRACT RESULTS. Let H be a real Hilbert space. A linear operator $A:D(A) \subset H \to H$ is said to satisfy Property I (cf. [2]) if the following conditions hold: (i) A is closed, densely defined, with a closed range $R(A)$, (ii) A has a non-trivial null-space $N(A)$, and $N(A) = N(A^*)$, where A^* is the adjoint of A, (iii) the restriction of A to $D(A) \cap R(A)$ has an inverse, denoted by $A^{-1}:R(A) \to R(A)$, which is assumed to be compact. An operator A satisfying Property I induces an orthogonal decomposition of the Hilbert space H as $H = R(A) \oplus N(A)$. Let us denote by Q and P the orthogonal projections onto $R(A)$ and $N(A)$, respectively, along this decomposition. So, for every $u \in H$, we have $u = Qu + Pu$. From (iii) above it follows that there exist positive constants c such that $-c^{-1}\|Au\|^2 \leq (Au,u)$, for all $u \in D(A)$. Namely $\|A^{-1}\|^{-1}$ is one such a constant c. Let us denote by $\alpha = \sup c$, that is, the supremum of all these constants; α may be $+\infty$. So

$$(1.1) \qquad -\frac{1}{\alpha}\|Au\|^2 \leq (Au,u), \quad \forall\, u \in D(A).$$

We remark that A is monotone, i.e., $(Au,u) \geq 0$, for $u \in D(A)$, if and only if $\alpha = +\infty$. It is easily seen that if A is self-adjoint, $A = A^*$, then α is the smallest positive eigenvalue of $-A$. However this fact is also valid for more general classes of operators. The number α can be characterized as the largest positive number such that $A^{-1} + \alpha^{-1}I$ is a monotone operator in $R(A)$.

Before stating our results let us recall some basic definitions. A bounded linear operator $T:H \to H$ in a Hilbert space H is said to be strictly monotone if there is a constant $c > 0$ such that

$$(1.2) \qquad (Tu,u) \geq c\|Tu\|^2, \quad \forall\, u \in H.$$

A linear operator $T:D(T) \subset H \to H$ is said to be anglebounded if there is a constant $c > 0$ such that

$$|(Tu,v)-(Tv,u)| \leq c(Tu,u)^{1/2}(Tv,v)^{1/2}, \quad \forall\, u,v \in D(T).$$

Bounded linear operators which are anglebounded satisfy (1.2). This can be easily seen using a theorem of Browder-Gupta [6] on the decomposition of anglebounded mappings. However there are bounded strictly monotone operators which are not anglebounded. Cf. Hess [16] and Brézis-Haraux [4]. Clearly, monotone self-adjoint operators are anglebounded. The following lemmas have proved to be very useful in our investigations.

Lemma 1. Let $A:D(A) \subset H \to H$ be an operator satisfying Property I. (i) When $0 < \alpha < \infty$, suppose that $A^{-1} + \alpha^{-1}I:R(A) \to R(A)$ is strictly monotone. If equality holds in (1.1), that is

$$- \frac{1}{\alpha}\|Au\|^2 = (Au,u) \tag{1.3}$$

for some $u \in D(A)$, then $u \in N(A) \oplus N(A+\alpha I)$.

ii) When $\alpha = +\infty$, suppose that A is angelbounded. If equality holds in (1.1), that is, $(Au,u) = 0$, for some $u \in D(A)$, then $u \in N(A)$.

Remark 1. If $A = A^*$, then $A^{-1} + \alpha^{-1}I$ is also self-adjoint, and so necessarily strictly monotone.

Proof of Lemma 1. Let $u = u_0 + u_1$, with $u_0 \in N(A)$ and $u_1 \in R(A)$, be an element in $D(A)$ for which (1.3) holds. If $u_1 = 0$, the proof is finished. Suppose that $u_1 \neq 0$. Then we obtain from (1.3) that

$$- \frac{1}{\alpha} \|Au_1\|^2 = (Au_1,u_1),$$

which can be written as

$$(A^{-1}x + \alpha^{-1}x, x) = 0 \tag{1.4}$$

with $x = Au_1$. Since $A^{-1} + \alpha^{-1}I$ is strictly monotone we obtain $A^{-1}x + \alpha^{-1}x = 0$, or $Au_1 + \alpha u_1 = 0$. This proves the lemma.

Remark 2. The additional assumption on A in order to have the conclusions of the above lemma is actually needed. Indeed, the mapping A in R^2 defined as a rotation by 90° satisfies $(Au,u) = 0$ for all $u \in R^2$, and $N(A) = \{0\}$. The previous lemma includes the case where A is self-adjoint. It is interesting to notice that in such a case $(A = A^*)$, a direct proof using the variational characterization of the first eigenvalue could be given. Observe that Lemma 1 does not establish that there always exist $u \in D(A)$, $u \neq 0$, such that $Au + \alpha u = 0$. This is proved next.

Lemma 2. Let A, $D(A) \subset H \to H$ be an operator satisfying Property I. Suppose that $0 < \alpha < \infty$ and that $A^{-1} + \alpha^{-1}I:R(A) \to R(A)$ is strictly monotone. Then $-\alpha$ is an eigenvalue of A.

Remark 3. It is easy to prove that the interval $(-\alpha,0)$ is in the resolvent set of A. So α is the smallest positive eigenvalue of $-A$. Lemma 2 which includes the case when $A = A^*$ establishes an assertion made in the Introduction.

Proof of Lemma 2. By definition of α there exists a sequence (x_n) in $R(A)$, with $\|x_n\| = 1$, such that $(A^{-1}x_n+\alpha^{-1}x_n, x_n) \to 0$. By taking subsequences, we may suppose that $x_n \rightharpoonup x$ and $A^{-1}x_n \longrightarrow A^{-1}x$, since $A^{-1}:R(A) \to R(A)$ is compact. ("$\rightharpoonup$" denotes weak convergence in H, while "$\longrightarrow$" denotes strong convergence in H). So

$$(1.6) \qquad 0 = \lim(A^{-1}x_n+\alpha^{-1}x_n,x_n) = (A^{-1}x,x) + \alpha^{-1},$$

which implies, in particular, that $x \neq 0$. From (1.6) and (1.5) we obtain $-\alpha^{-1}\|x\|^2 \leq (A^{-1}x,x) = -\alpha^{-1}$, which gives $\|x\| \geq 1$. On the other hand, $\|x\| \leq \liminf \|x_n\|=1$. Thus $\|x\| = 1$, and we get from (1.6) that $(A^{-1}x,x)+\alpha^{-1}(x,x)=1$ So, using Lemma 1, it follows that $u = A^{-1}x$ is an eigenfunction of A corresponding to the eigenvalue $-\alpha$. The proof of lemma 2 is complete.

2. PROOF OF THEOREM 1. In this section we apply the abstract results of the previous section to study the equation (0.1) in the Hilbert space $H = L^2(\Omega)$. First we prove Theorem 1 stated in the Introduction. The proof rests on a topological degree computation which requires a priori estimates. Let Q and P denote the orthogonal projections of $H = L^2(\Omega)$ onto $R(A)$ and $N(A)$,

respectively. Projecting the equation (0.1) onto $R(A)$ and $N(A)$, one obtains the following system

$$(2.1)\qquad \begin{cases} Qu + A^{-1}Q(Bu-h) = 0 \\ P(Bu-h) = 0, \end{cases}$$

which is easily seen to be equivalent to (0.1). For $t \in [0,1]$ we consider the mapping F_t from H into itself (we identify H with $R(A) \times N(A)$) defined by

$$(2.2)\qquad F_t(u) = \begin{bmatrix} Qu+tA^{-1}Q\{(1-t)\alpha u+t(Bu-h)\} \\ P\{(1-t)\alpha u+t(Bu-h)\} \end{bmatrix}.$$

Hence the original equation (0.1) is equivalent to $F_1(u) = 0$. Also observe that F_t, for each $t \in [0,1]$, is a compact perturbation of the identity, since $A^{-1}:R(A) \to R(A)$ is compact and $N(A)$ is finite dimensional. To compute the topological degree of F_t, we require the following a priori bound on the solutions of $F_t(u) = 0$, $t \in [0,1]$.

<u>Lemma 3</u>. <u>There exists a constant</u> $R > 0$ <u>such that</u> $F_t(u) = 0$ <u>implies</u> $\|u\| < R$, <u>for any</u> $t \in [0,1]$.

Before proving this lemma, we will complete the proof of Theorem 1, using this a priori estimate.

It follows from Lemma 3 that the topological degree $\deg(F_t, B_R, 0)$, where $B_R = \{u \in H: \|u\| < R\}$, is well defined and independent of $t \in [0,1]$. Hence F_t is an admissible homotopy for computing the degree of F_1. It is easily seen that F_0 is homotopic to the identity, so that $\deg(F_0, B_R, 0) = 1$. Therefore $\deg(F_1, B_R, 0) = 1$, which implies the existence of $u \in B_R$ such that $F_1(u) = 0$. It remains now to establish the a priori' estimate.

Remark 4. The proof of Theorem 1 could be done using Borsuk's theorem in a similar way as one of the authors did with Ni in [11]. J.P. Gossez also proposed another proof using a perturbation argument introduced by Hess in treating resonant problems. Here the perturbed equation would be

$$Au_n + (1 - \frac{1}{n})Bu_n + \frac{1}{n}\frac{\alpha}{2}u_n = h.$$

Proof of Lemma 3. We argue by contradiction and suppose the existence of sequences $(t_n) \subset [0,1]$ and $(u_n) \subset D(A)$ such that $F_{t_n}(u_n) = 0$ and $\|u_n\| \to \infty$. Let $v_n = u_n/\|u_n\|$, so that $\|v_n\| = 1$. Without loss of generality, passing to subsequences if necessary, we may assume that $t_n \to t^* \in [0,1]$ and $v_n \rightharpoonup v \in H$ as $n \to \infty$. Thus from (2.2) we obtain

$$(2.3)\quad \begin{cases} Qv_n + t_n A^{-1} Q\Big[(1-t_n)\alpha v_n + t_n \|u_n\|^{-1}(Bu_n - h)\Big] = 0 \\ P\Big[(1-t_n)\alpha v_n + t_n\|u_n\|^{-1}(Bu_n - h)\Big] = 0. \end{cases}$$

It follows from the assumption on the linear growth of g, inequality (0.2), that $(Bu_n)/\|u_n\|$ is bounded in $L^2(\Omega)$. So we may assume that it converges weakly. Hence, using the compact character of A^{-1} in (2.3) we see that Qv_n converges strongly (i.e., in the norm of $L^2(\Omega)$) to an element $\omega \in R(A)$. On the other hand, since $N(A)$ is finite dimensional, $Pv_n \to z \in N(A)$. Denoting by $v = \omega + z$, we then have that $v_n \to v$, strongly in $L^2(\Omega)$ and $\|v\| = 1$.

Now we define the functions

$$(2.4)\qquad g_1(x,s) = \begin{cases} k_+(x)s, & \text{if } s \geq 0 \\ k_-(x)s, & \text{if } s < 0 \end{cases}$$

$$g_2(x,s) = g(x,s) - g_1(x,s).$$

For simplicity, at the moment, we assume that the limit in (0.3) is uniform. The more general case is contained in Theorem 2. Then it follows that $g_2(x,s)$ is sublinear in s, that is, for any given $\varepsilon > 0$ there is an L^2-function $C_\varepsilon(x)$ such that

$$(2.5)\qquad |g_2(x,s)| \leq \varepsilon|s| + C_\varepsilon(x),\ s \in R,\ x \in \Omega.$$

Hence

$$\limsup_{n\to\infty} \|g_2(x,u_n)/\|u_n\|\| \le \varepsilon$$

which implies that $g_2(x,u_n)/\|u_n\| \to 0$ strongly in $L^2(\Omega)$. Therefore, writing $Bu_n = g_1(x,u_n) + g_2(x,u_n)$ and passing to the limit as $n \to \infty$ in (2.3) we obtain

$$(2.6) \qquad \begin{cases} Qv+t^*\Lambda^{-1}Q[(1-t^*)\alpha v + t^*g_1(x,v)] = 0 \\ P[(1-t^*)\alpha v + t^*g_1(x,v)] = 0 \end{cases}$$

Since $v \neq 0$, (2.6) implies that $t^* \neq 0$. (2.6) can be written as

$$(2.7) \qquad Av + t^*[(1-t^*)\alpha v+t^*g_1(x,v)] = 0.$$

It follows from (2.4) that $g_1(x,v) = k_v(x)v$ where $k_v(x) = k_+(x)$ if $v \geq 0$ and $k_v(x) = k_-(x)$ if $v < 0$. Denoting $m(x) = (1-t^*)\alpha + t^*k_v(x)$, we may rewrite (2.7) as

$$(2.8) \qquad Av = -t^*m(x)v.$$

From inequality (1.1) we derive $(Av, Av+\alpha v) \geq 0$. Using the expression of Av given in (2.8) we get

$$(2.9) \qquad t^* \int_\Omega m(x)[t^*m(x) - \alpha]v^2 \geq 0.$$

In view of (0.4) we see that $m(x) \geq 0$, $t^*m(x) - \alpha \leq 0$, and strict inequalities hold in both cases if $t^* < 1$. Hence, we derive from (2.9) that $t^* = 1$ and (2.9) is in fact equality to 0. So $(Av, Av+\alpha v) = 0$, which implies, by Lemma 1, that $v \in N(A) \oplus N(A+\alpha)$. And also (2.8) becomes

$$Av = -k_v(x)v. \tag{2.10}$$

Writing $v = v_o+v_1$, where $v_o \in N(A)$ and $v_1 \in N(A+\alpha)$, we derive from (2.10) that

$$(\alpha-k_v(x))v_1 = k_v(x)v_o. \tag{2.11}$$

Now, taking the inner product of (2.11) with v_o and v_1 we obtain $([\alpha-k_v(x)]v_1,v_o) = (k_v(x)v_o,v_o)$ and $([\alpha-k_v(x)]v_1,v_1) = (k_v(x)v_o,v_1)$. Using (0.4) we obtain successively that $(k_v(x)v_1,v_o) \le 0$ and $(k_v(x)v_o,v_1) \ge 0$. Altogether we get $(k_v(x)v_o,v_1) = 0$ and then it follows that

$$(k_v(x)v_o,v_o) = 0, \text{ and} \tag{2.12}$$

$$([\alpha-k_v(x)]v_1,v_1) = 0. \tag{2.13}$$

Let $A_o = \{x \in \Omega: v_o(x) \neq 0\}$ and $A_1 = \{x \in \Omega: v_1(x) \neq 0\}$. It follows from (2.12) and (2.13) that $k_v = 0$ in A_o and $k_v = \alpha$ in A_1. So $A_o \cap A_1 = \phi$. If $A_o = \phi$, then $v = v_1$ and (2.13) is a contradiction to hypothesis (0.6). If $A_o \neq \phi$, $v = v_o$ in A_o and so $k_v = k_{v_o}$ in A_o. In this case (2.12) yields a contradiction to (0.5). This completes the proof of Theorem 1.

We now establish a more general version of Theorem 1, which does not require the existence of the limits in (0.3). Let us first introduce some notation. Let

$$(2.14) \quad \ell_{\pm}(x) = \liminf_{u\to\pm\infty} \frac{g(x,u)}{u}, \quad k_{\pm}(x) = \limsup_{u\to\pm\infty} \frac{g(x,u)}{u}.$$

Clearly these limits always exist. The use of $k_{\pm}$ with a different meaning of the one used so far should not lead to any confusion. Instead of (0.4) we now assume:

$$(2.15) \quad 0 \le \ell_+(x) \le k_+(x) \le \alpha, \quad 0 \le \ell_-(x) \le k_-(x) \le \alpha.$$

Here the function $g:\bar{\Omega}\times R \to R$ is also assumed to be a Carathēodory function satisfying (0.2). So the limits in (2.14) are L^{∞}- functions. With these preliminaries we may state our next result.

Thoerem 2. *Let A satisfy Property I, and $N(A)$ be finite dimensional. Suppose that g is a Carathēodory function satisfying (0.2) and (2.15). Assume $\alpha < \infty$ and that $A^{-1} + \alpha^{-1}I$ is strictly monotone. If*

$$(2.16) \quad \ell_+(x)v^+(x)+\ell_-(x)v^-(x) \not\equiv 0, \quad \forall\, 0 \ne v \in N(A), \text{ and}$$

$$(2.17) \quad [\alpha-k_+(x)]v^+(x)+[\alpha-k_-(x)]v^-(x) \not\equiv 0, \ \forall\, 0\ne v \in N(A+\alpha I),$$

then equation (0.1) is solvable for every $h \in L^2(\Omega)$.

Remark 5. These pointwise assumptions are equivalent to the integral conditions as remarked in the Introduction.

Proof of Theorem 2. The proof parallels that of Theorem 1 up to the point where we introduce the function $g_1(x,v)$. Here we cannot proceed likewise, since the limits (0.4) do not exist. Still we can introduce one function that will play adequately the role of $g_1(x,v)$. Let us denote by $K(x)$ the weak limit in $L^2(\Omega)$ of $(Bu_n)/\|u_n\|$. As before $v_n = u_n/\|u_n\|$ converges strongly to $v \in D(A)$, $\|v\| = 1$. From (2.3) we derive

$$Av + t^*[(1-t^*)\alpha v + t^* K(x)] = 0. \tag{2.18}$$

The following lemma gives enough information about $K(x)$ to allows us to finish the proof of Theorem 2.

<u>Lemma 4. The function $K(x)$ satisfies the following set of inequalities</u>

$$\ell_+(x)v(x) \le K(x) \le k_+(x)v(x), \quad \text{a.e. in } \{v(x) > 0\} \tag{2.19}$$

$$k_-(x)v(x) \le K(x) \le \ell_-(x)v(x), \quad \text{a.e. in } \{v(x) > 0\} \tag{2.20}$$

$$K(x) = 0 \quad , \quad \text{a.e. in } \{v(x) = 0\}. \tag{2.21}$$

Let us use the lemma and finish the proof of Theorem 2. Let us denote by $k_v(x)$ the function defined by $K(x)/v(x)$ if $v(x) \neq 0$ and $k_v(x) = 0$ if $v(x) = 0$. The notation k_v is used here purposedly to show the analogy with the proof of Theorem 1. Then it follows from Lemma 4 that

$$\ell_+(x) \le k_v(x) \le k_+(x) \quad \text{if} \quad v(x) > 0, \text{ and} \tag{2.22}$$

$$\ell_-(x) \le k_v(x) \le k_-(x) \quad \text{if} \quad v(x) < 0. \tag{2.23}$$

Let us denote by $m(x) = (1-t^*)\alpha + t^* k_v(x)$. Then equation (2.18) becomes one of the form (2.8). With the aid of the inequalities (2.22) and (2.23) we can proceed as in Theorem 1. And then we derive $t^*=1$, $v \in N(A) + N(A+\alpha I)$ and equation (2.10), which gives

$$\alpha v_1 = k_v(x)v. \tag{2.24}$$

Setting $v = v_0+v_1$, with $v_0 \in N(A)$ and $v_1 \in N(A+\alpha I)$, we may rewrite this expression as $(\alpha-k_v(x))v_1 = k_v(x)v_o$, which is identical to the equation (2.11) in the proof of theorem 1. Then we proceed in a similar way to get a contradiction to (2.16), (2.17) or both.

Proof of Lemma 4. We have to proceed to a more careful analysis of the function $K(x)$. For that matter let us introduce the following notation: $\Omega_\pm = \{x\varepsilon\Omega: v(x) \gtrless 0\}$, $\Omega_o = \{x \varepsilon \Omega: v(x) = 0\}$, and $\Phi_n(x) = g(x,u_n(x))/\|u_n\|$. Observe that for $u_n(x) \neq 0$ we have

$$\Phi_n(x) = \frac{g(x,u_n(x))}{u_n(x)} \frac{u_n(x)}{\|u_n\|} = \frac{g(x,u_n(x))}{u_n(x)} v_n(x).$$

It is then easy to prove that:

(i) In Ω_+: $\liminf_{n\to\infty} \Phi_n(x) = \ell_+(x)v(x)$, $\limsup_{n\to\infty} \Phi_n(x) = k_+(x)v(x)$, (ii) In Ω_-: $\liminf_{n\to\infty} \Phi_n(x) = k_-(x)v(x)$, $\limsup_{n\to\infty} \Phi_n(x) = \ell_-(x)v(x)$ (iii) In Ω_o, we claim $\lim_{n\to\infty} \Phi_n(x) = 0$. Indeed, we may assume $|u_n(x)| > 1$. If $|u_n(x)| \leq 1$, then $g(x,u_n(x))$ is bounded and since $\|u_n\| \to \infty$, we get $\Phi_n(x) \to 0$. If $|u_n(x)| > 1$ we may write Φ_n as above. And then we observe using (0.2) that $|g(x,u_n(x))/u_n(x)| < \gamma + b(x)$. Since $v_n(x) \to 0$, when $x \varepsilon \Omega_o$, we prove our claim. Now lemma 4 follows readily from the pointwise limits above.

3. SECOND ORDER ELLIPTIC EQUATIONS. In this section we consider the case when A in equation (0.1) is a second

order self-adjoint elliptic operator. Of course, Theorem 1 and 2 can be used to give statements on the solvability of (0.1) in this case. However, if 0 is the first eigenvalue of A, then condition (0.5) in Theorem 1 [resp. condition (2.16) in Theorem 2] imply that both k_+ and k_- [resp. ℓ_+ and ℓ_-] are different from zero on subsets of Ω with positive measure. This is readily seen from the fact that the eigenfunctions in $N(A)$ have a definite sign, that is, if $0 \neq v \in N(A)$ then either $v > 0$ or $v < 0$ a.e. in Ω. In this section we propose to study existence questions when condition (2.16) is not satisfied. Namely we assume condition (3.3) for a result about "double resonance" (Theorem 3) and condition (3.4) for a result about "crossing of eigenvalues" (Theorem 4).

Let $\mathcal{L}$ be a second order differential operator of the form

$$\mathcal{L} = - \sum_{i,j=1}^{N} \frac{\partial}{\partial x_j}\left(a_{ij}(x)\frac{\partial}{\partial x_i}\right) + a_0(x)$$

which is assumed to be uniformly elliptic and symmetric. In terms of the coefficients this means

$$\sum_{i,j=1}^{N} a_{ij}(x)\xi_i\xi_j > 0, \; x \in \bar{\Omega}, \; 0 \neq \xi \in R^N, \quad a_{ij} = a_{ji}.$$

Ω is a bounded domain in R^N with a C^1 boundary $\partial\Omega$. We assume $a_{ij} \in C^1(\bar{\Omega})$ and $a_0 \in C^0(\bar{\Omega})$. Let us denote

[illegible] the first eigenvalue of the eigenvalue problem

$\mathcal{L}u = \lambda u$ in Ω and $u = 0$ on $\partial\Omega$. It is well known that λ_1 is a simple eigenvalue and that there is a corresponding smooth eigenfunction $\phi > 0$ in Ω, $\partial\phi/\partial\nu < 0$ on $\partial\Omega$, where $\partial/\partial\nu$ stands for the outward normal derivative. We consider the Dirichlet problem

$$(3.1) \qquad -\mathcal{L}u + \lambda_1 u + g(x,u) = h \quad \text{in} \quad \Omega,\ u = 0 \quad \text{on } \partial\Omega,$$

where $h \in L^p(\Omega)$, for some $p > N$ and $g:\bar{\Omega}\times R \to R$ is a Carathēodory function satisfying the linear growth condition (0.2). A typical example of (3.1) is when $\mathcal{L} = -\Delta$. Problem (3.1) may be set in the abstract framework introduced in the previous sections. We let $A = -\mathcal{L} + \lambda_1$ with domain $D(A) = H^2(\Omega) \cap H_0^1(\Omega)$. It is well known that A is self-adjoint with a closed range $R(A)$ and a one-dimensional subspace $N(A)$, which is the linear span of the function ϕ defined above. Here $\alpha = \lambda_2 - \lambda_1$, where λ_2 is the second eigenvalue of $\mathcal{L}$. We remark that all functions in $N(A)$ have definite sign, while the ones in $N(A+\alpha I)$ all change sign. This explains why condition (0.6) [resp. (2.17)] is more easily satisfied than condition (0.5) [resp. (2.16)]. Concerning the nonlinearity we assume that the following limits are ≥ 0 and finite a.e. in Ω

$$(3.2) \quad \ell_\pm(x) = \liminf_{u\to\pm\infty} \frac{g(x,u)}{u}, \quad k_\pm(x) = \limsup_{u\to\pm\infty} \frac{g(x,u)}{u}.$$

As remarked above, it follows in this case that

and ℓ_- are $\neq 0$ on subsets of Ω with positive measure. Let us then consider the case when this does not happen. Namely

(3.3) $\ell_+(x) \equiv 0;\ 0 \leq \ell_-(x),\ k_+(x),\ k_-(x) \leq \alpha;\ \ell_-(x) \not\equiv 0.$

In condition (3.3), one could have ℓ_-, k_- or k_+ equal to α on subsets of Ω with positive measure. So, problem (3.1) in this case exhibits double resonance. An analogous situation could be studied with the roles of ℓ_+ and ℓ_- interchanged. Later we discuss a result when (3.3) is replaced by

(3.4) $k_+(x) = \ell_+(x) \equiv 0;\ 0 \leq \ell_-(x),\ k_-(x) \leq \delta;\ \ell_-(x) \not\equiv 0,$

where $\delta > \alpha$. Under this assumption, problem (3.1) exhibits a "crossing of eigenvalues".

<u>Theorem 3</u>. <u>Let</u> $\mathcal{L}$ <u>satisfy the hypotheses above</u>. <u>Let us suppose that</u> g <u>satisfies condition</u> (3.3) <u>and that there exists an</u> L^2-<u>function</u> $c(x)$ <u>such that</u>

(3.5) $$g(x,u) \geq c(x),\quad u \geq 0,\quad x \in \Omega.$$

<u>Then</u>, <u>problem</u> (3.1) <u>has a solution</u> $u \in H^2 \cap H_0^1$ <u>for a given</u> $h \in L^p$, $p > N$, <u>provided</u>

(3.6) $$\int_{v>0} [\lambda_2-\lambda_1-k_+]v^2 + \int_{v<0} [\lambda_2-\lambda_1-k_-]v^2 > 0,\quad 0 \neq v \in N(\mathcal{L}+\lambda_2$$

<u>and</u>

(3.7) $$\int h\phi < \int g_+\phi,$$

<u>where</u> $\phi > 0$ <u>is an eigenfunction of</u> $\mathcal{L}$ <u>corresponding to the eigenvalue</u> λ_1, <u>and</u>

(3.8) $$g_+(x) = \liminf_{u\to+\infty} g(x,u).$$

Remark 6. Theorem 3 extends a result of Dancer [8]. An example of application of the previous theorem is the Dirichlet problem for $\Delta u + \lambda_1 u^+ - \lambda_2 u^- + \tilde{g}(x,u) = h$, where $\tilde{g}$ is a Carathéodory function with a sublinear growth, and satisfying condition (3.5). Observe that in the absence of the non-degeneracy condition (2.16), a condition of the Landesman-Lazer type (namely condition 3.7) is needed.

Proof of Theorem 3. Let $A = -\mathcal{L} + \lambda_1$, with $D(A) = H^2 \cap H_0^1$. We proceed as in Theorem 2 up to the point when (2.24) is obtained. Recall that the assumptions (2.16) and (2.17) were used only to show that (2.24) could not hold. Instead we use now (3.6) and (3.7) for that purpose. Indeed, if $v_1 \neq 0$, then (2.24) yields a contradiction to (3.6) in a similar manner as in the proof of Theorem 2. Alternatively, if $v_1 = 0$, then (2.24) gives $k_v(x)v = 0$, with $v = v_0 \neq 0$. Now, either $v > 0$ or $v < 0$ a.e. in Ω. If $v < 0$, then it follows from (2.23) that $\ell_-(x) \equiv 0$. This is a contradiction to (3.3). Thus, we conclude that $v > 0$ a.e. in Ω and $v \in N(A)$. Next we claim the contents of the following lemma.

Lemma 5. There exists n_o such that, for $n \geq n_o$, $v_n(x) > 0$ a.e. in Ω, where (v_n) is the sequence defined in the proof of Theorem 2.

Let us assume the lemma, and finish the proof

$$(3.9)\qquad Au_n + t_n[(1-t_n)\alpha u_n + t_n(Bu_n - h)] = 0.$$

We know that $t_n \to 1$, $v_n = u_n/\|u_n\| \to v$ in L^2 and a.e. in Ω. Taking the innerproduct of (3.9) with v and observing that $(Au_n, v) = (u_n, Av) = 0$ and $(v_n, v) \to \|v\|^2 > 0$, we have

$$(3.10)\qquad \int Bu_n v \leq \int hv$$

for large n. Next observe that the integrand $g(x, u_n(x))v(x)$ in the left side of (3.10) is bounded below by the L^1-function $c(x)v(x)$. This follows from Lemma 5 and hypothesis (3.5). Consequently passing to the limit in (3.10) and using Fatou's lemma one comes to a contradiction to (3.7). The proof of Theorem 3 is complete.

Remark 7. Observe that, assuming Lemma 5, the preceding proof holds in the abstract framework of Section 2. The differential structure of A and the regularity assumptions have not been used. They will be needed only to prove the lemma.

Proof of Lemma 5. In order to prove that $v_n > 0$ a.e. in Ω, we use the L^p theory for the linear Dirichlet problem. Using a standard bootstrap argument we prove that $v_n \in C^{1,\mu}(\bar{\Omega})$ and that $v_n \to v$ in $C^{1,\mu}$. Then the proof is finished, observing that $v > 0$ and that $\partial v/\partial \nu \leq -\varepsilon < 0$ on $\partial\Omega$ in view of the maximum principle.

Finally we present a result on the crossing of eigenvalues.

Theorem 4. Let $\mathcal{L}$ be as in Theorem 3. Let us suppose conditions (3.4) and (3.5). Then problem (3.1) has a solution $u \in H^2 \cap H_0^1$ for a given $h \in L^p$, $p > N$, provided that (3.7) holds. Moreover $u \in C^{1,\mu}(\bar{\Omega})$.

Remark 8. Observe that condition (3.4) implies (2.19) since $k_+(x) \equiv 0$ and all eigenfunctions in $N(-\mathcal{L} + \lambda_2)$ change sign. Condition (2.19) is a non-degeneracy condition at the second eigenvalue, just like (3.6) is. An example of application of Theorem 4 is the Dirichlet problem for $\Delta u + \lambda_1 u^+ - \rho u^- + \tilde{g}(x,u) = h$, where $\tilde{g}$ is a Carathéodory function with sublinear growth and ρ is any positive number $> \lambda_1$.

Proof of Theorem 4. Let us define the following Carathéodory function

$$(3.11) \qquad g_1(x,u) = \begin{cases} \alpha u & \text{if } u \le 0 \\ g(x,u), & \text{if } u \ge 1 \\ ug(x,1), & \text{if } 0 \le u \le 1 \end{cases}$$

and let B_1 be the corresponding Nemytskii mapping. Consider the following family of Dirichlet problems

$$(3.12) \qquad Au + (1-t)Bu + tB_1 u = h$$

with $t \in [0,1]$. For $t = 0$, (3.12) is our original problem, while for $t = 1$, it is a problem of the type considered in Theorem 3. Now we proceed as in Theorems 1 and 2 and transform (3.12) in a system in $R(A) \times N(A)$; we identity this cartesian product with L^2. For $t \in [0,1]$, let us define the mapping $\Phi_t : L^2 \to L^2$ by

$$\Phi_t(u) = \begin{cases} Qu + A^{-1}Q[(1-t)Bu + tB_1u - h] \\ P[(1-t)Bu+tB_1u-h]. \end{cases} \tag{3.13}$$

Clearly, u is a solution of (3.12) if and only if $\Phi_t(u) = 0$. Observe that Φ_t is a compact perturbation of the identity. To compute the topological degree of Φ_t we need the following a priori estimate.

Lemma 6. *There is* $R > 0$ *such that* $\Phi_t(u) = 0$ *implies* $\|u\| < R$, *for every* $t \in [0,1]$.

Assuming the lemma, the proof of Theorem 4 is finished in a standard way using the Leray-Schauder theory. Indeed, the lemma implies that $\deg(\Phi_0, B_R, 0) = \deg(\Phi_1, B_R, 0)$ and it follows from Theorem 3 that $\deg(\Phi_1, B_R, 0) \neq 0$. Here B_R denotes the ball of radius R in $L^2(\Omega)$. Thus, there exists u in the interior of B_R such that $\Phi_0(u) = 0$, i.e., u is a solution of (3.1).

Proof of Lemma 6. Assume by contradiction that there exist $\|u_n\| \to \infty$, $t_n \to t^*$ such that $\Phi_{t_n}(u_n) = 0$. That

$$(3.14) \qquad Au_n + (1-t_n)Bu_n + t_n B_1 u_n = h$$

Let $v_n = u_n/\|u_n\|$. As in the proof of Lemma 3, we conclude that $v_n \to v$ in L^2 and a.e. in Ω. We divide the rest of the proof in three steps.

<u>Step I</u>. We claim that $v \in N(A)$ and $v > 0$. Dividing (3.14) by $\|u_n\|$ and passing to the limit we derive

$$(3.15) \qquad Av + (1-t^*)k_v v - t^*\alpha v^- = 0,$$

where k_v is the function introduced in the proof of Theorem 2. The derivation of (3.15) is similar to the derivation of (2.18). In view of (2.22) and our hypothesis (3.4) it follows that $k_v = 0$ if $v > 0$. So we may rewrite (3.15) as

$$(3.16) \qquad Av - (1-t^*)k_v v^- - t^*\alpha v^- = 0.$$

If $v > 0$ a.e. in Ω, it follows from (3.16) that $v \in N(A)$, and the proof of step I is complete. If $v < 0$ on a subset of Ω with positive measure we proceed as follows. Taking the inner product of (3.16) with ϕ we obtain

$$(3.17) \qquad \int_{v<0} [(1-t^*)k_v + t^*\alpha] v\phi = 0.$$

Since $(1-t^*)k_v + t^*\alpha \geq 0$ and is > 0 if $t^* \neq 0$, we conclude that $t^* = 0$ and $k_v = 0$ in $v < 0$. So (3.16) becomes $Av = 0$, i.e. $v \in N(A)$. Thus either $v > 0$ a.e. in Ω and the proof of step I is complete or $v < 0$ a.e. in Ω. In the latter case, (3.17) implies

$k_v \equiv 0$, and as a consequence of (2.27) one obtains $\ell_-(x) \equiv 0$. This contradicts (3.4).

<u>Step II</u>. We claim that $v_n > 0$ for large n. With a standard bootstrap argument we prove that $v_n \in C^{1,\mu}(\bar{\Omega})$ and $v_n \to v$ in $C^{1,\mu}$. Again the maximum principle applies to give the result.

<u>Step III</u>. Taking the inner product of (3.14) with v, we obtain

$$\int [(1-t_n)Bu_n + t_n B_1 u_n] v = \int hv. \tag{3.16}$$

In view of (3.5) and step II above we see that the integrand in the left side of (3.16) is bounded below by an L^1-function. Taking limits and using Fatou's lemma we come to a contradiction to condition (3.7). The proof of the lemma is complete.

Work partially supported by CNPq.

References

[1] Aguinaldo, L. & Schmitt, K. - "On the boundary value problem $u'' + u = \alpha u^- + p(t)$, $u(0) = 0 = u(\pi)$. Proc. A.M.S. vol 68 (1978) p. 64-68.

[2] Amann, H. & Mancini, G. - "Some applications of monotone operator theory to resonance problems". Nonlinear Analysis, Theory, Methods and Applications vol 3 (1979) p. 815-830.

[3] Amann, H, Ambrosetti, A., Mancini, G. - "Elliptic Equations with Noninvertible Fredholm Linear Part and Bounded Nonlinearities". Math. Zeit. vol 158 (1978) p. 179-194.

[4] Brézis, H. & Haraux, A. - "Image d'une somme d'opérateur monotones et applications". Israel J. Math. Vol 23 (1976) p. 165-186.

[5] Brézis, H. & Nirenberg, L. - "Characterization of the Ranges of Some Nonlinear Operators and Applications to Boundary Value Problems". Ann. Sc. Norm. Sup. Pisa. Serie IV, Vol 5 (1978) p. 225-326.

[6] Browder, F.E. & Gupta, C.P. - "Monotone operators and nonlinear integral equations of Hammerstein type". Bull. A.M.S. Vol 75 (1969) p. 1347-1353.

[7] Castro, A. - "A two point boundary value problem with jumping nonlinearities". To appear.

[8] Dancer, E.N. - "On the Dirichlet problem for weakly nonlinear elliptic partial differential equations". Proc. Royal Soc. Edinburgh, vol 76 (1977) p. 283-30

[9] de Figueiredo, D.G. - "Semilinear Elliptic Equations at resonance: Higher Eigenvalues and Unbounded Nonlinearities". To appear.

[10] de Figueiredo, D.G. and Gossez, J.P. - "Nonlinear Perturbations of a Linear Elliptic Problem Near its First Eigenvalue". J. Diff. Eq. Vol 30 (1978) p. 1-19.

[11] de Figueiredo, D.G., Ni, W. M. - "Perturbations of Second Order Linear Elliptic Problem by Nonlinearities Without Landesman-Lazer Condition". Nonlinear Analysis, Theory, Methods & Applications. Vol 3 (1979) pp. 629-634.

[12] Fucik, S. & Krbec, M. - "Boundary value problem with bounded nonlinearity and general nullspace of linear part". Math. Z. Vol 155 (1977) pp 129-138.

[13] Gonçalves, J.V.A. - "On bounded nonlinear perturbations of an elliptic equations at resonance". To appear in Nonlinear Analysis, TMA.

[14] Hess, P. - "A remark on the preceding paper of Fucik and Krbec". Math. Z. Vol 155 (1977) p. 139-141.

[15] Hess, P. - "Nonlinear perturbations of linear elliptic and parabolic problems at resonance: existence of multiple Solutions". To appear.

[16] Hess, P. - "On nonlinear equations of Hammerstein type in Banach spaces". Proc. AMS Vol 31 (1971), 308-321.

[17] Landesman, E.M. & Lazer, A.C. - "Nonlinear perturbations of linear elliptic boundary value problem at resonance", J. Math. Mech. 19, 609-623 (1970).

[18] Mawhin, J. - "Landesman - Lazer's type problems for nonlinear equations". Conferenze del Seminario di Matematica dell'Università di Bari (1977).

Received April 1980; revised September 1980

SEMILINEAR ELLIPTIC EQUATIONS AT RESONANCE: HIGHER EIGENVALUES AND UNBOUNDED NONLINEARITIES

Djairo G. de Figueiredo[1,2]

Dept.o de Matematica
Universidade de Brasilia
Brasilia, Brasil

1. Let L be a uniformly strongly elliptic operator of order $2m$ with smooth coefficients acting on real-valued functions defined in a bounded domain Ω in R^N. Since we intend to study the Dirichlet problem we may then consider L as an operator in $L^2(\Omega)$ with domain $D(L) = H^{2m} \cap H_0^m(\Omega)$. Such an operator L is an unbounded closed linear operator in $L^2(\Omega)$ with finite dimensional nullspace $N(L)$. Assuming also that the coefficients of L are symmetric, as we do, it follows that L is self-adjoint and its spectrum is made up of real eigenvalues $\lambda_1 < \lambda_2 < \dots \lambda_n \to +\infty$, with corresponding finite dimensional eigenspaces $N(L-\lambda_i)$, $i = 1,2,\dots$; moreover $(L-\lambda)^{-1}$ which is a well defined operator from $R(L-\lambda)$ into $D(L) \cap R(L-\lambda)$ is compact. From now on we suppose that 0 is

---

[1]The author acknowledges a grant of the "Conselho Nacional do Desenvolvimento Cientifico e Tecnologico" (CNP_q) of Brasil for a stay at the Courant Institute of Mathematical Sciences where this research was conducted.

[2]Present address: Dept.o de Matematica, Universidade de Brasilia, Campus Universitario, Asa Norte, Brasilia, DF CEP 70910.

ISBN 0-12-186280-1

Originally published in: Recent Advances in Differential Equations, Ed. Roberto Conti. Academic Press, 1981, 89-99, © Academic Press, Inc. 1981
Reprint by Springer-Verlag Berlin Heidelberg 2013 with kind permission,
DOI: 10.1007/978-3-319-02856-9_10

a higher (i.e. $\lambda_1 < 0$) eigenvalue of L, and let α, be the smallest positive eigenvalue of $-L$.

We consider a nonlinearity given by a function $g : \Omega \times R \to R$ satisfying Caratheodory's conditions and the following basic assumption regarding its growth:

(g) there are a positive number $\gamma < \alpha$ and L^2-functions $b(x)$, $c(x)$ and $d(x)$ such that

$$|g(x,u)| \leq \gamma |u| + b(x), \quad u \in R, \quad x \in \Omega \tag{1}$$

$$ug(x,u) \geq -c(x)|u| - d(x), \quad u \in R, \quad x \in \Omega. \tag{2}$$

Theorem 1. *Assume* L *and* g *as above. Then there exists a* $u \in H^{2m}(\Omega) \cap H_0^m(\Omega)$ *satisfying*

$$Lu + g(x,u) = h, \tag{3}$$

for a given $h \in L^2(\Omega)$, *provided*

$$(h,v) < \int_{v>0} g_+(x)v(x)dx + \int_{v<0} g_-(x)v(x)dx, \quad \forall\, v \in N(L),\ v \neq 0$$

where

$$g_-(x) = \limsup_{u\to-\infty} g(x,u), \quad g_+(x) = \liminf_{u\to+\infty} g(x,u).$$

Observe that g in the above theorem could be an unbounded function with any linear growth $< \alpha$. This means that eventually we could have a situation like this

$$0 = \lim_{u\to+\infty} \frac{g(x,u)}{u} < \beta < \lim_{u\to-\infty} \frac{g(x,u)}{u} < \alpha$$

where β is as close to α as we wish. So we have here resonance at a higher eigenvalue (or the "nonlinearity jumping off an eigenvalue" in the terminology of Fučik); in this connection see the papers of Fučik [8] and Dancer [7]. We remark that in this paper the nonlinearity does not "cross" an eigenvalue; this follows from the fact that $\gamma < \alpha$ and (2) above.

The case of bounded nonlinearities has been extensively

studied in the literature starting with the pioneering work of Landesman and Lazer [11]; some interesting recent research appears in Fučik and Krbec [9] and Hess [10]. The case of nonlinearities with a sublinear growth and satisfying (2) above was investigated in Brezis and Nirenberg [6]. Unbounded nonlinearities g without condition (2) above lead to other difficulties, and to treat them some additional hypotheses have been imposed on g by several authors in recent papers, e.g. Ambrosetti and Prodi [4], Ambrosetti and Mancini [2],[3], Amann, Ambrosetti and Mancini [1], Berger and Podolak [5].

Instead of restricting ourselves to the case of differential operators we preferred to follow the work of Brezis and Nirenberg [6] and in Section 2 we consider more general equations of the type (3) given by

$$Au + Bu = h \tag{4}$$

where A is some (unbounded) linear operator in a Hilbert space H and B is some nonlinear operator. The problem taken up in [6] was the characterization of the $R(A+B)$. In this paper we pursue a more modest goal, but we are able to consider more general nonlinearities as far as growth at infinity is concerned.

2. We shall study the solvability of the equation (4) in $H = L^2(\Omega)$. As in [6] we assume the following condition on A:

Property I. $A : D(A) \subset H \to H$ *is a closed densely defined linear (not necessarily bounded) operator such that* $R(A) = N(A)^{\perp}$. *Consequently we have that* $N(A) = N(A^*)$, $H = R(A) \oplus N(A)$ *and* A *is a bijective mapping from* $D(A) \cap R(A)$ *onto* $R(A)$, *whose inverse we denote by* A^{-1}. *It is also assumed that* A^{-1} *is compact.*

As usual $N(A)$ denotes the nullspace of A, $R(A)$ its range and A^* the adjoint of A.

The following assertion is an immediate consequence of Property I: there exists α_o such that

$$\alpha_o \|Qu\| \leq \|Au\| , \qquad u \in D(A) \tag{5}$$

where Q is the orthogonal projection on R(A) and $||\cdot||$ is the norm of H. And from (5) it follows readily that

$$(Au,u) \geq -(1/\alpha_o) ||Au||^2, \qquad u \in D(A) \tag{6}$$

where (,) is the inner-product in H. Let us denote by α the largest positive constant α_o for which (6) holds:

$$(Au,u) \geq -(1/\alpha) ||Au||^2 , \qquad u \in D(A) . \tag{7}$$

The interest in this class of linear operators A lies mostly in the fact that $A = \pm L$ with $D(A) = H^{2m}(\Omega) \cap H_0^m(\Omega)$, where L is the elliptic operator of Section 1, satisfies Property I. In this case α is the smallest strictly positive eigenvalue of -L (resp. L) accordingly to A = L (resp. A = -L).

The nonlinear operator B is the Niemytskii mapping given by $Bu = g(x,u)$, where $g : \Omega \times R \to R$ is a function satisfying Carathéodory's conditions and the condition (g) above regarding its growth.

Inequality (1) implies that B is bounded and continuous from $L^2(\Omega)$ into $L^2(\Omega)$. As a consequence of (2) we have that

$$\begin{aligned} g_-(x) &\equiv \limsup_{u\to-\infty} g(x,u) \leq c(x) \\ g_+(x) &\equiv \liminf_{u\to+\infty} g(x,u) \geq -c(x) \end{aligned} \tag{8}$$

and so $-\infty \leq g_-(x) < +\infty$ and $-\infty < g_+(x) \leq +\infty$ for $x \in \Omega$. We note that when we write $x \in \Omega$ we actually mean $x \in \Omega$ (a.e.).

Now we state our main result.

<u>Theorem 2</u>. *Suppose that* A *satisfies Property* I *and that* N(A) *is finite dimensional. Assume also that* B *is a mapping defined by a function* g(x,u) *satisfying Carathéodory's conditions and condition* (g). *Conclusion:*

(a) *If*

$$(h,v) < \int_{v>0} g_+(x)v(x)dx + \int_{v<0} g_-(x)v(x)dx, \tag{9}$$

for all $0 \neq v \in N(A)$, *then* $h \in \text{Int } R(A+B)$.

(b) *If*

$$(h,v) \leq \int_{v>0} g_+(x)v(x)dx + \int_{v<0} g_-(x)v(x)dx \qquad (10)$$

for all $0 \neq v \in N(A)$, *then* $h \in \overline{R(A+B)}$.

Remarks: 1) The above theorem with some stronger assumption on the behavior of g at $\pm\infty$ was proved in [6; Theorem IV.5]. Namely it is assumed there, in addition to the hypothesis above, that one of the following conditions hold:

either (a) $g_-(x) \leq g_+(x)$, $x \in \Omega$,

or (b) $|g(x,u)| \leq \gamma_1|u| + b(x)$, $\gamma_1 > 0$, $x \in \Omega$, $u \in R$.

Our proof contains a new idea (a decomposition of g as stated in Lemma 2 below), and provides a simpler proof of the Brezis-Nirenberg theorem. The above theorem was originally proved by the author under an additional restriction on g enabling us a certain decomposition of g needed in Step 3' of the proof; we thank Ed Joer for suggesting us a modification of that decomposition so as to eliminate the need of our extra assumption.

2) It follows from the remarks after the statement of Property I that Theorem 1 is an immediate corollary of Theorem 2. Also the following result is a corollary of Theorem 2.

Theorem 3. *Let* L *be as in Section* 1 *and let* α *be the smallest positive eigenvalue of* L. *Suppose that* $g : \Omega \times R \to R$ *is a function satisfying Carathéodory's conditions and the following assumptions:*

$$|g(x,u)| \leq \gamma|u| + b(x), \quad u \in R, \quad x \in \Omega$$

$$ug(x,u) \leq c(x)|u| + d(x), \quad u \in R, \quad x \in \Omega$$

where $0 < \gamma < \alpha$, *is a constant and* b, c *and* d *are* L^2-*functions in* Ω. *Then, for given* $h \in L^2(\Omega)$, *the Dirichlet problem for*

$$Lu + g(x,u) = h$$

has a solution provided the inequality

$$(h,v) > \int_{v>0} \left[\limsup_{u\to+\infty} g(x,u)\right] v(x)\,dx + \int_{v<0} \left[\liminf_{u\to-\infty} g(x,u)\right] v(x)\,dx$$

holds for all $v \in N(L)$, $v \neq 0$.

Lemma 1. *Let* $g(x,u)$ *be a function satisfying Carathéodory's condition and condition* (g). *Then for each* $\gamma' > \gamma$ *there are positive constants* c_1 *and* c_2 *such that*

$$(Bu,u) \geq \frac{1}{\gamma'} \|Bu\|^2 - c_1\|u\| - c_2. \tag{11}$$

If, in addition, $g(x,u)u \geq 0$ *for all* $x \in \Omega$ *and all* $u \in R$, *then* (11) *holds with* $c_1 = 0$.

Proof. Using (1) and (2) we obtain

$$ug \geq \frac{1}{\gamma} |g|^2 - \frac{b}{\gamma} |g| - 2c|u| - 2d.$$

Now we use the inequality $pq \leq \varepsilon p^2 + (1/4\varepsilon)q^2$ with $\varepsilon = 1 - (\gamma/\gamma')$, $p = |g|$, $q = b$, and it follows from the above inequality that

$$ug \geq \frac{1}{\gamma'} |g|^2 - 2c|u| - \frac{1}{4\varepsilon\gamma} b^2 - 2d.$$

Integrating we then obtain (11). The second part of the lemma is a consequence of the fact that $c_1 = 2\|c\|$. Q.E.D.

Lemma 2. *Let* $g(x,u)$ *be a function satisfying Carathéodory's conditions and condition* (g). *Then for each* $R > 0$ *there is a decomposition*

$$g(x,u) = g_R(x,u) + q_R(x,u)$$

of g *by functions* g_R *and* q_R *satisfying Carathéodory's conditions and the following additional conditions:*

$$uq_R(x,u) \geq 0, \quad x \in \Omega, \quad u \in R,$$

q_R *satisfies inequality* (1) *with same; there is an* L^2*-function* $a_R(x)$ *such that*

$$|g_R(x,u)| \leq a_R(x), \quad x \in \Omega, \quad u \in R$$

$$g_{R,-}(x) \equiv \limsup_{u\to-\infty} g_R(x,u) = \begin{cases} -R & \textit{if} \quad g_-(x) \leq -R \\ g_-(x) & \textit{if} \quad g_-(x) > -R \end{cases}$$

and

$$g_{R,+}(x) \equiv \liminf_{u\to+\infty} g_R(x,u) = \begin{cases} R & \textit{if} \quad g_+(x) \leq R \\ g_+(x) & \textit{if} \quad g_+(x) > R \end{cases}$$

Proof. Given $R > 0$ define

$$\hat{\bar{g}}_R(x,u) = \begin{cases} \sup\{g(x,u),-R\} & \text{if} \quad u \leq -1 \\ \inf\{g(x,u),R\} & \text{if} \quad u \geq 1 \end{cases}$$

and

$$\hat{q}_R(x,u) = g(x,u) - \hat{g}_R(x,u) \qquad x \in \Omega, \quad |u| \geq 1.$$

Next we define $q_R(x,u)$ as $\hat{q}_R(x,u)$ if $|u| \geq 1$ and as $q_R(x,u) = |u|\hat{q}_R(x, u/|u|)$ if $|u| \leq 1$. And finally we define $g_R = g - q_R$. It is a straightforward calculation to check that all conditions of the lemma 2 are satisfied. Q.E.D.

Proof of Theorem 2. Step 1. For each positive integer n there exists $u_n \in D(A)$ such that

$$Au_n + \frac{1}{n} Pu_n + Bu_n = h, \tag{12}$$

where P is the orthogonal projection over $N(A)$. This is proved using the Leray-Schauder fixed point theorem, see step 1 in the proof of Theorem III.2 of [6].

Step 2. There exists a constant $c > 0$ such that

$$\|\frac{1}{n} Pu_n\| \leq c, \quad \text{for all} \quad n. \tag{13}$$

From now on all positive constants independent of n are designated by the same letter c; of course, they depend on A, B and h. To prove (13) we proceed as follows. Take the inner product of (12) with u_n and estimate using (7) and (11)

$$-\frac{1}{\alpha}\|Au_n\|^2+\frac{1}{n}\|Pu_n\|^2+\frac{1}{\gamma'}\|Bu_n\|^2 \le c+c\,\|u_n\|\,,$$

where $\gamma < \gamma' < \alpha$. To estimate Bu_n use the equation (12) to get

$$\left(\frac{1}{\gamma''}-\frac{1}{\alpha}\right)\|Au_n\|^2+\frac{1}{n}\|Pu_n\|^2-c\left\|\frac{1}{n}\,Pu_n\right\|^2 \le c+c\|u_n\|\,, \tag{14}$$

where $\gamma' < \gamma'' < \alpha$. The last term in the left side of (14) is absorbed by the second term for large n. Besides that we estimate $\|u_n\| \le \|Pu_n\|+\|Qu_n\|$ and use (5). All this in (14) gives

$$c\,\|Au_n\|^2+\frac{1}{n}\|Pu_n\|^2 \le c+c\,\|Pu_n\|+c\,\|Au_n\|\,.$$

Finally we use the fact that given $as^2 - bs$ with $a > 0$ there are numbers $0 < a' < a$ and $c > 0$ such that $as^2 - bs > a's^2 - c$ for all $s \in R$, and obtain

$$c\,\|Au_n\|^2+\frac{1}{n}\|Pu_n\|^2 \le c+c\,\|Pu_n\|\,. \tag{15}$$

Now omitting the first term in (15) we readily prove (13).

<u>Step 3</u>. Suppose that (Pu_n) contains a bounded subsequence. In view of (15) it follows that (Au_n) also contains a bounded sequence, and using the fact that A^{-1} is compact we obtain a convergent subsequence of (Qu_n). All this together with the finite dimensionality of $N(A)$ implies that (u_n) contains a convergent subsequence; we use the same notation for the subsequence and write that $u_n \to u \in H$. So $Bu_n \to Bu$ and $Au_n \to h-Bu$. Since A is closed, it follows that $u \in D(A)$ and $Au = h-Bu$. Thus we have proved that $h \in R(A+B)$ on the assumption that (Pu_n) contains a bounded subsequence.

<u>Step 3'</u>. On the other hand suppose that $\|Pu_n\| \to \infty$, and let us prove that such an assumption leads to a contradiction. First we observe that (15) and (5) imply that

$$\lim_{n\to\infty} \frac{\|Qu_n\|}{\|Pu_n\|} = 0 \,. \tag{16}$$

Let $v_n = u_n/\|Pu_n\|$. It follows from (16) that (v_n) is a bounded sequence. As in step 3 we may assume, passing to a subsequence, that $v_n \to v$. Again (16) implies that $\|v\| = 1$. It also follows from (15), using (12) and (13), that

$$\|Bu_n\|^2 \le c + c\,\|Pu_n\|$$

and so $\|Bu_n\| / \|Pu_n\| \to 0$. Dividing the equation (12) by $\|Pu_n\|$ and going to the limit we obtain that $Av_n \to 0$. Thus $v \in N(A)$.

Now let us denote by G_R and Q_R the Niemytskii operators defined respectively by the functions g_R and q_R of the decomposition of g in Lemma 2. So $Bu = G_Ru + Q_Ru$. Taking the inner product of (18) with u_n and using Lemma 1 to the operator Q_R we get

$$-\frac{1}{\alpha}\|Au_n\|^2 + \frac{1}{n}\|Pu_n\|^2 + \frac{1}{\gamma'}\|Q_Ru_n\|^2 + (G_Ru_n,u_n) \le (h,u_n) + c$$

where $\gamma < \gamma' < \alpha$. Using the equation (12) to estimate the third term in the left side of the above inequality we get

$$\frac{1}{n}\|Pu_n\|^2 + (G_Ru_n,u_n) \le (h,u_n) + c. \tag{17}$$

Dividing by $\|Pu_n\|$ and taking limits we obtain by Fatou's lemma that

$$\int_{v>0} g_{R,+}(x)v(x)\,dx + \int_{v<0} g_{R,-}(x)v(x)\,dx \le (h,v)\,. \tag{18}$$

Now letting $R \to +\infty$ and using (8) we obtain

$$\int_{v>0} g_+(x)v(x)\,dx + \int_{v<0} g_-(x)v(x)\,dx \le (h,v)$$

which contradicts (9). So on the assumption of (9) we proved that $h \in R(A+B)$. In fact we have to prove that $h \in \operatorname{Int} R(A+B)$, and this is proved next.

Step 4. Assume (9). We claim that there is a $\beta > 0$ such that

$$\ell(v) \equiv \int_{v>0} g_+ v + \int_{v<0} g_- v - (h,v) \geq \beta > 0 \qquad (19)$$

for all $v \in N(A)$, $\|v\| = 1$. Suppose this were not so. Then there would exist $v_n \in N(A)$, $\|v_n\| = 1$, $v_n \to v$ and $\ell(v_n) \to 0$. Using (8) we write

$$\ell(v_n) = \int_{v_n>0} (g_+ + c) v_n + \int_{v_n<0} (g_- - c) v_n - (h, v_n) - (c, v_n).$$

Taking limits and using Fatou's lemma

$$0 = \lim \ell(v_n) \geq \int_{v>0} (g_+ + c) v + \int_{v<0} (g_- - c) - (h,v) - (c,v)$$

which contradicts (9). Thus in view of (19) every $h + f$, for any $f \in H$ with $\|f\| < \beta$, would satisfy (9) and by steps 3 and 3', $h + f \in R(A+B)$. So part (i) of the teorem is proved.

Step 5. Assume (10). Passing to the limit in (17) as in step 3 but maintaining the term $\frac{1}{n} \|Pu_n\|$ we obtain

$$\limsup \frac{1}{n} \|Pu_n\| + \int_{v>0} g_{2,+} v + \int_{v<0} g_{2,-} v \leq (h,v).$$

Finally letting $R \to +\infty$ and using (10) we obtain that $\frac{1}{n} \|Pu_n\| \to 0$. Equation (12) then implies that $h \in \overline{R(A+B)}$.

Q.E.D.

REFERENCES

1. Amann, H., Ambrosetti, A. and Mancini, G., Elliptic equations with noninvertible Fredholm linear part and bounded nonlinearities. To appear.
2. Ambrosetti, A. and Mancini, G., Existence and multiplicity results for nonlinear elliptic problems with linear part at resonance. *J. Diff. Eq.*, to appear.

3. Ambrosetti, A. and Mancini, G., Theorems of existence and multiplicity for nonlinear elliptic problems with noninvertible linear part. *Ann. Scuola Norm. Sup. Pisa*, to appear.
4. Ambrosetti, A. and Prodi, G., On the inversion of some differentiable mappings with singularities between Banach spaces. *Ann. Mat. Pura App., IV, 93*, 231-247 (1972).
5. Berger, M. S. and Podolak, E., On nonlinear Fredholm operator equations. *Bull. A. M. S. 80*, 861-864 (1974).
6. Brezis, H. and Nirenberg, L., Characterizations of the ranges of some nonlinear operators and applications to boundary value problems. *Ann. Scuola Norm. Sup. Pisa, Ser. IV, V*, 115-175 (1978).
7. Dancer, E. N., On the Dirichlet problem for weakly nonlinear elliptic partial differential equations. *Proc. Royal Soc. Edinburgh 76A*, 283-300 (1977).
8. Fučik, S., Boundary value problems with jumping nonlinearities. *Časopis Pest. Mat. 101*, 69-87 (1976).
9. Fučik, S. and Krbec M., Boundary value problems with bounded nonlinearity and general null-space of the linear part. *Math. Z.*, to appear.
10. Hess, P., Nonlinear perturbations of linear elliptic and parabolic problems at resonance: existence of multiple solutions. To appear.
11. Landesman, E. A. and Lazer A. C., Nonlinear perturbations of linear elliptic boundary value problems at resonance. *J. Math. Mech. 19*, 609-623 (1970).

2. Ambrosetti, A., and Hess, P., "Theorems of existence and multiplicity for nonlinear elliptic boundary value problems," J. Math. Anal. Appl., Vol. 73, pp. 411-422, 1980.
3. Amann, H., "On the existence of positive solutions of nonlinear elliptic boundary value problems," Indiana Univ. Math. J., 21, 125-146 (1971).
4. Amann, H. and Laetsch, T., "Positive solutions of convex nonlinear eigenvalue problems," Indiana Univ. Math. J., 25, 259-270 (1976).
5. Brezis, H. and Turner, R. E. L., "On a class of superlinear elliptic problems," Comm. Partial Diff. Eqns., 2, 601-614 (1977).
6. Crandall, M. G., "On the existence problem for weakly nonlinear elliptic partial differential equations," Proc. Amer. Math. Soc., 21, 291-296 (1969).
7. Fabry, C., "Boundary value problems with superlinear nonlinearities," Nonlinear Anal., 1, 571-571 (1977).
8. Hess, P. and Kato, T., "On some linear and nonlinear eigenvalue problems with an indefinite weight function," Comm. Partial Diff. Eqns., 5, 999-1030 (1980).
9. Kazdan, J., "Nonlinear perturbations of linear elliptic and parabolic equations," preprint.
10. Kazdan, J. L. and Warner, F. W., "Remarks on some quasilinear elliptic equations," Comm. Pure Appl. Math., 28, 567-597 (1975).

J. Math. pures et appl.,
61, 1982, p. 41 à 63.

A PRIORI ESTIMATES AND EXISTENCE OF POSITIVE SOLUTIONS OF SEMILINEAR ELLIPTIC EQUATIONS

By D. G. de FIGUEIREDO, P.-L. LIONS and R. D. NUSSBAUM

Introduction

We are interested here in the existence of positive solutions of the problem:

$$(0.1)\qquad \begin{cases} -\Delta u = f(u) \quad \text{in } \Omega \subset \mathbb{R}^N; \\ u \in C^2(\overline{\Omega}), \qquad u=0 \quad \text{on } \partial\Omega, \qquad u>0 \quad \text{in } \Omega; \end{cases}$$

where Ω is some bounded regular domain and f is some given nonlinearity. More precisely we will study the question of *a priori* estimates of solutions of (0.1); and this will give information on the structure of the full set of solutions of (0.1).

Such problems arise in a variety of situations in the theory of nonlinear diffusion generated by nonlinear sources, in the theory of thermal ignition of gases, in the theory of gravitational equilibrium of stars and elsewhere; detailed references can be found in D. I. Joseph and T. S. Lundgren [15], or in I. M. Gelfand [12]. It is now well-understood (*see* H. Brezis and R. E. L. Turner [8], R. D. Nussbaum [23]) that *a priori* bounds in the sup norm of positive solutions of parametrized versions of (0.1) provide a great deal of information about existence of positive solution and the structure of the positive solution set, although the results in [15] show that even in some simple cases the detailed structure of the solution set can be surprisingly complicated.

We will study here the super-linear case, in which $f(u)$ is assumed to grow faster than linearly in u. Theorems of S. I. Pohozaev [27] and D. D. Joseph and T. S. Lundgren [15] show that if $f(u)$ grows faster than u^σ, where $\sigma=(N+2)/(N-2)$, then (0.1) may not have any positive solution and even if the equation does have positive solutions, *a priori* bounds may fail. (We will recall below the proof of the main result of [27] concerning the non existence of solutions of (0.1) if $f(u)$ grows faster than u^σ, where $\sigma=(N+2)/(N-2)$).

JOURNAL DE MATHÉMATIQUES PURES ET APPLIQUÉES. — 0021-7824/1982/41/$ 5.00

Originally published in: J. Math. Pures Appl. 61 (1982), 41-63,

DOI: 10.1007/978-3-319-02856-9_11

Thus the outstanding question is whether *a priori* bounds hold for positive solutions if $f(u)$ satisfies:

$$\lim_{u \to +\infty} f(u) u^{-\sigma} = 0, \qquad \text{where} \quad \sigma = \frac{N+2}{N-2}. \tag{0.2}$$

We give here *a priori* estimates and existence results under more restrictive assumptions than (0.2) (that we would like but we do not know how to avoid). These assumptions are given below, let us mention for the moment that they concern f and its primitive F and therefore are not just assumptions on the behavior of f (let us also mention that they require that f lies in a convex cone).

Previous work on this problem has been done by R. E. L. Turner [30], R. D. Nussbaum [23], H. Brezis and R. E. L. Turner [8] (*see* also H. Berestycki and P.-L. Lions [6] for a particular case). The key new ingredient in our approach is a theorem from a recent, beautiful paper by B. Gidas, Wei-Ming Ni and L. Nirenberg [13].

For simplicity we have restricted ourselves to an equation in variational form, and a result of A. Ambrosetti and P. H. Rabinowitz [3] provides a partial answer to the question of existence of solutions of (0.1) in this case. However, we could allow some functions of the form $f(x, u, |\nabla u|)$ to which the results of [3] would not apply. In any event, *a priori* bounds provide more information about the structure of the solution set of parametrized versions of (0.1) than do variational methods. Our results enable us to employ the apparatus of the fixed point index for compact mappings of a cone and global bifurcation theorems for cone mappings [23]. One drawback of our methods should be mentioned: our arguments depend strongly on symmetry properties of the Laplacian, and it is not clear at present whether these arguments can be extended to general, second-order, elliptic semilinear equations as in [8]. In fact, the reader will observe that some caution is necessary in even allowing an x-dependence in f.

The paper is organized as follows: in section I, we give our main results on *a priori* bounds for solutions of (0.1); while in section II, we apply those results to obtain various existence results.

After this work was completed, we were told by B. Gidas that he and J. Spruck have obtained very recently a result which complements Theorem 1.2 below.

The results given in this paper were announced in [21], [11].

The second author would like to thank J. Leray for very useful suggestions and remarks.

1. *A priori* bounds

We shall investigate the question of *a priori* bounds for positive solutions of the nonlinear elliptic boundary value problem:

$$\begin{cases} -\Delta u = f(u) \quad \text{in } \Omega, & u \in C^2(\overline{\Omega}); \\ u = 0 \quad \text{on } \partial\Omega, & u > 0 \quad \text{in } \Omega. \end{cases} \tag{1}$$

We will always assume that Ω is a bounded, open subset of $\mathbb{R}^N$ with a smooth boundary $\partial\Omega$ (at least of class $C^{2,\alpha}$, $0<\alpha<1$, as defined on p. 89 of [14]). On the nonlinearity f, we will always at least assume:

(2) $$f \in C(\mathbb{R}_+, \mathbb{R}) \text{ and } f \text{ is locally Lipschitzian;}$$

(3) $$\varliminf_{t\to+\infty} \frac{f(t)}{t} > \lambda_1,$$

where λ_1 is the first eigenvalue of $(-\Delta)$ acting on $H_0^1(\Omega)$. The assumption (3) is an assumption of superlinearity.

Our results on *a priori* bounds for solutions of (1) will be divided in three sections: we first study the case when Ω is convex, then the general case, and finally the case where Ω has some geometrical properties.

1.1. THE CASE OF A CONVEX DOMAIN.

THEOREM 1.1. — *Under assumptions* (2), (3), *if we assume in addition that* Ω *is convex and:*

(4) $$\lim_{t\to+\infty} f(t)\,t^{-\sigma}=0, \qquad \textit{with} \quad \sigma=\frac{N+2}{N-2} \quad (\sigma<\infty, \textit{ if } N=2)$$

and:

(5) $$\varlimsup_{t\to\infty} \frac{tf(t)-\theta F(t)}{t^2 f(t)^{2/N}} \leqq 0, \qquad \textit{for some} \quad 0 \leqq \theta < \frac{2N}{N-2},$$

where $F(t)=\int_0^t f(s)\,ds$ (*if* $N=2$, *this condition is vacuous*); *then we have:*

$$\| u \|_{L^\infty(\Omega)} \leqq C,$$

where C *depends only on* Ω *and on the behaviour of f in the limits arising in* (3)-(5).

Remark 1.1. — Let us first remark that (5) holds as soon as one of the two assumptions which follow holds:

(1) $tf(t)-\theta F(t)\leqq 0$ for some $\theta\in[0, 2N/(N-2))$ and for t large.

This is obviously equivalent to: $F(t)\,t^{-\theta}$ is non increasing for t large; and this is the case in particular if there exists $p\in(1, (N+2)/(N-2))$ such that:

$$(p+1)\varlimsup_{t\to+\infty} f(t)\,t^{-p} < \frac{2N}{N-2}\varliminf_{t\to+\infty} f(t)\,t^{-p} < +\infty;$$

(2) $\varlimsup_{t\to+\infty} f(t)\,t^{-N/(N-2)}=0.$

Let us also mention that the set of functions f satisfying (2)-(5) is (for $N\geqq 2$) a convex cone.

Finally let us remark that if $\varliminf_{t \to +\infty} f(t)\, t^{-N/(N-2)} > 0$ then (5) reduces to:

$$tf(t) - \theta F(t) \leqq 0 \quad \text{for some } \theta \in \left[0, \frac{2N}{N-2}\right), \qquad \text{for} \quad t \geqq t_0.$$

The latter condition, which is motivated by the Pohozaev identity (which we describe below), was introduced in section 4 of [23].

We believe that condition (5) is not necessary and more precisely we conjecture that under assumptions (2), (3), (4), and for a general regular bounded domain Ω, solutions of (0.1) are *a priori* bounded in $L^\infty(\Omega)$.

Proof of Theorem 1.1. — We shall divide the proof into several steps. We shall prove that (i) u and $f(u)$ are bounded in L^1_{loc}; (ii) u and ∇u are bounded in a neighborhood of the boundary $\partial\Omega$; (iii) ∇u is bounded in L^2 and (iv) u is bounded in the sup norm. In all that follows C will denote a constant depending only on f and Ω and possibly changing value in the course of the proof.

Step 1. — We prove that u, $f(u)$ are bounded in L^1_{loc}. More precisely, let v_1 be an eigenfunction corresponding to λ_1, i.e.:

$$(6) \qquad -\Delta v_1 = \lambda_1 v_1 \quad \text{in } \Omega, \qquad v_1 \in C^2(\overline{\Omega}), \qquad v_1 = 0 \quad \text{on } \partial\Omega,\ v_1 > 0 \quad \text{in } \Omega.$$

Then we want to prove that:

$$(7) \qquad \int_\Omega u v_1\, dx, \qquad \int_\Omega |f(u)|\, v_1\, dx \leqq C.$$

This is a standard result, and we prove it only for completeness (*see* also R. E. L. Turner [30], R. D. Nussbaum [23], or H. Brezis and R. E. L. Turner [8]). If we multiply (1) by v_1, integrate by parts twice and use (6), we obtain:

$$\int_\Omega f(u) v_1\, dx = \int_\Omega \lambda_1 u v_1\, dx.$$

Now, because of (3), there exists $\lambda > \lambda_1$ such that:

$$f(t) \geqq \lambda t, \qquad \text{for } t \geqq t_0.$$

Hence:

$$\lambda \int_\Omega u v_1\, dx \leqq \int_\Omega f(u) v_1\, dx + C = \lambda_1 \int_\Omega u v_1\, dx + C,$$

that is:

$$\int_\Omega u v_1\, dx \leqq \frac{C}{\lambda - \lambda_1} \qquad \text{and} \qquad \int_\Omega f(u) v_1\, dx \leqq \frac{C\lambda_1}{\lambda - \lambda_1}.$$

And we easily obtain (7).

Step 2. — We now prove that u, ∇u are bounded in a neighborhood of $\partial\Omega$. Indeed, applying the results of B. Gidas, Wei-Ming Ni and L. Nirenberg [13] and recalling that Ω is convex, we find that there exists $t_0>0$ such that: $u(x-tn(x))$ is nondecreasing for $t\in[0, t_0]$ and for x on $\partial\Omega$, where $n(x)$ is the unit outward normal to $\partial\Omega$ at x. In addition t_0 depends only on the geometry of Ω. Furthermore, by using the results of [13] one can show that there exists $\alpha>0$ (depending only on the convex region Ω) such that:

$$\begin{cases} u(x-t\nu) \text{ is nondecreasing for } t\in[0, t_0], \text{ for } \nu\in\mathbb{R}^N \\ \text{satisfying } |\nu|=1 \text{ and } (\nu . n(x))\geqq\alpha \text{ and for } x\in\partial\Omega. \end{cases}$$

The t_0 in the above condition may have to be decreased somewhat from its previous value. The existence of α above follows easily from [13] if all sectional curvatures at every point of $\partial\Omega$ are strictly positive (then the result is purely local); a little more care is necessary for the case of general convex regions (with $C^{2,\beta}$ boundaries).

The fact that $u(x-t\nu)$ is nondecreasing in t for x, t and ν as above implies that there exist positive numbers γ and ε (depending only on Ω) such that:

$$(8)\qquad \begin{cases} \text{for all } x\in\{z\in\Omega : d(z, \partial\Omega)<\varepsilon\} \text{ there exists a measurable set } I_x \text{ with:} \\ \text{(i) meas } (I_x)\geqq\gamma, \\ \text{(ii) } I_x\subset\{x\in\Omega : d(x, \partial\Omega)>\varepsilon/2\}, \\ (iii)\ u(\xi)\geqq u(x) \text{ for all } \xi \text{ in } I_x. \end{cases}$$

Actually, in our situation I_x is simply a piece of a cone with vertex at x.

Now, using (8), we can deduce that u and ∇u are bounded in a neighborhood of $\partial\Omega$. Indeed, by step 1 we have that for all x in $\{z\in\overline{\Omega} : d(z, \partial\Omega)<\varepsilon\}$:

$$C\geqq\int_{I_x} u(\xi)v_1(\xi)\,d\xi\geqq u(x)\gamma(\min_{\xi\in I_x} v_1(\xi)),$$

[Notice that the second part of the inequality follows from (i) and (iii).] Furthermore, in view of condition (ii) in (8) and the fact that $v_1(\xi)>0$ for $\xi\in\Omega$, there exists $\delta>0$ (depending only on Ω) such that:

$$\inf_{\xi\in I_x} v_1(\xi)\geqq\delta>0.$$

Thus we obtain:

$$(9)\qquad u(x)\leqq C \quad \text{for all } x \quad \text{in } \{z\in\Omega : d(z, \partial\Omega)<\varepsilon\}.$$

We now claim that elliptic regularity results imply:

$$|\nabla u(x)|\leqq C \quad \text{for all } x \quad \text{in } \{z\in\Omega : d(z, \partial\Omega)<\varepsilon/2\}.$$

To be more precise, de Giorgi-Nash type theorems (for example, Theorem 14.1 on p. 201 in [18]) imply that there exists a number α, $0<\alpha<1$, such that:

$$\|u\|_{C^{0,\alpha}(\Omega_1)}\leqq C, \qquad \Omega_1=\left\{x\in\Omega : \left(\frac{\varepsilon}{4}\right)<d(x, d\Omega)<\left(\frac{3\varepsilon}{4}\right)\right\}.$$

If r_1 and r_2 are any positive numbers with $\varepsilon/4<r_1<\varepsilon/2$ and $\varepsilon/2<r_2<(3\varepsilon/4)$ Schauder interior estimates (e. g., Theorem 6.2 on p. 85 of [14]) imply:

$$\|u\|_{C^{2,\alpha}(\Omega_2)}\leqq C, \qquad \Omega_2=\{x\in\Omega : r_1\leqq d(x, \partial\Omega)\leqq r_2\}.$$

Finally, Schauder boundary estimates (e. g., Theorem 6.6 on p. 93 of [14]) imply:

$$\|u\|_{C^{2,\alpha}(\Omega_3)}\leqq C, \qquad \Omega_3=\{x\in\Omega : d(x, \partial\Omega)\leqq r_2\}$$

and of course the latter inequality implies:

$$(10) \qquad |\nabla u(x)|\leqq C \quad \text{for all } x \quad \text{in } \left\{z\in\Omega : d(z, \partial\Omega)<\frac{\varepsilon}{2}\right\}.$$

Notice that inequality (9) when combined with (7) implies that $\|f(u)\|_{L^1(\Omega)}\leqq C$.

Step 3. – We next prove that ∇u is *a priori* bounded in L^2. We assume first that $N\geqq 3$. In this case the above claim is a consequence of the following two Lemmas. The first lemma below is a slight generalization of an identity due to S. I. Pohozaev [27].

LEMMA 1.1 (*see* S. I. Pohozaev). – *Let Ω be as above and suppose that $g : \overline{\Omega}\times\mathbb{R}\to\mathbb{R}$ is a continuous map and that $w\in C^2(\overline{\Omega})$ satisfies:*

$$\Delta w(x)+g(x, w(x))=0 \qquad \textit{for} \quad x\in\Omega,$$
$$w(x)=0 \qquad \textit{for} \quad x\in\partial\Omega.$$

If $y\in\mathbb{R}^N$ is a fixed vector and $n(x)$ denotes the unit outward normal to $\partial\Omega$ at x, then w satisfies:

$$(11) \quad \int_{\partial\Omega}(x-y).n(x)|\nabla w(x)|^2\,dS=2\,N\int_{\Omega}G(x, w)\,dx$$
$$+2\sum_{i=1}^{N}\int_{\Omega}(x_i-y_i)G_{x_i}(x, w)\,dx-(N-2)\int_{\Omega}g(x, w)\,w\,dx,$$

where $G(x, s)=\int_0^s g(x, t)\,dt$.

In particular one sees immediately that if Ω is star shaped with respect to y and if $g(x, t)=g(t)$ and g satisfies: $2\,NG(t)-(N-2)\,tg(t)\leqq 0$ for $t\geqq 0$, then the unique solution of (0.1) is $u=0$ (this result is due to Pohozaev [27]). Let us also remark that (5) appears to be (in some sense) a strong negation of: $2\,N/(N-2)\,G(t)\leqq tg(t)$ for $t\geqq 0$. Finally let us mention that (11) is the real motivation of the introduction of (5) (as it will be seen below).

LEMMA 1.2. — *If, for* $N \geqq 3$, $u \in H_0^1(\Omega)$ *satisfies:*

$$(12)\quad \left\{ \begin{array}{l} \displaystyle\int_\Omega |\nabla u|^2 dx = \int_\Omega u f(u)\, dx, \qquad \|f(u)\|_{L^1(\Omega)} \leqq C, \\ \displaystyle\left| (N-2) \int_\Omega |\nabla u|^2 dx - 2N \int_\Omega F(u)\, dx \right| \leqq C. \end{array} \right.$$

then $|\nabla u|_{L^2(\Omega)} \leqq C_1$.

We have proved in Step 2 that any positive solution u of (0.1) satisfies $\|f(u)\|_{L^1} \leqq C$ and $\sup_{x \in \partial\Omega} |\nabla u(x)| \leqq C$, so equation (11) implies that the final part of condition (12) is satisfied, and Lemma 1.2 gives the desired conclusion, at least for $N \geqq 3$.

Proof of Lemma 1.1 (*see* S. I. Pohozaev [27]). — If we multiply the defining equation for $w(x)$ by:

$$\sum_{i=1}^{N} (x_i - y_i) \left(\frac{\partial w}{\partial x_i} \right)$$

and integrate by parts, we find:

$$\begin{aligned} -\int_\Omega \sum_{i,j} (x_i - y_i) \frac{\partial w}{\partial x_i} \frac{\partial}{\partial x_j} \left(\frac{\partial w}{\partial x_j} \right) dx &= \int_\Omega |\nabla w|^2 dx - \int_{\partial\Omega} (x - y).n(x) |\nabla w|^2 dS \\ &\quad + \sum_{i,j} \int_\Omega (x_i - y_i) \left(\frac{\partial^2 w}{\partial x_i \partial x_j} \right) \left(\frac{\partial w}{\partial x_j} \right) dx \\ &= \sum_i \int_\Omega (x_i - y_i) \frac{\partial w}{\partial x_i} g(x, w)\, dx \\ &= -N \int_\Omega G(x, w)\, dx - \sum_i \int_\Omega (x_i - y_i) G_{x_i}(x, w)\, dx. \end{aligned}$$

But we have:

$$\sum_{i,j} \int_\Omega (x_i - y_i) \left(\frac{\partial^2 w}{\partial x_i \partial x_j} \right) \left(\frac{\partial w}{\partial x_j} \right) dx = \frac{1}{2} \int_{\partial\Omega} (x - y).(n(x)) |\nabla w|^2 dS - \frac{N}{2} \int_\Omega |\nabla w|^2 dx$$

and because $\int_\Omega |\nabla w|^2 dx = \int_\Omega w g(x, w)\, dx$, the proof is complete.

Proof of Lemma 1.2. — In view of Sobolev inequalitites, of Step 2 and of Holder inequalities, we have: for some $C > 0$:

$$\begin{aligned} \int_\Omega u f(u)\, dx = \int_\Omega |\nabla u|^2 dx &\geqq \frac{1}{C} \left[\int_\Omega u^{2N/(N-2)} dx \right]^{1-2/N} \\ &\geqq \frac{1}{C} \left[\int_\Omega u^{2N/(N-2)} dx \right]^{1-2/N} \left[\int_\Omega |f(u)|\, dx \right]^{2/N} \geqq \frac{1}{C} \int_\Omega u^2 |f(u)|^{2/N} dx, \end{aligned}$$

hence we have:

$$\int_\Omega u^2\,|f(u)|^{2/N}\,dx \leqq C\int_\Omega uf(u)\,dx.$$

Now in view of (12), we obtain:

$$\int_\Omega F(u)\,dx \leqq \frac{N-2}{2N}\int_\Omega uf(u)\,dx + C.$$

In order to deduce from the two above inequalitites a bound for $\int_\Omega uf(u)\,dx$, let us assume that f satisfies: $\exists\,\theta\in[0,\,2N/(N-2))$, $\forall\,\varepsilon>0$, $\exists\,t_\varepsilon>0$ such that:

$$tf(t)\leqq\theta F(t)+\varepsilon\,t^2\,|f(t)|^{2/N},\qquad \text{for}\quad t\geqq t_\varepsilon.$$

This yields:

$$\int_\Omega uf(u)\,dx\leqq\theta\int_\Omega F(u)\,dx+\varepsilon\int_\Omega u^2\,|f(u)|^{2/N}\,dx+C_\varepsilon$$

and in view of the above inequalities, we deduce:

$$\left(1-\frac{N-2}{2N}\theta-C\varepsilon\right)\int_\Omega F(u)\,dx\leqq C_\varepsilon,$$

that is the announced estimate:

$$\int_\Omega |\nabla u|^2\,dx=\int_\Omega uf(u)\,dx\leqq C.$$

To conclude, let us just remark that the assumption made on f is equivalent to (5).

It remains to consider the case $N=2$. In this case use condition (4) on f to select a number γ, $0<\gamma<1$, such that:

$$\overline{\lim_{t\to\infty}}\,\frac{tf(t)}{t^2\,f(t)^\gamma}=0.$$

Given $\varepsilon>0$, there exists $t_\varepsilon>0$ such that:

$$tf(t)\leqq\varepsilon\,t^2\,f(t)^\gamma\qquad\text{for}\quad t\geqq t_\varepsilon$$

Just as for $N\geqq 3$ we have.

$$\int_\Omega |\nabla u|^2\,dx=\int_\Omega uf(u)\,dx\leqq C_\varepsilon+\varepsilon\int_{u\geqq t_\varepsilon} u^2\,|f(u)|^\gamma\,dx$$

$$\leqq C_\varepsilon+\varepsilon\left[\int_\Omega |f(u)|\right]^\gamma\left[\int_\Omega u^{2/(1-\gamma)}\,dx\right]^{1-\gamma}\leqq C_\varepsilon+\varepsilon C\int_\Omega|\nabla u|^2\,dx.$$

Here C depends only on $\|f(u)\|_{L^1}$ and the constants in the Sobolev embedding theorem. For ε small enough that is $\varepsilon C<1$ we obtain L^2 bounds on ∇u for $N=2$.

Remark 1.2. – Lemma 1.2 is the only place where we use (5).

Step 4. – We conclude that $\|u\|_{L^\infty} \leqq C$. This claim comes from the combination of the H^1 bound obtained in Step 3 together with assumption (4). If (4) is a little bit strengthened, namely if we assume:

$$(4') \qquad \lim_{t \to +\infty} f(t)\, t^{-p}=0, \qquad \text{with} \quad p<\sigma=\frac{N+2}{N-2} \quad (\text{or } p<\infty, \text{ if } N=2),$$

then the conclusion follows from a simple bootstrap argument using the H^1 bound and classical regularity results of [1] (for example, $\|f(u)\|_{L^{p_1}} \leqq C$ for $p_1=(\sigma+1)/p$, thus $\|u\|_{W^{2,p_1}} \leqq C$, etc.). In the special case $N=2$, Step 3 and the Sobolev embedding Theorem imply that $\|u\|_{L^t} \leqq C$ for any $t<\infty$, and condition (4) on f and the regularity results of [1] imply L^∞ bounds.

In the general case [assumption (4) and $N \geqq 3$] we proceed as follows, adapting a device due to H. Brezis and T. Kato [7]. Let $p>1$: by (0.1) we have obviously:

$$p \int_\Omega |\nabla u|^2 u^{p-1}\, dx=\int_\Omega f(u)\, u^p\, dx;$$

or:

$$\frac{4p}{(p+1)^2} \int_\Omega |\nabla u^{(p+1)/2}|^2\, dx=\int_\Omega f(u)\, u^p\, dx.$$

Now, because of (4), we have: $\forall \varepsilon>0, \exists C_\varepsilon>0$:

$$f(t)\, t^p \leqq \varepsilon\, t^{p+\sigma}+C_\varepsilon.$$

Therefore we have:

$$\int_\Omega |\nabla u^{(p+1)/2}|^2\, dx \leqq C\varepsilon \int_\Omega u^{\sigma-1}\, u^{p+1}+C_\varepsilon.$$

Applying Sobolev inequality which introduces the exponent $q=(N(p+1))/(N-2)$ we find:

$$\left[\int_\Omega u^q\, dx\right]^{(N-2)/N} \leqq C\varepsilon \int_\Omega u^{(N-2)/N\, q+4/(N-2)}\, dx+C_\varepsilon,$$

and applying Hölder inequality we obtain:

$$\left[\int_\Omega u^q\, dx\right]^{(N-2)/N} \leqq C\varepsilon\left[\int_\Omega u^q\, dx\right]^{(N-2)/N}\left[\int_\Omega u^{2N/(N-2)}\, dx\right]^{2/N}+C_\varepsilon.$$

Now by Step 3 and anew Sobolev inequality:

$$\int_{\Omega} u^{2N/(N-2)}\, dx \leqq C.$$

And finally choosing ε small enough, we obtain:

$$\| u \|_{L^q(\Omega)} \leqq C \quad \text{for any } q < \infty.$$

In particular: $\| u \|_{L^r} \leqq C$, for $r > N/2\sigma$ and this implies $\| f(u) \|_{L^{r/\sigma}} \leqq C$, and $r/\sigma > N/2$. Now, in view of (1), this proves the desired *a priori* bound, namely:

$$\| u \|_{L^\infty} \leqq C.$$

Remark 1.3. – To shed some light on the above result it is of interest to look at the simple case $\Omega = [-1, 1]$ $(N=1)$. Then, under assumptions (2) and (3), the solutions of (1) are *a priori* bounded. Indeed, a simple O.D.E. argument shows that necessarily (u being positive) $u(-x) = u(x)$ and $u(x)$ decreases for $x \geqq 0$. Now, as in Step 1, we find that:

$$\| u \|_{L^1_{loc}},\ \| f(u) \|_{L^1_{loc}} \leqq C.$$

Now, by the above properties of u, this implies that $\| f(u) \|_{L^1} \leqq C$ and that $\| u'' \|_{L^1} \leqq C$. Since $u'(0)=0$, we deduce that $| u' | \leqq C$ and:

$$\| u \|_{L^\infty} \leqq C.$$

1.2 The case of a general domain.

Theorem 1.2. – *Assume conditions* (2)-(5) *on f and in addition suppose that:*

(14) $f(t)\, t^{-\sigma}$ *is nonincreasing for* $t \geqq 0$, *with* $\sigma = (N+2)/(N-2)$ (*if* $N=2$, *this condition disappears*). *Then we have:*

$$\| u \|_{L^\infty(\Omega)} \leqq C,$$

where C *depends only on* Ω *and on the behaviour of f in the limits arising in* (3)-(5).

Remark 1.4. – Let us recall that we conjecture that, under assumptions (2), (3), (4) only, the conclusion of Theorem 1.2 still holds.

Remark 1.5. – *In the case where we assume that f depends on x*, we need to assume:

(2′) $f \in C(\overline{\Omega} \times \mathbb{R}_+, \mathbb{R})$, $f(x, .)$ is Lipschitz continuous on any bounded interval [0, L], uniformly in x;

(3′) $\varliminf_{t \to +\infty} f(x, t)\, t^{-1} > \lambda_1$, uniformly in $x \in \overline{\Omega}$;

(4′) $\lim_{t \to +\infty} f(x, t)\, t^{-\sigma} = 0$ uniformly in x for $\sigma = (N+2)/(N-2)$;

(5′) $\varlimsup_{t \to +\infty} tf(x, t) - \theta \{ F(x, t) + (1/N)\, \mathrm{div}_x F(x, t) \} / t^2 f(x, t)^{2/N} \leqq 0$, uniformly in $x \in \overline{\Omega}$;

where:

$$F(x, t)=\int_0^t f(x, s)\,ds \qquad \text{and} \qquad \operatorname{div}_x F(x, t)=\sum_{i=1}^{N} \frac{\partial F}{\partial x_i}(x, t).$$

(15) $\left\{\begin{array}{l}\text{There exist } r, \delta > 0 \text{ such that } f(.,t)\in C^1(\Omega_r) \text{ (where } \Omega_r=\{x : d(x, \partial\Omega) < r\}) \text{ for all} \\ t \geqq 0 \text{ and } \nabla_x f(x, t).\theta \leqq 0 \text{ for all } x \text{ in } \Omega_r,\ t \geqq 0 \text{ and unit vectors } \theta \text{ such that} \\ |\theta-n(x)| < \delta.\end{array}\right.$

In the statement of (15) $n(x)$ is defined for $x\in\Omega$ to be $n(y)$, the unit outward normal vector at $y\in\partial\Omega$, where $y\in\partial\Omega$ is defined by $|y-x|=d(x, \partial\Omega)$. It can be shown (*see* the appendix of [14]) that under our assumptions on Ω there exists $\rho=\rho(\Omega) > 0$ such that for $x\in\Omega_\rho$ there exists a unique $y=y(x)\in\partial\Omega$ with $|y-x|=d(x, \partial\Omega)$ and that the map $x \to y(x)$ is C^1 on Ω_ρ.

Finally, if Ω is not convex or if Ω does not possess certain geometrical properties (*see* Section 1.3) we need to assume in addition:

(14′) $\left\{\begin{array}{l}\text{There exists } \varepsilon > 0 \text{ such that } f(x, t)\, t^{-\sigma} \text{ is a nonincreasing function of } t \text{ for } t \geqq 0 \text{ and} \\ \text{all } x \text{ such that } d(x, \partial\Omega) < \varepsilon\ (\sigma=(N+2)/(N-2)).\end{array}\right.$

Condition (14′) is vacuous if $N=2$.

Under assumptions (2′)-(5′), (15) and (14′) (if Ω is not convex) there exist *a priori* bounds in the L^∞ norm for all positive solutions of:

$$\Delta u(x)+f(x, u(x))=0, \qquad x\in\Omega,$$
$$u(x)=0, \qquad x\in\partial\Omega.$$

The proof is close enough to the proof of Theorem 1.2 below that we leave it to the reader.

Before going into the proof of Theorem 1.2, let us first say that the only part of the proof of Theorem 1.1 which uses the convexity of Ω is Step 2, where we apply the results of [13] to obtain (8). Here we are going to prove a variant of (8). We shall prove that there exist ε, γ and $C > 0$ such that:

(8′) $\left\{\begin{array}{l}\text{for all } x\in\Omega_\varepsilon=\{x\in\overline{\Omega},\ d(x, \partial\Omega) < \varepsilon\}, \text{ there exists a measurable set } I_x \text{ such that:} \\ \quad\text{(i) meas } (I_x) \geqq \gamma, \\ \quad\text{(ii) } I_x \subset \Omega-\Omega_{\varepsilon/2}, \\ \quad\text{(iii) } u(x) \leqq C\, u(\xi) \text{ for all } \xi \text{ in } I_x.\end{array}\right.$

To this end we shall follow Gidas, Ni and Nirenberg [13] and use a classical transform, the so-called Kelvin transform (*see* p. 66 of [14]). Of course once (8′) is proved, the remainder of the proof of Theorem 1.1 is still valid, and Theorem 1.2 is proved.

At this point, before proving (8′), we would like to describe briefly a version of a theorem in [13] which is adequate for our purposes. Let us first establish notation. Let $\mathcal{O}$ be a bounded open domain on $\mathbb{R}^N$ with a C^2 boundary, and for each $x\in\partial\mathcal{O}$, let $\eta(x)$ denote the unit, outward normal vector to $\mathcal{O}$ at x. If γ is a unit vector in $\mathbb{R}^N$, let T_α denote the hyperplane $\{x : (x.\gamma)=\alpha\}$ and let α^* be the largest value such that $T_\alpha\cap\mathcal{H}$ is nonempty. For $\alpha<\alpha^*$, let $\Sigma(\alpha)$ denote an open "cap", the part of $\mathcal{O}$ which is one the same side of T_α as T_{α^*}, so $\Sigma(\alpha)=\{x\in\mathcal{O} : (x.\gamma)>\alpha\}$. Let $\Sigma'(\alpha)$ denote the reflection of $\Sigma(\alpha)$ in T_α.

Let $v\in C^2(\mathcal{O})$ satisfy:

$$\begin{cases} -\Delta v=f(x, v), & v>0, \quad x\in\mathcal{O}, \\ v(x)=0, & x\in\partial\mathcal{O}; \end{cases}$$

where $f(x, t)$ satisfies (2').

Let $x_0\in T_{\alpha^*}\cap\partial\mathcal{O}$ and let us assume that:

(1) there exists $\alpha_1<\alpha^*$ such that for $\alpha_1\leqq\alpha\leqq\alpha^*$, one has $\Sigma'(\alpha)\subset\mathcal{O}$ where $\gamma=\eta(x_0)$ and $\Sigma'(\alpha)$ is not internally tangent to $\partial\mathcal{O}$ at any point Q not on T_α, and T_α is not orthogonal to $\partial\mathcal{O}$ at any point Q in $\partial\mathcal{O}$;

(2) there exist $r_1, r_2>0$ such that for any x in $\partial\mathcal{O}$ with $|x-x_0|\leqq r_1$ and any number $s\geqq 0$, the map $(t\mapsto f(x-t\eta(x_0), s))$ is nondecreasing for $0\leqq t\leqq r_2$ and $x+t\eta(x_0)\in\Sigma(\alpha_1)$ for $0\leqq t\leqq r_2$. Then Theorems 2.1 and 2.1' in [13] imply that $v(x-t\eta(x_0))$ is a nondecreasing function of t for $x\in\partial\mathcal{O}$ *such that* $|x-x_0|\leqq r_1$ and for $0\leqq t\leqq r_2$.

Now we prove Theorem 1.2, that is we prove (8'). To this end, let $x_0\in\partial\Omega$, and let B be a closed ball which intersects $\bar\Omega$ only at x_0. By translation and rotation of the axes we may assume that the center of the ball is at the origin and that $x_0=(r, 0, \ldots, 0)$. Let J denote reflection in the boundary of B, so $J(x)=r^2 x/|x|^2$. Since we assumed that Ω is regular, we may assume that $0<\alpha\leqq r\leqq\beta$, where α and β are independent of x_0. For notational convenience we can suppose $r=1$ and let $\mathcal{O}=J(\Omega)$. Notice that $\mathcal{O}$ has a nonempty open cap at $J(x_0)\in\partial\mathcal{O}$ which meets the conditions of [13] (*see* also above), and in fact the same is true for all points $x_1\in\partial\mathcal{O}$, which are close enough to x_0.

Next, we define for $y\in\mathcal{O}$:

$$|y|^{N-2}w(y)=u(x), \qquad \text{if} \quad y=\frac{x}{|x|^2} \quad (\text{and } x\in\Omega).$$

This is the "Kelvin transform" (p. 66 of [14]).

Obviously, $w\in C^2(\bar{\mathcal{O}})$, $w=0$ on $\partial\mathcal{O}$ and $w>0$ in $\mathcal{O}$. Furthermore a lengthy calculation shows that, since u satisfies (1), w satisfies:

$$(16) \quad -\Delta w(y)=|y|^{-N-2}f(|y|^{N-2}w(y)), \quad w\in C^2(\bar{\mathcal{O}}), \quad w>0 \text{ in } \mathcal{O}, \quad w=0 \text{ on } \partial\mathcal{O}.$$

But we assume (14), that is, $|y|^{-N-2}f(|y|^{N-2}t)\overset{\text{def}}{=}g(y, t)$ is nonincreasing in $|y|$, for any $t\geqq 0$; therefore all the conditions of [13] (*see* also above) are satisfied and we obtain:

$$(17)\quad \begin{cases} \forall x_0\in\partial\Omega, \exists \varepsilon_{x_0}, \gamma_{x_0}>0 \text{ such that for all } y \text{ in } B(J(x_0), \varepsilon_{x_0})\cap\mathcal{O} \text{ there exists a measurable} \\ \text{set } K_y \text{ such that:} \\ \text{(i) meas}(K_y)\geqq\gamma_{x_0}, \\ \text{(ii) } K_y\subset\{y\in\mathcal{O},\ d(y, \partial\mathcal{O})>\varepsilon_{x_0}/2\}, \\ \text{(iii) } w(y)\leqq w(\xi), \text{ for all } \xi \text{ in } K_y. \end{cases}$$

Now, since the mapping J is a C^1 diffeomorphism (and $\partial\Omega$ is compact) one can see with the aid of the change of variables formula in R^N that (17) implies (8'), and this proves Theorem 1.2.

1.3. A GEOMETRICAL ASSUMPTION. — Of course, the main difference between Theorems 1.1 and 1.2 is that we need, in the case of a general domain, the assumption (14). The goal of this section is to give a geometrical assumption on the domain Ω (generalizing the convexity assumption of section (11) which allows us to dispense with assumption (14).

We will assume:

(18) $\partial\Omega = \Gamma_1 \cup \Gamma_2$ where Γ_1, Γ_2 are closed and satisfy:

(i) at every point of Γ_1, all sectional curvatures of Γ_1 are bounded away from 0 by a positive constant $\alpha > 0$;

(ii) there exists a point x_0 in $\mathbb{R}^N$ such that $(x - x_0, n(x)) \leqq 0$, for all x in Γ_2.

Of course if Ω is strictly convex, then $\Gamma_1 = \partial\Omega$, $\Gamma_2 = \emptyset$ and (18) is satisfied. But if $\Omega = \Omega_1 - \overline{\Omega}_2$, with Ω_1 strictly convex, Ω_2 starshaped and $\overline{\Omega}_2 \subset \Omega_1$ then Ω satisfies (18) with $\Gamma_1 = \partial\Omega_1$, $\Gamma_2 = \partial\Omega_2$. Therefore, we see by this example that (18) may include domains which are not at all convex.

We then have:

THEOREM 1.3. — *Suppose that* $N \geqq 3$ (*we already know that condition* (14) *on* f *is unnecessary for* $N = 2$) *and suppose that assumptions* (2)-(4) *and* (18) *hold and that:*

$$(5'') \qquad \lim_{t \to +\infty} \frac{t f(t) - \theta F(t)}{t^2 f(t)^{2/(N+1)}} \leqq 0, \quad \textit{for some} \quad 0 \leqq \theta < \frac{2N}{N-2}.$$

Then we have $\| u \|_{L^\infty(\Omega)} \leqq C$, *where* C *depends only on* Ω *and on the behavior of* f *in the limits arising in* (3), (4) *and* (5″).

Remark 1.6. — Theorem 1.3 shows that condition (14) on f is unnecessary for domains with a certain geometrical property. The question is whether (14) is unnecessary for general regions Ω.

Proof of Theorem 1.3. — If we follow the proof of Theorem 1.1, as in Step 1 that:

$$(19) \qquad \int_\Omega |f(u)| \, \delta \, dx \leqq C, \qquad \int_\Omega u \, \delta \, dx \leqq C,$$

where δ denotes the function $\delta(x) = d(x, \partial\Omega)$. The same kind of argument used in Step 2 of the proof of Theorem 1 together with condition (i) of (18) yields that there exists $\varepsilon > 0$ and C (independent of u) such that:

$$u(x) \leqq C \quad \text{for all } x \text{ with } d(x, \Gamma_1) < \varepsilon.$$

By the same kind of regularity results used before in Step 2 one finds:

$$(20) \qquad |\nabla u(x)| \leqq C \quad \text{for all } x \text{ in } \Gamma_1$$

(the argument in this case is slightly different; one may have to work with $w=\zeta u$, where ζ is a function which vanishes at the "ends" of the strip $\{x : d(x, \Gamma_1) < \varepsilon\}$).

To prove, as in Step 3, that $|\nabla u|_{L^2}$ is bounded, recall that (11) gives:

$$(11') \quad 2N \int_\Omega F(u)\,dx - (N-2)\int_\Omega uf(u)\,dx = \int_{\Gamma_1} (x-x_0).n(x)\,|\nabla u|^2\,dS + \int_{\Gamma_2} (x-x_0).n(x)\,|\nabla u|^2\,dS.$$

Using (20) and part (ii) of (18) we obtain from (11') that:

$$\frac{2N}{N-2}\int_\Omega F(u)\,dx - \int_\Omega uf(u)\,dx \leqq C.$$

To conclude we argue as follows: we first recall that by a well-known inequality (*see* for example Lemma 2.2 of [8], or [20]) we have:

$$\left\|\frac{u}{\delta^\tau}\right\|_{L^q(\Omega)} \leqq C\,\|\nabla u\|_{L^2(\Omega)} \quad \text{if } 0 \leqq \tau \leqq 1, \quad \text{with} \quad q = \frac{2N}{N-2+2\tau}.$$

Combining this inequality, (19) and Hölder inequality, we find:

$$\int_\Omega uf(u)\,dx = \int_\Omega |\nabla u|^2\,dx \geqq \frac{1}{C}\left[\int_\Omega \frac{u^q}{\delta^{\tau q}}\,dx\right]^{2/q}$$
$$\geqq \frac{1}{C}\left[\int_\Omega \frac{u^q}{\delta^{\tau q}}\right]^{2/q}\left[\int_\Omega |f(u)|\,\delta\,dx\right]^{1-(2/q)}$$
$$\geqq \frac{1}{C}\int_\Omega \frac{u^2}{\delta^{2\tau}}\{|f(u)|\,\delta\}^{1-(2/q)}\,dx = \frac{1}{C}\int_\Omega u^2\,|f(u)|^{2\tau}\,dx,$$

if we choose q and τ such that: $0 \leqq \tau \leqq 1$, $q = 2N/(N-2+2\tau)$, $2\tau = 1-(2/q)$ i.e.:

$$\tau = \frac{1}{N+1}, \qquad q = 2\,\frac{N-1}{N+1}.$$

Hence, on one hand, we have:

$$\int_\Omega u^2\,|f(u)|^{2/(N+1)}\,dx \leqq C\int_\Omega uf(u)\,dx;$$

on the other hand, in view of (11'), (20) and part (ii) of (18), we obtain:

$$\int_\Omega F(u)\,dx \leqq \frac{N-2}{2N}\int_\Omega uf(u)\,dx + C.$$

In order to deduce from the two preceding inequalities a bound for $\int_\Omega uf(u)\,dx$, let us assume that there exists $\theta \in [0, 2N/(N-2))$ such that $\forall \varepsilon > 0$, $\exists t_\varepsilon > 0$ such that we have:

$$tf(t) \leqq \theta F(t) + \varepsilon t^2 |f(t)|^{2/(N+1)} \quad \text{for } t \geqq t_\varepsilon;$$

therefore this yields:

$$\int_\Omega uf(u)\,dx \leqq \theta \int_\Omega F(u)\,dx + \varepsilon \int_\Omega u^2 |f(u)|^{2/(N+1)}\,dx + C_\varepsilon$$

and in view of the preceding inequalities on the left hand side members we deduce:

$$\left(1 - \frac{N-2}{2N}\theta - C\varepsilon\right) \int_\Omega uf(u)\,dx \leqq C_\varepsilon,$$

and we obtain:

$$\int_\Omega |\nabla u|^2\,dx = \int_\Omega uf(u)\,dx \leqq C.$$

The remainder of the proof is then identical to the corresponding one in Theorem 1.1.

Remark 1.7. — Condition (5″) is slightly less general than (5). This is because of the presence of Γ_2: indeed we have only inequality (19) instead of estimates on $\| f(u) \|_{L^1}$ and $\| u \|_{L^1}$, since *a priori* we know nothing about u near Γ_2.

2. Existence results

2.1. EXISTENCE OF CONTINUA OF SOLUTIONS. — Our motivation for proving Theorems 1.1-1.3 is their usefulness in establishing results about the existence and structure of the set of positive solutions of (1) or of parametrized versions of (1). We shall derive these results from some general propositions concerning nonlinear mappings which take a cone in a Banach space into itself. In the following propositions we shall need to use the idea of the Leray-Schauder degree of a mapping (*see* J. Leray and J. Schauder [19] and L. Nirenberg [22] for the definition and elementary properties of the Leray-Schauder degree) and the fixed index of a compact map of a closed, convex set into itself. Recall that if C is a closed, convex subset of a Banach space Y, W is a relatively open subset of C (i.e., $W = \mathcal{O} \cap C$, where $\mathcal{O}$ is an open subset of Y), and $f : \overline{W} \to C$ is a compact map such that $f(x) \neq x$ for $x \in \overline{W} - W$, then $i_C(f, W)$, the fixed point index of f on W, can be defined as follows: Let $\mathcal{O}$ be a bounded, open subset of Y such that $\mathcal{O} \cap C = W$ and let $g : \overline{\mathcal{O}} \to C$ be a compact, continuous map such that $g \mid \overline{W} = f$ (the existence of g follows from a result of J. Dugundji). Then *define* $i_C(f, W)$ be the formula:

$$i_C(f, W) = \deg(I - g, \mathcal{O}, 0).$$

It is easy to show that $x-g(x) \neq 0$ for $x \in \partial\mathcal{C}$ and that this definition is independent of the particular set $\mathcal{C}$ above and the particular extension g of f. Properties of Leray-Schauder degree such as the homotopy and additivity properties carry over directly to the fixed point index.

We can now state some useful propositions. In the following recall that a cone C in a Banach space Y is a closed subset of Y such that: (i) if $x, y \in \mathrm{C}$ and $\alpha, \beta \geqq 0$, then $\alpha x + \beta y \in \mathrm{C}$ and (ii) if $x \in \mathrm{C}$ and $x \neq 0$, then $-x \notin \mathrm{C}$. A cone C induces a partial ordering on Y by $x \leqq y$ iff $y - x \in \mathrm{C}$. The next proposition is basically due to M. A. Krasnosel'skii ([18], [19]); the statement about the fixed point index was obtained by T. B. Benjamin [5], Appendix 1. Generalizations of Proposition 2.1 are given in [26], Section 1.

Proposition 2.1 (*see* [5], [16]). — *Let C be a cone in a Banach space X and* $\Phi : \mathrm{C} \to \mathrm{C}$ *a compact map such that* $\Phi(0) = 0$. *Assume that there exist numbers* $0 < r < \mathrm{R}$ *and a vector* $v \in \mathrm{C} - \{0\}$ *such that:* (i) $x \neq t\,\Phi(x)$ *for* $0 \leqq t \leqq 1$ *and* $\|x\| = r$ *and* (ii) $x \neq \Phi(x) + tv$ *for* $t \geqq 0$ *and* $\|x\| = \mathrm{R}$. *Then if* $\mathrm{U} = \{x \in \mathrm{C} : r < \|x\| < \mathrm{R}\}$ *and* $\mathrm{B}_\rho = \{x \in \mathrm{C} : \|x\| < \rho\}$, *one has:*

$$i_{\mathrm{C}}(\Phi, \mathrm{B}_{\mathrm{R}}) = 0, \qquad i_{\mathrm{C}}(\Phi, \mathrm{B}_r) = 1 \qquad \textit{and} \qquad i_{\mathrm{C}}(\Phi, \mathrm{U}) = -1.$$

In particular, Φ *has a fixed point in* U. *Condition* (i) *above is satisfied if there exists a bounded linear operator* $\mathrm{A} : \mathrm{X} \to \mathrm{X}$ *such that* $\mathrm{A}(\mathrm{C}) \subset \mathrm{C}$, A *has spectral radius strictly less than* 1 *and* $\Phi(x) \leqq \mathrm{A}(x)$ *for* $x \in \mathrm{C}$ *and* $\|x\| = r$.

Remark 2.1. — The conclusions of Proposition 2.1 remain valid if assumption: (ii) is replaced by the assumption that; (ii') there exists a compact map $\mathrm{F} : \overline{\mathrm{B}}_{\mathrm{R}} \times [0, \infty) \to \mathrm{C}$ such that $\mathrm{F}(x, 0) = \Phi(x)$ for $\|x\| = \mathrm{R}$, $\mathrm{F}(x, t) \neq x$ for $\|x\| = \mathrm{R}$ and $0 \leqq t < \infty$ and $\mathrm{F}(x, t) = x$ has no solution $x \in \overline{\mathrm{B}}_{\mathrm{R}}$ for $t \geqq t_0$. In fact the same proof used in [5] or [26] remains valid.

Our next proposition is proved for the Banach space case in Theorem 6.2 of [29], though the ideas go back to Leray and Schauder [19]. If one uses the fixed point index rather than degree theory, the same proof as in [29] gives the following result.

Proposition 2.2 (*see* [19], [29]). — *Let C be a closed convex set in a Banach space* X, $0 \in \mathrm{C}$, *and let* $\mathrm{F} : \mathrm{C} \times [0, \infty) \to \mathrm{C}$ *be a compact map such that* $\mathrm{F}(x, 0) = 0$ *for all* $x \in \mathrm{C}$. *Let* $\mathcal{S} = \{(x, t) : \mathrm{F}(x, t) = x\}$ *and let* $\mathcal{S}_0$ *denote the connected component of* $\mathcal{S}$ *which contains* (0, 0). *Then* $\mathcal{S}_0$ *is unbounded in* $\mathrm{C} \times [0, \infty)$.

The final abstract proposition we need was proved (in slightly less generality than stated here) for the Banach space case in [19], p. 63]. Again, essentially the same argument, using the fixed point index instead of Leray-Schauder degree (*see* [25], Lemma 3.4), yields

Proposition 2.3. — *Let C be a closed convex subset of a Banach space* X, U *a bounded relatively open subset of* C *and* $\mathrm{F} : \overline{\mathrm{U}} \times [a, b] \to \mathrm{C}$ *a compact map such that* $\mathrm{F}(x, t) \neq x$ *for* $(x, t) \in (\overline{\mathrm{U}} - \mathrm{U}) \times [a, b]$. *If* $\mathrm{F}_t(x) \overset{\mathrm{def}}{=} \mathrm{F}(x, t)$, *assume that* $i_{\mathrm{C}}(\mathrm{F}_a, \mathrm{U}) \neq 0$. *Then if* $\mathcal{S} = \{(x, t) \in \mathrm{U} \times [a, b] : \mathrm{F}(x, t) = x\}$, *there exists a connected component* D *of* $\mathcal{S}$ *such that* $\mathrm{D} \cap (\mathrm{U} \times \{a\})$ *and* $\mathrm{D} \cap (\mathrm{U} \times \{b\})$ *are nonempty.*

We now briefly show how the above abstract results, when combined with the *a priori* bounds of Section 1, yield a variety of results about (1) and parametrized versions of (1).

THEOREM 2.1. – *Let* $f:\mathbb{R}^+ \to \mathbb{R}^+$ *and* Ω *satisfy the hypotheses of either Theorem* 1.1, 1.2 *or* 1.3 *and assume, in addition:*

(21) $$\overline{\lim_{t\to 0^+}}\frac{f(t)}{t}<\lambda_1, \qquad f(0)=0.$$

Then equation (1) *has at least one positive solution u:*

(1) $$\begin{cases} -\Delta u=f(u) & \text{in } \Omega, \quad u\in C^2(\overline{\Omega}),\\ u=0 & \text{on } \partial\Omega, \quad u>0 \quad \text{on } \Omega.\end{cases}$$

Proof. – Let L denote the inverse of $(-\Delta)$ with Dirichlet boundary conditions and consider L as a map from $C(\overline{\Omega})\to C(\overline{\Omega})$. It is well-known that L is compact. Regularity theory and the maximum principle imply that solving (1) is equivalent to finding a nonzero solution $u\in C=\{u\in C(\overline{\Omega}): u(x)\geqq 0 \text{ for all } x \text{ in } \Omega\}$ of the equation:

$$u(x)=(\Phi u)(x)\stackrel{\text{def}}{=}L(f(u(x))).$$

Select $r>0$ and $\alpha<\lambda_1$ so that $f(t)\leqq \alpha t$ for $0\leqq t\leqq r$. Our assumptions imply that:

$$\Phi u\leqq \alpha L u, \qquad \|u\|\leqq r$$

and the spectral radius of αL is less than 1, so condition (i) in Proposition 2.1 is satisfied.

Now we use Remark 2.1 and define:

$$F(u,t)(x)\stackrel{\text{def}}{=}L(f(u(x)+t)).$$

The argument in Step 1 of the proof of Theorem 1.1 shows that the above equation has no positive solutions for $t\geqq t_0$. Furthermore, the *a priori* estimates for any positive solution are uniform in t, so there exists a constant $R>r$ such that the equation has no solution u in K with $\|u\|\geqq R$. Proposition 2.1 and Remark 2.1 now give the theorem. Note that we also obtain a result on the fixed point index of Φ.

COROLLARY 2.1. – *Let* $f:[\alpha,\beta]\times\mathbb{R}^+\to\mathbb{R}^+$ *be a continuous map such that* $f(\lambda, .)$ *satisfies the hypotheses of Theorem* 2.1 *uniformly in* $\lambda\in[\alpha,\beta]$. *Then there exists a closed, connected set* D *in* $[\alpha,\beta]\times C(\overline{\Omega})$ *such that:* (i) *for all* λ *in* $[\alpha,\beta]$ *there exists u in* $C(\overline{\Omega})$ *such that* $(\lambda,u)\in D$ *and* (ii) *for all* (λ,u) *in* D *we have:*

$(1)_\lambda$ $$\begin{cases} -\Delta u=f(\lambda,u) & \text{in } \Omega, \quad u\in C^2(\overline{\Omega}),\\ u=0 \quad \text{on } \partial\Omega & \text{and} \quad u>0 \quad \text{on } \Omega.\end{cases}$$

Proof. – Let L be defined as in the proof of Theorem 2.1 and define $F: C\times[\alpha,\beta]\to C$ by:

$$(F(u,\lambda))(x)=L(g)(x), \qquad g(y)=f(\lambda,u(y)).$$

The argument of Theorem 2.1 and the results of Section 1 show that there exist positive numbers r and R with:

$$F(u, \lambda) \neq u \quad \text{for} \quad \alpha \leqq \lambda \leqq \beta, \quad u \in C \quad \text{and} \quad \|u\| = r \quad \text{or} \quad \|u\| = R.$$

Furthermore, if $F_\lambda(u) = F(u, \lambda)$, we have seen that:

$$i_C(F_\lambda, U) = -1 \quad \text{for} \quad \alpha \leqq \lambda \leqq \beta, \quad U = \{u \in C : r < \|u\| < R\}.$$

Corollary 2.1 now follows easily from Proposition 2.3.

As another illustration, we have:

COROLLARY 2.2. – *Let $f : \mathbb{R}^+ \to \mathbb{R}^+$ and Ω satisfy the hypotheses of Theorem* 1.1, 1.2 *or* 1.3 *and assume in addition:*

(22) $f \in C^1(\mathbb{R}^+)$, $f'(t) > 0$ *for* $t > 0$ *and* $f'(t)$ *is nondecreasing for* $t > 0$,

(23) $f(0) > 0$ *and* $\lim_{t \to +\infty} f(t) t^{-1} = +\infty$.

Consider the following problem:

$$(24) \quad \begin{cases} -\Delta u = \lambda f(u) & \text{in } \Omega, \quad u \in C^2(\Omega), \\ \lambda > 0, \quad u = 0 & \text{on } \partial\Omega \quad \text{and} \quad u > 0 \quad \text{on } \Omega. \end{cases}$$

Define $\mathcal{S} = \{(0, 0)\} \cup \{(u, \lambda) : (u, \lambda)$ solves (24)$\}$ and define $\mathcal{S}_0$ to be the connected component of $\mathcal{S}$ (viewed as a subset of $C(\bar{\Omega}) \times [0, \infty)$) which contains $(0, 0)$. Then $\mathcal{S}_0$ is unbounded but $\{(u, \lambda) \in \mathcal{S} : \lambda \geqq \varepsilon\}$ is bounded in $C(\bar{\Omega}) \times [0, \infty)$ for each $\varepsilon > 0$. There exists Λ, $0 < \Lambda < \infty$, such that (24) has no solution for $\lambda > \Lambda$ and such that for $0 < \lambda < \Lambda$ (24) has at least two solutions u with $(u, \lambda) \in \mathcal{S}_0$. In addition, for $0 \leqq \lambda \leqq \Lambda$, there exists a minimal solution u_λ of (24) ($u_0 \overset{\text{def}}{=} 0$) and the map $\lambda \to u_\lambda \in C(\bar{\Omega})$ is continuous for $0 \leqq \lambda \leqq \Lambda$. If $f'(t)$ is strictly increasing (24) has exactly one solution for $\lambda = \Lambda$.

Remark 2.2. – The fact that there exists Λ such that (24) has no solution for $\lambda > \Lambda$ and at least one solution for $0 < \lambda < \Lambda$ and the fact that (24) has a minimum solution u_λ for $0 \leqq \lambda \leqq \Lambda$ [i.e., a nonnegative solution which is less than or equal to every other nonnegative solution v of (24)] such that $\lambda \to u_\lambda \in C(\bar{\Omega})$ is continuous for $0 \leqq \lambda \leqq \Lambda$) is well-known by now (*see* M. G. Crandall and P. H. Rabinowitz [10], compare also with D. S. Cohen [9], D. D. Joseph and T. S. Lundgren [15] and C. Bandle [4] and note that many of the basic ideas already appear in M. A. Krasnosel'skii's book [17]). Crandall and Rabinowitz [10] prove uniqueness of the solution of (24) for $\lambda = \Lambda$ if f is C^2 and $f''(t) > 0$ for $t \geqq 0$, but a somewhat more careful argument gives the result here. Crandall and Rabinowitz also show the existence of pairs of positive solutions for $0 < \lambda < \Lambda$.

The novelty of our result is the new information about $\mathscr{S}$, e. g. that the pairs of positive solutions can be selected to lie in a continuum $\mathscr{S}_0$.

Proof of Corollary 2.2. — Define Φ as in Theorem 2.1 and let $C=\{u\in C(\overline{\Omega}):u(x)\geqq 0 \text{ for all } x\in\Omega\}$. It is easy to see that $\mathscr{S}=\{(u, \lambda):\lambda\geqq 0, u\in C\}$ such that $u=\lambda\Phi(u)$ and $\mathscr{S}_0$ is the connected component of $\mathscr{S}$ containing (0, 0). Proposition 2.2 immediately implies that $\mathscr{S}_0$ is unbounded.

Assumption (23) on f implies that given $\varepsilon>0$ and $\lambda_2>\lambda_1$ there exists $t_\varepsilon>0$ such that:

$$\lambda f(t)t^{-1}\geqq\lambda_2 \quad \text{for} \quad t\geqq t_\varepsilon, \quad \lambda\geqq\varepsilon$$

(here λ_1 denotes the smallest eigenvalue of $-\Delta$). The results of Section 1 imply that there exists M_ε such that any solution of (24) with $\lambda\geqq\varepsilon$ satisfies $\sup_{x\in\Omega} u(x)\leqq M_\varepsilon$. Of course we have:

$$\lim_{\varepsilon\to 0^+} M_\varepsilon=+\infty.$$

If $\Phi'(u_\lambda)$ denotes the Fréchet derivative of Φ at the minimum solution u_λ, $0\leqq\lambda\leqq\Lambda$, and if $L_\lambda=\lambda\Phi'(u_\lambda)$, so:

$$(L_\lambda h)=\lambda L(g), \qquad g(y)=f'(u_\lambda(y))h(y), \qquad L=(-\Delta)^{-1},$$

then an abstract argument of Krasnosel'skii [17] (*see* also H. Amann [2] and Crandall and Rabinowitz [10]) implies that the spectral radius of L_λ is less than or equal to 1 (although the exact hypotheses of Krasnosel'skii's results are not satisfied). Since $\lambda\to u_\lambda$ is an increasing map in the partial ordering induced by C, it follows that $L_\lambda\leqq(\lambda/\mu)L_\mu$ for $0\leqq\lambda<\mu\leqq\Lambda$, and basic theory of positive linear operators implies that the spectral radius of L_λ is strictly less than that of L_μ and hence strictly less than 1. The implicit function theorem implies then that $\lambda\to u_\lambda$ is C^1 for $0\leqq\lambda<\Lambda$, and the *a priori* bounds on u_λ and monotonicity of $\lambda\to u_\lambda$ give continuity of $\lambda\to u_\lambda$ at Λ. For $0\leqq\lambda<\Lambda$, the above comments show that there exists $\rho=\rho(\lambda)>0$ such that:

$$i_C(\lambda F, \mathcal{O}_\lambda)=i_C(A_\lambda, \mathcal{O}_\lambda)=+1,$$

$$\mathcal{O}_\lambda=\{u\in C:\|u-u_\lambda\|<\rho_\lambda\}, \qquad A_\lambda u\overset{\text{def}}{=}u_\lambda+L_\lambda(u-u_\lambda).$$

Furthermore, we clearly have $\{u_\lambda:0\leqq\lambda\leqq\Lambda\}\subset\mathscr{S}_0$.

Given $\varepsilon>0$, let $\mathscr{S}_\varepsilon=\{(u, \lambda)\in\mathscr{S}_0:\lambda\geqq\varepsilon\}$, a bounded set. By using a topological Lemma of Whyburn (as in the proof of Theorem 6.2 in [29]), we see that there exists a bounded set $W\subset C(\overline{\Omega})\times[\varepsilon, \infty)$ such that $\mathscr{S}_\varepsilon\subset W$, W is open in $C(\overline{\Omega})\times[\varepsilon, \infty)$, and $\lambda\Phi(x)\neq x$ for $(x, \lambda)\in\overline{W}-W$. If $W_\lambda=\{x:(x, \lambda)\in W\}$, the homotopy property for the fixed point index implies that $i_C(\lambda\Phi, W_\lambda)$ is constant for $\lambda\geqq\varepsilon$, and because $\lambda\Phi$ has no fixed points for $\lambda>\Lambda$, the constant must be zero. If, for some λ with $\varepsilon\leqq\lambda<\Lambda$, $\lambda\Phi$ does not have a fixed point $u\neq u_\lambda$ with $(u, \lambda)\in\mathscr{S}_0$, the additivity property of the fixed point index implies:

$$i_C(\lambda\Phi, W_\lambda)=i_C(\lambda\Phi, \mathcal{O}_\lambda)=+1,$$

which contradicts the fact that $i_C(\lambda\Phi, W_\lambda)=0$. We conclude that there must be a second solution $u\neq u_\lambda$ with $(u, \lambda)\in\mathcal{S}_0$.

2.2. EXISTENCE OF ONE SOLUTION. – The goal of this section is to explain how the combination of some arguments in the proofs of Theorems 1.1-1.3 together with simple variational techniques enable us to give very general existence results (nearly optimal) even without having *a priori* estimates. We emphasize that, since we do not have *a priori* estimates under conditions such as those which follow, the proof we give does not give any information on the structure of the set of solutions of (1) but only gives the existence of one solution, under very general conditions:

THEOREM 2.3. – *Let $f:\mathbb{R}^+\to\mathbb{R}^+$ satisfy* (2) *and* (3) (*smoothness and superlinearity*). *We assume in addition that f and Ω satisfy:*

$$(26)\quad\begin{cases} \text{if } N=2, \text{ then } \lim_{t\to+\infty} f(t)e^{-t^\alpha}=0, \text{ for some } \alpha<2,\\ \text{if } N\geqq 3 \text{ and if } \Omega \text{ does not satisfy (18), then } f(t)\,t^{-(N+2)/(N-2)} \text{ is nonincreasing for } t\geqq 0\\ \text{and } \lim_{t\to+\infty} f(t)\,t^{-(N+2)/(N-2)}=0,\\ \text{if } N\geqq 3 \text{ and if } \Omega \text{ satisfies (18), then } \lim_{t\to+\infty} f(t)\,t^{-(N+2)/(N-2)}=0. \end{cases}$$

$$(21)\qquad \overline{\lim_{t\to 0}}\, f(t)\,t^{-1}<\lambda_1;\qquad f(0)=0.$$

Then, there exists a solution u of:

$$(1)\qquad \begin{cases} -\Delta u=f(u) & \text{in } \Omega, \quad u\in C^2(\overline{\Omega}),\\ u=0 & \text{on } \partial\Omega, \quad u>0 \quad \text{in } \Omega. \end{cases}$$

Remark 2.3. – This result generalizes the result of A. Ambrosetti and P. H. Rabinowitz [3]. Because of [20], it appears to be nearly optimal.

Remark 2.4. – If f belongs to a set F such that all g in F satisfy the assumptions of Theorem 2.2 in a uniform way, then the proof will show that it is possible to select a solution u of (1) bounded in $C^2(\overline{\Omega})$ for all f in F.

Proof of Theorem 2.2. – We will give the proof only in the third case of (26), that is, when $N\geqq 3$ and Ω satisfies (18); the other cases are treated by similar arguments. The proof is divided into three steps: (1) we first define a sequence of approximate problems, (2) we prove H^1 bounds on the sequence of solutions, and (3) we pass to the limit.

Step 1. – Because of (3), there exists a sequence $t_n\geqq 1$, $t_n\to+\infty$, and a constant $\alpha>0$ such that:

$$\frac{f(t)}{t}\geqq\lambda_1+\alpha \qquad \text{for} \quad t\geqq t_n.$$

Select a fixed constant γ with $1<\gamma<(N+2)/(N-2)$, define $\mu_n=f(t_n)\,t_n^{-1}$ and define a modified function $f_n(t)$ by $f_n(t)=0$ for $t\leqq 0$, $f_n(t)=f(t)$ for $0\leqq t\leqq t_n$ and

$f_n(t)=f(t_n)+\mu_n(t-t_n)^\gamma$ for $t\geqq t_n$. Notice that f_n satisfies the hypotheses of Theorem 1.3. Notice also that $\lim_{t\to+\infty} f_n(t)t^{-1}=+\infty$ and that if $F_n(t)=\int_0^t f_n(s)\,ds$, there exists θ_n with $0\leqq\theta_n<1/2$ and s_n such that:

$$F_n(t)\leqq\theta_n\, tf_n(t)\quad\text{for } t\geqq s_n.$$

A nearly immediate application of [3] or of arguments in Section 2 of [10] gives:

LEMMA 2.1. – *Let* $S^n(v)=\int_\Omega 1/2\,|\nabla v|^2-F_n(v)\,dx$ *for* $v\in H_0^1(\Omega)$. *Then there exists* $w\in H_0^1(\Omega)$ *such that* $S^n(w)\leqq 0$ *for all n. Let* $J=\{g\in C([0,1];H_0^1(\Omega)):g(0)=0,\ g(1)=w\}$ *and set* $b^n=\inf_{g\in J}\sup_{t\in[0,1]} S^n(g(t))$.

Then $b^n>0$ *and there exists* $u^n\in C^2(\overline{\Omega})$ *such that:*

$$(1)_n\qquad \begin{cases} -\Delta u^n=f_n(u^n) & \text{in } \Omega,\\ u^n=0 \quad \text{on } \partial\Omega, & u^n>0,\end{cases}$$

and $S^n(u^n)=b^n$.

Proof of Lemma 2.1. – It suffices to verify that S^n satisfies all hypotheses of Theorem 2.1 in [3], for instance the Palais-Smale condition. However, we have defined f_n so that $F_n(t)\leqq\theta_n tf_n(t)$ for $t\geqq s_n$, $0\leqq\theta_n<1/2$, and exactly the arguments in Lemma 3.6 of [4] apply here. Notice that if v_1 is the positive eigenvector of $-\Delta$ and R is chosen large enough (independent of n), w can be taken equal to $R v_1$.

Step 2. – We now obtain H^1 bounds on $\{u_n\}$. It is clear that $0\leqq b^n\leqq C$. In addition using the proof of Theorem 1.3 and the fact that f_n satisfies (26) uniformly in n, we obtain:

$$|Du_n(x)|\leqq C,\quad\text{for all } x \text{ in } \Gamma_1\ (\textit{see}\ (18)).$$

Applying (12′) (*see* the proof of Theorem 1.3), we deduce that:

$$(N-2)\int_\Omega|\nabla u^n|^2\,dx=2N\int_\Omega F_n(u^n)\,dx-\int_{\partial\Omega}(x-x_0,n(x))\,|\nabla u^n|^2\,dS$$

and because of (18):

$$\frac{1}{2}\int_\Omega|\nabla u^n|^2\,dx-\frac{N}{N-2}\int_\Omega F_n(u^n)\,dx\geqq -C.$$

But:

$$0\leqq b^n=\frac{1}{2}\int_\Omega|\nabla u^n|^2\,dx-\int_\Omega F_n(u^n)\,dx\leqq C,$$

therefore $\int_\Omega |\nabla u^n|^2\,dx \leqq C$ and this implies that $\| u^n \|_{H^1} \leqq C$, and also that $\| u^n \|_{L^\infty} \leqq C$ (by the same argument as in Section 1).

Step 3. – If we observe that:

$$\int_\Omega f_n(u^n)\,v_1\,dx = \lambda_1 \int_\Omega u^n\,v_1\,dx$$

and that (21) holds for f_n (uniformly in n), we deduce that there exists $\alpha > 0$ such that:

$$\| u^n \|_{L^\infty(\Omega)} \geqq \alpha.$$

It is then obvious to pass to the limit.

REFERENCES

[1] S. Agmon, A. Douglis and L. Nirenberg, *Estimates Near the Boundary for Solutions of Elliptic Partial Differential Equations Satisfying General Boundary Conditions*, I (*Comm. Pure Appl. Math.*, Vol. 12, 1959, pp. 623-727).

[2] H. Amann, *Fixed point equations and nonlinear eigenvalue problems in ordered Banach spaces* (*S.I.A.M.*, Vol. 18, 1976, pp. 620-709).

[3] A. Ambrosetti and P. H. Rabinowitz, *Dual Variational Methods in Critical Point Theory and Applications* (*J. Functional Analysis*, Vol. 14, 1973, pp. 369-381).

[4] C. Bandle, *Existence Theorems, Qualitative Results and a priori bounds for a Class of Nonlinear Dirichlet Problems* (*Arch. Rat. Mech. Anal.*, Vol. 58, 1975, pp. 219-238).

[5] T. B. Benjamin, *A Unified Theory of Conjugate Flows* (*Phil. Trans. Royal Soc.*, Vol. 269 A, 1971, pp. 587-643).

[6] H. Berestycki and P. L. Lions, *Une méthode locale pour l'existence de solutions positives de problèmes semilinéaires dans* R^n (*J. Analyse Mathématique*, Vol. 38, 1980, pp. 144-187).

[7] H. Brezis and T. Kato, *Remarks on the Schrodinger Operator With Singular Complex Potentials* (*J. Math. pures et appl.*, Vol. 58, 1978, p. 137-151).

[8] H. Brezis and R. E. L. Turner, *On a Class of Superlinear Elliptic Problems* (*Comm. in P.D.E.*, Vol. 2, 1977, pp. 601-614).

[9] D. S. Cohen, *Positive Solutions of Nonlinear Eigenvalue Problems: Applications to Nonlinear Reactor Dynamics* (*Arch. Rat. Mech. Anal.*, Vol. 26, 1967, pp. 305-315).

[10] M. G. Crandall and P. H. Rabinowitz, *Some Continuation and Variational Methods for Positive Solutions of Nonlinear Elliptic Eigenvalue Problems* (*Arch. Rat. Mech. Anal.*, Vol. 58, 1975, pp. 207-218).

[11] D. de Figueiredo, P. L. Lions and R. D. Nussbaum, *Estimations* a priori *pour les solutions positives de problèmes elliptiques semi-linéaires* (*C. R. Acad. Sc.*, Paris, série A, 1980).

[12] I. M. Gelfand, *Some Problems in the Theory of Quasilinear Equations* (*Amer. Math. Soc. Transl.*, Vol. 29, 1963, pp. 295-381).

[13] B. Gidas, Wei-Ming Ni and L. Nirenberg, *Symmetry and Related Properties* via *the Maximum Principle* (*Comm. Math. Phys.*, Vol. 68, 1979, pp. 209-243).

[14] D. Gilbarg and N. S. Trudinger, *Elliptic Partial Differential Equations of Second Order*, Springer Verlag, New York, 1977.

[15] D. D. Joseph and T. S. Lundgren, *Quasilinear Dirichlet Problems Driven by Positive Sources* (*Arch. Rat. Mech. Anal.*, Vol. 49, 1972, pp. 241-269).

[16] M. A. Krasnosel'skii, *Fixed Points of Cone-Compressing of Cone-Extending Operators* (*Soviet Math. Dokl.*, Vol. 1, 1960, pp. 1285-1288).

[17] M. A. KRASNOSEL'SKII, *Positive Solutions of Operator Equations*, Noordhoff, Groningen, the Netherlands, 1964.

[18] O. LADYZHENSKAYA and N. URALTSEVA, *Linear and Quasilinear Elliptic Equations*, Translated by Scripta Technica, Académic Press, New York, 1968.

[19] J. LERAY and J. SCHAUDER, *Topologie et équations fonctionnelles* (*Ann. Scient. Éc. Norm. Sup.*, Vol. 51, 1934, pp. 45-78).

[20] J. L. LIONS and G. MAGENES, *Nonhomogeneous Boundary Value Problems and Applications*, Vol. I, Springer, Berlin, 1970.

[21] P. L. LIONS, A priori *Estimates and Existence Results for Some Semilinear Elliptic Problems*, M.R.C. Technical Summary Report, 1975, Madison, Wisconsin, 1979.

[22] L. NIRENBERG, *Topics in Non Linear Functional Analysis*, N.Y.U. Courant Institute Lecture Notes, New York, 1974.

[23] R. D. NUSSBAUM, *Positive Solutions of Nonlinear Elliptic boundary Value Problems* (*J. Math. Anal. Appl.*, Vol. 51, 1975, pp. 461-482).

[24] R. D. NUSSBAUM, *A Global Bifurcation Theorem With Applications to Functional Differential Equations* (*J. Functional Analysis*, Vol. 19, 1975, pp. 319-338).

[25] R. D. NUSSBAUM, *Uniqueness and Nonuniqueness for Periodic Solutions of* $x'(t) = -g(x(t-1))$ (*J. Differential Equations*, Vol. 34, 1979, pp. 24-54).

[26] R. D. NUSSBAUM, *Periodic Solutions of Some Nonlinear Autonomous Functional Differential Equations, II* (*J. Differential Equations*, Vol. 14, 1973, pp. 368-394).

[27] S. I. POHOZAEV, *Eigenfunctions of* $\Delta u + \lambda f(u) = 0$ (*Soviet Math. Dokl.*, Vol. 6, 1965, pp. 1408-1411).

[28] P. H. RABINOWITZ, *Variational Methods for Nonlinear Eigenvalue Problems* (in *Eigenvalues and Nonlinear Problems*, C.I.M.E., Ediz. Cremonese, Rome, 1979.

[29] P. RABINOWITZ, *Some Aspects of Nonlinear Eigenvalue Problems* (*Rocky Mountain J.*, Vol. 3, 1973, pp. 161-202).

[30] R. E. L. TURNER, A Priori *Bounds for Positive Solutions of Nonlinear Elliptic Equations in two Variables* (*Duke Math. J.*, Vol. 41, 1974, p. 759-774).

(Manuscrit reçu en septembre 1980.)

D. G. DE FIGUEIREDO,
Univ. Brasilia,
Brasilia,
Brazil.

P. L. LIONS,
Ceremade,
Université Paris-IX - Dauphine,
place de-Lattre-de-Tassigny,
75775 Paris Cedex 16,
France.

R. D. NUSSBAUM,
Rutgers University,
Dept. of Mathematics,
New Brunswick,
N.J. 08903,
U.S.A.

Nonlinear Analysis, Theory, Methods & Applications, Vol. 8, No. 6, pp. 655–665, 1984.
Printed in Great Britain.
0362-546X/84 \$3.00 + .00

ON THE SUPERLINEAR AMBROSETTI–PRODI PROBLEM

DJAIRO G. DE FIGUEIREDO*
Department of Mathematics, University of Brasília, 70910 Brasília, D.F. Brazil

(*Received* 15 *June* 1983; *received for publication* 8 *September* 1983)

Key words and phrases: Semilinear elliptic boundary value problem, monotone iteration method, mountain pass theorem.

0. INTRODUCTION

LET Ω BE a smooth bounded domain in R^N. We consider the semilinear elliptic boundary value problem

$$-\Delta u = g(x,u) + f(x) \quad \text{in } \Omega, \qquad u = 0 \quad \text{on } \partial\Omega, \tag{1}$$

where $f(x)$ is some given C^α-function in Ω, and the nonlinearity g satisfies the smoothness condition below, besides other conditions that will be timely introduced as we proceed:

$$g\colon \bar{\Omega}\colon R \to R \quad \text{is a } C^1\text{-function.} \tag{2}$$

Problem (1) is said to be of the Ambrosetti–Prodi type if

$$\limsup_{s\to-\infty} \frac{g(x,s)}{s} < \lambda_1, \tag{3a}$$

and

$$\lambda_1 < \liminf_{s\to+\infty} \frac{g(x,s)}{s}, \tag{3b}$$

where the inequalities hold uniformly in Ω, and the limits could assume value $-\infty$ or $+\infty$, respectively, on the whole of Ω or on subsets of positive measure. Here λ_1 denotes the first eigenvalue of the eigenvalue problem $-\Delta u = \lambda u$ in Ω, $u = 0$ on $\partial\Omega$. Let ϕ be the eigenfunction corresponding to λ_1 which is >0 in Ω and $\int \phi^2 = 1$. And let N be the subspace of $C^\alpha(\bar{\Omega})$ generated by ϕ, and $N^\perp = \{u \in C^\alpha(\bar{\Omega})\colon \int u\phi = 0\}$. Consequently any $f \in C^\alpha(\bar{\Omega})$ can be uniquely written as $f = t\phi + h$, with $t \in R$ and $h \in N^\perp$. Using this decomposition we shall look at the parametrized form of (1):

$$-\Delta u = g(x,u) + t\phi + h \quad \text{in } \Omega, \qquad u = 0 \quad \text{on } \partial\Omega. \tag{1_t}$$

Conditions (3a) and (3b) are implied by the more restrictive conditions (4a) and (4b), respectively:

$$\limsup_{s\to-\infty} g'_s(x,s) < \lambda_1, \tag{4a}$$

$$\lambda_1 < \liminf_{s\to+\infty} g'_s(x,s), \tag{4b}$$

where g'_s denotes the partial derivative of g with respect to s.

* Work completed at the Mathematics Research Center, University of Wisconsin-Madison, Wisconsin, under a Guggenheim fellowship.

Originally published in: Nonlinear Anal.-TMA 8 (1984), 655-665,

DOI: 10.1007/978-3-319-02856-9_12

Viewing the future use of variational methods we suppose

$$\lim_{s\to+\infty}\frac{g(x,s)}{s^\sigma}=0,\qquad 1<\sigma\leqslant\frac{N+2}{N-2}\quad \text{if } N\geqslant 3, \tag{5}$$

$$1<\sigma<\infty\qquad \text{if } N=2.$$

No assumption on the growth of g as $s\to-\infty$ will be necessary. Also by force of some desirable Palais–Smale condition for a functional associated with our problem (1) we found it necessary to assume the following technical condition, but only in the case $N\geqslant 3$ and $(N+1)/(N-1)<\sigma\leqslant(N+2)/(N-2)$:

$$\liminf_{s\to+\infty}\frac{sg(x,s)-\theta G(x,s)}{s^2 g(x,s)^{2/(N+1)}}\geqslant 0,\quad \text{for some } \theta>2, \tag{6}$$

where $G(x,s)=\int_0^s g(x,\xi)\,d\xi$.

THEOREM 1. Assume (2), (3b), (4a), (5) and (6). Then for each given $h\in N^\perp$ there exists $t_0\in R$ such that (1_t) has at least two $C^{2,\alpha}$ solutions for $t<t_0$.

The first result of this sort was proved by Ambrosetti and Prodi [1], under much stronger assumptions, but yielding a sharper conclusion. In their work as well as in the subsequent work of other authors, e.g. Berger and Podolak [2], Amann and Hess [3], Fučik [4], the nonlinearity g is assumed to have linear growth. Kazdan and Warner [5] relaxed this condition but obtained only one solution. Dancer [6] obtains a second solution, provided $\sigma\leqslant(N+1)(N-1)$ in (5). He uses a topological degree argument, and a crucial point in his proof is obtaining *a priori* estimates on the solutions of (1_t). Under his restriction on σ these estimates are readily obtained using a technique due to Brézis and Turner [7]. For nonlinearities g growing faster than $(N+1)(N-1)$ this is not a simple matter, and as far as we know estimates are not available yet. We found it easier to obtain a second solution in this case using a variational method. Here also there is a delicate point which is the establishment of the Palais–Smale condition for some associated functional. We required a technical condition (6), which we believe to be not too restrictive. For example, pure powers like $g(x,s)=a(x)s^p$ for large s and $1<p<(N+2)/(N-2)$ do satisfy (6). More generally the condition introduced by Ambrosetti and Rabinowitz [8] in similar situations also gives (6). For the early history of the Ambrosetti–Prodi problem see the survey paper [9]. We should also mention the work of Berestycki and Lions [17] on the superlinear Ambrosetti–Prodi problems with convex nonlinearities g.

We mention that the steps in the proof of theorem 1 will make clear that a theorem of Ambrosetti and Rabinowitz [8] on the existence of positive solutions for some semilinear elliptic equations can be slightly extended. For completeness let us state it in case of the Laplacian.

THEOREM 2. [Ambrosetti–Rabinowitz]. Let $g\colon \Omega\times R^+\to R^+$ be a Carathéodory function satisfying conditions (3b), (5) and (6) above as well as $g(x,0)=0$ and

$$\limsup_{s\to 0}\frac{g(x,s)}{s}<\lambda_1.$$

Then the boundary value problem

$$-\Delta u = g(x, u) \quad \text{in } \Omega, \qquad u = 0 \quad \text{on } \partial\Omega$$

has a nontrivial positive solution u in H_0^1. If g is C^α then $u \in C^{2,\alpha}(\bar{\Omega})$.

The above theorem should be compared with theorem 2.3 in de Figueiredo, Lions and Nussbaum [10]. Although theorem 2 above applies to more general second order elliptic operators and nonlinearities g depending on x, its technical condition (6) seems more restrictive than the requirements made in theorem 2.3.

The paper is divided in three parts. In the first section we prove that there exists a $t_2 \in R$ such that for all $t \leq t_2$, problem (1_t) has a minimal *negative* solution $u_t \in C^{2,\alpha}(\overline{\Omega})$ such that the eigenvalue problem

$$-\Delta v - g_s'(x, u_t)v = \mu v \quad \text{in } \Omega, \qquad v = 0 \quad on\ \partial\Omega,$$

has a positive first eigenvalue μ_1. In the second part we take an appropriate truncation of the nonlinearity g and construct a functional whose critical points are precisely the weak solutions of (1_t). It is then shown that the minimal solution u_t found in the previous section is a local minimum of this functional. Under our assumptions the functional is shown to satisfy the Palais–Smale condition. The mountain pass theorem of [8] is applied to get a second solution of (1_t) in H_0^1. Its regularity in the case $\sigma < (N+2)/(N-2)$ is seen by the standard way using a bootstrap argument. The case $\sigma = (N+2)/(N-2)$ requires a great deal more of work. We prove it in Section 3 using an argument due to Brézis and Kato [11].

The novelty of the present work is the treatment of superlinear nonlinearities g with growth at infinite "touching" the critical exponent $(N+2)/(N-2)$, in the sense of condition (5).

1. EXISTENCE OF A MINIMAL NEGATIVE SOLUTION

Since some of the assertions made in the present section are true under weaker assumptions than the ones placed in theorem 1, we shall state them separately. Also this will make clear the role of (4a) which we believe it could be replaced by (3a), as well as the role of (6) which eventually could be either weakened or dropped altogether.

LEMMA 3. Assume (3a) and (3b). Then there exists a number τ, independent of $h \in N^\perp$, such that (1_t) has no solution for $t > \tau$.

Proof. It follows from (3a) and (3b) that there are numbers $C > 0$ and $\underline{\mu} < \lambda_1 < \bar{\mu}$ such that

$$g(x, s) \geq \underline{\mu}s - C \quad \text{and} \quad g(x, s) \geq \bar{\mu}s - C \tag{7}$$

for all $x \in \overline{\Omega}$ and all $s \in R$. Take the inner product of (1_t) with ϕ, integrate by parts, and estimate $\int g(x, u)\phi$ using (7). This will give an upper bound on the values of t for which (1_t) has a solution. ■

In this section we use the method of monotone iteration to get a minimal solution of (1_t). We say that $u \in C^{2,\alpha}(\bar{\Omega})$ is a subsolution of (1_t) if

$$-\Delta \underline{u} \leq g(x, \underline{u}) + t\phi + h \quad \text{in } \Omega, \qquad \underline{u} \leq 0 \text{ on } \partial\Omega. \tag{8}$$

A supersolution is defined likewise by changing in (8) the inequalities $\leq$ for $\geq$. It is well known, see for instance [12, 13], that if (1_t) has a subsolution $\underline{u}$ and a supersolution $\bar{u}$ such that $\underline{u} \leq \bar{u}$, then (1_t) has a solution u_t such that $\underline{u} \leq u_t \leq \bar{u}$ and if u is any other solution with $\underline{u} \leq u \leq \bar{u}$ then $u_t \leq u$. This solution is obtained by an iterative method.

LEMMA 4. Assume (3a) and (3b). Given $h \in N^{\perp}$ and a compact interval $[a, b]$, there is a function $w \in C^{2,\alpha}(\bar{\Omega})$ with $w = 0$ on $\partial\Omega$ such that: (i) w is a subsolution of (1_t) for each $t \in [a, b]$, (ii) $w \leq v$ for all v which are supersolutions of (1_t) with $t \in [a, b]$. In particular, w bounds from below all solutions of (1_t) for $t \in [a, b]$.

Proof. The unique solution w of the linear problem

$$-\Delta w = \mu w - C + a\phi + h \quad \text{in } \Omega, \qquad u = 0 \quad \text{on } \partial\Omega$$

where μ and C were introduced in (7), satisfies all the requirements of the lemma. ∎

LEMMA 5. Assume (3a) and (3b). Given $h \in N^{\perp}$, there is a $t_1 \in R$ and a function W in $C^{2,\alpha}(\bar{\Omega})$ which is < 0 in Ω, $= 0$ in $\partial\Omega$ and which is a supersolution for all problems (1_t) with $t \leq t_1$.

Proof. Let $\varepsilon > 0$ and $p > N$ be given. As a consequence of the Sobolev imbedding $W^{2,p}(\Omega) \subset C^{1,\alpha}(\bar{\Omega})$ and the strong maximum principle, we see that there are numbers α, $K > 0$ such that

$$-K < v - \alpha\phi < 0 \quad \text{in } \Omega \text{ for } v \in W^{2,p}(\Omega) \text{ with } v = 0 \text{ on } \partial\Omega, \text{ and } \|v\|_{W^{2,p}} \leq \varepsilon. \tag{9}$$

By the L^p-regularity theory, the operator $-\Delta$ with Dirichlet boundary condition has a bounded inverse from L^p to $W^{2,p}$, i.e. there is a constant $C > 0$ such that

$$\|v\|_{W^{2,p}} \leq C\|\zeta\|_{L^p} \quad \text{for} \quad -\Delta v = \zeta \text{ in } \Omega, \qquad v = 0 \quad \text{on } \partial\Omega. \tag{10}$$

Let

$$m = \max\{|g(x, s) + h(x)| : x \in \Omega, -K \leq s \leq \varepsilon\} \tag{11}$$

and take subdomains Ω_1 and Ω_2 of Ω such that $\Omega_1 \subset \bar{\Omega}_1 \subset \Omega_2 \subset \bar{\Omega}_2 \subset \Omega$ and the Lebesgue measure of $\Omega\backslash\Omega_1$ is smaller than $[\varepsilon/Cm]^p$, where C and m have been introduced in (10) and (11) respectively. Next define a C^{∞} function ζ which is 0 in Ω_1, $= m$ in $\Omega\backslash\Omega_2$ and assumes values between 0 and m in $\Omega_2\backslash\Omega_1$. Let v be the unique solution of the Dirichlet problem $-\Delta v = \zeta$ in Ω, $v = 0$ on $\partial\Omega$. Using (10) and the choice of Ω_1 we have

$$\|v\|_{W^{2,p}} \leq C\|\zeta\|_{L^p} \leq Cm|\Omega\backslash\Omega_1|^{1/p} \leq \varepsilon. \tag{12}$$

We claim that $W = v - \alpha\phi$ is a supersolution of (1_t) provided we choose t sufficiently large negatively. Indeed, if $x \in \Omega\backslash\Omega_2$

$$-\Delta W = \zeta - \alpha\lambda_1\phi = m - \alpha\lambda_1\phi \geq g(x, v - \alpha\phi) + h(x) - \alpha\lambda_1\phi$$

where the inequality follows from (9), (11) and (12). If $x \in \Omega_2$, use the fact that $\phi(x)$ is bounded away from 0 by a positive constant for $x \in \Omega_2$ and obtain $\bar{t}$ such that $\zeta(x) \geq g(x, v - \alpha\phi) + h(x) + \bar{t}\phi$. Then, for $x \in \Omega_2$

$$-\Delta W = \zeta - \alpha\lambda_1\phi \geq g(x, v - \alpha\phi) + h(x) + (\bar{t} - \alpha\lambda_1)\phi.$$

Thus the lemma is proved with $t_1 = \min[\bar{t} - \alpha\lambda_1, -\alpha\lambda_1]$.

Remark. Kazdan and Warner [5] under the assumptions of lemma 5 have proved by a similar argument the existence of a supersolution which happens to be *not* negative in Ω. It is apparent that problem (1_t) could have a supersolution for $t > t_1$, although it is not necessarily negative.

Using the method of monotone iteration it follows

COROLLARY 6. Assume (3a) and (3b). Given $h \in N^\perp$, there is a $t_1 \in R$ such that problem (1_t) has a negative solution $u_t \in C^{2,\alpha}$ for each $t \leq t_1$. Moreover u_t is minimal, i.e. given any solution u of (1_t) we have $u_t \leq u$.

Remark. Existence of a negative solution in the Ambrosetti–Prodi problem is implicit in the work of Lazer and McKenna [14]. It was essentially noted in [9], but to our knowledge was first explicitly proved by Solimini [15]. That author [15] and Ambrosetti [16] have made a relevant use of this fact to prove multiplicity results (existence of 3 solutions) in certain Ambrosetti–Prodi problems. We remark that all these authors work with nonlinearities g having linear growth, essentially

$$g(x, u) = \mu^+ u^+ - \mu^- u^- + k(x, u)$$

where $\mu^\pm = \lim_{\mu \to \pm\infty} g'(u)$ and $k(x, u)$ is bounded. Their interesting proof of the existence of a negative solution apparently does not extend to our general case.

LEMMA 7. Assume (3a) and (3b). Let u_t be the minimal solution of (1_t). Then the first eigenvalue μ_1 of the eigenvalue problem

$$-\Delta v - g_s'(x, u_t)v = \mu v \quad \text{in } \Omega, \qquad v = 0 \quad \text{on } \partial\Omega \tag{13}$$

is ≥ 0.

Proof. We follow Dancer [6] in this argument. Suppose by contradiction that $\mu_1 < 0$ and let $v > 0$ be the corresponding eigenfunction normalized by $\int v^2 = 1$. Look at $u_t + \delta v$ and write

$$-\Delta(u_t + \delta v) = g(x, u_t) + t\phi + h + \delta[g_s'(x, u_t)v + \mu_1 v]$$

and using the expression

$$g(x, u_t + \delta v) - g(x, u_t) = \int_0^1 g_s'(x, u_t + r\delta v)\delta v \, dr$$

we obtain

$$-\Delta(u_t + \delta v) = g(x, u_t + \delta v) + t\phi + h + \delta v\left[\int_0^1 [g_s'(x, u_t) - g_s'(x, u_t + r\delta v)] \, dr + \mu_1\right].$$

By the continuity of g_s' the expression inside the brackets is <0 for $|\delta|$ small. So if we take $\delta<0$ and small we get

$$-\Delta(u_t+\delta v)\geq g(x,u_t+\delta v)+t\phi+h$$

which says that $u_t+\delta v$ is a supersolution of (1_t). By lemma 4 and using the monotone iteration method we would obtain a solution u of (1_t) with $u\leq u_t+\delta v<u_t$ in Ω, contradicting corollary 6.

For future purposes we need $\mu_1>0$. We can prove that this is so under a slightly stronger assumption than one in lemma 7.

LEMMA 8. Assume (4a) and (3b). Given $h\in N^{\perp}$, there is a $t_2\in R$ such that the first eigenvalue μ_1 of problem (13) for $t\leq t_2$ is positive.

Proof. Let us assume by contradiction that there are $t_n\to-\infty$ such that the first eigenvalue of (13) with $t=t_n$ is 0. So there is $v_{t_n}>0$ in Ω, $\int v_{t_n}^2=1$ such that

$$-\Delta v_{t_n}=g_s'(x,u_{t_n})v_{t_n}.$$

To simplify notation let us denote $v_n=v_{t_n}$, $u_n=u_{t_n}$. Then

$$\int|\nabla v_n|^2=\int g_s'(x,u_n)v_n^2\leq C\int v_n^2=C$$

where $C=\sup\{g_s'(x,s);x\in\bar{\Omega},s\leq 0\}$ which exists in view of assumption (4a). So we may assume that $v_n\rightharpoonup v_0$ in H_0^1, $v_n\to v_0$ in L^2 and a.e., and that $v_n(x)\leq h(x)$ for some $h\in L^2$. Clearly $\|v_0\|_{L^2}=1$. Then

$$-\int\Delta v_n\cdot\phi=\int g_s'(x,u_n)v_n\phi=\int_{u_n<s_0}+\int_{u_n>s_0}\tag{14}$$

where $s_0<0$ and $\underline{\mu}<\lambda_1$ (used next) are such that

$$g_s'(x,s)\leq\underline{\mu}\quad\text{for } x\in\bar{\Omega},\quad s\leq s_0.$$

From (14) we obtain

$$\lambda_1\int v_n\phi\leq\underline{\mu}\int v_n\phi+\int g_s'(x,u_n)v_n\phi\chi_{[u_n\geq s_0]}\tag{15}$$

Now we remark that a simple modification of lemma 5 yields the following stronger conclusion: given any $\beta<0$ there is a t_β such that problem (1_t) for $t\leq t_\beta$ has a supersolution $W\leq\beta\phi$. This implies that $\chi_{[u_n\geq s_0]}$ tends to 0 a.e. By Fatou's lemma we obtain from (15)

$$\lambda_1\int v_0\phi\leq\underline{\mu}\int v_0\phi$$

which is impossible. ■

Remark. It follows from lemma 8 that

$$\int|\nabla v|^2-\int g_s'(x,u_t)v^2\geq\mu_1\int v^2,\quad \mu_1>0.$$

On the other hand we know that there are constants c, $c_0 > 0$ such that

$$\int |\nabla v|^2 - \int g_s'(x, u_t)v^2 \geq c \int |\nabla v|^2 - c_0 \int v^2$$

which is Garding's inequality. Using the previous inequality we can easily see that $c_0 = 0$.

2. EXISTENCE OF THE SECOND SOLUTION

Now fix $t \leq t_2$ and let w be the subsolution associated with problem (1_t), by lemma 4. We now modify our function g as follows. Let $\tilde{g}(x, s)$ be the new function such that:

(i) $\tilde{g}(x, s) = g(x, s)$ for $x \in \Omega$, $s \geq w(x)$.

(ii) $\tilde{g}(x, s) \geq \underline{\mu} s - C$ and $\tilde{g}(x, s) \geq \bar{\mu} s - C$ for $x \in \Omega$ and $s \in R$, where C, $\underline{\mu}$ and $\bar{\mu}$ are the constants in (7).

(iii) $\tilde{g}(x, s) \geq 0$ for $x \in \Omega$ and all s smaller than a certain constant s_1.

(iv) $\tilde{g}(x, s)$ has linear growth at $-\infty$, i.e. $\tilde{g}(x, s)$ is like cs, $c < 0$, at $-\infty$.

(v) $\tilde{g}(x, s)$ is a C^1 function.

In view of (ii) we see that all eventual solutions of the modified problem (1_t) with g replaced by $\tilde{g}$ are bounded from below by $w(x)$, see lemma 4. So the solutions of the modified problem are the same as the ones of the original problem. Thus from now on we assume that g has properties (i)–(v) above. Let us now look for solutions of (1_t) as critical points of the functional

$$J(u) = \frac{1}{2}\int |\nabla u|^2 - \int G(x, u) - \int fu$$

defined in H_0^1.

LEMMA 9. The functional J satisfies the Palais–Smale condition.

Proof. Case 1. $N \geq 3$, $(N + 1)/(N - 1) < \sigma \leq (N + 2)/(N - 2)$. Let (u_n) be a sequence in H_0^1 such that $|J(u_n)| \leq C$ and $J'(u_n) \to 0$. We claim (u_n) contains a convergent subsequence in H_0^1. From $J'(u_n) \to 0$ we have for all $v \in H_0^1$

$$\left| \int \nabla u_n \nabla v - \int g(x, u_n) v - \int fv \right| \leq \varepsilon_n \|v\|_{H^1}, \quad \varepsilon_n \to 0. \tag{16}$$

Let $s_2 < s_1$ be such that $G(x, s) \leq 0$ for $s < s_2$, and define

$$w_n = \begin{cases} u_n - s_2 & \text{if } u_n \leq s_2 \\ 0 & \text{if } u_n \geq s_2. \end{cases}$$

Taking $v = w_n$ in (16):

$$\int |\nabla w_n|^2 - \int g(x, u_n) w_n \leq \int f w_n + \varepsilon_n \|w_n\|_{H^1} \tag{17}$$

and since $g(x, s) \geq 0$ for $s \leq s_2$ we obtain that $\|w_n\|_{H^1} \leq C$. It then follows from (17) that

$$-\int_{u_n \leq s_2} g(x, u_n)\, u_n \leq C + C\|u_n\|_{H^1} \tag{18}$$

where we have used the fact that g has linear growth at $-\infty$. Next from $|J(u_n)| \leq C$ it follows

$$\int |\nabla u_n|^2 \leq C + 2\int G(x, u_n) + 2\int f u_n \leq C + 2\int_{u_n \leq s_3} G(x, u_n) + C\|u_n\|_{H^1}$$

where $s_3 > 0$ is chosen from (6) in such a way that

$$sg(x, s) - \theta G(x, s) \geq -\varepsilon s^2 g(x, s)^{2/N+1}, \quad x \in \bar{\Omega}, \quad s \geq s_3$$

for some given $\varepsilon > 0$. Then

$$\int |\nabla u_n|^2 \leq C + C\|u_n\|_{H^1} + \frac{2}{\theta}\int_{u_n \geq s_3} u_n g(x, u_n) + \frac{2\varepsilon}{\theta}\int_{u_n \geq s_3} u_n^2 g(x, u_n)^{2/(N+1)}. \tag{19}$$

Using (18) the first integral in the right side of (19) can be taken over all Ω, which is then estimated using (16) with $v = u_n$:

$$\int |\nabla u_n|^2 \leq C + C\|u_n\|_{H^1} + \frac{2}{\theta}\int |\nabla u_n|^2 + \frac{2\varepsilon}{\theta}\int_{u_n \geq s_3} u_n^2 g(x, u_n)^{2/(N+1)}$$

or

$$\int |\nabla u_n|^2 \leq C + C\varepsilon \int_{u_n \geq s_3} u_n^2 g(x, u_n)^{2/(N+1)}. \tag{20}$$

Now we claim that

$$\int_{u_n \geq s_3} g(x, u_n)\phi \leq C. \tag{21}$$

Indeed, it follows from (16) with $v = \phi$ that

$$-\varepsilon_n \|\phi\|_{H^1} \leq \lambda_1 \int u_n \phi - \int g(x, u_n)\phi - \int f\phi \leq (\lambda_1 - \bar{\mu})\int u_n \phi + C$$

which implies that $\int u_n \phi$ and $\int g(x, u_n)\phi$ are bounded above. This gives (21) in view of hypothesis (iii) on g. The integral in (20) is estimated using Hölder's inequality by

$$\left(\int_{u_n \geq s_3} g(x, u_n)\phi\right)^{2/(N+1)} \left(\int_{u_n \geq s_3} \phi^{-2/(N-1)} u_n^{2(N+1)/(N-1)}\right)^{(N-1)/(N+1)}$$

which is then bounded by $c\|\nabla u_n\|_{L^2}^2$ using the Hardy–Sobolev inequality [7]. This shows that $\|\nabla u_n\|_{L^2}$ is uniformly bounded. So we may assume that $u_n \rightharpoonup u$ in H_0^1 and $u_n \to u$ in L^2. Now we claim that $u_n \to u$ in H_0^1. In the case $\sigma < (N+2)/(N-2)$ this follows readily from the fact that J' is an operator of the form identity minus a compact operator. Although this is not true in the case $\sigma = (N+2)/(N-2)$ we can still prove that $u_n \to u$ as follows. Let $\varepsilon > 0$ be given. Then there is an s_4 such that $g(x, s) \leq \varepsilon s^\sigma$ for $s \geq s_4$. Now using (16) with $v = u_n - u$ we have, with $C_1 \geq \|u_n\|_{H^1}, \|u\|_{H^1}$,

$$\int |\nabla(u_n - u)|^2 \leq -\int \nabla u \cdot \nabla(u_n - u) + \int f(u_n - u) + 2C_1\varepsilon_n + \int g(x, u_n)(u_n - u).$$

The first three terms in the above estimate go to zero as $n \to \infty$. We estimate the last term by

$$C\|u_n - u\|_{L^2} + \varepsilon \int_{u_n \geq s_4} u^\sigma |u_n - u|.$$

Taking this last integral over the whole of Ω with $|u_n|$ replacing u_n, and using Hölder's inequality we have finally

$$\int |\nabla(u_n - u)|^2 \leq -\int \nabla u \,.\, \nabla(u_n - u) + C\|u_n - u\|_{L^2} + 2C_1\varepsilon_n + C_2\varepsilon$$

where C_2 is a constant independent of ε. This proves the claim.

Case 2. $N \geq 3$, $\sigma \leq (N+1)/(N-1)$. This case is much simpler. Taking (16) with $v = u_n$ and using property (iii) of g we obtain

$$\int |\nabla u_n|^2 \leq C + C\|u_n\|_{H^1} + \int_{u_n \geq s_4} g(x, u_n)u_n. \tag{22}$$

The integral on the right side is then estimated using the Hardy–Sobolev inequality as above, leading to a uniform bound for the L^2-norm of the gradient of u_n. The convergence of a subsequence follows as in case 1.

Case 3. $N = 2$. We proceed as in case 2.

Proof of theorem 1 *completed*. In order to apply the mountain pass theorem it remains to check that

(i) there is $v \in H_0^1$ such that $J(v) < J(u_t)$,
(ii) there is an $r > 0$ such that $\inf\{J(u) : \|u - u_t\| = r\} > J(u_t)$.

To see (i) we take $v = R\phi$:

$$J(R\phi) = \tfrac{1}{2}\lambda_1 R^2 - \int G(x, R\phi) - R\int f\phi.$$

From (7), $G(x, s) \geq \tfrac{1}{2}\bar{\mu}s^2 - C$ and we obtain

$$J(R\phi) \leq \tfrac{1}{2}(\lambda_1 - \bar{\mu})R^2 - CR - C$$

which gives the claim by taking R sufficiently large.

As for (ii) we proceed as follows. Using the fact that u_t is a critical point of J we have:

$$J(u_t + v) - J(u_t) = \frac{1}{2}\int |\nabla v|^2 - \int [G(x, u_t + v) - G(x, u_t) - g(x, u_t)v]. \tag{23}$$

By Taylor's formula, the expression into the brackets is $\tfrac{1}{2}g_s'(x, u_t)v^2 + r(x, v)$ where $|r(x, v)| \leq \varepsilon(x, v)|v|^2$ if $|v| < 1$ and $\varepsilon(x, v) \to 0$ as $v \to 0$. On the other hand $|r(x, v)| \leq c|v|^{\sigma+1}$ for $|v| > 1$. Altogether

$$|r(x, v)| \leq \varepsilon|v|^2 + c|v|^{\sigma+1}$$

which gives

$$\int |r(x, v)|\,dx \leq \varepsilon\|v\|_{L^2}^2 + c\|v\|_{L^{\sigma+1}}^{\sigma+1} \leq \varepsilon\|v\|_{H^1}^2 + c\|v\|_{H^1}^{\sigma+1}.$$

Estimating (23) we obtain

$$J(u_t + v) - J(u_t) \geq \frac{1}{2}\int |\nabla v|^2 - \frac{1}{2}\int g_s'(x, u_t)v^2 - \varepsilon\|v\|_{H^1}^2 - c\|v\|_{H^1}^{\sigma+1}$$

which in view of lemma 8 (and the remark right after it) is bounded below by $c\|v\|_{H^1}^2 - c\|v\|_{H^1}^{\sigma+1}$. So if $\|v\|_{H^1} = r$ and r is small the quadratic term dominates and we have claim (ii). The final step is the regularity of this solution obtained by the mountain pass theorem. In the case $\sigma < (N+2)/(N-2)$ this follows readily via a standard bootstrap argument. In the case $\sigma = (N+2)/(N-2)$, this follows from theorem 10 proved in the next section. ■

3. A REGULARITY RESULT

In the present section we establish a regularity result for H_0^1 solutions of semilinear elliptic equations of second order. The growth of the nonlinearity $g(x,u)$ as $u \to \infty$ prevents a direct use of bootstrap arguments. The main ideas of the proof below are borrowed from Brézis and Kato [11] where they treat the linear Schrödinger equation.

THEOREM 10. Let $g: \Omega \times R \to R$ be a Carathéodory function such that

$$|g(x,s)| \leq c|s|^\sigma + c(x)$$

where $\sigma = (N+2)/(N-2)$, c is a constant and $c(x)$ is an $L^{2N/(N-2)}$ function. If $u \in H_0^1(\Omega)$ is a solution of

$$-\Delta u = g(x,u) \quad \text{in } \Omega, \qquad u = 0 \quad \text{on } \partial\Omega \tag{24}$$

then $u \in L^p$ for some $p > 2N/(N-2)$.

Remark. Once we know $u \in L^p$ for some $p > 2N/(N-2)$ then a bootstrap argument shows that $u \in C^{1,\alpha}$. If more regularity of g is required (for example C^1 in the situation of the previous section) then $u \in C^{2,\alpha}$.

Proof. Let $\chi \in C_0^\infty(R)$ with compact support containing a neighborhood of zero. Then g decomposes as $g = g_1 + g_2$ where $g_1(x,s) = \chi(s)g(x,s)$ and $g_2(x,s) = (1-\chi(s))g(x,s)$. Consequently $g(x,u(x))$, where $u \in H_0^1$ is the given solution of (24), can be written as

$$g(x,u(x)) = a(x)u(x) + b(x)$$

where $a(x) \in L^{N/2}$ and $b(x) \in L^{2N/(N-2)}$. Now we use the following lemma from Brézis and Kato [11, p. 139]: "Given $a(x) \in L^{N/2}$ and $\varepsilon > 0$ there is a constant $k_\varepsilon > 0$ such that

$$\int |a(x)|u^2 \leq \varepsilon \int |\nabla u|^2 + k_\varepsilon \int u^2, \quad \forall u \in H_0^1.\text{"} \tag{25}$$

Let k_1 be the constant in (25) corresponding to $\varepsilon = 1$. It is easy to see that, for $k > k_1$, the problem

$$-\Delta v + kv = a(x)v \quad \text{in } \Omega, \qquad v = 0 \quad \text{on } \partial\Omega$$

has only the trivial solution $v = 0$. So the problem

$$-\Delta v + kv = a(x)v + b(x) + ku(x) \quad \text{in } \Omega, \qquad v = 0 \quad \text{on } \partial\Omega \tag{26}$$

has a *unique* solution v, which happens to be $v = u$. The question now is to prove that the unique solution $v \in H_0^1$ of the problem

$$-\Delta v + kv = a(x)v + d(x) \quad \text{in } \Omega, \qquad v = 0 \quad \text{on } \partial\Omega$$

where $d(x)$ is an $L^{2N/(N-2)}$ function, is in some L^p with $p > 2N/(N-2)$. For this matter we proceed as in [11] by truncation of $a(x)$ and subsequent truncation of the solution of the truncated problem. A slight change in the argument of [11] is necessary since there $d(x)$ is assumed L^∞. Finally we come to $v \in L^p$ where $p = [2N/(N-2)][N/(N-2)]$. ■

Note. After the completion of this work we learned from Professor Brézis that the two truncations are not really needed if one does a truncation of $a(x)$ different from the one in [11]. This will appear in one of his forthcoming articles. Also Professor Ni has mentioned to me that Professor K. C. Chang has obtained a result comparable with our theorem 1.

Acknowledgement—The author thanks Professor Antonio Ambrosetti for his hospitality at the Scuola Internazionale Superiore di Studi Avanzati (Trieste Italia), where part of this work was done.

REFERENCES

1. AMBROSETTI A. & PRODI G., On the inversion of some differentiable mappings with singularities between Banach spaces, *Annali. Math. pura appl. Ser. IV*, **93**, 231–247 (1972).
2. BERGER M. S. & PODOLAK E., On the solutions of a nonlinear Dirichlet problem, *Indiana Univ. Math. J.* **24**, 837–846 (1975).
3. AMANN H. & HESS P., A multiplicity result for a class of elliptic boundary value problems, *Proc. R. Soc. Edinb.* **84A**, 145–151 (1979).
4. FUČIK S., Remarks on a result by A. Ambrosetti and G. Prodi, *Boll. Un. mat. Ital.* **11**, 259–267 (1975).
5. KAZDAN J. L. & WARNER F. W., Remarks on some quasilinear elliptic equations, *Communs pure appl. Math.* **XXVIII**, 567–597 (1975).
6. DANCER E. N., On the ranges of certain weakly nonlinear elliptic partial differential equations, *J. Math. pures appl.* **57**, 351–366 (1978).
7. BRÉZIS H. & TURNER R. L., On a class of superlinear elliptic problems, *Communs P.D.E.* **2**, 601–614 (1977).
8. AMBROSETTI A. & RABINOWITZ P., Dual variational methods in critical point theory and applications, *J. funct. Analysis* **14**, 349–381 (1973).
9. DE FIGUEIREDO, D. G., Lectures on boundary value problems of the Ambrosetti–Prodi type, *Atas do 12° Seminário Brasileiro de Análise*, São Paulo (October 1980).
10. DE FIGUEIREDO D. G., LIONS P. -L. & NUSSBAUM R., *A priori* estimates and existence of positive solutions of semilinear elliptic equations, *J. Math. pures appl.* **61**, 41–63 (1982).
11. BRÉZIS H. & KATO T., Remarks on the Schrödinger operator with singular complex potentials, *J. Math. pures appl.* **58**, 137–151 (1979).
12. AMANN H., On the existence of positive solutions of nonlinear elliptic boundary value problems, *Indiana Univ. Math. J.* **21**, 125–146 (1971).
13. SATTINGER D. H., *Topics in Stability and Bifurcation Theory*, Springer Lecture Notes in Mathematics, Vol. 309 (1973).
14. LAZER A. C. & MCKENNA P. J., On the number of solutions of a nonlinear Dirichlet problem, to appear.
15. SOLIMINI S., Differential problems with symmetries: equations with jumping nonlinearities, to appear.
16. AMBROSETTI A., Elliptic equtions with jumping nonlinearities, to appear.
17. BERESTYCKI H. & LIONS P. L., Sharp existence results for a class of semilinear elliptic problems, *Bol. Soc. Bras. Mat.* **12**, 9–20 (1981).

[illegible]

[illegible]

[illegible]

REFERENCES

[illegible]

COMM. IN PARTIAL DIFFERENTIAL EQUATIONS, 9(7), 699-717 (1984)

A VARIATIONAL APPROACH TO SUPERLINEAR ELLIPTIC PROBLEMS

Djairo G. de Figueiredo

Departamento de Matemática
Universidade de Brasília
Brasília, Brazil

Sergio Solimini

Scuola Internazionale Superiore di Studi Avanzati
Trieste, Italy

ABSTRACT

This paper contains a variational treatment of the Ambrosetti-Prodi problem, including the superlinear case. The main result extends previous ones by Kazdan-Warner, Amann-Hess, Dancer, K. C. Chang and de Figueiredo. The required abstract results on critical point theory of functionals in Hilbert space are all proved using Ekeland's variational principle. These results apply as well to other superlinear elliptic problems provided an ordered pair of a sub- and a supersolution is exhibited.

 0360-5302/84/0907-0699$3.50/0

Originally published in: Comm. Partial Differential Equations 9 (1984), 699-717,

DOI: 10.1007/978-3-319-02856-9_13

INTRODUCTION.

The main objective of the present paper is to give a variational treatment of the superlinear Ambrosetti-Prodi problem and to prove the existence of two solutions for certain values of the parameter. In this process we produce interesting and simple proofs of some abstract results using Ekeland's variational principle [1]. Some of these abstract results, namely Propositions 3 and 5, are known in the literature, where they are proved via the deformation lemma, c.f. Hofer [2] and Rabinowitz [3]. Our results on the superlinear Ambrosetti-Prodi problem extend previous theorems of Dancer [4], K. C. Chang [5] and one of the authors [6].

Let Ω be a smooth bounded domain in R^N, $N \geq 2$. Let $g : \Omega \times R \to R$ be a function satisfying the following conditions:

(g1) g is a C^1-function,

(g2) $\limsup_{s\to-\infty} \frac{g(x,s)}{s} < \lambda_1 < \liminf_{s\to+\infty} \frac{g(x,s)}{s}$

(g3) $\lim_{s\to+\infty} \frac{g(x,s)}{s^\sigma} = 0$, where $1 < \sigma \leq (N+2)/(N-2)$ if $N \geq 3$ or $1 < \sigma < \infty$ if $N = 2$

(g4) $\liminf_{s\to+\infty} \frac{sg(x,s) - \theta G(x,s)}{s^2 g(x,s)^{2/(N+1)}} \geq 0$, if $N \geq 3$ and $(N+1)/(N-1) < \sigma \leq (N+2)/(N-2)$ where $\theta > 2$ and $G(x,s) = \int_0^s g(x,\xi)d\xi$.

The statements about the limits above are supposed to hold

uniformly for $x \in \Omega$. The limits in (g2) could assume value $-\infty$ or $+\infty$, respectively, on the whole of Ω or on subsets of positive measure. In (g2), λ_1 denotes the first eigenvalue of the eigenvalue problem $-\Delta u = \lambda u$ in Ω, $u = 0$ on $\partial\Omega$.

We denote by ϕ the eigenfunction corresponding to λ_1 and such that $\phi > 0$ in Ω and $\int \phi^2 = 1$. Let $\phi^\perp = \{u \in C^\alpha(\bar{\Omega}) : \int u\phi = 0\}$. Any function $f \in C^\alpha(\bar{\Omega})$ can be uniquely decomposed as $f = t\phi + h$, where $t \in R$ and $h \in \phi^\perp$.

We shall look at the parametrized family of Dirichlet problems:

(P_t) $$-\Delta u = g(x,u) + t\phi + h \quad \text{in } \Omega, \quad u = 0 \quad \text{on } \partial\Omega ,$$

and prove the following result.

Theorem 1. *Assume (g1), (g2), (g3) and (g4). Then given $h \in \phi^\perp$, there exists a $\bar{t} \in R$ (depending on h), such that problem (P_t) has at least two $C^{2,\alpha}$-solutions if $t < \bar{t}$, and no solution if $t > \bar{t}$. Moreover, there is at least one solution for $t = \bar{t}$ provided we assume the following additional condition on g, in the case that $N \geq 3$ and the σ in (g3) is $> (N+1)/(N-1)$:*

(g5) *there is a function $a : \bar{\Omega} \to R$, with $a(x) > 0$ if $x \in \bar{\Omega}$, such that*

$$\lim_{s \to +\infty} \frac{g(x,s)}{s^\alpha} = a(x)$$

for some $(N+1)/(N-1) < \alpha < (N+2)/(N-2)$.

Remarks. 1) Condition (g2) characterizes (P_t) as a boundary value problem of the Ambrosetti Prodi type. 2) Condition (g3), together with a convenient modification of the nonlinearity $g(x,s)$ for $s < 0$, allows us to treat (P_t) by variational techniques. In this way we obtain its solutions as critical points of an associated Euler-Lagrange functional. 3) Condition (g4), in conjunction with (g2) and (g3) are used to prove that such a functional satisfies the Palais-Smale condition. 4) Condition (g5) implies an a priori bound on the solutions of (P_t) for h fixed and t in a compact set. Observe that in the cases $N = 2$ and $N \geq 3$ with $1 < \sigma \leq (N+1)/(N-1)$ an a priori bound on solutions follows from the other hypotheses of the theorem. We remark that (g5) is a strong restriction; once it is assumed, then (g3) and (g4) are automatically fulfilled for the corresponding ranges of N and σ. The a priori bounds mentioned above are consequences of similar results for positive solutions of certain second order elliptic equations. Indeed, just observe that (P_t) has a minimal solution u_t and that any other solution u may be written as $u = u_t + v$, where v is a positive solution of an appropriate second order elliptic equation. The bound on the v's follows from the work of Gidas-Spruck [7] in the case when (g5) is assumed. In the other cases the estimates of Brézis-Turner [8] will suffice.

The paper is organized as follows. In the first part we recall known results and prepare our differential problem (P_t) to be treated by variational methods. In the second part we state and prove some abstract results, which will be used in the next section. Finally in the third part, we show how the abstract theorems of the previous section are used to prove results on the existence of two solutions for some second order elliptic problems. There the basic assumption is that these problems have a strict subsolution ω and a strict supersolution W, with $\omega \leq W$. In particular these results imply Theorem 1. The solutions so obtained are in H_0^1. However in view of the regularity assumptions on g and Ω made above, their regularity follows either by standard bootstrap arguments or more sophisticate ones due to Brézis-Kato [9], see also [6].

The contents of this paper were presented by the first author at the AMS Summer Institute on Nonlinear Functional Analysis and Applications in Berkeley (July 1983). The preparation of this paper was partially sponsored by the United States Army under Contact No. DAAG29-80-C-0041 while the first author was visiting the Mathematics Research Center.

1. VARIATIONAL FORMULATION.

We start with the following result which is essentially due to Kazdan-Warner [10].

Lemma 2. Assume (g1) and (g2), and let $h \in \phi^{\perp}$ be given. Then (i) there exists a $t_1 \in R$ such that (P_t) has no solution if $t > t_1$. (ii) For each given $t \in R$ there exists a

$C^{2,\alpha}$-function ω_t, which is a strict subsolution of (P_t) and bounds from below all eventual supersolutions of (P_t). (iii) There exists a $t_2 \in R$ such that (P_{t_2}) has a strict supersolution $w_{t_2} \in C^{2,\alpha}(\bar{\Omega})$.

From now on we assume $h \in \phi^1$ be given and fixed. The set J of the t's such that (P_t) has a supersolution is a halfline from $-\infty$. Let $\bar{t}$ be the supremum of J. We assert that if $t < \bar{t}$ then (P_t) has a strict supersolution. Indeed, a supersolution of problem $(P_{\tilde{t}})$, with $t < \tilde{t} < \bar{t}$, is a strict supersolution of problem (P_t).

Hereafter let $t < \bar{t}$ be fixed. Our goal is to prove that (P_t) has two solutions. We remark at this point that the method of monotone iteration yields readily the existence of one solution. This was the approach taken by Kazdan-Warner [10], and also used by Amann-Hess [11], Dancer [4], Berestycki [12] and one of the authors [6] to obtain the first solution (the minimal one) of the Ambrosetti-Prodi problem. The second solution was obtained by degree arguments, in the case of [4], [11] and [12], and by a variational argument, in the case of [6]. In the present paper we proceed solely by variational techniques, even for the first solution. In this way we are in the spirit of the work of Hofer [2].

In order to formulate variationally our problem, we recall [see Lemma 2] that all eventual solutions of (P_t) are bounded from below by ω_t. We remark that ω_t is the solution of the

Dirichlet problem

$$-\Delta\omega_t = \mu\omega_t - C + t\phi + h \quad \text{in} \quad \Omega, \quad \omega_t = 0 \quad \text{on} \quad \partial\Omega ,$$

where $\mu < \lambda_1$ and $C > 0$ are chosen in such a way that $g(x,s) \geq \mu s - C$ for all $x \in \Omega$ and $s \in \mathbb{R}$. [Such a lower bound on g exists as a consequence of (g2), cf. [6]]. So ω_t depends only on this lower bound for $g(x,s)$ and consequently a modification of g maintaining it produces the same ω_t. Thus we modify the nonlinearity $g(x,s)$ for $s < \min \omega_t$, so as to preserve the inequality $g(x,s) \geq \mu s - C$. The solutions of (P_t) with this modified nonlinearity are the same as the ones of the original problem. In addition, we do this modification in such a way that the new g is still a C^1-function and moreover $g(x,s) \geq 0$ for large negative s and g has linear growth as $s \to -\infty$. With this new nonlinearity the functional Φ below is well defined in H_0^1 and satisfies the Palais-Smale condition, cf. [6]:

$$\Phi(u) = \int \frac{1}{2} |\nabla u|^2 - G(x,u) - (t\phi + h)u .$$

2. ABSTRACT RESULTS.

Let H be a real Hilbert space, with inner product $\langle \cdot , \cdot \rangle$ and norm $\|\cdot\|$. A functional $\Phi : H \to \mathbb{R}$ of class C^1 is said to satisfy the Palais Smale condition if every sequence (u_n) in H, such that $\Phi(u_n)$ is bounded and $\Phi'(u_n) \to 0$, is precompact. The following result was a proposition proved by Hofer in [2] using the deformation lemma adapted for closed convex subsets of a Hilbert space. We provide below a different proof.

Proposition 3. Let $\Phi \in C^1(H,R)$ satisfy the Palais Smale condition. Let C be a closed convex subset of H. Suppose th $K \equiv I - \Phi'$ maps C into C and that Φ is bounded below in C. Then there is $u_0 \in C$ such that $\Phi'(u_0) = 0$ and $\inf_C \Phi = \Phi(u_0)$.

Proof. By Ekeland's variational principle, given $\varepsilon > 0$ there is an ε-minimum of Φ, that is, $u_\varepsilon \in C$ such that $\Phi(u_\varepsilon) \leq \inf_C \Phi + \varepsilon$ and

$$(1) \qquad \Phi(u_\varepsilon) \leq \Phi(u) + \varepsilon\|u - u_\varepsilon\|, \quad \forall\, u \in C\,.$$

Take in (1) $u = (1 - t)u_\varepsilon + tKu_\varepsilon$, with $0 \leq t \leq 1$, and use Taylor's formula to expand $\Phi(u_\varepsilon + t(Ku_\varepsilon - u_\varepsilon))$ about u_ε. We then obtain

$$t\|\Phi'(u_\varepsilon)\|^2 \leq \varepsilon t\|\Phi'(u_\varepsilon)\| + o(t)$$

from which follows that $\|\Phi'(u_\varepsilon)\| \leq \varepsilon$, making $t \to 0$. Finally use the Palais-Smale condition to conclude that there exists $u_0 \in C$ such that $u_\varepsilon \to u_0$, for some sequence $\varepsilon \to 0$. The continuity of Φ and Φ' completes the proof. □

Remark. As a matter of fact one does not need the continuity of Φ' for the validity of the previous Proposition. To prove tha $\Phi'(u_0) = 0$ one uses the fact that $\Phi(u_0) \leq \Phi(u)$ and then appl Taylor's formula, as before. Or just observe that u_0 is an ε-minimum for every $\varepsilon > 0$.

An operator $T : H \to H$ is said to be of type S^+ if for every sequence $u_n \rightharpoonup u$ ($\rightharpoonup$ means weak convergence) such that $\limsup (Tu_n, u_n - u) \leq 0$, it follows that $u_n \to u_0$.

Proposition 4. Let $\Phi \in C^1(H,R)$ be such that Φ' is of type S^+. Suppose that Φ is bounded below in a ball $\bar{B}$. Then there exists $v_0 \in \bar{B}$ such that $\Phi(v_0) = \inf_{\bar{B}} \Phi$ and $\Phi'(v_0) = \lambda v_0$, with $\lambda \leq 0$.

Proof. By Ekeland's variational principle, given $\varepsilon > 0$ there exists $v_\varepsilon \in \bar{B}$ such that $\Phi(v_\varepsilon) \leq \inf_{\bar{B}} \Phi + \varepsilon$ and

(2) $$\Phi(v_\varepsilon) \leq \Phi(v) + \varepsilon \| v - v_\varepsilon \|, \quad \forall\, v \in \bar{B} .$$

Now let $u \in \bar{B}$ be arbitrary and take in (2) $v = v_\varepsilon + t(u - v_\varepsilon)$, with $0 \leq t \leq 1$. Use Taylor's formula to expand $\Phi(v_\varepsilon + t(u - v_\varepsilon))$ about v_ε and obtain, making $t \to 0$:

(3) $$0 \leq \langle \Phi'(v_\varepsilon), u - v_\varepsilon \rangle + \varepsilon \| u - v_\varepsilon \|, \quad \forall\, u \in \bar{B} .$$

Next taking in (3) $u = v_0$, where v_0 is the weak limit of v_ε for some sequence $\varepsilon \to 0$, and making $\varepsilon \to 0$, one obtains $\limsup_{\varepsilon \to 0} \langle \Phi'(v_\varepsilon), v_\varepsilon - v_0 \rangle \leq 0$. Since Φ' is of type S^+, it follows that $v_\varepsilon \to v_0$. We obtain readily that $\Phi(v_0) = \inf_{\bar{B}} \Phi$. On the other hand, from (3) it follows that $0 \leq \langle \Phi'(v_0), u - v_0 \rangle$ for all $u \in \bar{B}$. This implies that there exists $\lambda \leq 0$ such that $\Phi'(v_0) = \lambda v_0$. □

The Mountain Pass Theorem of Ambrosetti-Rabinowitz [13] states that a functional $\Phi \in C^1(H,R)$, which satisfies the Palais-Smale condition and such that

(4) $$\inf\{\Phi(u) : \|u\| = r\} > \max\{\Phi(0), \Phi(e)\} \equiv a ,$$

where $0 < r < \|e\|$, has necessarily a critical point u_0 at the level

$$c \equiv \inf_{\gamma \in \Gamma} \max_{t \in [0,1]} \Phi(\gamma(t)) \tag{5}$$

where $\Gamma = \{\gamma \in C^0([0,1],H) : \gamma(0) = 0, \gamma(1) = e\}$. Clearly $c >$ a, which implies that $u_0 \neq 0$ and $u_0 \neq e$. What happens however if the inequality in (4) is replaced by equality? In this direction, Pucci-Serrin [14] observed that if there are numbers r and R such that $0 < r < R < \|e\|$, and

$$\inf\{\Phi(u) : r \leq \|u\| \leq R\} \geq \max\{\Phi(0), \Phi(e)\} \equiv a \tag{6}$$

then c as defined in (5) is a critical value for Φ. Moreover if $c = a$ then there is a critical point at every sphere $\partial B_\rho = \{u \in H : \|u\| = \rho\}$ for $r < \rho < R$. See also Willem [15] for an interesting proof of this fact. Rabinowitz in [3] proved a stronger form of the dual Mountain Pass Theorem, which provides an improved form of the Pucci-Serrin result. In his result, Rabinowitz replaces condition (6) by the much weaker condition $\inf\{\Phi(u) : \|u\| = r\} \geq \Phi(0) > \Phi(e)$ with $\|e\| > r$, and proves that there is a critical point at the level

$$b \equiv \sup_{U \in O} \inf_{u \in \partial U} \Phi(u) ,$$

where $O = \{U \subset H \mid U \text{ is open, } 0 \in U, e \in \bar{U}\}$. Moreover if $b = \Phi(0)$ then there is a critical point at ∂B_r. The proposition below is essentially this result, but its proof differs from the one given in [3].

<u>Proposition 5.</u> <u>Let</u> $\Phi \in C^1(H,R)$ <u>satisfy the Palais Smale condition. Suppose that</u>

(7) $$\inf\{\Phi(u) : \|u\| = r\} \geq \max\{\Phi(0), \Phi(e)\} ,$$

<u>where</u> $0 < r < \|e\|$. <u>Then</u> Φ <u>has a critical point</u> $u_0 \neq 0$.

<u>Proof</u>. It suffices to consider the case where we have equality in (7), since the case of inequality is the usual Mountain Pass Theorem. (i) First suppose that $\Phi(0) < \Phi(e)$. Then either e is a local minimum (and we are through) or there is an e' near e such that $\Phi(e') < \inf\{\Phi(u) : \|u\| = r\}$. In the latter case we use the usual Mountain Pass Theorem to conclude. (ii) Next suppose that $\Phi(0) = \Phi(e)$. As in (i) above either e is a local minimum or there is e' near e with $\Phi(e') < \Phi(0)$. (iii) So we may always suppose from now one that $\Phi(0) > \Phi(e)$. Also we may assume that

$$\Phi(0) = \inf_{\bar{B}_r} \Phi .$$

Otherwise we would apply the usual Mountain Pass Theorem to conclude. (iv) Now let $0 < \rho < r$. We claim that either there is an $\alpha > \Phi(0)$ such that $\inf\{\Phi(u) : \|u\| = \rho\} \geq \alpha$ or there is a critical point of Φ on ∂B_ρ. If the first alternative occurs we apply the usual Mountain Pass Theorem and the proof is complete. To prove the claim, let us suppose that there is no such an α. Then there are $u_n \in \partial B_\rho$ such that

$$\Phi(u_n) \leq \Phi(0) + \frac{1}{n}$$

Consider the ring $R = \{u \in H : 0 < \rho - \eta \leq \|u\| \leq \rho + \eta < r\}$ for an appropriate $\eta > 0$. Now use Ekeland variational principle for

Φ in the ring. Thus there are $v_n \in R$ such that

$$\Phi(v_n) \leq \phi(u_n), \quad \|v_n - u_n\| \leq \frac{1}{n}, \tag{8}$$

$$\Phi(v_n) \leq \Phi(u) + \frac{1}{n}\|u - v_n\|, \quad \forall u \in R.$$

For large n, v_n is in the interior of R. So take in (8) $u = v_n + tu$, where $t \in R$ is small and u is arbitrary with $\|u\| = 1$. Using Taylor's formula to expand $\Phi(v_n + tu)$ about v_n and letting $t \to 0$ we obtain $\|\Phi'(v_n)\| \leq \frac{1}{n}$. Finally using the Palais-Smale condition we conclude that $v_n \to u_0 \in \partial B_\rho$ and u_0 is a critical point of Φ.

Remark. It has been observed by Ekeland that the usual Mountain Pass Theorem can be proved using his variational principle, instead of the deformation lemma.

3. APPLICATION TO ELLIPTIC PROBLEMS.

Since the abstract results presented in the previous section allow us to treat a larger class of second order elliptic problems we start anew the abstract setting of the differential problem. We observe that for the use of variational methods and the existence of H_0^1-solutions a great deal less of regularity is required from both the nonlinearity g and the linear part of the equation. We also remark that the basic assumption is the existence of an ordered pair of a subsolution and a supersolution. So besides the Ambrosetti-Prodi problem stated in the introduction we can treat other problems where such a pair is produced.

Consider the second order elliptic operator

$$Lu = - \sum_{i,j=1}^{N} D_j(a_{ij}(x)D_i u) + c(x)u$$

where $a_{ij} \in L^\infty(\Omega)$, $c \in L^{N/2}(\Omega)$ and $c(x) \geq 0$ a.e. Ellipticity here means that there is a constant $c > 0$ such that $\sum a_{ij}(x)\xi_i\xi_j \geq c|\xi|^2$. Let $a : H_0^1 \times H_0^1 \to R$ be the bilinear form associated to L and defined by

$$a[u,v] = \int \sum a_{ij}(x)D_i u D_j v + c(x)uv \ .$$

We assume that $g : \Omega \times R \to R$ is a Carathéodory function such that

$$|g(x,s)| \leq c|s|^\sigma + d(x)$$

where $\sigma = (N+2)/(N-2)$ and $d \in L^{2N/(N+2)}$ if $N \geq 3$, or $d \in L^p$ and $1 < p, \sigma < \infty$ if $N = 2$. We then consider the Dirichlet problem

$$(9) \qquad \begin{cases} a[u,v] = \int g(x,u)v, & \forall\, v \in H_0^1 \\ u \in H_0^1 \end{cases}$$

Let $\Phi : H_0^1 \to R$ be the corresponding Euler-Lagrange functional defined by

$$(10) \qquad \Phi(u) = \frac{1}{2}\, a[u,u] - \int G(x,u)$$

where $G(x,s) = \int_0^s g(x,\xi)d\xi$. We renorm H_0^1 with the norm $\|u\| =$

$a[u,u]^{1/2}$; denote the corresponding inner product by $\langle\cdot,\cdot\rangle$: $\langle u,v\rangle = a[u,v]$. Clearly Φ is C^1 and its derivative at u is given by

$$\langle\Phi'(u),v\rangle = \langle u,v\rangle - \int g(x,u)v, \quad \forall\, v \in H_0^1 .$$

We then see that the critical points of Φ are precisely the solutions of (9). Let us define the map $K : H_0^1 \to H_0^1$ by

$$\langle Ku,v\rangle = \int g(x,u)v, \quad \forall\, v \in H_0^1 .$$

<u>Theorem 6</u>. <u>Assume conditions on</u> g <u>that guarantee that</u> Φ <u>defined in (10) satisfies the Palais-Smale condition. Suppose that there exist a subsolution</u> $\omega \in H_0^1$ <u>and a supersolution</u> $W \in H_0^1$ <u>of (9) such that</u> $\omega \leq W$; <u>that is,</u>

(11) $$a[\omega,v] \leq \int g(x,\omega)v \quad \text{and} \quad a[W,v] \geq \int g(x,W)v,$$
$$\forall\, v \in H_0^1,\ v \geq 0.$$

<u>Assume also that</u> $g(x,s)$ <u>is a nondecreasing function of</u> s <u>for</u> $\omega(x) \leq s \leq W(x)$. <u>Then there exists</u> $u_0 \in H_0^1$ <u>such that</u> $u_0 \in [\omega,W]$, $\Phi(u_0) = \inf_{[\omega,W]} \Phi$ <u>and</u> $\Phi'(u_0) = 0$ <u>[Consequently</u> u_0 <u>is a solution of (9)].</u>

<u>Remarks</u>. 1) We use the notation $[\omega,W]$ to denote the set of $u \in H_0^1$ such that $\omega(x) \leq u(x) \leq W(x)$ for $x \in \Omega$ (a.e.).

2) The set of assumptions on g prescribed at the Introduction are sufficient to guarantee that Φ satisfies the Palais-Smale condition.

<u>Proof</u>. (i) $C = [\omega,W]$ is a closed and convex subset of H_0^1. (ii) K maps C into C. Indeed for $u \in [\omega,W)$, we have

$g(x,w) \leq g(x,u) \leq g(x,W)$, which implies $a[w,v] \leq a[Ku,v] \leq a[W,v]$ for all $v \in H_0^1$, $v \geq 0$. By the Maximum Principle (see Gilbarg-Trudinger [16, p. 168]) it follows that $w \leq Ku \leq W$. (iii) Φ is bounded below in C. Indeed, if $u \in [w,W]$ then $|u(x)| \leq |w(x)| + |W(x)|$, which implies that $\|u\|_{L^{2N/(N-2)}} \leq C$. So $\int |G(x,u)| \leq C$, which proves the claim. (iv) Finally the conclusion of the theorem follows readily from Proposition 3. □

Corollary 7. Same hypotheses of Theorem 6, except the assumption that g is nondecreasing in s for $w(x) \leq s \leq W(x)$. Assume however that g is locally lipschitzian and w and W are continuous in $\bar{\Omega}$. Then the same conclusions of Theorem 6 hold.

Proof. In view of the hypotheses, there is a positive constant M such that the function $g_M(x,s) \equiv g(x,s) + Ms$ is nondecreasing in s for $w(x) \leq s \leq W(x)$. The Dirichlet problem

$$(12) \qquad \begin{cases} a_M[u,v] = \int g_M(x,u)v, & \forall v \in H_0^1 \\ u \in H_0^1 \end{cases}$$

where $a_M[u,v] = a[u,v] + M \int uv$, has the same solutions as problem (9). Moreover the functional associated to (12) is the same as Φ. And we may apply Theorem 6 to problem (12). □

Theorem 8. Assume conditions on g that guarantee that Φ satisfies the Palais-Smale condition and Φ' is of type S^+. Suppose that g is locally lipschitzian and the coefficients of L are such that the results of the L^p-regularity theory hold true. Assume also that $w \in C^{1,\alpha}$ is a strict subsolution and

$W \in C^{1,\alpha}$ is a strict supersolution, with $w \leq W$. Then there exists $u_0 \in [w,W]$ such that $\Phi'(u_0) = 0$ and u_0 is a local minimum of Φ. That is, there exists $\varepsilon > 0$, such that

$$\Phi(u_0) = \inf_{B_\varepsilon(u_0)} \Phi$$

Remark. The Ambrosetti-Prodi problem stated in the Introduction satisfies all the hypotheses of the above theorem. As a matter of fact the functional Φ corresponding to the Ambrosetti-Prodi problem is weakly lower semicontinuous. So Proposition 4 is not needed in this case. The conclusion follows from more elementary considerations.

Proof. From Theorem 6 it follows that there is $u_0 \in [w,W]$ such that $\Phi'(u_0) = 0$ and

$$\Phi(u_0) = \inf_{[w,W]} \Phi . \tag{13}$$

It can be shown that $u_0 \in C^{1,\alpha}$; the first step in the proof of this fact uses arguments from Brezis-Kato [9] to show that $u_0 \in L^p$ for $p > 2N/(N-2)$, see [6]. Then a bootstrap argument is used. Suppose by contradiction that u_0 is not a local minimum. Then for every $\varepsilon > 0$ there is $u_\varepsilon \in \overline{B_\varepsilon(u_0)}$ such that $\Phi(u_\varepsilon) < \Phi(u_0)$. By Proposition 4 there is $v_\varepsilon \in \overline{B_\varepsilon(u_0)}$ such that

$$\Phi(v_\varepsilon) = \inf_{B_\varepsilon(u_0)} \Phi \leq \Phi(u_\varepsilon) < \Phi(u_0) \text{ , and} \tag{14}$$

$$\Phi'(v_\varepsilon) = \lambda_\varepsilon(v_\varepsilon - u_0), \text{ with } \lambda_\varepsilon \leq 0 . \tag{15}$$

By the same arguments used to establish that $u_0 \in C^{1,\alpha}$, we prove that $v_\varepsilon \in C^{1,\alpha}$. The strong maximum principle and the fact that w and W are not solutions imply that $v_\varepsilon \in [w,W]$ for small ε. Then (14) contradicts (13). □

Corollary 9. In addition to all the hypotheses of Theorem 8, assume that there is $v \in H_0^1$ such that $\|v\| > \max\{\|w\|,\|W\|\}$ and $\Phi(v) \leq \min\{\Phi(w),\Phi(W)\}$. Then the Dirichlet problem (9) has two solutions.

Remark. The existence of such a v is trivial in the case that Φ is not bounded below in H_0^1. This is always the case for superlinear problems.

Proof. From Theorem 8 it follows the existence of the first solution, which is a local minimum of Φ. The second solution is a consequence of Proposition 5. □

REFERENCES

[1] Ekeland, I. - "Non convex minimization problems". Bull. AMS 1 (1979), pp. 443-474.

[2] Hofer, H. - "Variational and Topological Methods in Partially Ordered Hilbert Spaces". Math. Ann. 261 (1982), pp. 493-514.

[3] Rabinowitz, P. - "Some aspects of critical point theory". MRC Tech. Rep. #2465 (Jan. 1983).

[4] Dancer, E. N. - "On the ranges of certain weakly nonlinear elliptic partial differential equations". J. Math. Pures et

Appl. 57 (1978), pp. 351-366.

[5] Chang, K. C. - personal communication.

[6] de Figueiredo, D. G. - "On the superlinear Ambrosetti-Prodi problem". MRC Tech. Rep. #2522 (May 1983).

[7] Gidas, B. and Spruck, J. - "A priori bounds for positive solutions of nonlinear elliptic equations". Comm. PDE 6 (1981), pp. 883-901.

[8] Brézis, H. and Turner, R.E.L. - "On a class of superlinear elliptic problems". Comm. PDE 2 (1977), pp. 601-614.

[9] Brézis, H. and Kato, T. - "Remarks on the Schrödinger operator with singular complex potentials". J. Math. Pures et Appl. 58 (1979), pp. 137-151.

[10] Kazdan, J. and Warner, F. W. - "Remarks on some quasilinear elliptic equations". Comm. Pure Appl. Math. XXVIII (1975), pp. 567-597.

[11] Amann, H. and Hess, P. - "A multiplicity result for a class of elliptic boundary value problems". Proc. Royal Soc. Edinburgh 84A (1979), pp. 145-151.

[12] Berestycki, H. - "Le nombre de solutions de certains problemes semilinéaires elliptiques". J. Fct. Anal.

[13] Ambrosetti, A. and Rabinowitz, P. - "Dual variational methods in critical point theory and applications". J. Fctl. Anal. 14 (1973), pp. 349-381.

[14] Pucci, P. and Serrin, J. - "A mountain pass theorem". To appear.

[15] Willem, M. - "Lectures on Critical Point Theory". Trabalho de Matemática nº199. Universidade de Brasilia (Feb. 1983).

[16] Gilbarg, D. and Trudinger, N. S. - "Elliptic Partial Differential Equations". Springer Verlag (1977).

Received January 1984

Reprinted from JOURNAL OF DIFFERENTIAL EQUATIONS
All Rights Reserved by Academic Press, New York and London

Vol. 60, No. 1, October 1985
Printed in Belgium

Radial Solutions for a Dirichlet Problem in a Ball

D. G. COSTA* AND D. G. DE FIGUEIREDO†

Departamento de Matematica, Universidade de Brasilia,
Caixa Postal 15301, Brasilia 70910, DF, Brazil

Received February 10, 1984; revised August 21, 1984

The Ambrosetti–Prodi boundary value problem with an asymptotically linear nonlinearity is considered. Under general conditions on the nonlinearity it is shown that there exist positive and negative solutions. In the case when the domain is a ball in R^n and the nonlinearity "crosses" the first n eigenvalues, corresponding to radial eigenfunctions, it is proved that there are at least $n+1$ radial solution.

Let $\Omega \subset R^N$ be a bounded smooth domain. Let $\lambda_1 < \lambda_2 < \cdots < \lambda_k < \cdots$ denote the eigenvalues of $-\Delta$ on Ω with Dirichlet boundary condition and $\phi_1 > 0$ an eigenfunction corresponding to λ_1. We consider the Ambrosetti–Prodi problem

$$-\Delta u = f(x, u) + h(x) + t\phi_1(x) \quad \text{in } \Omega, \qquad u = 0 \quad \text{on } \partial\Omega, \tag{1_t}$$

where $h \in C(\bar{\Omega})$, $\int h\phi_1 = 0$, and $f\colon \bar{\Omega} \times R \to R$ is a continuous function such that there exist the limits

$$\lim_{s \to \pm\infty} \frac{f(x, s)}{s} = f_\pm \qquad \text{uniformly for } x \in \bar{\Omega}, \text{ with} \tag{2}$$

$$f_- < \lambda_1 < f_+, \qquad f_+ \neq \lambda_j. \tag{3}$$

We assume without loss of generality that $f(x, 0) \equiv 0$.

In Section 1 of the present paper we shall show that for $t < 0$ sufficiently large problem (1_t) possesses a large positive solution and a large negative solution. We remark that for a special class of problems (1_t) existence of a negative solution was first observed by Lazer–McKenna [3]. Subsequently, Ambrosetti [1] and Solimini [5], again for a special class of problems (1_t) proved the existence of both a negative and a positive solution for $t < 0$

* Partially supported by CNPq/Brazil.
† Guggenheim Fellow, 1983.

0022-0396/85 $3.00

Originally published in: J. Differential Equations 60 (1985), 80-89,
© 1985 Academic Press, Inc.
Reprint by Springer-Verlag Berlin Heidelberg 2013 with kind permission,
DOI: 10.1007/978-3-319-02856-9_14

sufficiently large. In [2], one of the authors proved existence of a negative solution for a general class of Ambrosetti–Prodi problems including the superlinear case. At present we do not know whether a positive solution exists in the case of nonlinearities f which grow more rapidly than linear.

In Section 2 we consider the case when Ω is a ball and f is a C^1 function which does not depend explicitely on n. Let us denote by $\mu_1 < \mu_2 < \cdots$ the eigenvalues of $-\Delta$ acting on the radial functions of $H_0^1(\Omega)$. (It is well known that $\mu_1 = \lambda_1$). We assume there that the limits below exist and satisfy

$$f_- \equiv \lim_{s \to -\infty} f'(s) < \mu_1, \qquad \mu_n < f_+ \equiv \lim_{s \to +\infty} f'(s) < \mu_{n+1}, \quad n > 1. \tag{4}$$

It is then proved that for $t < 0$ sufficiently large (1_t) possesses a radial positive solution U_t and a radial negative solution u_t. It is also proved that given any $c \neq 0$ there exists a radial solution u of the equation in (1_t) such that $u - c \in H_0^1(\Omega)$ and $u - U_t$ has at least n "nodal lines." More precisely there are at least n concentric spheres where the function $u - U_t$ vanishes.

In Section 3 we prove under the same assumptions of the previous section that problem (1_t) has at least $n+1$ radial solutions for $t < 0$ sufficiently large. Here we use a shooting argument in a manner similar to the work of Lazer–McKenna [4]. In their paper they treated a two-point boundary value problem, which is the o.d.e. analogue of (1_t), with $-\Delta u$ replaced by $-u''$. Our analysis would correspond in their case to looking for solutions which are symmetric with respect to the middle point of the interval. Without this assumption of symmetry they can actually obtain $2n$ solutions. We believe that this is also true in the p.d.e. case because we are in fact crossing also a number of eigenvalues that correspond to non-radial eigenfunctions. As a matter of fact we prove in Section 4 that this is the case when $N = 3$. Finally we observe that in the case when Ω is a region between two concentric spheres there are at least $2n$ solutions for problem (1_t), $t < 0$ large.

1. Existence of Large Positive and Negative Solutions

In this section we let $\Omega \subset R^N$ be an arbitrary bounded smooth domain. Let $\lambda_1 < \lambda_2 < \cdots < \lambda_k < \cdots$ denote the eigenvalues of $-\Delta$ on Ω with Dirichlet boundary condition and $\phi_1 > 0$ an eigenfunction corresponding to λ_1. We consider the Ambrosetti–Prodi problem

$$-\Delta u = f(x, u) + h(x) + t\phi_1(x) \quad \text{in } \Omega, \qquad u = 0 \quad \text{on } \partial\Omega, \tag{1_t}$$

where $h \in C(\bar{\Omega})$, $\int h\phi_1 = 0$, and $f\colon \bar{\Omega} \times R \to R$ is a continuous function such that $f(x, 0) \equiv 0$ and there exist the limits

$$\lim_{s \to \pm\infty} \frac{f(x,s)}{s} = f_\pm, \qquad \text{uniformly for } x \in \bar{\Omega}, \text{ with} \tag{2}$$

$$f_- < \lambda_1 < f_+, \qquad f_+ \neq \lambda_j. \tag{3}$$

We shall show that for $t < 0$ sufficiently large problem (1_t) possesses a large positive solution and a large negative solution. To that end consider the related problems

$$-\Delta v = f_\pm v + \xi(x, v + \alpha\phi_1) + h \quad \text{in } \Omega, \qquad v = 0 \quad \text{on } \partial\Omega \ (\alpha \in R), \tag{$1_\alpha^\pm$}$$

where $\xi(x, s) = f(x, s) - (f_+ s^+ - f_- s^-)$, so that

$$\lim_{|s| \to \infty} \frac{\xi(x,s)}{s} = 0 \qquad \text{uniformly for } x \in \bar{\Omega}. \tag{4}$$

Denote by $S^\pm(\alpha)$ the set of solutions of $(1_\alpha^\pm)$ and let $M^\pm(\alpha) = \sup\{\|v\|_{C^1}\colon v \in S^\pm(\alpha)\}$.

LEMMA 1. $\lim_{|\alpha| \to \infty} M^\pm(\alpha)/\alpha = 0$.

Proof. We shall consider only the case of $M^+(\alpha)$, the other case being entirely similar. Any solution $v \in S^+(\alpha)$ is a solution of the equation $v = K_+[\xi(\cdot, v + \alpha\phi_1) + h]$, where $K_+\colon L^2(\Omega) \to L^2(\Omega)$ is the resolvent operator $K_+ = (-\Delta - f_+)^{-1}$ (which exists in view of (3)). As it is well known, $K_+\colon C^0(\bar{\Omega}) \to C^1(\bar{\Omega})$ is a bounded operator and, therefore, any $v \in S^+(\alpha)$ satisfies

$$\|v\|_{C^1} \leqslant c_+ \|\xi(\cdot, v + \alpha\phi_1) + h\|_{C^0},$$

or yet, since one has by (4) that $|\xi(x, s)| \leqslant \varepsilon|s| + c_\varepsilon$ for all $x \in \bar{\Omega}$, $s \in R$, where $\varepsilon > 0$ can be chosen arbitrarily (and c_ε depends only on ε),

$$\|v\|_{C^1} \leqslant c_+[\varepsilon \|v\|_{C^0} + \varepsilon |\alpha| \|\phi_1\|_{C^0} + c_\varepsilon + \|h\|_{C^0}].$$

Choosing $\varepsilon > 0$ so that $1 - c_+\varepsilon > \frac{1}{2}$ one obtains

$$\|v\|_{C^1} \leqslant 2c_+\varepsilon |\alpha| \|\phi_1\|_{C^0} + 2c_+(c_\varepsilon + \|h\|_{C^0}),$$

from which the claim follows readily. ∎

THEOREM 1. *Suppose that* (2) *and* (3) *hold. Then there exists* $t_0 < 0$ *such that for all* $t < t_0$ *problem* (1_t) *has a positive solution* U_t *and a negative solution* u_t *satisfying*

$$\lim_{t \to -\infty} \left\| \frac{U_t}{t} - \frac{\phi_1}{\lambda_1 - f_+} \right\|_{C^1} = \lim_{t \to -\infty} \left\| \frac{u_t}{t} - \frac{\phi_1}{\lambda_1 - f_-} \right\|_{C^1} = 0. \tag{5}$$

Proof. Let $\eta > 0$ be such that $w + \phi_1 > 0$ for all w satisfying $\|w\|_{C^1} \leqslant \eta$, $w|\partial\Omega = 0$. By Lemma 1 there exists $\alpha_0 > 0$ such that $M^+(\alpha) \leqslant \eta\alpha$ for all $\alpha \geqslant \alpha_0$, that is,

$$\|v\|_{C^1} \leqslant \eta\alpha \tag{6}$$

for any solution v of (1_α^+), $\alpha \geqslant \alpha_0$. For any such v, in view of the choice of η and the fact that $\|v/\alpha\|_{C^1} \leqslant \eta$, we have that

$$U \equiv v + \alpha\phi_1 = \alpha\left(\frac{v}{\alpha} + \phi_1\right) > 0. \tag{7}$$

So U is a positive solution of the problem

$$-\Delta u = f_+(u - \alpha\phi_1) + \xi(x, u) + h + \alpha\lambda_1\phi_1 \quad \text{in } \Omega, \qquad u = 0 \text{ on } \partial\Omega,$$

for $\alpha \geqslant \alpha_0$, that is, U is a positive solution of (1_t) for $t = \alpha(\lambda_1 - f_+) \leqslant \alpha_0(\lambda_1 - f_+) \equiv t_0^+$. And, in view of (6) and (7), U/t satisfies the inequality

$$\left\| \frac{U}{t} - \frac{\phi_1}{\lambda_1 - f_+} \right\|_{C^1} = \left\| \frac{v}{t} \right\|_{C^1} \leqslant \frac{\eta}{f_+ - \lambda_1}.$$

Similarly, we prove the existence of a negative solution u_t of (1_t) for $t \leqslant -\alpha_0(\lambda_1 - f_-) \equiv t_0^-$ satisfying

$$\left\| \frac{u_t}{t} - \frac{\phi_1}{\lambda_1 - f_-} \right\|_{C^1} \leqslant \frac{\eta}{\lambda_1 - f_-}.$$

The proof is complete by letting $t_0 = \min\{t_0^+, t_0^-\}$ and noting that the chosen $\eta > 0$ can be taken arbitrarily small. ▮

2. THE CASE WHEN Ω IS A BALL

We now let Ω be the unit ball $B = B_1(0) \subset R^N$ and consider the Ambrosetti–Prodi problem

$$-\Delta u = f(u) + h + t\phi_1 \quad \text{in } B, \qquad u = 0 \quad \text{on } \partial B, \tag{8_t}$$

where $h \in C(\bar{B})$ is a given radial function with $\int h\phi_1 = 0$ and $f\colon R \to R$ is a C^1 function satisfying $f(0) = 0$,

$$\lim_{s \to -\infty} f'(s) \equiv f_- < \mu_1, \qquad \mu_n < \lim_{s \to +\infty} f'(s) \equiv f_+ < \mu_{n+1}, n \geqslant 1. \tag{9}$$

Here we are denoting by $\mu_1 = \lambda_1 < \mu_2 < \cdots$ the eigenvalues of $-\Delta$ acting on the radial functions of $H_0^1(B)$, that is, the eigenvalues of the problem

$$-u'' - \frac{N-1}{r} u' \equiv \mu u, \qquad 0 < r < 1,\ u'(0) = u(1) = 0. \tag{10}$$

As it is well known, these eigenvalues are given by $\mu_j = \nu_j^2$, where the ν_j's are the positive zeros of the Bessel function $J_{(N-2)/2}$, and corresponding eigenfunctions are given by $\phi_j(r) = r^{-(N-2)/2} J_{(N-2)/2}(\nu_j r)$.

THEOREM 2. *Under assumption* (9), *there exists* $t_1 < 0$ *such that for all* $t \leqslant t_1$ *problem* (8_t) *has a radial positive solution* U_t *and a radial negative solution* u_t *satisfying* (5).

Proof. In view of Theorem 1 there exists $t_0 < 0$ such that for all $t \leqslant t_0$ problem (8_t) has a positive solution U_t and a negative solution u_t satisfying (5). It remains to show that both U_t and u_t are radial functions for $t \leqslant t_0$ sufficiently large. Indeed, letting $z_t = \partial U_t/\partial\omega$ denote any angular derivative of U_t, z_t satisfies $-\Delta z_t = f'(U_t) z_t$ in B, $z_t = 0$ on ∂B. Therefore if we assume that $z_t \not\equiv 0$, z_t is an eigenfunction of the above problem corresponding to the eigenvalue $1 = \lambda_j\,(f'(U_t))$. But then the fact that $U_t(x) \to +\infty$ for all $x \in B$ and the Lebesgue dominated convergence theorem imply that $f'(U_t) \to f_+$ in L^p, $p \geqslant 1$, and hence that $1 = \lambda_j(f'(U_t)) \to \lambda_j(f_+) = \lambda_j/f_+$ as $t \to -\infty$, which is a contradiction. Similarly we show that u_t is a radial function for $t \leqslant t_0$ sufficiently large. ∎

Now we make the change of (dependent) variables $u = w + U_t$ to rewrite problem (8_t) as

$$-\Delta w = f(w + U_t) - f(U_t) \qquad \text{in} \quad B, \qquad w = 0 \quad \text{on } \partial B,$$

or yet as

$$-\Delta w = f_+ w + g(w + U_t) - g(U_t) \qquad \text{in} \quad B, \qquad w = 0 \quad \text{on } \partial B, \tag{11_t}$$

where $g(s) \equiv f(s) - f_+ s$. Notice that, in view of (9), $g\colon R \to R$ is a C^1 function satisfying

$$\lim_{s \to -\infty} g'(s) = f_- - f_+, \qquad \lim_{s \to +\infty} g'(s) = 0. \tag{12}$$

On the other hand, given $0 \neq \bar{c} \in R$, let us define the function

$$\bar{v}(r) = \frac{\bar{c}}{r^{(N-2)/2}} \frac{J_{(N-2)/2}(\sqrt{f_+}\, r)}{J_{(N-2)/2}(\sqrt{f_+})}, \qquad r = |x|.$$

Then it is easy to see that $\bar{v}$ is a solution of the problem

$$-\Delta \bar{v} = f_+ \bar{v} \quad \text{in } B, \qquad \bar{v} = \bar{c} \quad \text{on } \partial B, \tag{13}$$

and, since $\nu_n < \sqrt{f_+} < \nu_{n+1}$ by (9), $\bar{v}$ vanishes on the n concentric spheres $r = r_j = \nu_j / \sqrt{f_+}$ $(j = 1, ..., n)$ in B.

THEOREM 3. *Suppose* (9) *holds and let* $0 \neq \bar{c} \in R$ *be given. Then there exists* $\bar{t} = \bar{t}(\bar{c}) < 0$ *such that for all* $t \leqslant \bar{t}$ *problem*

$$-\Delta u = f(u) + h + t\phi_1 \quad \text{in} \quad B, \qquad u = \bar{c} \quad \text{on } \partial B, \tag{$14_{t,\bar{c}}$}$$

has a radial solution $\bar{u}$ *with the property that* $\bar{u} - U_t$ *vanishes on* n *concentric spheres in* B.

Proof. We want to find a radial solution w of (11_t), with the boundary condition replaced by $w = \bar{c}$ on ∂B, and such that w vanishes on n concentric spheres in B. Since the function $\bar{v}$ has this latter property and its n zeros in $(0, 1)$ (as a function of r) are all simple, it suffices to find a solution z of the problem

$$-\Delta z = f_+ z + g(z + \bar{v} + U_t) - g(U_t) \quad \text{in } B, \qquad z = 0 \quad \text{on } \partial B, \tag{15_t}$$

with $\|z\|_{C^0} \leqslant \bar{\varepsilon}$, $\bar{\varepsilon} = \bar{\varepsilon}(\bar{c})$ sufficiently small, for then $\bar{u} = z + \bar{v} + U_t$ is a solution of $(14_{t,\bar{c}})$ with the desired property that $\bar{u} - U_t = z + \bar{v}$ vanishes on n concentric spheres in B. (We point out that the same argument used in the proof of Theorem 2 shows that the solutions of (15_t) are radial for $t < 0$ sufficiently large). So Lemma 2 below concludes the proof.

LEMMA 2. *Given* $\bar{\varepsilon} > 0$ *there exists* $\bar{t} = \bar{t}(\bar{\varepsilon})$ *such that for all* $t \leqslant \bar{t}$ *problem* (15_t) *has a unique solution* z *satisfying* $\|z\|_{C^0} \leqslant \bar{\varepsilon}$.

Proof. We observe that solving (15_t) is equivalent to solving the operator equation

$$z = KG_t(z) = T_t(z),$$

where $K: L^2(B) \to H_0^1(B)$ is the resolvent operator $K = (-\Delta - f_+)^{-1}$ and G_t is the Nemytskii operator associated with the function $g(z + \bar{v}(r) + U_t(r)) - g(U_t(r))$. We also observe that the regularity theory for elliptic equations implies that K maps $L^p(B)$ continuously into $W^{2,p}(B)$ and (12) implies that

G_t maps $L^p(B)$ continuously into itself, so that T_t maps $L^p(B)$ into $W^{2,p}(B)$. Given the ball $B_{\bar{\varepsilon}}(0)$ in $C^0(\bar{B})$, we shall show that for $t<0$ sufficiently large we have $T_t(B_{\bar{\varepsilon}}(0)) \subset B_{\bar{\varepsilon}}(0)$ and T_t is a contraction. Indeed, fixing $p>N/2$ and using the Sobolev imbedding theorem, we have

$$\|T_t(z)\|_{C^0} \leqslant c\|T_t(z)\|_{W^{2,p}} = c\|KG_t(z)\|_{W^{2,p}} \leqslant \text{const. } \|G_t(z)\|_{L^p},$$

and $\|G_t(z)\|_{L^p} \to 0$ as $t \to -\infty$ uniformly for $z \in B_{\bar{\varepsilon}}(0)$, since we can estimate $\|G_t(z)\|_{L^p} \leqslant \delta(t)\ \|z+\bar{v}\|_{C^0}$, where $\delta(t) \to 0$ as $t \to -\infty$. (In computing $\|G_t(z)\|_{L^p}$ apply the mean value theorem for the function g and use the Lebesgue dominated convergence theorem, keeping in mind that $z+\bar{v}+U_t \to +\infty$ pointwisely in B and that $g'(s) \to 0$ as $s \to +\infty$.) Therefore, $T_t(B_{\bar{\varepsilon}}(0)) \subset B_{\bar{\varepsilon}}(0)$ for $t<0$ sufficiently large.

Similarly, for arbitrary $z_1, z_2 \in B_{\bar{\varepsilon}}(0)$, we have the estimate $\|T_t(z_1) - T_t(z_1)\|_{C^0} \leqslant \text{const.} \|G_t(z_1) - G_t(z_2)\|_{L^p} \leqslant \text{const.} \delta(t) \|z_1 - z_2\|_{C^0}$, where $\delta(t) \to 0$ as $t \to -\infty$, so that $T_t \colon B_{\bar{\varepsilon}}(0) \to B_{\bar{\varepsilon}}(0)$ is a contraction for $t<0$ sufficiently large. The proof is complete. ∎

Remark. It should be pointed out that we stated and proved Theorem 3 as a p.d.e. result, ignoring for the time being its natural 1-dimensional character (i.e., search of radial solutions in a ball). The reason for that is the fact that indeed, for an arbitrary bounded smooth domain Ω, Theorem 3 has the following analogue which we now describe.

For given $0 \neq c \in R$, let v_c denote the solution of the problem $-\Delta v = f_+ v$ in Ω, $v=c$ on $\partial\Omega$ (i.e., $v_c = -\Delta(-\Delta - f_+)^{-1}c$). Suppose that for some $\bar{c} \neq 0$, $\bar{v} = v_{\bar{c}}$ satisfies the condition that $\nabla\bar{v}(x) \neq 0$ whenever $\bar{v}(x)=0$, that is, the graph of $\bar{v}$ is transversal to $\Omega \times \{0\}$ at the points of $M \times \{0\}$, where $M = \{x \in \Omega \mid \bar{v}(x) = 0\}$ is the "null manifold" of $\bar{v}$. Let m be the number of components of M. Then, for $t<0$ sufficiently large (and under hypotheses (9)), there exists a solution u of the problem

$$-\Delta u = f(u) + h + t\phi_1 \quad \text{in } \Omega, \qquad u=c \quad \text{on } \partial\Omega,$$

such that $u-U_t$ has at least m "nodal lines", that is, the null manifold of $u-U_t$ has at least m components.

3. Existence of Many Solutions

In this section we consider again the case when Ω is the unit ball $B = B_1(0) \subset R^N$ and prove under the same assumption (9) of the previous section that the Ambrosetti–Prodi problem (8_t) has at least $n+1$ radial solutions for $t<0$ sufficiently large. For that matter we shall use a shooting argument as in [5].

It should be remarked that only from now on is that the one-dimensional character of problem (8_t) will play an important role in proving the existence of other solutions besides $U_t > 0$ and $u_t < 0$, when $n \geqslant 2$ in (9).

So, we start by rewriting (8_t) as

$$-u'' - \frac{N-1}{r} u' = f(u) + h + t\phi, \quad 0 < r < 1, \quad u'(0) = u(1) = 0, \qquad (16_t)$$

which in turn, again through the change of dependent variables $u(r) = w(r) + U_t(r)$, can be rewritten as

$$-w'' - \frac{N-1}{r} w' = f_+ w + g(w + U_t) - g(U_t), \qquad 0 < r < 1, \qquad (17a_t)$$

$$w'(0) = w(1) = 0, \qquad (17b)$$

where we recall that $g(s) \equiv f(s) - f_+ s$.

LEMMA 3. *The initial value problem*

$$-w'' - \frac{N-1}{r} w' = f_+ w + g(w + U_t(r)) - g(U_t(r)), \; 0 < r < 1,$$

$$w(0) = a, \; w'(0) = 0,$$

has a unique solution $w(r) = w(r;\, a)$ *which is defined for* $0 \leqslant r \leqslant 1$ *and depends continuously on* $a \in R$.

Proof. Since $-w'' - ((N-1)/r)w' = -(1/r^{N-1})\,(r^{N-1}w')'$, it can be seen that the initial value problem in question is equivalent to the following Volterra integral equation

$$w(r) = a + \int_0^r K(r, \sigma)\, F(\sigma, w(\sigma))\, d\sigma,$$

where $F(\sigma, s) \equiv f_+ s + g(s + U_t(\sigma)) - g(U_t(\sigma))$ and $K(r, \sigma) = (\sigma/(N-2))\,[(\sigma/r)^{N-2} - 1]$, when $N \geqslant 3$, or $K(r, \sigma) = \sigma \ln(\sigma/r)$ when $N = 2$. Since the kernel $K(r, \sigma)$ is nice, the result follows from the standard theory of Volterra equations. ∎

THEOREM 4. *Under assumption* (9), *there exists* $t_2 < 0$ *such that for all* $t \leqslant t_2$ *problem* (17_t) *has at least n distinct nontrivial solutions* $w_0, w_1, ..., w_{n-1}$ *with the property that* $w_j(r)$ *has exactly j simple zeros in the open interval* $(0, 1)$ *and* $w_j(0) < 0$, $j = 0, ..., n-1$. (*Therefore, for* $t \leqslant t_2$ *problem* (8_t) *has at least the* $n + 1$ *distinct radial solutions* U_t, $w_j + U_t$, $j = 0, ..., n-1$.)

Proof. From Theorem 2 we already know 2 solutions for problem (17_t), namely $w \equiv 0$ (corresponding to the positive solution U_t of (8_t)) and $w_0 = u_t - U_t < 0$ (corresponding to the negative u_t of (8_t), provided $t \leqslant t_1$. On the other hand, from Theorem 3 we have a solution $\bar{w} = \bar{u} - U_t$ of equation $(17a_t)$, for $t \leqslant \bar{t}$, such that $\bar{w}(r)$ has n simple zeros in the open interval $(0, 1)$ and $\bar{w}(0) < 0$ (take $\bar{c}$ in $(14_{t,\bar{c}})$ so that the function $\bar{v}$ in (13) satisfies $\bar{v}(0) < 0$). Therefore, in view of Lemma 3, we can use the idea of Lemma 2.3 in [4] to show that, for each $0 < j < n$, problem (17_t) has a solution $w_j(r)$ with exactly j simple zeros in $(0, 1)$ and such that $w_j(0) < 0$.

4. The Case $N = 3$

In the case that Ω is the unit ball $B = B_1(0)$ in R^3 we are able to improve Theorem 4 and obtain $2n$ radial solutions for the Ambrosetti–Prodi problem (8_t), $t < 0$ sufficiently large. Namely, we have

Theorem 5. *Let $N = 3$ and suppose* (9) *holds. Then, for all $t < 0$ sufficiently large, problem (17_t) has at least $2n - 1$ distinct nontrivial solutions w_j, $j = 0, ..., n-1$, and $\tilde{w}_j$, $j = 1, ..., n-1$, with the property that w_j and $\tilde{w}_j$ have exactly j simple zeros in $(0, 1)$ and $w_j(0) < 0$, $\tilde{w}_j(0) > 0$. (Therefore, for $t < 0$ sufficiently large, problem (8_t) has at least the following $2n$ distinct radial solutions: U_t, $w_j + U_t$, $j = 0, ..., n-1$, $\tilde{w}_j + U_t$, $j = 1, ..., n-1$.)*

Proof. By making the change of variables $v(r) = rw(r)$ and letting $V_t(r) \equiv rU_t(r)$, we transform (17_t) into the problem

$$-v'' = f_+ v + \tilde{g}(r, v + V_t) - \tilde{g}(V_t), \qquad 0 < r < 1,\ v(0) = v(1) = 0, \quad (18_t)$$

where $\tilde{g}(r, v) \equiv rg(v/r)$. Note that (9) (hence (12)) implies that $\tilde{g}$ is continuous on $[0, 1] \times R$ (by defining $\tilde{g}(0, v) = (f_+ - f_-)v^-$), $\tilde{g}$ is of class C^1 on $(0, 1) \times R$, $\tilde{g}_v$ is continuous on $[0, 1] \times (R \backslash 0)$ and satisfies $\lim_{v \to -\infty} \tilde{g}_v(r, v) = f_- - f_+$, $\lim_{v \to +\infty} \tilde{g}_v(r, v) = 0$ uniformly for $r \in [0, 1]$. Therefore, problem (18_t) can be treated in a similar manner as the problem in [4]. It then follows that (18_t) has solutions v_j, $j = 0, ..., n-1$, and $\tilde{v}_j$, $j = 1, ..., n-1$, such that v_j and $\tilde{v}_j$ have exactly j simple zeros in $(0, 1)$ and $v_j'(0) < 0$, $\tilde{v}_j'(0) > 0$. Going back to our original (dependent) variable we first claim that if $v(r)$ is a solution of (18_t), then $w(r) = v(r)/r$ is a solution of (17_t). Indeed, the only nonobvious property to check is that $w(r)$ satisfies the boundary condition $w'(0) = 0$. For that we observe that any solution $v(r)$ of (18_t) satisfies $v''(0) = 0$, so that using l'Hospital's rule we obtain

$$\lim_{r \downarrow 0} w'(r) = \lim_{r \downarrow 0} \frac{rv'(r) - v(r)}{r^2} = \lim_{r \downarrow 0} \frac{rv''(r)}{2r} = \frac{v''(0)}{2} = 0.$$

Consequently $w_j(r) = v_j(r)/r$ and $\tilde{w}_j(r) = \tilde{v}_j(r)/r$ are solutions of (17_t) with the stated properties. ∎

Remark. We observe that in the case Ω is the region between two concentric spheres in R^N, say $\varepsilon < r < 1$, we again obtain $2n$ radial solutions for the corresponding Ambrosetti–Prodi problem. Indeed, we are led to the o.d.e. problem $(17a_t)$, $w(\varepsilon) = w(1) = 0$, or yet, making the change of variable $v(r) = r^{(N-1)/2} w(r)$, to the self-adjoint problem

$$-v'' + \frac{C_N}{r^2} v = f_+ v + \tilde{g}(r, v + V_t) - \tilde{g}(r, V_t), \qquad \varepsilon < r < 1, \; v(\varepsilon) = v(1) = 0,$$

where $C_N = (N-1)(N-3)/4$, and $V_t(r)$ and $\tilde{g}(r, v)$ are as before.

Acknowledgment

The present research was done when both authors were visiting the Mathematics Research Center of the University of Wisconsin, whose hospitality they gratefully acknowledge.

References

1. A. Ambrosetti, Elliptic equations with jumping nonlinearities, to appear.
2. D. G. de Figueiredo, "On the Superlinear Ambrosetti–Prodi Problem," MRC Tech. Rep., No. 2522, May 1983.
3. A. C. Lazer and P. J. McKenna, On the number of solutions of a nonlinear Dirichlet problem, *J. Math. Anal. Appl.* **84** (1981), 282–294.
4. A. C. Lazer and P. J. McKenna, On a conjecture related to the number of solutions of a nonlinear Dirichlet problem, to appear.
5. S. Solimini, Existence of a third solution for a class of b.v.p. with jumping nonlinearities, *Nonlinear Anal. TMA* **7** (8) (1983), 917–927.

Printed by the St. Catherine Press Ltd., Tempelhof 41, Bruges, Belgium

On Pairs of Positive Solutions for a Class of Semilinear Elliptic Problems

DJAIRO G. DE FIGUEIREDO
& PIERRE-LOUIS LIONS

Introduction. The question of existence of positive solutions for semilinear elliptic problems of the type

$$-\Delta u = f(u) \quad \text{in } \Omega, \qquad u = 0 \quad \text{on } \partial\Omega, \tag{1}$$

depends very strongly on the behavior of the function $f: R^+ \to R$ at 0 and at $+\infty$. Here Ω denotes a smooth bounded domain in R^N, $N \geq 2$, and f is always assumed to be locally Lipschitzian. We distinguish two special classes of such problems, which have been extensively studied in recent years: the sublinear problems and the superlinear ones. See, for instance, the review papers [1], [2], [3]. The sublinear problems are characterized by the inequalities

$$\liminf_{s\to 0+} \frac{f(s)}{s} > \lambda_1 \quad \text{and} \quad \limsup_{s\to +\infty} \frac{f(s)}{s} < \lambda_1 . \tag{2}$$

Here $\lambda_1 = \lambda_1(\Omega)$ denotes the first eigenvalue of $-\Delta$ in $H_0^1(\Omega)$. The superlinear problems are characterized by the inequalities

$$\limsup_{s\to 0+} \frac{f(s)}{s} < \lambda_1 \quad \text{and} \quad \liminf_{s\to +\infty} \frac{f(s)}{s} > \lambda_1 . \tag{3}$$

In the present paper we propose to discuss problems which are sublinear at 0 and superlinear at $+\infty$. Namely,

$$\liminf_{s\to 0+} \frac{f(s)}{s} > \lambda_1 \quad \text{and} \quad \liminf_{s\to +\infty} \frac{f(s)}{s} > \lambda_1 . \tag{4}$$

We allow these limits to be $+\infty$. For instance, this is the case for the first limit in (4) if $f(0) > 0$. And the second limit in (4) is $+\infty$ if f grows more rapidly than linearly. We remark that these types of problems have been studied in the case of $f(0) > 0$ and f convex by several authors, for instance, Gelfand [4], Joseph-Lundgren [5], Crandall-Rabinowitz [6], Bandle [7], Mignot-Puel [8]. Also some problems of this type are discussed in Lions [2]. In this paper we treat a much larger class of nonlinearities. The dominating theme throughout the paper is that

Indiana University Mathematics Journal ©, Vol. 34, No. 3 (1985)

Originally published in: Indiana Univ. Math. J. 34 (1985), 591-606,

DOI: 10.1007/978-3-319-02856-9_15

of a supersolution of (1). The main highlights being: (i) the question of existence of supersolutions, (ii) their actual construction, (iii) how they can be obtained from solutions of the symmetrized problem by the use of Schwarz symmetrization techniques, (iv) existence theorems for problem (1) that possesses a supersolution.

In Section 1 we present results on the existence of two positive solutions for problem (1), one of the basic assumptions being the existence of a *strict* supersolution. Here topological degree techniques and variational methods are used to treat different classes of nonlinearities. Condition (4) puts no restriction on the growth of f at $+\infty$. To obtain the first solution of (1) is no serious problem. Indeed, if we know that a supersolution w of (1) exists, it follows from (4) that a subsolution $\underline{u}$ can be found, with $\underline{u} \leq w$. And then the monotone iteration method applies to yield a solution of (1), which is in fact the minimal solution. However, if we are interested in deciding whether or not there are more solutions, we need some sort of compactness in our problem. More precisely, if we plan on treating the problem by topological degree arguments, a priori bounds on the solutions are in order. If we are to use variational methods, then a Palais-Smale condition is required on the Euler-Lagrange functional associated with (1). In both cases these properties will be consequences of the conditions assumed on f as far as its growth at $+\infty$ is concerned. We discuss these points in Section 1. More general results are obtained in Section 4.

The existence of a supersolution of (1) is connected with how far below the line $\lambda_1 s$ the graph of $f(s)$ goes. We propose to discuss this matter in detail in Section 2, showing how to actually construct a supersolution in some examples. In Section 3 we use the idea of Schwarz symmetrization to obtain statements about the existence of a supersolution of (1) from the knowledge of a solution of (1) in the case of a ball.

In Section 4 we invoke some results of [9] and prove a fairly general theorem on the existence of pairs of positive solutions. In this way we are able to treat a class of nonlinearities f more general than those of Section 1. Under these less restrictive assumptions on f it is not clear that one has a priori bounds on the solutions of (1). Also the corresponding Euler-Lagrange functional apparently does not satisfy the Palais-Smale condition. However we are able to proceed via an appropriate truncation of the nonlinearity f.

1. Existence theorems. In this section we establish results on the existence of positive solutions for the Dirichlet problem

$$-\Delta u = f(u) \quad \text{in } \Omega, \qquad u = 0 \quad \text{on } \partial\Omega, \tag{1}$$

where Ω is a smooth bounded domain in R^N, $N \geq 2$, and the nonlinearity f satisfies different sets of conditions, as specified in the sequel. The following conditions will be always assumed:

$$f : R^+ \to R \text{ is locally Lipschitzian} \tag{f1}$$

$$\liminf_{s \to 0+} \frac{f(s)}{s} > \lambda_1 \tag{f2}$$

(f3) $$\liminf_{s\to+\infty} \frac{f(s)}{s} > \lambda_1$$

(f4) $$\lim_{s\to+\infty} \frac{f(s)}{s^\sigma} = 0, \text{ with } 1 < \sigma \le (N+2)/(N-2), \text{ if } N \ge 3 \text{ or}$$

$$1 < \sigma < \infty, \text{ if } N = 2.$$

We remark however that condition (f4) is not needed when we want just to state the existence of *one* solution. Its importance appears when we have to prove the existence of *two* solutions, since in this case a priori bounds on the solutions or a Palais-Smale condition is needed. Besides the above stated conditions on f, the following ones will be useful in treating the problem by topological degree arguments. The following conditions have to be assumed *only* in the case $N \ge 3$:

(f5) $$\limsup_{s\to+\infty} \frac{sf(s) - \theta F(s)}{s^2 f(s)^{2/N}} \le 0, \text{ with } \theta \in [0, 2N/(N-2)),\ F(s) = \int_0^s f$$

(f6) Ω is convex or $f(s)s^{-p}$ is nonincreasing for $s > 0$,

with $p = (N+2)/(N-2)$.

Assumptions (f4), (f5) and (f6) are used in Theorem 1 below only to assert that there exist a priori bounds on the positive solutions of (1), or more generally, a priori bounds on positive solutions of a certain parametrized family of Dirichlet problems. And for that matter we rely on the results of de Figueiredo-Lions-Nussbaum [9]. So Theorem 1 is true if (f4), (f5) and (f6) are replaced by another set of conditions that insures the existence of such bounds. We remark however that these assumptions are not too restrictive. For instance, (f5) is satisfied if σ in (f4) is $\le N/(N-2)$ or if $f(s) = cs^p$, with $c > 0$ and $1 < p < (N+2)/(N-2)$. It has been established in [9] that more general conditions on Ω and f also yield a priori bounds on the positive solutions of (1). Therefore we see that large classes of nonlinearities may be allowed. For the question of a priori bounds on positive solutions of (1) we refer also to the work of Brézis-Turner [10] and Gidas-Spruck [11].

A function $w \in C^{2,\alpha}(\bar{\Omega})$ is said to be a *supersolution* of (1) if $-\Delta w \ge f(w)$ in Ω and $w \ge 0$ on $\partial\Omega$. A supersolution which is not a solution is said to be strict. Now we state our first existence result.

Theorem 1. *In addition to (f1)–(f6), assume that (1) possesses a strict supersolution w. Then (1) has two solutions $0 < u_1 < u_2$, and u_1 is the minimal solution.*

Remarks. 1) Conditions (f1)–(f6) are not sufficient to insure the existence of a positive solution of (1). Indeed, take any f such that $f(s) \ge \alpha s$, for $s \ge 0$ and $\alpha > \lambda_1$. Clearly 0 is the only possible solution of (1). Hence we see that the graph of $f(s)$ has to cross the straight line $\lambda_1 s$. And the question is to determine "how much" it has to be below $\lambda_1 s$, so that a positive solution of (1) exists. It is well known that if $f(s) = 0$ for some $s > 0$ then a positive solution of (1) exists under hypotheses (f1) and (f2). We shall see that the graph of f does not have to go below that far. So the interesting case is $f(s) > 0$, for all $s > 0$.

2) The limits in (f2) and (f3) can be $+\infty$. This will be the case, for instance, if $f(0) > 0$ and $f(s)$ behaves like cs^p for large s, with $c > 0$ and $p > 1$.

3) The requirement that the supersolution be strict is essential, as we show next. Let $g : R^+ \to R^+$ be a continuous strictly convex function, with $g(0) > 0$, and consider the eigenvalue problem:

$$(2)_\lambda \qquad -\Delta u = \lambda g(u) \quad \text{in } \Omega, \qquad u = 0 \quad \text{on } \partial\Omega.$$

It is known (see for instance, Crandall-Rabinowitz [6]) that there exists $0 < \lambda^* < \infty$ such that $(2)_\lambda$ has a minimal solution u_λ for $0 < \lambda < \lambda^*$, and no solution for $\lambda > \lambda^*$. We recall that this branch of minimal solutions is obtained by a continuation argument using the implicit function theorem. Moreover if g is such that one can prove an a priori bound on this branch of minimal solutions for $0 \le \lambda < \lambda^*$, then such a branch actually reaches $\lambda = \lambda^*$. That is, $(2)_\lambda$ has also a minimal solution u_{λ^*} for $\lambda = \lambda^*$. One can in fact prove that such a u_{λ^*} is the unique solution of $(2)_{\lambda^*}$. Now we claim that u_{λ^*} is the only supersolution of $(2)_{\lambda^*}$. Indeed, suppose that v is a strict supersolution of $(2)_{\lambda^*}$. Then $v \ge u_{\lambda^*}$ and we have

$$-\Delta(v - u_{\lambda^*}) \ge \lambda^*(g(v) - g(u_{\lambda^*})) \ge \lambda^* g'(u_{\lambda^*})(v - u_{\lambda^*})$$

where one of the inequalities above is strict in a set of positive measure, as a consequence of $v \not\equiv u_{\lambda^*}$ and the strict convexity of g. Consequently, the first eigenvalue of the eigenvalue problem

$$[-\Delta - \lambda^* g'(u_{\lambda^*})]u = \mu u \quad \text{in } \Omega, \qquad u = 0 \quad \text{on } \partial\Omega$$

is positive. However, this implies, via the continuation argument, that there exists $\lambda > \lambda^*$ such that $(2)_\lambda$ has a solution. This contradicts the maximality of λ^*.

Proof of Theorem 1. Let us denote by X the Banach space of C^1-functions in $\bar{\Omega}$ which are 0 on $\partial\Omega$, endowed with the usual C^1-norm. Define f for negative s as $f(s) = f(0)$. The solutions of (1) with this extended f are the same as the solutions of the original problem (1). For in either case the solutions are all positive. Let us define the one-parameter family of functions

$$(3) \qquad f_\mu(s) = \mu f(s) + (1 - \mu)(\lambda s^+ + 1), \qquad 0 \le \mu \le 1$$

where λ is some fixed number $> \lambda_1$ and s^+ denotes the function which is s for $s \ge 0$ and 0 for $s < 0$. In view of assumptions (f1), (f3), (f4), (f5), (f6), there is a constant $c_1 > 0$ such that

$$(4) \qquad \|u\|_X \le c_1$$

for all eventual solutions of the problems

$(5)_\mu$ $$-\Delta u = f_\mu(u) \quad \text{in } \Omega, \qquad u = 0 \quad \text{on } \partial\Omega, \quad 0 \le \mu \le 1.$$

Now let $k > 0$ be such that $f(s) + ks$ is increasing for $s \in [0,\|w\|_{L^\infty}]$. Let $K_\mu = (-\Delta + k)^{-1}(f_\mu(\cdot) + k\cdot)$. More precisely, let us define $K_\mu : X \to X$ as follows $K_\mu v = u$ where u is the solution of the Dirichlet problem

(6) $$-\Delta u + ku = f_\mu(v) + kv \quad \text{in } \Omega, \qquad u = 0 \quad \text{on } \partial\Omega.$$

The mapping K_μ so defined is compact. From the Schauder estimates it follows that there is a constant $c_2 > 0$ such that

(7) $$\|K_\mu v\|_X \le c_2, \qquad \forall\, v \in X, 0 \le v \le w.$$

As a consequence of (f2) we see that there is $\varepsilon > 0$ such that $\varepsilon\phi_1$ is a subsolution for *all* problems $(5)_\mu$. Here ϕ_1 denotes a positive eigenfunction corresponding to $\lambda_1(\Omega)$. Also we may take $\varepsilon\phi_1 < w$ in Ω. It follows from the maximum principle that any solution u of $(5)_\mu$ such that $u \ge \varepsilon\phi_1$ in Ω satisfies the strict inequalities $u > \varepsilon\phi_1$ in Ω and $\partial u/\partial\nu < \varepsilon(\partial\phi_1/\partial\nu)$ on $\partial\Omega$. Now consider the bounded open set

$$\mathcal{O} = \left\{ u \in X : \|u\|_X < c_1 + c_2 + 1; u > \varepsilon\phi_1 \text{ in } \Omega; \frac{\partial u}{\partial\nu} < \varepsilon\frac{\partial\phi_1}{\partial\nu} \text{ on } \partial\Omega \right\},$$

where c_1 and c_2 are the constants defined in (4) and (7), respectively, and $\varepsilon\phi_1$ is the subsolution defined above. By the foregoing remarks it follows that $0 \notin (I - K_\mu)(\partial\mathcal{O})$. So the degree $d(I - K_\mu,\mathcal{O},0)$ is independent of $\mu \in [0,1]$. Clearly the degree $d(I - K_0,\mathcal{O},0) = 0$ since $(5)_0$ has no solution. Hence

$$d(I - K_1,\mathcal{O},0) = 0.$$

Now let us consider the following open subset of $\mathcal{O}$:

$$\mathcal{O}' = \left\{ u \in \mathcal{O} : u < w \text{ in } \Omega; \frac{\partial n}{\partial\nu} > \frac{\partial w}{\partial\nu} \text{ on } \partial\Omega \right\},$$

and we *claim* that $d(I - K_1,\mathcal{O}',0) = 1$. Once this is proved, it follows that $d(I - K_1,\mathcal{O} \setminus \bar{\mathcal{O}}',0) = -1$. So $(5)_1$, which is the same as (1), has two solutions $\tilde{u}_1 \in \mathcal{O}'$ and $u_2 \in \mathcal{O} \setminus \bar{\mathcal{O}}'$, which are not necessarily ordered. So we have to proceed further in order to complete the proof of the theorem. Let $v = \min(w,u_2)$. It follows that v is in $W_0^{1,\infty}(\Omega)$, $\varepsilon\phi_1 \le v$ and

$$-\Delta v \ge f(v), \qquad \text{in } \mathcal{D}'(\Omega).$$

So the monotone iteration method yields the existence of a solution u_1 of (1), with $0 < u_1 < u_2$ and u_1 is the minimal solution of (1). To finish the proof we have to prove the above claim. To do that observe that K_1 maps $\mathcal{O}'$ into $\mathcal{O}'$. Let $u_0 \in \mathcal{O}'$ and consider the constant mapping $C : \mathcal{O}' \to \mathcal{O}'$ defined by $C(u) = u_0$. By the convexity of $\mathcal{O}'$, it follows that $I - K_1$ is homotopic to $I - C$ in $\mathcal{O}'$ and $d(I - K_1,\mathcal{O}',0) = d(I - C,\mathcal{O}',0)$. But this last degree is trivially equal to 1. The proof is complete. □

Our next result provides the existence of two positive solutions for (1) under a different set of assumptions on the nonlinearity f. We shall drop conditions (f5) and (f6), and replace them by (f7) below. Hence a priori bounds on the solutions are not available any longer. Condition (f7) is needed *only* when $N \geq 3$ and the σ in (f4) is $> (N+1)/(N-1)$.

$$\text{(f7)} \qquad \liminf_{s\to+\infty} \frac{sf(s) - \theta F(s)}{s^2 f(s)^{2/(N+1)}} \geq 0, \text{ with } \theta > 2 \text{ and } F(s) = \int_0^s f.$$

Theorem 2. *In addition to (f1)–(f4) and (f7), assume that problem (1) possesses a strict supersolution w. Then (1) has two positive solutions.*

Remark. Assumptions (f4) and (f7) are used to assure that the Euler-Lagrange functional associated with (1) satisfies the Palais-Smale condition. At this point we rely on the work in [12]. We remark that (f7) is satisfied, for instance, if one assumes the condition introduced by Ambrosetti and Rabinowitz in [13]. Namely, that there are numbers $\theta > 2$ and $s_0 > 0$ such that $\theta F(s) \leq sf(s)$ for $s \geq s_0$. So functions f behaving at $+\infty$ like cs^p, with $c > 0$ and $1 < p < (N+2)/(N-2)$ do satisfy (f7). Hence we see that there are classes of functions which satisfy both (f5) and (f7). However it is apparent that there are functions satisfying one of these conditions but not the other.

Proof. Define f for negative s as $f(s) = f(0)$. Let us consider the functional $\Phi : H_0^1 \to R$ defined by

$$\Phi(u) = \frac{1}{2}\int |\nabla u|^2 - F(u), \qquad F(s) = \int_0^s f.$$

In view of hypotheses (f1), (f3), (f4) and (f7), this functional is of class C^1 and satisfies the Palais-Smale condition. The critical points of Φ are precisely the H_0^1-solutions of (1). In view of (f4) these solutions in fact belong to $C^{2,\alpha}(\bar{\Omega})$. This follows readily by a bootstrap argument in the case when σ in (f4) is $< (N+2)/(N-2)$. The case when $\sigma = (N+2)/(N-2)$ is treated by more sophisticated arguments due to Brézis-Kato [14], see also [12]. Now we use the fact that (1) possesses an ordered pair of a sub- and a supersolution: $\varepsilon\phi_1 < w$ for $\varepsilon > 0$ sufficiently small. Then it is known (see [15]) that there is a point u_1 in the interval $[\varepsilon\phi_1, w]$, which is a local minimum of Φ. Now, either (i) u_1 is a *strict* local minimum of Φ, or (ii) it is not. Hence at this point the proof bifurcates to cover separately these two possibilities. Let us first assume (i). Define the function $g : \Omega \times R \to R$ by

$$g(x,s) = f(s^+ + u_1(x)) - f(u_1(x)).$$

The functional $\Psi : H_0^1 \to R$ defined by

$$\Psi(v) = \frac{1}{2}\int |\nabla v|^2 - \int G(x,v),$$

where $G(x,s) = F(s^+ + u_1(x)) - f(u_1(x))s^+ - F(u_1(x))$, satisfies the Palais-Smale condition. Its critical points are the H_0^1-solutions of

$$-\Delta v = f(v^+ + u_1) - f(u_1),$$

and it follows that they are ≥ 0 a.e. (Ω). So a nonzero critical point v of Ψ provides a second nontrivial solution of problem (1): $u_2 = v + u_1$. To show that Ψ has a nonzero critical point it is enough to prove that 0 is a strict local minimum of Ψ. (The Mountain Pass Theorem then completes the proof.) For that matter we observe that there is $\varepsilon > 0$ such that

$$(*) \qquad \Psi(v) > 0 \qquad \text{for all } 0 < \|v\|_{H^1} \leq \varepsilon.$$

This is a simple consequence of the identity:

$$\Psi(v) = \frac{1}{2}\int |\nabla v^-|^2 + \Phi(v^+ + u_1) - \Phi(u_1),$$

and the fact that u_1 is a strict local minimum of Φ. Next we claim that there is $\alpha > 0$ such that $\Psi(v) \geq \alpha$ for $\|v\|_{H^1} = \varepsilon$. In fact, suppose that this is not the case. Then there exists v_n, with $\|v_n\|_{H^1} = \varepsilon$, and $\Psi(v_n) \to 0$. Since $v_n \rightharpoonup v$ we get $\Psi(v) \leq 0$. In view of $(*)$ above we see that $v = 0$. So passing to a subsequence we see that $\int G(x,v_n) \to 0$, which implies that $\Psi(v_n) \geq (1/4)\varepsilon^2$ for n sufficiently large. But this contradicts the fact that $\Psi(v_n) \to 0$.

Now we proceed to consider alternative (ii). That is, suppose that given $\varepsilon_0 > 0$ such that $\Phi(u) \geq \Phi(u_1)$ for $\|u - u_1\| \leq \varepsilon_0$, there is an $0 < \varepsilon < \varepsilon_0$ such that

$$\inf\{\Phi(u) : \|u - u_1\|_{H^1} = \varepsilon\} = \Phi(u_1).$$

Let (ω_n) be such that $\|\omega_n - u_1\|_{H^1} = \varepsilon$, $\Phi(\omega_n) \to \Phi(u_1)$ and $\omega_n \rightharpoonup \omega$. It follows from (f4) that Φ is weakly lower semicontinuous. So $\Phi(\omega_n) \to \Phi(\omega)$, which gives $\Phi(\omega) = \Phi(u_1)$ showing that ω is a minimum of Φ in the ball $\|u - u_1\|_{H^1} \leq \varepsilon_0$. We see readily that $\omega \neq u_1$. Otherwise, supposing $\omega = u_1$, we come to a contradiction passing to the limit in the identity below:

$$\Phi(\omega_n) = \frac{1}{2}\int |\nabla(\omega_n - u_1)|^2 + \int \nabla\omega_n \cdot \nabla u_1 - \frac{1}{2}\int |\nabla u_1|^2 - \int F(\omega_n).$$

(Recall that the first term in the right side of the previous equality is $\varepsilon^2/2$.) Consequently ω is another critical point of Φ, since $0 < \|\omega - u_1\| \leq \varepsilon < \varepsilon_0$. □

Remark. If $f(0) > 0$ the above proof can be considerably shortened. Indeed, at the point where it bifurcates we have only to apply the strong form of the Mountain Pass Theorem, as proved by Rabinowitz [16]; see also a simple proof in [15]. The critical point so obtained is clearly nontrivial. We also remark that the above proof in the case of alternative (ii) can also be done using the Ekeland variational principle as in [15].

2. The case when Ω is a ball. Suppose that Ω is the open ball

$$B_\rho = \{x \in R^N : \|x\| < \rho\}.$$

In this section we show how to construct supersolutions of problem (1), and discuss some examples. Let us begin by considering the following linear Dirichlet problem:

$$(8) \qquad -\Delta v = \mu v + 1 \quad \text{in } B_1, \qquad v = 0 \quad \text{on } \partial B_1,$$

where $0 \leq \mu < \lambda_1(B_1)$. Let us denote by v_μ the solution of (8). We know that v_μ is positive, radially symmetric and its maximum M_μ is assumed at the center of the ball. That is

$$M_\mu = v_\mu(0) = \max\{v_\mu(x) : x \in B_1\}.$$

These functions v_μ and their corresponding maxima, M_μ, play an important role in the questions of existence of solutions for semilinear problems like (1). Problem (8) can be solved explicitly. Since $v(x)$ is a function only of the radius $r = |x|$, (denote it again by $v(r)$), it has to be the solution of the following two-point boundary value problem

$$(9) \qquad -v'' - \frac{N-1}{r} v' = \mu v + 1 \quad \text{in } (0,1), \qquad v'(0) = v(1) = 0.$$

In the case $N = 2$ this equation becomes

$$v'' + \frac{1}{r} v' + \mu v = -1$$

which is the Bessel's equation of zero order. So the solution of (9) is

$$v_\mu(r) = \frac{1}{\mu}\left[\frac{J_0(\sqrt{\mu}\, r)}{J_0(\sqrt{\mu})} - 1\right]$$

where J_0 denotes the Bessel function of zero order. Recalling that $J(0) = 1$, we obtain a simple expression M_μ in this case. Namely

$$M_\mu = \frac{1}{\mu}\left[\frac{1}{J_0(\sqrt{\mu})} - 1\right].$$

In the case $N \geq 3$, we make the substitution $\omega(r) = v(r) r^p$, where

$$p = (N-2)/2.$$

So ω is a solution of

$$\omega'' + \frac{1}{r}\omega' + \left(\mu - \frac{p^2}{r^2}\right)\omega = -r^p, \qquad \omega(0) = \omega(1) = 0,$$

which is given readily in terms of Bessel's functions of order p. Thus the solution of (9) is

$$v_\mu(r) = \frac{1}{\mu}\left[\frac{J_p(\sqrt{\mu}\, r)}{r^p J_p(\sqrt{\mu})} - 1\right], \qquad p = (N-2)/2.$$

Recalling that $r^{-p}J_p(r) \to [2^p\Gamma(p+1)]^{-1}$ as $r \to 0$, we obtain

$$M_\mu = \frac{1}{\mu}\left[\frac{\mu^{p/2}}{2^p\Gamma(p+1)J_p(\sqrt{\mu})} - 1\right), \qquad p = (N-2)/2,$$

which of course is valid for all $N \geq 2$. In the case $N = 3$, the expressions of v_μ and M_μ simplify to more elementary functions. Namely,

$$v_\mu(r) = \frac{1}{\mu}\left[\frac{\sin(\sqrt{\mu}\, r)}{r \sin\sqrt{\mu}} - 1\right], \qquad M_\mu = \frac{1}{\mu}\left[\frac{\sqrt{\mu}}{\sin\sqrt{\mu}} - 1\right].$$

From the former expressions we see the following properties of M_μ which will be useful in the sequel. Recalling that

$$r^{-2}[r^{-p}J_p(r)2^p\Gamma(p+1) - 1] \to [2(2p+2)]^{-1} \qquad \text{as } r \to 0,$$

we conclude that

$$M_0 = (2N)^{-1}.$$

It is easily seen that the first eigenvalue $\lambda_1(B_1)$ of $-\Delta\phi = \lambda\phi$ in B_1, $\phi = 0$ on ∂B_1 is the square of the first positive zero ν_p of $J_p(r)$: $\lambda_1(B_1) = \nu_p^2$. Consequently $M_\mu \to +\infty$ as $\mu \to \lambda_1(B_1)$. Since $M_\mu < M_{\mu'}$ for $0 \leq \mu < \mu' < \lambda_1(B_1)$ we have a pretty good idea of the graph of M_μ.

Proposition 3. *Suppose that the function $f: R^+ \to R^+$ satisfies the following condition: there are numbers $s_0 > 0$ and $0 \leq \mu < \lambda_1(B_1)$ such that*

(f8) $$f(s) \leq \mu s + M_\mu^{-1} s_0, \ \forall 0 \leq s \leq s_0.$$

Then problem (1) with $\Omega = B_1$ has a strict supersolution.

Remark. A special case of (f8) is

$(\text{f8})_0$ $$f(s) \leq 2Ns_0, \ \forall 0 \leq s \leq s_0, \text{ for some } s_0 > 0.$$

Proof. The solution w of the linear Dirichlet problem

(8′) $$-\Delta w = \mu w + C \quad \text{in } B_1, \qquad w = 0 \quad \text{on } \partial B_1$$

is such that $0 \leq w \leq M_\mu C$, for any given constant $C > 0$. We see that w is a supersolution of (1) provided $\mu s + C \geq f(s)$ for all $0 \leq s \leq M_\mu C$. But this follows from (f8) taking $C = M_\mu^{-1} s_0$. □

Remark. In the case when $\Omega = B_\rho$, the analogue of condition (f8) is: there are numbers $s_0 > 0$ and $0 \leq \mu < \lambda_1(B_\rho)$ such that

$$f(s) \leq \mu s + \rho^{-2} M_{\mu\rho^2}^{-1} s_0, \qquad \forall\, 0 \leq s \leq s_0.$$

Remark. The idea of using linear Dirichlet problems like (8′) to obtain a supersolution of (1) has been used before by several authors. A special case of Proposition 3 for increasing functions f was obtained by Bandle [23] using this method.

Examples. 1) $f(s) = \lambda e^s$. Consider the nonlinear eigenvalue problem

$$-\Delta u = \lambda e^u \quad \text{in } B_1, \qquad u = 0 \quad \text{on } \partial B_1. \tag{10}$$

It is known that there is a $\bar{\lambda} > 0$, such that (10) has solutions for $\lambda < \bar{\lambda}$ and no solution if $\lambda > \bar{\lambda}$. What can be said about the value of $\bar{\lambda}$? In the case when $N = 2$, problem (10) can be solved explicitly. This was already proved by Liouville in 1853, see Bandle [17]. In this case it is seen that $\bar{\lambda} = 2$. Proposition 3 above provides lower bounds for $\bar{\lambda}$ in any dimension N. For instance, $(f8)_0$ in the case of $N = 2$, gives $\bar{\lambda} \geq 4/e = 1.4715178\ldots$. Using (f8) we can improve this bound and get

$$\bar{\lambda} \geq \max\left\{\frac{\ell n(1 + \mu M_\mu)}{M_\mu} : 0 \leq \mu < \lambda_1(B_1)\right\} \tag{11}$$

which in the case of $N = 2$ gives $\bar{\lambda} \geq 1.804169\ldots$. In the case of $N = 3$, $(f8)_0$ gives $\bar{\lambda} \geq 6/e = 2.2072767\ldots$, while estimate (11) gives $\bar{\lambda} \geq 2.865171\ldots$.

2) $f(s) = \lambda(1 + \alpha s)^\beta$, α, $\beta > 0$. We see then that the problem

$$-\Delta u = \lambda(1 + \alpha u)^\beta \quad \text{in } B_1, \qquad u = 0 \quad \text{on } \partial B,$$

has a solution for each $\lambda \leq \bar{\lambda}$ where

$$\bar{\lambda} = \max\left\{\frac{1}{\alpha M_\mu}[(1 + \mu M_\mu)^{1/\beta} - 1] : 0 \leq \mu < \lambda_1(B_1)\right\}.$$

3. The case of a general domain Ω. In this section we use Schwarz symmetrization for two purposes. First, we state a condition on f (depending only on the volume of Ω) which insures the existence of a strict supersolution of (1) for a general domain Ω. Second, we prove that the existence of a solution of (1) for the ball with the same volume as Ω implies the existence of a solution for general Ω. This result is useful in constructing strict supersolutions of (1) as we shall see.

Let us begin by reviewing the basic facts about symmetrization, which will be used in the sequel. More about symmetrization can be seen in Hardy-Littlewood-Pólya [18], Pólya-Szegö [19] and Bandle [17]. Let $u : R^N \to R$ be an L^1-function. We define the Schwarz symmetrization u^* (also called the symmetric decreasing rearrangement) of u as the radially symmetric, positive and nonincreasing function $u^* : R^N \to R$ such that

$$\text{meas}\{x \in R^N : |u(x)| > t\} = \text{meas}\{x \in R^N : u^*(x) > t\}, \qquad \forall\, t > 0.$$

And one sees that

$$u^*(x) = \inf[t > 0 : \text{meas}\{x \in R^N : |u(x)| > t\} \leq \omega_N |x|^N],$$

where ω_N is the volume of the unit ball in R^N.

The following properties of u^* will be needed here

(I) If $u \in L^p(R^N)$, $1 \le p \le \infty$, then $u^* \in L^p(R^N)$ and $\|u\|_{L^p} = \|u^*\|_{L^p}$.

(II) If $u \in H^1(R^N)$, then $u^* \in H^1(R^N)$, and $\|\nabla u^*\|_{L^2} \le \|\nabla u\|_{L^2}$.

The proof of (I) is given in [18] and [19]. An elementary proof of (II) can be seen in [20]. Now let Ω be a smooth bounded domain in R^N. Let us denote by Ω^* the ball centered at 0 and with the same volume as Ω. Then, it follows from (I) and (II) that

(III) $\lambda_1(\Omega^*) \le \lambda_1(\Omega)$.

The next result is proved in Lions [21]:

(IV) Consider the following two linear Dirichlet problems:

$$-\Delta u = \lambda u + 1 \quad \text{in } \Omega, \qquad u = 0 \quad \text{on } \partial\Omega;$$

$$-\Delta v = \lambda v + 1 \quad \text{in } \Omega^*, \qquad v = 0 \quad \text{on } \partial\Omega^*$$

where $0 \le \lambda < \lambda_1(\Omega^*)$. Then $u^* \le v$ a.e. (Ω^*).

(V) If u is a Lipschitz continuous function in Ω, then u^* is Lipschitz continuous in Ω^*. See [17]. The next result is due to Talenti [22]. A simple proof can be seen in [21]. We remark that in fact a more general result is true, as proved in both [21] and [22].

(VI) Let h be a Lipschitz continuous function in Ω. Consider the following two Dirichlet problems:

$$-\Delta u = h \quad \text{in } \Omega, \qquad u = 0 \quad \text{on } \partial\Omega;$$

$$-\Delta v = h^* \quad \text{in } \Omega^*, \qquad v = 0 \quad \text{on } \partial\Omega^*.$$

Then $u^* \le v$ a.e. (Ω^*).

Now we give a simple sufficient condition on f for the existence of a strict supersolution of (1).

Proposition 4. *Consider (1) for a smooth bounded domain Ω in R^N, and let Ω^* be the ball in R^N with the same volume as Ω. Suppose that the function f satisfies the following condition: there exist numbers $s_0 > 0$ and $0 \le \mu < \lambda_1(\Omega^*)$ such that*

$$f(s) \le \mu s + \rho^{-2} M_{\mu\rho^2}^{-1} s_0, \qquad \forall\, 0 \le s \le s_0,$$

where ρ is the radius of the ball Ω^*. Then the problem (1) has a strict supersolution.

Proof. Proceed as in the proof of Proposition 3, using now (I) and (IV) above. □

Now let us show how a solution of (1) can be obtained from a solution of

(12) $\quad -\Delta v = f(v) \quad \text{in } \Omega^*, \qquad v > 0 \quad \text{in } \Omega^*, \qquad v = 0 \quad \text{on } \partial\Omega^*.$

Theorem 5. *Let $f: R^+ \to R^+$ be a nondecreasing function satisfying hypotheses (f1) and (f2). Suppose that (12) has a solution v. Then problem (1) has a minimal solution $\underline{u}$ and $\underline{u}^* \leq \underline{v}$, where $\underline{v}$ is the minimal solution of (12).*

Proof. Let us denote by $\hat{\phi}$ the eigenfunction corresponding to $\lambda_1(\Omega^*)$, which is > 0 and $\int \hat{\phi}_1^2 = 1$. Choose $\varepsilon > 0$ and $\delta > 0$ such that $u_0 = \varepsilon\phi_1$ is a subsolution of (1), $v_0 = \delta\hat{\phi}_1$ is a subsolution of (12) and

$$u_0^* \leq v_0. \tag{13$_0$}$$

Now we define iteratively the sequences (u_n), (v_n) by

$$-\Delta u_{n+1} = f(u_n) \quad \text{in } \Omega, \qquad u_{n+1} = 0 \quad \text{on } \partial\Omega,$$

$$-\Delta v_{n+1} = f(v_n) \quad \text{in } \Omega^*, \qquad v_{n+1} = 0 \quad \text{on } \partial\Omega^*.$$

In view of (VI), $u_1^* \leq w_1$, where w_1 is the solution of

$$-\Delta w_1 = f(u_0^*) \quad \text{in } \Omega^*, \qquad w_1 = 0 \quad \text{on } \partial\Omega^*.$$

So $-\Delta w_1 \leq f(v_0) = -\Delta v_1$, which implies by the maximum principle $v_1 \geq w_1$. Consequently

$$u_1^* \leq v_1. \tag{13$_1$}$$

By induction, it follows that

$$u_n^* \leq v_n. \tag{13$_n$}$$

But, since problem (12) has a solution, it follows that (v_n) converges to $\underline{v}$ in the $C^{2,\alpha}$-norm. Now from (I) it follows that $\|u_n\|_{L^\infty} = \|u_n^*\|_{L^\infty} \leq \|v_n\|_{L^\infty}$. So (u_n) also converges to a function $\underline{u}$ which is the minimal solution of (1). And from (13) we obtain $\underline{u}^* \leq \underline{v}$. □

The following results show how Theorem 5 can be used to establish the existence of strict supersolutions of (1).

Corollary 6. *Let $f_1: R^+ \to R^+$ be an increasing function satisfying hypotheses (f1) and (f2), such that $f_1(s) > f(s)$ for s in some interval $(0,\varepsilon)$, and $f_1(s) \geq f(s)$ for all $s \in R^+$. Suppose that the problem*

$$-\Delta v = f_1(v) \quad \text{in } \Omega^*, \qquad v = 0 \quad \text{on } \partial\Omega^*$$

has a solution. Then (1) possesses a strict supersolution w.

Proof. Indeed, w is the solution of the problem

$$-\Delta w = f_1(w) \quad \text{in } \Omega, \qquad w = 0 \quad \text{on } \partial\Omega$$

given by Theorem 5. □

Corollary 7. *Let f satisfy all the hypotheses of Theorem 5. Moreover f is C^1. Suppose that the first eigenvalue of $-\Delta\omega - f'(v)\omega = \lambda\omega$ in Ω^*, $\omega = 0$ on $\partial\Omega^*$ is positive. Then problem (1) has a strict supersolution.*

Proof. By the implicit function theorem the problem

$$-\Delta z = f(z) + \varepsilon z \quad \text{in } \Omega^*, \qquad z = 0 \quad \text{on } \partial\Omega^*$$

has a solution z. So by Theorem 5, the analogous problem in Ω has also a solution, u. Clearly u is a strict supersolution of (1). □

We have seen above that the existence of a solution in the ball Ω^* implies the existence of a minimal positive solution in Ω. In fact, using the full generality of [22]—see also [21]—we may improve this argument as follows. Assume that f is nondecreasing, (f1) holds and that there exists v solution of:

$$(12) \qquad -\Delta v = f(v) \quad \text{in } \Omega^*, \qquad v > 0 \quad \text{in } \Omega^*, \qquad v = 0 \quad \text{on } \partial\Omega^*.$$

Then let $a_{ij}(x) = a_{ji}(x) \in L^\infty(\Omega)$ satisfy: $(a_{ij}(x)) \geq I_N$ a.e. in Ω and consider the following problem:

$$-\frac{\partial}{\partial x_i} a_{ij}(x) \frac{\partial u}{\partial x_j} = f(u) \quad \text{in } \Omega, \qquad u > 0 \quad \text{in } \Omega, \qquad u = 0 \quad \text{on } \partial\Omega.$$

If

$$\liminf_{t \to 0+} \frac{f(t)}{t} > \lambda_1\left(-\frac{\partial}{\partial x_i}\left(a_{ij}\frac{\partial}{\partial x_j}\right), H_0^1(\Omega)\right),$$

then the preceding problem has a minimal solution u such that:

$$u^* \leq \underline{v},$$

where $\underline{v}$ is the minimal solution of (12).

4. Existence results with neither a priori bounds nor Palais-Smale. In this section we consider once more problem (1) under the basic assumption that a strict supersolution exists. Again conditions (f1) through (f4) are assumed. However we shall assume neither condition (f5) nor (f7). Consequently one does not know whether the solutions of (1) are a priori bounded. Also the corresponding Euler-Lagrange functional does not clearly satisfy the Palais-Smale condition. Thus the methods of Section 1 do not apply directly. We found it useful to rely on the results of [9]. It is proved there that the bound

$$(14) \qquad \|\nabla u\|_{L^\infty(\partial\Omega)} \leq k_0, \qquad \text{for all solutions of (1)},$$

holds if in addition to (f1), (f3) and (f4) one assumes condition (f6). Also we recall (cf. [9]) that the condition below in conjunction with (f1), (f3) and (f4) gives inequality (17) used in the proof of Theorem 8 presented next.

(f9) $\partial\Omega = \Gamma_1 \cup \Gamma_2$, where Γ_1 and Γ_2 are closed and satisfy

(i) at every point of Γ_1 all sectional curvatures of Γ_1 are bounded away from 0 by a positive constant;
(ii) there is a point x_0 in R^N such that $(x - x_0, n(x)) \leq 0$ for all $x \in \Gamma_2$, where $n(x)$ denotes the outward unit normal to $\partial\Omega$ at x.

Theorem 8. *In addition to (f1)–(f4), assume that (1) possesses a strict supersolution w. Suppose that one of the conditions (f6) or (f9) holds. Then (1) has two solutions* $0 < u_1 < u_2$.

Proof. i) We first observe that (f3) implies that there exist $\mu > \lambda_1$ and $R > 0$ such that

$$f(s) \geq \mu s, \qquad \text{for all } s \geq R. \tag{15}$$

Also it follows from (f4) that given $\varepsilon > 0$ there is $C_\varepsilon > 0$, such that

$$f(s) < \varepsilon s^{(N+2)/(N-2)} + C_\varepsilon, \qquad \text{for all } s \geq 0. \tag{16}$$

If (f6) is assumed we have from [9] that (14) holds, with k_0 depending only on μ and R given in (15). Using Pohozaev's identity (cf. [9]) we see that there exists a constant $k_1 > 0$ *depending only on* μ and R, such that

$$\left| \frac{N-2}{2N} \int_\Omega |\nabla u|^2 - \int_\Omega F(u) \right| \leq k_1. \tag{17}$$

If we assume (f9) we also obtain (17) from the results of [9]. So in any case we have that (17) holds true for all solutions of (1).

ii) By the results of Section 1 there is $u_1 \in H_0^1$ in the interval $[0,w]$ which is a local minimum of

$$\Phi(v) = \frac{1}{2} \int_\Omega |\nabla v|^2 - \int_\Omega F(v).$$

Consider the functional $\Psi : H_0^1 \to R$ defined by

$$\Psi(v) = \frac{1}{2} \int_\Omega |\nabla v|^2 - \int_\Omega G(x,v)$$

where $G(x,s) = F(s^+ + u_1(x)) - f(u_1(x))s^+ - F(u_1(x))$. Let $C_0 = \|w\|_{L^\infty} + 1$. It follows from (f3) that there is a $v_0 \in C^1(\bar{\Omega})$, depending only on C_0, μ and R, such that $v_0 > 0$ in Ω and $\Psi(v_0) \leq 0$. Let

$$C_1 = \|u_1 + v_0\|_{L^\infty} + 1 \quad \text{and} \quad b = \max\{\Psi(tv_0) : 0 \leq t \leq 1\}.$$

Then b depends only on f on the interval $[0,C_2]$ where $C_2 = \max(C_0,C_1)$ and on μ and R.

iii) Now if v is a critical point of Ψ, we see that $v \geq 0$ and $u = v + u_1$ is a solution of (1). If we suppose that $\Psi(v) \in [0,b]$, then it follows from (17) that

$$\|u\|_{L^\infty} \leq C_3$$

where C_3 depends on k_1, b, C_ε and f restricted to the interval $[0,C_0]$. Consequently C_3 depends only on μ, R, C_ε and f restricted to the interval $[0,C_2]$. Therefore we modify f outside of the interval $[0,\max(C_2,C_3)]$, keeping the same μ, R and C_ε. Moreover this modification is done in such a way that the new f satisfies the assumptions which enable us to use the results of Section 1. Thus the critical point v of the modified Ψ obtained by mountain passing satisfies $\Psi(v) \in [0,b]$. It is then readily seen that $u = v + u_1$ is the second solution of (1) we are looking for. □

References

1. H. Amann, *Fixed point equations and nonlinear eigenvalue problems in ordered Banach spaces*, SIAM Review **18** (1976), 620–709.
2. P.-L. Lions, *On the existence of positive solutions of semilinear elliptic equations*, SIAM Review **24** (1982), 441–467.
3. D. G. de Figueiredo, *Positive solutions of semilinear elliptic equations*, Lecture Notes in Math. **957** (pp. 34–87), Springer-Verlag, Berlin, 1982.
4. I. M. Gelfand, *Some problems in the theory of quasilinear equations*, Amer. Math. Soc. Transl. Ser. 2, **29** (1963), 295–381.
5. D. D. Joseph & T. S. Lundgren, *Quasilinear Dirichlet problems driven by positive sources*, Arch. Rational Mech. Anal. **49** (1973), 241–269.
6. M. G. Crandall & P. H. Rabinowitz, *Some continuation and variational methods for positive solutions of nonlinear elliptic eigenvalue problems*, Arch. Rational Mech. Anal. **58** (1975), 207–218.
7. C. Bandle, *Existence theorems, qualitative results and a priori bounds for a class of nonlinear Dirichlet problems*, Arch. Rational Mech. Anal. **58** (1973), 349–381.
8. F. Mignot & J.-P. Puel, *Sur une classe de problèmes nonlinéaires avec nonlinearité positive, croissante, convexe*, Comm. Partial Differential Equations **5** (1980), 791–836.
9. D. G. de Figueiredo, P.-L. Lions & R. D. Nussbaum, *A priori estimates and existence results for positive solutions of semilinear elliptic equations*, J. Math. Pures Appl. **61** (1982), 41–63.
10. H. Brézis & R. E. L. Turner, *On a class of superlinear elliptic problems*, Comm. Partial Differential Equations **2** (1977), 601–604.
11. B. Gidas & J. Spruck, *A priori bounds for positive solutions of nonlinear elliptic equations*, Comm. Partial Differential Equations **6** (1981), 883–901.
12. D. G. de Figueiredo, *On the superlinear Ambrosetti-Prodi problem*, Nonlinear Anal. Theory; Methods and Applications **16** (1984), 655–665.
13. A. Ambrosetti & P. H. Rabinowitz, *Dual variational methods in critical point theory and applications*, J. Funct. Anal. **14** (1973), 349–381.
14. H. Brézis & T. Kato, *Remarks on the Schrödinger operator with singular complex potentials*, J. Math. Pures Appl. **58** (1979), 137–151.
15. D. G. de Figueiredo & S. Solimini, *A variational approach to superlinear elliptic problems*, Comm. Partial Differential Equations **9** (1984), 699–717.
16. P. H. Rabinowitz, *Some aspects of critical point theory*, MRC Technical Summary Report #2465 (January 1983).
17. C. Bandle, *Isoperimetric Inequalities and Applications*, Pitman Publishing Ltd., London, 1980.
18. G. H. Hardy, J. E. Littlewood & G. Pólya, *Inequalities*, Cambridge University Press, London, 1967.
19. G. Pólya & G. Szegö, *Isoperimetric Inequalities in Mathematical Physics*, Princeton University Press, 1951.
20. E. H. Lieb, *Existence and uniqueness of the minimizing solution of Choquard's non linear equation*, Stud. Appl. Math. **57** (1977), 93–105.
21. P.-L. Lions, *Quelques remarques sur la symmetrization de Schwarz*, in "Nonlinear Partial Differential Equations and Their Applications," Collège de France Seminar, Vol. 1, Pitman, London, 1981.
22. G. Talenti, *Elliptic equations and rearrangements*, Ann. Scuola Norm. Sup. Pisa Cl. Sci. (4) **3** (1976), 697–718.

23. C. Bandle, *Bounds for the solutions of boundary value problems*, J. Math. Anal. Appl. **54** (1976), 706–716.

The preparation of this paper was partially sponsored by the United States Army under Contract No. DAAG 29-80-C-0041 while the first author was visiting the Mathematics Research Center with the support of the John Simon Guggenheim Memorial Foundation.

de Figueiredo: Universidade de Brasília—70910 Brasília-DF, Brasil
Lions: Université Paris—IX, Dauphine—75775 Paris Cedex 16, France

Received May 10, 1984

Proceedings of Symposia in Pure Mathematics
Volume 45 (1986), Part 1

Positive Solutions for Some Classes of Semilinear Elliptic Problems

DJAIRO G. DE FIGUEIREDO

Introduction. Let us consider the Dirichlet problem

$$-\Delta u = f(u) \quad \text{and} \quad u > 0 \quad \text{in } \Omega, \qquad u = 0 \quad \text{on } \partial\Omega. \tag{0.1}$$

where Ω is a bounded smooth domain in R^N, $N \geqslant 2$, and $f\colon R^+ \to R$ is a function satisfying the condition below, besides other requirements that will be timely introduced as we proceed.

$$f\colon [0, \infty) \to R \text{ is locally lipschitzian.} \tag{$f1$}$$

The following quantities will play an important role in the analysis conducted in this paper:

$$I = \inf_{s>0} \frac{f(s)}{s}, \qquad S = \sup_{s>0} \frac{f(s)}{s}. \tag{0.2}$$

Clearly $I \in R \cup \{-\infty\}$ and $S \in R \cup \{+\infty\}$. It is clearly seen that a necessary condition for existence of a solution of (0.1) is

$$I \leqslant \lambda_1 \leqslant S. \tag{0.3}$$

(This is proved via a multiplication of the equation in (0.1) by an eigenfunction $\phi_1 > 0$ of $(-\Delta, H_0^1)$ corresponding to the first eigenvalue λ_1.) Also by this very procedure we see that if either $I = \lambda_1$ or $S = \lambda_1$, then u is necessarily a positive multiple of ϕ_1 and $f(s) = \lambda_1 s$ for s in some interval $[0, a]$. Consequently the interesting cases are when

$$I < \lambda_1 < S. \tag{0.4}$$

That is, (0.4) is a necessary condition for existence of a solution of (0.1) which is not trivial. By trivial we mean $u = k\phi_1$ for some positive constant k.

Condition (0.4) is achieved in classes I and II below by stating hypotheses on the behavior of $f(s)/s$ at 0 and at $+\infty$. For classes III and IV we shall need

---

1980 *Mathematics Subject Classification*. Primary 35J20, 35J25, 47H15.

Originally published in: Nonlinear Functional Analysis and its Applications. Proc. Sympos. Pure Math. 45, Part 1, Amer. Math. Soc. (1986), 371-379,

DOI: 10.1007/978-3-319-02856-9_16

assumptions on $f(s)/s$ at intermediate points between 0 and ∞. Of course, the necessary condition (0.4) is not sufficient for the existence of solutions of (0.1) in most cases as simple examples show. The object of this paper is to exhibit some further conditions on the nonlinearity f that insure the existence of solutions for (0.1). This will be done for classes III and IV.

Class I (Sublinear):

$$\liminf_{s\to 0}\frac{f(s)}{s}>\lambda_1 \quad\text{and}\quad \limsup_{s\to+\infty}\frac{f(s)}{s}<\lambda_1.$$

Class II (Superlinear):

$$\limsup_{s\to 0}\frac{f(s)}{s}<\lambda_1 \quad\text{and}\quad \liminf_{s\to+\infty}\frac{f(s)}{s}>\lambda_1.$$

Class III (Sub-super!):

$$\liminf_{s\to 0}\frac{f(s)}{s}>\lambda_1 \quad\text{and}\quad \liminf_{s\to+\infty}\frac{f(s)}{s}>\lambda_1.$$

Class IV (Super-sub!):

$$\limsup_{s\to 0}\frac{f(s)}{s}<\lambda_1 \quad\text{and}\quad \limsup_{s\to+\infty}\frac{f(s)}{s}<\lambda_1.$$

It is well known that problems (0.1) in class I have at least one solution. This can be proved by the method of monotone iterations (see, for instance, [**1, 16**] or the review paper [**2**]).

Problems (0.1) in class II do not have a solution in general. The most famous case is Pohožaev's example [**3**]

$$-\Delta u=u^{\sigma}\quad\text{and}\quad u>0\quad\text{in }\Omega,\qquad u=0\quad\text{on }\partial\Omega. \tag{0.5}$$

where $\sigma=(N+2)/(N-2)$, $N\geqslant 3$, and Ω is a star-shaped bounded domain in R^N. It is also known [**4**] that if $\sigma<(N+2)/(N-2)$, $N\geqslant 3$, then (0.5) possesses a solution. And more generally, also [**4**], if

$$\lim_{s\to+\infty}\frac{f(s)}{s^{\sigma}}=0,\quad\text{where } \sigma=\frac{N+2}{N-2},\, N\geqslant 3, \tag{0.6}$$

and if there exist numbers $s_0>s$ and $\theta>2$ such that

$$sf(s)\geqslant\theta F(s)\quad\text{for } s\geqslant s_0,\, F(s)=\int_0^s f, \tag{0.7}$$

then problem (0.1) has a solution. Conditions (0.6) and (0.7) are used to prove that the Euler–Lagrange functional associated to (0.1), namely

$$\Phi(u)=\frac{1}{2}\int|\nabla u|^2-\int F(u). \tag{0.8}$$

satisfies the Palais–Smale condition in $H_0^1(\Omega)$. However it has been proved in [**5**] that for convex domains Ω existence of solutions of (0.1) holds without assumption (0.7), assuming only (0.6) and the fact that the problem (0.1) is in class II. This is due to fairly general a priori estimates of the L^{∞} norms of ∇u on $\partial\Omega$ for solutions of u of (0.1), obtained in [5] by the use of the Gidas–Ni–Nirenberg techniques [6]. So existence holds in class II even if Φ (apparently) does not satisfy the Palais–Smale condition. Nevertheless the aforementioned results still require condition (0.6). Observe however that the Euler–Lagrange functional (0.8) is still defined in $H_0^1(\Omega)$ if one only assumes

$$|f(s)| \leqslant cs^\sigma + c, \qquad \sigma = (N+2)/(N-2), n \geqslant 3. \tag{0.9}$$

The investigation of problems of this type has been conducted by Brézis–Nirenberg [12] and solutions of (0.1) have been obtained under various additional conditions of the nonlinearity f. All previous remarks were made in case $N \geqslant 3$. Similar statements can be made in the case $N = 2$, where even exponential growth of f is allowed [5].

Problems (0.1) in class III also do not have a solution in general. As observed above the graph of $f(s)$ has to be somewhere below the straight line $\lambda_1 s$. But this requirement only does not suffice. We can see that problems like

$$-\Delta u = \lambda e^u \quad \text{in } \Omega, \qquad u = 0 \quad \text{on } \partial\Omega \tag{0.10}$$

may have no solution for some λ's such that $\lambda < \lambda_1/e$. (That is, $\inf\{\lambda e^s/s: s > 0\} = \lambda e < \lambda_1$.) So the graph of $f(s)$ has to have a "fairly good" crossing of the line $\lambda_1 s$. Clearly if $f(s)$ vanishes at some point s_0, this is enough to guarantee the existence of a solution for (0.1). (The method of monotone iteration readily gives the solution. Here s_0 is a supersolution of (0.1).) However the graph of $f(s)$ does not have to go below $\lambda_1 s$ that far. In fact, problem (0.10) has a solution if $\lambda > 0$ is small enough. In §1 we shall see a sufficient condition on f that provides a strict supersolution for (0.1), and consequently a solution, obtained again by the method of monotone iterations. We see then that the existence of one solution for (0.1) in the case of class III does not involve the behavior of f at $+\infty$. However if we are willing to discuss the existence of further solutions, the behavior of f at $+\infty$ becomes very useful in two respects. First, if we are planning to use topological degree methods, conditions on f at $+\infty$ will be required in such a way as to obtain a priori bounds on the solutions of (0.1). Second, if we choose to use variational methods, again conditions on f at $+\infty$ will be necessary to define the Euler–Lagrange functional (0.8) in H_0^1. Also some further conditions on f at $+\infty$ will insure that Φ satisfies the Palais–Smale condition, which will permit us to use the usual critical point theory, namely here the Mountain Pass Theorem [4]. All this will be discussed in §1.

Problems in class IV clearly do not have a solution in general, since the necessary condition (0.4) may not be satisfied. In §2 we show that the solutions of a given problem (0.1) in this class are a priori bounded in L^∞. So a truncation on the nonlinearity f may be performed and the truncated problem has the same solutions as the original problem. With this truncated f, the Euler–Lagrange functional is well defined in H_0^1, bounded below, and satisfies the Palais–Smale condition. Clearly 0 is a local minimum of this functional, which also has a global minimum. It could very well happen that 0 is this global minimum. In fact this will be the case if one does not insure that $f(s)$ goes sufficiently enough above $\lambda_1 s$. This will be achieved, for instance, by assuming a condition on the primitive F, which will guarantee that the functional is negative somewhere. Consequently the global minimum is negative and it is achieved at a point $u \neq 0$.

1. Problems (0.1) in class III. The condition imposed on $f(s)/s$ near 0 implies immediately that $\underline{u} = \varepsilon\phi_1$, for $\varepsilon > 0$ sufficiently small, is a strict subsolution. In order to use the method of monotone iteration one shall insure the existence of a supersolution $\bar{u} \geqslant \underline{u}$. For that matter one uses the solution of a linear problem

$$(1.1) \qquad -\Delta v = \mu v + C \quad \text{in } \Omega, \qquad v = 0 \quad \text{on } \partial\Omega,$$

$0 \leqslant \mu < \lambda_1(\Omega)$ and C is a positive constant. Since other domains besides Ω will be appearing in this exposition we use the notation $\lambda_1(\Omega)$ to denote the first eigenvalue of the Laplacian in $H_0^1(\Omega)$. The solution v of (1.1) is positive in Ω and its maximum M depends on μ, C, and Ω. If $\mu < \lambda_1(\Omega^*)$, where Ω^* is the ball with the same volume as Ω, centered at 0, then it follows that $M \leqslant \omega(0)$, where ω is the solution of

$$(1.2) \qquad -\Delta\omega = \mu\omega + C \quad \text{in } \Omega^*, \qquad \omega = 0 \quad \text{on } \partial\Omega^*.$$

The proof of this statement uses the facts that $v^* \leqslant \omega$ in Ω^* and $\|v\|_{L^\infty} = \|v^*\|_{L^\infty}$, where v^* represents the decreasing symmetric rearrangement of v. For proofs of these facts see P. L. Lions **[7]**, Talenti **[8]**, or Bandle **[9]**. Observe also that $\lambda_1(\Omega^*) \leqslant \lambda_1(\Omega)$. So a bound for M can be written out explictly by solving (1.2), which indeed reduces to an ordinary differential equation. If we denote by M_ρ the maximum, $z(0)$, of the solution z of

$$(1.3) \qquad -\Delta z = \rho z + 1 \quad \text{in } B_1(0), \qquad z = 0 \quad \text{on } \partial B_1(0),$$

where $0 \leqslant \rho < \lambda_1(B_1(0))$, then $M \leqslant CR^2 M_{\mu R^2}$, where R is the radius of the ball Ω^*. In this way the following result was established in **[10]**.

PROPOSITION 1.1. *If there exist numbers $s_0 > 0$ and $0 \leqslant \mu < \lambda_1(\Omega^*)$ such that*

$$(1.4) \qquad f(s) \leqslant \mu s + s_0/R^2 M_{\mu R^2}, \qquad 0 \leqslant s \leqslant s_0,$$

then problem (0.1) *possesses a strict supersolution $\bar{u}$.*

Existence of a strict supersolution for problem (0.1) can also be inferred from informations on the symmetrized problem. For instance, see **[10]**.

PROPOSITION 1.2. *Suppose that there exists a function f_1: $R^+ \to R$ which is nondecreasing, $f_1(s) \geqslant f(s)$ and $f_1(s) > f(s)$ for s in some interval* $(0, a)$. *Assume that the problem*

$$(1.5) \qquad -\Delta v = f_1(v) \quad \text{in } \Omega^*, \qquad v = 0 \quad \text{on } \partial\Omega^*,$$

has a positive solution v. Then problem (0.1) *in class* III *possesses a strict supersolution.*

PROOF. It suffices to prove that the problem

$$(1.6) \qquad -\Delta u = f_1(u) \quad \text{in } \Omega, \qquad u = 0 \quad \text{on } \partial\Omega$$

has a solution. In fact, one may assume that the solution v given in the statement of the proposition is a minimal solution. Also the solution u of (1.6) obtained next will be a minimal solution. To prove the existence of such a solution we proceed by monotone iterations. Take a subsolution of (1.6) given by $u_0 = \varepsilon\phi_1$, and a subsolution of (1.5) given by $v_0 = \delta\hat{\phi}_1$, where $\hat{\phi}_1 > 0$ is a first eigenfunction of $-\Delta$ on $H_0^1(\Omega^*)$. Choose $\varepsilon > 0$ small enough such that $u_0^* \leqslant v_0$. Then proceed through the two following iteration schemes:

$$-\Delta u_n = f_1(u_{n-1}) \quad \text{in } \Omega, \qquad u_n = 0 \quad \text{on } \partial\Omega,$$
$$-\Delta v_n = f_1(v_{n-1}) \quad \text{in } \Omega^*, \qquad v_n = 0 \quad \text{on } \partial\Omega^*.$$

It is then seen that $u_n^* \leqslant v_n$ in Ω^*. Since v_n converges to v, it follows that u_n also converges to a solution of (1.6). □

Now we assume the following basic hypotheses on the behavior of f at $+\infty$:

$$(f2) \qquad \lim_{s\to+\infty} \frac{f(s)}{s^\sigma} = 0, \quad \text{where } \sigma = \frac{N+2}{N-2}, N \geqslant 3.$$

From now on (in this section) we restrict ourselves to the case $N \geqslant 3$. The case $N = 2$ can be treated likewise, and in view of the Sobolev imbedding theorems in R^2 the nonlinearity f can be allowed to grow even exponentially (see [5]).

Suppose in addition to $(f2)$ that Ω is convex and the technical condition

$$(f3) \qquad \limsup_{s\to+\infty} \frac{s(f) - \theta F(s)}{s^2 f(s)^{2/N}} \leqslant 0, \qquad 0 \leqslant \theta < \frac{2N}{N-2},$$

holds. It has been proved in [5] that the solutions of a problem (0.1) in class III are a priori bounded in L^∞, and consequently in C^1. As established in [5] the convexity hypotheses in Ω can be relaxed at the expense of some additional assumptions on f. The following result holds.

Proposition 1.3. *Assume $(f2)$ and $(f3)$, and that Ω is convex. Suppose that a problem* (0.1) *in class* III *possesses a strict supersolution $\bar{u}$. Then it has at least two solutions u_1, u_2 and $u_1 \leqslant u_2$.*

Proof. Let us prove the result in the simpler case when $f(0) > 0$. The general case is proved in [10]. Extend the function f for $s < 0$ as $f(s) = f(0)$, and consider the problem

$$(1.7) \qquad -\Delta u = f(u) \quad \text{in } \Omega, \qquad u = 0 \quad \text{on } \partial\Omega.$$

The solutions of (1.7) can be proved to be all positive and in fact they are precisely the solutions of (0.1). The a priori estimates mentioned above hold for all solutions of the following parametrized family of problems.

$$(1.8) \qquad -\Delta v = tf(v) + (1-t)(\lambda v^+ + 1) \quad \text{in } \Omega, \qquad v = 0 \quad \text{on } \partial\Omega,$$

where $0 \leqslant t \leqslant 1$ and λ is some fixed number greater than $\lambda_1(\Omega)$ (v^+ denotes the positive part of v). Hence there exists a number $R > 0$ independent of t such that (1.8) has no solution in $\|u\|_{C^1} \geqslant R$. In order to apply topological degree theory we take our problem to a functional-theoretical setting. Let X be the Banach space of C^1 functions in Ω which vanish on $\partial\Omega$. The equations

$$-\Delta u = tf(v) + (1-t)(\lambda v^+ + 1) \quad \text{in } \Omega, \qquad u = 0 \quad \text{on } \partial\Omega$$

define mappings $T_t\colon X \to X$ by $T_t v = u$, $0 \leqslant t \leqslant 1$. From the elliptic theory these mappings are compact. The question of solvability of (0.1) is equivalent to the problem of obtaining a solution $u \in X$ of $u - T_1 u = 0$. From the a priori bounds it follows that

$$\deg(I - T_t, U_R, 0) = \text{const}, \tag{1.9}$$

where $U_R = \{u \in X\colon \|u\|_X < R\}$. Since

$$-\Delta v = \lambda v^+ + 1 \quad \text{in } \Omega, \qquad v = 0 \quad \text{on } \partial\Omega$$

has no solution, it follows that the degree in (1.9) is zero. On the other hand since the problem possesses a strict subsolution $\underline{u}$ and a strict supersolution $\bar{u}$, with $\underline{u} \leqslant \bar{u}$, we can construct an open set $\mathcal{O} \subset U_R$ such that $\deg(I - T_1, \mathcal{O}, 0) = 1$. So $\deg(I - T_1, U_R \setminus \bar{\mathcal{O}}, 0) = -1$, and we obtain two solutions $v_1 \in \mathcal{O}$, $v_2 \in U_R \setminus \bar{\mathcal{O}}$. On the basis of these two solutions one can indeed show that existence of two ordered solutions. □

Suppose now that the condition

$$(f4) \qquad \liminf_{s \to +\infty} \frac{sf(s) - \theta F(s)}{sf(s)^{2/(N+1)}} \geqslant 0, \qquad \theta > 2,$$

holds. It has been proved in [**11**] that under hypotheses $(f2)$ and $(f4)$ the Euler–Lagrange functional associated with (0.1) satisfies the Palais–Smale condition. Then the following result [**10**] holds true.

PROPOSITION 1.4. *Assume* $(f2)$ *and* $(f4)$. *Suppose that a problem* (0.1) *in class* III *possesses a strict supersolution* $\bar{u}$. *Then it has at least two solutions.*

PROOF. The existence of a strict subsolution $\underline{u}$ and a strict supersolution $\bar{u}$, with $\underline{u} \leqslant \bar{u}$, implies that the Euler–Lagrange functional (0.8) has a local minimum u_1, which is in fact in the interval $[\underline{u}, \bar{u}]$. Clearly u_1 is a solution of (0.1). Next define the functional

$$\Psi(v) = \frac{1}{2}\int |\nabla v|^2 - \int G(x, v),$$

where $G(x, s) = F(s^+ + u_1(x)) - f(u_1(x))s^+ - F(u_1(x))$. It is seen that 0 is a local minimum of Ψ and that there exists $v_0 \in H_0^1$ such that $\Psi(v_0) < 0$. The hypotheses on the growth of f imply that Ψ satisfies the Palais–Smale condition. Apply then the strong form of the Mountain Pass Theorem (which holds for the case of nonstrict minimum) to obtain another critical point $v_1 \neq 0$ of Ψ. Then $u_2 = v_1 + u_1$ is the other solution of (0.1). □

The next result is also proved in [**10**]. The proof relies on certain estimates on the gradient of u on $\partial\Omega$ which allow one to perform appropriate truncations on f. In this way one can obtain existence of pairs of solutions for (0.1) without either one of the technical conditions $(f3)$ or $(f4)$.

Proposition 1.5. *Assume* $(f2)$ *and that* Ω *is convex. Suppose that problem* (0.1) *in class* III *possesses a strict supersolution* $\bar{u}$. *Then it has at least two solutions.*

2. Problems (0.1) in class IV. We start this section by proving that, as a consequence of the behavior of f at $+\infty$, the solutions of a given problem (0.1) in class IV are a priori bounded in L^∞. This will permit a truncation of the nonlinearity f and the subsequent treatment of the problem by variational methods.

Proposition 2.1. *Let f be a function satisfying* $(f1)$ *and in the class* IV. *Then all (eventual) solutions of* (0.1) *are a priori bounded in* L^∞.

Proof. (i) First we prove an a priori bound in H_0^1. As a consequence of the behavior of f at $+\infty$, there are constants $0 < \mu < \lambda_1$ and $C_\mu > 0$ such that

$$f(s) \leqslant \mu s + C_\mu \quad \text{for all } s \geqslant 0. \tag{2.1}$$

Multiplying the equation in (0.1) by u and integrating by parts we get

$$\lambda_1 \int u^2 \leqslant \int |\nabla u|^2 = \int f(u)u \leqslant \mu \int u^2 + C_\mu \int u. \tag{2.2}$$

We have used (2.1) to obtain the second inequality in (2.1). It follows from (2.2) that the solutions of (0.1) are a priori bounded in L^2, and then in H_0^1. Moreover this bound depends only on μ and C_μ.

(ii) Now we write $f(s) = f^+(s) - f^-(s)$ and then the equation in (0.1) can be written as

$$-\Delta u + \frac{f^-(u)}{u} u = f^+(u). \tag{2.3}$$

(Assuming, as we shall do, that $\liminf_{s\to 0}(f(s)/s) > -\infty$, we see that, for each given solution u of (0.1), the coefficient of μ in (2.3) is continuous and bounded in Ω. Maybe this assumption would not be necessary in the arguments we present in the sequel.) For each solution u of (0.1) let us denote the coefficient of u in (2.3) by c_u. Clearly $c_u \geqslant 0$ in Ω. Also for each given solution u of (0.1) let us denote by ω_u the solution of the problem

$$-\Delta\omega_u = f^+(u) \quad \text{in } \Omega, \qquad \omega_u = 0 \quad \text{on } \partial\Omega. \tag{2.4}$$

Clearly ω_u is a nonnegative $C^{2,\alpha}$ function. From (2.3) and (2.4) follows that

$$-\Delta(u - \omega_u) + c_u(u - \omega_n) = -c_u\omega_u \leqslant 0 \tag{2.5}$$

and the maximum principle gives $u \leqslant \omega_u$. Once we have this last inequality we may proceed by a bootstrap argument in (2.4) to get the required L^∞ bound on the solutions of (0.1). Indeed, by part (i) of this proof and Sobolev imbedding, the solutions of (0.1) are a priori bounded in L^p for $p = 2N/(N-2)$. From (2.4) the corresponding ω_u are a priori bounded in $W^{2,p}$ and consequently in an $L^{p'}$ with $p' > p$. The inequality $u \leqslant \omega_u$ gives the same for u, and we proceed... □

In view of Proposition 2.1, given a problem (0.1) in class IV we may assume that f is bounded. This is achieved by a truncation of f preserving condition (2.1).

The solutions of the truncated problem are precisely the same as the original problem. For this modified (but equivalent) problem the Euler–Lagrange functional Φ defined in (0.8) is well defined in H_0^1, bounded from below, and coercive. Consequently it possesses a global minimum u_1. By the hypothesis on $f(s)$ at $s = 0, u = 0$ is a local minimum of Φ. Thus a priori u_1 could be 0. This will indeed be the case if, for instance, the graph of $f(s)$ does not cross the line $\lambda_1 s$ for some $s > 0$. A necessary and sufficient condition for $u_1 \neq 0$ is the existence of a point $v_0 \neq 0$ where $\Phi(v_0) \leqslant 0$. So the following result holds.

PROPOSITION 2.2. *Let f be a function satisfying* ($f1$) *and in class* IV. *Suppose that there exists* $0 \neq v_0 \in H_0^1$ *such that* $\Phi(v_0) \leqslant 0$. *Then problem* (0.1) *has two solutions.*

It suffices, in addition to the above remarks, to observe that the condition $\Phi(v_0) \leqslant 0$ for some $v_0 \neq 0$ implies that there exists another minimum of Φ, $u_1 \neq 0$. Then use the Mountain Pass Theorem (see Corollary 2.16 of **[13]** or previous results in **[14]**).

Next we present a sufficient condition in terms of the primitive F of f which insures that there is a $v_0 \neq 0$ such that $\Phi(v_0) < 0$.

PROPOSITION 2.3. *Suppose that there is an* $s_0 > 0$ *such that*

$$(2.6)\quad F(s_0) \geqslant \frac{1}{2} s_0^2 \frac{1}{R^2} \min\left\{ \frac{(1+t)^2}{t^2} \frac{(1+t)^N - 1}{2-(1+t)^N} : 0 < t < 2^{1/N} - 1 \right\},$$

where R is the radius of the largest ball contained in Ω. Then the functional Φ is negative for some $u \in H_0^1$.

We may assume that the ball centered at 0 with radius R is contained in Ω. Defining, for each $t \in [0,1]$,

$$u_t(x) = \begin{cases} s_0 & \text{if } \|x\| < \dfrac{R}{1+t}, \\ s_0\left[1 - \dfrac{1+t}{tR}\left(\|x\| - \dfrac{R}{1+t}\right)\right] & \text{if } \dfrac{R}{1+t} < \|x\| \leqslant R, \\ 0 & \text{if } x \in \Omega \setminus B_R(0), \end{cases}$$

the result of Proposition 2.3 follows by a simple calculation.

REMARK. It should be remarked that a problem (0.1) in class IV may have two solutions even if $\Phi(u) > 0$ for all $u \neq 0$. This would mean that the global minimum of Φ is 0. So the argument above would not work. For these cases some other arguments have to be produced! See Remark 1.12 in **[15]**.

REFERENCES

1. H. Amann, *Fixed point equations and nonlinear eigenvalue problems in ordered Banach spaces*, SIAM Rev. **18** (1976), 620–709.

2. D. G. de Figueiredo, *Positive solutions of semilinear elliptic equations*, Lecture Notes in Math., vol. 947, Springer-Verlag, 1982, pp. 34–87.

3. S. I. Pohožaev, *Eigenfunctions of* $\Delta u + \lambda f(u) = 0$, Soviet Math. Dokl. **6** (1965), 1408–1411.

4. A. Ambrosetti and P. R. Rabinowitz, *Dual variational methods in critical points theory and applications*, J. Funct. Anal. **14** (1973), 349–381.

5. D. G. de Figueiredo, P.-L. Lions and R. D. Nussbaum, *A priori estimates and existence results for positive solutions of semilinear elliptic equations*, J. Math. Pures Appl. **61** (1982), 41–63.

6. G. Gidas, W.-M. Ni and L. Nirenberg, *Symmetry and related properties via the maximum principle*, Comm. Math. Phys. **68** (1979), 209–243.

7. P.-L. Lions, *Quelques remarques sur la symmetrization de Schwarz*, Nonlinear Partial Differential Equations and Their Applications, College de France Seminar, Vol. 1, Pitman, London, 1981, pp. 308–319.

8. G. Talenti, *Elliptic equations and rearrangements*, Ann. Scuola Norm. Sup. Pisa (4) **3** (1976), 697–718.

9. C. Bandle, *Isoperimetric inequalities and applications*, Pitman, London, 1980.

10. D. G. de Figueiredo and P.-L. Lions, *On pairs of positive solutions for a class of semilinear elliptic problems*, MRC Tech. Summary Rep. 2660, 1984.

11. D. G. de Figueiredo, *On the superlinear Ambrosetti–Prodi problem*, Nonlinear Anal. **8** (1984), 655–665.

12. H. Brézis and L. Nirenberg, *Positive solutions of nonlinear elliptic equations involving critical Sobolev exponents*, Comm. Pure Appl. Math.

13. P. R. Rabinowitz, *Some aspects of critical point theory*, MRC Tech. Summary Rep. 2645, Jan. 1983.

14. ______, *Variational methods for nonlinear eigenvalue problems*, Eigenvalues of Nonlinear Problems (G. Prodi. ed.), C.I.M.E., Edizioni Cremonese, Rome, 1974, pp. 141–195.

15. P.-L. Lions, *On the existence of positive solutions of semilinear elliptic equations*, SIAM Rev. **24** (1982), 441–457.

16. H. Berestycki and P.-L. Lions, *Some applications of the method of super and subsolutions*, Bifurcation and Nonlinear Eigenvalue Problems, Lecture Notes in Math., vol. 782, Springer-Verlag, 1982, pp. 16–41.

UNIVERSIDADE DE BRASÍLIA, BRAZIL

SIAM J. MATH. ANAL.
Vol. 17, No. 4, July 1986

006

A MAXIMUM PRINCIPLE FOR AN ELLIPTIC SYSTEM AND APPLICATIONS TO SEMILINEAR PROBLEMS*

DJAIRO G. DE FIGUEIREDO[†] AND ENZO MITIDIERI[‡]

Abstract. The Dirichlet problem in a bounded region for elliptic systems of the form

$$(*)\qquad -\Delta u=f(x,u)-v, \qquad -\Delta v=\delta u-\gamma v$$

is studied. For the question of existence of positive solutions the key ingredient is a maximum principle for a linear elliptic system associated with $(*)$. A priori bounds for the solutions of $(*)$ are proved under various types of growth conditions on f. Variational methods are used to establish the existence of pairs of solutions for $(*)$.

Key words. elliptic systems, maximum principle, a priori estimates, positive solutions, monotone iteration method, mountain pass theorem

AMS(MOS) subject classifications. Primary 35J65, 35J50, 47H15

Introduction. In this paper we propose to discuss the elliptic system

$$(0.1)\qquad -\Delta u=f(x,u)-v, \qquad -\Delta v=\delta u-\gamma v \quad \text{in } \Omega,$$

where Ω is a bounded smooth domain in R^N, $N\geqq 2$, subject to Dirichlet boundary conditions $u=v=0$ on $\partial\Omega$. The solutions (u,v) of this problem represent steady state solutions of reaction diffusion systems of interest in biology, namely systems of the form

$$(0.2)\qquad u_t=D_1\Delta u+f(u)-v, \qquad v_t=D_2\Delta v+\varepsilon(u-\gamma v)$$

where D_1, D_2, ε and γ are positive constants, and one looks for solutions $u(t,x)$, $v(t,x)$ defined in $(0,\infty)\times\Omega$, subject to Dirichlet boundary conditions on $(0,\infty)\times\partial\Omega$. The type of nonlinearities which are of importance in the applications will be described in the Examples I and II below. System (0.2) shows that both species may diffuse. In this sense it is an extension of the well-known FitzHugh–Nagumo system, which serves as a model for nerve conduction, cf. [5] or Hastings [7]. We also mention Koga–Kuramoto [10], where the complete system (0.2) appears and steady state solutions are discussed. There is an extensive bibliography in this subject. We mention three additional papers, which are more closely related to the investigation presented here, namely Rothe–de Mottoni [13], Rothe [14] and Lazer–McKenna [11].

In the applications the constants γ and δ, which appear in system (0.1), are taken to be *positive*. So we shall make this assumption throughout this paper. It follows then that the second equation in (0.1) can be solved for v in terms of u. Let us denote by B its solution operator under Dirichlet boundary conditions. That is, given u, we define

*Received by the editors July 6, 1984, and in revised form February 10, 1985. This work was sponsored by the U. S. Army under contract DAAG29-80-C-0041.

† Departamento de Matemática, Universidade de Brasília, Brasília, Brazil. This research was partially done at the Scuola Internazionale Superiore di Studi Avanzati in Trieste, when the first author held a Guggenheim Fellowship.

‡ Istituto di Matematica, Università Degli Studi di Trieste, 34100, Trieste, Italy.

Originally published in: SIAM J. Math. Anal. 17 (1986), 836-849,

DOI: 10.1007/978-3-319-02856-9_17

Bu as the solution of the problem $-\Delta v+\gamma v=\delta u$ in Ω, $v=0$ on $\partial\Omega$. Thus our problem becomes the one of finding u such that

$$-\Delta u+Bu=f(x,u) \quad \text{in } \Omega, \qquad u=0 \quad \text{on } \partial\Omega. \tag{0.3}$$

We observe that the left side of (0.3) contains a local (differential) operator $-\Delta$, and a nonlocal (integral) operator B. This fact gives rise to quite interesting questions. It is essential at the outset to understand the operator $-\Delta+B$. In §1 we study its spectral properties and establish a maximum principle for solutions of linear equations like

$$-\Delta u+Bu-\lambda u=g(x) \quad \text{in } \Omega, \qquad u=0 \quad \text{on } \partial\Omega, \tag{0.4}$$

where the real parameter λ is restricted to certain ranges depending on γ, δ and the region Ω. In §2 we establish a priori bounds for solutions of (0.3) under the main assumption that the nonlinearity f at ∞ is below the smallest eigenvalue of the operator $-\Delta+B$; this assumption will be stated precisely as condition (f2) and it characterizes a class of systems which are here called sublinear. The two examples below, which were treated by previous authors [9], [11], [13] and [14], are included in the classes studied in the present paper. Their results are therefore sharpened as far as ranges of the parameters involved and signs of the solutions.

Example I. $f(u)=\lambda u-g(u)$, where λ is a real parameter larger than the first eigenvalue of the operator $-\Delta+B$, and g is a function behaving like u^3, but not necessarily odd; cf. [11], [13], [14].

Example II. $f(u)=u(u-a)(1-u)$, where a is such that $0<a<1/2$. This is the sort of nonlinearity arising in the FitzHugh–Nagumo equations, [5], [9].

The a priori bounds obtained in §2 will be needed in an essential way to perform appropriate truncations of the nonlinearity f, so the problem could be treated by variational methods. This will be done in §5.

In §3 we discuss a class of systems whose model nonlinearity is the one given by Example I. Using the results of §1 we are able to establish the existence of a positive and a negative solution. This result complements a previous one by Lazer and McKenna [11], who proved the existence of two nontrivial solutions by topological degree arguments. Their method, however, does not yield the signs of the solutions obtained. The maximum principle for equations like (0.4) becomes very useful in this respect.

In §4 we sketch a result on the existence of positive solutions for a superlinear elliptic system. Results similar to the ones known for the scalar case hold true in view of the aforementioned maximum principle. The question of the a priori bounds for positive solutions of superlinear elliptic systems may be a hard one. If the growth of the nonlinearity at $+\infty$ is at most like $(N+1)/(N-1)$, for $N\geq 3$, then the results of Brézis–Turner [2] extend readily. The range $[(N+1)/(N-1),\ (N+2)/(N-2)]$ poses serious difficulties. The methods used in de Figueiredo–Lions–Nussbaum [3] to treat the scalar case rely on the results of Gidas–Ni–Nirenberg [6], which are not available as yet for the type of systems studied here. We remark that Troy [15] has extended some of the results in [6] to systems. However, Troy's systems do not include the ones we are concerned with. Also in §4 we prove a nonexistence result basing it on our extension to systems of the well-known Pohozaev's identity.

In §5 we consider a class of systems whose model nonlinearity is the one given in Example II. Using the Mountain Pass Theorem of Ambrosetti–Rabinowitz [1], we establish Theorem 5.1 on the existence of two nontrivial solutions for such systems, extending a previous result of Klaasen–Mitidieri [9]. This result shows clearly the relevance of the volume of Ω and of the parameters γ and δ on the existence questions.

It also exhibits the importance of a large positive parameter λ on the existence of two positive solutions for the system

$$-\Delta u=\lambda f(x,u)-v, \qquad -\Delta v=\delta u-\gamma v, \quad \text{in } \Omega$$

subject to Dirichlet boundary conditions, and the nonlinearity f is of the type given by Example II. This relates to the scalar case studied in Rabinowitz [12].

The contents of this paper are as follows:

1. The operator $-\Delta+B$.
2. A priori bounds for solutions of sublinear elliptic systems.
3. Existence of positive solutions.
4. Remarks on a superlinear system.
5. Existence of two nontrivial solutions for a class of sublinear systems.

1. The operator $-\Delta+B$. Consider the linear Dirichlet problem

$$-\Delta v+\gamma v=\delta u \quad \text{in } \Omega, \qquad v=0 \quad \text{in } \partial\Omega, \tag{1.1}$$

where $\Omega\subset R^N$ is a bounded and smooth domain, γ and δ are positive constants. Let us denote by B its solution operator: $v=Bu$. It is well known that

$$B:L^2(\Omega)\to H^2(\Omega)\cap H_0^1(\Omega); \quad B:L^p(\Omega)\to W^{2,p}(\Omega); \quad B:C^\alpha(\overline{\Omega})\to C^{2+\alpha}(\overline{\Omega}).$$

Let us define the operator

$$T\equiv -\Delta+B:L^2(\Omega)\to L^2(\Omega), \quad \text{with } D(T)=H^2(\Omega)\cap H_0^1(\Omega).$$

Clearly T is symmetric, that is, $(Tu_1,u_2)=(u_1,Tu_2)$ for all $u_1,u_2\in D(T)$, where $(\ ,\ )$ denotes the L^2 inner product. Using the L^2 regularity theory, one can prove that T is a closed operator. Let us denote by $0<\lambda_1<\lambda_2\leqq\lambda_3\leqq\cdots$ the eigenvalues of $-\Delta$ under Dirichlet boundary conditions, and by ϕ_k the corresponding eigenfunctions. Then it is easily verified that

$$\hat{\lambda}_k=\lambda_k+\frac{\delta}{\gamma+\lambda_k}, \qquad k=1,2,\cdots \tag{1.2}$$

are eigenvalues of T. Moreover the same ϕ_k's defined above are their corresponding eigenfunctions. Since (ϕ_k) is a complete orthonormal set in L^2, it is readily shown that the $\hat{\lambda}_k$'s are the *only* eigenvalues of T. We shall prove in the sequel that in fact the *spectrum* $\sigma(T)$ of T *consists precisely of these eigenvalues*. For each λ in the resolvent set $\rho(T)$ of T, let us denote by $T_\lambda=(T-\lambda I)^{-1}$ its corresponding resolvent operator.

LEMMA 1.1 (a representation formula of the resolvent operator for some values of λ). *Suppose that the real numbers a and b satisfy the following conditions*

$$a>-\lambda_1, \quad \gamma+b>-\lambda_1, \quad b\neq 0, \quad \textit{and} \tag{1.3}$$

$$b\gamma+\delta=ab. \tag{1.4}$$

Then $\lambda=-a-b$ *is in the resolvent set* $\rho(T)$ *and*

$$T_\lambda=\left[1-b(\gamma+b-\Delta)^{-1}\right](a-\Delta)^{-1}. \tag{1.5}$$

Proof. With $\lambda=-a-b$, one can write

$$T-\lambda I=(a-\Delta)+b\left[\frac{b\gamma+\delta}{b}-\Delta\right](\gamma-\Delta)^{-1}.$$

Using condition (1.4) above, one obtains

$$T-\lambda I=(a-\Delta)\left[1+b(\gamma-\Delta)^{-1}\right]=(a-\Delta)(\gamma-\Delta)^{-1}(\gamma+b-\Delta).$$

Finally using condition (1.3), it follows that

$$T_\lambda=(\gamma+b-\Delta)^{-1}(\gamma-\Delta)(a-\Delta)^{-1},$$

which readily gives (1.5). □

Remark 1.1. A calculation shows that λ, taken in the ranges indicated below, are representable as $\lambda=-a-b$, with a and b satisfying (1.3) and (1.4):

(i) All $\lambda\leqq-\gamma-2\sqrt{\delta}$. These λ's correspond to $b>0$.

(ii) If $\gamma+\lambda_1>\sqrt{\delta}$, there are some additional values of λ, namely $2\sqrt{\delta}-\gamma\leqq\lambda<\lambda_1+\delta/(\gamma+\lambda_1)$. These λ's correspond to b negative in the range $-\lambda_1-\gamma<b<-\delta/(\gamma+\lambda_1)$.

Remark 1.2 (*monotonicity of the sequence* $\hat{\lambda}_k$). We observe that $\gamma+\lambda_1>\sqrt{\delta}$ implies that $\hat{\lambda}_1<\hat{\lambda}_2\leqq\hat{\lambda}_3\leqq\cdots$. Of course one does not have in general such a monotonicity of the eigenvalues $\hat{\lambda}_k$. Clearly $\gamma+\lambda_1>\sqrt{\delta}$ is not a necessary condition, since it in fact implies the stronger statement that the function $s\mapsto s+\delta/(\gamma+s)$ is monotonically increasing in the whole halfline $[\lambda_1,\infty)$. A necessary and sufficient condition for this monotonicity involves also the second eigenvalue λ_2, namely $\delta<(\gamma+\lambda_1)(\gamma+\lambda_2)$.

COROLLARY 1.2 (compactness of T_λ). *For all* $\lambda\in\rho(T)$, *the resolvent operator* T_λ *is compact.*

Proof. For any $\lambda,\mu\in\rho(T)$ one has the resolvent equation

$$T_\mu-T_\lambda=(\mu-\lambda)T_\mu T_\lambda.$$

So if T_λ is compact for some λ, then it is compact for all λ's in the resolvent set. By the previous lemma T_λ is compact for $\lambda\leqq-\gamma-2\sqrt{\delta}$. □

The following result is an immediate consequence of Lemma 1.1 and Remark 1.1 above.

COROLLARY 1.3 (positiveness of T_λ for some values of λ). *If* $\gamma+\lambda_1>\sqrt{\delta}$, *then* T_λ *is positive for all* $2\sqrt{\delta}-\gamma\leqq\lambda<\hat{\lambda}_1$.

Remark 1.3. The positiveness of T_λ is a *maximum principle* for the equation

$$-\Delta v+Bv-\lambda v=u \quad \text{in } \Omega, \qquad v=0 \quad \text{on } \partial\Omega.$$

It says that if $u\in L^2$ and $u\geqq0$ a.e., then $v\geqq0$ a.e. In fact, it follows from the representation formula (1.5) that a *strong maximum principle* holds. Namely, if $u\in C^0(\Omega)$ and $u\geqq0$ in Ω, then $v>0$ in Ω and the outward normal derivative $(\partial v/\partial\nu)<0$. (Recall that Ω is being assumed to be smooth. So the interior sphere condition is satisfied.)

Remark 1.4. If $\gamma>2\sqrt{\delta}$, then the condition $\gamma+\lambda_1>\sqrt{\delta}$ is automatically satisfied, and Corollary 1.3 says that in this case T_λ is positive for λ in an interval which contains 0. In general one cannot expect that T_0 be positive. Indeed, if $\gamma=\delta=1$, then Corollary 1.3 says that T_λ is positive for $1\leqq\lambda<\hat{\lambda}_1$.

PROPOSITION 1.4. *The spectrum* $\sigma(T)$ *of* T *consists of precisely the eigenvalues* $\hat{\lambda}_k$.

Proof. We have seen above that the point spectrum $P\sigma(T)=\{\hat{\lambda}_k: k=1,2,\cdots\}$. Let $\lambda\notin P\sigma(T)$. Then $T-\lambda I$ is one-to-one. If we show that $T-\lambda I$ is onto, it follows by the Closed Graph Theorem that $\lambda\in\rho(T)$. Thus we claim that the equation $Tu-\lambda u=v$ has a solution u for each given $v\in L^2$. Taking $\mu\in\rho(T)$, we see that this equation is equivalent to $Tu-\mu u=(\lambda-\mu)u+v$, or

$$u=(\lambda-\mu)T_\mu u+T_\mu v. \tag{1.6}$$

By the Fredholm alternative (1.6) is solvable iff the homogeneous equation $u = (\lambda-\mu)T_\mu u$ has only the solution $u=0$. But this is actually the case, since this homogeneous equation is equivalent to $Tu=\lambda u$. Recall that $\lambda \notin P\sigma(T)$. □

Remark 1.5. The above proposition follows also from general results in functional analysis. Namely, since T is a self-adjoint operator, it follows that its residual spectrum $R\sigma(T)$ is empty. Next, since $-\Delta-\lambda$ is Fredholm for every $\lambda\in C$, it follows that $-\Delta+B-\lambda$ is also Fredholm for all $\lambda\in C$. Consequently the continuous spectrum $C\sigma(T)$ is also empty.

Remark 1.6 (*a useful inequality*). Let $\tilde{\lambda}$ denote the smallest of the eigenvalues $\hat{\lambda}_k$. We have seen above that $\tilde{\lambda}=\hat{\lambda}_1$ if $\gamma+\lambda_1>\sqrt{\delta}$. We assert that

$$(1.7)\qquad (Tu,u)\geqq\tilde{\lambda}\|u\|^2_{L^2},\qquad \forall u\in D(T).$$

Indeed, since (ϕ_k) is a complete orthonormal set in L^2, we can write $u=\sum\alpha_k\phi_k$ where $\alpha_k=(u,\phi_k)$. So

$$(Tu,u)=\sum\alpha_k(Tu,\phi_k)=\sum\alpha_k(u,T\phi_k)=\sum\alpha_k^2\hat{\lambda}_k,$$

from which the claim follows. A similar argument shows that

$$(1.8)\qquad \int|\nabla u|^2+(Bu,u)\geqq\tilde{\lambda}\|u\|^2_{L^2},\qquad \forall u\in H_0^1.$$

Remark 1.7 (*uncoupling of systems and maximum principles*). The usual maximum principle for systems, as well as the maximum principle proved here, seems to be related with the possibility of uncoupling the elliptic system. To make precise our observation, let us look at the linear elliptic system

$$(1.9)\qquad -\Delta u=au+bv+f(x),\qquad -\Delta v=cu+dv+g(x)$$

subject to Dirichlet boundary conditions: $u=v=0$ on $\partial\Omega$, where Ω is some bounded domain in R^N, and a, b, c and d are real constants. Suppose that $b\neq 0$ and $c\neq 0$; otherwise the problem trivializes. The uncoupling of system (1.9) is possible if the matrix of the coefficients

$$M=\begin{bmatrix} a & b\\ c & d\end{bmatrix}$$

has two distinct eigenvalues, μ_1 and μ_2. Such a condition is equivalent to

$$(1.10)\qquad (a-d)^2+4bc>0.$$

Of course this is the case if b and c both have the same sign. However, to infer the signs of u and v from the signs of the corresponding functions in the uncoupled system, one requires that both b and c be positive. This gives the usual maximum principle for systems. On the other hand, if b and c have opposite signs, the uncoupling is still possible provided a and d "compensate" for the negativeness of bc. Through some calculations one can prove the following result, which essentially gives our maximum principle.

PROPOSITION 1.5. *In addition to* (1.10) *assume that* $bc<0$, $c(a-d)>0$, $\mu_1<\hat{\lambda}_1$ *and* $\mu_2<\hat{\lambda}_1$. *Then if* $f\geqq 0$, $g\geqq 0$ *and* $cf\geqq(a-\mu_1)g$, *it follows that the solutions u and v of* (1.9) *are positive in* Ω.

2. A priori bounds for solutions of sublinear elliptic systems. Let us consider the elliptic system

$$(2.1)\qquad -\Delta u=f(x,u)-v,\qquad -\Delta v=\delta u-\gamma v\quad \text{in } \Omega,$$

where Ω is a bounded smooth domain in R^N, subject to Dirichlet boundary conditions. We always assume that γ and δ are *positive constants*. The nonlinearity f is subject to the following conditions:

(f1) $f:\bar{\Omega}\times R\to R$ is locally Lipschitzian,

(f2) $\limsup_{|s|\to\infty}(f(x,s)/s)<\tilde{\lambda}$ (uniformly in Ω), where $\tilde{\lambda}$ denotes the smallest eigenvalue of the operator $-\Delta+B$ studied in §1.

Condition (f2) characterizes system (2.1) as being *sublinear*.

Examples. 1) $f(u)=\lambda u-h(u)u$, where h is a C^1 function such that $h(0)=0$, $h'(s)s>0$ for all $s\neq 0$ and $\liminf_{s\to\pm\infty}h(s)>\lambda$, (for instance $h(s)=s^2$). This is the case considered in [11] and [14].

2) $f(u)=u(u-a)(1-u)$, where $0<a<1$. This is the type of nonlinearity that appears in the FitzHugh–Nagumo equations; cf. [5], [9].

Remark 2.1. By a solution of (2.1) we mean a classical solution. That is, a pair of functions (u,v) which are in $C^2(\Omega)\cap C^0(\bar{\Omega})$ and which are 0 on $\partial\Omega$. We observe that if, $u,v\in H_0^1(\Omega)\cap C^0(\bar{\Omega})$ satisfy (2.1) in the distribution sense, then by a bootstrap argument it follows that $u,v\in C^{2,\alpha}(\bar{\Omega})$. We remark that in general one cannot drop the hypothesis that u and v are in $C^0(\bar{\Omega})$ in order to be able to bootstrap. However, this would be possible provided one assumes some growth condition on f.

In order to obtain the a priori bound for the solutions of (2.1), we shall assume either one of the conditions below.

(f3) $\lim_{|s|\to\infty}(f(x,s)/|s|^p)=0$, where $1<p<(N+2)/(N-2)$, if $N\geqq 3$, and $1<p<\infty$, if $N=2$,

(f4) $\limsup_{|s|\to\infty}(f(x,s)/s)<-\delta/\gamma$,

where the limits are uniform in Ω.

Remark 2.2. In the scalar case (i.e. $-\Delta u=f(x,u)$) condition (f4) corresponds to $f(x,s)<0$ for $s>\beta>0$ and $f(x,s)>0$ for $s<-\beta$, where β is some real number.

PROPOSITION 2.1. *Under hypotheses* (f1), (f2) *and* (f3), *the solutions of* (2.1) *are a priori bounded in* L^∞.

Proof. It follows from (f2) that there exist $0<\mu<\tilde{\lambda}$ and $M>0$ such that

$$f(x,s)\leqq\mu s+M,\quad \text{for } 0\leqq s<\infty,\qquad f(x,s)\geqq\mu s-M\quad\text{for } -\infty<s<0. \tag{2.2}$$

The second equation in (2.1) can be solved for v in terms of u. And in this way system (2.1) is equivalent to the equation

$$-\Delta u+Bu=f(x,u), \tag{2.3}$$

using the notation of §1. So we need only to prove bounds on u. The corresponding bounds on v are obtained immediately from the second equation in (2.1). Multiplying (2.3) by u, integrating by parts and using (1.8), we obtain

$$\tilde{\lambda}\int u^2\leqq\int|\nabla u|^2+\int(Bu)u=\int f(x,u)u. \tag{2.4}$$

Next we estimate the last term in (2.4) using (2.2)

$$\int f(x,u)u\leqq\mu\int u^2+M\int|u| \tag{2.5}$$

which implies $\int u^2\leqq C$. (We shall use the same C to denote different constants.) Using (2.4) and (2.5) again and recalling that B is a bounded linear operator in L^2, we conclude that $\int|\nabla u|^2\leqq C$. It follows from (f3) that given $\varepsilon>0$ there exists $C_\varepsilon>0$ such that

$$|f(x,s)|\leqq\varepsilon|s|^p+C_\varepsilon.$$

Finally using this inequality and invoking L^p estimates and the Sobolev imbedding theorem, we conclude that there exist a constant $\hat{C}$ such that $\|u\|_{L^\infty} \leqq \hat{C}$. □

Remark 2.3. We emphasize that the dependence of $\hat{C}$ on f is through the constants μ, M and C_ε. So if we change f for $|s| \geqq \hat{C}$ maintaining μ, M and C_ε, the new equation (2.3) with this modified f has the same solutions of the original equation (2.3). This fact will be used in §5.

The following result was proved by Rothe [14] and Lazer and McKenna [11] under less general hypotheses on f. The main idea in the proof below is taken from those papers.

PROPOSITION 2.2. *Under hypotheses* (f1), (f2) *and* (f4), *the solutions of* (2.1) *are a priori bounded in* L^∞.

Proof. (i) We first claim that for $u \in C^0(\overline{\Omega})$, with $u = 0$ on $\partial\Omega$, one has

$$\frac{\delta}{\gamma}\min u \leqq (Bu)(x) \leqq \frac{\delta}{\gamma}\max u, \qquad x \in \Omega. \tag{2.6}$$

Indeed we know that $v = Bu$ satisfies the equation

$$v = \frac{1}{\gamma}\Delta v + \frac{\delta}{\gamma}u. \tag{2.7}$$

Let us prove the first inequality in (2.6). If $v \geqq 0$ that inequality is trivially true. So let us assume that for $x_1 \in \Omega$ we have $v(x_1) = \min v < 0$. Then $\Delta v(x_1) \geqq 0$ and (2.7) implies that $v(x_1) \geqq (\delta/\gamma)u(x_1)$, from which the first inequality in (2.6) follows readily. In a similar way we prove the second inequality in (2.6).

(ii) It follows from (f4) that there exist positive constants k and m such that

$$\frac{f(x,s)}{s} \leqq -k < -\frac{\delta}{\gamma}, \qquad |s| \geqq m. \tag{2.8}$$

We claim that $\|u\|_{L^\infty} \leqq m$ for all solutions u of (2.1). Indeed, suppose by contradiction that $\|u\|_{L^\infty} = M > m$ for some solution u. It follows from (2.6), using the first equation in (2.1) that

$$\frac{\delta}{\gamma}\min u \leq \Delta u + f(x,u) \leq \frac{\delta}{\gamma}\max u. \tag{2.9}$$

If there is $x_0 \in \Omega$ such that $u(x_0) = M$, we obtain from (2.9) and (2.8) that

$$-\frac{\delta}{\gamma}M \leqq f(x_0, u(x_0)) \leqq -ku(x_0) = -kM,$$

which is impossible. In a similar way we arrive to a contradiction if $u(x) = -M$ for some $x \in \Omega$. □

Next we discuss the question of bounds for positive solutions of the system (2.1). As remarked before we need only to obtain bounds on u, and then corresponding bounds on v follow readily.

PROPOSITION 2.3. *In addition to* (f1) *assume the following condition*:

(f5) *there exists a constant* $m > 0$ *such that* $f(x,s) = 0$ *for* $s \geqq m$.

Then all nonnegative solutions u *of* (2.3) *are bounded above by* m.

Proof. Given a solution u of (2.1), define the function ω as $\omega(x) = u(x) - m$ for $u(x) > m$ and $\omega(x) = 0$ for $u(x) \leqq m$. Such a ω belongs to $H_0^1(\Omega)$. So it follows from (2.3) that

$$\int |\nabla\omega|^2 + (Bu, \omega) = \int f(x,u)\omega. \tag{2.10}$$

In view of (f5) and the fact that $(Bu)(x)>0$ for $x\in\Omega$, we conclude from (2.10) that $\int|\nabla\omega|^2=0$, which implies $\omega=0$. □

Remark 2.4. This proposition will be used as follows. Suppose that the function f is such that there is an $m>0$ for which $f(x,m)=0$. Then we consider system (2.1) with f replaced by a new function $\tilde{f}$ defined as f for $s\leqq m$ and as 0 for $s>m$. If for this new system we could find a nonnegative solution u, then by the proposition above such a u would be indeed a solution of the original system.

Remark 2.5. Now if $f(x,s)=f(s)$ satisfies (f2), $f(0)\geqq 0$ and $f(s)>0$ for $0<s<\varepsilon$, then either there is an $m>0$ such that $f(m)=0$ or f satisfies (f3). In the first case we treat the problem as in the previous remark. In the second case we proceed as in Proposition 2.1 and obtain an a priori bound on positive solutions.

Remark 2.6. Similar statements can be made for nonpositive solutions u.

Remark 2.7. *A sufficient condition for all (eventual) nontrivial solutions of* (2.1) *to be positive*. Assume that $\gamma+\lambda_1>\sqrt{\delta}$ and that $f(x,u)\geqq\alpha u$ for all u, where $-\gamma+2\sqrt{\delta}\leqq\alpha<\lambda_1$. Then *the nontrivial solutions* u of (2.1) *are positive in* Ω. From (2.3) we obtain $-\Delta u+Bu\geqq\alpha u$, and the result follows readily by Corollary 1.3.

Remark 2.8. The previous condition applied to Example 2 gives interesting conclusions. Indeed, we can in this case compute explicitly the value of m in (2.8). Then truncate f outside $|s|\geqq m$ in such a way that the new f has derivative equals to $-a$ for $|s|>m$. By Proposition 2.2 the solutions of (2.1) with this new f are the same as the solutions of the original equation. Moreover, from the way the truncation is done, it follows (by a straightforward calculation) that now $f(u)\geqq -au$, (where we are denoting also by f the truncated function) provided $\delta/\gamma<a$. So the previous sufficient condition applies. Summarizing, *the solutions of* (2.1), *in the case of Example* 2, *are positive if*

$$\frac{\delta}{\gamma}<sa\leqq\gamma-2\sqrt{\delta}\,. \tag{2.11}$$

Observe that, if (2.11) is assumed, then the condition $\gamma+\lambda_1>\sqrt{\delta}$ is automatically satisfied, cf. Remark 1.4. We remark that no solution of (2.3) in this example can be nonpositive (i.e. $u\leqq 0$ in Ω). In fact the solutions in general change sign.

3. Existence of positive solutions. We consider again system (2.1) of the previous section or its equivalent expression in the form of equation (2.3). In this section we examine the question of existence of a positive solution under an additional condition on the nonlinearity f at 0. In order to simplify the presentation in the sequel, we suppose that f does not depend on x. The case when f depends also on x can also be treated by the method used here; under appropriate conditions on f similar results may be obtained. So we assume the next condition.

(f6) $\liminf_{s\to 0}(f(s)/s)>\hat{\lambda}_1$.

Examples. Condition (f6) is satisfied, for instance, if (i) $f(0)>0$, or (ii) $f(s)$ is C^1 and $f'(0)>\lambda_1$. A special case of (ii) was considered in [11].

THEOREM 3.1. *Assume that* $\gamma+\lambda_1>\sqrt{\delta}$. *In addition to conditions* (f1), (f2) *and* (f6), *suppose that* f *is* C^1 *for* $s\geqq 0$ *and*

$$\inf\{f'(s):0\leqq s<\beta\}\geqq-\gamma+2\sqrt{\delta}\,, \tag{3.1}$$

where $\beta\leqq+\infty$ *is the first positive zero of* $f(s)$. *Then equation* (2.3) *has a positive solution* u, *or equivalently, system* (2.1) *has a pair* (u,v) *of positive solutions.*

Remark 3.1. The hypothesis $\gamma+\lambda_1>\sqrt{\delta}$ in Theorem 3.1 implies that $\hat{\lambda}=\lambda_1$. Recall also that under this hypothesis $-\gamma+2\sqrt{\delta}<\lambda_1$, and so we can make use of Corollary 1.3. The condition on the differentiability of f can be relaxed and in consequence (3.1) has to be replaced by an appropriate one-sided Lipschitz condition.

Proof of Theorem 3.1. (i) It follows from (f6) that there exist $\nu > \hat{\lambda}_1$ and $s_0 > 0$ such that $f(s) \geqq \nu s$ for $0 \leqq s \leqq s_0$. Thus $\varepsilon\phi_1$ is a subsolution of (2.3) for all ε such that $0 < \varepsilon < \varepsilon_0 = s_0/\max\phi_1$.

(ii) If $\beta < \infty$ then $\omega(x) \equiv \beta$ in Ω is a supersolution of (2.3). If $\beta = +\infty$ we construct a supersolution ω for (2.3) as follows. It follows from (f2) that there exist $-\gamma + 2\sqrt{\delta} < \mu < \hat{\lambda}_1$ and $C > 0$ such that $f(s) \leqq \mu s + C$. We then take ω as the solution of $-\Delta\omega + B\omega = \mu\omega + C$ in Ω, $\omega = 0$ on $\partial\Omega$. In view of Corollary 1.3, $\omega > 0$ in Ω and $\varepsilon > 0$ can be chosen in such a way that $\varepsilon\phi_1 < \omega$ in Ω.

(iii) So (2.3) possesses an ordered pair of a sub- and a supersolution. Now in order to apply the method of monotone iteration, it is still required that (a) $T_\lambda = (-\Delta + B - \lambda I)^{-1}$ be a positive operator for some real number λ, and (b) the function $s \to f(s) - \lambda s$, for the same λ, be nondecreasing in the interval $[0, \max\omega]$. These two requirements are accomplished if one chooses $\lambda = -\gamma + 2\sqrt{\delta}$. Indeed, (a) then follows by Corollary 1.3 and (b) follows from (3.1). Therefore the method of monotone iteration can be applied and one obtains a solution of (2.3) in the interval $[\varepsilon\phi_1, \omega]$.

Remark 3.2. It should be remarked that besides (f2) no growth condition is required on f.

Remark 3.3. A statement similar to Theorem 3.1 holds true for the existence of negative solutions of (2.1). In this case, condition (3.1) is replaced by

$$\inf\{f'(s) : \beta' < s \leqq 0\} \geqq -\gamma + 2\sqrt{\delta} \tag{3.1$^-$}$$

where $-\infty \leqq \beta' < 0$ is the first negative zero of $f(s)$. In order to prove such a result, we can reduce it to the situation of Theorem 3.1 by the substitution $z = -u$.

Example. $f(u) = au - u^3$ with $a > 0$. In this case $\beta = \sqrt{a}$, and $\min\{f'(s) : 0 \leqq u \leqq \beta\} = -2a$. So conditions (f6) and (3.1) are satisfied if $\hat{\lambda}_1 < a \leqq \gamma/2 - \sqrt{\delta}$. We then see that in this example there are values of a for which (2.3) has a positive solution provided

$$\hat{\lambda}_1 < \gamma/2 - \sqrt{\delta}. \tag{3.2}$$

Clearly this is the case for instance if γ is large. This is also the case if $\gamma > (\sqrt{3} + 1)\sqrt{\delta}$ and Ω is a sufficiently large ball. Indeed, for large balls λ_1 is essentially zero and this last inequality implies readily condition (3.2). Clearly in this example there is also a negative solution, namely $-u$, where u is the positive solution.

Comparison with the results of Lazer–McKenna. In [11] the following system is studied

$$-k\Delta u = \lambda u - h(u)u - v, \qquad -\Delta v + v = u \quad \text{in } \Omega \tag{3.3}$$

subject to Dirichlet boundary conditions. Under certain conditions on k, λ and h it is proved that

$$-k\Delta u + (1 - \Delta)^{-1}u = \lambda u - h(u)u, \tag{3.4}$$

which is an equivalent form of (3.3), has exactly three solutions. In [11] a topological degree argument is used, which does not give the sign of the two nontrivial solutions. Under essentially the same hypotheses, our Theorem 3.1 says that one of these solutions is positive and the other is negative. Our precise result is the following. We state only the one corresponding to the existence of a positive solution. A similar one can be drawn for the existence of a negative solution.

COROLLARY 3.2. *Under the assumptions below, equation* (3.4) *has a positive solution*:

$$1+\lambda_1 > 1/\sqrt{k}, \tag{3.5}$$

$$h \in C^1(R,R), \quad h(0)=0 \quad h'(s)s>0 \quad \forall s \neq 0, \tag{3.6}$$

$$k\lambda_1 + \frac{1}{1+\lambda_1} < \lambda, \tag{3.7}$$

$$\sup\{h'(s)s+h(s): 0 \leqq s \leqq \beta\} \leqq \lambda + k - 2\sqrt{k}, \tag{3.8}$$

where β is the only positive solution of $h(s)=\lambda$. (Observe that β could be $+\infty$.)

Remark 3.4. If $h'(s)$ is nondecreasing, then $\beta<\infty$ and (3.8) simplify to $\beta h'(\beta) \leqq k-2\sqrt{k}$. So positive solutions of (3.4) exist if the *diffusion rate k is large*.

4. Remarks on a superlinear system. Consider the elliptic system

$$-\Delta u = f(u) - v, \qquad -\Delta v = \delta u - \gamma v \quad \text{in } \Omega, \tag{4.1}$$

subject to Dirichlet boundary conditions, with $\gamma, \delta > 0$ and $-\gamma + 2\sqrt{\delta} < 0$. Assume the following conditions on the nonlinearity f:

(f1)′ $f: R^+ \to R^+$ locally Lipschitzian,
(f7) $\liminf_{s\to+\infty}(f(s)/s) > \hat{\lambda}_1$,
(f8) $\limsup_{s\to 0}(f(s)/s) < \hat{\lambda}_1$,
(f9) $\lim_{s\to+\infty}(f(s)/s^\sigma)=0$ where $1<\sigma\leqq(N+1)/(N-1)$, if $N \geqq 3$ and $1<\sigma<\infty$, if $N=2$.

As seen in the previous section, (4.1) is equivalent to

$$-\Delta u + Bu = f(u). \tag{4.2}$$

Under the hypotheses above we may proceed as in the scalar case (cf. Brézis–Turner [2]) and we prove that (4.2) has a positive solution. Condition (f9) is used to get a priori bounds for the positive solutions of (4.2). We do not know how to proceed in order to obtain such bounds in the case when $(N+1)/(N-1)<\sigma<(N+2)/(N-2)$ and $N \geqq 3$. The results of [3] for the scalar case are not immediately extended to this case. For that purpose, the first step would be to see how the results of Gidas–Ni–Nirenberg [6] look (if at all!) in this case. We remark that the extension obtained by Troy [15] does not cover the type of systems studied in this paper.

Remark 4.1. The condition $-\gamma+2\sqrt{\delta}<0$ is used in order to guarantee that the operator $T_0 \equiv (-\Delta+B)^{-1}$ is positive. If this condition is not satisfied, but one has $\gamma+\lambda_1>\sqrt{\delta}$, everything still works provided $\hat{\lambda}_1$ in the right sides of assumptions (f7) and (f8) is replaced by $\hat{\lambda}_1 - \gamma + 2\sqrt{\delta}$.

Nonexistence of positive solutions in the case when $f(u)=u^p$, for $p \geqq (N+2)/(N-2)$ and $N \geqq 3$. As in the scalar case this is proved using an identity of the Pohozaev type. The function $f(u)=u^p$ for $u \geqq 0$ is extended as $f(u)=0$ for $u<0$. Then it follows from Remark 2.7 that all eventual solutions u and v of (4.1) are positive in Ω, provided we assume that $-\gamma+2\sqrt{\delta}<0$. Consequently the nonexistence of nontrivial solutions for system (4.1) (in *star-shaped* domains Ω) with such an f follows readily from the two lemmas below.

LEMMA 4.1. *Let u and v be solutions of* (4.1). *Then the following identity holds*

$$2N\int F(u) - (N-2)\int uf(u) - 2\int uv - \frac{2}{\delta}\int|\nabla v|^2 = \oint (x\cdot\nu)\left[|\nabla u|^2 - \frac{1}{\delta}|\nabla v|^2\right] \tag{4.3}$$

where $F(s)=\int_0^s f$ and $\int$ denotes (volume) integral over Ω and $\oint$ (surface) integral over $\partial\Omega$. Here ν denotes the outward unit normal.

LEMMA 4.2. *Let u and v be solutions of* (4.1). *Assume that* $-\gamma+2\sqrt{\delta}<0$. *Then* $u-(1/\sqrt{\delta})v$ *is positive in* Ω *and*

$$\frac{\partial u}{\partial \nu}<\frac{1}{\sqrt{\delta}}\frac{\partial v}{\partial \nu}<0 \quad on\ \partial\Omega.$$

To conclude this section, we prove the two lemmas above.

Proof of Lemma 4.1. First we use the general form of Pohozaev's identity for solutions of the $-\Delta u=g(x,u)$ in Ω and $u=0$ on $\partial\Omega$; see [3]. This identity will be applied separately to the first and second equations in (4.1). Observe that for the first equation, $g(x,s)=f(s)-v(x)$, and for the second equation, $g(x,s)=\delta u(x)-\gamma s$. Then we obtain the following two identities

$$2N\int[F(u)-uv]-2\int(x\cdot\nabla v)u-(N-2)\int[f(u)-v]u=\oint(x\cdot\nu)|\nabla u|^2, \tag{4.4}$$

$$2N\int\left[\delta uv-\frac{1}{2}\gamma v^2\right]+2\delta\int(x\cdot\nabla u)v-(N-2)\int[\delta u-\gamma v]v=\oint(x\cdot\nu)|\nabla v|^2. \tag{4.5}$$

(If one prefers to ignore [3], identities (4.4) and (4.5) may be obtained in the standard way Pohozaev's identities are proved. Use the multiplier $x\cdot\nabla u$ in the first equation of (4.1) and $x\cdot\nabla x$ in the second.) It follows from the divergence theorem that

$$\int(x\cdot\nabla v)u+\int(x\cdot\nabla u)v=-N\int uv. \tag{4.6}$$

Next dividing (4.5) through by δ, subtracting the result from (4.4) and using (4.6), we obtain

$$2N\int F(u)-(N-2)\int uf(u)-4\int uv+\frac{2\gamma}{\delta}\int v^2=\oint(x\cdot\nu)\left[|\nabla u|^2-\frac{1}{\delta}|\nabla v|^2\right]. \tag{4.7}$$

Now it follows from the second equation in (4.1) that

$$\int|\nabla v|^2-\delta\int uv-\gamma\int v^2. \tag{4.8}$$

Taking (4.8) into (4.7), we obtain the identity (4.3). □

Proof of Lemma 4.2. It follows from $-\gamma+2\sqrt{\delta}<0$ that there exists a real number k such that $\sqrt{\delta}<k<\gamma-\sqrt{\delta}$. Using (4.1), it is easy to check that

$$(-\Delta+k)\left(u-\frac{1}{\sqrt{\delta}}v\right)\geqq 0 \quad \text{in } \Omega$$

from which the assertion of the lemma follows. Observe that we know that all (eventual) solutions of (4.1) would be positive. □

5. Existence of two nontrivial solutions for a class of sublinear systems. Let us once more consider system (2.1) under conditions (f1), (f2), (f3) or (f4). As in previous sections we discuss, instead of system (2.1), its equivalent form given by equation (2.3). In this section we propose to treat the question of existence of solutions of (2.3) by a variational argument. So we look for the critical points of the functional

$$\Phi(u)=\frac{1}{2}\int|\nabla u|^2+\frac{1}{2}(Bu,u)-\int F(x,u) \tag{5.1}$$

where $F(x,s)=\int_0^s f(x,\xi)\,d\xi$. Although this functional is well defined in H_0^1 if we assume (f3), this is not the case if (f4) is assumed instead. Observe that both (f2) and (f4) restrict f only in one direction. So some truncation has to be done. The existence of a priori bounds on the solutions of (2.3) in either case ((f3) or (f4) assumed), as proved in §2, allows us to truncate the nonlinearity f in such a way that the functional Φ is *well defined* in H_0^1 and it is *bounded from below*. Indeed, if case (f3) is assumed, we choose an appropriate $\tilde{C}>\hat{C}$ and do this truncation for $|s|\geqq\tilde{C}$ (see Proposition 2.1 and Remark 2.3) preserving μ, M and C_ε and in such a way that $\lim_{|s|\to\infty}(f(x,s)/s)=l$ where $0<l<\tilde{\lambda}$. In case we assume (f4) the truncation is done for $|s|\geqq m$ (see Proposition 2.2), and in such a way that $\lim_{|s|\to\infty}(f(x,s)/s)=-k$, where the constant k is given in (2.8). The truncation so done has the very essential feature that the new equation (2.3) with this truncated function has the same solutions as the solutions of the original equation (2.3).

One sees immediately that $\Phi: H_0^1(\Omega)\to R$ is C^1 and

$$(\Phi'(u),\omega)_{H^1}=\int\nabla u\cdot\nabla\omega+\int(Bu)\omega-\int f(x,u)\omega. \tag{5.2}$$

So the critical points of Φ are the H_0^1 solutions of (2.3). By a bootstrap argument it follows that these solutions are in fact in $C^{2,\alpha}(\bar{\Omega})$.

LEMMA 5.1. *The functional Φ defined above satisfies the Palais–Smale condition.*

Proof. (i) In view of Poincaré's inequality we may consider H_0^1 endowed with the inner product $(u,\omega)_{H^1}=\int\nabla u\cdot\nabla\omega$. It is well known that the nonlinear operator $\tilde{f}: H_0^1\to H_0^1$ defined by $(\tilde{f}(u),\omega)_{H^1}=\int f(x,u)\omega$, $\forall\,\omega\in H_0^1$, is compact. (Recall that f has linear growth in view of the truncation.) On the other hand the (linear) operator $\tilde{B}: H_0^1\to H_0^1$ defined by $(\tilde{B}u,\omega)_{H^1}=\int(Bu)\omega$ is also compact. This follows readily from the compact imbedding of H_0^1 in L^2. Consequently $\Phi'=I+\tilde{B}-\tilde{f}$, that is, Φ' is of the form identity + compact operator. Thus to prove the Palais–Smale condition, it is enough to show that any sequence $(u_n\in H_0^1)$ such that $|\Phi(u_n)|\leqq C$ and $\Phi'(u_n)\to 0$ in H_0^1 possesses a subsequence (denoted again by u_n) such that $\|u_n\|_{H^1}\leqq C$.

(ii) It follows from $\Phi'(u_n)\to 0$ that given $\varepsilon_n\downarrow 0$ there exists a subsequence of (u_n) (denoted again by u_n) such that

$$\left|\int\nabla u_n\cdot\nabla\omega+\int(Bu_n)\omega-\int f(x,u_n)\omega\right|\leqq\varepsilon_n\|\omega\|_{H^1}. \tag{5.3}$$

Now using (5.3) with $\omega=u_n$ and estimating with the help of (1.8), we get

$$\tilde{\lambda}\int|u_n|^2\leq\int f(x,u_n)u_n+\varepsilon_n\|u_n\|_{H^1}. \tag{5.4}$$

From the properties of the truncated f we obtain from (5.4)

$$\int u_n^2\leqq C+C\varepsilon_n\|u_n\|_{H^1}. \tag{5.5}$$

Next from $|\Phi(u_n)|\leqq C$ we infer that

$$\int|\nabla u_n|^2\leq\int(Bu_n)u_n+2\int|F(x,u_n)|+C \tag{5.6}$$

and finally using the properties of the truncated f, we obtain from (5.6) and (5.5) that $\int|\nabla u_n|^2\leqq C+C\varepsilon_n\|u_n\|_{H^1}$ which proves that $\|u_n\|_{H^1}\leqq C$. □

Remark 5.1. It follows immediately from the previous remarks that system (2.1) has *at least one solution* under hypotheses (f1), (f2), and (f3) or (f4). Indeed, since Φ is C^1 functional, bounded below and satisfying the Palais–Smale condition, it follows that it has a global minimum u_1, $\Phi(u_1)=\inf\{\Phi(u):u\in H_0^1\}$. One cannot expect in general the existence of more solutions. Indeed if $f(u)=\lambda u$ with $\lambda<\tilde{\lambda}$, equation (2.3) in this case has only the trivial solution! So some additional assumption is necessary.

Now we treat a problem which is superlinear at 0, in the sense that the condition below holds:

(f10) f is differentiable at 0, $f(x,0)=0$, and $f'(x,0)<\tilde{\lambda}$.

Example 2 in §2 satisfies condition (f10).

THEOREM 5.2. *Assume conditions* (f1), (f2), (f3) *or* (f4), *and* (f10). *In addition suppose that there exists* $\xi>0$ *such that*

$$F(\xi)\geqq F(s)\quad \forall 0\leqq s\leqq \xi, \tag{5.7}$$

$$\frac{2F(\xi)}{\xi^2}>\min\left\{\frac{1}{R^2}\frac{(1+t)^2}{t^2}\frac{(1+t)^N-1}{2-(1+t)^N}+\frac{\delta}{\gamma}\frac{(1+t)^N}{2-(1+t)^N}:0<t<2^{1/N}-1\right\} \tag{5.8}$$

where R denotes the radius of the largest ball contained in Ω. Then equation (2.3) *has at least two nontrivial solutions.*

Remark 5.2. Condition (5.8) is the analogue of a condition introduced by one of the authors (D.G.F.) in [4] for the scalar case. We remark that if there is a $\xi>0$ such that $F(\xi)>0$ then the condition is satisfied (for example if Ω is a large ball and δ is very small). The special case of Example 2 was studied by Klaasen and Mitidieri [9]. Condition (5.8) follows readily from their conditions: (i) Ω to be a large ball, and (ii) $\gamma/\delta>9/(2a^2-5a+2)$.

Proof. It suffices to prove that there exists $\tilde{u}\in H_0^1$ such that $\Phi(\tilde{u})<0$. Once this is done, we see that the global minimum u_1 of Φ is a nontrivial solution since $\Phi(u_1)=\inf\Phi<0$. The second solution is obtained immediately by an application of the Mountain Pass Theorem of Ambrosetti–Rabinowitz [1], since 0 is a strict local minimum in view of assumption (f10). In order to see that there are points in H_0^1 where the functional Φ is negative we consider the functions u_t below. We may assume that the ball centered at 0 with radius R is contained in Ω, where R is the radius of the largest ball contained in Ω. Defining

$$u_t(x)=\begin{cases}\xi & \text{if } \|x\|\leqq R/(1+t),\\ \xi\left[1-\frac{1+t}{tR}\left(\|x\|-\frac{R}{1+t}\right)\right] & \text{if } \frac{R}{1+t}\leqq\|x\|\leqq R,\\ 0 & \text{if } x\in\Omega\setminus B_R(0),\end{cases}$$

the result follows by a calculation from conditions (5.7) and (5.8). □

REFERENCES

[1] A. AMBROSETTI AND P. H. RABINOWITZ, *Dual variational methods in critical point theory and applications*, J. Funct. Anal., 14 (1973), pp. 349–381.

[2] H. BRÉZIS AND R. L. TURNER, *On a class of superlinear elliptic problems*, Comm. Partial Differential Equations, 2 (1977), pp. 601–614.

[3] D. G. DE FIGUEIREDO, P.-L. LIONS AND R. NUSSBAUM, *A priori estimates and existence of positive solutions of semilinear elliptic equations*, J. Math. Pures Appl., 61 (1982), pp. 41–63.

[4] D. G. DE FIGUEIREDO, *Positive solutions for some classes of semilinear elliptic problems*, to appear.

[5] R. FitzHugh, *Impulses and physiological states in theoretical models of nerve membrane*, Biophysic. J., 1 (1961), pp. 445–466.

[6] B. Gidas, W. M. Ni and L. Nirenberg, *Symmetry and related properties via the maximum principle*, Comm. Math. Phys., 68 (1979), pp. 209–243.

[7] S. P. Hastings, *Some mathematical problems from neurobiology*, Amer. Math. Monthly, 82 (1975), pp. 881–895.

[8] G. Klaasen and W. Troy, *Standing wave solutions of a system of reaction-diffusion equations derived from FitzHugh–Nagumo equations*, SIAM J. Appl. Math., 44 (1984), pp. 96–110.

[9] G. Klaasen and E. Mitidieri, *Standing wave solutions for a system derived from the FitzHugh–Nagumo equations for nerve conduction*, to appear.

[10] S. Koga and Y. Kuramoto, *Localized patterns in reaction-diffusion systems*, Progr. Theoret. Phys., 63 (1980), pp. 106–121.

[11] A. C. Lazer and P. J. McKenna, *On steady state solutions of a system of reaction-diffusion equations from biology*, Nonlinear Anal., Theory, Meth. Appl., 6 (1982), pp. 523–530.

[12] P. H. Rabinowitz, *Pairs of positive solutions of nonlinear elliptic partial differential equations*, Indiana Univ. Math. J., 23 (1973), pp. 173–186.

[13] F. Rothe and P. de Mottoni, *A simple system of reaction-diffusion equations describing morphogenesis: Asymptotic behavior*, Ann. Mat. Pura Appl., 122 (1979), pp. 141–157.

[14] F. Rothe, *Global existence of branches of stationary solutions for a system of reaction diffusion equations from biology*, Nonlinear Anal., Theory, Meth. Appl., 5 (1981), pp. 487–498.

[15] W. C. Troy, *Symmetry properties in systems of semilinear elliptic equations*, J. Differential Equations, 42 (1981), pp. 400–413.

JOURNAL OF MATHEMATICAL ANALYSIS AND APPLICATIONS **123**, 170–180 (1987)

On the Uniqueness of Solution for a Class of Semilinear Elliptic Problems

D. G. COSTA, D. G. DE FIGUEIREDO, AND J. V. A. GONÇALVES

Universidade de Brasilia Departamento de Matemática,
Caixa Postal 153101, 70910 Brasilia, DF-Brasil

Submitted by E. Stanley Lee

Received August 1, 1985

INTRODUCTION

We shall discuss here the uniqueness of solution of the Dirichlet problem

$$-\Delta u = f(u) + \rho h(x) \quad \text{in } \Omega, \qquad u = 0 \quad \text{on } \partial\Omega, \tag{1}$$

for large values of the real parameter ρ. Here Ω is a bounded smooth domain in $\mathbb{R}^N$, h is a given real-valued function in $L^2(\Omega)$ and the nonlinearity $f: \mathbb{R} \to \mathbb{R}$ is a C^1 function satisfying additional conditions that will be timely posed as we go along. By a solution of (1) we *always* mean a strong solution, that is, $u \in H^2(\Omega) \cap H_0^1(\Omega)$. Let us explain at the outset why such an uniqueness result is to be expected in certain cases. Through the change of variables $v = u/\rho$, problem (1) becomes

$$-\Delta v = \rho^{-1} f(\rho v) + h \quad \text{in } \Omega, \qquad v = 0 \quad \text{on } \partial\Omega. \tag{2}$$

Let us assume that the limits $f_\pm = \lim_{s \to \pm\infty} f'(s)$ exist as real numbers. Thus the problem

$$-\Delta v = f_+ v^+ - f_- v^- + h \quad \text{in } \Omega, \qquad v = 0 \quad \text{on } \partial\Omega, \tag{3}$$

may be viewed as the limiting problem to (2), that is, "problem (2) with $\rho = \infty$"; this will be made precise shortly. If adequate hypotheses on f_+, f_-, and on h are made, so as to have existence and uniqueness of solution for problem (3), then it is reasonable to expect that existence and uniqueness of solution will also hold for problem (2) with large ρ's. In fact we shall start by treating the question of uniquenes separately from that of existence and uniqueness; it turns out that, under some mild nondegeneracy conditions and some a priori bounds, problem (2) has at most

0022-247X/87 $3.00

Originally published in: J. Math. Anal. Appl. 123 (1987), 170-179,

DOI: 10.1007/978-3-319-02856-9_18

one solution (for large ρ's) provided the same is true for the limiting problem (3). This is the contents of Theorem 1.

Let us denote by $0 < \lambda_1 < \lambda_2 < \lambda_3 < \cdots$, the distinct eigenvalues of $-\Delta$ on Ω under Dirichlet boundary condition. It is well known that λ_1 is a simple eigenvalue and its corresponding one-dimensional eigenspace N_1 is generated by a positive eigenfunction ϕ_1, that is $\phi_1(x) > 0$ for $x \in \Omega$ and $\phi_1(x) = 0$ for $x \in \partial\Omega$. The other eigenvalues λ_k, $k \geqslant 2$, are multiple in general but their corresponding eigenspaces N_k are finite-dimensional. We observe that, for $k \geqslant 2$, no eigenfunction $\psi \in N_k$ has a definite sign. Now we single out two conditions on the nonlinearity f, which are assumed throughout this paper, except in Section 4: (a) $f: \mathbb{R} \to \mathbb{R}$ is a C^1 function, and (b) the limits below exist as real numbers:

$$f_{\pm} = \lim_{s \to \pm\infty} f'(s).$$

Theorem 1 is presented for didactical reasons. It isolates a set of three conditions (two conditions of nondegeneracy and an a priori estimate) which are precisely what is needed to assert the uniqueness of solution of problem (2) from that of the limiting problem (3).

THEOREM 1. *Assume two nondegeneracy conditions and an a priori estimate as follows*:

(ND1) $h(x) \neq 0$ *for* $x \in \Omega$ *(a.e.)*,

(ND2) *for any pair of subsets A, B of Ω, with positive measure, such that $\Omega = A \cup B$ (a.e.) the Dirichlet problem below has only the trivial solution*

$$-\Delta\omega = (f_+ \chi_A + f_- \chi_B)\omega \quad in\ \Omega, \qquad \omega = 0 \quad on\ \partial\Omega,$$

where χ_A denotes the characteristic function of the set A,

(E) *there are positive numbers M and $\bar{\rho}$ such that $\|v\|_H{}^2 \leqslant M$ for all solutions v of problem* (2) *with $\rho \geqslant \bar{\rho}$.*

Then the uniqueness of solution of problem (3) *implies the same for problem* (2).

The theorem although interesting poses immediately the questions of when does the limiting problem (3) have a unique solution, and when do conditions (ND2) and (E) hold true. In other words, which further conditions on f and h are needed to ensure the validity of those hypotheses. We have some answers which are contained in the following theorems. We shall consider different cases depending on the positioning of f_+ and f_- with respect to the eigenvalues λ_k's. Here we do not consider jumping nonlinearities or crossing of eigenvalues (i.e., we assume that there are no eigenvalues strictly between f_+ and f_-), but we allow f_+ or f_- or both to be equal to eigenvalues. As we shall see shortly the consideration of different cases according to the following conditions is quite natural:

$(\mathrm{H1})_0$ $\quad f_+, f_- \leqslant \lambda_1$ and $(f_+, f_-) \neq (\lambda_1, \lambda_1)$,

$(\mathrm{H1})_k, k \geqslant 1$ $\quad \lambda_k \leqslant f_+, f_- \leqslant \lambda_{k+1}$ and $(f_+, f_-) \neq (\lambda_i; \lambda_i)$, $i = k, k+1$,

and the corresponding nonreasonance conditions:

$(\mathrm{H2})_0$ $\quad \int h\phi_1 < 0$ if $f_+ = \lambda_1$, or $\int h\phi_1 > 0$ if $f_- = \lambda_1$,

$(\mathrm{H2})_1$ $\quad h \notin N_2^{\perp}$ if f_+ or f_- is λ_2, and $\int h\phi_1 > 0$ if $f_+ = \lambda_1$ or $\int h\phi_1 < 0$ if $f_- = \lambda_1$,

$(\mathrm{H2})_k, k \geqslant 2$ $\quad h \notin N_i^{\perp}$ if f_+ or f_- is λ_i, $i = k, k+1$.

Notation. $N_i^{\perp}$ denotes the orthogonal complement of N_i in $L^2(\Omega)$. The integrals are taken over the whole of Ω. When we write (H1) we mean either one of the hypotheses above: $(\mathrm{H1})_0$ or $(\mathrm{H1})_k$, $k \geqslant 1$.

THEOREM 2. *Assume* (H1) *and the corresponding* (H2). *Then the limiting problem* (3) *has a unique solution and* (ND2) *holds, provided the following condition is satisfied*

(H3) $\quad h \in L^p(\Omega)$ *for some* $p > N$, $\qquad$ *if* f_+ *or* f_- *is* λ_1.

Remark. (H3) is vacuous in the case $(\mathrm{H.})_k$ for $k \geqslant 2$. Theorem 2 is an existence and uniqueness result. Condition (H3) is assumed *only* for existence of solutions of (3).

To obtain the a priori bound we need an additional condition on f, but again only in the case when f_+ or f_- is λ_1. Namely, for some $s_0 > 0$,

$(\mathrm{H4})_0$ $\quad f(s)/s \leqslant \lambda_1$ for $s \geqslant s_0\, [s \leqslant -s_0]$ if $f_+ = \lambda_1\, [f_- = \lambda_1]$.

$(\mathrm{H4})_1$ $\quad f(s)/s \geqslant \lambda_1$ for $s \geqslant s_0\, [s \leqslant -s_0]$ if $f_+ = \lambda_1\, [f_- = \lambda_1]$.

THEOREM 3. *Assume either one of the three following sets of conditions*:

(i) $(\mathrm{H1})_0, (\mathrm{H}_2)_0, (\mathrm{H3}), (\mathrm{H4})_0$,

(ii) $(\mathrm{H1})_1, (\mathrm{H2})_1, (\mathrm{H3}), (\mathrm{H4})_1$,

(iii) $(\mathrm{H1})_k, (\mathrm{H2})_k, k \geqslant 2$.

Then the a priori estimate (E) *holds true.*

In Section 1 we prove Theorem 1. In Section 2 we study the limiting equation (3) and prove Theorem 2. In Section 3 we discuss a priori bounds for the solutions of (2) under hypotheses $(\mathrm{H.})_k$, $k \geqslant 1$, and prove Theorem 3 under hypotheses (ii) and (iii). The case (i) is contained in Section 4 where the a priori bounds are obtained under more general conditions, namely f_+ and f_- are $\leqslant \lambda_1$, but one of them or both could be $-\infty$. Existence of solutions of (1) under he hypotheses considered here is well known. For completeness in Section 5 we explain how the a priori bounds obtained previously give quickly such an existence by topological degree arguments. The following statement is an immediate corollary of the above theorems.

THEOREM 4. *In addition to* (ND1) *assume either one of the three sets of conditions in Theorem* 3. *Then problem* (1) *has one and only one solution for large* ρ.

Remarks. (1) The theorems above pose no restriction upon h, other than (ND1), if f_+ and f_- are not eigenvalues. In that case conditions $(\mathrm{H2})_k$, $k \geqslant 0$, (H3), $(\mathrm{H4})_0$ and $(\mathrm{H4})_1$ would disappear from the statements.

(2) The case when f_+ or f_- is equal to the eigenvalue λ_1 is much more involved. Here the fact that the eigenfunctions corresponding to λ_1 have definite sign causes some difficulty. Conditions $(\mathrm{H2})_0$ and prt of condition $(\mathrm{H2})_1$ are conditions of the so called Landesman–Lazer type, [1].

(3) The novelty here is the treatment of uniqueness. Existence of at least one solution has been well studied in the literature. We mention Brézis–Nirenberg [2], Kazdan–Warner [3], where extensive references to additional work in this area could be found. In particular, existence under the set of assumptions (i) has been discussed by the Figueiredo–Gossez [4], under more general conditions. Also, existence under the set of assumptions (ii) and (iii) was fully discussed in Berestycki-de Figueiredo [5].

1. PROOF OF THEOREM 1

We shall need the following result:

LEMMA 1. *Let* (α_n) *be a sequence of positive real numbers, with* $\alpha_n \to +\infty$. *Let* $\omega_n, \omega \in L^2(\Omega)$, *with* $\omega_n \to \omega$ *in the* L^2-*sense. Then*

$$\alpha_n^{-1} f(\alpha_n \omega_n) \to f_+ \omega^+ - f_- \omega^- \quad \text{in } L^2. \tag{4}$$

Remarks. A similar result is proved in Gallouet–Kavian [6] under slightly stronger assumptions. We shall use the notation: $l(\omega) = f_+ \omega^+ - f_- \omega^-$.

Proof of Lemma 1. We shall use the fact that $f_n \equiv \alpha_n^{-1} f(\alpha_n \omega_n)$ converges to $l(\omega)$ in L^2 if and only if every subsequence of f_n possesses a further subsequence converging to $l(\omega)$ in L^2. Therefore, without loss of generality, we may assume that $\omega_n \to \omega$ a.e. and that $|\omega_n| \leqslant g$ with $g \in L^2$. It follows readily from the Lebesgue Convergence Theorem that $f_n \chi_{\{\omega \neq 0\}} \to l(\omega)$ in L^2. On the other hand, from the linear growth $|f(s)| \leqslant a\,|s| + b$ we have $|f_n| \leqslant a\,|\omega_n| + \alpha_n^{-1} b$, hence $f_n \chi_{\{\omega = 0\}} \to 0$ in L^2. ∎

Proof of Theorem 1. Suppose by contradiction that there is a sequence $\rho_n \to +\infty$, such that problem (2) with $\rho = \rho_n$ has two distinct solutions $v_n, \bar{v}_n$. Now we claim that the sequences (v_n), $(\bar{v}_n)$ both converge in H^1 to the unique solution v_∞ of (3). Indeed, it follows from (E) that given any subsequence of (v_n) there is a a further subsequence, denoted again by (v_n), which converges in H^1 to some function $\hat{v} \in H^1$. Using Lemma 1 we obtain

$$\int \nabla \hat{v} \cdot \nabla \psi = \int l(v)\psi + \int h\psi, \qquad \forall \psi \in \mathscr{D}(\Omega), \tag{5}$$

that is, $\hat{v}$ is a weak solution of (3). By the elliptic regularity it follows that $\hat{v} \in H^2(\Omega)$, and consequenctly (3) possesses a solution. [Observe that although uniqueness of solution of (3) is assumed in Theorem 1, its existence is not.] Since such a solution is unique, by hypothesis, we have that $\hat{v} = v_\infty$. Therefore the whole sequence (v_n) converges to v_∞. The same is true for $(\bar{v}_n)$. Next let us denote by $\omega_n = v_n - \bar{v}_n$. Using the mean value theorem we have

$$-\Delta \omega_n = f'(\theta_n)\omega_n \quad \text{in } \Omega, \qquad \omega_n = 0 \quad \text{on } \partial\Omega, \tag{6}$$

where $\theta_n(x)$ is some number between $\rho_n v_n(x)$ and $\rho_n \bar{v}_n(x)$ for each $x \in \Omega$ (a.e.). It is not clear (and it may not be true) that the function $x \mapsto f'(\theta_n(x))$ is measurable; nevertheless the function $x \mapsto f'(\theta_n(x))\,\omega_n(x)$ is! Passing to subsequences, if necessary, we can assume that both (v_n) and $(\bar{v}_n)$ converge a.e. to v_∞. Let us denote by $\Omega_+ = \{x \in \Omega : v_\infty(x) > 0\}$, $\Omega_- = \{x \in \Omega : v_\infty(x) < 0\}$, and $\Omega_0 = \{x \in \Omega : v_\infty(x) = 0\}$. It follows from the fact that $h \neq 0$ a.e. in Ω, that Ω_0 has measure zero (use Lemma A4 of [7, p. 53]). So $\Omega = \Omega_+ \cup \Omega_-$ a.e. Denoting by $\tilde{\omega}_n = \omega_n / \|\omega_n\|_{H^1}$ we obtain from (6),

$$-\Delta \tilde{\omega}_n = f'(\theta_n)\,\tilde{\omega}_n \quad \text{in } \Omega, \qquad \tilde{\omega}_n = 0 \quad \text{on } \partial\Omega. \tag{7}$$

On the other hand, passing to a subsequence, if necessary, we may assume that there is $\tilde{\omega} \in H_0^1(\Omega)$ such that $\tilde{\omega}_n \rightharpoonup \tilde{\omega}$ in H^1, $\tilde{\omega}_n \to \tilde{\omega}$ in L_2 and a.e. Hence

$$f'(\theta_n)\tilde{\omega}_n \to (f_+ \chi_{\Omega_+} + f_- \chi_{\Omega_-})\tilde{\omega} \equiv m(x)\tilde{\omega} \qquad \text{a.e. in } \Omega.$$

Thus by the Lebesgue Convergence Theorem we obtain that $\tilde{\omega}$ is an H_0^1 solution of $-\Delta\tilde{\omega} = m(x)\tilde{\omega}$, and consequently by elliptic regularity a strong solution. An easy argument involving this equation and (3) with $v = v_\infty$ leads to $|\Omega_-|\,|\Omega_+| > 0$. Hence condition (ND2) with $A = \Omega_+$ and $B = \Omega_-$ gives $\tilde{\omega} = 0$ a.e. in Ω. But, on the other hand, from (7) we have

$$1 = \int |\nabla\tilde{\omega}_n|^2 = \int f'(\theta_n)\tilde{\omega}_n^2$$

which gives $1 = \int m(x)\tilde{\omega}^2$, contradicting $\tilde{\omega} = 0$ a.e. in Ω. ∎

2. The Limiting Equation

This section is devoted to the proof of Theorem 2.

I. *Uniqueness*

Suppose by contradiction that v_1 and v_2 are two distinct solutions of the limiting equation (3). Then $0 \neq v = v_2 - v_1$ is a solution of the problem

$$-\Delta v = m(x)v \quad \text{in } \Omega, \qquad v = 0 \quad \text{on } \partial\Omega, \tag{8}$$

where the weight function $m(x)$ is defined by

$$m(x) = \begin{cases} \dfrac{l(v_2(x)) - l(v_1(x))}{v_2(x) - v_1(x)} & \text{if } \ v_2(x) \neq v_1(x) \\ \bar{\lambda} & \text{if } \ v_2(x) = v_1(x) \end{cases}$$

with $\bar{\lambda} = (\lambda_k + \lambda_{k+1})/2$ in the case of hypothesis $(\text{H1})_k$, when $k \geqslant 1$, and $\bar{\lambda}$ is any fixed real number $< \lambda_1$, in the case of hypothesis $(\text{H1})_0$. Observe that $m \in L^\infty(\Omega)$ and we claim that

$$\lambda_k \leqslant m(x) \leqslant \lambda_{k+1}, \qquad k \geqslant 1, \qquad \text{or} \qquad m(x) \leqslant \lambda_1 \quad \text{in } \Omega \quad \text{(a.e.)}, \tag{9}$$

and all the inequalities above are strict on subsets of Ω with positive measre. Indeed, if we assume by contradiction that $m(x) = \lambda_j$ in Ω (a.e.) for some $j \geqslant 1$, then it would result the three following facts: (i) both v_1 and v_2 have the same definite sign all over Ω (a.e.), (ii) f_+ or f_- is equal to λ_j, and (iii) v_1 and v_2 satisfy

$$-\Delta v_i = \lambda_j v_i + h \quad \text{in } \Omega, \qquad v_i = 0 \quad \text{on } \partial\Omega, \ i = 1, 2. \tag{10}$$

Consequently $h \in N_j^\perp$, which contradicts hypothesis (H2). To complete the uniqueness proof we now use comparison of eigenvalues. Let $\mu_j(m)$ be the eigenvalues of $-\Delta v = \mu m(x) v$ in Ω, $v=0$ on $\partial\Omega$. From (8) it follows that $\mu_p(m)=1$ for some positive integer p. On the other hand, from the inequalities (9) and the assertion right after them, we obtain

$$\mu_p(\lambda_k) > \mu_p(m) > \mu_p(\lambda_{k+1}) \qquad \text{for } k \geqslant 1, \quad \text{or} \quad \mu_p(m) > \mu_p(\lambda_1).$$

Hence, using the fact that $\mu_p(m)=1$, we have that $\lambda_p/\lambda_k > 1 > \lambda_p/\lambda_{k+1}$, $k \geqslant 1$, or $1 > \lambda_p/\lambda_1$, which is impossible in either case. ∎

Remark. We observe that in the case of hypotheses $(\mathrm{H1})_k$ and $(\mathrm{H2})_k$, with $k \geqslant 2$, the uniqueness of solution of (3) follows even when $h=0$. [That is, $\omega=0$ is the only solution of (3) in this case.] Indeed, in this case, the equation in (3) could be written as $-\Delta v = (f_+ \chi_{v>0} + f_- \chi_{v<0}) v$ and $m(x) = f_+ \chi_{v>0} + f_- \chi_{v<0}$ satisfies the inequalities $\lambda_k \leqslant m(x) \leqslant \lambda_{k+1}$, $k \geqslant 2$. Moreover, both inequalities are strict in sets of positive measure, because, otherwise we would have $-\Delta v = \lambda_k v$ in Ω (a.e.) with v possessing a definite sign, which is impossible! This has been observed before by Dancer [8].

II. *Existence*

Existence follows readily from usual topological degree arguments, as soon as we have established an a priori bound on the solutions of the parametrized family of Dirichlet problems:

$$-\Delta v = (1-\theta)\, l(v) + \theta \bar{\lambda} v + h \quad \text{in } \Omega, \qquad v=0 \quad \text{on } \partial\Omega,$$

where $0 \leqslant \theta \leqslant 1$ and $\bar{\lambda}$ is the real number defined in the uniqueness part proved above. However this is a special case of Lemmas 2 and 4, when $f(s) = \bar{\lambda} s$.

3. A priori Bounds under Hypotheses $(\mathrm{H.})_k$, $k \geqslant 1$

Viewing later uses of topological degree arguments we prove an a priori bound on the solutions of the parametrized family of problems

$$-\Delta v = \theta \rho^{-1} f(\rho v) + (1-\theta)\, l(v) + h \quad \text{in } \Omega. \qquad v=0 \quad \text{on } \partial\Omega, \quad (\mathrm{P}_{\rho,\theta})$$

where $\rho \geqslant 1$, $\theta \in [0,1]$.

LEMMA 2. *Assume hypotheses* $(\mathrm{H1})_k$ *and* $(\mathrm{H2})_k$ *for some* $k \geqslant 1$. *In addition assume* (H3) *and* $(\mathrm{H4})_1$, *whenever* f_+ *or* f_- *is equal to* λ_1. *Then there is a constant* $C>0$ *independent of* ρ *and* θ *such that* $\|v\|_{H^1} \leqslant C$ *for all eventual solutions* v *of* $(P_{\rho,\theta})$.

Proof. Suppose by contradiction that there are sequences (v_n), (ρ_n), (θ_n) such that $\|v_n\|_{H^1} \to \infty$, $\rho_n \to \bar{\rho} \in [1, \infty]$, $\theta_n \to \bar{\theta} \in [0, 1]$, and

$$-\Delta v_n = \theta_n \rho_n^{-1} f(\rho_n v_n) + (1-\theta_n)\, l(v_n) + h \quad \text{in } \Omega, \qquad v_n = 0 \quad \text{on } \partial\Omega. \tag{11}$$

Let $\omega_n = v_n / \|v_n\|_{H^1}$ and suppose without loss of generality that $\omega_n \rightharpoonup \omega$ in H_0^1, $\omega_n \to \omega$ in L^2 and a.e. in Ω. Dividing through the Eq. (11) by $\|v_n\|_{H^1}$ and passing to the limit we obtain that ω is a solution of $-\Delta\omega = l(\omega)$ in Ω with $\omega = 0$ on $\partial\Omega$. By the results of Section 2 we conclude that either (i) $\omega = 0$, or (ii) $\omega = k\phi_1$, with $k > 0$ [or $k < 0$] if f_+ [or f_-] is equal to λ_1. We next show that either possibility leads to a contradiction. Indeed from (11) it follows

$$1 = \int |\nabla \omega_n|^2 = \theta_n \rho_n^{-1} \|v_n\|_{H^1}^{-1} \int f(\rho_n v_n)\omega_n + (1-\theta_n) \int l(\omega_n)\omega_n$$
$$+ \int h\, \|v_n\|_{H^1}^{-1} \omega_n$$

which is impossible for large n, in the case when $\omega = 0$. Now suppose that $f_+ = \lambda_1$ and $\omega = k\phi_1$, with $k > 0$. The other case is treated similarly. In view of (H3) we obtain through a botstrapping argument that $\omega_n \to \omega$ in $C^{1,\alpha}$. Hence $v_n > 0$ for large n. On the other hand it follows from (11), under the present hypotheses, that

$$0 = \theta_n \int [f(\rho_n v_n) - \lambda_1 \rho_n v_n]\, \rho_n^{-1} \phi_1 + \int h\phi_1.$$

Using $(H4)_1$ and Fatou's lemma, it follows that the lim inf of the first integral above is $\geqslant 0$, which contradicts $(H2)_1$. ∎

Lemma 2 is precisely Theorem 3 under the set of conditions (ii) or (iii).

4. A priori Bounds under Hypotheses $(H.)_0$

As promised in the Introduction we now obtain a priori bounds on the solutions of (2) under a hypothesis weaker than (b), in the case $f_\pm \leqslant \lambda_1$, namely, (b′) the limits $f_\pm = \lim_{s \to \pm\infty} f'(s)$ exist as real numbers or $-\infty$. Nevertheless, the simple proof below requires a stronger condition on h, namely $h \in L^\infty(\Omega)$.

(i) First, suppose that $\lim_{s \to -\infty} f'(s) = -\infty$. Then given $M > 0$ there exists $s_M > 0$ such that $f(s)s^{-1} < -M$ for $s < -s_M$. Hence there is $C_M > 0$, such that

$$f(s) \geqslant -Ms - C_M, \qquad \forall s \leqslant 0; \tag{12}$$

one may choose C_M such that $\|h\|_{L^\infty} \leqslant C_M$. And now we claim

LEMMA 3. *Suppose that* $\lim_{s \to -\infty} f'(s) = -\infty$. *Then there exists a constant* $C > 0$, *depending only on* M *and* C_M *given in the estimate* (12), *such that*

$$-C \leqslant v(x), \qquad \forall x \in \Omega, \tag{13}$$

for all eventual solutions of (2) *with* $\rho \geqslant 1$.

Remark. The bound (13) depends on f through (12). We may truncate f for $s < -C$ in such a way that the truncated $\tilde{f}$ is such that $\tilde{f}'(s) \to l < -M$, as $s \to -\infty$, where l is some real number. And then we see that problem (2) with $\tilde{f}$ replacing f has the same solutions as the original problem (2).

Proof of Lemma 3. Let $\omega \in C^{1,\alpha}(\bar{\Omega})$ be the unique solution of

$$-\Delta\omega = -M\omega - C_M + h \quad \text{in } \Omega, \qquad \omega = 0 \quad \text{on } \partial\Omega.$$

By the choice of C_M, we see that $\omega < 0$ in Ω. We claim that $\omega(x) \leqslant v(x)$, $x \in \Omega$, for all eventual solutions v of (2). Once this is done the proof will be completed by taking $-C = \inf \omega$. Thus suppose by contradiction that there is a solution v of (2) such that $\omega(x) > v(x)$ for some $x \in \Omega$. Let $z = v - \omega$. So there is a $z_0 \in \Omega$ where

$$0 > z(x_0) = \min\{z(x) : x \in \Omega\} \qquad \text{and} \qquad -\Delta z(x_0) \leqslant 0. \tag{14}$$

On the other hand we have

$$-\Delta z = -\Delta v + \Delta\omega = \rho^{-1} f(\rho v) + M\omega + C_M \tag{15}$$

and we prove that the right side of (15) evaluated at x_0 is >0, thus contradicting (14). Indeed, since $v(x_0) < \omega(x_0) < 0$ it follows from (12) that

$$\rho^{-1} f(\rho v(x_0)) \geqslant -Mv(x_0) - \rho^{-1} C_M > -M\omega(x_0) - C_M,$$

which finishes the proof.

(ii) In a similar way we prove that if $\lim_{s \to +\infty} f'(s) = -\infty$, then there is a constant $C > 0$ such that

$$v(x) \leqslant C, \qquad \forall x \in \Omega,$$

for all eventual solutions of (2). Consequently we may assume without loss of generality that both f_+ and f_- are finite. And now we state the a priori bound for the solutions of $(P_{\rho,\theta})$, in this case. The proof is identical to that of Section 3.

LEMMA 4. *In addition to* $f_\pm \in [-\infty, \lambda_1]$ *and* $h \in L^\infty(\Omega)$, *assume* $(\mathrm{H2})_0$ *and* $(\mathrm{H4})_0$. *Then there is a constant* $C > 0$ *independent of* ρ *and* θ *such that* $\|v\|_{H^1} \leqslant C$ *for all eventual solutions* v *of* $(P_{\rho,\theta})$.

Lemma 4 contains Theorem 3 under the set of conditions (i).

5. FINAL REMARKS

Lemma 2 says that the degree associated with Eq. (2) is the same as the degree associated with Eq. (3). The a priori bound proved in the existence part of Section 2 says that the degree associated with Eq. (3) is the same as the one associated with the linear problem

$$-\Delta v = \bar{\lambda} v + h \quad \text{in } \Omega, \qquad v = 0 \quad \text{on } \partial\Omega, \tag{16}$$

where $\bar{\lambda}$ is a fixed real number $\neq \lambda_k$ for all $k = 1, 2,\ldots$. But it is well known that the degree associated with problem (16) is $(-1)^\sigma$ where $\sigma = \sum_{\lambda_k < \bar{\lambda}} \dim N_k$. Consequently problem (2) possesses a solution. Of course, the foregoing remarks can be formalized by transforming the problems into functional equations and using all the apparatus of the theory of topological degree.

Another way of proving the uniqueness of solution of problem (2) for large ρ is to show that, under the set of conditions (H1), (H2) [and also the other ones in the case that f_+ or f_- is λ_1], all the eventual solutions of (2) for large ρ are nondegenerate and its index is precisely the degree associated with problem (2).

REFERENCES

1. E. M. LANDESMAN AND A. C. LAZER, Nonlinear perturbations of linear elliptic boundary value problems at resonance, *J. Math. Mech.* **19** (1970), 609–623.
2. H. BRÉZIS AND L. NIRENBERG, Characterizations of the ranges of some nonlinear operators and applications to boundary value problems, *Ann. Scuola Norm. Sup. Pisa* **V** (1978), 225–326.
3. J. L. KAZDAN AND F. W. WARNER, Remarks on quasilinear elliptic equations, *Comm. Pure Appl. Math.* **XXVIII** (1975), 567–597.
4. D. G. DE FIGUEIREDO AND J.-P. GOSSEZ, Nonlinear perturbations of a linear elliptic problem near its first eigenvalue, *J. Differential Equations* **30** (1978), 1–19.

5. H. Berestycki and D. G. de Figueiredo, Double resonance in semilinear elliptic problems, *Comm. Partial Differential Equations* **6** (1981), 91–120.
6. T. Gallouet and O. Kavian, Resultats d'existence et de non existence pour certains problèmes demi-linéaires a l'infini, *Ann. Fac. Sci. Toulouse* **II** (1981), 201-246.
7. D. Kinderlehrer and G. Stampacchia, "An Introduction to Variational Inequalities and Their Applications," Academic Press, New York, 1980.
8. E. N. Dancer, On the Dirichlet problem for weakly nonlinear elliptic partial differential equations, *Proc. Roy. Soc. Edinburg* **76** (1977), 283–300.

ROCKY MOUNTAIN
JOURNAL OF MATHEMATICS
Volume 18, Number 2, Winter 1988

ON SUPERLINEAR ELLIPTIC PROBLEMS WITH NONLINEARITIES INTERACTING ONLY WITH HIGHER EIGENVALUES

DJAIRO G. DE FIGUEIREDO

1. Introduction. Consider the Dirichlet problem

$$-\Delta u = g(x,u) \text{ in } \Omega, \quad \mu = 0 \text{ on } \partial\Omega, \tag{1}$$

where Ω is a bounded smooth domain in $R^N, N \geq 2$. For the purposes of this section we shall assume that $g : \overline{\Omega} \times R \to R$ is continuous, where $\overline{\Omega}$ denotes the closure of Ω. Grosso modo problem (1) is said to be *superlinear at* $+\infty$ if

$$\lim_{s\to+\infty} \frac{g(x,s)}{s} = +\infty. \tag{2}$$

In [**1**] Ambrosetti and Rabinowitz proved the existence of a solution for problem (1) under a set of conditions which we discuss next. First, growth assumptions at $\pm\infty$ on g were made so as to guarantee that the Euler-Lagrange functional

$$\Phi(u) = \frac{1}{2}\int |\nabla u|^2 - \int G(x,u), \quad G(x,s) = \int_0^s g(x,\xi)d\xi \tag{3}$$

is well defined in $H_0^1(\Omega)$. (Integrals $\int$ are supposed to be taken over the whole of Ω, unless indicated otherwise.) The critical points of Φ are then the H_0^1 solutions of (1). Also, in [**1**], the following condition is assumed in order to ensure that Φ satisfies the Palais-Smale condition:

$$\begin{array}{l}\text{There are numbers } \theta \in (0,1/2) \text{ and } s_0 > 0 \text{ such that} \\ \quad 0 < G(x,s) \leq \theta s g(x,s) \quad \text{for } |s| \geq s_0, x \in \Omega.\end{array} \tag{4}$$

We remark that condition (4) implies that the function $g(x,s)$ is superlinear in both directions, that is, at $\pm\infty$. Namely, (4) implies

$$\lim_{s\to\pm\infty} \frac{g(x,s)}{s} = +\infty.$$

Received by the editors on March 20, 1986.

Originally published in: Rocky Mountain J. Math. 18 (1988), 287-303,

DOI: 10.1007/978-3-319-02856-9_19

An adequate condition at $s = 0$ guarantees that 0 is a local minimum of Φ, and a nontrivial critical point of Φ is then obtained in [**1**] using the Mountain Pass Theorem. In this way the authors of [**1**] have introduced in the literature a simple but extremely useful tool to treating problems in Nonlinear Partial Differential Equations.

Another instance of a superlinear problem appears in the so called Ambrosetti-Prodi problem

$$(5) \qquad -\Delta u = g(x,u) + t\varphi + h \text{ in } \Omega, \quad u = 0 \text{ on } \partial\Omega,$$

where $\varphi > 0$ is an eigenfunction associated with the first eigenvalue λ_1 of $(-\Delta, H_0^1(\Omega))$, t is a real parameter and $\int h\varphi = 0$. In this problem, the nonlinearity g satisfies condition (2), but at $-\infty$ one has

$$\lim_{s\to-\infty} \frac{g(x,s)}{s} < \lambda_1.$$

This problem was studied in [**2, 4,** and **5**] by variational methods. Indeed, for $t < 0$ and large in absolute value, problem (5) possesses a solution which is a local minimum of the corresponding Euler-Lagrange functional. And again the Mountain Pass Theorem is invoked to get a second solution for problem (5).

The natural question now is: Suppose again that problem (1) is superlinear at $+\infty$, namely (2) is satisfied, but at $-\infty$ one has

$$(6) \qquad \lim_{s\to-\infty} \frac{g(x,s)}{s} = \lambda \quad \text{with } \lambda_k < \lambda < \lambda_{k+1},$$

where λ_k and λ_{k+1} are two consecutive eigenvalues of $(-\Delta, H_0^1(\Omega))$. As we shall see, the existence of one solution for this type of problem is not hard to establish. Nevertheless it may not be true that the associated Euler-Lagrange functional has a local minimum. This precludes the use of the Mountain Pass Theorem, as before, in order to obtain a second solution. However, the setting invites the application of the Generalized Mountain Pass Theorem, due to Rabinowitz [**8**]. Our present work on this problem was motivated by Ruf-Srikanth [**10**] which treated the nonlinearity $g(x,u) = \lambda u + (u^+)^p + t\varphi + h$, where λ is as in (6), t, φ, and h are as in problem (5), and $p > 1$ is restricted in a usual way, see below. In this paper we show that a result similar to the one in [**10**] can be established for a larger class of nonlinearities. Namely, we consider the following parametrized form of (1)

$$(7)_t. \qquad -\Delta u = \lambda u + f(x,u) + t\varphi + h \text{ in } \Omega, \quad u = 0 \text{ on } \partial\Omega,$$

where $\lambda_k < \lambda < \lambda_{k+1}, t$ is a real parameter, $\varphi > 0$ is an eigenfunction associated with the first eigenvalue of $(-\Delta, H_0^1(\Omega)), h \in C^\nu(\overline{\Omega}), 0 < \nu \le 1, \int h\varphi = 0$. In order to prove existence of two solutions for $(7)_t$ we assume that the nonlinearity f satisfies the following set of conditions:

(F1) $f : \overline{\Omega} \times R \to R$ is a C^1 function.

(F2) There exists $0 < \alpha < 1$ such that $\lim_{s\to-\infty} f(x,s)|s|^{-\alpha} = 0$.

(F3) $\lim_{s\to-\infty} f_s'(x,s) = 0$.

(F4) There are $0 < \theta < 1/2$ and $s_0 > 0$ such that $0 < F(x,s) \le \theta s f(x,s)$, for $s > s_0$ and all $x \in \Omega$.

(F5) $\lim_{s\to+\infty} f(x,s)s^{-\sigma} = 0$, where $\sigma \le (N+2)/(N-2)$ if $N \ge 3$ or $1 < \sigma < \infty$ if $N = 2$.

(F6) $f_s'(x,s) \ge -\mu$ where $\mu < \lambda - \lambda_k$.

(F7) $\sigma\theta < \min\{\frac{1}{1+\alpha}, \frac{N+2}{2N}\}$.

THEOREM. *Under assumptions* (F1) *through* (F7), *there exists a* $\hat{t} > 0$ *such that, for all* $t \ge \hat{t}$, *problem* $(7)_t$ *has (at least) two solutions.*

REMARKS ON THE ABOVE ASSUMPTIONS ON THE NONLINEARITY. (1) All the above limits are supposed to be uniform with respect to $x \in \Omega$.

(2) Conditions (F1), (F2) and (F5) imply that there is a constant $c > 0$ such that $|f(x,s)| \le C|s|^\sigma + C$. Consequently, the Euler-Lagrange functional associated with problem $(7)_t$ is well defined in $H_0^1(\Omega)$.

(3) Condition (F4) implies that $F(x,s) \ge Cs^{1/\theta} - C$ for $s > 0$, which in its turn implies that f is superlinear at $+\infty$: $f(x,s) \ge Cs^{(1/\theta)-1} - C$, for $s > 0$. This last inequality, together with (F5), implies that $\sigma > (1/\theta) - 1$. It then follows that

$$(*) \qquad 1/\theta < 2N/(N-2) \Rightarrow (N-2)/2N < \theta < 1/2 \text{ for } N \ge 3.$$

(4) Either condition (F2) or (F3) implies

$$\lim_{s\to-\infty} f(x,s)s^{-1} = 0,$$

which implies that problem (7) is not resonant at $-\infty$. And condition (F4) implies

$$\lim_{s\to+\infty} f(x,s)s^{-1} = +\infty$$

(5) Condition (F3) does not imply (F2). Indeed, if $f(x,s)$ is equal to $s|\ln s|^{-1}$ for $s < -1$, then f satisfies (F3) but not (F2). It is also easy to see that (F2) puts no restriction on the growth of $f'_s(x,s)$ at $-\infty$; so (F2) and (F3) are independent.

(6) Condition (F2) is verified, for instance, if $f(x,s)$ is bounded for $s < 0$. We recall that the case treated in [**10**] is $f(x,s) \equiv 0$ for $s < 0$, which gives also (F3) trivially.

(7) Condition (F7) is not as restrictive as it may appear at first sight. For instance, if $f(x,s)$ is a *pure power* at $+\infty$, say s^σ with $1 < \sigma < (N+2)/(N-2)$ if $N \geq 3$, then one could take $\theta = 1/(\sigma+1)$. Thus $\sigma\theta = 1-\theta$, and from (*) above we see that $\sigma\theta < (N+2)/2N$. So (F7) would be satisfied for all $\alpha \leq (N-2)/(N+2)$. If $N = 2$ and again the case of a pure power we observe that $\sigma\theta = \sigma/(\sigma+1) < 1$, and in this case one could take $\alpha < 1/\sigma$. Summarizing we see that, in the case when f is a pure power at ∞, condition (F7) establishes a *balance* between the growth σ at $+\infty$ and the growth α at $-\infty$. So (F7) seems a reasonable condition to be assumed.

(8) We do not know if conditions (F2) and (F6) are indeed necessary. (F2) enters in an apparently essential way to get the PS condition in §2. And(F6) permits one to do required "linking"in the Generalized Mountain Pass Theorem, done in §4.

REMARK. It is natural to ask if this problem has more than two solutions. In the ODE case a result of Ruf-Srikanth [**11**] shows that (for $f(x,u) = (u^+)^p, h = 0$) the problem has $2k+2$ solutions. This result has been extended to the PDE case when Ω is a ball and for a large class of nonlinearities by Padua [**7**]. Both the ODE case and the case when Ω is a ball use bifurcation arguments that are not applicable to the case of a general domain. It is also interesting to remark that if f is asymptotically linear at $+\infty$, then the work of Lazer and McKenna [**6**] shows that there are more solutions.

The Palais-Smale condition. In this section we show that the Euler-Lagrange functional associated with an equation from a certain class of superlinear elliptic problems satisfies the Palais-Smale condition (in short, the PS condition). We recall that a C^1 functional $\Phi : H_0^1 \to R$ satisfies the PS condition if every sequence (u_n) in H_0^1 with $|\Phi(u_n)| \leq$ const and $\Phi'(u_n) \to 0$ contains a convergent subsequence. If Φ' is of the form identity + compact, then, in order to check PS, it suffices to prove that $||u_n||_{H^1} \leq$ const. Let us consider problem (1) and let us assume the following conditions on the nonlinearity g:

(G1) $g : \overline{\Omega} \times R \to R$ is continuous.

(G2) There are $\lambda \neq \lambda_j$ for $j = 1, 2, \dots$ and $0 \leq \alpha < 2$ such that

$$\lim_{s \to -\infty} \frac{g(x,s) - \lambda s}{|s|^\alpha} = 0$$

(G3) $\lim_{s \to +\infty} g(x,s)s^{-\sigma} = 0$, where $\sigma \leq (N+2)/(N-2)$ if $n \geq 3$ or $1 < \sigma < \infty$ if $N = 2$.

(G4) There are $0 < \theta < 1/2$ and $s_0 > 0$ such that $0 < G(x,s) \leq \theta s g(x,s)$, for $s > s_0$ and all $x \in \Omega$.

(G5) $\sigma\theta < \min\{\frac{1}{1+\alpha}, \frac{N+2}{2N}\}$.

REMARKS ON THE ABOVE CONDITIONS. (1) All limits in the above conditions are supposed to hold uniformly with respect to $x \in \Omega$.

(2) Conditions (G1), (G2) and (G3) imply that there is a constant $c > 0$ such that $|g(x,s)| \leq C|s|^\sigma + C$. Consequently the Euler-Lagrange functional Φ in (3) is well defined in $H_0^1(\Omega)$.

(3) Conditions (G1) and (G2) imply that there is a constant C such that

$$|g(x,s)s - 2G(x,s)| \leq C + C|s|^{1+\alpha}, \quad s < 0, \tag{8}$$

$$|g(x,s)s - \lambda s^2| \leq C + C|s|^{1+\alpha}, \quad s < 0. \tag{9}$$

(4) Condition (G2) implies that

$$\lim_{s \to -\infty} g(x,s)s^{-1} = \lambda$$

and condition (G4) implies that

$$\lim_{s \to +\infty} g(x,s)s^{-1} = +\infty.$$

LEMMA 1. *If $g(x,s)$ satisfies conditions* (G1) *through* (G5), *then the functional Φ, as defined in* (3), *satisfies the Palais-Smale condition.*

PROOF. Let (u_n) be a sequence in $H_0^1(\Omega)$ for which

$$|\Phi(u_n)| = \left|\frac{1}{2}\int |\nabla u_n|^2 - \int G(x, u_n)\right| \leq C, \tag{10}$$

$$|\langle \Phi'(u_n), v\rangle| = \left|\int \nabla u_n \nabla v - \int g(x, u_n)v\right| \leq \varepsilon_n ||v||_{H^1}, \tag{11}$$

where $\langle\ ,\ \rangle$ denotes the inner product in $H_0^1, \varepsilon_n \to 0$ as $n \to \infty$ and v is any element in H_0^1. As we remarked above to prove that Φ satisfies the PS condition it suffices to prove that $||u_n||_{H^1}$ is bounded uniformly with respect to n. Taking $v = u_n$ in (11) and using (10) we obtain

$$\int [g(x, u_n)u_n - 2G(x, u_n)] \leq C + \varepsilon_n ||u_n||_{H^1}. \tag{12}$$

The integral in (12) is broken into three parts. The first is estimated using (8) as follows:

$$\left|\int_{u_n<0} [g(x, u_n)u_n - 2G(x, u_n)]\right| \leq C + C\int |u_n^-|^{1+\alpha},$$

where $u^- = \max\{0, -u\}$. The second integral taken over $0 \leq u_n \leq s_0$ is obviously bounded uniformly with respect to n. And the third one is estimated using (G4):

$$\int_{u_n>s_0} g(x, u_n)u_n - 2G(x, u_n)) \geq \left(\frac{1}{\theta} - 2\right)\int_{u_n>s_0} G(x, u_n).$$

Altogether, one obtains the estimate

$$\int_{u_n>s_0} G(x, u_n) \leq C + \varepsilon_n ||u_n||_{H^1} + C||u_n^-||_{L^{1+\alpha}}^{1+\alpha}. \tag{13}$$

Next let us take $v = u_n^-$ in (11) to obtain

$$\left|\int |\nabla u_n^-|^2 - \int_{u_n<0} g(x, u_n)u_n\right| \leq \varepsilon_n ||u_n^-||_{H^1}. \tag{14}$$

Using (8) and (9), it follows from (14) that

$$\left| \int |\nabla u_n^-|^2 - 2\int_{u_n<0} G(x,u_n) \right| \le C + \varepsilon_n \|u_n^-\|_{H^1} + C\|u_n^-\|_{L^{1+\alpha}}^{1+\alpha}, \tag{15}$$

$$\left| \int |\nabla u_n^-|^2 - \lambda \int |u_n^-|^2 \right| \le C + \varepsilon_n \|u_n^-\|_{H^1} + C\|u_n^-\|_{L^{1+\alpha}}^{1+\alpha}. \tag{16}$$

Now we consider the following alternatives: either (i) $\|u_n^-\|_{H^1} \le$ const. or (ii) $\|u_n^-\|_{H^1} \to \infty$, passing to a subsequence if necessary. In case (i) holds, estimate (15) implies that $|\int_{u_n<0} G(x,u_n)| \le$ const. Consequently it follows from (10) and (13) that

$$\|u_n^+\|_{H^1} \le C + \varepsilon_n \|u_n^+\|_{H^1},$$

which implies that $\|u_n^+\|_{H^1}$ is bounded. So $\|u_n\|_{H^1} \le C$ and the proof of the lemma would be complete in case (i). Next let us consider case (ii) and prove that it cannot hold, completing in this way the proof of the lemma. It follows from (10) using (13) and (15) that

$$\int |\nabla u_n^+|^2 \le C + C\varepsilon_n \|u_n^-\|_{H^1} + C\|u_n^-\|_{L^{1+\alpha}}^{1+\alpha}. \tag{17}$$

Now we obtain from (11) that

$$\left| \int \nabla u_n^- \nabla v - \lambda \int u_n^- v \right| \le I_1 + I_2 + I_3 + \varepsilon_n \|v\|_{H^1}, \tag{18}$$

where the integrals I_1, I_2 and I_3 are explicitly written below and then estimated.

$$I_1 \equiv \left| \int \nabla u_n^+ \nabla v \right| \le \|u_n^+\|_{H^1} \|\nu\|_{H^1}$$

$$I_2 \equiv \left| \int_{u_n>0} g(x,u_n) v \right| \le (C + C\|u_n^+\|_{L^{p\sigma}}^{\sigma}) \|v\|_{L^q},$$

where we have used (G3) and $p = 2N/(N+2), q = 2N/(N-2)$ if $N \ge 3$. If $N = 2$ we take $1 < p \le 1/\sigma\theta$; this can be done because (G5) in this case implies $\sigma\theta < 1$.

$$I_3 \equiv \left| \int_{u_n<0} g(x,u_n) v + \lambda u_n^- v \right| \le \int_{u_n<0} |\delta(x,u_n)||v|,$$

where $\delta(x,s) = g(x,s) - \lambda s$. Using (G2) we obtain the estimate

$$I_3 \leq (C + C||u_n^-||_{H^1}^{\alpha})||v||_{H^1}.$$

Altogether, the left side of (18) is estimated by

$$(C + ||u_n^+||_{H^1} + C||u_n^+||_{L^{p\sigma}}^{\sigma} + C||u_n^-||_{H^1}^{\alpha})||v||_{H^1}.$$

Our purpose now is to show that the expression in parenthesis goes to zero when divided by $||u_n^-||_{H^1}$. Firstly, we see that (17) implies

$$(19) \qquad ||u_n^+||_{H^1}/||u_n^-||_{H^1} \to 0 \quad \text{as } n \to \infty.$$

Secondly we claim that

$$(20) \qquad ||u_n^+||_{L^{p\sigma}}^{\sigma}/||u_n^-||_{H^1} \to 0 \quad \text{as } n \to \infty.$$

We remark that this is the only point in the present proof where condition (G5) is used. It follows from the fact that $G(x,s) \geq Cs^{1/\theta} - C$, for $s > 0$, and from estimate (13) that

$$(21) \qquad \int (u_n^+)^{1/\theta} \leq C + \varepsilon_n||u_n||_{H^1} + C||u_n^-||_{L^{1+\alpha}}^{1+\alpha}.$$

Recalling the statement at (19) and that the $L^{1+\alpha}$ norm is bounded by the H^1 norm, we obtain from (21) that the $L^{1/\theta}$ norm of $u_n^+/||u_n^-||_{H^1}^{\theta(1+\alpha')}$ goes to zero, where α' is any number with $\alpha < \alpha' < 1$. Our claim (20) is that the $L^{p\sigma}$ norm of $u_n^+/||u_n^-||^{1/\sigma}$ goes to zero. So this will be achieved if $p\sigma \leq 1/\theta$ and $1/\sigma \geq \theta(1+\alpha')$. But this is guaranteed by(G5). So (20) is established. Next let us define $w_n = u_n^-/||u_n^-||_{H^1}$ and recall that we are treating case (ii), that is $||u_n^-||_{H^1} \to \infty$. Let us assume that $w_n \to w_0$ weakly in H_0^1 and strongly in L^2. It follows from (16) that $\int w_n^2 \to \lambda^{-1}$. So $||w_0||_{L^2}^2 = \lambda^{-1}$ and $w_0 \not\equiv 0$. Now, dividing (18) through by $||u_n^-||_{H'}$ and passing to the limit using (19) and (20), we get

$$\int \nabla w_0 \nabla v - \lambda \int w_0 v = 0 \text{ for all } v \in H_0^1.$$

This means that w_0 is an eigenfunction ($\neq 0$) corresponding to λ, which is impossible.

3. Existence of a first solution of $(7)_t$**.** In this section we consider problem $(7)_t$ as stated in the introduction. In order to prove the existence of one solution of $(7)_t$ we do not have to assume all conditions (F1) through (F7) stated in the Introduction. In fact the following result is true.

LEMMA 2. *Suppose that* $f(x,s)$ *is continuous and*
(F8) $\lim_{s\to-\infty} f(x,s)s^{-1} = 0$
Then there exists $\bar{t} > 0$ *such that* $(7)_t$ *has a negative solution* u_t *for all* $t \geq \bar{t}$. *Moreover,*

$$u_t \leq \frac{Ct}{\lambda_1 - \lambda}\varphi, \tag{22}$$

where C *is a positive constant independent of* t.

PROOF. We use an argument similar to the one in [**3**; Theorem 1]. Let us define

$$\tilde{f}(x,s) = \begin{cases} f(x,s) & \text{if } s < 0 \\ f(x,0) & \text{if } s \geq 0. \end{cases}$$

We claim that the Dirichlet problem

$$-\Delta u = \lambda u + \tilde{f}(x,u) + t\varphi + h \text{ in } \Omega, \quad u = 0 \text{ on } \partial\Omega \tag{23$_t$}$$

has a negative solution for $t > 0$ large. Such a solution is obviously a solution of the original problem $(7)_t$. Let $z_0 \in C^{2+\nu}(\overline{\Omega})$ be the solution of

$$-\Delta z_0 = \lambda z_0 + h \text{ in } \Omega, \quad z_0 = 0 \text{ on } \partial\Omega.$$

Now let us consider the Dirichlet problem

$$-\Delta w = \lambda w + \tilde{f}(x, w + z_0 - t\beta\varphi) \text{ in } \Omega, \quad w = 0 \text{ on } \partial\Omega, \tag{24$_t$}$$

where $\beta = 1/(\lambda - \lambda_1)$. It follows from (F8) that

$$\lim_{|s|\to\infty} \tilde{f}(x,s)s^{-1} = 0$$

which implies that, given $\varepsilon > 0$, there is a constant C_ε such that $|\tilde{f}(x,s)| \leq \varepsilon|s| + C_\varepsilon$, for all $s \in R$ and all $x \in \Omega$. In view of this, it is then well known that problem $(24)_t$ has a solution w_t for each t and

$$||w_t||_{C^1} \leq C||\tilde{f}(x, w_t + z_0 - t\beta\varphi||_{C^0},$$

where the constant C does not depend on t and the norms considered here are maximum norms. Using the above estimate on $\tilde{f}$ we then obtain

$$||w_t||_{C^1} \leq \varepsilon C||w_t + z_0 - t\beta\varphi||_{C^0} + C.$$

Consequently we find, by taking $\varepsilon > 0$ small enough, that $||w_t||_{C^1} \leq C + C\varepsilon|t|$. So $||w_t/t||_{C^1} \to 0$. Then, using the fact that the eigenfunction φ is positive in Ω and its outer normal derivative at the boundary is strictly negative, we conclude that $w_t + z_0 - t\beta\varphi < 0$ in Ω for $t > 0$ and sufficiently large. Finally it is easy to check that $u_t = w_t + z_0 - t\beta\varphi$ is a solution of $(23)_t$ which is negative in Ω, and the estimate (22) is also clear.

Next we show that if one assumes (F3) instead of (F8), then u_t can be taken in such a way that u_t is a nondegenerate critical point of the Euler Lagrange functional associated with $(7)_t$ and its Morse index is k. That is a great deal of information about this critical point. Yet we use it here merely in a rather mild way. We prefer to work with a translation of $(7)_t$ as follows. Let $u = v + u_t$ in problem (7_t) for $t \geq \bar{t}$. Then v satisfies the equation

$$-\Delta v = \lambda v + f(x, v + u_t) - f(x, u_t) \text{ in } \Omega, \quad v = 0 \text{ on } \partial\Omega. \tag{25}$$

Clearly, $v = 0$ is a solution of this equation, and we shall look for other solutions of $(25)_t$ as critical points of the functional

(26)

$$\Psi_t(v) = \frac{1}{2}\int |\nabla v|^2 - \frac{\lambda}{2}\int v^2 - \int F(x, v+u_t) + \int F(x, u_t) + \int f(x, u_t)v.$$

LEMMA 3. *Let us assume* (F1), (F3) *and* (F5). *Then there exists* $\hat{t} \geq \bar{t}$ *such that* $v = 0$ *is a nondegenerate critical point of* Ψ_t *with Morse index* k, *for all* $t \geq \hat{t}$.

REMARKS. (1) We are assuming that $\lambda_1 < \lambda_2 \leq \cdots \leq \lambda_k < \lambda < \lambda_{k+1}$. So there are k eigenvalues of $(-\Delta, H_0^1(\Omega))$, counting multiplicities, which are less than λ. To each λ_j corresponds an eigenfunction $\varphi_j \in H_0^1(\Omega)$. Let V be the finite dimensional subspace of $H_0^1(\Omega)$ generated by $\varphi_1, \ldots, \varphi_k$, and let $W = V^\perp$. So $H_0^1(\Omega) = V \oplus W$.

(2) In view of the hypotheses, the functional Ψ_t is well defined in H_0^1 and it is of class C^2. Their first and second derivatives are represented

as follows:

$$\langle \Psi_t'(u), v\rangle = \int \nabla u \nabla v - \lambda \int uv - \int f(x, u+u_t)v + \int f(x, u_t)v,$$

$$\langle \Psi_t''(u)v_1, v_2\rangle = \int \nabla v_1 \nabla v_2 - \lambda \int v_1 v_2 - \int f_s'(x, u+u_t)v_1 v_2,$$

for all u, v, v_1, v_2 in $H_0^1(\Omega)$.

PROOF OF LEMMA 3. It suffices to prove that there exists $\hat{t} \geq \bar{t}$ such that, for each $t \geq \hat{t}$, there is a positive constant $c(t)$ depending only on t such that

$$\langle \Psi_t''(0)v, v\rangle \leq -c(t)||v||_{H^1}^2, \quad v \in V, \tag{27}$$

and

$$\langle \Psi_t''(0)w, w\rangle \geq c(t)||w||_{H^1}^2, \quad w \in W. \tag{28}$$

Instead of proving (27) and (28) it suffices to show that

$$\langle \Psi_t''(0)v, v\rangle < 0, \quad \text{for all } v \in V \text{ with } ||v||_{H^1} = 1, \tag{29}$$

and

$$\langle \Psi_t''(0)w, w\rangle > 0, \text{ for all } w \in W \text{ with } ||w||_{H^1} = 1. \tag{30}$$

Indeed (29) clearly implies (27) in view of the compactness of the unit ball in V. To see that (30) implies (28), let us assume by contradiction that there exists a sequence $w_n \in W$ with $||w_n||_{H^1} = 1, w_n \to w_0$ weakly in H_0^1 and strongly in L^2, such that $\langle \Psi_t''(0)w_n, w_n\rangle \downarrow 0$, that is

$$\lim_{n\to\infty} \left\{ \int |\nabla w_n|^2 - \lambda \int w_n^2 - \int f_s'(x, u_t)w_n^2 \right\} = 0. \tag{31}$$

On one hand (31) gives

$$1 - \lambda \int w_0^2 - \int f_s'(x, u_t)w_0^2 = 0$$

which implies $w_0 \neq 0$. On the other hand, (31) also yields the inequality

$$\int |\nabla w_0|^2 - \lambda \int w_0^2 - \int f_s'(x, u_t) w_0^2 \leq 0$$

which is a contradiction to (30) with $w = w_0/||w_0||_{H^1}$. Now let us prove (30). The relation (29) is proved likewise. Suppose, by contradiction, that there is a sequence $t_n \to +\infty$ and $w_n \in W$ with $||w_n||_{H^1} = 1$ and $\langle \Psi''_{t_n}(0) w_n, w_n \rangle \leq 0$. Then

$$\int |\nabla w_n|^2 - \lambda \int w_n^2 \leq \int f_s'(x, u_{t_n}) w_n^2. \tag{32}$$

The left side of (32) can be estimated using $\lambda_{k+1} \int w_n^2 \leq \int |\nabla w_n|^2$, and we are led to

$$1 - \frac{\lambda}{\lambda_{k+1}} \leq \int f_s'(x, u_{t_n}) w_n^2. \tag{33}$$

The sequence w_n can be chosen in such a way that $w_n \to w_0$ weakly in H_0^1, strongly in L^2, a.e. in Ω and $|w_n(x)| \leq k(x)$ where $k(x)$ is some $L^2(\Omega)$ function. It follows from the estimate (22) in Lemma 2 that the integrand in (33) converges a.e. to 0. Since u_{t_n} is a negative function, this same integrand is bounded uniformly by an L^2 function. So applying the Lebesgue Dominated Theorem to the expression (33) we are led to a contradiction.

4. Existence of a second solution. In this section we show that the functional Ψ_t defined in (26) possesses a nontrivial critical point for $t \geq \hat{t}$. In this way one obtains a second solution of $(7)_t$ in addition to the solution u_t found in §3. And this will complete the proof of the Theorem. So in this section we assume all the hypotheses of the Theorem. The proof below uses the Generalized Mountain Pass Theorem, see [8] or [9; Theorem 4.1], and the lemmas in the sequel are the verification of the hypotheses of the theorem applied to our functional Ψ_t.

LEMMA 4. *The functional ψ_t satisfies the* PS *condition.*

PROOF. The function $g(x,s) = \lambda s + f(x, s + u_t(x)) - f(x, u_t(x))$ satisfies all conditions of Lemma 1, as consequences of (F1), (F2), (F4), (F5) and (F7). The θ in (G4) should be slightly larger than the one in (F4).

From now on let us fix $t \geq \hat{t}$.

LEMMA 5. *There are positive constants ρ and δ such that*

$$\Psi_t(w) \geq \delta \quad \text{for } w \in W \text{ with } \|w\|_{H^1} = \rho. \tag{34}$$

PROOF. Since Ψ_t is of class C^2 it follows that, given $\varepsilon > 0$, there is $\rho > 0$ such that $|||\Psi''(u) - \Psi''(0)||| \leq \varepsilon$ if $\|u\| \leq \rho$, where $|||\cdot|||$ denotes the norm in the space $L(H_0^1, H_0^1)$ of linear bounded operators in H_0^1. Now take $\varepsilon > 0$ such that $\varepsilon < c(t)/2$. We claim that the corresponding ρ does the job. Indeed, given any $\overline{w} \in W$ with $\|\overline{w}\|_{H^1} = 1$, define the function $\theta : R \to R$ by

$$\theta(s) = \Psi_t(s\overline{w}).$$

So $\theta'(s) = \langle \Psi_t'(s\overline{w}), \overline{w}\rangle$, $\theta''(s) = \langle \Psi_t(s\overline{w})\overline{w}, \overline{w}\rangle$. Using Taylor's formula, $\theta(s) = \theta(0) + \theta'(0)s + \frac{1}{2}\theta''(\xi)s^2$, where ξ is a number between 0 and s, we have

$$\Psi_t(sw) = \frac{1}{2}\langle \Psi_t''(\xi\overline{w})\overline{w}, \overline{w}\rangle s^2.$$

Now take $s = \rho$ and write

$$\Psi_t(\rho\overline{w}) = \frac{1}{2}\langle \Psi_t''(0)\rho\overline{w}, \rho\overline{w}\rangle + \frac{1}{2}\langle (\Psi_t''(\xi\overline{w}) - \Psi_t''(0))\rho\overline{w}, \rho\overline{w}\rangle.$$

Using Lemma 3 we then estimate

$$\Psi_t(\rho\overline{w}) \geq \frac{1}{2}c(t)\|\rho\overline{w}\|_{H^1}^2 - \frac{1}{2}\varepsilon\|\rho\overline{w}\|_{H^1}^2 \geq \frac{1}{4}c(t)\|\rho\overline{w}\|_{H^1}^2.$$

So the lemma is proved with $\delta = \frac{1}{4}c(t)p^2$ and the above ρ.

LEMMA 6. *Given $\varepsilon > 0$ there exists an element $e \in W$ with $\|e\|_{H^1} = \varepsilon$ such that the set*

$$\{x \in \Omega : v(x) + e(x) > 1\}$$

has positive measure for each $v \in V$ with $||v||_{H^1} \leq 1$.

PROOF. Suppose by contradiction that, for each $e \in W$ with $||e||_{H^1} = \varepsilon$, there is a $v_e \in V$ with $||v_e||_{H^1} \leq 1$ and such that

$$v_e(x) + e(x) - 1 \leq 0 \text{ a.e. in } \Omega.$$

This implies that

$$e(x) \leq 1 - v_e(x) \leq \text{k a.e. in } \Omega.$$

Similarly, we get $-e(x) \leq$ const. a.e. in Ω. So $e \in L^\infty(\Omega)$, for all $e \in W$. This would give that $H_0^1(\Omega) \subset L^\infty(\Omega)$, which is false for all $N \geq 2$.

Now we claim that there is a constant $\eta > 0$ such that

$$\inf_{\substack{v \in V \\ ||v||_{H^1}=1}} \int [(v + e - 1)^+]^2 \geq \eta > 0 \tag{35}$$

Indeed, otherwise there would exist a sequence $v_n \in V$ with $||v_n||_{H^1} = 1$ such that $[v_n + e - 1]^+ \to 0$ in L^2. Since V is finite dimensional we may assume that $v_n \to v_0$ strongly in both H_0^1 and L^2 and $||v_0||_{H^1} = 1$. Hence $[v_0 + e - 1]^+ = 0$ which contradicts the statement of Lemma 6.

LEMMA 7. *There exist $R > 0$ and $\varepsilon > 0$ such that*

$$\Psi_t(u) \leq 0 \textit{ for all } u \in \partial Q$$

where $Q = (\overline{B}_r \cap V) \oplus \{re : 0 \leq r \leq R\}$, where $\overline{B}_R$ is a closed ball in H_0^1 centered at 0 and e, with $||e||_{H^1} = \varepsilon$, is the function found in Lemma 6.

PROOF. The set ∂Q is composed of three sets: $\Gamma_1 = \overline{B}_R \cap V, \Gamma_2 = \{u \in H_0^1 : u = v + se$, with $v \in V, ||v||_{H^1} = R, 0 \leq s \leq R\}$, and $\Gamma_3 = \{u \in H_0^1 : u = v + Re$, with $v \in V, ||v||_{H^1} \leq R\}$. The proof of the lemma is divided in three parts, each one showing that $\Psi_t \leq 0$ in each of the three sets above.

(i) Here we show even more. Namely $\Psi_t \leq 0$ on V. Indeed, for $v \in V$, one uses the estimate $\int |\nabla v|^2 \leq \lambda_k \int v^2$ and Taylor's formula $F(x, v+u_t) - F(x, u_t) + f(x, u_T)v = \frac{1}{2} f'_s(x, \xi(x))v^2$ where $\xi(x)$ is some number between $u_t(x)$ and $u_t(x) + v(x)$ to obtain the estimate

$$\Psi_t(v) \leq \frac{1}{2}(\lambda_k - \lambda) \int v^2 - \frac{1}{2} \int f'_s(x, \xi(x))v^2.$$

Next, using (F6), we obtain

$$\Psi_t(v) \leq \frac{1}{2}(\lambda_k - \lambda + \mu) \int v^2 \leq 0.$$

(ii) Here we take $u \in \Gamma_2$, and show that $\Psi_t(u) \leq 0$ for any choice of R. The number $\varepsilon > 0$ will be chosen at this step. We could estimate the functional as in (i):

$$\Psi_t(u) \leq \frac{1}{2}\Big(\big(1 - \frac{\lambda}{\lambda_k}\big) \int |\nabla u|^2 + \frac{s^2}{2}\Big(\int |\nabla e|^2 - \lambda \int e^2\Big) + \frac{\mu}{2} \int |v + se|^2$$

or

$$\Psi_t(u) \leq \frac{R^2}{2}\left(\Big(1 - \frac{\lambda - \mu}{\lambda_k}\Big) + \int |\nabla e|^2 - (\lambda - \mu) \int e^2\right). \tag{36}$$

So $\Psi_t(u) \leq 0$ if one chooses $\varepsilon > 0$ such that

$$\varepsilon^2 + \frac{\lambda - \mu}{\lambda_{k+1}}\varepsilon^2 \leq \frac{\lambda - \mu}{\lambda_k} - 1. \tag{37}$$

Indeed, since $||e||_{H^1} = \varepsilon$, we obtain

$$\int |\nabla e|^2 - (\lambda - \mu) \int e^2 \leq \Big(1 + \frac{\lambda - \mu}{\lambda_{k+1}}\Big) \int |\nabla e|^2 \leq \frac{\lambda - \mu}{\lambda_k} - 1,$$

which, applied in (36), gives that $\Psi_t(u) < 0$.

(iii) Now we show how to select $R > 0$ in such a way that $\Psi_t(u) \leq 0$ for $u \in \Gamma_3$. Using the fact that $\int |\nabla v|^2 \leq \lambda_k \int v^2 < \lambda \int v^2$ we obtain the estimate

$$\begin{aligned}\Psi_t(u) \leq &\frac{R^2}{2}\left(\int |\nabla e|^2 - \lambda \int e^2\right) \\ &- \int (F(x, v + Re + u_t) - F(x, u_t) - f(x, u_t)(v + Re))\end{aligned}$$

which could be estimated further by

$$\Psi_t(u) \le C + CR + CR^2 - \int_{v+Re+u_t>0} F(x, v + Re + u_t),$$

where the constants C do not depend on R; observe that $|F(s)| \le C|s|^{1+\alpha} + C$ if $s < 0$. Now, by the superlinearity of f at $+\infty$, given any $K > 0$ there exist $s_K > 0$ such that $f(x, s) \ge 2Ks$ for $s > s_K$. So there is a constant C_K such that $F(x, s) \ge C_K + Ks^2$ for $s > 0$. Then

$$\Psi_t(u) \le C + CR + CR^2 - C_K - KR^2 \int \left(\left(\frac{v}{R} + e + \frac{u_t}{R}\right)^+\right)^2.$$

Now we pick $R > 0$ such that $u_t(x)/R \ge -1$ for all $x \in \Omega$; this can be done because of the continuity of u_t. So

$$(\frac{v}{R} + e + \frac{u_t}{R})^+ \ge (\frac{v}{R} + e - 1)^+,$$

and then

$$\Psi_t(y) \le C + CR + CR^2 - C_K KR^2 \int \left(\left(\frac{v}{R} + e - 1\right)^+\right)^2.$$

Applying (35) we get finally that

$$\Psi_t(u) \le C + CR + CR^2 - C_K - K\eta R^2.$$

So the result is achieved if K is chosen in such a way that $K\eta > C$.

Acknowledgements. The author thanks Alan Lazer for helpful discussions as well as the Mathematics Department of the University of Miami for a stimulating scientific atmosphere and hospitality during his visit.

REFERENCES

1. A. Ambrosetti and P.H. Rabinowtiz, *Dual variational methods in critical point theory and applications.* J. Fctl. Anal. **14** (1973), 349-381.

2. K.C. Chang, *Variational methods and sub- and super-solutions,* Scientia Sinica, **XXVI** (1983), 1256-1265.

3. D.G. Costa and D.G. de Figueiredo, *Radial solutions for a Dirichlet problem in a ball,* J. Diff. Eq. **60** (1985), 80-89.

4. D.G. de Figueiredo, *On the superlinear Ambrosetti-Prodi Problem,* Nonlinear Analysis TMA **8** (1984), 655-665.

5. ——— and S. Solimini, *A variational approach to superlinear elliptic problems,* Comm. in PDE **9** (1984), 699-717.

6. A.C. Lazer and P.J. McKenna, *Multiple point theory and boundary value problems with nonlinearities crossing multiple eigenvalues* II, to appear.

7. J.C.N. de Padua, *Multiplicity results for a superlinear Dirichlet problem,* Doctoral Dissertation, University of Brasilia, 1986.

8. P.H. Rabinowitz, *Some critical point theorems and applications to semilinear elliptic partial differential equations,* Ann. Sci. Norm. Sup. Pisa, ser IV **5** (1978), 215-223.

9. ———, *Some aspects of critical point theory,* MRC Rep. No. 2465 (Jan. 1983).

10. B. Ruf and P.N. Srikanth, *Multiplicity results for superlinear elliptic problems with partial interference with the spectrum,* SISSA preprint.

11. ——— and ———, *Multiplicity results for ODE's with nonlinearities crossing all but a finite number of eigenvalues,* SISSA report.

DEPARTAMENTO DE MATEMATICA, UNIVERSIDADE DE BRASILIA, 70910 BRASILIA DF, BRAZIL

Math. Ann. 281, 589–610 (1988)

Mathematische Annalen

Nonresonance Below the First Eigenvalue for a Semilinear Elliptic Problem*

Djairo G. De Figueiredo[1] and Jean-Pierre Gossez[2]

[1] Departamento de Matematica, Universidade de Brasilia, 70910 Brasilia D.F., Brasil
[2] Département de Mathématique, C.P.214, Université Libre de Bruxelles, Bd. du Triomphe, Campus Plaine, B-1050 Bruxelles, Belgium

1. Introduction and Statement of Results

Let Ω be a bounded open subset in $\mathbb{R}^N$ with smooth boundary $\partial\Omega$ and let $f(x,s)$ be a real-valued Caratheodory function on $\Omega\times\mathbb{R}$. We consider the semilinear Dirichlet problem

$$\left.\begin{array}{ll} -\Delta u=f(x,u)+h(x) & \text{in } \Omega\,, \\ u=0 & \text{on } \partial\Omega\,. \end{array}\right\} \tag{1.1}$$

The nonlinearity f generates a potential

$$F(x,s)=\int_0^s f(x,t)dt$$

which, in this paper, will always be assumed to lie asymptotically as $s\to\pm\infty$ at the left side of the first eigenvalue λ_1 of $-\Delta$ on $H_0^1(\Omega)$, i.e.

$$\limsup_{s\to\pm\infty}\frac{2F(x,s)}{s^2}\leqq\lambda_1\,, \tag{1.2}$$

with some uniformity with respect to x. We are interested in the additional conditions to be imposed on f or F in order that (1.1) be solvable *for any h*. Such conditions are sometimes referred to as *nonresonance conditions*.

A classical result in that direction, which goes back to Hammerstein [7], says that if f is continuous, satisfies a linear growth condition and if for some number $\mu<\lambda_1$,

$$\limsup_{s\to\pm\infty}\frac{2F(x,s)}{s^2}\leqq\mu \tag{1.3}$$

uniformly for $x\in\Omega$, then (1.1) is solvable for any h. This was extended by Mawhin-Ward-Willem [9] who assume that f grows at most as $|s|^\sigma$ for some

* Part of this work was carried out with the support of the CNPq and the FNRS

Originally published in: Math. Ann. 281 (1988), 589-610,

Springer-Verlag Berlin Heidelberg 2013, DOI: 10.1007/978-3-319-02856-9_20

$\sigma<(N+2)/(N-2)$ ($\sigma<\infty$ if $N=1$ or 2) and that for some function $\alpha\in L^{\infty}(\Omega)$ with $\alpha(x)\leqq\lambda_1$ a.e. in Ω and $\alpha(x)<\lambda_1$ on a subset of positive measure,

$$\limsup_{s\to\pm\infty}\frac{2F(x,s)}{s^2}\leqq\alpha(x) \tag{1.4}$$

uniformly for a.e. $x\in\Omega$.

We are mainly interested in this paper in the situation where (1.4) does not hold, e.g. when

$$\limsup_{s\to\pm\infty}\frac{2F(x,s)}{s^2}=\lambda_1 \tag{1.5}$$

for a.e. $x\in\Omega$. The new assumption which will eventually yield the solvability of (1.1) bears on the way the limits in (1.5) are approached. Roughly speaking, it consists in requiring that the quotient $2F(x,s)/s^2$ stays below λ_1 for sufficiently many values of s as $s\to\pm\infty$. This is expressed in terms of the density of certain sets at infinity. Density conditions of the same type but bearing on the quotient $f(x,s)/s$ were introduced in [5] (see in this respect Proposition 4.10 below).

Let us state our results. Take an exponent p with $p>2N/(N+2)$ if $N\geqq 2$, $p\geqq 1$ if $N=1$. We assume that the Caratheodory function $f(x,s)$ and its potential $F(x,s)$ satisfy:

(f_1) there exist a constant A and a function $B(x)$ in $L^p(\Omega)$ such that

$$|f(x,s)|\leqq A|s|+B(x)$$

for a.e. $x\in\Omega$ and all $s\in\mathbb{R}$;

(F_1) inequality (1.2) holds uniformly for a.e. $x\in\Omega$, i.e. for any $\varepsilon>0$ there exists $b_\varepsilon(x)$ in $L^1(\Omega)$ such that

$$F(x,s)\leqq\frac{\lambda_1}{2}s^2+\varepsilon s^2+b_\varepsilon(x)$$

for a.e. $x\in\Omega$ and all $s\in\mathbb{R}$.

To state our last assumption on F, we need the following

Definition 1.1. Let E be a measurable subset of $\mathbb{R}$. We say that E has *positive density at* $+\infty$ if

$$\liminf_{r\to+\infty}\frac{\mu_1(E\cap[0,r])}{\mu_1([0,r])}>0.$$

Here μ_1 denotes the Lebesgue measure on $\mathbb{R}$. Of course, one can give a similar definition at $-\infty$ by considering the intersection of E with the interval $[r,0]$ and letting $r\to-\infty$.

It will also be convenient to say that a measurable subset A of a measurable set B is a *full* subset of B if $B\backslash A$ has measure zero.

The last assumption imposed on F in our first result is now the following, where we denote $\mathbb{R}\backslash\{0\}$ by $\mathbb{R}_0$:

(F_2) there exist a full subset $\Omega'\subset\Omega$ and $\eta>0$ such that

$$\bigcap_{x\in\Omega'}\left\{s\in\mathbb{R}_0;\ \frac{2F(x,s)}{s^2}\leqq\lambda_1-\eta\right\}$$

has positive density at both $+\infty$ and $-\infty$.

Theorem 1.2. *Assume* (f_1), (F_1), *and* (F_2). *Then, for any* $h \in L^p(\Omega)$, *there exists* $u \in H_0^1(\Omega) \cap W^{2,p}(\Omega)$ *solution of* (1.1), *which minimizes the associated functional*

$$\phi(u) = 1/2 \int_\Omega |\nabla u|^2 - \int_\Omega F(x,u) - \int_\Omega hu \tag{1.6}$$

on $H_0^1(\Omega)$.

Theorem 1.2 applies in particular to the case where $f(x,u)$ does not depend on x, i.e. $f(x,u) = f(u)$.

Assumption (F_2) is global with respect to Ω. It can be replaced by a local one, at the expense of strengthening a little bit the density condition at infinity. Let us introduce the following variant of (F_2):

$(F_3)_+$ there exist an open set $\omega_+ \subset \Omega$ and a corresponding full subset $\omega'_+ \subset \omega_+$ with the following property: for any $0 \leqq \upsilon < 1$ there exists $\eta > 0$ such that

$$\liminf_{r \to \infty} \frac{\mu_1(E(\omega'_+, \lambda_1 - \eta) \cap [\upsilon r, r])}{\mu_1([\upsilon r, r])} > 0, \tag{1.7}$$

where

$$E(\omega'_+, \lambda_1 - \eta) = \bigcap_{x \in \omega'_+} \left\{ s \in \mathbb{R}_0;\ \frac{2F(x,s)}{s^2} \leqq \lambda_1 - \eta \right\};$$

$(F_3)_-$ there exist an open set $\omega_- \subset \Omega$ and a corresponding full subset $\omega'_- \subset \omega$ with the following property: for any $0 \leqq \upsilon < 1$ there exists $\eta > 0$ such that

$$\liminf_{r \to -\infty} \frac{\mu_1(E(\omega'_-, \lambda_1 - \eta) \cap [r, \upsilon r])}{\mu_1([r, \upsilon r])} > 0.$$

Condition (F_2) at $+\infty$ corresponds to $(F_3)_+$ with $\omega_+ = \Omega$ and the single choice $\upsilon = 0$. Observe also that if for some open set $\omega_+ \subset \Omega$ and some number $\mu < \lambda_1$,

$$\limsup_{s \to +\infty} \frac{2F(x,s)}{s^2} \leq \mu$$

uniformly for a.e. $x \in \omega_+$, then $(F_3)_+$ holds. Indeed, for a full subset $\omega'_+ \subset \omega_+$, the set $E(\omega'_+, (\mu + \lambda_1)/2)$ then contains a half line. Similar observations can of course be made for $(F_3)_-$.

Theorem 1.3. *Assume* (f_1), (F_1), $(F_3)_+$, *and* $(F_3)_-$. *Then for any* $h \in L^p(\Omega)$, *there exists* $u \in H_0^1(\Omega) \cap W^{2,p}(\Omega)$ *solution of* (1.1), *which minimizes* ϕ *on* $H_0^1(\Omega)$.

The main part of the proof of these two theorems consists in showing that the functional ϕ is coercive on $H_0^1(\Omega)$. Conditions (F_2) and $(F_3)_\pm$ are applied in direct connection with two propositions relative to inverse images. Given a Lipschitz function u and a Borel set B in its range, these propositions provide an uniform estimate from below on the measure of $u^{-1}(B)$ in terms of the measure of B and the Lipschitz constant of u. These propositions are proved by rearrangement methods. In this respect, the main (technical) difficulty to go from the global assumption (F_2) to the local one $(F_3)_\pm$ comes from the fact that the classical Polya-Szegö theorem about the Schwarz rearrangement of a nonnegative Lipschitz function is no longer true if the function does not vanish on the boundary. We use instead the Steiner rearrangement when dealing with such functions.

Section 2 is devoted to these two propositions. Since they may have some independent interest, we have tried to present them in a rather sharp form, which is stronger than what is really needed for their application in this paper. Section 3 contains the proof of Theorems 1.2 and 1.3 as well as various remarks. Several examples are given in Sect. 4, in particular when equality (1.5) holds. We also show there that if a nonlinearity satisfies our assumptions in a nontrivial way, i.e. with (1.5), then a certain crossing of the eigenvalue λ_1 must necessarily occur (cf. Proposition 4.2).

2. Two Propositions About Inverse Images

The following proposition enters the proof of Theorem 1.2.

Proposition 2.1. *There exists a constant $c=c(N)>0$ such that*

$$\mu_N(u^{-1}(B)) \geqq c\frac{\mu_1(B)^N}{\mathrm{Lip}(u)^N} \tag{2.1}$$

for any open set $\Omega \subset \mathbb{R}^N$ with finite measure, any nonconstant Lipschitz function u on $\bar{\Omega}$ which vanishes on $\partial\Omega$ and any Borelian set B in the range of u.

Here μ_1 and μ_N denote the Lebesgue measures on $\mathbb{R}$ and $\mathbb{R}^N$ respectively and $\mathrm{Lip}(u)$ is the Lipschitz constant of u on Ω.

Proof of Proposition 2.1. Let Ω, u, and B be as in the proposition. We first consider the case where the function u is nonnegative. Let then u^* be the Schwarz rearrangement of u. It is defined on the corresponding rearranged domain Ω^*, which is the ball in $\mathbb{R}^N$ centered at zero with measure $\mu_N(\Omega)$. It is known that the range of u^* coincides with that of u and that

$$\mu_N(u^{*-1}(B)) = \mu_N(u^{-1}(B)); \tag{2.2}$$

moreover u^* is Lipschitzian on $\bar{\Omega}^*$ and one has the Polya-Szegö estimate

$$\mathrm{Lip}(u^*) \leqq \mathrm{Lip}(u). \tag{2.3}$$

The proof of (2.3) is given in [3, 10] when Ω is bounded. It carries over with little change to the present situation if one observes that our conditions imply that $u(x)\to 0$ as $|x|\to\infty$, $x\in\Omega$.

We claim that for any radius R of the ball Ω^*,

$$\mu_1(B) = \int_{u^{*-1}(B)\cap R} \left|\frac{\partial u^*}{\partial r}\right| dr, \tag{2.4}$$

where r denotes the distance to the origin. This is clear when B is an interval since, along R, u^* is nonincreasing and is the integral of its derivative. The general case of a Borelian set B then follows since both sides of (2.4) are measures on the range of u. Relation (2.4) implies that for any radius R of Ω^*,

$$\mu_1(u^{*-1}(B)\cap R) \geqq \frac{\mu_1(B)}{\mathrm{Lip}(u^*)}. \tag{2.5}$$

We now consider the radially symmetric set $u^{*-1}(B)$ and let A denote the ball centered at zero of radius $\mu_1(u^{*-1}(B)\cap R)$. To compare the N-measures of $u^{*-1}(B)$ and of A, we write

$$\mu_N(u^{*-1}(B))=\int\Bigg(\int_{u^{*-1}(B)\cap R_\theta} r^{N-1}dr\Bigg)d\theta\,,$$

where R_θ is the radius through the point θ of the unit sphere. An application of Lemma 2.2 below to the inner integral gives

$$\mu_N(u^{*-1}(B))\geqq\mu_N(A)\,.$$

Consequently, using (2.5), we obtain

$$\mu_N(u^{*-1}(B))\geqq\omega_N\frac{\mu_1(B)^N}{\mathrm{Lip}(u^*)^N}\,,$$

where ω_N denotes the N-measure of the unit ball in $\mathbb{R}^N$. The conclusion (2.1) with $c=\omega_N$ then follows by combining the previous inequality with (2.2) and (2.3).

To deal with the case of an arbitrary function u, we write

$$u=u_+-u_-\,.$$

This provides a decomposition of u into two nonnegative Lipschitz functions u_+ and u_- which vanish on $\partial\Omega$ and satisfy $\mathrm{Lip}(u_+)\leqq\mathrm{Lip}(u)$ and $\mathrm{Lip}(u_-)\leqq\mathrm{Lip}(u)$. Moreover, writing

$$B=B^-\cup B^0\cup B^+\,,$$

where $B^-=B\cap\mathbb{R}_0^-$, $B^0=B\cap\{0\}$, and $B^+=B\cap\mathbb{R}_0^+$, we have

$$\mu_N(u^{-1}(B))\geqq\mu_N(u_-^{-1}(-B^-))+\mu_N(u_+^{-1}(B^+))\,.$$

Applying (2.1) to u_- and to u_+, we deduce

$$\mu_N(u^{-1}(B))^{1/N}\geqq\frac{\omega_N^{1/N}}{\mathrm{Lip}(u)}(\mu_1(B^-)^N+\mu_1(B^+)^N)^{1/N}\,.$$

Consequently

$$\mu_N(u^{-1}(B))^{1/N}\geqq\frac{\omega_N^{1/N}c_0}{\mathrm{Lip}(u)}\mu_1(B)\,,$$

where the constant $c_0>0$ comes from the equivalence between the N-norm and the 1-norm on $\mathbb{R}^2$. Q.E.D.

Lemma 2.2. *Let $f:[a,b]\to\mathbb{R}$ be nondecreasing. Take $D\subset[a,b]$ measurable and let D' be the interval with left point a and length $\mu_1(D)$. Then*

$$\int_{D'} f\leqq\int_D f\,. \tag{2.6}$$

Proof. Letting $A=D\setminus(D\cap D')$ and $B=D'\setminus(D\cap D')$, we have the disjoint unions

$$D=(D\cap D')\cup A\,,$$

$$D'=(D\cap D')\cup B\,.$$

Introducing this in (2.6), we are reduced to proving

$$\int_B f \leqq \int_A f .$$

Since $\mu_1(A)=\mu_1(B)$ and $\sup B \leqq \inf A$, we have

$$\int_B f \leqq \mu_1(B) f(\sup B) \leqq \mu_1(A) f(\inf A) \leqq \int_A f . \quad \text{Q.E.D.}$$

Remark 2.3. It follows from (2.1) that for any $q>N$ and any open set Ω with finite measure, there exists $c_1=c_1(N,q,\mu_N(\Omega))$ such that

$$\mu_N(u^{-1}(B)) \geqq c_1 \frac{\mu_1(B)^q}{\operatorname{Lip}(u)^q} \tag{2.7}$$

for any u and B as in Proposition 2.1. Indeed (2.1) implies

$$c_2 \frac{\mu_1(B)}{\operatorname{Lip}(u)} \leqq 1 ,$$

where $c_2=c^{1/N}\mu_N(\Omega)^{-1/N}$. Rewriting (2.1) in the form

$$\mu_N(u^{-1}(B)) \geqq \frac{c}{c_2{}^N}\left(c_2 \frac{\mu_1(B)}{\operatorname{Lip}(u)}\right)^N$$

and observing that the Nth power of a number between 0 and 1 is greater than its qth power, we deduce (2.7). We also remark that an estimate like (2.7) cannot hold with an exponent $q<N$. This can be seen by considering the functions

$$u_k(x)=\begin{cases} 1-k|x| & \text{for} \quad |x| \leqq 1/k, \\ 0 & \text{for} \quad 1/k \leqq |x| < 1 \end{cases}$$

on $\Omega=$ the unit ball in $\mathbb{R}^N$ and the Borelian set $]0,1[$.

Estimate (2.7) was announced in [5]. The preceding remark shows that the exponent N in (2.1) is sharp.

Remark 2.4. The result of Proposition 2.1 is valid for an arbitrary open set Ω provided the vanishing of u on $\partial\Omega$ is understood to hold also at infinity, i.e. $u(x)\to 0$ as $|x|\to\infty$, $x\in\Omega$. Estimate (2.3) remains true under these conditions, with a proof easily adapted from [3, 10].

In the proof of Theorem 1.3, we will need a result analogous to Proposition 2.1 for functions which do not necessarily vanish on the boundary. For such functions, even nonnegative, the Lipschitz character is not necessarily preserved by taking the Schwarz rearrangement (cf. Example 2.5 below; cf. also Remark 2.9). We will see, however, that the conclusion of Proposition 2.1 essentially remains valid.

Example 2.5. Let $\Omega=B(0,1)\subset\mathbb{R}^2$ and consider $u(x,y)=(x^2+y^2)^{1/2}$. Then $u^*=0$ on $\partial B(0,1)$ and, for $0<t<1$, $u^*>t$ in $B(0,r)$ where $\pi r^2=\pi-\pi t^2$. Letting $t\to 0$, one easily concludes that u^* is not Lipschitzian on $\bar{\Omega}^*$.

Proposition 2.6. *Let $\Omega\subset\mathbb{R}^N$ be an open parallelipiped. Then there exists a constant $c=c(N,\Omega)>0$ such that*

$$\mu_N(u^{-1}(B)) \geqq c \frac{\mu_1(B)^N}{\operatorname{Lip}(u)^N} \tag{2.8}$$

for any nonconstant Lipschitz function u on $\bar{\Omega}$ and any Borelian set B in the range of u.

Observe that without loss of generality, we can suppose the faces of the parallelipiped Ω to be parallel to the coordinate hyperplanes. This will be assumed from now on, without further notice.

The proof of Proposition 2.6 uses the Steiner rearrangement of u. It will be denoted by $\tilde{u}$. It is obtained from u by N successive one-dimensional Schwarz rearrangements with respect to each variable $x_1, \ldots, x_N$ (cf. [11, 8]). The function $\tilde{u}$ is defined on the corresponding rearranged domain $\tilde{\Omega}$, which here is simply the parallelipiped Ω centered at zero. It enjoys properties similar to those of u^*: the range of $\tilde{u}$ coincides with that of u,

$$\mu_N(\tilde{u}^{-1}(B)) = \mu_N(u^{-1}(B)), \tag{2.9}$$

$\tilde{u}$ is Lipschitzian on $\tilde{\Omega}$ and one has a Polya-Szegö type estimate:

$$\mathrm{Lip}(\tilde{u}) \leqq 3^N \,\mathrm{Lip}(u). \tag{2.10}$$

The proof of (2.10) can be adapted from that of a similar estimate sketched in [8; p. 82] for periodic functions.

The following preliminary result will be used in the proof of Proposition 2.6.

Lemma 2.7. *For any measurable function u on Ω there exists α_u such that*

$$\mu_N\{x \in \Omega;\ u(x) \leqq \alpha_u\} \geqq \mu_N(\Omega)/2 \tag{2.11}$$

and

$$\mu_N\{x \in \Omega;\ u(x) \geqq \alpha_u\} \geqq \mu_N(\Omega)/2\,. \tag{2.12}$$

Proof. Consider the distribution functions of u:

$$d_u(t) = \mu_N\{x \in \Omega;\ u(x) > t\},$$

$$d^*_u(t) = \mu_N\{x \in \Omega;\ u(x) \geqq t\},$$

and define

$$\alpha_u = \inf\{t;\ d_u(t) \leqq \mu_N(\Omega)/2\}\,.$$

Since d_u is right continuous, one has

$$d_u(\alpha_u) \leqq \mu_N(\Omega)/2\,,$$

which gives (2.11). We claim that (2.12) holds, i.e.

$$d^*_u(\alpha_u) \geqq \mu_N(\Omega)/2\,. \tag{2.13}$$

Indeed, by the definition of α_u, $d_u(t) > \mu(\Omega)/2$ for all $t < \alpha_u$ and consequently $d^*_u(t) > \mu(\Omega)/2$ for all $t < \alpha_u$. Since d^*_u is left continuous, we get (2.13). Q.E.D.

Proof of Proposition 2.6. Take u and B as in Proposition 2.6 and let α_u be such that (2.11) and (2.12) hold. We decompose B into a disjoint union

$$B = B^- \cup B^0 \cup B^+\,,$$

where $B^- = B \cap [\min u, \alpha_u[$, $B^0 = B \cap \{\alpha_u\}$ and $B^+ =]\alpha_u, \max u]$. If we can show that for some constant $c = c(N, \Omega) > 0$, one has

$$\mu_N(u^{-1}(B^-)) \geqq c \frac{\mu_1(B^-)^N}{\mathrm{Lip}(u)^N}$$

and

$$\mu_N(u^{-1}(B^+)) \geqq c \frac{\mu_1(B^+)^N}{\mathrm{Lip}(u)^N},$$

then, by an argument similar to that used in the second part of the proof of Proposition 2.1, we conclude that (2.8) holds (with the new constant $cc_0{}^N$).

We are thus reduced in this way to proving estimate (2.8) for functions u and Borelian sets B which satisfy in addition either

$$\mu_N\{x \in \Omega;\ u(x) \leqq \inf B\} \geqq \mu_N(\Omega)/2 \tag{2.14}$$

or

$$\mu_N\{x \in \Omega;\ u(x) \geqq \sup B\} \geqq \mu_N(\Omega)/2. \tag{2.15}$$

Let us first consider the case where (2.14) holds. Let $\tilde{u}$ be the Steiner rearrangement of u. Then $\max \tilde{u}$ is attained at 0 while $\min \tilde{u}$ is attained at each of the corners of the parallelipiped $\tilde{\Omega}$. Moreover, by (2.14),

$$\mu_N\{x \in \tilde{\Omega};\ \tilde{u}(x) \leqq \inf B\} \geqq \mu_N(\Omega)/2. \tag{2.16}$$

Let x_0 be the corner of the parallelipiped $\tilde{\Omega}$ with positive coordinate and let $y_0 = x_0/2^{1/N}$. We claim that

$$\tilde{u}(y_0) \leqq \inf B. \tag{2.17}$$

Indeed, if (2.17) does not hold, then $\tilde{u}$ is $> \inf B$ on a neighbourhood of the parallelipiped centered at 0 with corner y_0, which contradicts (2.16). We now deduce from (2.17) that $\tilde{u}$ is $\leqq \inf B$ on $(y_0 + \mathbb{R}^N{}_+) \cap \tilde{\Omega}$, i.e. on the shaded parallelipiped in the following figure:

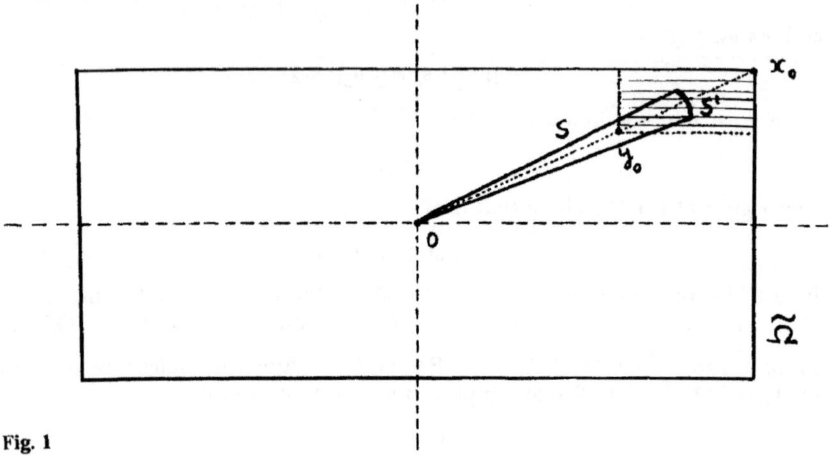

Fig. 1

We can then construct a small sector S around the radius $[0, x_0]$, whose cap S' entirely lies in this shaded parallelipiped. The $(N-1)$ measure of S' can be estimated from below by a positive number depending only on $\tilde{\Omega}$.

We are now in a position to adapt the argument of the proof of Proposition 2.1. We claim that for any radius R of the above sector S,

$$\mu_1(B) = \int_{\tilde{u}^{-1}(B) \cap R} \left| \frac{\partial \tilde{u}}{\partial r} \right| dr, \tag{2.18}$$

where r denotes the distance to the origin. This follows as in the proof of Proposition 2.1, with the additional observation that since $\tilde{u}$ is $\leqq \inf B$ on S', B entirely lies in the image of R by $\tilde{u}$. It follows from (2.18) that

$$\mu_1(\tilde{u}^{-1}(B) \cap R) \geqq \frac{\mu_1(B)}{\mathrm{Lip}(\tilde{u})} \tag{2.19}$$

for any radius R of S.

We now consider the set $\tilde{u}^{-1}(B) \cap S$ and compute its N-measure by the formula

$$\mu_N(\tilde{u}^{-1}(B) \cap S) = \int \left(\int_{\tilde{u}^{-1}(B) \cap R_\theta} r^{N-1} dr \right) d\theta,$$

where R_θ is the radius of S through the point θ of the unit sphere and where the outer integral is extended on that part of the unit sphere homothetic to the cap S'. Applying Lemma 2.2 to the inner integral, we deduce from (2.19) that

$$\mu_N(\tilde{u}^{-1}(B)) \geqq \frac{\mu_1(B)^N}{\mathrm{Lip}(\tilde{u})^N} c_3, \tag{2.20}$$

where c_3 is the N-measure of the sector of radius 1 through the cap S'. Of course, c_3 is estimated from below by a positive number depending only on $\tilde{\Omega}$. Estimate (2.8) now follows by combining (2.20) with (2.9) and (2.10).

Estimate (2.8) is thus proved if u and B satisfy (2.14). Now if u and B satisfy (2.15), then $-u$ and $-B$ satisfy (2.14), so that (2.8) still holds in this case. Q.E.D.

Remark 2.8. An estimate like (2.8) does not hold with a constant c depending only on N and the measure of the parallelipiped Ω. This can be seen by considering the domains

$$\Omega_k = \{(x, y) \in \mathbb{R}^2;\ |x| < k \text{ and } |y| < 1/k\},$$

the functions

$$u_k(x, y) = \begin{cases} -1 & \text{for} \quad x \leqq -1, \\ x & \text{for} \quad -1 \leqq x \leqq 1, \\ +1 & \text{for} \quad x \geqq 1, \end{cases}$$

and the Borel set $]-1, 1[$.

Remark 2.9. An estimate like (2.8) does not hold in an arbitrary bounded open set Ω, even if u vanishes on a significant part of $\partial\Omega$. This can be seen by considering the domain.

$$\Omega = \{(x, y) \in \mathbb{R}^2;\ 0 < x < 1 \text{ and } -x^2 < y < x^2\},$$

the function $u(x, y) = 1 - x$ and the Borel sets $]1 - 1/k, 1[$. One can also see here that both the Schwarz and the Steiner rearrangements of u are not Lipschitzian. Indeed $u^*(0, 0) = 1$ and $u^* \leqq 1 - t$ outside $B(0, r)$ where $\pi r^2 = 2t^3/3$; letting $t \to 0$ yields the conclusion for u^*. On the other hand, $\tilde{u}(0, 0) = 1$ and $\tilde{u}(0, y) = 1 - |y|^{1/2}$; letting $y \to 0$ yields the conclusion for $\tilde{u}$.

3. Proof of the Theorems

Proof of Theorem 1.2. It will be divided into several steps.

Step 1. We consider the functional ϕ defined by (1.6). It is easily verified that ϕ is weakly lower semicontinuous on $H_0^1(\Omega)$ and we will show that ϕ is coercive on $H_0^1(\Omega)$, i.e. that $\phi(u) \to +\infty$ as $\|u\| \to \infty$. Here $\| \;\|$ denotes the $H_0^1(\Omega)$ norm. Assuming this for a moment, we deduce that the minimum of ϕ is achieved at some point u in $H_0^1(\Omega)$. Since ϕ is a C^1 functional on $H_0^1(\Omega)$ whose critical points are the weak solutions of (1.1), we obtain that u solves (1.1). Finally, a standard bootstrap argument yields that any $H_0^1(\Omega)$ solution of (1.1) belongs to $W^{2,p}(\Omega)$.

Step 2. To prove that ϕ is coercive on $H_0^1(\Omega)$, we assume by contradiction the existence of a sequence u_n such that $\|u_n\| \to \infty$ and

$$\phi(u_n) \leqq c \tag{3.1}$$

for some constant c. Write

$$v_n = \frac{u_n}{\|u_n\|}.$$

Then, for a subsequence, $v_n \to v$ weakly in $H_0^1(\Omega)$, strongly in $L^2(\Omega)$ and a.e. in Ω.

We first show that $v \not\equiv 0$. For that purpose observe that, by (3.1),

$$\frac{1}{2} \int_\Omega |\nabla v_n|^2 - \int_\Omega \frac{F(x, u_n)}{\|u_n\|^2} - \int_\Omega \frac{h v_n}{\|u_n\|} \leqq \frac{c}{\|u_n\|^2}, \tag{3.2}$$

which implies, by (F_1),

$$\frac{1}{2} \int_\Omega |\nabla v_n|^2 - \left(\frac{\lambda_1}{2} + \varepsilon \right) \int_\Omega v_n^2 - \int_\Omega \frac{b_\varepsilon}{\|u_n\|^2} - \int_\Omega \frac{h v_n}{\|u_n\|} \leqq \frac{c}{\|u_n\|^2}. \tag{3.3}$$

If $v \equiv 0$, then the last four terms go to zero as $n \to \infty$. This leads to a contradiction since each v_n has norm 1 in $H_0^1(\Omega)$.

We claim that v is an eigenfunction corresponding to λ_1. Indeed, letting first $n \to \infty$ in (3.3) and then $\varepsilon \to 0$, we get

$$\int_\Omega |\nabla v|^2 \leqq \lambda_1 \int_\Omega v^2,$$

which yields the claim.

We also deduce from (3.2), by using Poincaré's inequality, that

$$\limsup \int_\Omega \left(\lambda_1 v_n^2 - \frac{2F(x, u_n)}{\|u_n\|^2} \right) \leqq 0. \tag{3.4}$$

Write $\|u_n\| = t_n$. By the mean value theorem,

$$\int_\Omega \left(\frac{2F(x, t_n v_n)}{t_n^2} - \frac{2F(x, t_n v)}{t_n^2} \right) = \int_\Omega \frac{2}{t_n^2} f(x, t_n v + \theta(t_n v_n - t_n v))(t_n v_n - t_n v), \tag{3.5}$$

where $\theta = \theta(n, x) \in [0, 1]$, and we observe, using (f_1), that the right-hand side of (3.5) goes to zero as $n \to \infty$. It thus follows from (3.4) that

$$\limsup \int_\Omega \left(\lambda_1 v^2 - \frac{2F(x, t_n v)}{t_n^2} \right) \leqq 0. \tag{3.6}$$

This will be used in Step 4 to reach the final contradiction.

Step 3. We will show that for some $\delta > 0$ and all n sufficiently large,

$$\mu_N \left\{ x \in \Omega ; \frac{2F(x, t_n v(x))}{t_n^2 v(x)^2} \leqq \lambda_1 - \eta \right\} \leqq \delta, \tag{3.7}$$

where η is the positive number entering assumption (F_2).

Since v is an eigenfunction corresponding to λ_1, we have either $v > 0$ on all Ω or $v < 0$ on all Ω. We consider the first case. The second case can be treated similarly, by using that part of assumption (F_2) relative to $-\infty$.

We apply Proposition 2.1 to the open set Ω, the function $t_n v$ and the Borelian set

$$B_n = E(\Omega', \lambda_1 - \eta) \cap [0, t_n \max v],$$

where $E(\Omega', \lambda_1 - \eta)$ is the set entering assumption (F_2):

$$E(\Omega', \lambda_1 - \eta) = \bigcap_{x \in \Omega'} \left\{ s \in \mathbb{R}_0 ; \frac{2F(x, s)}{s^2} \leqq \lambda_1 - \eta \right\}.$$

This gives

$$\mu_N \{ x \in \Omega ; t_n v(x) \in B_n \} \geqq c \frac{\mu_1(B_n)^N}{\mathrm{Lip}(t_n v)^N} \tag{3.8}$$

for some constant $c > 0$. The left-hand side of (3.8) is equal to

$$\mu_N \{ x \in \Omega' ; t_n v(x) \in B_n \} \tag{3.9}$$

and the set in (3.9) is clearly included in the set in (3.7). Moreover, assumption (F_2) tells us that for some $\delta_1 > 0$ and all n sufficiently large,

$$\mu_1(B_n) \geqq \delta_1 \, t_n \max v.$$

Finally,

$$\mathrm{Lip}(t_n v) = t_n \, \mathrm{Lip}(v),$$

where $\mathrm{Lip}(v)$ is a constant. Collecting these informations, we deduce (3.7) from (3.8).

Step 4. We will now deduce a contradiction from (3.6) and (3.7). Let χ_n be the characteristic function of the set in (3.7). From (3.6) we have

$$\liminf \int_\Omega \left(\lambda_1 - \frac{2F(x, t_n v)}{t_n^2 v^2} \right) v^2 \chi_n + \liminf \int_\Omega \left(\lambda_1 - \frac{2F(x, t_n v)}{t_n^2 v^2} \right) v^2 (1 - \chi_n) \leqq 0.$$

We apply the definition of χ_n to the first term and Fatou's lemma together with (F_1) to the second term to get

$$\eta \liminf \int_\Omega v^2 \chi_n \leqq 0 .$$

This implies, for a subsequence,

$$\lim \int_\Omega v^2 \chi_n = 0 .$$

Thus $v^2 \chi_n \to 0$ in $L^1(\Omega)$ and consequently in measure.

Recall that χ_n is the characteristic function of a varying set of measure $\geqq \delta > 0$. Choose $\varepsilon > 0$ so small that

$$\mu_N\{x \in \Omega;\ v^2(x) < \varepsilon\} < \frac{\delta}{3} . \tag{3.10}$$

The existence of such an ε can be seen as follows: in the contrary, for each $k = 1, 2, \ldots$,

$$\mu_N \left\{ x \in \Omega;\ v^2(x) < \frac{1}{k} \right\} \geqq \frac{\delta}{3} ;$$

calling A_k the set involved in this inequality and A the intersection of those A_k, we get $\mu_N(A) \geqq \delta/3$ and $v = 0$ a.e. in A; this contradicts the unique continuation property of the eigenfunction v. We then deduce from (3.10) that

$$\mu_N\{x \in \Omega;\ v^2(x) \geqq \varepsilon\} \geqq \mu_N(\Omega) - \delta/3$$

and consequently

$$\mu_N\{x \in \Omega;\ \chi_n(x) v^2(x) \geqq \varepsilon\} = \mu_N(\{x \in \Omega;\ v^2(x) \geqq \varepsilon\} \cap \{x \in \Omega;\ \chi_n(x) = 1\}) \geqq \frac{2\delta}{3} .$$

This contradicts the convergence in measure of $\chi_n v^2$ towards zero. Q.E.D.

Proof of Theorem 1.3. It is identical to that of Theorem 1.2 except for Step 3. The claim here is that for some $\eta > 0$ and some $\delta > 0$,

$$\mu_N \left\{ x \in \Omega;\ \frac{2F(x, t_n v(x))}{t_n^2 v(x)^2} \leqq \lambda_1 - \eta \right\} \geqq \delta \tag{3.11}$$

for all n sufficiently large. As before we can consider only the case where v is > 0 on Ω.

Replacing if necessary the open set ω_+ of assumption $(F_3)_+$ by a smaller one, we can always suppose that ω_+ is a parallelipiped. Since the eigenfunction v is not constant on ω_+, we have, on ω_+,

$$0 \leqq \min v < \max v .$$

Let us write $\min v = a$ and $\max v = b$. We apply Proposition 2.6 to the open set ω_+, the function $t_n v$ and the Borelian set

$$B_n = E(\omega'_+, \lambda_1 - \eta) \cap [a t_n, b t_n] ,$$

where η is associated to the number $\upsilon = a/b$ according to assumption $(F_3)^+$. This gives

$$\mu_N\{x \in \omega_+;\ t_n v(x) \in B_n\} \geqq c \frac{\mu_1(B_n)^N}{\mathrm{Lip}(t_n v)^N} \tag{3.12}$$

for some positive constant $c > 0$.

Estimate (3.11) can then be deduced from (3.12) as in the proof of Theorem 1.2. Q.E.D.

Remark 3.1. A natural question concerns the introduction in this problem of strong nonlinearities, as in [4]. This is possible under assumption (1.4), as was observed in [2], where this question is also studied with the operator $-\Delta$ replaced by the pseudo-laplacian $-\Delta_p$. The difficulty to include strong nonlinearities in our approach lies at the end of Step 2, to derive (3.6) from (3.4). The same difficulty arises when one tries to relax the linear growth assumption for f into the more natural growth condition in $|s|^\sigma$.

Remark 3.2. The method of this paper does not apply to the study of the Neumann problem

$$\begin{cases} -\Delta u = f(x,u) + h(x) & \text{in } \Omega, \\ \partial u/\partial n = 0 & \text{on } \partial\Omega \end{cases}$$

near its first eigenvalue $\lambda_1 = 0$. This is due to the fact that the corresponding eigenfunctions are constant. A similar difficulty arises in the study of the periodic solutions of Liénard equation (cf. [6]).

Remark 3.3. It is easily seen that Theorems 1.2 and 1.3 still hold if assumptions (F_2) and $(F_3)_\pm$ are weakened in the following way. It suffices in (F_2) to assume that for some nonnegative function $d(x)$ in $L^1(\Omega)$, the set

$$\bigcap_{x \in \Omega'} \left\{ s \in \mathbb{R}_0;\ \frac{2F(x,s)}{s^2} \leqq \lambda_1 - \eta + \frac{d(x)}{s} \right\}$$

has a positive density at both $+\infty$ and $-\infty$. In $(F_3)_+$, one can replace the set $E(\omega'_+, \lambda_1 - \eta)$ by the larger set

$$\bigcap_{x \in \omega'_+} \left\{ s \in \mathbb{R}_0;\ \frac{2F(x,s)}{s^2} \leqq \lambda_1 - \eta + \frac{d(x)}{s} \right\},$$

where the nonnegative function $d(x)$ lies in $L^1(\omega_+)$ and possibly depends on υ. Of course, a similar variant can be introduced for $(F_3)_-$.

Remark 3.4. Assumption $(F_3)_+$ of Theorem 1.3 can also be weakened in the following way, which involves the verification of (1.7) for only one choise of υ. Denote by v the positive normalized (i.e. $\|v\| = 1$) eigenfunction associated to λ_1. It suffices to require the existence of a full subset ω'_+ of an open parallelipiped $\omega_+ \subset \Omega$, a number υ with

$$\frac{\min_{\omega_+} v}{\max_{\omega_+} v} < \upsilon < 1$$

and a number $\eta > 0$ such that (1.7) holds. Using Lemma 4.5 of Sect. 4 as well as the notations introduced there, we see that this will be satisfied if there exist a full subset ω'_+ of an open parallelipiped $\omega_+ \subset \Omega$ and a number $\eta > 0$ such that

$$\frac{\underline{d}(E(\omega'_+, \lambda_1 - \eta), 0)}{\bar{d}(E(\omega'_+, \lambda_1 - \eta), 0)} > \frac{\min_{\omega_+} v}{\max_{\omega_+} v}.$$

A similar observation holds of course for $(F_3)_-$.

Remark 3.5. Results similar to Theorems 1.2 and 1.3 also hold for the problem

$$\begin{cases} Lu = f(x,u) + h(x) & \text{in } \Omega, \\ D^\alpha u = 0 & \text{on } \partial\Omega \text{ for } \quad |\alpha| \leqq m-1, \end{cases}$$

where L is a differential operator of the form

$$Lu = \sum_{|\alpha|, |\beta| \leqq m} (-1)^{|\beta|} D^\beta (a_{\alpha\beta}(x) D^\alpha u)$$

with the following assumptions: $a_{\alpha\beta} = a_{\beta\alpha}$; $a_{\alpha\beta} \in C^{|\alpha| + |\beta|}(\bar{\Omega})$ for $|\alpha| + |\beta| > 0$; $a_{\alpha\beta} \in L^\infty(\Omega)$ for $|\alpha| = |\beta| = 0$; for some $\gamma > 0$,

$$\sum_{|\alpha| = |\beta| = m} a_{\alpha\beta}(x) \xi^\alpha \xi^\beta \geqq \gamma |\xi|^{2m}$$

for a.e. $x \in \Omega$ and all $\xi \in \mathbb{R}^N$; the eigenfunctions of L associated to the first eigenvalue λ_1 have the unique continuation property, i.e. the only function $v \in H_0^m(\Omega)$ satisfying $Lv = \lambda_1 v$ and vanishing on a set of positive measure is $v \equiv 0$. Assumption $(F_3)_\pm$ must however be reinforced to take into account the fact that the eigenfunctions corresponding to λ_1 may not be of constant sign on Ω. One requires that the open sets ω_+ and ω_- be equal.

4. Examples

We first give an example of a nonlinearity $f(x,s) = f(s)$ which satisfies the assumptions of Theorem 1.2 together with (1.5).

Example 4.1. Take a sequence

$$8 < a_1 < 2a_1 < a_2 < 2a_2 < \ldots < a_k < 2a_k < \ldots$$

such that $\lim a_k / a_{k+1} = 0$ and consider the set

$$E = \bigcup_{k=1}^{\infty} [2a_k, a_{k+1}].$$

Letting

$$\zeta(r) = \frac{\mu_1(E \cap [0,r])}{\mu_1([0,r])},$$

we see that $\zeta(r)$ is increasing in each interval $[2a_k, a_{k+1}]$, decreasing in each interval $[a_k, 2a_k]$ and $\zeta(2a_k) = 1/2\zeta(a_k)$; consequently

$$\liminf_{r\to\infty} \zeta(r) = \liminf_{k\to\infty} \zeta(2a_k)$$
$$= 1/2 \liminf_{k\to\infty} \zeta(a_k) \geqq 1/2 \liminf_{k\to\infty} \frac{a_k - 2a_{k-1}}{a_k} = 1/2$$

and

$$1 \geqq \limsup_{r\to\infty} \zeta(r) = \limsup_{k\to\infty} \zeta(a_k)$$
$$\geqq \limsup_{k\to\infty} \frac{a_k - 2a_{k-1}}{a_k} = 1\,.$$

Take $\alpha > 0$ and define a function ψ on $\mathbb{R}^+$ in the following way: put

$$\psi = \begin{cases} \alpha & \text{on} \quad E\,, \\ 0 & \text{on} \quad [0, 3/2a_1]\,, \\ 0 & \text{at each point} \quad 3/2a_k\,, \end{cases}$$

and connect these values in a smooth almost piecewise linear oscillating manner, as described in the following figure:

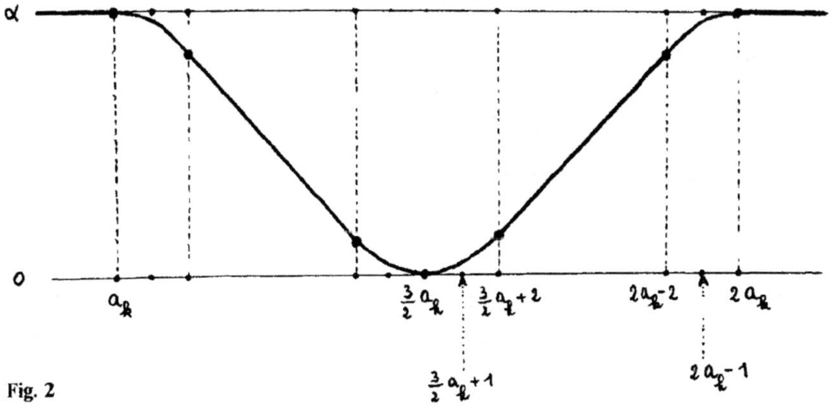

Fig. 2

This function is extended to all $\mathbb{R}$ by taking, for instance, $\psi(s) = +\psi(-s)$ for $s \in \mathbb{R}^-$. We then define the potential F by

$$F(s) = \lambda_1 \frac{s^2}{2} - \psi(s)\frac{s^2}{2}\,.$$

Clearly (F_1) and (1.5) hold, and, by the computation of $\liminf \zeta(r)$ above, the set

$$\left\{ s \in \mathbb{R}_0 : \frac{2F(s)}{s^2} \leqq \lambda_1 - \alpha \right\} = E \cup (-E)$$

has positive density at both $+\infty$ and $-\infty$. It remains to verify (f_1). We have

$$\frac{f(s)}{s} = \lambda_1 - \psi(s) - \frac{s}{2}\psi'(s)\,. \tag{4.1}$$

A simple calculation gives

$$\limsup_{s\to\infty}\frac{f(s)}{s}=\limsup_{s\to-\infty}\frac{f(s)}{s}=\lambda_1+\frac{3\alpha}{2},$$

$$\liminf_{s\to\infty}\frac{f(s)}{s}=\liminf_{s\to-\infty}\frac{f(s)}{s}=\lambda_1-3\alpha,$$

which yields (f_1).

We observe in this example that the quotient $f(s)/s$ does not lie asymptotically at the left side of λ_1 as $s\to\pm\infty$. Some sort of crossing of λ_1 occurs. However it is not of the Ambrosetti-Prodi type, which would mean

$$\limsup_{s\to-\infty} f(s)/s<\lambda_1<\liminf_{s\to+\infty} f(s)/s.$$

It turns out that the crossing exhibited in the above example must always occur under our assumptions if (1.5) holds. This is a consequence of the following

Proposition 4.2. *Let $f:\mathbb{R}^+\to\mathbb{R}$ be continuous and denote by F the associated potential. Suppose that*

$$\limsup_{s\to\infty}\frac{2F(s)}{s^2}=\lambda_1$$

and that, for some $\eta>0$, the set

$$\left\{s\in\mathbb{R}_0^+;\ \frac{2F(s)}{s^2}\leqq\lambda_1-\eta\right\}$$

has positive density at infinity. Then

$$\limsup_{s\to\infty}\frac{f(s)}{s}>\lambda_1.$$

Proof. We write $2F(s)/s^2$ as $\lambda_1-\psi(s)$ so that

$$\liminf_{s\to\infty}\psi(s)=0 \tag{4.2}$$

and the set

$$\{s\in\mathbb{R}_0^+;\ \psi(s)\geqq\eta\} \tag{4.3}$$

has positive density at infinity. By (4.1) we have to show that

$$\liminf_{s\to\infty}\left(\psi(s)+\frac{s}{2}\psi'(s)\right)<0. \tag{4.4}$$

By (4.2) there exists a sequence $b_k\to\infty$ such that $\psi(b_k)=\varepsilon_k\to 0$. Without loss of generality, one can always assume that ψ takes a value $\geqq\eta$ at the left side of b_1 and that $\varepsilon_k<\eta$ for all k. Call a_k the largest $s<b_k$ such that $\psi(s)=\eta$. Thus $\psi(a_k)=\eta$ and $\psi<\eta$ on $]a_k,b_k[$. Define

$$\beta_k=\inf\{\psi(s)+s/2\psi'(s);\ s\in[a_k,b_k]\}.$$

Thus, on $[a_k, b_k]$,

$$\psi(s)+s/2\psi'(s)\geqq\beta_k,$$

i.e.

$$(s^2\psi(s))'\geqq 2s\beta_k,$$

which gives, after integration,

$$\varepsilon_k-\beta_k\geqq\frac{a_k^2}{b_k^2}(\eta-\beta_k). \tag{4.5}$$

We claim that for some constant $c>0$ and all k sufficiently large,

$$\beta_k\leqq -c, \tag{4.6}$$

which will of course yield the conclusion (4.4).

To prove (4.6) we first observe that no subsequence of β_k converges to zero. Indeed, suppose by contradiction that there exists such a subsequence (still denoted by β_k). Then (4.5) implies that

$$\frac{a_k}{b_k}\to 0. \tag{4.7}$$

Since the set (4.3) does not intersect the union of the intervals $]a_k, b_k[$, Lemma 4.3 below shows that (4.7) contradicts the assumption of positive density for (4.3). We now claim that no subsequence of β_k satisfies an inequality like

$$\beta_k\geqq d>0 \tag{4.8}$$

for some constant d. Indeed, suppose by contradiction that there exists such a subsequence (still denotes by β_k). Then either $\beta_k>\eta$ for an infinite number of values of k, or $\beta_k<\eta$ for an infinite number of values of k, or $\beta_k=\eta$ for an infinite number of values of k. We look separately at these values of k. In the first case (4.5) implies

$$\frac{a_k^2}{b_k^2}\geqq\frac{\beta_k-\varepsilon_k}{\beta_k-\eta}>1$$

since $\varepsilon_k<\eta$. This is contradictory since $a_k\leqq b_k$. In the second case, (4.5) implies

$$\frac{a_k^2}{b_k^2}\leqq\frac{\varepsilon_k-\beta_k}{\eta-\beta_k},$$

where the right-hand side becomes <0 since (4.8) holds. This is again contradictory. In the third case, (4.5) implies

$$\beta_k\leqq\varepsilon_k$$

which, by (4.8), is also contradictory.

We have thus a sequence β_k such that no subsequence goes to zero and no subsequence satisfies (4.8). This implies that (4.6) holds. Q.E.D.

Lemma 4.3. *Suppose that the measurable set $G\subset\mathbb{R}^+$ does not intersect a sequence of disjoint intervals $]a_k, b_k[$ where a_k and $b_k\to\infty$. If G has positive density at infinity,* *then there exists a constant c such that*

$$\frac{b_k}{a_k} \leqq c \tag{4.9}$$

for all k.

Proof. The quotient

$$\frac{\mu_1(G \cap [0,r])}{\mu_1([0,r])}$$

is multiplied by a_k/b_k when r crosses the hole $]a_k, b_k[$. It follows that if for a subsequence, $a_k/b_k \to 0$, then G does not have positive density at infinity. Q.E.D.

We now construct a nonlinearity $f(x,s)$ which satisfies the assumptions of Theorem 1.3 together with (1.5).

Example 4.4. Take ψ as in Example 4.1 and consider, for $0 < \eta \leqq \alpha/2$, the set

$$E_\eta = \{s \in \mathbb{R}^+ ;\ \psi(s) \geqq \eta\}.$$

We claim that for any $0 \leqq \upsilon < 1$, there exists $\eta > 0$ such that

$$\liminf_{r \to \infty} \frac{\mu_1(E_\eta \cap [\upsilon r, r])}{\mu_1([\upsilon r, r])} > 0. \tag{4.10}$$

For that purpose we first observe that the holes in E_η are intervals $]c_k, d_k[$ centered at $3/2a_k$, with length at most equal to

$$\frac{a_k \eta - 4\eta + 2\alpha}{\alpha}.$$

Letting

$$\zeta_\eta(r) = \frac{\mu_1(E_\eta \cap [0,r])}{\mu_1([0,r])}$$

and computing as in Example 4.1, we obtain

$$\begin{aligned}
\liminf_{r \to \infty} \zeta_\eta(r) &= \liminf_{k \to \infty} \zeta_\eta(d_k) \\
&= \liminf_{k \to \infty} \frac{c_k}{d_k} \zeta_\eta(c_k) \geqq \liminf_{k \to \infty} \frac{c_k}{d_k} \zeta_\eta(a_k) \\
&\geqq \liminf \frac{c_k}{d_k} \frac{a_k - 2a_{k-1}}{a_k} \geqq \frac{3\alpha - \eta}{3\alpha - \eta}
\end{aligned}$$

and

$$\limsup_{r \to \infty} \zeta_\eta(r) \geqq \limsup_{r \to \infty} \zeta(r) = 1.$$

Consequently, given any υ with $0 \leqq \upsilon < 1$, we can choose η such that

$$\frac{\liminf \zeta_\eta(r)}{\limsup \zeta_\eta(r)} > \upsilon$$

i.e., with the notations of Lemma 4.5 below,

$$\frac{\underline{d}(E_\eta, 0)}{\overline{d}(E_\eta, 0)} > \upsilon .$$

The claim (4.10) is then a consequence of this Lemma 4.5. We can now construct the desired example. Choose two balls B_+ and B_- in Ω and define

$$F(x,s)=\begin{cases} \lambda_1 \dfrac{s^2}{2} & \begin{array}{l}\text{if } x \notin B_+ \cup B_- \\ \text{or if } x \in B_+ \text{ and } s<0 \\ \text{or if } x \in B_- \text{ and } s \geqq 0,\end{array} \\ \lambda_1 \dfrac{s^2}{2} - \psi(s)\dfrac{s^2}{2} & \begin{array}{l}\text{if } x \in B_+ \text{ and } s>0 \\ \text{or if } x \in B_- \text{ and } s<0.\end{array} \end{cases}$$

Clearly (F_1) and (1.5) hold, and the calculation in Example 4.1 shows that (f_1) holds. Moreover, (4.10) shows that $(F_3)_+$ and $(F_3)_-$ are satisfied on B_+ and B_-, respectively.

Lemma 4.5. *Let $G \subset \mathbb{R}^+$ be measurable and denote, for $0 \leqq \upsilon < 1$,*

$$\underline{d}(G,\upsilon) = \liminf_{r\to\infty} \frac{\mu_1(G \cap [\upsilon r, r])}{\mu_1([\upsilon r, r])},$$

$$\overline{d}(G,\upsilon) = \limsup_{r\to\infty} \frac{\mu_1(G \cap [\upsilon r, r])}{\mu_1([\upsilon r, r])}.$$

Assume $\underline{d}(G,0)>0$. Then $\underline{d}(G,\upsilon)>0$ for all $0 \leqq \upsilon < 1$ such that

$$\upsilon < \frac{\underline{d}(G,0)}{\overline{d}(G,0)}.$$

Proof. Writing

$$\frac{\mu_1(G\cap[\upsilon r, r])}{(1-\upsilon)r} = \frac{\mu_1(G\cap[0,r])}{(1-\upsilon)r} - \frac{\mu_1(G\cap[0,\upsilon r])}{(1-\upsilon)r},$$

we obtain

$$\underline{d}(G,\upsilon) \geqq \frac{1}{1-\upsilon}\underline{d}(G,0) - \frac{\upsilon}{1-\upsilon}\overline{d}(G,0),$$

which yields the conclusion. Q.E.D.

We now turn to a nonlinearity $f(x,s)=f(s)$ for which the global result (Theorem 1.2) applies while the local one (Theorem 1.3) does not. This illustrates the fact that the density requirement at infinity is stronger in Theorem 1.3 than in Theorem 1.2.

Example 4.6. Take a sequence

$$8 < b_1 < 4b_1 < b_2 < 4b_2 < \dots$$

such that $\lim b_k/b_{k+1}=0$. This implies, as in Example 4.1, that the set

$$G=\bigcup_{k=1}^{\infty}[4b_k, b_{k+1}]$$

has positive density at $+\infty$. Take $\alpha>0$ and define a function φ on $\mathbb{R}^+$ in the following way: put

$$\varphi=\begin{cases}\alpha & \text{on } G,\\ 0 & \text{on } [0,2b_1],\\ 0 & \text{on } \bigcup_{k=1}^{\infty}[2b_k,3b_k],\end{cases}$$

and connect these values by a construction similar to that in Example 4.1, as described in the following figure:

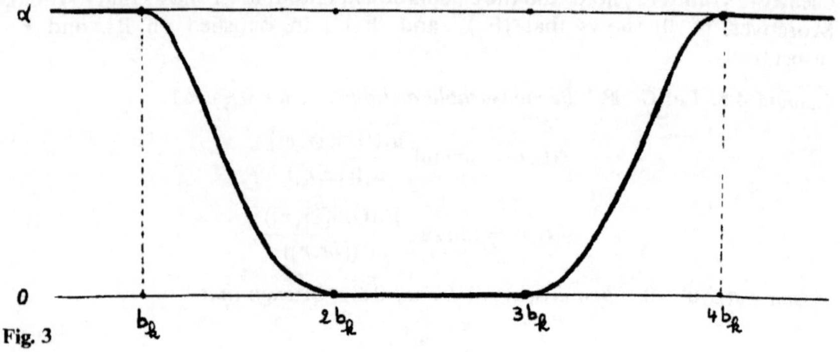

Fig. 3

Then, for any $\eta>0$, the set

$$G_\eta=\{s\in\mathbb{R}^+;\ \varphi(s)\geqq\eta\}$$

does not intersect

$$\bigcup_{k=1}^{\infty}[2b_k,3b_k],$$

and consequently

$$\underline{d}(G_\eta, 2/3)=0.$$

A potential $F(s)$ can then be constructed from this function φ, as in Example 4.1, which satisfies (f_1), (F_1), (1.5), (F_2) but does not satisfy $(F_3)_\pm$.

Our next two examples concern nonlinearities whose potential oscillates below λ_1 but which fall outside our present results.

Example 4.7. Consider

$$F(s)=\frac{3}{8}\lambda_1 s^2+\frac{1}{8}\lambda_1 s^2 \sin s. \tag{4.11}$$

Clearly (F_1), (1.5), and (F_2) hold. However, (f_1) is not verified since

$$f(s)=\frac{3}{4}\lambda_1 s+\frac{1}{4}\lambda_1 s\sin s+\frac{1}{8}\lambda_1 s^2\cos s.$$

Example 4.8. Consider

$$F(s)=\frac{3}{8}\lambda_1 s^2+\frac{1}{8}\lambda_1 s^2\sin\log(1+\log(1+|s|)). \tag{4.12}$$

Clearly (F_1) and (1.5) hold, as well as (f_1) since

$$\limsup_{s\to\infty}\frac{f(s)}{s}=\lambda_1,\qquad \liminf_{s\to\infty}\frac{f(s)}{s}=\frac{\lambda_1}{2}. \tag{4.13}$$

However (F_2) does not hold. This can be verified directly by using Lemma 4.3. It also follows from Proposition 4.2 and (4.13).

Observe that a little less damping in the oscillations in (4.12) leads to a nonlinearity which satisfies the assumptions of Theorem 1.2:

Example 4.9. Consider

$$F(s)=\frac{3}{8}\lambda_1 s^2+\frac{1}{8}\lambda_1 s^2\sin\log(1+|s|).$$

Clearly (F_1) and (1.5) hold, as well as (f_1) since

$$\lambda_1<\limsup_{s\to\infty}\frac{f(s)}{s}<\frac{9}{8}\lambda_1,$$

$$\frac{3}{8}\lambda_1<\liminf_{s\to\infty}\frac{f(s)}{s}<\frac{1}{2}\lambda_1.$$

A direct computation show that (F_2) also holds.

To conclude this section, we show that a density condition at infinity for $f(s)/s$ implies a strict inequality at infinity for $2F(s)/s^2$. This was observed in a particular case in [1]. See also [2].

Proposition 4.10. *Let $f:\mathbb{R}^+\to\mathbb{R}$ be continuous with*

$$\limsup_{s\to\infty}\frac{f(s)}{s}\leqq\lambda_1. \tag{4.14}$$

Suppose that for some $\eta>0$, the set

$$E=\left\{s\in\mathbb{R}_0;\ \frac{f(s)}{s}\leqq\lambda_1-\eta\right\}$$

has positive density at $+\infty$. Then the associated potential F satisfies

$$\limsup_{s\to+\infty}\frac{2F(s)}{s^2}<\lambda_1.$$

Proof. By (4.14) and the definition of E, we have

$$F(s) \leqq \int_{[0,s]\cap E} (\lambda_1 t - \eta t) + \int_{[0,s]\cap E^c} (\lambda_1 t + \varepsilon t + b_\varepsilon)$$

where the constant b_ε is associated to $\varepsilon > 0$ in the detailed expression of (4.14) and where E^c denotes the complement of E. Lemma 2.2 implies

$$\int_{[0,s]\cap E} \eta t \geqq \int_0^{\mu_1(E\cap[0,s])} \eta t = \eta/2\mu_1(E\cap[0,s])^2,$$

where, by the positive density at infinity, the last term is $\geqq \eta\delta s^2/2$ for some $\delta > 0$ and all s sufficiently large. Consequently, taking $\varepsilon = \eta\delta/4$, we obtain

$$F(s) \leqq \lambda_1 \frac{\delta^2}{2} - \frac{\eta\delta}{4}\frac{s^2}{2} + sb_\varepsilon,$$

which yields the conclusion. Q.E.D.

Acknowledgements. We wish to thank P. L. Lions for suggesting the use of the Steiner rearrangement and providing us with a variant of Example 2.5 and A. Anane for some interesting comments about the density condition at infinity, in particular Lemma 4.5.

References

1. Abdoral, O.: Ph.D. Thesis, Universidade de Brasilia (in preparation)
2. Anane, A.: Ph.D. Thesis, Université Libre de Bruxelles, 1987
3. Bandle, C.: Isoperimetric inequalities and applications. London: Pitman 1980
4. De Figueiredo, D., Gossez, J.-P.: Nonlinear perturbations of a linear elliptic problem near its first eigenvalue. J. Differ. Equations **30**, 1-19 (1978)
5. De Figueiredo, D., Gossez, J.-P.: Conditions de non-résonance pour certains problèmes elliptiques semi-linéaires. C.R. Acad. Sc. Paris **302**, 543–545 (1986)
6. Gossez, J.-P., Omari, P.: Nonresonance with respect to the Fučik spectrum for periodic solutions of second order ordinary differential equations (to appear)
7. Hammerstein, A.: Nichtlineare Integralgleichungen nebst Anwendungen. Acta Math. **54**, 117–176 (1930)
8. Kawohl, B.: Rearrangements and convexity of level sets in PDE (Lect. Notes Math., Vol. 1150). Berlin-Heidelberg-New York: Springer 1985
9. Mawhin, J., Ward, J. Jr., Willem, M.: Variational methods and semilinear elliptic equations. Arch. Rat. Mech. Anal. **95**, 269–277 (1986)
10. Mossino, J.: Inégalités isopérimétriques et applications en physique. Paris: Herman 1984
11. Polya, G., Szegö, G.: Isoperimetric inequalities in mathematical physics. Princeton: Princeton University Press 1951

Received August 18, 1987

MAXIMUM PRINCIPLES FOR LINEAR ELLIPTIC SYSTEMS (*)

by DJAIRO G. DE FIGUEIREDO (in Campinas)
and ENZO MITIDIERI (in Trieste) (**)

SOMMARIO. - *In questo lavoro si ottengono alcune condizioni sufficienti (necessarie) per la validità del principio di massimo per sistemi ellittici di tipo cooperativo e non.*

SUMMARY. - *In this paper we obtain some sufficient (necessary) conditions for the validity of the maximum principle for cooperative and non-cooperative elliptic systems.*

0. Introduction. In this paper we shall prove some results concerning the maximum principle for weakly-coupled elliptic systems of the form

$$(0.1)\qquad \begin{aligned} \mathcal{L}(D)U &= A(x)U + F \quad \text{in} \quad \Omega \subset \mathbf{R}^N \\ U &= 0 \quad \text{on} \quad \partial\Omega \end{aligned}$$

where $\Omega \subset \mathbf{R}^N (N \geq 1)$ is a smooth bounded domain, $\mathcal{L} = [L_1(D), L_2(D), \dots, L_n(D)]$ is a diagonal-matrix of second order elliptic operators, $A(x) = (a_{ij}(x))$ is a $n \times n$ coefficient matrix and F is a given n-vector function defined in Ω. By a maximum principle we mean the statement of the positiveness of the solution U of (0.1) (that is, $U \geq 0$ in Ω) when the given function $F \geq 0$ in Ω.

The classical maximum principle of Protter-Weinberger (see [8]) for weakly-coupled elliptic systems of the form (0.1) holds if the following conditions on $A(x)$ are satisfied:

(*) Pervenuto in Redazione il 3 giugno 1991.

(**) Indirizzo degli Autori: D.G. de Figueiredo: Departemento de Matematica – Universidade de Campinas – Unicamp – Campinas (Brazil); E. Mitidieri: Dipartimento di Matematica e Informatica – Università degli Studi di Udine – Via Zanon, 6 – 33100 Udine (Italy) e Dipartimento di Scienze Matematiche – Università degli Studi di Trieste – Piazzale Europa, 1 – 34100 Trieste (Italy).

Originally published in: Rendiconti dell'Istituto di Matematica dell'Univ. Trieste, Vol. 22 (1990), 36-66. © Rendiconti dell'Istituto di Matematica dell' Università di Trieste (RIMUT). The original publication is available at http://rendiconti .dmi.units.it/
Reprint by Springer-Verlag Berlin Heidelberg 2013 with kind permission,
DOI: 10.1007/978-3-319-02856-9_21

(i) $a_{ij}(x) \geq 0 \quad \forall i \neq j \quad x \in \Omega$

(ii) $\sum_{j=1}^{n} a_{ij}(x) \leq 0 \quad \forall i \quad x \in \Omega.$

A system (0.1) for which (i) above holds is said to be *cooperative.* It is clear that since (ii) is assumed, the above result cannot contain the maximum principle for the single equation. Indeed a maximum principle for $-\Delta u = a(x)u + f$ holds, for example, if $a(x) < \lambda_1$ in Ω. Here λ_1 denotes the first eigenvalue of $(-\Delta, H_0^1(\Omega))$. Hence, it is natural to ask if it is possible to refine the above mentioned result for systems by relaxing condition (ii) somehow. In this paper we show that this is actually possible. Let us briefly sketch the contents of the present paper.

In section 1 we treat a general class of cooperative systems like (0.1) with variable coefficients. We start by giving a proof of the above mentioned result, see Theorem 1.1. Our proof is completely different from the "parabolic" proof contained in [8]. Then we prove our Theorem 1.2, which is a noticeable improvement of the previous theorem. The case when the elliptic operators L_k are all equal to the same selfadjoint elliptic operator is considered. We then obtain the right generalization of the scalar case (single equation) stated above.

In Section 2 we consider the case when all the entries of the matrix A are constant and the operators L_k are all equal to a self adjoint operator L. We then obtain a necessary and sufficient condition for the existence of a maximum principle in terms of the characteristic polynomial of A.

In Section 3 we analyse a 2×2 noncooperative system. Using the idea of embedding such a system into 3×3 cooperative system we are able to prove maximum principles of noncooperative systems. In this way we recapture a result obtained before by the authors in [5]. We also treat fourth order elliptic operators, and in particular we obtain a maximum principle for the biharmonic operator under the so-called Navier boundary conditions.

Some of the results contained in this paper were announced at the "Conference on Reaction Diffusion equations" held in Edinburgh in June 1988.

1. Maximum principle for cooperative elliptic systems with variable coefficients. Let us first fix the notation and state some conditions that will be used throughout this section.

Consider the following set of second order elliptic operators with real valued coefficients defined in some bounded domain Ω in $\mathbf{R}^N$,

$$L_k(D) = -\sum_{ij} b_{ij}^k(x) D_i D_j + \sum_i b_i^k(x) D_i, \quad k = 1, \dots, n$$

where $b_{ij}^k = b_{ji}^k$ and

$$\sum_{ij} b_{ij}^k(x) \xi_i \xi_j \geq \lambda(x) |\xi|^2, \quad x \in \Omega, \ \xi \in \mathbf{R}^N, \ \forall k$$

for some function $\lambda(x) > 0$, $x \in \Omega$. So each L_k is elliptic as defined in Gilbarg-Trudinger [6]. No regularity on the coefficients is needed. We shall assume that there is a constant $M > 0$ such that

$$\text{(1.1)} \qquad \frac{|b_i^k(x)|}{\lambda(x)} \leq M, \quad x \in \Omega, \ \forall i, \ \forall k.$$

Let $A = [a_{kj}(x)]$ be a $n \times n$ matrix whose entries are real-valued functions defined in Ω. We assume that the off-diagonal entries are nonnegative (cooperativeness):

$$\text{(1.2)} \qquad a_{kj}(x) \geq 0, \quad x \in \Omega, \quad k \neq j.$$

Let $F(x) = (f_1(x), \dots, f_n(x))$ be a given n-vector whose components are real-valued functions defined in Ω.

Let us consider the elliptic system

$$\text{(1.3)}_k \qquad L_k(D) u_k = \sum_j a_{kj}(x) u_j + f_k \quad k = 1, \dots, n$$

or, in short

$$\text{(1.3)} \qquad \mathcal{L}(D) U = AU + F$$

where $\mathcal{L}(D)$ denotes the diagonal operator-matrix $(L_1(D), \dots, L_n(D))$ and U is the solution n-vector $(u_1(x), \dots, u_n(x))$.

We say that an n-vector $V(x) = (v_1(x), \ldots, v_n(x))$ is ≥ 0 if all its components are nonnegative functions.

In this section we shall discuss the nonnegativeness of solutions of system (1.3).

More precisely, we investigate when $F \geq 0$ implies that a solution of the Dirichlet problem

$$(1.4) \qquad \begin{array}{lll} \mathcal{L}(D)U & = AU + F & \text{in } \Omega \\ U & = 0 & \text{on } \partial\Omega \end{array}$$

is such that $U \geq 0$ in Ω.

By a solution we mean a classical solution, that is a function U defined in $\bar{\Omega}$ which is continuous in $\bar{\Omega}$ and belongs to $C^2(\Omega)$. Here we do not discuss existence questions. We assume that a classical solution of (1.4) is given and we prove its nonnegativeness.

The first result is the following (see [8] for a different proof):

THEOREM 1.1. *Suppose in addition to the above assumptions that*

$$(1.5) \qquad a_k(x) \equiv \sum_j a_{kj}(x) \leq 0 \quad x \in \Omega, \quad k = 1, \ldots, n.$$

Then $F \geq 0$ in Ω implies that $U \geq 0$, being U a solution to (1.4).

REMARK 1.1. Condition (1.5) and the hypothesis that the off-diagonal entries of A are ≥ 0 imply that $a_{kk}(x) \leq 0 \ \forall x \in \Omega, k = 1, \ldots, n.$

We shall prove next that if the conclusion of Theorem 1.1 holds under the stronger assumption

$$(1.5) \qquad a_k(x) < 0, \text{ for } x \in \Omega \text{ and } k = 1, \ldots, n$$

then Theorem 1.1 is true.

Indeed, let $\alpha = \alpha(x_1)$ be a positive C^2 real-valued function which will be chosen later. let $v_k(x)$ $(k = 1, \ldots, n)$ be defined by $u_k(x) = \alpha(x_1)v_k(x)$. Through some calculations we come to

$$L_k u_k = \alpha L_k v_k - 2\alpha' \sum_i b_{1i}^k D_i v_k - \alpha'' b_{11}^k v_k + \alpha' b_1^k v_k .$$

Thus the v_k's satisfy the system

$$L_k v_k - \frac{2\alpha'}{\alpha}\sum_i b_{i1}^k D_i v_k = \left(\frac{\alpha''}{\alpha} b_{11}^k - \frac{\alpha'}{\alpha} b_1^k\right) v_k + \sum_j a_{kj} v_j + \frac{f_k}{\alpha},$$

or

$$(1.6) \qquad \hat{\mathcal{L}}(D)V = \hat{A}V + \frac{F}{\alpha}$$

where $\mathcal{L}$ and $\hat{\mathcal{L}}$ have the same principal part and b_i^k is replaced by

$$\tilde{b}_i^k = b_i^k - \frac{2\alpha'}{\alpha} b_{i1}^k .$$

Also A and $\tilde{A}$ have the same off-diagonal terms and a_{kk} is replaced by

$$\tilde{a}_{kk} = a_{kk} + \frac{\alpha''}{\alpha} b_{11}^k - \frac{\alpha'}{\alpha} b_1^k .$$

Now we choose the function $\alpha(x_1)$ in such a way that $\tilde{a}_{kk} < 0$ in Ω, and consequently $\tilde{a}_k < 0$ in Ω.

Using the ellipticity of L_k and condition (1.1) we have

$$(1.7) \qquad \frac{\alpha''}{\alpha} b_{11}^k - \frac{\alpha'}{\alpha} b_1^k \le \frac{\alpha''}{\alpha}\lambda(x) - \frac{\alpha'}{\alpha} b_1^k \le \lambda(x)\left[\frac{\alpha''}{\alpha} - M\frac{\alpha'}{\alpha}\right]$$

supposing that $\alpha > 0$, $\alpha' \le 0$ and $\alpha'' \le 0$.

Our claim is attained by showing that an α as such can be chosen to make the right hand side of (1.7) strictly negative. Since Ω is bounded, we may assume that Ω is contained in the strip $\{x, : 0 < x_1 < K\}$ for some $K \ge 0$. Thus, choosing

$$\alpha(x_1) = -e^{\beta x_1} + e^{\beta K} + 1$$

where $\beta > M$, we see that α fulfills all the above requirements. Clearly $U \ge 0$ if and only if $V \ge 0$.

Proof of Theorem 1.1. By the previous remark we assume (1.5)' instead of (1.5). Using the notation in (1.5) we write system $(1.3)_k$ as

$$(1.8)_k \qquad L_k(D)u_k = a_k(x)u_k + \sum_{j\neq k} a_{kj}(x)[u_j - u_k] + f_k(x)$$

Suppose by contradiction that U is not positive. This means that at least one of its components is not positive. Without loss of generality, assume that it is u_1. Hence, there is $x^1 \in \Omega$ such that

$$u_1(x^1) = \min_{x\in\Omega} u_1(x) < 0 \,. \tag{1.9}$$

Since $L_1(D)u_1(x^1) \leq 0$, it follows that, for at least one $j \neq 1$, we have

$$a_{1j}(x^1)[u_j(x^1) - u_1(x^1)] < 0 \,.$$

Without loss of generality assume that this happens for $j = 2$. So,

$$u_2(x^1) < u_1(x^1) \tag{1.10}$$

and let $x^2 \in \Omega$ be such that

$$u_2(x^2) = \min_{x\in\Omega} u_2(x) < 0 \,. \tag{1.11}$$

Using the second equation in $(1.8)_k$ we infer that

$$a_{2j}(x^2)[u_j(x^2) - u_2(x^2)] < 0$$

for some $j \neq 2$. Clearly $j \neq 1$ in view of (1.9), (1.10) and (1.11). As in the previous case we may assume that

$$u_3(x^2) < u_2(x^2)$$

and we proceed as before using the third equation in $(1.8)_k$. Observe that $u_3(x^2) < u_1(x^2)$. Repeating the above argument after using the $(n-1)^{th}$ equation, we come to a point $x^n \in \Omega$ where

$$u_n(x^n) = \min_{x\in\Omega} u_n(x) < 0$$

and $u_n(x^n) < u_j(x^n)$ for $j \neq n$.

Finally, using the nth equation in $(1.8)_k$ we come to a contradiction.

◇

Next we obtain a maximum principle for the solutions of (1.4) under a weaker hypothesis than (1.5).

DEFINITION 1.1. The operator $\mathcal{L} - A$, satisfies property (ψ) if there exists a C^2-function $\psi : \bar{\Omega} \to \mathbf{R}^n$ such that

(i) $\psi(x) > 0$ for $x \in \bar{\Omega}$

(ii) $\mathcal{L}(\psi) \geq A\psi$ in Ω.

REMARK 1.2. If property (ψ) holds for $\mathcal{L}(D) - A$ where $A \geq B$ (i.e. $a_{ij}(x) \geq b_{ij}(x)$ $i, j, \dots, n$) then it also holds for $\mathcal{L}(D) - B$.

REMARK 1.3. If the coefficients of the matrix A are bounded i.e. $a_{ij} \in L^\infty(\Omega)$ then a sufficient condition in order that $\mathcal{L}(D) - A$ satisfies property (ψ) is the solvability of the following system

$$(*) \qquad \begin{cases} \mathcal{L}(\phi) = A_\infty \phi + F & \text{in } \Omega \\ \phi = 0 & \text{on } \partial\Omega \\ \phi > 0 & \text{in } \Omega \end{cases}$$

where $F = (F_1, \dots, F_n)$ with $F_j = \sum_{j=1}^n a_{ij}^\infty$, and $a_{ij}^\infty = ||a_{ij}||_{L^\infty}$.

Indeed if $(*)$ has a solution then, the operator $\mathcal{L} - A_\infty$ satisfies property (ψ) with $\psi = \phi + 1$ (i.e. $\psi_i = \phi_i + 1, i = 1, \dots, n$) and then by the preceding Remark $\mathcal{L} - A$ satisfies the same property.

REMARK 1.4. In the special case $L_1(D) = L_2(D) = L(D)$ with L self-adjoint (see [6]).

$$(1.12) \qquad L(D)u = -\sum_{ij} D_i(a_{ij} D_j u + b_i(.u)) + \sum_i b_i D_i u$$

where the coefficients are continuous functions defined in a domain $\Omega' \supset \bar{\Omega}$ we can give a simple sufficient condition as follows:

Let $\lambda_1(\Omega)$ be the first eigenvalue of $(L(D), H_0^1(\Omega))$. Suppose that

$$a_k(x) \leq \Lambda < \lambda_1(\Omega) \quad x \in \Omega\ .$$

Let $\hat{\Omega}$ be, such that $\Omega' \supset \hat{\Omega} \supset \bar{\Omega}$, and $\Lambda \leq \lambda_1(\hat{\Omega}) < \lambda_1(\Omega)$, where $\lambda_1(\Omega)$ denotes the first eigenvalue of $(L(D), H_0^1(\hat{\Omega}))$. Let ϕ be a positive eigenfunction corresponding to $\lambda_1(\hat{\Omega})$. Then we see that the operators $\mathcal{L} - A$ satisfies property (ϕ) with $\psi = (\phi, \dots, \phi)$.

THEOREM 1.2. *If the operators $\mathcal{L} - A$ satisfies property (ψ) then $F \geq 0$ in Ω implies $U \geq 0$, being U a solution to* (1.4).

COROLLARY 1.1. *Let $\mathcal{L}$ be as in Remark* 1.4. *Assume that $a_k(x) \leq \Lambda < \lambda_1(\Omega), \forall x \in \Omega$. Then $F \geq 0$ in Ω implies $U \geq 0$ in Ω, being U a solution of* (1.4).

REMARK 1.5. Theorem 1.2 above contains Theorem 1.1. Indeed, if $a_k(x) \leq 0$, then it is sufficient to consider $\psi = (1, 1, \ldots, 1)$.

Proof of Theorem 1.2. Let us introduce the functions v_k, $k = 1, \ldots, n$ by

$$u_k(x) = v_k(x)\psi_k(x) \quad x \in \Omega,$$

and put $h_k(x) = L_k(\psi_k) - \sum_{j=1}^n a_{kj}\psi_j(x)$.

Then

$$L_k u_k = \psi_k L_k v_k + v_k L_k \psi_k - 2\sum_{1j} b_{ij}^{(k)} D_i\psi_k D_j v_k$$

which implies

$$L_k v_k - \frac{2}{\psi_k}\sum_{ij} b_{ij}^{(k)} D_i\psi_k D_j v_k = \frac{v_k}{\psi_k}[-h_k - \sum_{j=1}^n a_{kj}\psi_j] +$$

$$+ \sum_{j=1}^n \frac{a_{kj} v_j \psi_j}{\psi_k} + \frac{f_k}{\psi_k}.$$

Thus, system (1.1) is equivalent to

$$\tilde{L}_k v_k = \sum_{j=1} c_{kj} v_j + \frac{f_k}{\psi_k} \quad k = 1, \ldots, n \tag{1.13}$$

where $\tilde{L}_k$ and L_k have the same principal part and

$$\tilde{b}_i^k = b_i^k - \frac{2}{\psi_k}\sum_j b_{ij}^k D_j\psi_j.$$

$$c_{kj} = \begin{cases} -\frac{1}{\psi_k}[h_k + \sum_{i=1}^{n} a_{ki}\psi_i - a_{kk}\psi_k] \text{ if } j = k \\ \frac{a_{kj}\psi_j}{\psi_k} \; j \neq k \end{cases}$$

For system (1.13) we have

$$\tilde{c}_k = \sum_j c_{kj} = -\frac{h_k}{\psi_k} \leq 0$$

and so Theorem 1.1 may be applied to conclude the proof.

◇

REMARK 1.6. Assume that each L_k is strictly elliptic, that is $\lambda(x) \geq \lambda_0$ for some positive constant λ_0 and all $x \in \Omega$. Suppose that the a_k's are L^∞ functions and let $M > 0$ be such that $|a_k(x)| \leq M, k = 1, \ldots, n$ $x \in \Omega$. It follows from a result of Protter and Weinberger [8; p. 73] that if the domain is contained in a sufficiently narrow slab bounded by two parallel planes, or if M is sufficiently small, then $\mathcal{L} - A$ satisfies property (ψ). To see how narrow has the slab to be or how small M, look at the footnote on page 74 of [6].

REMARK 1.7 (a strong maximum principle). Assume that the diagonal entries a_{kk} are L^∞ functions, and that Ω is smooth. Then on the hypotheses of Theorem 1.2, if, for some k, f_k is positive on a set of positive measure, it follows that $u_k > 0$ in Ω and the outward normal derivative $\frac{\partial u_k}{\partial \nu} < 0$. To see that let M be a positive constant such that $M \geq |a_{kk}(x)|$ for all $x \in \Omega$ and apply the usual strong maximum principle for scalar equations to the equation

$$(L_k + M)u = \sum_{j \neq k} a_{kj}u_j + (a_{kk} + M)u_k + f_k \,.$$

REMARK 1.8. If $f : \Omega \to \mathbf{R}^n$ is a given C^1 function with cooperative Jacobian J_f and $f(0) = 0$, then it follows from Theorem 1.1 and the mean value theorem (we apply the mean value theorem componentwise) that if there exists a function $\psi : \Omega \to \mathbf{R}^n$ such that

(i) $\psi > 0$ in $\bar{\Omega}$

(ii) $\mathcal{L}(\psi) \geq f(\psi)$ in Ω

then for every $F : \Omega \to \mathbf{R}^n$ with $F \geq 0$ in Ω the solutions to

$$\begin{aligned} \mathcal{L}(u) &= f(U) + F \quad \text{in } \Omega \\ U &= 0 \quad\quad\quad\quad \text{on } \partial\Omega \end{aligned}$$

are non negative.

2. A maximum principle for a cooperative system with constant coefficients.

In this section we study system (1.4) of the previous section in the case when

$$L_1(D) = L_2(D) = \ldots = L_n(D) = L$$

with L self-adjoint i.e. L is of the form

$$L(D) = -\sum_{ij} D_i(a_{ij} D_j + b_i) + \sum_i b_i D_i$$

the coefficients $a_{ij}, b_i \in C(\Omega)$ and A is a constant matrix with nonnegative off-diagonal entries. Namely

$$\begin{aligned} \mathcal{L}(D)U &= AU + F, \quad \text{in } \Omega \\ U &= 0 \quad\quad\quad\quad \text{on } \partial\Omega \end{aligned} \tag{2.1}$$

For such a system we prove the following result, which is a strong sharpening of Theorem 1.2. We shall denote by λ_1 the principal eigenvalue of $(L, H_0^1(\Omega))$.

THEOREM 2.1. *Let $p_1(\lambda), \ldots, p_n(\lambda)$ be the characteristic polynomials of the first n-principal minors of A, namely*

$$(a_{11}), \begin{pmatrix} a_{11} & a_{12} \\ a_{21} & a_{22} \end{pmatrix}, \begin{pmatrix} a_{11} & a_{12} & a_{13} \\ a_{21} & a_{22} & a_{23} \\ a_{31} & a_{32} & a_{33} \end{pmatrix}, \ldots A.$$

Then

$$p_i(\lambda_1) > 0 \quad i = 1, \ldots, n \tag{2.2}$$

is a necessary and sufficient condition for

$$(2.3) \qquad F > 0 \Rightarrow U > 0 ,$$

where U is a solution to (3.1).

REMARK 2.1. $F > 0$ means $F_j(x) \geq 0$ and $F_j(x) \not\equiv 0$ for all j. $F \geq 0$ means $F_j(x) \geq 0$ for all j. Conditions (2.2) in the above theorem is sufficient for $F \geq 0 \Rightarrow U \geq 0$.

REMARK 2.2. We see as a consequence of Lemma 3.1 below, that condition (2.2) implies that the characteristic polynomial $p(\lambda)$ of any principal minor is positive at λ_1, i.e. $p(\lambda_1) > 0$. This fact dissipates a possible suspicion raised by the apparent asymmetry of condition (2.2).

To fix terminology used here we recall that a principal minor is a minor obtained from the original matrix by dropping lines and columns of the same order. For instance, the second and fourth lines and the second and fourth columns. All othe rminors are called secondary.

DEFINITION 2.1. A matrix of the form

$$M_n = \begin{bmatrix} a_{11} & -a_{12} & \dots & -a_{1n} \\ -a_{21} & a_{22} & & -a_{2n} \\ \dots & \dots & & \dots \\ -a_{n1} & -a_{n2} & & a_{nn} \end{bmatrix}$$

where $a_{ij} \geq 0$ for all $i, j = 1, \dots, n$ is called an M-matrix. Now we list some facts about these matrices and, for sake of completeness, we include the proofs.

LEMMA 2.1. *Suppose that all principal minors of M_n, including M_n itself, have positive determinants.*

Then the $(n+1) \times (n+1)$ matrix below has non positive determinant

$$z_{n+1} = \begin{vmatrix} z_1 & t_2 \ \dots \ t_{n+1} \\ z_2 & \\ \vdots & M_n \\ z_{n+1} & \end{vmatrix}$$

where $z_i \leq 0, i = 1, \dots, n+1$, $t_i \leq 0$ *for* $i = 2, \dots, n+1$. *Moreover, if* $z_i < 0$ *for all* i, *then* $\det Z_{n+1} < 0$.

Proof. By induction. The result is clearly true for 1×1 M-matrices. Now, assume the result is true for $(n-1) \times (n-1)$ M-matrices. Let M_n be a given $n \times n$ M-matrix satisfying the hypothesis of the Lemma. Then,

$$\det Z_{n+1} = z_1 \det M_n + \sum_{j=2}^{n} (-1)^{j+1} \det \hat{Z}_{1,j} \tag{2.4}$$

where $Z_{1,j}$ is the minor obtained from Z_{n+1} by omitting line j, $2 \leq j \leq n+1$, and the first column. The matrix $Z_{1,j}$ has its $(j-1)^{th}$ column formed by nonpositive numbers. We transpose this column with the first one and we obtain a matrix $\hat{Z}_{ij}$ of the type Z_n whose determinant is nonpositive by induction hypothesis. Since

$$\det Z_{1,j} = (-1)^{j-2} \det \hat{Z}_{1,j}$$

we get the result from (2.4).

◇

LEMMA 2.2. *Let* M_n *be an* M*-matrix and suppose that its* n*-first principal minors have positive determinants. Then*

(i) all other principal minors have positive determinants

(ii) $(-1)^{i+j} \det M_{ij} \geq 0$ *when* M_{ij} *is the secondary minor determined from* M_n *by dropping the* i^{th} *line and the* j^{th} *column,* $i \neq j$.

Proof. By induction. For $n = 2$ we obtain

$$a_{11}a_{22} - a_{12}a_{21} > 0 \quad a_{11} > 0$$

by hypothesis, which implies $a_{22} > 0$. Now, we assume the result true for $(n-1) \times (n-1)$ M-matrices, and let M_n be a $n \times n$ M-matrix satisfying the hypothesis of the lemma. Let us prove (ii). Assume $i < j$. The case $i > j$ can be proved similarly. Observe that the ith column of M_{ij} is formed by non-positive numbers: we transpose it with the first column. The $(j-i)^{th}$ line of M_{ij} is formed by nonpositive numbers: we transpose it with the first line. We then obtain a matrix $\hat{M}_{ij}$ of the type Z_{n-1}. So, by Lemma 2.1, det $\hat{M}_{ij} \leq 0$. On the other hand

$$\det M_{ij} = (-1)^{i-1}(-1)^{j-2} \det \hat{M}_{ij} .$$

So the result follows. Now let us prove (i). In view of the induction hypothesis it remains to prove that the principal minors containing a_{nn} have positive determinants. To do that it suffices to prove that $a_{nn} > 0$. For that matter we write

$$\det M_n = a_{nn} \det M_{n-1} - \sum_{j=2}^{n} a_{nj}(-1)^{n+j} \det M_{nj} .$$

By part (ii) each term ($j = 2, \ldots, n$) of the above summation is ≥ 0. Since det $M_n > 0$ and det $M_{n-1} > 0$ it follows that $a_{nn} > 0$.

◊

LEMMA 2.3. *Let M be an M-matrix. If all solutions X of*

$$MX = Y \tag{2.5}$$

for $Y > 0$ are such that $X > 0$, then all principal minors of M have positive determinant. (Here $X > 0$ means $X_j > 0$ for each j).

Proof. The ith equation in (2.5) gives

$$a_{ii}x_i = \sum_{i \neq j} a_{ij}x_j + y_1 .$$

Since $a_{ij} \geq 0$ for all i, j and $x_i > 0$, $y_i > 0$ it follows $a_{ii} > 0$. Now we proceed by induction. Assume the result has been proved to a system (2.5) with $n - 1$ equations.

Let us prove it for a system with n equations. From such a system we can write n system with $n - 1$ equations and $n - 1$ unknown using the following procedure: drop the ith equation and write the other $n - 1$ equations in the form

$$a_{jj}x_j - \sum_{k \neq i} a_{jk}x_k = a_{ji}x_i + y_j . \tag{2.6}$$

The matrix of system (2.6) is an M-matrix. So by the induction hypothesis all $j \times j$ ($j \leq n-1$) principal minors of M have positive determinant. It remains to show that det $M > 0$. We first observe that det $M \neq 0$. Indeed, if det $M = 0$, then there would exist $X_0 \neq 0$, such that $MX_0 = 0$. Then if $X > 0$ is a solution to (2.5) for a given $Y > 0$, then $X + tX_0$ for any $t \in \mathbf{R}$ is also a solution. Clearly for a convenient choice, at t we can get $X + tX_0$ not > 0. Now, by Cramer's rule

$$(2.7) \qquad x_1 \det M = \det \begin{vmatrix} y_1 & -a_{12} & \dots & -a_{1n} \\ y_2 & a_{22} & & -a_{2n} \\ \dots & & & \\ y_n & -a_{n2} & & a_{nn} \end{vmatrix}$$

By Lemma 2.1 it follows that the determinant in the right hand side of (2.7) is positive. Consequently det $M > 0$.

◇

To prove Theorem 2.1 we need some preliminaries.

Let E be a Banach space ordered by a closed convex cone K. An operator $P : E \to E$ is said to be positive if $P(K) \subset K$. A set $\mathcal{P}$ of positive bounded linear operators in E is said to be *positive-numbers-like* if

(i) $I \in \mathcal{P}$

(ii) $P_1 P_2 = P_2 P_1 \forall P_1, P_2 \in \mathcal{P}$

(iii) $||P_1 + P_2|| = ||P_1|| + ||P_2||, \quad ||P_1 P_2|| = ||P_1|| \, ||P_2|| \forall \mathcal{P}, \mathcal{P}_2 \in \mathcal{P}$,

where $||P||$ denotes the norm of the operators P, and I is the identity operator.

REMARK 2.3. Let $\mathcal{P}$ be a positive-numbers-like set of operators. Then, if and $\mathcal{P} \in \mathcal{P}$ and $||P|| < 1$, it follows that $(I - P)^{-1}$ is a positive bounded lienar operator and

$$||(I - P)^{-1}|| = (I - ||P||)^{-1} .$$

Indeed,

$$|| \lim_n \sum_{j=1}^{n} P^j || = \lim_n || \sum_{j=1}^{n} P^j || = \lim_n \sum_{j=1}^{n} ||P||^j .$$

REMARK 2.4. Let $\mathcal{P}$ be as in the previous remark. Then the set formed by all polynomials

$$\sum_{|\alpha|\leq m} a_\alpha P_1^{\alpha_1} \dots P_k^{\alpha_K} \dots , \quad a_\alpha \geq 0 ,$$

for any set of operators $P_1, \dots, P_k$ in $\mathcal{P}$, is a positive-numbers-like set of operators.

REMARK 2.5. Let $\mathcal{P}$ be as in Remark 3 above. Then, the set made up by operators of form $(I - P_1)^{-1} P_2$, with $P_1, P_2 \in \mathcal{P}$ and $||P_1|| < 1$, is also a positive-numbers-like set of operators.

REMARK 2.6. A set $\mathcal{P}$ of positive bounded linear operators with the property that there exists $\phi \in K$ such that $P\phi = ||P||\phi$ for all $P \in \mathcal{P}$, is a positive-numbers-like set of operators.

EXAMPLE 2.1. Let $E = L^2(\Omega)$ where Ω is a bounded domain in $\mathbf{R}^N$. Let K be the cone of a.e. non-negative L^2 -functions. Let L and λ_1 be as defined in the beginning of this section.

For each real number $a < \lambda_1$, consider the operator $B_a : L^2(\Omega) \to L^2(\Omega)$ defined by $B_a u = v$ where

$$\begin{cases} (L - a)v &= u \quad \text{in } \Omega \\ \qquad v &= 0 \quad \text{on } \partial\Omega \end{cases} .$$

The norm of B_a is $(\lambda_1 - a)^{-1}$ and $B_a\phi = (\lambda_1 - a)^{-1}\phi$, where ϕ is the first eigenfunction of $(L, H_0^1(\Omega))$.

The set

$$\mathcal{P} = \{I\} \cup \{B_a : a < \lambda_1\}$$

is a positive-numbers-like set of operators in $L^2(\Omega)$.

Now we prove a result on the positivity of an inverse matrix of positive-numbers-like operators. Consider the following matrix

$$\mathbf{P} = \begin{bmatrix} I & -P_{12} & -P_{13} & \dots & -P_{1n} \\ -P_{21} & I & -P_{23} & & -P_{2n} \\ -P_{31} & -P_{32} & I\dots & & -P_{3n} \\ \dots & \dots & \dots & & \dots \\ -P_{n1} & -P_{n2} & -P_{n3} & & I \end{bmatrix} \tag{2.8}$$

where P_{ij} are positive-numbers-like operators in a Banach space E with a cone K. We consider the equation

$$\mathbf{P}U = F \tag{2.9}$$

where $U = (u_1, \ldots, u_n)$ and $F = (f_1, \ldots, f_n)$ with $u_j, f_j \in E$.

We now state conditions on the operators P_{ij} which ensure that a solution $U \in K^n$ when $F \in K^n$. For that matter consider the following matrix of scalars associated with the matrix $\mathbf{P}$.

$$\|\mathbf{P}\| = \begin{bmatrix} 1 & -\|P_{12}\| & -\|P_{13}\| & \ldots & -\|P_{1n}\| \\ -\|P_{21}\| & 1 & -\|P_{23}\| & \ldots & -\|P_{2n}\| \\ -\|P_{31} & -\|P_{32}\| & 1 & \ldots & -\|P_{3n}\| \\ -\|P_{n1}\| & -\|P_{n2}\| \ldots & -\|P_{n3}\| & \ldots & 1 \end{bmatrix} \tag{2.10}$$

we have the following result.

THEOREM 2.2. *Let* $\mathbf{P}$ *be a matrix as in (2.8). Suppose that the* n *first principal minors of the matrix* $\|\mathbf{P}\|$ *defined in (2.10) have positive determinants.*

Then $U \in K^n$ if $F \in K^n$, for all solutions U of (2.9).

REMARK 2.7 By Lemma 3.1 it follows that all principal minors of $\|\mathbf{P}\|$ have positive determinants.

Proof. By induction. For $n = 2$ we have

$$u_1 - P_{12} u_2 = f_1$$

$$-P_{21} u_1 + u_2 = f_2 .$$

Applying P_{21} to the first equation and adding the result to the second equation we obtain

$$(I - P_{12} P_{21}) u_2 = P_{21} f_1 + f_2 . \tag{2.11}$$

By hypothesis $\|P_{12}\| \; \|P_2\| < 1$, which implies, in view of Remark 2.3, that $(I - P_{12} P_{21})^{-1}$ is a positive operator.

So $u_2 \in K$ and $u_1 = f_1 + P_{12} u_2 \in K$.

Next, assume that the result is true for systems with $j \leq n-1$ equations; let us prove that it is true for $j = n$.

Applying $P_{i1}, 2 \leq i \leq n$, to the first equation and adding the ith equation we obtain a system with $n-1$ equation, whose matrix is

$$\begin{bmatrix} I - P_{12}P_{21} & -P_{23} - P_{13}P_{21} \dots & -P_{2n} - P_{1n}P_{21} \\ -P_{32} - P_{12}P_{31} & I - P_{13}P_{31} \dots & -P_{3n} - P_{1n}P_{31} \\ \dots & \dots & \\ -P_{n2} - P_{12}P_{n1} & -P_{n3} - P_{13}P_{n1} \dots & I - P_{1n}P_{n1} \end{bmatrix}$$

and the right hand side is in K^n.

$$(f_2 + p_{12}f_1, \dots, f_n + p_{1n}f_1) \in K^n .$$

It follows from the hypothesis through Remark 2.3, that

$$Q_{ij} = (I - P_{1j}P_{j1})^{-1}, \quad j = 2, \dots, n$$

are positive operators.

So we are reduced to a system with $n-1$ equations

$$\mathbf{P}_1 \begin{bmatrix} u_2 \\ \\ \\ u_n \end{bmatrix} = \begin{bmatrix} g_2 \\ \cdot \\ \cdot \\ \cdot \\ g_n \end{bmatrix} \tag{2.12}$$

where

$$g_j = P_{j1}(f_j + P_{1j}f_1)$$

and the matrix $\mathbf{P}_1$ is given by

$$\begin{bmatrix} I & -Q_{12}(P_{23}+P_{13}P_{21})\dots & -Q_{12}(P_{2n}+P_{1n}P_{21}) \\ -Q_{13}(P_{32}+P_{12}P_{31}) & I & -Q_{13}(P_{3n}+P_{1n}P_{31}) \\ \cdot & & \\ \cdot & & \\ \cdot & & \\ \cdot & & \\ -Q_{1n}(P_{n2}+P_{12}P_{n1}) & Q_{1n}(P_{n3}+P_{13}P_{n1})\dots & I \end{bmatrix}$$

This matrix is like matrix $\mathbf{P}$. So, by the induction hypothesis, the positiveness of $u_2, \dots, u_n$ will follow, if one proves that the principal minors of the matrix $||\mathbf{P}_1||$, defined analogously to $||\mathbf{P}||$, have positive determinants. But this is the same as to show that this is true for the matrix below

$$\mathbf{C} = \begin{bmatrix} 1 - p_{12}p_{21} & -p_{23} - p_{13}p_{21} \cdots & -p_{2n} - p_{1n}p_{21} \\ -p_{32} - p_{12}p_{31} & 1 - p_{13}p_{31} \cdots & -p_{3n} - p_{1n}p_{31} \\ \cdots & \cdots & \cdots \\ -p_{n2} - p_{12}p_{n1} & -p_{n3} - p_{13}p_{n1} & 1 - p_{1n}p_{n1} \end{bmatrix}$$

where $p_{ij} = ||P_{ij}||$.

However, this is equivalent to show that the n first principal minors of the matrix

$$\mathbf{C}_1 = \begin{bmatrix} 1 & -p_{12} & -p_{-1} \\ 0 & 1 \ldots\ldots\ldots\ldots & \\ 0 & & \\ \cdot & & \mathbf{C} \\ 0 & & \\ 0 & & \\ 0 & 1 & \end{bmatrix}$$

are positive.

Observe the matrix $\mathbf{C}_1$ is obtained from the matrix $||\mathbf{P}||$ by multiplying the first line by p_{i1} and adding the ith line to it.

This procedure does not change the determinants of the first n principal minors of the matrix $\mathbf{C}_1$.

The proof is complete.

◇

Proof of Theorem 2.1.

Necessity.

Multiplying each equation in system (2.1) by ϕ, the first positive eigenfunction of $(L, H_0^1(\Omega))$, and integrating by parts we obtain the following linear algebraic system

$$(\lambda_1 I - A)X = Y \tag{2.13}$$

where $X = (x_1, \ldots, x_n)$, $x_i = \int_\Omega u_i\phi_i$ and $Y = (y_1, \ldots, y_n)$, $y_i = \int_\Omega f_i\phi$.

The matrix $\lambda_1 I - A$ is an M-matrix. So, an application of Lemma (2.1) concludes the proof.

Sufficiency.

System (2.1) can be written as

$$u_1 - a_{12}B_1u_2 \ldots - a_{1n}B_1u_n = B_1f_1$$

$$-a_{21}B_2u_1 + u_2 \ldots - a2_nB_2u_n = B_2f_2$$

$$-a_{n1}B_nu_1 - a_{n2}B_nu_2 \ldots + u_n = B_nf_n$$

where $B_i = B_{a_{ii}}$ (see the definition of B_a in example 2.1). Recall that the hypothesis of the theorem implies $a_{ii} < \lambda_i \forall i = 1, \ldots, n$.

The matrix of the above system is like **P** in Theorem 2.2. So to use Theorem 2.2 in the present situation it suffices to verify that the matrix of scalars

$$\begin{bmatrix} 1 & -a_{12}(\lambda_1 - a_{11})^{-1} \ldots & -a_{1n}(\lambda_1 - a_n)^{-1} \\ -a_{21}(\lambda_1 - a_{22})^{-1} & 1 & -a_{2n}(\lambda_1 - a_{22})^{-1} \\ -a_{n1}(\lambda_1 - a_{nn})^{-1} & -a_{n2}(\lambda_1 - a_{nn})^{-1} & 1 \end{bmatrix}$$

has its n first principal minors with positive determinants. But by hypothesis this is the case since these minors are

$$1,\ (\lambda_1 - a_{11})^{-1}(\lambda_1 - a_{22})^{-1}p_2(\lambda), \ldots, (\lambda_1 - a_{11})^{-1} \ldots (\lambda_1 - a_{nn})^{-1}p_n(\lambda).$$

This concludes the proof.

◇

REMARK 2.8. A useful consequence of Theorem 2.1 is that it furnishes a sufficient condition for the validity of condition (ψ) in the case of systems with non-constant coefficients, see Remark 1.3.

We conclude this Section by proving a positivity result connected with the classical eigenvalue problem for systems. See [7] for general results in this direction.

We assume that $\mathcal{L}$ is as in (2.1) and A is a constant $n \times n$ cooperative matrix.

THEOREM 2.3. *Let $p_1(\lambda),\dots,p_n(\lambda)$ be the characteristic polynomials of the first n principal minors of A.*

Let

$$v^* := \max\{v : p_j(v) = 0, \quad j = 1,\dots,n\},$$

and assume that $a_{kk} > 0$ for at least one $k(1 \le k \le n)$.

Then $v^* > 0$ and $\mu_1(A) := \frac{\lambda_1}{v^*}$ is an eigenvalue of

$$(2.14) \qquad \begin{cases} \mathcal{L}\emptyset &= \mu A\emptyset \quad \text{in } \Omega \\ \emptyset &= 0 \quad \text{on } \partial\Omega \end{cases}$$

with non negative eigenfunction $\overline{\emptyset}$.

Moreover, if $\mu < \mu_1(A)$, then $F \ge 0$ in Ω implies $U \ge 0$ in Ω, being U a solution of

$$(2.15) \qquad \begin{cases} \mathcal{L}U &= \mu AU + F \quad \text{in } \Omega \\ U &= 0 \quad \text{on } \partial\Omega \end{cases}$$

Proof. Let us denote by $p_{1,\mu},\dots,p_{n,\mu}$ the characteristic polynomials of the first n principal minors of μA.

Clearly we have

$$p_{j,\mu}(\lambda) = \mu^j p_j(\frac{\lambda}{\mu}) \quad j = 1,\dots,n.$$

Now we apply Theorem 2.1 to the equation (2.15). If $\mu < \frac{\lambda_1}{v^*}$ then $p_{j,\mu}(\lambda) > 0$ for all $j = 1,\dots,n$.

So Theorem 2.1 implies that a maximum principle holds for (2.15). Next suppose that $\mu = \frac{\lambda_1}{v^*}$. Then for some $j, p_{j,\mu}(\lambda_1) = 0$. This means that λ_1 is an eigenvalue of the matrix

$$\mu A_j = \mu \begin{pmatrix} a_{11} & \dots & a_{1j} \\ a_{j1} & \dots & a_{j1} \end{pmatrix}$$

Let $e \in \mathbf{R}^J$ be a corresponding eigenvector

$$\mu A_j e = \lambda_1 e.$$

We claim that the coordinates $e_i (i = 1, \ldots, j)$ of e are all positive. Indeed this follows from Perron-Frobenius theorem applied to the matrix $\mu(A_j + mI)$ where m is a positive constant such that $m > |a_{ii}|$, $i = 1, \ldots, J$. To complete the proof take $\overline{\varnothing} = (\phi_1 e_1, \ldots, \phi_1 e_j, \ldots, 0 \ldots 0)$. It is easy to see that $\mathcal{L}\overline{\varnothing} = \mu A \overline{\varnothing}$.

◊

3. Embedding of a 2×2 noncooperative system into a 3×3 cooperative system.

In this section, we shall use the results of the two previous sections to obtain maximum principles for a class of noncooperative systems with variable coefficients, as well as stronger results for such systems in the case of constant coefficients. Since we restrict ourselves to a 2×2 system with the same differential operator in the left side, we use the simpler notation below

$$Lu = au + bv + f \tag{3.1a}$$

$$Lv = cu + dv + g\,, \tag{3.1b}$$

where a, b, c, d, f and g are given functions defined in Ω.

Let us start with the simpler case when $b \leq 0$ and $c \leq 0$ in Ω. Although the system is noncooperative in this case, it can be changed into a cooperative system for the unknowns u and $-v$. Namely

$$Lu = au - b(-v) + f$$

$$L(-v) = (-c)u + d(-v) - g\,.$$

So using Corollary 1.1 we obtain the following result.

THEOREM 3.1. *Suppose that L is self-adjoint and λ_1 is the first eigenvalue of $(L(D), H_0^1(\Omega))$. Assume also that*

(i) $b(x), c(x) \leq 0$ in Ω

(ii) $\max\{\sup_{x\in\Omega}[a(x) - b(x)]; \sup_{x\in\Omega}[-c(x) + d(x)]\} < \lambda_1$.

Then, if $f \geq 0$ and $g \leq 0$ in Ω, it follows that $u \geq 0$, $v \leq 0$ in Ω, being (u, v) the solution to (3.1).

Next we consider the more challenging case which corresponds to $b(x) \leq 0$, $b(x) \not\equiv 0$, $c(x) \geq 0$ in Ω. Our main purpose is to find out conditons on the coefficients of the system (3.1) which will give us a maximum principle. The idea now is first to embed the given 2×2 (non-cooperative) system into a 3×3 (cooperative) system. Then we apply the results already established for cooperative systems in order to obtain information about the given system. We introduce a new unknown $w = \varepsilon u + \delta v$, where $\varepsilon \neq 0$, $\delta \neq 0$ are real parameters to be determined as we proceed. So we have an additional equation

$$(3.1c) \qquad -\Delta w = (a\varepsilon + c\delta) u + (b\varepsilon + d\delta) v + \varepsilon f + \delta g .$$

Since $-\varepsilon u - \delta v + w = 0$ we may rewrite system (3.1) as follows

$$-\Delta u = (a - r\varepsilon) u + (b - r\delta) v + rw + f$$
$$-\Delta v = cu + dv + g$$
$$-\Delta w = (a\varepsilon + c\delta - s\varepsilon) u + (b\varepsilon + d\delta - s\delta) v + sw + \varepsilon f + \delta g ,$$

where the real-valued functions $r(x)$ and $s(x)$ are to be determined later. In order to get cooperativeness, we observe initially that in the first equation we should have $r \geq 0$ and $b - r\delta \geq 0$ in view of the hypothesis $b \leq 0$ and $h \not\equiv 0$, we infer that $\delta < 0$ and $r \geq 0$, $r \not\equiv 0$. On the other hand, aiming at a maximum principle we should have $\varepsilon f + \delta g \geq 0$, which implies that $\varepsilon > 0$. We may without loss of generality take $\varepsilon = 1$. So we obtain the following system:

$$(3.2a) \quad -\Delta u = (a - r) u + (b - r\delta) v + rw + f$$
$$(3.2b) \quad -\Delta v = cu + dv + g$$
$$(3.2c) \quad -\Delta w = (a - s + c\delta) u + (b + d\delta - s\delta) v + sw + f + \delta g$$

which will be cooperative if

(3.3)

$$r(x) \geq 0; \quad b(x) - r(x)\delta \geq 0; \quad \frac{b(x) + d(x)\delta}{\delta} \leq s(x) \leq a(x) + c(x)\delta$$

for all x in Ω. These conditions can be satisfied by a proper choice of δ $r(x)$ and $s(x)$ provided that there is a $\delta < 0$ such that

$$c(x)\delta^2 + (a(x) - d(x))\delta - b(x) \leq 0, \quad \forall x \in \Omega . \tag{3.4}$$

We summarize the foregoing arguments in the

PROPOSITION 3.1. *The noncooperative system* (3.1) $[b(x) \leq 0,$ $b(x) \not\equiv 0,\ c(x) \geq 0]$ *can be embedded into a cooperative system (3.2) if there exists a* $\delta < 0$ *such that* (3.4) *holds.*

REMARK 3.1. Embedding in the previous proposition means that: if (u, v, w) is a solution to (3.2), then (u, v) is a solution of (3.1), and moreover any solution of (3.1) is obtained in this way.

REMARK 3.2. The existence of δ as required in the above proposition is equivalent to the following conditions on the entries of the matrix A:

$$[a(x) - d(x)]^2 + 4b(x)c(x) \geq 0, \quad \forall x \in \Omega . \tag{3.5}$$

$$d(x) < a(x), \quad \forall x \in \Omega, \tag{3.6}$$

$$\sup_{x\in\Omega} \delta_-(x) \leq \inf_{x\in\Omega} \delta_+(x) , \tag{3.7}$$

where $\delta_-(x)$ and $\delta_+(x)$ are the two roots of the quadratic (3.4). By (3.5) and (3.6) these roots are < 0.

REMARK 3.3. If the entries of the matrix A are constants, then the existence of δ as stated above is equivalent to $(a - d)^2 + 4bc \geq 0$ (which is also equivalent to saying that the matrix A has real eigenvalues) and $d < a + (-bc)^{1/2}$.

Now we obtain a maximum principle for system (3.1) by applying Corollary 1.1 to system (3.2). We should then require that δ, r and s be chosen to satisfy the additional conditions

$$a(x) + b(x) - r(x)\delta < \lambda_1, \quad c(x) + d(x) < \lambda_1 \tag{3.8}$$

$$a(x) + c(x)\delta + b(x) + d(x)\delta - s(x)\delta < \lambda_1 , \tag{3.9}$$

for all x in Ω. The less restrictive choice of s in (3.9) is obtained form (3.3), namely $s(x) = [b(x) + d(x)\delta]/\delta$. And we get from (3.9)

$$a(x) + c(x)\delta < \lambda_1 . \tag{3.10}$$

In this way we see that the best choice of δ to attend (3.10) is

$$\delta_- = \sup_{x+\Omega} \delta_-(x)$$

where $\delta_-(x)$ is defined right after (3.7). Next we choose $r(x) = \frac{b(x)}{\delta_-}$ and the first requirement in (3.8) is attained if $a(x) < \lambda_1$, for all x in Ω. Summarizing, we have the following maximum principle

THEOREM 3.2. *Consider system* (3.1) *with* $b(x) \leq 0, b(x) \not\equiv 0$ *and* $c(x) \geq 0$ *in* Ω. *Assume condition* (3.4) *of Proposition 3.1. Suppose in addition that*

$$\begin{aligned} &a(x) < \lambda_1; \\ &c(x) + d(x) < \lambda_1; \quad a(x) + c(x)\delta_- < \lambda - 1, \quad \forall x \in \Omega . \end{aligned} \tag{3.11}$$

Then, if $f(x) \geq 0, g(x) \geq 0$ *and* $f(x) + \delta_- g(x) \geq 0$ *in* Ω, *it follows* $u \geq 0, v \geq 0$ *in* Ω.

For recent results on the maximum principle for non cooperative systems with variable coefficients see [3].

THE MAXIMUM PRINCIPLE OF [5]. In [5] the authors of the present paper considered the following system in Ω

$$-\Delta u = \lambda u - v + f \tag{3.12 a}$$

$$-\Delta v = \delta u - \gamma v , \tag{3.12 b}$$

subject to Dirichlet boundary conditions on $\partial\Omega$. Here δ, γ are positive constants. We showed that $f \geq 0$ in Ω implies $u, v \geq 0$ in Ω, provided

$$\sqrt{\delta} < \gamma + \lambda_1 , \tag{3.13}$$

$$-\gamma + 2\sqrt{\delta} \leq \lambda < \lambda_1 + \frac{\delta}{\lambda_1 + \gamma}. \tag{3.14}$$

We remark that this result is not a consequence of Theorem 3.2. Indeed, although the condition (3.4) gives the first half of the inequality in (3.14), the conditions (3.11) in the notation of the system (3.12) are

$$\lambda < \lambda_1, \quad \delta - \gamma < \lambda_1,$$

where the first inequality is more restrictive than (3.14), and the second is not comparable with (3.13) for general values of δ.

However we can derive the maximum principle in [5] from our Theorem 2.1. In fact, in the case when the coefficients of (3.1) are constant, the parameters r and s in the system (3.2) can be chosen also as constants. So in this case, with the further assumption that $L = -\Delta$, a maximum principle for (3.2) holds if (1) the system is cooperative, i.e.

$$r > 0, \quad b - r\delta \geq 0, \tag{3.16}$$

$$\frac{b + d\delta}{\delta} \leq s \leq a + c\delta, \quad \text{and} \tag{3.17}$$

(ii) the conditions of Theorem 2.1 are verified, namely

$$a - r < \lambda_1. \tag{3.18}$$

$$\begin{vmatrix} \lambda_1 - a + r & -b + r\delta \\ -c & \lambda_1 - d \end{vmatrix} > 0 \tag{3.19}$$

$$\begin{vmatrix} \lambda_1 - a + r & -b + r\delta & -r \\ -c & \lambda_1 - d & 0 \\ -a + s - c\delta & -b - d\delta + s\delta & \lambda_{1-s} \end{vmatrix} > 0 \tag{3.20}$$

Now we proceed to choose the parameters δ, r and s in such a way that conditions (3.16)–(3.20) are satisfied. First choose $r = b/\delta$. Therefore condition (3.16) is satisfied, and (3.18) and (3.19) are verified if

$$a - \frac{b}{\delta} < \lambda_1, \quad d < \lambda_1. \tag{3.21}$$

Also, condition (3.20) can be written as

$$\left(\lambda_1 - s + \frac{b}{\delta}\right)\left(\lambda_1 - a - \frac{bc}{\lambda_1 - d}\right) > 0 .$$

Now we choose $s < \lambda_1 + b/\delta$. [Such a choice can be done attending (3.17), because $(b + d\delta)/\delta < \lambda_1 + b/\delta$]. So the above inequality becomes

$$a < \lambda_1 - \frac{bc}{\lambda_1 - d} . \tag{3.22}$$

Finally it remains to choose δ. Take as δ the smallest root of

$$c\delta^2 + (a - d)\delta - b = 0 .$$

And then check that

$$\frac{-bc}{\lambda_1 - d} \le \frac{b}{\delta}, \text{ provided } \sqrt{-bc} < \lambda_1 - d .$$

Summarizing, we conclude that the parameters δ, r and s can be chosen to attend conditions (3.16)-(3.20) if

$$d + 2\sqrt{-bc} \le a \tag{3.23}$$

$$d < \lambda_1, \quad a < \lambda_1 - \frac{bc}{\lambda_1 - d} \tag{3.24}$$

$$\sqrt{-bc} < \lambda_1 - d . \tag{3.25}$$

THEOREM 3.3. *Consider system* (3.1) *with constant matrix A and* $L = -\Delta$. *Assume conditions* (3.23)-(3.25). *Then* $f \ge 0, g = 0$ *in* Ω *implies* $u, v \ge 0$ *in* Ω.

REMARK 3.4. The above theorem gives the maximum principle [5]. For others results connected with [5] see [3,4,10].

On the biharmonic equation. Let us consider the fourth order elliptic equation

$$(\Delta + a(x))(\Delta + b(x))v = \mu v + f \text{ in } \Omega\ , \tag{3.26}$$

subject to the so-called Navier boundary conditions, namely

$$v = 0, \quad \Delta v = 0 \quad \text{on } \partial\Omega\ . \tag{3.27}$$

We discuss now conditions on the functions a and b and the real parameter μ that will lead to a maximum principle for the problem above. For that matter we introduce a new dependent variable $u = -(\Delta + b)v$, and transform the above equation into the system below

$$-\Delta u = au + \mu v + f$$

$$-\Delta v = u + bv$$

subject to Dirichlet boundary conditions $u = v = 0$ on $\partial\Omega$. Then applying previous results we have the following statements.

A) If $\mu \geq 0$, the system is cooperative and ther is a maximum principle for (3.26)-(3.27) if $a(x) + \mu < \lambda_1$ and $1 + b(x) < \lambda_1$, for all x in Ω.

B) If $\mu \geq 0$ and a and b are constants, there is a maximum principle if

$$a < \lambda_1 \text{ and } (\lambda_1 - a)(\lambda_1 - b) > \mu\ .$$

C) If $\mu < 0$, the system above is noncooperative and we apply Theorem 3.2 to obtain a maximum principle for (3.26)-(3.27) provided

$$a(x) \geq b(x) + 2\sqrt{-\mu},\ a(x) < \lambda_1 \text{ and } 1 + b(x) < \lambda_1 \text{ for all } x \text{ in } \Omega\ .$$

D) If $\mu < 0$ and the coefficients a and b are constant, we apply Theorem 3.3 and the conditions are

$$a \geq b + 2\sqrt{-\mu},\ b < \lambda_1,\ a < \lambda_1 - \frac{\mu}{\lambda_1 - b},\ \sqrt{-\mu} < \lambda_1 - b\ .$$

E) Suppose now that $a = b = 0$. So in (3.26) we have precisely the biharmonic operator Δ^2. Then from B) above we have a maximum principle provided

$$0 \leq \mu < \lambda_1^2 .$$

On the other hand if $\mu < 0$, D) above does not give any information about the biharmonic operator. Nevertheless we have some partial results if the coefficients are constant and $\partial\Omega$ has some regularity property. Let us consider the simplest case

$$(3.28) \qquad \begin{cases} \Delta^2 v + \delta^2 v = f & \text{in } \Omega \\ \quad\; v = \Delta v = 0 & \text{on } \partial\Omega \end{cases}$$

where $\delta \in \mathbf{R}$ and f is a given positive function defined on Ω. Observe that the biharmonic operator subject to Navier boundary conditions has eigenvalues λ_K^2, $k = 1, 2, \ldots$, where λ_K are the eigenvalues of $(-\Delta, H_0^1(\Omega))$.

By putting $B = (-\Delta)^{-1}$ and supposing that $\delta^2 < \lambda_1$, it follows that (3.28) is equivalent to

$$(3.29) \qquad v = (I + \delta^2 B^2)^{-1} B^2 (f) .$$

Since $(I + \delta^2 B^2)^{-1} = T \cdot (I + \delta^2 B^2)$ where $T = (I - \delta^4 B^4)^{-1}$, it follows that, in order to prove that $v \geq 0$ in Ω, it is sufficient to prove that the operator $B^2 - \delta^2 B^4$ is positive, hence in particular that $B - \delta^2 B^3$ is positive.

PROPOSITION 3.2. *Let $\Omega \subset \mathbf{R}^N$ be a bounded $C^{1,1}$ domain. Then there exists a constant $K > 0$ such that*

(i) $K < \lambda_1^2$
(ii) if $\delta^2 \leq K$ then $f \geq 0 \Rightarrow v \geq 0$ in Ω
(iii) $K \leq 1 + \frac{\lambda_1 \lambda_2}{\lambda_1^2 + \lambda_2^2}$.

Proof. We shall denote by $G(x, y)$ the Green function of $-\Delta$ subject to Dirichlet boundary conditions, that is $G(\cdot, y)(y \in \Omega)$ is the solution to the problem

$$(3.30) \qquad -\Delta_x G(x, y) = \delta_y(x) \quad x \in \Omega$$

$$G(x,y)=0 \quad x\in\partial\Omega\,.$$

Let $\lambda\in\mathbf{R}^+$ and consider

$$g_\lambda(x)=(Bf)(x)-\lambda(B^3f)(x) \quad x\in\Omega\,.$$

We see that the solution of (3.28) is $v=(TB)g_{\delta^2}$ where $T=(I-\delta^4B^4)^{-1}$.

So in order to show that v is positive, it is sufficient to prove that there exists $\lambda\in\mathbf{R}^+$ such that $g_\lambda(x)\geq 0$ $(x\in\Omega)$ (this will imply that $g_{\delta^2}(x)\geq 0$ if $\delta^2\leq\lambda$). By using the integral representation we see that

$$g_\lambda(x)=\int_\Omega G(x,y)f(y)\,dy-\lambda\int_{\Omega^3}G(x,b)G(a,b)G(a,y)f(y)\,da\,db\,dy\,.$$

Hence, in order to show that $g_\lambda\geq 0$ it is sufficient to prove that the function

$$H(x,y)=\Big(\int_{\Omega^2}G(x,b)G(a,b)G(a,y)\,da\,db\Big)\,(G(x,y))^{-1}$$

is uniformly bounded in Ω.

Indeed if the last statement is true, i.e.

$$H(x,y)\leq\overline{K}\quad\forall x,y\in\Omega \tag{3.31}$$

then if we take $\lambda\leq(\overline{K})^{-1}$ we are done. In fact if $\lambda\leq(\overline{K})^{-1}$ then

$$(\overline{K})^{-1}\int_{\Omega^2}G(x,b)G(a,b)G(a,y)\,da\,db\leq G(x,y)$$

which implies (since $f\geq 0$) that

$$\lambda\int_{\Omega^3}G(x,b)G(a,b)G(a,y)f(y)\,da\,db\,dy\leq\int_\Omega G(x,y)f(y)\,dy$$

i.e. $g_\lambda(x)>0$ in Ω.

To prove that (3.31) holds we can use the results of Zhao [11] in a special situation.

Namely by [11] we know that if $\Omega\subset\mathbf{R}^N$ is $C^{1,1}$ domain then there exists $A>0$ such that:

$$\forall x,y,z\in\Omega\quad \frac{G(x,z)G(y,z)}{G(x,y)}\leq A[\,|x-z|^{2-N}+|y-z|^{2-N}]\,,$$

hence,

$$H(x,y) = \int_{\Omega^2} G(x,b)G(a,b)G(a,y)G(x,y)^{-1}da\,db =$$

$$= \int_{\Omega^2} G(x,b)G(y,b)G(x,y)^{-1}\cdot G(a,b)\cdot G(a,y)\cdot G(y,b)^{-1}da\,db \leq$$

$$\leq A^2 M(x,y)$$

where

$$M(x,y) = \int_{\Omega^2} (|x-b|^{2-N} + |y-b|^{2-N})(|a-b|^{2-N} + |a-y|^{2-N})\,da\,db\,,$$

Since $M(\cdot,\cdot)$ is continuous we conclude that

$$\sup_{x,y\in\Omega} H(x,y) \leq A^2 \max_{x,y\in\overline{\Omega}} M(x,y) = K$$

This concludes the proof of (ii).

To prove (i) it is sufficient to restrict our attention to $\langle\phi_1\rangle$ (the eigenspace generate by ϕ_1). Indeed if $B - KB^3$ is a positive operator then $B\phi_1 - KB^3\phi_1 > 0$ i.e. $K < \lambda_1^2$. The same remark can be used to prove (iii), in fact (ii) is the necessary condition on $B - KB^3$ in order to be positive on $\langle\phi_1, \phi_2\rangle$.

REMARK 3.5. Proposition above generalizes results of Sweers [9] and Bonnet [1] which hold respectively if $\Omega \subset \mathbf{R}^N$ is ball and if Ω is of class C^2. We observe that in general we are not able to find the "best constant" K. Some results in this direction are contained in [2].

REMARK 3.6. It is possible to prove by using the results of Zhao [11] in the full generality that a maximum principle still holds for equation (3.26) in a bounded $C^{1,1}$ domain, without requiring that the coefficients satisfy condition C).

References

[1] Bonnet A., *A positivity related property for the biharmonic operator*, (Preprint).

[2] Caristi G. and Mitidieri E., *Maximum principles for a class of non cooperative elliptic systems*, In press on "DELFT Progress Reports" (1989).

[3] Cosner G. and Schaefer P.W., *Sign-definite solutions in some linear elliptic systems*, to appear on Proc. Roy. Soc. Edinburgh.

[4] Cosner G. and Schaefer P.W., *Sign-definite solutions in elliptic systems with constant coefficients*, Proc. Int. Conf. on Theory and Applications of Differential Equations, Ohio U. (1988).

[5] de Figueiredo and Mitidieri E., *A maximum principle for an elliptic system and applications to semilinear problems*, SIAM J. on Math. Anal. *17* (1986) 836-849.

[6] Gilbarg D. and Trudinger N.S., *Elliptic partial differential equations of second order*, Springer-Verlag, Berlin and New York (1977).

[7] Hess P., *On the eigenvalue problem for weakly coupled elliptic systems*, Arch. for Rational mech. Anal. *81* (1983), 151-159.

[8] Protter M.H. and Weinberger H.F., *Maximum principles in differential equations*, Prentice-Hall, Englewood Cliffs, NJ (1967).

[9] Sweers G., *A strong maximum principle for a non cooperative elliptic system* (P.H.D. thesis Delft T.H.) (1988) – to appear on SIAM J. on Math. Anal.

[10] Weinberger H.F., *Some remarks on invariant sets for systems*, Pitman Research Notes in Mathematics Series *175* (1988), 189-207.

[11] Zhao Z., *Green function for Schroedinger operator and conditioned Feynmann-Kac gauge*, J. Math. Anal. and Appl. *116* (1986), 309-334.

J. reine angew. Math. 421 (1991), 1—22

Journal für die reine und angewandte Mathematik

On a superlinear Sturm-Liouville equation and a related bouncing problem

By *D. G. Figueiredo* at Campinas and *B. Ruf* at Milano

Introduction

In this paper we study the solvability of the Neumann problem

$$(1)\qquad \begin{cases} -u'' = \lambda u + g(x,u) + h(x) & \text{in} \quad (0,1), \\ u'(0) = u'(1) = 0 \end{cases}$$

where $g : [0,1] \times \mathbb{R} \to \mathbb{R}$ is a continuous function and $h = h(x)$ is some given function in $L^2(0,1)$. The basic assumptions on g concern its behavior at both $+\infty$ and $-\infty$, namely

$$(2)\qquad \lim_{s\to-\infty} \frac{g(x,s)}{s} = 0, \quad \lim_{s\to+\infty} \frac{g(x,s)}{s} = +\infty .$$

This type of hypothesis lends a very strong asymmetry to problem (1), and it implies the fact that the nonlinearity crosses all but a finite number of eigenvalues of the eigenvalue problem

$$(3)\qquad -v'' = \mu v \quad \text{in} \quad (0,1), \quad v'(0) = v'(1) = 0 .$$

Our research focuses on the important role of the parameter λ as regard to the existence of solutions. If $\lambda < 0 = \lambda_1$, where λ_1 denotes the first eigenvalue of the problem (3), the problem (1) is of the Ambrosetti-Prodi type, and it follows from the results of Amann-Hess [1], Dancer [4], deFigueiredo-Solimini [6] that (1) admits at least two solutions for certain h's and no solution for others.

Here we consider the case when $\lambda \geqq 0$, which presents completely different features.

Originally published in: J. Reine Angew. Math. 421 (1991), 1-22,

DOI: 10.1007/978-3-319-02856-9_22

The following additional condition (Ambrosetti-Rabinowitz [2]) will be required to prove that the functional associated to problem (1) satisfies the (PS) condition, see Lemma 1.4: there are numbers $0 < \theta < \frac{1}{2}$ and $s_0 > 0$, such that

$$(4) \qquad 0 < G(x,s) \leqq \theta s g(x,s), \quad \forall x \in (0,1), \quad \forall s \geqq s_0,$$

where $G(x,s) = \int_0^s g(x,\xi)\,d\xi$.

Also the condition below will be imposed in the case when $\lambda = 0$:

$$(5) \qquad g(x,s) > 0, \quad \forall (x,s) \in (0,1) \times \mathbb{R}, \quad \lim_{s \to -\infty} g(x,s) = 0.$$

1. Theorem. *Assume* (2), (4) *and* (5). *Then for* $\lambda = 0$ *equation* (1) *has a solution for a given* h *in* L^2 *if and only if* $\int h\,dx < 0$.

2. Theorem. *Assume* (2) *and* (4). *Then for* $0 < \lambda < \pi^2/4$ *equation* (1) *has a solution for all* h *in* L^2.

A corresponding result to Theorem 1 (with Dirichlet boundary conditions instead) has been obtained by Kannan-Ortega [7]. Their proof is based on fixed point techniques, while here variational methods will be applied. For results related to Theorem 2 (again under Dirichlet boundary conditions) we refer to Castro-Shivaji [3], Ruf-Solimini [10], where the existence of many solutions is proved for the case $h(x) = \text{const} \ll 0$, and to deFigueiredo [5], Padua [8], Ruf-Srikanth [11], where multiple solutions for $h = te_1$, with $t \gg 0$, (e_1 is a first positive eigenfunction of $-d^2/dx^2$ under Dirichlet boundary conditions) are obtained. In all these works Dirichlet boundary conditions are assumed. Observe that our results above give a complete information on the whole range of the mapping

$$u \mapsto u'' + \lambda u + g(x,u).$$

Our results apply to the case $g(x,u) = \exp(u)$, which has been studied recently in the literature.

The proofs for Theorems 1 and 2 rely on variational methods. That is, we consider the functional

$$I(u) = \frac{1}{2}\int_0^1 |u'|^2 - \frac{\lambda}{2}\int_0^1 u^2 - \int_0^1 G(x,u) - \int_0^1 hu.$$

It is easily seen that I is a C^1 functional from H^1 into $\mathbb{R}$, where $H^1 = H^1(0,1)$ denotes the usual Sobolev space. Furthermore, if $u \in H^1$ is a critical point of I, i.e. $I'(u)[v] = 0$ for all $v \in H^1$, then u is a (weak) solution of (1). Hence the problem of finding solutions for (1) is transferred to looking for critical points of the functional I.

We recall that for $\lambda < 0$ the mentioned Ambrosetti-Prodi type result is obtained by proving (for suitable h) the existence of a strict local minimum of I (thus yielding a first solution). Then one applies the mountain-pass theorem of Ambrosetti-Rabinowitz [2] to obtain a second solution. For $\lambda \geqq 0$ we encounter the difficulty that the existence of a local minimum cannot be proven anymore. However, we will show that we still have a mountain-pass solution provided that $\lambda < \gamma = \frac{\pi^2}{4}\left(\text{and } \int_0^1 h < 0 \text{ if } \lambda = 0\right)$.

In the course of the proof turns up very naturally (and quite surprisingly) the following characterization of the value γ.

3. Theorem. *Let* $B = \left\{u \in H^1 : \int_0^1 u^2 + \|u\|_{L^\infty}^2 = 1\right\}$ *and* $E_1 = \left\{u \in H^1 : \int_0^1 u = 0\right\}$. *Then*

$$\gamma = \inf_{B \cap E_1} \int |u'|^2 \tag{6}$$

and one has: $\gamma = \frac{\pi^4}{4}$. *Furthermore, the minimizers of* (6) *are given by*

$$u_1(x) = \pm\sqrt{2}\left(\cos\frac{\pi}{2}x - \frac{2}{\pi}\right),$$

$$u_2(x) = \pm\sqrt{2}\left(\cos\frac{\pi}{2}(1-x) - \frac{2}{\pi}\right).$$

To prove this theorem we will associate an Euler-Lagrange equation to (6). Since the constraint B is not differentiable we have to recur to methods of convex analysis, in particular, we shall use the subdifferential. Finally, we point out that the solutions u_1, u_2 can be interpreted as the stationary solutions for a bouncing problem: u_1 (and u_2) describe the motion of a harmonic oscillator on which a constant force (gravity) is acting, and which bounces against a rigid obstacle. We have not found in the literature any association of variational problems of the form (6) to bouncing problems. It is interesting to note that there is no unique minimizer to (6).

1. The variational setting

To prove Theorems 1 and 2 we first of all note that

$$\begin{cases} \text{For any } u \in H^1, \text{ with } \sup_{x\in(0,1)} u > 0, \text{ one has} \\ I(\varrho u) \to -\infty, \quad \text{for } \varrho \to +\infty. \end{cases} \tag{1.1}$$

Indeed, since

$$\frac{I(\varrho u)}{\varrho^2} = \frac{1}{2}\int |u'|^2 - \frac{\lambda}{2}\int u^2 - \int \frac{G(x, \varrho u)}{\varrho^2} - \frac{1}{\varrho}\int hu \tag{1.2}$$

it suffices to show that

$$\lim_{\varrho\to+\infty} \int \frac{G(x,\varrho u)}{\varrho^2} = +\infty. \tag{1.3}$$

From hypothesis (2) we see that, given any $M > 0$ and any $\varepsilon > 0$, there exist positive constants c_M and c_ε such that

$$g(x,s) \geqq Ms - c_M, \quad \forall s > 0 \tag{1.4}$$

and

$$|g(x,s)| \leqq \varepsilon|s| + c_\varepsilon, \quad \forall s < 0. \tag{1.5}$$

It then follows that

$$G(x,s) \geqq \frac{1}{2}Ms^2 - c_M s, \quad \forall s > 0 \tag{1.6}$$

and

$$|G(x,s)| \leqq \frac{\varepsilon}{2}s^2 + c_\varepsilon|s|, \quad \forall s < 0. \tag{1.7}$$

Now let $u \in H^1$ be fixed as in (1.1). Given any $K > 0$, choose M and ε such that

$$\frac{M}{2}\int_{u>0} u^2 - \frac{\varepsilon}{2}\int_{u<0} u^2 > K.$$

So we can estimate

$$\int \frac{G(x,\varrho u)}{\varrho^2} \geqq \frac{M}{2}\int_{u>0} u^2 - \frac{c_M}{\varrho}\int_{u>0} u - \frac{\varepsilon}{2}\int_{u<0} u^2 - \frac{c_\varepsilon}{\varrho}\int_{u<0} |u|$$

which implies that

$$\liminf_{\varrho\to+\infty} \int \frac{G(x,\varrho u)}{\varrho^2} \geqq K.$$

Since K is arbitrary, the statement in (1.3) follows readily.

Next we assert that (denoting by $\mathbb{1}$ the constant function equal 1 on $(0,1)$).

$$\begin{cases} \text{for } \lambda > 0, \text{ or } \lambda = 0 \text{ and } \int h < 0 \\ I(\varrho(-\mathbb{1})) \to -\infty \qquad \text{as } \varrho \to +\infty. \end{cases} \tag{1.8}$$

Indeed, using (1.2) with $u = -\mathbb{1}$, the result in (1.8), for the case $\lambda > 0$, follows from the fact that

$$\lim_{\varrho\to+\infty} \int \frac{G(x,-\varrho)}{\varrho^2} = 0,$$

which is a consequence of hypothesis (2); use (1.7).

In the other case, that is $\lambda = 0$ and $\int h < 0$, we have

$$I(\varrho(-\mathbb{1})) = -\int G(x, -\varrho) + \varrho \int h.$$

So the result in (1.8) follows because

$$\liminf_{\varrho \to +\infty} \int \frac{G(x, -\varrho)}{\varrho} \geqq 0, \tag{1.9}$$

which is a consequence of hypothesis (5).

The aim now is to find a surface of codimension 1 in H^1 on which I is bounded below, and which "separates" the cases (1.1) and (1.2). This will enable us to apply the mountain-pass theorem.

A suitable surface is given by the boundary of the negative cone in H^1: Let

$$N = \{u \in H^1 : u(x) < 0, \forall x \in [0,1]\}$$

(since $H^1(0,1) \hookrightarrow C(0,1)$ we may assume that u is continuous), and let ∂N denote the boundary of N. The set ∂N can be characterized as follows. Let $E_1 = \left\{u \in H^1 : \int_0^1 u = 0\right\}$ and let the mapping $Q : E_1 \to \partial N$ be given by $u \mapsto u - \sigma(u)\mathbb{1}$, where $\sigma(u) := \sup_{(0,1)} u \geqq 0$. The mapping Q is continuous, since for

$$u, y \in E_1 : |\sigma(u) - \sigma(y)| = |\sigma(u^+) - \sigma(y^+)| \leqq \|u^+ - y^+\|_{L^\infty} \leqq \|u - y\|_{L^\infty} \leqq C\|u - y\|_{H^1}.$$

Here $u^+(x) = \max(u(x), 0)$. The inverse of Q is the projection $P : \partial N \to E_1$,

$$z \mapsto Pz = z - \int z.$$

Hence Q is a homeomorphism, and we have

Lemma 1.1. *The set ∂N is an unbounded C^0-manifold (modelled globally over E_1) which separates H^1 into two components.*

Next we want to show that the functional I restricted to ∂N is bounded from below. For this, we introduce the set $B = \{u \in H^1; |||u|||^2 := \|u\|_{L^2}^2 + \|u\|_{L^\infty}^2 = 1\}$, and define

$$\gamma := \inf_{B \cap E_1} \int_0^1 |u'|^2 dx. \tag{1.10}$$

In the next Lemma we show that $\gamma > 0$ and that the infimum is assumed. A more detailed study of (1.10) is done in section 2.

Lemma 1.2. *There exists $\bar{u} \in B \cap E_1$ such that*

$$\int_0^1 |\bar{u}'|^2 = \inf_{B \cap E_1} \int_0^1 |u'|^2 = \gamma > 0.$$

Proof. Let $\{u_n\} \subset B \cap E_1$ be a minimizing sequence: $\int_0^1 |u_n'|^2 \to \lambda$. Since $\|u_n\|_{L^2}^2 \leqq 1$ we have $\|u_n\|_{H^1}^2 \leqq C$, for all $n \in \mathbb{N}$, and hence we can find a subsequence (again denoted by $\{u_n\}$) such that $u_n \to \bar{u}$ in L^2 and L^∞, and $u_n \rightharpoonup \bar{u}$ in H^1. Hence we find $\int_0^1 |u_n'|^2 \to \int_0^1 |\bar{u}'|^2 = \gamma$. Since $\bar{u} \in B \cap E_1$ we have $\int_0^1 \bar{u} = 0$ and hence $\gamma > 0$. □

We now prove

Proposition 1.3. *Let $\lambda < \gamma$, where γ is given by* (1.10). *Then $I|\partial N$ is bounded below.*

Proof. For $y \in \partial N$ we write $y = u - \sigma(u)\mathbb{1}$, with $u = Py$, where $P: \partial N \to E_1$ is the orthogonal projection onto E_r, as in Lemma 1.1. Then, using the characterization of γ, we have

$$(1.11) \qquad I(y) = \frac{1}{2}\int |u'|^2 - \frac{\lambda}{2}\int u^2 - \frac{\lambda}{2}\sigma^2(u) - \int G(x,y) - \int hy$$

$$\geqq \frac{\gamma - \lambda}{2} |||u|||^2 - \int G(x,y) - \|h\|_{L^2}\|y\|_{L^2}.$$

Now let $\varepsilon > 0$ be such that $\varepsilon < \gamma - \lambda$. Using (1.7) we estimate

$$(1.12) \qquad \left|\int G(x,y)\right| \leqq \frac{\varepsilon}{2}\|y\|_{L^2}^2 + c_\varepsilon \int |y|.$$

Finally using the fact that $\|y\|_{L^2} \leqq |||u|||$, it follows from (1.11) and (1.12) that I is indeed bounded below in ∂N. □

In order to apply the mountain-pass theorem we need to verify that the functional $I: H^1 \to \mathbb{R}$ satisfies the Palais-Smale condition:

(PS) Every sequence $\{u_n\} \subset H^1$ which satisfies $|I(u_n)| \leqq c$, for all $n \in \mathbb{N}$, and $I'(u_n) \to 0$ in H^{-1} as $n \to +\infty$, contains a convergent subsequence in the norm of H^1.

Lemma 1.4. (α) *Assume* (2), (4) *and* (5). *Then the functional $I: H^1 \to \mathbb{R}$ satisfies the* (PS) *condition, when $\lambda = 0$ and $\int h < 0$.*

(β) *Assume* (2) *and* (4). *Then the functional I satisfies the* (PS) *condition when $0 < \lambda < \pi^2/4$.*

Proof. Let $\{u_n\} \subset H^1$ be a sequence such that

$$\left|\frac{1}{2}\int |u_n'|^2 - \frac{\lambda}{2}\int u_n^2 - \int G(x, u_n) - \int h u_n\right| \leqq c \tag{1.13}$$

and

$$|\int u_n' v' - \lambda \int u_n v - \int g(x, u_n) v - \int h v| \leqq \varepsilon_n \|v\|, \quad \forall v \in H^1 \tag{1.14}$$

where c is a positive constant, $\varepsilon_n \to 0$ as $n \to \infty$, and $\|\cdot\|$ denotes the H^1 norm, namely

$$\|u\|^2 = \|u\|_{L^2}^2 + \|u'\|_{L^2}^2 .$$

(i) We claim that u_n is bounded, and consequently it can be proved easily that it contains a convergent (in the norm H^1) subsequence. Suppose by contradiction that $\|u_n\| \to \infty$. Let $z_n = u_n/\|u_n\|$.

We can assume, by passing to a subsequence, that, for some $z_0 \in H^1$, z_n converges to z_0, strongly in C^0 and weakly in H^1. Let

$$\Omega^+ = \{x \in (0,1) : z_0(x) > 0\} .$$

From (1.14) divided by $\|u_n\|$ and with $v = z_0^+$ we obtain

$$\int \frac{g(x, u_n)}{\|u_n\|} z_0^+ \leqq \int z_n'(z_0^+)' - \lambda \int z_n z_0^+ - \frac{1}{\|u_n\|}\int h z_0^+ + \frac{\varepsilon_n}{\|u_n\|}\|z_0^+\| . \tag{1.15}$$

Now in order to apply Fatou's lemma we observe that, in view of (1.4) and (1.5), $g(x, u_n)/\|u_n\|$ is bounded from below by a constant. Also it follows from (1.4) that, for all $x \in \Omega^+$

$$\liminf \frac{g(x, u_n(x))}{\|u_n\|} = +\infty .$$

Applying this to (1.15) we conclude that $z_0^+ = 0$. So $z_0 \leqq 0$ in Ω.

(ii) Now we claim that $z_0 = 0$. Suppose by contradiction that

$$\Omega^- = \{x \in (0,1) : z_0(x) < 0\}$$

is not empty. Using (1.14) divided by $\|u_n\|$ and with $v \in \mathscr{D}(\Omega^-)$, i.e., v is a C^∞ function with compact support in the open set Ω^-, we obtain

$$|\int z_n' v' - \lambda \int z_n v| \leqq \left|\int \frac{g(x, u_n)}{\|u_n\|} v\right| + \frac{1}{\|u_n\|}|\int h v| + \frac{\varepsilon_n}{\|u_n\|}\|v\| . \tag{1.16}$$

Since z_n converges uniformly to z_0, it follows that for n sufficiently large (n independent of x) $u_n(x) < 0$ in supp v.

Passing to the limit in (1.16) and using the Lebesgue dominated convergence theorem we obtain

$$\int z_0' v' - \lambda \int z_0 v = 0, \quad \forall v \in \mathscr{D}(\Omega^-). \tag{1.17}$$

From now on the proof splits in two cases. If $\lambda > 0$, we see from (1.17) that z_0 is a first eigenfunction of $-z'' = \lambda v$ in an interval $J \subset \Omega^-$, where J is any of the intervals that compose Ω^-. But this is impossible since (regardless of the type of boundary conditions that arise at the end points of J) this would imply $\lambda > \pi^2$. If $\lambda = 0$, it follows from (1.17) that $z_0 = \text{const} < 0$ in the whole of Ω. Going back to (1.14) with $v = 1$, we obtain

$$\left| \int g(x, z_n \|u_n\|) + \int h \right| \leqq \varepsilon_n \|v\|$$

or

$$\lim \int g(x, z_n \|u_n\|) = -\int h > 0$$

which contradicts hypothesis (5).

(iii) Now from the facts that $1 = \|z_n\|^2 = \|z_n'\|_{L^2}^2 + \|z_n\|_{L^2}^2$ and $z_n \to z_0 \equiv 0$ in L^2 it follows that $\|z_n'\|_{L^2} \to 1$.

Next we show that $\|z_n'\|_{L^2} \to 0$, thus coming to the final contradiction.

(iv) Dividing (1.14) by $\|u_n\|$ and taking $v = z_n$ we obtain

$$\int |z_n'|^2 - \lambda \int z_n^2 \leqq \int \frac{g(x, u_n)}{\|u_n\|} z_n + \frac{1}{\|u_n\|} \int h z_n + \frac{\varepsilon_n}{\|u_n\|} \|z_n\|. \tag{1.18}$$

So if we prove that

$$\limsup \int \frac{g(x, u_n)}{\|u_n\|} z_n \leqq 0 \tag{1.19}$$

the result follows readily from (1.18). To prove (1.19) we proceed as follows. From (1.13) and (1.14) we get

$$\left| \int \left[G(x, u_n) - \frac{1}{2} g(x, u_n) u_n \right] \right| \leqq \frac{1}{2} \left| \int h u_n \right| + C + \varepsilon_n \|u_n\|.$$

Using hypothesis (4) we obtain, for each $\varepsilon > 0$, that

$$\left(\frac{1}{2} - \theta \right) \int_{u_n \geqq s_0} u_n g(x, u_n) \leqq \varepsilon \int |u_n|^2 + C + \frac{1}{2} \|h\|_{L^2} \|u_n\|_{L^2} + \varepsilon_n \|u_n\|$$

(we have used (1.5) and (1.7) and the continuity of g), from which follows that

$$\limsup \int_{u_n \geqq s_0} \frac{g(x, u_n)}{\|u_n\|} z_n \leqq 0.$$

This implies (1.19), using again (1.5) and (1.7). □

Remark. The above proof can be modified at step (ii) to give that I satisfies (PS) for all $\lambda > 0$ on hypotheses (2) and (4). Indeed, using once more (1.13) and (1.14) one can show that for the sequence $\{u_n\}$ considered in step (i) one has

$$\int g(u_n)/\|u_n\| \xrightarrow[n\to\infty]{} 0.$$

This allows to conclude that z_0 (the limit of the convergent subsequence of $u_n/\|u_n\|$) satisfies the equation

$$-z'' - \lambda z = 0, \quad z'(0) = z'(1) = 0.$$

Since $z_0 \leqq 0$ (as shown in step (i)) and $\lambda > 0$, this is impossible; hence the result follows.

We are now in the position to complete the proof of Theorems 1 and 2 (up to the determination of the value γ).

1.5. *Proof of Theorems* 1 *and* 2. a) In Theorem 1 the condition $\int h < 0$ is necessary: in fact, integration of (1) yields $0 > -\int_0^1 g(x, u) = \int_0^1 h$.

b) To prove the existence of a solution, we apply the mountain-pass theorem of Ambrosetti-Rabinowitz [2]. By Proposition 1.3 we have

$$m = \inf_{\partial N} I(u) > -\infty.$$

We now choose (by (1.1) and (1.8)) a $s_0 \in \mathbb{R}^+$ such that

$$I(s\mathbb{1}) < m, \quad \text{for all} \quad s \in \mathbb{R}\setminus[-s_0, s_0].$$

Then we define the set of paths

$$\Gamma = \{\gamma : [0,1] \to H^1, \text{ continuous}, \gamma(0) = -s_0\mathbb{1}, \gamma(1) = s_0\mathbb{1}\}$$

and set

$$d = \inf_{\gamma\in\Gamma} \max_{u\in\gamma} I(u).$$

Note that $d \geqq m$, since $\gamma([0,1]) \cap E_1 \neq \emptyset$, $\forall \gamma \in \Gamma$. By the mountain-pass theorem d is a critical value and hence we have found a solution of equation (1). □

2. A bouncing problem

In the Lemma 1.2 we have shown that

$$\gamma = \inf_{B \cap E_1} \int_0^1 |u'|^2 \tag{2.1}$$

is positive, where $B = \{u \in H^1 : |||u|||^2 := \|u\|_{L^2}^2 + \|u\|_\infty^2 = 1\}$ and $E_1 = \left\{u \in H^1 : \int_0^1 u = 0\right\}$.

In order to determine the value of γ we will associate an Euler-Lagrange equation to (2.1). Note that B is not a differentiable manifold, but since the mapping $u \mapsto \frac{1}{2}\|u\|_\infty^2 =: \psi(u)$ is convex there exists its subdifferential $\partial\psi$. We will show that the Euler-Lagrange equation corresponding to (2.1) is

$$\int u'v' = \gamma \int uv + \gamma\langle \mu_u, v\rangle + \alpha \int v, \quad \text{for all} \quad v \in H^1, \tag{2.2}$$

where μ_u is an element of $\partial\psi(u)$, and α is some real number. By analysing more closely this equation we will be able to determine a minimizer of (2.1), which is given by

$$\gamma = \frac{\pi^2}{4}, \quad u(x) = \sqrt{2}\left(\sin\frac{\pi}{2}x - \frac{2}{\pi}\right), \tag{2.3}$$

$$\alpha = \frac{\pi}{\sqrt{2}}.$$

As we have already mentioned, the function $u(x)$ can be viewed as the stationary motion of a harmonic oscillator on which the constant force $\alpha = \pi/\sqrt{2}$ is acting, and which bounces against a rigid obstacle at $u_0 = \min u(x) = -\frac{2^{3/2}}{\pi}$.

We first recall the definition of the subdifferential.

Definition 2.1. Let E be a Banach space, and let $\psi : E \to \mathbb{R}$ be a continuous and convex functional. Then the *subdifferential* $\partial\psi(u)$ of ψ at a point $u \in E$ is a subset of E' whose elements are characterized as follows:

$$\mu_u \in \partial\psi(u) \Leftrightarrow \psi(u+v) - \psi(u) \geqq \langle \mu_u, v\rangle, \tag{2.4}$$

for all $v \in E$ (here $\langle\cdot,\cdot\rangle$ denotes the dual pairing between E and E'). An element $\mu_u \in \partial\psi(u)$ is called a *subgradient* of ψ at u.

One has the following characterization of $\partial\psi(u)$ for a certain class of convex functionals.

Lemma 2.2. *Let $\psi : E \to \mathbb{R}$ be convex and continuous. Assume that $\psi(x) \geqq 0$ for all $x \in E$, and*

$$\psi(tx) = t^2\psi(x), \quad \forall t \geqq 0. \tag{2.5}$$

Then $\mu \in \partial\psi(u)$ if and only if

$$\langle \mu, u\rangle = 2\psi(u) \tag{2.6}$$

and

$$\langle \mu, x\rangle \leqq \langle \mu, u\rangle, \quad \forall x \in G, \tag{2.7}$$

where $G = \{x \in E : \psi(x) \leqq \psi(u)\}$.

Proof. First assume that $\mu \in \partial\psi(u)$, that is

$$\psi(u+v) \geqq \psi(u) + \langle \mu, v\rangle, \quad \forall v \in E. \tag{2.8}$$

Choosing $v = t(x-u)$, $0 \leqq t \leqq 1$, $x \in G$, we have

$$\psi(u) \geqq \psi(tx + (1-t)u) = \psi(u+v)$$

which together with (2.8) yields (2.7).

Now, taking $v = tu$, $0 < |t| < 1$, in (2.8) we obtain

$$(1+t)^2\psi(u) \geqq \psi(u) + t\langle \mu, u\rangle$$

or

$$(2t+t^2)\psi(u) \geqq t\langle \mu, u\rangle .$$

Dividing by t (and attending to its sign) and letting t go zero we obtain (2.6).

Next assume (2.6) and (2.7), and let us prove that

$$\langle \mu, v\rangle - \langle \mu, u\rangle + \psi(u) \leqq \psi(v), \quad \forall v \in E. \tag{2.9}$$

If $\psi(v) = 0$, we get by (2.5) and (2.7) that $\langle \mu, v\rangle = 0$ and so the claim follows trivially. If $\psi(v) > 0$, let $z = \psi(u)^{1/2}\psi(v)^{-1/2}v$, which lies in G. So by (2.7)

$$\begin{aligned}\langle \mu, v\rangle &\leqq \psi(v)^{1/2}\psi(u)^{-1/2}\langle \mu, v\rangle\\ &= 2\psi(u)^{1/2}\psi(v)^{1/2},\end{aligned}$$

and the left side of (2.9) is estimated by

$$2\psi(u)^{1/2}\psi(v)^{1/2} - 2\psi(u) + \psi(u),$$

which is clearly less or equal to $\psi(v)$. □

To derive the Euler-Lagrange equation for our functional we will need the following lemma.

Lemma 2.3. *Let E be a Banach space and assume that $\Phi \in C^1(E, \mathbb{R})$ and*

$$\langle \Phi'(x), x \rangle \neq 0, \quad \forall x \in E \setminus \{0\}.$$

Let $\psi : E \to \mathbb{R}$ satisfy the hypothesis in Lemma 2.2. Let $y \in E$ be such that

$$\Phi(y) = \inf_{\psi(u)=1} \Phi(u).$$

Then there exists a $\lambda \in \mathbb{R}$ such that

$$\Phi'(y) \in \lambda \partial \psi(y). \tag{2.10}$$

Proof. First note that $\lambda = 0$ if and only if $\Phi'(y)[v] = 0$ for all $v \in E$. Suppose now that this is not the case. Let $G = \{x \in E : \psi(x) \leqq \psi(y) = 1\}$. Assume by contradiction that $\frac{1}{\lambda}\Phi'(y)$ is not a subgradient of ψ at y for all $\lambda \in \mathbb{R} \setminus \{0\}$. It follows from Lemma 2.2 that the set of G is not contained in either of the halfspaces

$$H^+ = \{x \in E; \langle \Phi'(y), x \rangle \geqq \langle \Phi'(y), y \rangle\},$$
$$H^- = \{x \in E; \langle \Phi'(y), x \rangle \leqq \langle \Phi'(y), y \rangle\}.$$

Now we recall that $\langle \Phi'(y), y \rangle \neq 0$. So there exists $\lambda \in \mathbb{R}$ such that

$$\langle \frac{1}{\lambda}\Phi'(y), y \rangle = 2\psi(y) = 2.$$

From now on we work with this λ.

By Lemma 2.2 there exists $x \in G$ such that

$$\langle \frac{1}{\lambda}\phi'(y), x \rangle > \langle \frac{1}{\lambda}\Phi'(y), y \rangle. \tag{2.11}$$

We now claim that there exists a $w \in \partial G$ such that

$$\langle \frac{1}{\lambda}\phi'(y), w \rangle < \langle \frac{1}{\lambda}\Phi'(y), y \rangle.$$

In fact, if this were not the case we would have $\partial G \subset H^+$. By (2.11) there exists a point $\bar{x} \in \operatorname{Int} G$ which is in H^+. So $G \subset H^+$, which is impossible.

To complete the proof we look at the plane Π in E containing the triangle $[y, x, w]$. Consider the lines $[y, x]$ and $[y, w]$. By the convexity of G the boundary ∂G (which passes through y and w) lies in one of the sectors formed by the lines $[y, x]$ and $[y, w]$, while the line $\{x \in \Pi; \langle \Phi'(y), x\rangle = \langle \Phi'(y), y\rangle\}$ lies in the interior of the other. Therefore there exists a sequence $\{w_n\} \subset \partial G$ such that $w_n \to y$, $\frac{w_n - y}{\|w_n - y\|} \to z$, and $\langle \Phi'(y), z\rangle < 0$.

On the other hand we have

$$\Phi(y) \leqq \Phi(w_n) = \Phi(y) + \langle \Phi'(y), w_n - y\rangle + o(w_n - y),$$

and hence

$$0 \leqq \langle \Phi'(y), \frac{w_n - y}{\|w_n - y\|}\rangle + \frac{o(w_n - y)}{\|w_n - y\|}.$$

This implies $\langle \Phi'(y), z\rangle \geqq 0$, contradicting the above inequality. □

Corollary 2.4. *If $\Phi : E \to \mathbb{R}$ and $\psi : E \to \mathbb{R}$ are homogeneous of degree 2, then the value λ in* (2.10) *satisfies*

$$\lambda = \Phi(y),$$

where y is as in Lemma 2.3.

Proof. Differentiating $\Phi(tu)$ with respect to t and setting $t = 1$ we have

$$\langle \Phi'(u), u\rangle = 2\Phi(u).$$

On the other hand we observe that

$$\psi((1+t)u) \geqq \psi(u) + t\langle \mu_u, u\rangle$$

implies

$$(1+t)^2 \psi(u) \geqq \psi(u) + t\langle \mu_u, u\rangle$$

and hence $2t + 2t^2 \geqq t\langle \mu_u, u\rangle$, $\forall t \in \mathbb{R}$, which yields $\langle \mu_u, u\rangle = 2$.

By Lemma 2.2 we conclude. □

We now apply Lemma 2.3 to the functional

$$\Phi(u) = \int_0^1 |u'|^2$$

on the space $E_1 = \left\{u \in H^1; \int_0^1 u = 0\right\}$ and constrained to the set $B_1 = \{u \in E_1; |||u||| = 1\}$. Let $\psi(u) = \frac{1}{2}\|u\|_{L^\infty}^2$ and denote $\tilde{\psi} = \psi | E_1$. By Lemma 2.3 and Corollary 2.4 we obtain that the minimizer $y \in E_1$ of

$$\gamma = \inf_{u \in E_1 \cap B} \int_0^1 |u'|^2$$

satisfies

(2.12) $$\int_0^1 y'v' = \gamma \int_0^1 yv + \gamma \langle \tilde{\mu}_y, v \rangle, \quad \forall v \in E_1,$$

for some $\tilde{\mu}_y \in \partial\tilde{\psi}(y) \subset E_1'$. By Lemma 2.2 we have

$$\langle \tilde{\mu}_x, x \rangle \leqq \langle \tilde{\mu}_y, y \rangle, \quad \forall x \in E_1 \quad \text{with} \quad \tilde{\psi}(x) \leqq \tilde{\psi}(y).$$

By a known separation theorem of convex analysis we can extend $\tilde{\mu}_y$ to a functional μ_y on H^1 which satisfies

$$\langle \mu_y, x \rangle \leqq \langle \mu_y, y \rangle, \quad \forall x \in H_1, \quad \text{with} \quad \psi(x) \leqq \psi(y).$$

Again by Lemma 2.2, this implies that $\mu_y \in \partial\psi(y) \subset (H^1)'$. Hence we can write (2.12) as

(2.13) $$\int_0^1 y'v' = \gamma \int_0^1 yv + \gamma \langle \mu_y, v \rangle + \alpha \int v, \quad \forall v \in H^1,$$

where

(2.14) $$\alpha = -\gamma \langle \mu_y, \mathbb{1} \rangle.$$

Equivalently to (2.13) we write

(2.15) $$-y'' = \gamma y + \gamma \mu_y + \alpha.$$

We have thus proved

Proposition 2.5. *The Euler-Lagrange equation corresponding to* (2.1) *is given by* (2.15).

To determine the solutions of (2.15) we first derive some properties of the subgradients $\mu_y \in \partial\psi(y)$.

Lemma 2.6. *Let* $\psi(u) = \frac{1}{2}\|u\|_\infty^2$, $u \in H^1(0,1)$, *and let* $K_u = \{x \in [0,1]; |u(x)| = \|u\|_\infty\}$. *Then*

1) $\operatorname{supp} \mu_u \subset K_u, \quad \forall \mu_u \in \partial\psi(u)$,

2) $\langle \mu_u, u \rangle = \|u\|_\infty^2$,

3) $\langle \mu_u, v \rangle \geqq 0$ ($\leqq 0$, *resp.*) *for* $v \in H^1$, $v \geqq 0$, *with* $\operatorname{supp} v \subset \mathscr{U}_\delta(x_M)$ ($\operatorname{supp} v \subset \mathscr{U}_\delta(x_m)$, *resp.*), *where* x_M (x_m) *is such that* $u(x_M) = \|u\|_\infty$ ($u(x_m) = -\|u\|_\infty$), *and* $\mathscr{U}_\delta$ *is a sufficiently small neighborhood of* x_M (x_m).

Proof. 1) Let $v \in C_0^\infty((0,1)\backslash K_u)$. Then, for $t \in \mathbb{R}$ with t sufficiently small, we get

$$\frac{1}{2}\|u\|_\infty^2 = \frac{1}{2}\|u + tv\|_\infty^2 \geqq \frac{1}{2}\|u\|_\infty^2 + \langle \mu_u, tv \rangle$$

and hence $\langle \mu_u, v \rangle = 0$.

2) We have for all $t \in \mathbb{R}$

$$\frac{1}{2}(1+t)^2\|u\|_\infty^2 = \frac{1}{2}\|u+tu\|_\infty^2 \geqq \frac{1}{2}\|u\|_\infty^2 + t\langle \mu_u, u\rangle$$

and hence

$$t\|u\|_\infty^2 + \frac{t^2}{2}\|u\|_\infty^2 \geqq t\langle \mu_u, u\rangle .$$

Since this holds for all $t \in \mathbb{R}$ the claim follows.

3) Let $\operatorname{supp} v \subset \mathscr{U}_\delta(x_M)$, $v(x) \geqq 0$. By the definition of μ_u we have

$$\frac{1}{2}\|u-v\|_\infty^2 \geqq \frac{1}{2}\|u\|_\infty^2 + \langle \mu_u, -v\rangle .$$

If $\mathscr{U}_\delta$ is sufficiently small we derive

$$0 \leqq \frac{1}{2}\|u\|_\infty^2 - \frac{1}{2}\|u-v\|_\infty^2 \leqq \langle \mu_u, v\rangle .$$

The other inequality is obtained similarly. □

We now derive some properties of μ_y, where y is a solution of (2.15) (not necessarily a minimizer).

Lemma 2.7. *Suppose that* $y \in E_1$ *solves* (2.15) *for some* $\gamma \in \mathbb{R}$ *and* $\alpha \in \mathbb{R}$. *Then*

1) $\gamma > 0$,

2) $\operatorname{supp} \mu_y$ *does not contain intervals.*

Proof. 1) Choosing $v = y$ in (2.12) we obtain

$$0 < \int_0^1 |y'|^2 = \gamma\left(\int_0^1 |y|^2 + \langle \mu_y, y\rangle\right) = \gamma |||y|||^2 .$$

2) Suppose to the contrary that $\operatorname{supp} \mu_y$ contains an interval I with (for example) $y(x) = \|y\|_\infty$, $x \in I$. Letting $v \in C_0^\infty(I)$, $v \geqq 0$, $v \not\equiv 0$, we obtain from (2.12)

$$0 = \gamma\|y\|_\infty \int_I v + \gamma\langle \mu_y, v\rangle + \alpha \int_I v .$$

One verifies easily (recalling the proof of Lemma 2.6, part 3)) that $\langle \mu_y, v\rangle > 0$, and hence $\gamma\|y\|_\infty < -\alpha$. On the other hand, choosing $v \equiv 1$ in (2.13) we obtain, since $y \in E_1$

$$0 = \gamma\langle \mu_y, 1\rangle + \alpha ,$$

and hence by Lemma 2.2 and Lemma 2.6, part 2:

$$-\alpha \leqq \gamma \langle \mu_y, \frac{y}{\|y\|} \rangle = \gamma \|y\|_\infty . \quad \square$$

Next we determine the precise form of the subgradients μ_y:

Lemma 2.8. *Let* $\psi(u) = \frac{1}{2}\|u\|_\infty^2$, $u \in H^1(0,1)$, *and let* $K_u = \{x \in [0,1]; |u(x)| = \|u\|_\infty\}$. *Then* $\partial\psi(y) = \{\mu_y\} \subset (H^1)'$ *is given by*:

1) *If* $K_y = \{x_0\}$, *with* $y(x_0) = \pm \|y\|_\infty$, *then* $\partial\psi(y) = \{\pm \|y\|_\infty \delta_{x_0}\}$.

2) *If* $K_y = \{x_1, \ldots, x_n\} \subset [0,1]$, *with* $y(x_i) = -\|y\|_\infty$, *for* $1 \leqq i \leqq m (\leqq n)$, $y(x_i) = \|y\|_\infty$, *for* $m+1 \leqq i \leqq n$, *then*

$$\partial\psi(y) = \left\{ \sum_{i=1}^{m} \|y\|_\infty \beta_i(-\delta_{x_i}) + \sum_{i=m+1}^{n} \|y\|_\infty \beta_i \delta_{x_i} ; 0 \leqq \beta_i \leqq 1 , \sum_{i=1}^{n} \beta_i = 1 \right\}.$$

Proof. 1) Assume that $K_y = \{x_0\}$. Then

$$\begin{aligned}\frac{1}{2}\|y+v\|_\infty^2 &\geqq \frac{1}{2}|y(x_0) + v(x_0)|^2 \\ &\geqq \frac{1}{2}|y(x_0)|^2 + y(x_0)v(x_0) \\ &= \frac{1}{2}\|y\|_\infty^2 \pm \|y\|_\infty v(x_0) .\end{aligned}$$

Hence $\mu_y = \pm \|y(x_0)\| \delta_{x_0} \in \partial\psi(y)$.

Let now μ_y be an arbitrary element of $\partial\psi(y)$. Since $\operatorname{supp} \mu_y \subset \{x_0\}$ it follows that $\langle \mu_y, v \rangle = f(v(x_0))$. Since $f(v(x_0))$ is linear in v, it follows that $f(v(x_0)) = a \cdot v(x_0)$, for some $a \in \mathbb{R}$. By Lemma 2.6, part 2 we have $\|y\|_\infty^2 = \langle \mu_y, y \rangle = ay(x_0) = a(\pm \|y\|_\infty)$.

2) We again show first that any functional μ_y of the form

$$\sum_{i=1}^{m} \|y\|_\infty \beta_i(-\delta_{x_i}) + \sum_{i=m+1}^{n} \|y\|_\infty \beta_i \delta_{x_i}$$

with $\sum \beta_i = 1$, $0 \leqq \beta_i \leqq 1$, is an element of $\partial\psi(u)$. In fact, we have

$$\begin{aligned}\frac{1}{2}\|u+v\|_\infty^2 &= \frac{1}{2} \sum_{i=1}^{n} \beta_i \|u+v\|_\infty^2 \\ &\geqq \frac{1}{2} \sum_{i=1}^{n} \beta_i |u(x_i) + v(x_i)|^2\end{aligned}$$

$$= \frac{1}{2} \sum_{i=1}^{n} \beta_i |u(x_i)|^2 + \sum_{i=1}^{n} \beta_i u(x_i) v(x_i)$$

$$= \frac{1}{2} \|u\|_\infty^2 + \sum_{i=1}^{m} \beta_i \|u\|_\infty (-v(x_i)) + \sum_{i=m+1}^{n} \beta_i \|u\|_\infty v(x_i)$$

$$= \frac{1}{2} \|u\|_\infty^2 + \langle \sum_{i=1}^{m} \beta_i \|u\|_\infty (-\delta_{x_i}) + \sum_{i=m+1}^{n} \beta_i \|u\|_\infty \delta_{x_i}, v \rangle .$$

On the other hand, if $\mu_y \in \partial\psi(y)$ and hence $\operatorname{supp} \mu_y \subset \{x_1, \dots, x_n\}$, it follows that

$$\langle \mu_y, v \rangle = f(v(x_1), \dots, v(x_n)) .$$

Choosing v with $\operatorname{supp} v \subset \mathcal{U}_\delta(x_1)$ (such that $\mathcal{U}_\delta(x_1) \cap \{x_2, \dots, x_n\} = \emptyset$) we find, using the linearity of f:

$$f(v(x_1), \dots, v(x_n)) = a_1 v(x_1), \quad \operatorname{supp} v \in \mathcal{U}_\delta(x_1)$$

and similarly for $i = 2, \dots, n$. Writing $v \in H^1$ as $v(x) = \sum_{i=1}^{n} z_i(x)$ with

$$z_i(x_i) = v(x_i), \quad \operatorname{supp} z_i \subset \mathcal{U}_\delta(x_i) ,$$

it follows that

$$f(v(x_1), \dots, v(x_n)) = \sum_{i=1}^{n} a_i v(x_i) .$$

By Lemma 2.6, part 3 we have $a_1, \dots, a_m \leqq 0$ and $a_{m+1}, \dots, a_n \geqq 0$, and hence by Lemma 2.6, part 2:

$$\|y\|_\infty^2 = \langle \mu_y, y \rangle = \sum_{i=1}^{n} a_i y(x_i)$$

$$= \sum_{i=1}^{m} a_i(-\|y\|_\infty) + \sum_{i=m+1}^{n} a_i \|y\|_\infty ,$$

which implies $\sum_{i=1}^{m} (-a_i) + \sum_{i=m+1}^{n} a_i = \|y\|_\infty$. Hence we have the desired result. □

We now arrive at the main result of this section:

Theorem 2.9. *The value* γ *in* (2.1) *is:*

$$\gamma = \frac{\pi^2}{4} .$$

The corresponding minimizers in $B \cap E_1$ are

$$
\begin{aligned}
y_1(x) &= \sqrt{2}\left(\cos\frac{\pi}{2}x - \frac{2}{\pi}\right),\\
y_2(x) &= \sqrt{2}\left(\cos\frac{\pi}{2}(1-x) - \frac{2}{\pi}\right),\\
y_3(x) &= -y_1(x),\\
y_4(x) &= -y_2(x).
\end{aligned}
\tag{2.16}
$$

The functions $y_1(x)$, $y_2(x)$ solve equation (2.15) *with $\alpha = \pi/\sqrt{2}$, while $y_3(x)$, $y_4(x)$ solve* (2.15) *with $\alpha = -\pi/\sqrt{2}$.*

Proof. Assume that y solves (2.15), and assume first that $\operatorname{supp}\mu_y = \{x_0\}$, $x_0 \in [0,1]$, with $y(x_0) = \|y\|_\infty$. By Lemma 2.6 we have $\mu_y = \|y\|_\infty \delta_{x_0}$, and hence (2.15) reads, using (2.14).

$$-y'' = \gamma y + \gamma\|y\|_\infty \delta_{x_0} - \gamma\|y\|_\infty .$$

On $(0,1)\backslash\{x_0\}$ we have therefore

$$-y'' = \gamma(y - \|y\|_\infty),$$

which implies that y is convex in $(0, x_0)$ and in $(x_0, 1)$. Also, we see that for $x \to x_0$ follows $y''(x) \to 0$. Furthermore (assuming that $x_0 > 0$), we know that $y'(0) = 0$, and hence $y(x)$ on $(0, x_0)$ is given by

$$y(x) = \varrho \cos\frac{\pi}{2x_0}x + c .$$

The value γ corresponding to this y is $\gamma = \pi^2/4x_0^2$, which is clearly minimal for $x_0 = 1$. The constants ϱ and c are now easily determined to yield y_3. The other solutions y_1, y_2, y_4 are obtained by symmetry arguments.

It remains to show that $\gamma > \pi^2/4$ if the corresponding μ_y in (2.14) is supported in more than one point. There are two cases:

(i) all points $x_i \in \operatorname{supp}\mu_y$ are such that $y(x_i) = \|y\|_\infty$;

(ii) $\operatorname{supp}\mu_y$ contains two points x_i, x_{i+1} such that $y(x_i) = -\|y\|_\infty$, $y(x_{i+1}) = \|y\|_\infty$.

Case (i). In this case, we have $\langle \mu_y, \mathbb{1}\rangle = \|y\|_\infty$, since for $s > 0$ and $-\varepsilon < s < 0$

$$
\begin{aligned}
\psi(y + s\mathbb{1}) - \psi(y) &= \frac{1}{2}\|y + s\mathbb{1}\|_\infty^2 - \frac{1}{2}\|y\|_\infty^2\\
&= s\|y\|_\infty + \frac{s^2}{2} \geq \langle \mu_y, s\mathbb{1}\rangle .
\end{aligned}
$$

Hence $\alpha = -\gamma\|y\|_\infty$ by (2.14). We now consider equation (2.15) on an interval (x_i, x_{i+1}), where x_i, x_{i+1} are two consecutive points in supp μ_y:

$$\begin{cases} -y'' = \gamma(y - \|y\|_\infty), & x \in (x_i, x_{i+1}), \\ y(x_i) = y(x_{i+1}) = \sup_{(x_i, x_{i+1})} y = \|y\|_\infty . \end{cases}$$

This yields $y(x) = -\varrho \sin[\pi(x - x_i)/(x_{i+1} - x_i)] + \|y\|_\infty$, and hence

$$\gamma = \pi^2/(x_{i+1} - x_i) \geqq \pi^2 > \pi^2/4 .$$

Case (ii). We recall that our contention is to prove that

$$\frac{\int_0^1 |y'|^2}{\int_0^1 |y|^2 + \|y\|_\infty^2} > \frac{\pi^2}{4}, \tag{2.17}$$

We may assume $\|y\|_\infty = 1$. Let us call $2\ell = x_{i+1} - x_i$. The enumerator of (2.17) can be estimated as follows

$$\int_0^1 |y'|^2 \geqq \int_{x_i}^{x_{i+1}} |y'|^2 \geqq \frac{2}{\ell}. \tag{2.18}$$

To prove (2.18) we see that if $\bar{y}$ is the minimizer of

$$\left\{ \int_{x_i}^{x_{i+1}} |y'|^2 : y \in H^1, y(x_i) = -1, y(x_{i+1}) = 1 \right\}$$

then $\bar{y}$ satisfies $\int \bar{y}' v' = 0$ for all $v \in C_0^1[x_i, x_{i+1}]$. And consequently $\bar{y} = \frac{1}{\ell}\left(x - \frac{x_i + x_{i+1}}{2}\right)$. This gives (2.18). To estimate the denominator of (2.17) we first prove that

$$\int_{x_i}^{x_{i+1}} y^2 \leqq 2\ell - \frac{8\ell}{3\pi}. \tag{2.19}$$

Let us assume (2.19) and complete the proof. The left side of (2.17) can be estimated from below by

$$\frac{2/\ell}{1 - 8\ell/3\pi + 1} = \frac{1}{\ell\left(1 - \frac{4\ell}{3\pi}\right)}.$$

The minimum value of the above expression occurs at $\ell = 1/2$, and calculating that value we see that it is larger than $\pi^2/4$. It remains to prove (2.19) in order to complete the proof. Observe that y satisfies

$$\begin{cases} -y'' = \gamma y + \alpha & \text{in} \quad (x_i, x_{i+1}), \\ y(x_i) = -1, & \qquad y(x_{i+1}) = 1 \end{cases} \tag{2.20}$$

where, according to Lemma 2.8:

$$\alpha = -\gamma\langle \mu_y, \mathbb{1}\rangle = -\gamma\|y\|_\infty\left(\sum_{i=1}^{m}\beta_i - \sum_{i=m+1}^{n}\beta_i\right).$$

So $-\gamma \leqq \alpha \leqq \gamma$. In order to simplify the analysis of (2.20), we introduce

$$z(t) = y\left(t + \frac{x_i + x_{i+1}}{2}\right)$$

and $\beta = \alpha/\gamma$. So (2.20) becomes

$$\text{(2.21)}\qquad \begin{cases} -z'' = \gamma(z+\beta) & \text{in } (-\ell,\ell), \\ z(-\ell) = -1, & z(\ell) = 1 \end{cases}$$

with $-1 \leqq \beta \leqq 1$.

The solution of (2.21) is

$$\text{(2.22)}\qquad z(t) = \varrho\sin(\sqrt{\gamma}t + \theta) - \beta$$

where the constants ϱ and θ are determined by the boundary conditions:

$$\text{(2.23)}\qquad \varrho\sin(\sqrt{\gamma}\ell + \theta) - \beta = -1;\quad \varrho\sin(\sqrt{\gamma}\ell + \theta) - \beta = 1.$$

From these expressions we obtain

$$\text{(2.24)}\qquad \varrho\sin\theta\cos\sqrt{\gamma}\ell = \beta;\quad \varrho\cos\theta\sin\sqrt{\gamma}\ell = 1.$$

From (2.21) we see that the maximum of the derivative $z'(t)$ occurs at a point in $[-\ell,\ell]$ and from the expression (2.22) we conclude that

$$\|z'\|_\infty = \max|z'(t)| = |\varrho\sqrt{\gamma}|.$$

So in order to maximize z' we have to maximize ϱ as a function of $\beta\in[-1,1]$. From (2.24) we obtain

$$\varrho^2 = \frac{1}{\sin^2\sqrt{\gamma}\ell} + \frac{\beta^2}{\cos^2\sqrt{\gamma}\ell}.$$

This implies $\beta = +1$ (or $\beta = -1$) and it follows from (2.23) that $\theta = \sqrt{\gamma}\ell$ (or $\theta = -\sqrt{\gamma}\ell$, resp.). Let us consider the first possibility (the other is treated in a similar way). Using then (2.24) we get

$$\text{(2.25)}\qquad \varrho = \frac{2}{\sin 2\sqrt{\gamma}\ell} = \frac{2\sqrt{\gamma}\ell}{\sin 2\sqrt{\gamma}\ell}\,\frac{1}{\sqrt{\gamma}\ell}.$$

In this case we have $2\sqrt{\gamma\ell} \leqq \pi/2$. Thus we can estimate ϱ in (2.25) by

$$\varrho \leqq \frac{\pi}{2} \frac{1}{\sqrt{\gamma\ell}}.$$

Altogether we obtain the estimate of z':

$$\|z'\|_\infty \leqq \frac{\pi}{2\ell}.$$

Now let $t_0 \in (-\ell, \ell)$ be the point where $z(t_0) = 0$, and let us define the function

$$r(t) = \begin{cases} -1, & -\ell \leqq t \leqq t_0 - \dfrac{2\ell}{\pi}, \\ \pi(t - t_0)/2\ell, & |t - t_0| \leqq \dfrac{2\ell}{\pi}, \\ 1, & t_0 + \dfrac{2\ell}{\pi} \leqq t \leqq \ell. \end{cases}$$

Clearly

$$\int_{-\ell}^{\ell} |z(t)|^2 \leqq \int_{-\ell}^{\ell} |r(t)|^2 = \left(2\ell - \frac{4\ell}{\pi}\right) + \frac{\pi^2}{4\ell^2} \int_{t_0 - \frac{2\ell}{\pi}}^{t_0 + \frac{2\ell}{\pi}} (t - t_0)^2 = 2\ell - \frac{8\ell}{3\pi},$$

which proves (2.19).

Bibliography

[1] *H. Amman, P. Hess*, A multiplicity result for a class of elliptic boundary value problems, Proc. Roy. Soc. Edinb. **84** (1979), 145–151.

[2] *A. Ambrosetti, P.H. Rabinowitz*, Dual variational methods in critical point theory and applications, J. Funct. Anal. **14** (1973), 349–381.

[3] *A. Castro, R. Shivaji*, Multiple solutions for a Dirichlet problem with jumping non-linearities, Trends in theory and practice in Nonlinear Analysis, Amsterdam 1985.

[4] *E.N. Dancer*, On the ranges of certain weakly nonlinear elliptic partial differential equations, J. Math. Pures et Appl. **57** (1978), 352–366.

[5] *D.G. de Figueiredo*, On superlinear elliptic problems with nonlinearities interacting only with higher eigenvalues, Rocky Mountain J. Math. **18** (1988), 287–303.

[6] *D.G. de Figueiredo, S. Solimini*, A variational approach to superlinear elliptic problems, Comm. PDE **9** (1984), 699–717.

[7] *R. Kannan, R. Ortega*, Existence of solutions of $x'' + x + g(x) = p(t)$, $x(0) = 0 = x(\pi)$, Proc. A.M.S. **96** (1986), 67–70.

[8] *J.C.N. Padua*, Multiplicity results for a superlinear Dirichlet problem, preprint.

[9] *P.H. Rabinowitz*, Variational Methods for nonlinear eigenvalue problems (The Varenna Lectures), Eigenvalues of Nonlinear problems, CIME, edizioni Cremonese Roma (1979), 140–195.

[10] *B. Ruf, S. Solimini*, On a superlinear Sturm-Liouville problem with arbitrarily many solutions, SIAM J. Math. Anal. 17 (1986), 761–771.

[11] *B. Ruf, P. N. Srikanth*, Multiplicity results for ODE's with nonlinearities crossing all but a finite number of eigenvalues, Nonlin. Anal. TMA 10 (1966), 157–163.

IMECC, Universidade Estadual de Campinas, Caixa Postal, 6065, 13081 Campinas-SP, Brasil
Dipartimento di Matematica, Università di Milano, Via Saldini 50, I-20133 Milano

Eingegangen 5. Januar 1990

Reprinted from JOURNAL OF DIFFERENTIAL EQUATIONS
All Rights Reserved by Academic Press, New York and London

Vol. 96, No. 1, March 1992
Printed in Belgium

The Exact Number of Solutions for a Class of Ordinary Differential Equations through Morse Index Computation*

D. G. COSTA

Departamento de Matemática, Universidade de Brasilia, 70910 Brasilia D.F., Brazil

D. G. DE FIGUEIREDO

IMECC-UNICAMP, Caixa Postal 6065, 13081 Campinas S. P., Brazil

AND

P. N. SRIKANTH

Tata Institute of Fundamental Research, P.O. Box 1234, Indian Institute of Science Campus, Bangalore 560012, India

Received April 3, 1990

We consider a class of second order ordinary differential equations with *jumping nonlinearities* that *cross* the first k eigenvalues, under zero boundary conditions and, by means of a Morse Index computation, show the existence of *exactly* $2k$ solutions. © 1992 Academic Press, Inc.

1. INTRODUCTION

In this paper we consider the question of determining the exact number of solutions of the two point boundary value problem

$$-u'' = f(u) - t \sin x \quad \text{in} \quad (0, \pi), \qquad u(0) = u(\pi) = 0, \tag{1}$$

for large values of the real parameter t. Here the nonlinearity $f: \mathbf{R} \to \mathbf{R}$ is a C^1 function such that the limits

$$f_{\pm} = \lim_{s \to \pm\infty} f'(s)$$

exist as real numbers. Through the change of variables $v = u/t$, problem (1) becomes

$$-v'' = t^{-1} f(tv) - \sin x \quad \text{in} \quad (0, \pi), \qquad v(0) = v(\pi) = 0, \tag{2}$$

*Work partially supported by CNPq/Brasil.

0022-0396/92 $3.00

Originally published in: J. Differential Equations 96 (1992), 185-199,

DOI: 10.1007/978-3-319-02856-9_23

and, therefore, it is natural to consider the following problem

$$-v'' = f_+ v^+ - f_- v^- - \sin x \qquad \text{in} \quad (0, \pi), \qquad v(0) = v(\pi) = 0, \qquad (3)$$

as the limiting problem to (2), that is, "problem (2) with $t = \infty$." Now, if suitable hypotheses are made on f_+ and f_-, then it is reasonable to expect that, for large t's, problem (2) has the same number of sulutions as problem (3). In fact one can consider different cases depending on the positioning of f_+ and f_- with respect to the eigenvalues $\lambda_k = k^2$, $k = 1, 2, \ldots$, of the problem $-v'' = \lambda v$ in $(0, \pi)$, $v(0) = v(\pi) = 0$. The case when there are no eigenvalues strictly between f_+ and f_- was considered in [2], where the question of uniqueness was treated for a corresponding (more general) elliptic problem with Dirichlet boundary conditions. Here we consider the case of jumping nonlinearities or crossing of eigenvalues. More precisely, we assume all along this paper the following condition:

$$\text{(H1)} \quad f_- < \lambda_1 < \lambda_2 < \cdots < \lambda_k < f_+ < \lambda_{k+1},$$

and prove the following theorem.

THEOREM 1. *Assume* (H1). *Then problem* (3) *has exactly* $2k$ *solutions.*

In fact we prove that these $2k$ solutions are nondegenerate. As a consequence, by local inversion type arguments, cf. [8, Thm 1], one obtains the following result.

COROLLARY 1. *Assume* (H1). *Then problem* (1) *has exactly* $2k$ *solutions for large* $t > 0$.

The novelty here is the *exactness* of the number of solutions of problems (1) and (3). Existence of at least $2k$ solutions was proved by Lazer and McKenna [4] using the shooting method and also by Ruf [7] using bifurcation arguments. Of course, existence is also established here by using bifurcation theory as is [7]. The corresponding elliptic problem in dimensions higher than 1 was considered by Adimurthi and Srikanth [1] for the case $k = 1$ and by Solimini [7] for the case $k = 1, 2$ and for $k = 3$ with f_+ close to λ_3.

The paper is organized as follows. In Section 2 we transform problem (3) in an equivalent one, namely problem $(*)_\lambda$, which is more suitable to be treated by bifurcation techniques. In Section 3 we give a first description of the bifurcation diagram. And, finally, in Section 4 we show that all solutions of problem $(*)_\lambda$ are nondegenerate, completing the description of the bifurcation diagram. This allows us to obtain the exactness stated in Theorem 1.

2. Properties of the Solutions of Problem (3)

In order to simplify the notation we now write

$$f_- = \alpha, \qquad f_+ = \lambda, \qquad b_\lambda(v) = \lambda v^+ - \alpha v^-, \qquad \text{and} \qquad \phi(x) = \sin x,$$

so problem (3) is rewritten as

$$-v'' = b_\lambda(v) - \phi \quad \text{in} \quad (0, \pi), \qquad v(0) = v(\pi) = 0. \tag{4}$$

Now we proceed to change problem (4) into one which is more suitable to be treated by bifurcation theory. The idea of transforming problem (4) into problem $(*)_\lambda$ below is due to A. C. Lazer. It is clear that $v_\lambda(x) = \phi(x)/(\lambda - 1)$ is a solution of problem (4) for all $\lambda > 1$, and, in fact, it is the only positive solution. Making the change of variables $v = w/(\lambda - 1)$, we obtain the equivalent problem

$$-w'' = b_\lambda(w) - (\lambda - 1)\phi \quad \text{in} \quad (0, \pi), \qquad w(0) = w(\pi) = 0,$$

which has the positive solution $w_0 = \phi$ for all $\lambda > 1$. Finally we rewrite the above equation around this solution by letting $y = w - \phi$;

$$-y'' = b_\lambda(y + \phi) - \lambda\phi \quad \text{in} \quad (0, \pi), \qquad y(0) = y(\pi) = 0, \tag*{$(*)_\lambda$}$$

which has $y = 0$ as a solution for all $\lambda > 1$. We observe that, as far as the number of solutions is concerned, problems $(*)_\lambda$ and (3) are completely equivalent for all $\lambda > 1$.

We recall that if we denote by K the operator defined by $Kz = y$ if and only if $-y'' = z$, $y(0) = y(\pi) = 0$, then

$$K(C[0, \pi]) \subset C_0^2[0, \pi] = \{y \in C^2[0, \pi] \mid y(0) = y(\pi) = 0\}.$$

Thus problem $(*)_\lambda$ is equivalent to the functional equation in the Banach space $X = C_0^1[0, \pi]$:

$$y = \lambda Ky + k(\lambda, y),$$

where $k(\lambda, y) = (\lambda - \alpha)\, K((y + \phi)^-)$. Here $K: X \to X$ and $k: \mathbf{R} \times X \to X$ are compact operators with $k(\lambda, y) = o(\|y\|)$ as $y \to 0$ in X, for uniformly bounded λ-intervals. Therefore, we are in position to use the results from bifurcation theory [3, 5, 6]. As usual we denote by S_k, $k \geqslant 1$, the set of functions y in X, which have exactly $k - 1$ zeros in the open interval $(0, \pi)$, and all their zeros, including $x = 0$ and $x = \pi$, are simple. Also we let

$$S_k^+ = \{y \in S_k \mid y'(0) > 0\}, \qquad S_k^- = -S_k^+,$$

and observe that these are open sets in X. From Rabinowitz's Global Bifurcation Theorem [6] we obtain, for each $k \geqslant 1$, a branch C_k of nontrivial solutions (λ, y), $y \not\equiv 0$, bifurcating at $(\lambda_k, 0)$ from the line $\mathbf{R} \times \{0\}$ of trivial solutions. From the Existence and Uniqueness Theorem for ODE's, we obtain that $C_k \subset (\mathbf{R} \times S_k) \cup \{(\lambda_k, 0)\}$. Moreover, one has that

$$C_k = C_k^+ \cup C_k^-, \qquad \text{with} \quad C_k^+ \cap C_k^- = \{(\lambda_k, 0)\} \qquad \text{and} \qquad C_k^\pm \subset S_k^\pm,$$

and both C_k^+ and C_k^- are unbounded in $\mathbf{R} \times X$.

Next we present a series of facts about problem $(*)_\lambda$ which will be helpful in describing the bifurcation diagram.

(A) Each nontrivial solution y of $(*)_\lambda$ when $\lambda > 1$ has $p \geqslant 0$ positive bumps and $n \geqslant 1$ negative bumps, where $|p - n| = 0$ or 1. To see that $n \geqslant 1$ we write $(*)_\lambda$ as

$$-y'' = \lambda y + (\lambda - \alpha)(y + \phi)^- \tag{5}$$

which shows that y cannot be $\geqslant 0$ and $\not\equiv 0$ in the whole interval $[0, \pi]$. So, in particular, $(*)_\lambda$ does not have positive solutions.

(B) If $y \not\equiv 0$ is a solution of $(*)_\lambda$, then y is *strictly concave* at a positive bump and *strictly convex* at a negative bump. This is clear for $y > 0$, and, for $y < 0$, it follows readily from (5), since

$$-y'' = \lambda y \quad \text{if} \quad y \geqslant -\phi, \qquad \text{and} \qquad -y'' = \alpha y + (\alpha - \lambda)\phi \quad \text{if} \quad y < -\phi.$$

(C) All positive bumps of a solution y of $(*)_\lambda$ have length equal to $\pi/\sqrt{\lambda}$ and the negative bumps have length $\geqslant \pi/\sqrt{\lambda}$. In fact, at a positive bump (a, b) the solution y is a multiple of $\sin \sqrt{\lambda}\,(x - a)$, $a < x < b$. Now, let (a, b) be a negative bump; multiply (5) by $\xi(x) = \sin [\pi(x - a)/(b - a)]$ and integrate in (a, b) to obtain

$$\frac{\pi^2}{(b-a)} \int_a^b y\xi = \lambda \int_a^b y\xi + (\lambda - \alpha) \int_a^b (y + \phi)^- \xi,$$

which gives the result.

(D) If $\lambda = 1$, the solutions of $(*)_\lambda$ are given by the continuum $y = s\phi$ with $s \geqslant -1$.

(E) If $\lambda > 1$, the function $\hat{y} = (\alpha - \lambda)(1 - \alpha)^{-1}\phi$ is the *only negative solution*. Moreover, any other solution y of $(*)_\lambda$ has the property that $y \geqslant \hat{y}$. Indeed, suppose that $y < \hat{y}$ in an interval (a, b) with $y(a) = \hat{y}(a)$ and $y(b) = \hat{y}(b)$. Let $z = y - \hat{y}$. It is easy to check that $-z'' = \alpha z$ in (a, b) with $z(a) = z(b) = 0$, which implies $z = 0$, a contradiction. In order to see that $\hat{y}$

is the *only* negative solution we use a simple argument of comparison of eigenvalues.

(F) Let $0 \leqslant a < b \leqslant \pi$ be such that $y(a) = y(b) = 0$, $y(x) < 0$ and $(y+\phi)^- \not\equiv 0$ in (a, b). Let μ_1 and ψ_1 be, respectively, the first eigenvalue and the first eigenfunction of the eigenvalue problem

$$-\psi'' = \mu a_\lambda(y+\phi)\psi \quad \text{in} \quad (a, b), \qquad \psi(a) = \psi(b) = 0,$$

where

$$a_\lambda(y+\phi) = \begin{cases} \lambda & \text{if} \quad y+\phi > 0 \\ \alpha & \text{if} \quad y+\phi < 0. \end{cases}$$

Then we claim that $\mu_1 > 1$. Indeed, multiplying Eq. (5) by ψ_1 and integrating by parts, we obtain

$$\mu_1 \int_a^b a_\lambda(y+\phi)y\psi_1 = \lambda \int_a^b y\psi_1 + (\lambda - \alpha) \int_a^b (y+\phi)^- \psi_1,$$

which can be written as

$$(\mu_1 - 1)\left\{\lambda \int_{I_1} y\psi_1 + \alpha \int_{I_2} y\psi_1\right\} + (\lambda - \alpha) \int_{I_2} \phi\psi_1 = 0,$$

where $I_1 = \{x \in (a, b) \mid (y+\phi)(x) > 0\}$, $I_2 = \{x \in (a, b) \mid (y+\phi)(x) < 0\}$. Since the first and second integrals above are negative and the third one is positive, it follows that indeed $\mu_1 > 1$.

(G) It follows from (F) and a argument on comparison of eigenvalues that, if we replace (a, b) by a subinterval in the above eigenvalue problem, then the first eigenvalue of this new eigenvalue problem is still strictly greater than 1.

(H) As a consequence of (F) and (G), it follows that if we divide (a, b) into disjoint parts where $(y+\phi) \geqslant 0$ and $(y+\phi) < 0$, then the part where $(y+\phi) < 0$ is connected. Indeed, suppose it is not connected. Then there exists a part $(a_0, b_0) \subset (a, b)$, where $-y'' = \lambda y$, $(y+\phi)(a_0) = (y+\phi)(b_0) = 0$, that is, we have

$$\begin{cases} -(y+\phi)'' = \lambda(y+\phi) - (\lambda - \lambda_1)\phi & \text{in} \quad (a_0, b_0) \\ (y+\phi)(a_0) = (y+\phi)(b_0) = 0, \end{cases}$$

with $y+\phi > 0$ in (a_0, b_0). Therefore, writing $w = y + \phi$, we obtain

$$\langle -w'' - \lambda w, w \rangle = -(\lambda - \lambda_1) \int_{a_0}^{b_0} \phi w < 0,$$

which implies that the first eigenvalue with weight $a_\lambda = \lambda$ on (a_0, b_0) is strictly less than 1, a contradiction to (G).

From the above properties we obtain immediately

PROPOSITION 1. *The first branch C_1 is described by*

$$C_1^+ = \{1\} \times \{s\phi \mid s \geq 0\},$$

$$C_1^- = \{1\} \times \{s\phi \mid -1 \leq s \leq 0\} \cup \left\{\left(\lambda, \frac{\alpha - \lambda}{1 - \alpha}\phi\right) \mid \lambda > 1\right\}.$$

In the next section we will obtain some information on the other branches.

3. A Preliminary Description of the Bifurcation Diagram

We start by showing that the branches C_k actually *bend* to the right of $\lambda_k = k^2$.

LEMMA 1. *If $y \in S_k$ is a solution of $(*)_\lambda$ then $\lambda \geq \lambda_k$.*

Proof. There are two possibilities: (i) $(y+\phi)^- \equiv 0$. It is then readily seen that $-y'' = \lambda y$. So $y \not\equiv 0$ if and only if $y = s\phi_k$ with s in some interval $[a_k, b_k]$, $a_k < 0 < b_k$; (ii) $(y+\phi)^- \not\equiv 0$. Then, a solution of $(*)_\lambda$ satisfies the eigenvalue problem

$$-y'' = m_\lambda(x) y \quad \text{in} \quad (0, \pi), \qquad y(0) = y(\pi) = 0,$$

where

$$m_\lambda(x) = \frac{b_\lambda(y+\phi) - \lambda\phi}{y} = \begin{cases} \lambda, & \text{if} \quad y + \phi > 0 \\ \alpha + \dfrac{\alpha - \lambda}{y}\phi, & \text{if} \quad y + \phi < 0. \end{cases} \tag{6}$$

So $\mu_j(m_\lambda) = 1$ for some $j \geq 1$. Moreover, since $y \in S_k$, we have necessarily $j = k$, that is, $\mu_k(m_\lambda) = 1$. It follows from (6) that $m_\lambda(x) \leq \lambda$ with strict inequality on a set of positive measure. Therefore, we obtain $\mu_k(m_\lambda) > \mu_k(\lambda) = \lambda_k/\lambda$, which completes the proof. ∎

Next we show that a given branch $C_k^\pm$ extends to $+\infty$ in the λ-direction. For that matter we need an a priori bound.

LEMMA 2. *Given $M > 1$ there exists a constant $C > 0$ depending on M such that $\|y\|_{C^1} \leq C$ for all solutions of $(*)_\lambda$ with $1 < \lambda \leq M$.*

Proof. If $0 \leqslant a < b \leqslant \pi$ we have from Eq. $(*)_\lambda$ that

$$y'(a) - y'(b) = \int_a^b [b_\lambda(y+\phi) - \lambda\phi]\,dx. \tag{7}$$

So it suffices to show that $\|y\|_{C^0} \leqslant C$. From Remark (E) above the negative bumps are all bounded from below. To bound the positive bumps, it suffices to show that if a is a zero of y then $y'(a)$ is bounded. However, this follows from (7) by taking b equal to the point where $y'(b) = 0$, in the negative bump adjacent to a. ∎

It follows directly from Rabinowitz's Global Bifurcation Theorem, in view of Lemmas 1 and 2, that the following result is true.

PROPOSITION 2. *For each* $k \geqslant 2$, *we have* $\text{Proj}_{\mathbf{R}} C_k^{\pm} = [\lambda_k, \infty)$. *In particular, problem* $(*)_\lambda$ *has at least* $2k$ *solutions if* $\lambda > \lambda_k$, $k \geqslant 1$.

4. All Solutions of $(*)_\lambda$ Are Nondegenerate

In this section we shall prove the key result of this paper, namely Theorem 2 below, a direct consequence of which is Theorem 1.

THEOREM 2. *Let* $y \in S_k$ *be a solution of* $(*)_\lambda$ *for some* $\lambda > \lambda_k$. *Then* y *is nondegenerate and its Morse Index is* $k-1$.

For the development of the present section we shall work in the framework of the Hilbert space $H_0^1 = H_0^1[0, \pi]$. Clearly, the space X of Section 2 is contained in H_0^1. Let us denote by $K: H_0^1 \to H_0^1$ the (solution) operator defined by $K\rho = y$ if and only if $y \in H^2 \cap H_0^1$ and $-y'' = \rho$.

Then, problem $(*)_\lambda$ becomes the functional equation

$$y - K[b_\lambda(y+\phi) - \lambda\phi] = 0$$

in H_0^1. The operator $S: y \mapsto K[b_\lambda(y+\phi) - \lambda\phi]$ is compact and Fréchet differentiable at a solution y of $(*)_\lambda$, and its derivative at y is the bounded linear operator on H_0^1 given by

$$S'(y): \psi \mapsto K[a_\lambda(y+\phi)\psi],$$

where

$$a_\lambda(y+\phi) = \begin{cases} \lambda, & \text{if } \; y+\phi > 0 \\ \alpha, & \text{if } \; y+\phi < 0. \end{cases}$$

In terms of differential equations, $\eta = S'(y)\psi$ if and only if $\eta \in H^2 \cap H_0^1$ and $-\eta'' = a_\lambda(y+\phi)\psi$. The operator $S'(y)$ is also compact and its eigenvalues are the inverses of the eigenvalues of problem

$$-\psi'' = \mu a_\lambda(y+\phi)\psi \quad \text{in} \quad (0, \pi), \qquad \psi(0) = \psi(\pi) = 0. \tag{8}$$

It is well known that (8) has a sequence of eigenvalues $0 < \mu_1 < \mu_2 < \cdots$, $\mu_j \to +\infty$ (here we are using the fact that $\alpha \geqslant 0$). Let us denote by $\psi_1, \psi_2, \ldots$ the corresponding eigenfunctions, by E_j the subspace of H_0^1 generated by $\psi_1, \ldots, \psi_j$ and by $E_j^\perp$ its orthogonal complement in H_0^1, which coincides with the subspace generated by $\psi_{j+1}, \psi_{j+2}, \ldots$.

A solution y of $(*)_\lambda$ is *nondegenerate* if the operator $S'(y)$ is invertible, which is equivalent to saying that $\mu_j \neq 1$ for all j. The *Morse Index* of a nondegenerate solution y of $(*)_\lambda$ is the number of eigenvalues μ_j of (8) which are <1. Let A be the quadratic form on H_0^1 defined by

$$A(\psi) = \int_0^\pi |\psi'|^2 - \int_0^\pi a_\lambda(y+\phi)\psi^2. \tag{9}$$

Using the orthonormality in H_0^1 of the ψ_j's, we obtain for $\psi = \sum_{j=1}^\infty \beta_j \psi_j$ that

$$A(\psi) = \sum_{j=1}^\infty \mu_j^{-1}(\mu_j - 1)\beta_j^2. \tag{10}$$

So, y has Morse Index n if and only if the quadratic form in (10) has exactly the n first coefficients negative.

LEMMA 3. *Let $y \in S_k$ be a solution of $(*)_\lambda$. Then the number of eigenvalues μ_j which are $\leqslant 1$ is at most $k-1$.*

Proof. With the notation in the proof of Lemma 1 we have $a_\lambda(y+\phi) \leqslant m_\lambda(x)$, with strict inequality on a set of positive measure. So, we obtain $\mu_k(a_\lambda(y+\phi)) > \mu_k(m_\lambda) = 1$, which gives the result. ∎

LEMMA 4. *Let $y \in S_k$ be a solution of $(*)_\lambda$. Suppose that there are exactly n eigenvalues μ_j which are <1 (We are not saying that $\mu_{n+1} > 1$. So we are not assuming that y is not degenerate.) Then, n is the maximum dimension of a subspace F of H_0^1, where A is <0.*

Proof. Suppose, by contradiction, that the dimension of F is $n+1$. Let $e_1, \ldots, e_{n+1}$ be a basis for F. Then, there are scalars β_{ji} $(j = 1, \ldots, n+1; i = 1, 2, \ldots)$ such that

$$e_j = \sum_{i=1}^\infty \beta_{ji} \psi_i.$$

Let $c_1, ..., c_{n+1}$ be scalars not all zero such that

$$\sum_{j=1}^{n+1} c_j \sum_{i=1}^{n} \beta_{ji}\psi_i = 0.$$

Then $\psi = \sum_{j=1}^{n+1} c_j e_j \in E_n^\perp$. From (10) we obtain $A(\psi) \geqslant 0$, which contradicts the hypothesis that $A(\psi) < 0$.

THEOREM 3. *Let $y \in S_k$ be a solution of $(*)_\lambda$. Then there exists a subspace F of H_0^1 with dimension $k-1$, where A is <0.*

Let us assume Theorem 3 and give the proof of Theorem 2.

Proof of Theorem 2. It follows from Lemma 4 and Theorem 3 that the number of eigenvalues $\mu_j < 1$ is at least $k-1$. Next, using Lemma 3 we obtain that there are exactly $k-1$ eigenvalues <1 and 1 is not an eigenvalue. So y is nondegenerate and its Morse Index is $k-1$. ∎

Thus it remains to prove Theorem 3. Its proof will follow readily from Propositions 3 and 4 below, whose proofs will be given in the next section.

PROPOSITION 3. *Let y be a solution of $(*)_\lambda$.*

(i) *Suppose $0 < a < b$ with $y > 0$ in $(0, a)$ and $y < 0$ in (a, b). Let b' be the point where y assumes its minimum in (a, b). Then there exists a function $\psi \in H_0^1$ with* supp $\psi \subset [0, b')$ *and $A(\psi) < 0$.*

(ii) *Suppose $0 \leqslant a < b < c < d \leqslant \pi$ with $y > 0$ in (b, c), $y < 0$ in (a, b) and in (c, d), and assume that $(y+\phi)^- \not\equiv 0$ on (b', c'), where b' [resp. c'] is the point where y attains its minimum in (a, b) [resp. (c, d)]. Then, for any $\varepsilon > 0$ sufficiently small, there exist orthogonal functions $\psi^{(b)}, \psi^{(c)} \in H_0^1$ such that* supp $\psi^{(b)}$, supp $\psi^{(c)} \subset [b' + \varepsilon, c' - \varepsilon]$ *and A is negative definite on the subspace $\langle \psi^{(b)}, \psi^{(c)} \rangle$ generated by $\psi^{(b)}$ and $\psi^{(c)}$.*

Now, if all positive interior bumps of a solution y are in the situation of Proposition 3 then we are finished. On the other hand, if there exist positive interior bumps (b, c) where $(y+\phi)^- \equiv 0$ on (b', c'), we need a different argument by means of Proposition 4 below.

PROPOSITION 4. *Let $0 \leqslant a < b < c < d \leqslant \pi$, b' and c' be as in Proposition 3 and assume that $(y+\phi)^- \equiv 0$ on (b', c'). Then, for any $\varepsilon > 0$ sufficiently small, there exist orthogonal functions $\psi^{(b)}, \psi^{(c)} \in H_0^1$ such that* supp $\psi^{(b)}$, supp $\psi^{(c)} \subset [b' - 2\varepsilon, c' - \varepsilon/2]$ *and A is negative definite on the subspace $\langle \psi^{(b)}, \psi^{(c)} \rangle$.*

Finally, we come to the proof of Theorem 3.

Proof of Theorem 3. Let $y \in S_k$ be a solution of $(*)_\lambda$. Since y has $k-1$ interior zeros, we have the following possibilities:

(i) If $k-1$ is an even number, then either $(\text{i})_1$ y has $(k-1)/2$ interior positive bumps and no boundary positive bump, or $(\text{i})_2$ y has $(k-1)/2-1$ interior positive bumps and 2 boundary positive bumps.

(ii) If $k-1$ is an odd number, then y has $(k-1-1)/2$ positive interior bumps and 1 positive boundary bump. In all cases, it follows from Propositions 3 and 4 that there are $k-1$ linearly independent functions in H_0^1 such that A is negative definite on the space generated by them. ∎

5. Proofs of Propositions 3 and 4

We start with two preliminary technical lemmas.

Lemma 5. *Let $\beta > 0$ be a real number and $g(x)$ a continuously differentiable function on a given interval $[0, T]$. Then the solution of*

$$w'' + \beta w = g(x) \quad \text{in } [0, T], \qquad w(0) = w_0, \qquad w'(0) = w_1$$

is given by

$$w(x) = \int_0^x \frac{\sin\sqrt{\beta}\,\xi}{\sqrt{\beta}}\, g(x-\xi)\, d\xi + w_0 \cos\sqrt{\beta}\, x + w_1 \frac{\sin\sqrt{\beta} x}{\sqrt{\beta}}$$

and its derivative is given by

$$w'(x) = \int_0^x \frac{\sin\sqrt{\beta}\,\xi}{\sqrt{\beta}}\, g'(x-\xi)\, d\xi + \frac{\sin\sqrt{\beta}\, x}{\sqrt{\beta}}\,[g(0) - \beta w_0] + w_1 \cos\sqrt{\beta}\, x.$$

Corollary. *If in addition to the hypotheses of the above lemma one has*

(i) $g(x) > 0$, $g'(x) > 0$ *in* $[0, T]$,

(ii) $w_0 \geqslant 0$, $w_1 \geqslant 0$,

(iii) $g(0) - \beta w_0 \geqslant 0$,

(iv) $T < \pi/(2\sqrt{\beta})$,

then $w(x) > 0$ *and* $w'(x) > 0$ *in* $(0, T)$.

Lemma 6. *Let $0 \leqslant a < b < c < d \leqslant \pi$ be such that $y(x) > 0$ in (b, c) and $y(x) < 0$ in $(a, b) \cup (c, d)$, where y is a solution of $(*)_\lambda$. Assume $\pi/2 < b$ and* *let b' [resp. c'] be the point where y attains its minimum in (a, b) [resp. (c, d)]. Then, we have*

$$c+(b-b')<c'$$

and

$$y(c+x) \leqslant y(b-x) \qquad \textit{for} \quad 0<x<b-b',$$
$$y'(c+x) \leqslant -y'(b-x) \qquad \textit{for} \quad 0<x<b-b'.$$

Proof. Let us introduce the notation

$$y_c(x)=y(c+x), \qquad y_b=y(b-x), \qquad \phi_c(x)=\phi(c+x), \qquad \phi_b(x)=\phi(b-x).$$

As a consequence of $\pi/2<b$ we obtain that

$$\phi_b(x)>\phi_c(x), \qquad \phi'_b(x)>\phi'_c(x), \qquad |\phi'_b(x)|<|\phi'_c(x)|, \tag{11}$$

for $0<x\leqslant\pi-c$. Now, y_b and y_c are solutions of the two following initial value problems in $(0, \pi-c)$,

$$-z''=b_\lambda(z+\phi_b)-b_\lambda(\phi_b), \qquad z(0)=0, \qquad z'(0)=\theta, \tag{12}$$
$$-z''=b_\lambda(z+\phi_c)-b_\lambda(\phi_c), \qquad z(0)=0, \qquad z'(0)=\theta, \tag{13}$$

where $\theta=y'(c)$. With the notation just introduced, the contention of the lemma is that

$$y_c(x)\leqslant y_b(x) \qquad \text{and} \qquad y'_c(x)\leqslant y'_b(x) \qquad \text{for} \quad 0<x<b-b'.$$

From now on we break the proof into several parts.

(i) The function $y_c(x)+\phi_c(x)$ is decreasing in $(0, c'-c)$ and is >0 at $x=0$. Let $x_0\in(0, c'-c)$ be the (unique) point where it vanishes; if there is no such point take $x_0=c'-c$. From the fact that

$$0<y_c(0)+\phi_c(0)<y_b(0)+\phi_b(0)$$

and from (11) above, it follows that the two functions y_b and y_c coincide in the interval $[0, x_0]$. Observe that they are solutions of the same initial value problem

$$-z''=\lambda z \qquad \text{in} \quad [0, x_0], \qquad z(0)=0, \qquad z'(0)=\theta.$$

If $x_0=c'-c$ we have that $y_b=y_c$ in the whole of $[0, c'-c]$ and, consequently, $c'-c=b-b'$ and $b-a=d-c=\pi/\sqrt{\lambda}$. Also we have that $y_b(x)=A\sin(\sqrt{\lambda}x)$, with $A<0$. So, in the case that $x_0=c'-c$, the proof is complete. Therefore, let us assume that $x_0 < c' - c$ and proceed with the proof.

(ii) Let $r = \min(c' - c, b - b')$. Assume first that $y_b + \phi_b \geqslant 0$ in $[0, b - b']$. Then, in the interval (x_0, r), the functions y_b and y_c satisfy the following equations, with the same initial data at $x = x_0$,

$$-y_b'' = \lambda y_b$$

$$-y_c'' = \alpha(y_c + \phi_c) - \lambda\phi_c = \alpha y_c + (\alpha - \lambda)\phi_c.$$

Therefore, letting $w = y_b - y_c$, we obtain

$$w'' + \lambda w = (\alpha - \lambda)(y_c + \phi_c), \qquad w(x_0) = w'(x_0) = 0, \tag{14}$$

and we are in the conditions of applying the Corollary of Lemma 5, since the hypothesis

$$r - x_0 < \pi/(2\sqrt{\lambda})$$

holds true by virtue of the fact that y_b satisfies the equation $-y_b'' = \lambda y_b$ in $(0, r)$ with $y_b(0) = 0$, $y_b'(0) < 0$, and $y_b'(x) \leqslant 0$ for x in $(0, r)$. Hence, at $x = r$, we obtain that $y_c(r) < y_b(r)$ and $y_c'(r) < y_b'(r)$. But this implies that $r = b - b'$ since, otherwise (if $r = c' - c < b - b'$), we would have

$$y_c'(c' - c) = 0 < y_b'(c' - c),$$

contradicting the fact that y_b is decreasing at $c' - c$. So, in this case, the proof is finished. It remains to consider the case in which the function $y_b + \phi_b$ is negative somewhere in the interval $(0, b - b')$.

(iii) By Remark (H) it is clear that if $y_b + \phi_b$ has zeros in $(0, b - b')$ then there exist at most two of them, say x_1, x_2 with $x_1 < x_2 \leqslant b - b'$ (In fact, there can exist at most two zeros in the *whole* negative part under consideration.) Observe that $y_b + \phi_b > 0$ on $(x_0, x_1) \cup (x_2, b - b')$ (assuming such an x_2 exists) and $y_b + \phi_b < 0$ on (x_1, x_2). We shall prove that the contention of the lemma holds in the interval (x_0, r), where $r = \min(c' - c, b - b')$. Then, as done earlier, we will conclude that $b - b' \leqslant c' - c$.

If $x_1 \leqslant r$ we see that in the interval (x_0, x_1) the situation is exactly as in part (ii) involving even the same equations. Now, if $x_2 \leqslant r$ then, in the interval (x_1, x_2), the functions y_b and y_c satisfy the equations

$$-y_b'' = \alpha y_b + (\alpha - \lambda)\phi_b,$$

$$-y_c'' = \alpha y_c + (\alpha - \lambda)\phi_c.$$

Therefore, letting $w = y_b - y_c$, we obtain

$$w'' + \alpha w = (\lambda - \alpha)(\phi_b - \phi_c) \tag{15}$$

with $w(x_1) > 0$ and $w'(x_1) > 0$. Now, we claim that $w'(x) > 0$ for all x in $(x_1, x_2]$. Indeed suppose that this is not the case. Then, there exists a first point x_0' in $(x_1, x_2]$ with $w'(x_0') = 0$. Consider the function $z = w'$ in the interval (x_0, x_0'). It follows from (14) and (15) that

$$z'' + a_\lambda z = \begin{cases} (\alpha - \lambda)(y_c' + \phi_c') & \text{in } (x_0, x_1) \\ (\lambda - \alpha)(\phi_b' - \phi_c') & \text{in } (x_1, x_0') \end{cases} \tag{16}$$

and observe that $z(x_0) = z(x_0') = 0$ and $z(x) > 0$ in (x_0, x_0'). Since $z > 0$, $(y_c + \phi_c)' < 0$, $(\phi_b - \phi_c)' > 0$ and $\lambda - \alpha > 0$, it follows from (16) that

$$\langle z'' + a_\lambda z, z \rangle = (\alpha - \lambda) \int_{x_0}^{x_1} (y_c + \phi_c)' z + (\lambda - \alpha) \int_{x_1}^{x_0'} (\phi_b - \phi_c)' z > 0,$$

a fact which can be translated into a statement about functions on (a, b) to lead to a contradiction to (16). ∎

Proof of Proposition 3. (i) Let $\alpha < e < b'$ be chosen such that $y + \phi > 0$ in $(0, e)$. Let ψ_0 be the first eigenfunction of

$$-\psi_0'' = \mu \psi_0 \quad \text{in } (0, e), \qquad \psi_0(0) = \psi_0(e) = 0,$$

and define

$$\psi = \begin{cases} \psi_0 & \text{in } (0, e) \\ 0 & \text{elsewhere.} \end{cases}$$

Since $\mu = \pi^2/e^2 < \pi^2/a^2 = \lambda$, it follows readily that $A(\psi) < 0$.

(ii) Without loss of generality we may assume that $\pi/2 < M$, where M is the point in (b, c) where y attains its maximum (If $\pi/2 > M$ we use the change of variables $x \to \pi - x$ to reduce to the above case; also, the *symmetric* case $\pi/2 = M$ can be handled by the arguments below or those of Proposition 4 to follow.) Moreover, since we are assuming $(y + \phi)^- \not\equiv 0$ on (b', c'), it follows that necessarily $(y + \phi)^- \not\equiv 0$ on (c, c').

Indeed, if $(y + \phi)^- \equiv 0$ on (c, c') then we should have $(y + \phi)^- \not\equiv 0$ on (b', b), so that letting $b_- < b$ be the smallest number such that $(y + \phi)^- \equiv 0$ on (b_-, b) and defining $c_+ = c + b - b_-$ we would obtain $y(b - x) = y(c + x) = A \sin \sqrt{\lambda} x$, $A < 0$, for all $x \in [0, b - b_-]$. But then, we could not have $\phi(c_+) < \phi(b_-)$ (since this would imply, by continuity, that $(y + \phi)(c_+ + \xi) < (y + \phi)(b_- - \xi)$ for $\xi > 0$ small and, hence, since

$(y+\phi)^-(b_- - \xi) \neq 0$ for $\xi > 0$ small, we would obtain $(y+\phi)^-(c_+ + \xi) \neq 0$ for $\xi > 0$ small). Therefore, it follows that $\phi(c_+) \geqslant \phi(b_-)$ which contradicts our assumption that $\pi/2 < M$.

Now, let $e = b - \varepsilon$, $e' = c + \varepsilon$ with $\varepsilon > 0$ being sufficiently small so that $y + \phi > 0$ in $I_\varepsilon = (e, e')$, let ψ_0 be the first eigenfunction of

$$-\psi_0'' = \mu\psi_0 \quad \text{in} \quad I_\varepsilon, \qquad \psi_0(e) = \psi_0(e') = 0,$$

and define

$$\psi^{(1)} = \begin{cases} \psi_0 & \text{in } I_\varepsilon \\ 0 & \text{elsewhere.} \end{cases}$$

As in (i) we see that $A(\psi^{(1)}) < 0$. Next, define

$$\psi^{(2)} = \begin{cases} y' & \text{in } (b', c') \\ 0 & \text{elsewhere,} \end{cases}$$

and note that from Eq. $(*)_\lambda$ we obtain $-y''' = a_\lambda(y+\phi)(y'+\phi') - \lambda\phi'$, which used in the quadratic form A gives

$$A(\psi^{(2)}) = \int_{b'}^{c'} a_\lambda(y+\phi)\,|y'|^2 + \int_{b'}^{c'} [a_\lambda(y+\phi) - \lambda]\,\phi' y' - \int_{b'}^{c'} a_\lambda(y+\phi)\,|y'|^2,$$

hence

$$A(\psi^{(2)}) = \int_{[b',b] \cap [y+\phi<0]} (\alpha - \lambda)\phi' y' + \int_{[c,c'] \cap [y+\phi<0]} (\alpha - \lambda)\phi' y'.$$

Now, in view of the fact that $(y+\phi)^- \not\equiv 0$ on $[c, c']$, the second integral above is negative. Therefore, we obtain $A(\psi^{(2)}) < 0$ as a consequence of

$$\left| \int_{[b',b] \cap [y+\phi<0]} \phi' y' \right| \leqslant \int_{[0, b-b'] \cap [y_b + \phi_b < 0]} |\phi_b'|\,|y_b'|$$
$$< \int_{[0, c'-c] \cap [y_c + \phi_c < 0]} |\phi_c'|\,|y_c'|,$$

where the last inequality follows from Lemma 6. Also, since $\psi^{(1)}$ and $\psi^{(2)}$ are orthogonal, we obtain that A is negative definite on the subspace $\langle \psi^{(1)}, \psi^{(2)} \rangle$. In particular this implies that $\mu_1^{(b',c')}(a_\lambda) < \mu_2^{(b',c')}(a_\lambda) < 1$, (where $\mu_i^J(m)$ denotes the eigenvalues on the interval J with weight m) and, hence, $\mu_1^{(b'+\varepsilon, c'-\varepsilon)}(a_\lambda) < \mu_2^{(b'+\varepsilon, c'-\varepsilon)}(a_\lambda) < 1$ for all $\varepsilon > 0$ sufficiently small. For any such fixed $\varepsilon > 0$, we may therefore choose orthogonal functions $\psi^{(b)}, \psi^{(c)} \in H_0^1$ such that supp $\psi^{(b)}$, supp $\psi^{(c)} \subset [b' + \varepsilon, c' - \varepsilon]$, and

$A(\psi^{(b)}) < 0$, $A(\psi^{(c)}) < 0$. Hence, A is negative definite on $\langle \psi^{(b)}, \psi^{(c)} \rangle$. The proof of Proposition 3 is complete. ∎

Proof of Proposition 4. Since $(y+\phi)^- \equiv 0$ on (b', c'), the function $z = y'$ is a solution of $-z'' = a_\lambda z = \lambda z$ in (b', c') with $z(b') = z(M) = z(c') = 0$ and $z > 0$ on (b', M), $z < 0$ on (M, c'). Therefore, for all $\varepsilon > 0$ sufficiently small, we have $\nu_{1b} := \mu_1^{(b'-2\varepsilon, M-\varepsilon)}(a_\lambda) < \mu_1^{(b', M)}(a_\lambda) = 1$ and $\nu_{1c} := \mu_1^{(M-\varepsilon, c'-\varepsilon/2)}(a_\lambda) < \mu_1^{(M,c')}(a_\lambda) = 1$. We let $\psi^{(b)} > 0$ [resp. $\psi^{(c)} > 0$] be a first eigenfunction on $(b' - 2\varepsilon, M - \varepsilon)$ [resp. $(M - \varepsilon, c' - \varepsilon/2)$], with weight a_λ, corresponding to the eigenvalue ν_{1b} [resp. ν_{1c}]. Again, it follows that A is negative definite on $\langle \psi^{(b)}, \psi^{(c)} \rangle$. ∎

Concluding Remark. We point out the importance of the method of obtaining the exactness result in this paper, through the use of Morse Index computations. As a byproduct, we showed that $(*)_\lambda$ possesses solutions of every admissible Morse Index. Namely, for $\lambda_k < \lambda < \lambda_{k+1}$, problem $(*)_\lambda$ has

1 solution of Morse Index 0 (the negative solution $\hat{y}$),
2 solutions of Morse Index 1 (those in S_2^+ and S_2^-),
⋮
2 solutions of Morse Index $k-1$ (those in S_k^+ and S_k^-),
1 solution of Morse Index k (the trivial solution $y \equiv 0$).

REFERENCES

1. ADIMURTHI AND P. N. SRIKANTH, On exact number of solutions at infinity for Ambrosetti–Prodi class of problems, *Boll. Un. Mat. Ital.* **6** (1984), 15–24.
2. D. G. COSTA, D. G. DE FIGUEIREDO, AND J. V. GONÇALVES, On the uniqueness of solutions for a class of semilinear elliptic problems, *J. Math. Anal. Appl.* **123** (1987), 170–180.
3. M. G. CRANDALL AND P. H. RABINOWITZ, Bifurcation from simple eigenvalues, *J. Funct. Anal.* **8** (1971), 321–340.
4. A. C. LAZER AND P. J. MCKENNA, On a conjecture related to the number of solutions of a nonlinear Dirichlet problem, *Proc. Roy. Soc. Edinburgh Sect. A* **95** (1983), 275–283.
5. P. H. RABINOWITZ, Nonlinear Sturm–Liouville problems for second order ordinary differential equations, *Comm. Pure Appl. Math.* **23** (1970), 939–962.
6. P. H. RABINOWITZ, Some global results for nonlinear eigenvalue problems, *J. Funct. Anal.* **7** (1971), 487–513.
7. B. RUF, Remarks and generalizations related to a recent multiplicity result of A. Lazer and P. J. McKenna, *Nonlinear Anal.* TMA **9** (1985), 1325–1330.
8. S. SOLIMINI, Some remarks on the number of solutions of some nonlinear elliptic problems, *Ann. Inst. H. Poincaré Anal. Non Linéaire* **2** (1985), 143–156.

Printed by Catherine Press, Ltd., Tempelhof 41, B-8000 Brugge, Belgium [illegible] Hence [illegible]. The proof of Proposition [illegible] is complete. ∎

Proof of Proposition [illegible] the function [illegible] and [illegible] on [illegible]. Therefore, for [illegible] sufficiently [illegible]. Let [illegible] be a first [illegible] eigenfunction on [illegible], which [illegible] corresponds to the eigenvalue [illegible]. From [illegible] it follows that [illegible] positive [illegible]. ∎

Concluding Remark. We point out the importance of the method of obtaining the existence result in this paper, through the use of Morse index computations. As a byproduct, we proved that [illegible] of every admissible Morse index. Namely, [illegible]

[illegible] solutions [illegible] index of the negative solution [illegible]

[illegible] of Morse index [illegible] in [illegible] and [illegible]

[illegible] Solutions of Morse index [illegible]

[illegible] of Morse index [illegible] the trivial solution [illegible]

References

1. [illegible]
2. [illegible]
3. [illegible]
4. [illegible]
5. [illegible]
6. [illegible]
7. [illegible]
8. [illegible]

COMMUN. IN PARTIAL DIFFERENTIAL EQUATIONS, 17(1&2), 339-346 (1992)

STRICT MONOTONICITY OF EIGENVALUES AND UNIQUE CONTINUATION

Djairo G. de Figueiredo*

Departamento de Matemática
Universidade de Campinas
13081 Campinas, São Paulo, Brazil

Jean-Pierre Gossez†

Département de Mathématique
Université Libre de Bruxelles
1050 Bruxelles, Belgique

Abstract

In this paper, we show that the strict monotonicity of the eigenvalues of an uniformly elliptic operator of second order is equivalent to a unique continuation property.

1 Introduction

This note is primarily concerned with the eigenvalue problem

$$Lu = \mu m u \text{ in } \Omega, \; u = 0 \text{ on } \partial\Omega, \tag{1}$$

where

$$Lu = -\Sigma_{i,j=1}^{N} D_j(a_{ij}(x) D_i u) + a_0(x) u \tag{2}$$

is an uniformly elliptic operator in a bounded domain $\Omega \subset \mathbf{R}^N$, with coefficients satisfying $a_{ij} = a_{ji} \in L^\infty(\Omega)$ and $0 \leq a_0$. The weight function $m(x)$ is assumed to be $\not\equiv 0$ and both a_0 and $m(x)$ are assumed to lie in $L^r(\Omega)$ for

*Supported in part by CNPq.
†Supported in part by FNRS

Originally published in: Comm. Partial Differential Equations 17 (1992), No. 1-2, 339-346, © 1992 Marcel Dekker, Inc.
Reprint by Springer-Verlag Berlin Heidelberg 2013 with kind permission,
DOI: 10.1007/978-3-319-02856-9_24

some $r > N/2$. As it is well known (cf. [Ma-Mi], [DF]), the eigenvalues μ in (1) form in general two sequences, one of positive eigenvalues, the other one of negative eigenvalues, which can be written respectively as

$$\mu_1(m) < \mu_2(m) \leq \dots,$$

$$\mu_{-1}(m) > \mu_{-2}(m) \geq \dots.$$

Here no negative (resp. positive) eigenvalue appears if $m(x) \geq 0$ (resp. ≤ 0) a.e. in Ω, and repetition is made accordingly to the multiplicity. Moreover one has the variational characterization

$$\frac{1}{\mu_k(m)} = \sup_{F_k} \inf \left\{ \int_\Omega mu^2; u \in F_k, a[u,u] = 1 \right\} \qquad (3)$$
$$\frac{1}{\mu_{-k}(m)} = \inf_{F_k} \sup \left\{ \int_\Omega mu^2; u \in F_k, a[u,u] = 1 \right\},$$

where F_k varies over all k-dimensional subspaces of $H_0^1(\Omega)$ and where $a[u,v]$ denotes the Dirichlet form associated with the differential operator (2).

We are interested in the monotone dependence of these eigenvalues with respect to the weight. Clearly (3) implies that if two weights m and $\hat{m}$ satisfy $m \leq \hat{m}$ a.e., then $\mu_j(m) \geq \mu_j(\hat{m})$ for all $j \in \mathbf{Z}_0$. Our purpose in this note is to show that strict monotonicity holds if and only if some unique continuation property is satisfied by the corresponding eigenfunctions. Applications of this strict monotonicity to nonlinear problems have been made recently in [Co-DF-Gs],[DF-Go], [DF-Mi], [Go-Om]. For completeness we have included at the end of this note some remarks on unique continuation.We compare there the unique continuation property that we consider with one of the more usual notions of unique continuation.

We thank Carlos Kenig for various discussions and much information on unique continuation.

2 Strict Monotonicity of Eigenvalues

We will say that a family of functions enjoys the *unique continuation property*, in short (U.C.P.), if no function, besides possibly the zero function, vanishes on a set of positive measure. Here we use the notation $\leq\not\equiv$ to mean inequality a.e. together with strict inequality on a set of positive measure.

Proposition 1 *Let m and $\hat{m}$ be two weights with $m \lneqq \hat{m}$ and let $j \in \mathbf{Z}_0$. If the eigenfunctions associated to $\mu_j(m)$ enjoy the U.C.P., then $\mu_j(m) > \mu_j(\hat{m})$.*

Proposition 2 *Let m be a weight and let $j \in \mathbf{Z}_0$. If the eigenfunctions associated to $\mu_j(m)$ do not enjoy the U.C.P., then there exists a weight $\hat{m}$ with $m \lneqq \hat{m}$, such that, for some $i \in \mathbf{Z}_0$ with $\mu_i(m) = \mu_j(m)$, one has $\mu_i(m) = \mu_i(\hat{m})$.*

Remark 1 The conclusion of Proposition 1 also holds if it is the eigenvalues associated to the larger weight which enjoy the U.C.P.. Similarly, in Proposition 2, one can construct a weight smaller than the given one with the desired property. This follows from the observation that $\mu_j(m) = -\mu_{-j}(-m)$.

PROOF: (Proposition 1). Assume first that $j > 0$. Since the extrema in (3) are achieved (cf. [DF]), there exists $F_j \subset H_0^1(\Omega)$ of dimension j such that

$$\frac{1}{\mu_j(m)} = \inf\left\{\int_\Omega mu^2; u \in F_j \text{ and } a[u,u] = 1\right\}. \tag{4}$$

Pick $u \in F_j$ with $a[u,u] = 1$. Either u achieves the infimum in (4) or not. In the first case, u is an eigenfunction associated to $\mu_j(m)$ (cf. [DF]) and so, by the U.C.P.,

$$\frac{1}{\mu_j(m)} = \int_\Omega mu^2 < \int_\Omega \hat{m}u^2.$$

In the second case

$$\frac{1}{\mu_j(m)} < \int_\Omega mu^2 \le \int_\Omega \hat{m}u^2.$$

Thus, in any case,

$$\frac{1}{\mu_j(m)} < \int_\Omega \hat{m}u^2.$$

It then follows, by a simple compactness argument, that

$$\frac{1}{\mu_j(m)} < \inf\left\{\int_\Omega \hat{m}u^2; u \in F_j \text{ and } a[u,u] = 1\right\}.$$

This yields the desired inequality

$$\frac{1}{\mu_j(m)} < \frac{1}{\mu_j(\hat{m})}$$

The argument is similar when $j < 0$, say $j = -k$, except that one starts from another variational formula, namely

$$\frac{1}{\mu_{-k}} = \sup_{F_{k-1}} \inf \left\{ \int_\Omega mu^2 ; u \perp F_{k-1} \text{ and } a[u,u] = 1 \right\},$$

where the orthogonality is meant with respect to the Dirichlet form $a[u,v]$ (cf. [DF]). QED

PROOF: (Proposition 2) Assume first that $j > 0$ and denote by u an eigenfunction associated to $\mu_j(m)$ which vanishes on a set of positive measure. Take i such that $\mu_j(m) = \mu_i(m) < \mu_{i+1}(m)$. Define

$$\hat{m}(x) = \begin{cases} m(x) & \text{if } u(x) \neq 0 \\ m(x) + \varepsilon & \text{if } u(x) = 0 \end{cases}$$

where $\varepsilon > 0$ is chosen so that

$$\mu_i(m) < \mu_{i+1}(\hat{m}), \tag{5}$$

which is possible by the continuous dependence of the eigenvalues with respect to the weight (cf. [DF]). Clearly

$$Lu = \mu_i(m) mu = \mu_i(m) \hat{m} u,$$

which shows that $\mu_i(m)$ is an eigenvalue for the weight $\hat{m}$, i.e., $\mu_i(m) = \mu_l(\hat{m})$ for some $l \in \mathbf{Z}_0$. Let us choose the largest l such that this equality holds. It follows from (5) that $l < i+1$. Moreover, by the monotone dependence, $\mu_i(\hat{m}) \leq \mu_i(m)$, which implies $l \geq i$. We then conclude that $l = i$, and so $\mu_i(m) = \mu_i(\hat{m})$. The same argument applies when $j < 0$, after observing that i above cannot be equal to -1 (since the eigenfunctions associated to $\mu_{-1}(m)$ have a definite sign, cf. [DF]). QED

3 Some remarks on unique continuation

There is an extensive literature on unique continuation. We refer to the recent work of Jerison-Kenig and Garofalo-Lin, among others, cf. [Je-Ke], [Ke], [Ga-Li], and the references therein. The unique continuation property as defined above differs from the two more usual notions of unique continuation, which we now define for completeness. A family of functions enjoys the *strong unique continuation property*, S.U.C.P. for short, if no function, besides possibly the zero function, has a zero of infinite order. A function $u \in L^2(\Omega)$ has a *zero of infinite order at* $x_0 \in \Omega$ if for each $n \in \mathbf{N}$,

$$\int_{|x-x_0| \leq R} u^2 = O(R^n)$$

as $R \to 0$. A family of functions enjoys the *weak unique continuation property*, for short W.U.C.P., if no function, besides possibly the zero function, vanishes in an open subset of Ω.

We now discuss the relationship between U.C.P. and S.U.C.P. for the family of $H^1_{loc}(\Omega)$ solutions of

$$Lu + V(x)u = 0 \tag{6}$$

where L is the differential operator (with $a_0(x) = 0$) defined above and the potential $V(x) \in L^{N/2}_{loc}(\Omega)$. Here we are assuming $N \geq 3$. As before the coefficients $a_{ij} \in L^\infty(\Omega)$.

Proposition 3 *Suppose that $u \in H^1_{loc}(\Omega)$ is a solution of (6), i.e.*

$$\int_\Omega \sum a_{ij}(x) D_i u D_j v + \int_\Omega V(x) uv = 0, \quad \forall v \in C_0^\infty(\Omega)$$

or in a shorter notation

$$\int_\Omega \langle A(x)\nabla u, \nabla v\rangle + \int_\Omega V(x)uv = 0 \quad \forall v \in C_0^\infty(\Omega). \tag{7}$$

If $u = 0$ on a set E of positive measure, then u has a zero of infinite order.

Lemma 1 *Let B_r and B_{2r} be two concentric balls contained in Ω. Then*

$$\int_{B_r} |\nabla u|^2 \leq \frac{c}{r^2} \int_{B_{2r}} u^2 \tag{8}$$

where the constant c does not depend on r.

PROOF: (Lemma 1) Take $\varphi \in C_0^\infty(\Omega)$, with supp$\varphi \subset B_{2r}$, $\varphi(x) \equiv 1$ for $x \in B_r$ and $|\nabla\varphi| \leq 2/r$. Using $\varphi^2 u$ as a test function in (7) we get

$$\int_\Omega \langle A(x)\nabla u, \nabla u\rangle \varphi^2 = -2\int_\Omega \langle A(x)\nabla u, \nabla\varphi\rangle \varphi u - \int_\Omega V(x)\varphi^2 u^2. \tag{9}$$

Using Schwarz inequality we can estimate the first integral in the right side of (9) by

$$\varepsilon \int_\Omega \langle A(x)\nabla u, \nabla u\rangle \varphi^2 + k_\varepsilon \int_\Omega \langle A(x)\nabla\varphi, \nabla\varphi\rangle u^2.$$

Also, by a result in [Br-Ka] we can estimate the second integral in the right side of (9) by

$$\varepsilon \int_\Omega |\nabla(\varphi u)|^2 + c_\varepsilon \int_\Omega |\varphi u|^2.$$

Using these estimates in (9) and the ellipticity condition we have

$$\int_{B_{2r}} \varphi^2 |\nabla u|^2 \leq \frac{c}{r^2} \int_{B_{2r}} u^2 + 2\varepsilon \int_{B_{2r}} |\nabla\varphi|^2 u^2 + 2\varepsilon \int_{B_{2r}} \varphi^2 |\nabla u|^2 + c_\varepsilon \int_{B_{2r}} |\varphi u|^2$$

which gives immediately (8). QED

PROOF: (Proposition 3) We know that almost every point of E is a point of density of E. Let $x_0 \in E$ be such a point. This means that

$$\lim_{r\to 0} \frac{|E \cap B_r|}{|B_r|} = 1$$

where B_r denotes the ball of radius r centered at x_0 and $|S|$ denotes the Lebesgue measure of a set S. So, given $\varepsilon > 0$ there is an $r_0 = r_0(\varepsilon)$ such that

$$\frac{|E^c \cap B_r|}{|B_r|} < \varepsilon \quad \text{for } r \le r_0$$

where E^c denotes the complement of the set E. Taking r_0 smaller, if necessary, we can assume $B_{r_0} \subset \Omega$. Since $u = 0$ on E we have

$$\int_{B_r} u^2 = \int_{B_r \cap E^c} u^2 \le \left[\int_{B_r \cap E^c} |u|^{\frac{2N}{N-2}}\right]^{\frac{N-2}{N}} |B_r \cap E^c|^{\frac{2}{N}}$$

which can be estimated by

$$c\varepsilon^{\frac{2}{N}} r^2 \int_{B_r} |\nabla u|^2 + c\varepsilon^{\frac{2}{N}} \int_{B_r} u^2,$$

where c is independent of ε and of r as $r \to 0$. By (8) above we then obtain

$$\int_{B_r} u^2 \le c\varepsilon^{\frac{2}{N}} \int_{B_{2r}} u^2 \quad \text{for } r \le r_0. \tag{10}$$

Now let us introduce the following function of r

$$f(r) = \int_{B_r} u^2$$

and let us fix $n \in \mathbf{N}$. Choose $\varepsilon > 0$ such that $c\varepsilon^{\frac{2}{N}} = 2^{-n}$. Observe that consequently r_0 depends on n. Then (10) can be written as

$$f(r) \le 2^{-n} f(2r), \quad \text{for } r \le r_0. \tag{11}$$

Iterating (11) we get

$$f(\rho) \le 2^{-kn} f(2^k \rho), \quad \text{if } 2^{k-1}\rho \le r_0. \tag{12}$$

Now given $0 < r < r_0(n)$ choose $k \in \mathbf{N}$ such that

$$2^{-k} r_0 \le r < 2^{-(k-1)} r_0.$$

From (12) we then obtain

$$f(r) \le 2^{-kn} f(2^k r) \le 2^{-kn} f(2r_0).$$

Since $2^{-k} \leq r/r_0$ we finally obtain

$$f(r) \leq \left(\frac{r}{r_0}\right)^n f(2r_0)$$

which shows that x_0 is a zero of infinite order. QED

Remark 2 S.U.C.P. holds in (6) if $Lu = -\Delta u$ and $V \in L_{loc}^{N/2}$, cf. [Je-Ke], or more generally if $a_{ij} \in C^\infty$ and $V \in L_{loc}^{N/2}$, cf. [So]. Also S.U.C.P. holds in (6) if a_{ij} are Lipschitz continuous and V belongs to a certain class of potentials, which is a subset of $L_{loc}^{N/2}$ and which contains L_{loc}^∞, cf. [Ga-Li]. In general Hölder continuity of the coefficients a_{ij} does not suffice for WUCP in (6), cf. [Mi]. As far as (1) is concerned, we conclude in particular the strict monotonicity of eigenvalues if a_{ij} are Lipschitz continuous and a_0 and m are in $L^\infty(\Omega)$.

References

[Br-Ka] H. Brézis, T. Kato, Remarks on the Schrödinger operator with singular complex potentials,*J. Math. Pures et Appl.* **58**, 137-151 (1979).

[Co-DF-Gs] D. G. Costa, D. G. deFigueiredo, J. V. Gonçalves, On the uniqueness of solution for a class of semilinear elliptic problems,*J. Math. Anal. Appl.* **123**, 170-180 (1987).

[DF] D. G. deFigueiredo, Positive solutions of semilinear elliptic equations, *Lecture Notes in Mathematics* **957**, 34-87 (1982).

[DF-Go] D. G. deFigueiredo, J.-P. Gossez, Conditions de non-résonance pour certains problèmes elliptiques semi-lineaires, *C.R.Acad.Sci. Paris* **302**, 543-545 (1986).

[DF-Mi] D. G. deFigueiredo, O. H. Miyagaki, Semilinear elliptic equations with the primitive of the nonlinearity away from the spectrum , to appear in Nonlinear Analysis, TMA.

[Ga-Li] N. Garofalo,F.-H. Lin, Unique continuation for elliptic operators: a geometric-variational approach, *Comm. Pure Appl. Math.* **XL**, 347-366 (1987).

[Go-Om] J.-P. Gossez, P. Omari, A necessary and sufficient condition of nonresonance for a semilinear Neumann problem,, to appear in Proc.Amer.Math.Soc.

[Je-Ke] D. Jerison, C. .E. Kenig, Unique continuation and absence of positive eigenvalues for Schrödinger operators, *Ann. Math.* **121**, 463-494 (1985).

[Ke] C. E. Kenig, Restriction theorems, Carleman estimates uniform Sobolev inequalities and unique continuation , Lectures Notes in Mathematics **1384**, 69-90 (1989).

[Ma-Mi] A. Manes, A. M. Micheletti, Un'estensione della teoria variazionale classica degli autovalori per operatori ellittici del secondo ordine, *Boll. UMI* **7**, 285-301 (1973).

[Mi] K. Miller, Nonunique continuation for uniformly parabolic and elliptic equations in self-adjoint divergence forms with Hölder continuous coefficients, *Arch. Rat. Mech. Anal.* ,**54**, 105-117 (1974).

[So] C. Sogge, Strong uniqueness theorems for second order elliptic differential equations,*Am. J. Math.*, **112**, 943-984 (1990).

Received June 1991

COMMUN. IN PARTIAL DIFFERENTIAL EQUATIONS, 17(5&6), 923-940 (1992)

POSITIVE SOLUTIONS OF SEMILINEAR ELLIPTIC SYSTEMS

Ph. Clément

Department of Technical Mathematics and Informatics
University of Technology, Delft
The Netherlands

D.G. de Figueiredo

Departemento de Matematica
Universidade de Campinas
Unicamp-Campinas, Brazil

E. Mitidieri

Dipartimento di Matematica e Informatica
Università degli Studi di Udine
Udine, Italy

ABSTRACT

We investigate the existence of positive solutions of a Dirichlet problem for the system

$$-\Delta u = f(v), -\Delta v = g(u)$$

in a bounded convex domain Ω of $\mathbb{R}^N$ with smooth boundary. In particular L^∞ a priori bounds are obtained in the same spirit as in De Figueiredo - Lions - Nussbaum [7].

Originally published in: Comm. Partial Differential Equations 17 (1992), No. 5-6, 923-940, © 1992 Marcel Dekker, Inc.
Reprint by Springer-Verlag Berlin Heidelberg 2013 with kind permission,
DOI: 10.1007/978-3-319-02856-9_25

§1. INTRODUCTION.

In this paper we investigate the existence of positive solutions for the nonlinear elliptic system:

$$(S) \qquad \begin{cases} -\Delta u &= f(v) \quad \text{in } \Omega, \ u = 0 \text{ on } \partial\Omega \\ -\Delta v &= g(u) \quad \text{in } \Omega, \ v = 0 \text{ on } \partial\Omega, \end{cases}$$

where Ω is a bounded domain of $\mathbb{R}^N$, $N \geq 2$, with boundary $\partial\Omega$ sufficiently smooth. Such problems have been intensively studied in the literature, see P.L. Lions [11] and the references therein for a good review of results and problems. Recently Cosner [2] considered the problem of existence of positive solutions for semilinear elliptic systems which are not necessarily variational. His main results, which are based on the cooperativity (quasimonotonicity) of the nonlinearities, are extensions to systems of the fundamental paper of Brézis-Turner [1] concerning a single equation. In view of recent advances in this area (see De Figueiredo, Lions, Nussbaum [7], Gidas-Spruck [10]), it is natural to ask whether it is possible to obtain the corresponding results for systems. In this paper we show that this program can be developed for problems like (S).
In particular we are interested in the case where

$$f(0) = g(0) = 0, \text{ and where } f, g \text{ are } C^1 \text{ nondecreasing functions.}$$

The linearization of (S) leads to a cooperative system

$$(LS) \qquad \begin{cases} -\Delta\tilde{u} = f'(v)\tilde{v} & \text{in } \Omega, \ \tilde{u} = 0 \text{ on } \partial\Omega \\ -\Delta\tilde{v} = g'(u)\tilde{u} & \text{in } \Omega, \ \tilde{v} = 0 \text{ on } \partial\Omega. \end{cases}$$

Then it is known that the trivial solution $(0,0)$ is exponentially stable for the corresponding system if and only if $f'(0)g'(0) < \lambda_1^2$ where λ_1 denotes the lowest eigenvalue of $(-\Delta, H_0^1(\Omega))$. As well-known λ_1 is positive and we shall denote by φ_1 the corresponding positive eigenfunction normalized by $\max\varphi_1 = 1$. The existence of nontrivial solutions of (S) is established in Theorem 3.1 by using a degree argument together with a priori estimates. The a priori bounds are obtained in Section 2, Theorem 2.1. The method of proof is closely related to the method used in [7]. However it necessitates a version for systems of "moving hyper-planes argument" due to [9] and of an extension of Rellich's identity due to [12]. In Sections 4 and 5 we consider the special case where $f(v) = v$, i.e. the case of the biharmonic equation with Navier boundary conditions. In Section 4, we prove the existence of solutions by using a variational argument.

Finally in Section 5, we rewrite $g(u)$ as $h(u) + \lambda u$ where $h(0) = h'(0) = 0$, and consider λ as a parameter. We construct branches of positive solutions bifurcating from the line of trivial solutions $(\lambda, 0)$, $\lambda \in \mathbb{R}$. Then we obtain positive solutions even for negative (small) λ, s.

§2. A PRIORI ESTIMATES.

In this section we prove a priori bounds for the positive solutions of the system (S). We shall follow an approach similar to the one used in [7] for the case of a single equation. In this way we need a version fo the Gidas-Ni-Nirenberg results [9], which has been obtained by Troy [16]. For that matter we have to assume that f and g are nondecreasing C^1 functions. Moreover for later use in Section 3 and 5, we need to consider the question of a priori bounds of positive solutions of (S) where the functions f and g depend on a parameter $t \in I$, where I is an index set. Let us state precisely the conditions needed for the estimates.

$(f1)$ $f_t, g_t : \mathbb{R}^+ \to \mathbb{R}$ are C^1 and $f_t'(s) \geq 0$, $g_t'(s) \geq 0$ for all $s \geq 0$, $t \in I$.

$(f2)$ There are numbers a', $b' \in]0, \infty]$ such that $a'b' > \lambda_1^2$ where λ_1 is the first eigenvalue of $(-\Delta, H_0^1(\Omega))$, and there are numbers a_t, $b_t \in]0, \infty[$ such that $\liminf_{s \to +\infty} \frac{f_t(s)}{s} = a_t$, $\liminf_{s \to +\infty} \frac{g_t(s)}{s} = b_t$, and $a_t b_t \geq a'b'$ for all $t \in I$.

$(f3)$ The functions f_t, g_t are uniformly bounded with respect to t on bounded intervals, that is for every $s_0 > 0$, we have

$$\sup_{t \in I} \sup_{0 \leq s \leq s_0} f_t(s) < \infty \quad \text{and} \quad \sup_{t \in I} \sup_{0 \leq s \leq s_0} g_t(s) < \infty.$$

Moreover there are numbers $\alpha, \beta \in]0, \infty[$ such that

$$\lim_{s \to 0} \frac{f_t(s)}{s^p} = \alpha \quad \text{and} \quad \lim_{s \to \infty} \frac{g_t(s)}{s^q} = \beta,$$

uniformly in $t \in I$ where $1 \leq p, q < \infty$ satisfy

$$\frac{1}{p+1} + \frac{1}{q+1} > \frac{N-2}{N} \text{ when } N \geq 3.$$

Theorem 2.1. Suppose that the family of functions (f_t), (g_t) satisfies $(f1)$, $(f2)$, $(f3)$ and that Ω is convex and $\partial\Omega \in C^3$. Then there is a constant $C > 0$, independent of t such that

$$\|u\|_{L^\infty} \leq C \text{ and } \|v\|_{L^\infty} \leq C \text{ for all positive solutions } (u, v) \text{ of } (S).$$

Remark. The proof is divided in several steps. All constants introduced in the proof will be denoted by the same letter c. We emphasize that they do not depend on t, nor on the particular positive solutions (u,v). We also remark that our estimates are also valid for non convex domains provided some additional technical assumptions are made. This is proved as much as in the same lines of [7]. In order to avoid cumbersome notations, we shall omit the index t for the functions f_t, g_t.

Proof. Step 1. We start by proving that the projections of the solutions (u,v) of the system (S) over φ_1, the positive eigenfunction of $(-\Delta, H_0^1(\Omega))$ normalized by $\max_{x\in\Omega}\varphi_1(x) = 1$, are bounded. Indeed, let a' and b' be as in $(f2)$, that is

$$0 < a' < a_t, \quad 0 < b' < b_t \quad \text{and } a'b' > \lambda_1^2.$$

Then, it follows from $(f2)$ and $(f3)$ that there exists a constant c depending only on a', b' such that

$$f(s) \geq a's - c \quad g(s) \geq b's - c, \quad \text{for all } s \geq 0. \tag{2.1}$$

Multiplying the equations of (S) by φ_1, integrating by parts and using the inequalities in (2.1) we obtain

$$\lambda_1 \int u\varphi_1 = \int f(v)\varphi_1 \geq a' \int v\varphi_1 - c\int \varphi_1$$
$$\lambda_1 \int v\varphi_1 = \int g(u)\varphi_1 \geq b' \int u\varphi_1 - c\int \varphi_1$$

These expressions lead to

$$\int u\varphi_1 \leq \frac{1}{a'b' - \lambda_1^2}(\lambda_1 + a')c\int \varphi_1$$

and a similar inequality for $\int v\varphi_1$. This shows that a constant $c > 0$ can be chosen such that

$$\int u\varphi_1 \leq c \text{ and } \int v\varphi_1 \leq c$$

for all (u,v) solutions of (S). It also follows from the above inequalities that there is a constant $c > 0$ such that

$$\int |f(v)|\varphi_1 < c \quad \text{and} \quad \int |g(u)|\varphi_1 < c$$

which imply that $f(v)$ and $g(u)$ are bounded in $L^1_{loc}(\Omega)$.

Step 2. Next we use the fact that f and g are nondecreasing in order to apply the results of [9] as extended to systems by [16]. This step is done in a similar way to the work developed in [7]. We then conclude that there is an $\varepsilon > 0$ such that

$$u(x), v(x) \leq c \ , \quad \text{for every } x \in \{u \in \Omega : d(y, \partial\Omega) < \varepsilon\}.$$

Also as in [7] we have by elliptic regularity that

$$|\nabla u(x)|, |\nabla v(x)| \leq c, \text{ for every } x \in \{u \in \Omega : d(y, \partial\Omega) < \varepsilon/2\}.$$

For the next step we need a version of Rellich identity, which is a special case of an identity proved in [12].

Lemma. Let u and v be in $C^2(\bar{\Omega})$, where Ω is a C^1 domain in $\mathbb{R}^N$, satisfying $u = v = 0$ on $\partial\Omega$. Then we have,

$$\int_\Omega [\Delta u \cdot (x, \nabla v) + \Delta v \cdot (x, \nabla u)] = (N-2) \int_\Omega \nabla u \cdot \nabla v + \oint_{\partial\Omega} (x, n) \frac{\partial u}{\partial n} \frac{\partial v}{\partial n} \tag{2.2}$$

where n denotes the exterior normal.

Step 3. We prove now that there is a constant $c > 0$ such that

$$\int v^{p+1} < c, \quad \int u^{q+1} < c.$$

Indeed, applying (2.2) for solutions u and v of (S) and observing that

$$\int f(v) \cdot (x, \nabla v) = -N \int F(v)$$

$$\int g(u) \cdot (x, \nabla u) = -N \int G(u)$$

where $F(v) = \int_0^v f(s)ds$ and $G(u) = \int_0^u g(s)ds$, we obtain using the boundary estimates of Step 2:

$$|N \int [F(v) + G(u)] - (N-2) \int \nabla u \cdot \nabla v| \leq c. \tag{2.3}$$

Next observe that from the equations (S) we have $\int \nabla u \cdot \nabla v = \int v f(v) = \int u \, g(u)$. To proceed with the estimates we now split the second term in (2.3) in order to balance properly the expression involving f and its primitive F with the one involving g and G. For that matter, choose $\theta \in]0,1[$ such that

$$\frac{1}{p+1} > \theta \frac{N-2}{N} \quad \text{and} \quad \frac{1}{q+1} > (1-\theta) \frac{N-2}{N}$$

which is possible by $(f3)$. So (2.3) can be written as

$$(2.4) \quad |\int F(v) - \theta\frac{N-2}{N}\int vf(v) + \int G(u) - (1-\theta)\frac{N-2}{N}\int ug(u)| \le c.$$

It follows from $(f3)$ that given $\varepsilon > 0$ there is a constant $c > 0$ such that, for all $s \ge 0$,

$$(\alpha - \varepsilon)s^p - c \le f(s) \le (\alpha + \varepsilon)s^p + c$$

and then

$$\frac{\alpha - \varepsilon}{p+1}s^{p+1} - cs \le F(s) \le \frac{\alpha + \varepsilon}{p+1}s^{p+1} + cs.$$

Using these inequalities we can estimate from below the two first terms in (2.4) as follows

$$\frac{\alpha-\varepsilon}{p+1}\int v^{p+1} - \theta\frac{N-2}{N}(\alpha+\varepsilon)\int v^{p+1} - c\int v =$$

$$= \alpha\{\frac{1}{p+1} - \theta\frac{N-2}{N}\}\int v^{p+1} - \varepsilon\{\frac{1}{p+1} + \theta\frac{N-2}{N}\}\int v^{p+1} - c\int v.$$

A similar expression is obtained for u, by estimating the third and fourth terms of (2.4). Thus in view of the choise of θ it follows that for $\varepsilon > 0$ sufficiently small,

$$\int v^{p+1} + \int u^{q+1} < c\,(1 + \int v + \int u),$$

from which the estimates announced at the beginning of Step 3 follow.

Step 4. We prove that there is a constant $c > 0$ such that $\|v\|_{L^\infty} \le c$, $\|u\|_{L^\infty} \le c$. This follows from a bootstrap argument using the Lemma below. So the proof of Theorem 2.1 is complete by application of the Lemma 2.2.

Lemma 2.2. Let $\bar{p} \ge p$, and suppose that there is a constant $c > 0$ such that

$$\|v\|_{L^{\bar{p}+1}} + \le c$$

for all solutions (u, v) of (S). Then these solutions are also bounded in $L^{\hat{p}+1}$ where

$$(2.5) \quad \hat{p} \ge \bar{p} + (p+1)^2(q+1)\eta \quad \text{with} \quad \eta = \frac{1}{p+1} + \frac{1}{q+1} - \frac{N-2}{N} (> 0).$$

Proof. By elliptic regularity and $(f3)$ solutions (u, v) (S) satisfy u is bounded in $W^{2,r}$ with $r = \frac{\bar{p}+1}{p}$. If $\frac{p}{\bar{p}+1} - \frac{2}{N} \le 0$, by Sobolev imbedding u is bounded in L^s, for every $s > 0$ and from the second equation in (S), it follows that v is also bounded in L^s, for every $s > 0$, hence in any $L^{\hat{p}+1}$ where $\hat{p}$ satisfies (2.5). If $\frac{p}{\bar{p}+1} - \frac{2}{N} > 0$, by Sobolev imbedding u is bounded in $L^{\bar{q}+1}$ where

(2.6) $$\frac{1}{\bar{q}+1} = \frac{p}{\bar{p}+1} - \frac{2}{N}.$$

Note that

$$\frac{p}{\bar{p}+1} - \frac{2}{N} \leq \frac{p}{p+1} - \frac{2}{N} = 1 - \frac{2}{N} - \frac{1}{p+1} < \frac{1}{q+1}$$

by $(f3)$. Hence $\frac{\bar{q}+1}{q} > 1$.
Since $g(u)$ is bounded in $L^{\frac{\bar{q}+1}{q}}$ we get v bounded in $W^{2,s}$ with $s = \frac{\bar{q}+1}{q}$. So by Sobolev imbedding v is bounded in $L^{\hat{p}+1}$ with

$$\frac{1}{\hat{p}+1} = \frac{q}{\bar{q}+1} - \frac{2}{N}.$$

Hence $\frac{1}{\hat{p}+1} = \frac{pq}{\bar{p}+1} - \frac{2q}{N} - \frac{2}{N}$ which implies

$$\frac{1}{\bar{p}+1} - \frac{1}{\hat{p}+1} = \frac{1-pq}{\bar{p}+1} + \frac{2}{N}(q+1) \geq \frac{1-pq}{p+1} + \frac{2}{N}(q+1) = (q+1)\eta.$$

So $\hat{p} = \bar{p} \geq (\bar{p}+1)(\hat{p}+1)(q+1)\eta$ which gives the result. ◻

§3. EXISTENCE OF POSITIVE SOLUTIONS FOR SYSTEM (S).

In this section we use degree theory to establish the existence of a positive solution (u,v) for system (S) under a set of conditions which, in particular, insure the a priori estimates of the previous section.

Theorem 3.1. Suppose that $f,g : \mathbb{R}^+ \to \mathbb{R}^+$ satisfy conditions $(f1)$, $(f2)$, $(f3)$, $f(0) = g(0) = 0$ and $f'(0)g'(0) < \lambda_1^2$. Suppose Ω convex and $\partial\Omega \in C^3$. Then system (S) has at least one positive solution $(u,v) \in C^{2,\alpha}(\bar{\Omega})$, for every $\alpha \in (0,1)$.

Remark. An example where all assumptions on the functions f and g are satisfied is $f(v) = v^p$ and $g(u) = u^q$ with $\frac{1}{p+1} + \frac{1}{q+1} > \frac{N-2}{N}$, $N \geq 3$. Note that a previous result of Cosner ([2], Theorem 2, p. 1435) requires the stronger assumption

$$1 < \max(p,q) < \frac{N}{N-2}.$$

Moreover it is proved in ([12], Proposition 3.1) that if $\frac{1}{p+1} + \frac{1}{q+1} \leq \frac{N-2}{N}$, $N \geq 3$, then system (S) has no positive solution.

Proof. Let $r_0 > 0$ be such that

$$f'(s)g'(t) < \lambda_1^2$$

for all $0 \le s,t \le r_0$. Let $\gamma = \max\{f'(s) : 0 \le s \le r_0\}$ and $\eta = \max\{g'(t) : 0 \le t \le r_0\}$. Clearly $\gamma\eta < \lambda_1^2$. It also follows that, for any $0 < r \le r_0$,

$$f(s) \le \gamma s \quad \text{and} \quad g(s) \le \eta s, \text{ for all } 0 \le s \le r. \tag{3.1}$$

Let X denote the Banach space $C(\bar{\Omega})$ endowed with the supnorm. The inverse of $-\Delta$ subject to Dirichlet boundary conditions defines a compact linear map $B : X \to X$. By the maximum principle $B(K) \subset K$, where K denotes the cone of functions in X which are nonnegative in $\bar{\Omega}$. So establishing the existence of a positive solution (u,v) for system (S) is equivalent to obtaining a nonzero solution of

$$u = \phi(u) := Bf(Bg(u))$$

in the cone K, since we then take $v = Bg(u)$. Elliptic regularity theory implies then that u and v are $C^{2,\alpha}(\bar{\Omega})$ for every $\alpha \in (0,1)$. The result follows by applying the theory of fixed point index for compact mapping of a cone, see [13]. More specifically we apply Proposition 2.1 of [7] which fits properly in the present situation. For that matter we next prove that numbers $0 < r < R$ and $t_0 > 0$ can be chosen such that

A) There is a bounded linear operator $A : X \to X$, such that $A(K) \subset K$, with spectral radius strictly less than one, and such that $\phi(u) \le Au$ for all $u \in K$, with $\|u\|_X = r$.

B) There is a compact map $\Psi : \bar{B}_R \times [0,\infty[\to K$, where $B_R = \{u \in K; \|u\|_X < R\}$, such that

 (i) $\Psi(u,0) = \phi(u)$ for all $u \in K$.
 (ii) $\Psi(u,t) \ne u$ for $0 \le t < \infty$, $u \in K$ with $\|u\|_X = R$.
 (iii) $\Psi(u,t) \ne u$ for $t \ge t_0$, and all $u \in \bar{B}_R$.

Proof of (A). Let $A = \gamma\eta B^2$, where γ and η have been defined above. All we have to prove is that we can choose $r > 0$ such that $\phi(u) \le Au$ for $u \in K$, $\|u\|_X = r$, since the other conditions are clearly satisfied. For any $r \le r_0$, it follows from (3.1) that $Bg(u) \le \eta Bu$ for $u \in K$, $\|u\|_X = r$. Let $c > 0$ be such that $\|Bu\|_X \le c\|u\|_X$. Then

$$\|Bg(u)\|_X \le c\eta r \qquad \text{for } u \in K, \|u\|_X = r.$$

Choosing now $0 < r < r_0$, such that $c\eta r \le r_0$, we see from (3.1) that

$$f(Bg(u)) \le \gamma Bg(u) \le \gamma\eta Bu$$

which proves $\phi(u) \le Au$, for $u \in K$, $\|u\|_X = r$.

Proof of (B). Define

$$\Psi(u,t) = Bf(Bg(u+t))$$

and we show that $R > 0$ can be chosen such that conditions (ii) and (iii) are fullfilled. Indeed, $\Psi(u,t) = u$ means

$$-\Delta u = f(v), \; -\Delta v = g(u+t) \tag{3.2}$$

with u, v satisfying Dirichlet boundary conditions. Using the argument in Step 1 of the proof of the Theorem 2.1 we conclude that the existence of positive solution (u,v) for (3.2) implies that t has to be bounded. So there exsits $t_0 > 0$ such that $\Psi(u,t) \neq u$ for $u \in K$ and $t \geq t_0$, proving (iii). Finally we observe that the a priori estimates of the previous section hold uniformly for $t \in [0, t_0]$. So there exists $R > 0$ such that (ii) above holds.

§4. THE NAVIER PROBLEM FOR THE BIHARMONIC EQUATION. A VARIATIONAL APPROACH.

In this section we consider the question of existence of nontrivial solutions for the biharmonic equation subject to Navier boundary conditions. Namely

$$\Delta^2 u = g(u) \quad \text{in } \Omega \,, \quad u = \Delta u = 0 \text{ on } \partial\Omega. \tag{4.1}$$

This problem is a special case of (S) when $f(v) = v$.

We now state our assumptions on g.

$(g1)$ $g : \mathbb{R} \to \mathbb{R}$ is locally lipschitzian.

$(g2)$ $|g(s)| \leq c|s|^\ell + c$ for all $s \in \mathbb{R}$, where

$$1 \leq \ell < \frac{N+4}{N-4} \text{ if } N \geq 5,$$

$$1 \leq \ell < \infty \text{ if } N \geq 2, 3, 4.$$

$(g3)$ $g(0) = 0$ and $\limsup_{s\to 0} \frac{g(s)}{s} < \lambda_1^2$.

$(g4)$ There are $s_0 > 0$ and $0 \leq \theta < \frac{1}{2}$ such that $0 < G(s) \leq \theta s g(s)$, for all $|s| \geq s_0$ where $G(s) = \int_0^s g(t)dt$.

By using a variational approach, we obtain the following result.

Theorem 4.1. Assume $(g1)$, $(g2)$, $(g3)$ and $(g4)$. Let Ω be a domain with C^2 boundary. Let σ be any number larger than one if $N = 2,3,4$ and $\sigma = \frac{2N}{N+4}$ if

$N \geq 5$. Then there exists a function $u \in W^{2,2}(\Omega) \cap W_0^{1,2}(\Omega)$, not identically zero, such that $\Delta u \in W^{2,\sigma}(\Omega) \cap W_0^{1,\sigma}(\Omega)$ and $g(u) \in L^\sigma(\Omega)$ satisfying (4.1).

Remark. It is known (see [3], [4], [8], [15]) that the solution to the linear problem

$$(P) \qquad \begin{aligned} \Delta^2 u &= \lambda u + f \quad \text{in } \Omega \\ u = \Delta u &= 0 \quad \text{on } \partial\Omega, \end{aligned}$$

where f is a given positive function, does not need to be positive when $\lambda < 0$. However it has been shown in [3], [4], [8], [15] that if Ω is sufficiently smooth, there exists $\lambda^* < 0$ such that the solution u to problem (P) is positive for $\lambda \in (\lambda^*, 0)$. Notice that in that case (P) is equivalent to the noncooperative system

$$\begin{aligned} -\Delta u &= v \quad \text{in } \Omega \quad u = 0 \quad \text{on} \quad \partial\Omega \\ -\Delta v &= \lambda u + f \quad \text{in } \Omega \quad v = 0 \quad \text{on} \quad \partial\Omega \end{aligned}$$

This observations will be used in Section 5, Theorem 5.1.

Proof. We consider the Hilbert space $H = W^{2,2}(\Omega) \cap W_0^{1,2}(\Omega)$ with inner product $\langle u, v\rangle_H = \int \Delta u \Delta v$, and the functional $I(u) = \frac{1}{2}\langle u, u\rangle_H - \int G(u)$, $u \in H$. Conditions $(g1)$ and $(g2)$ imply that I is well-defined in H and it is of class C^1, with

$$\langle I'(u), z\rangle_H = \int \Delta u \Delta z - g(u)z, \quad \text{for all } z \in H.$$

First we show that a critical point of I is a solution of (4.1) in the sense of Theorem 4.1. We only treat the case $N \geq 5$, the other case being simpler. Let σ' be the conjugate exponent to σ, i.e. $\sigma' = \frac{2N}{N-4}$. Since $\sigma' > 2$, if follows that $H \supset X := W^{2,\sigma'}(\Omega) \cap W_0^{1,\sigma'}(\Omega)$. So if u is a critical point of I, we have

$$(4.2) \qquad \int \Delta u \Delta z = \int g(u) z \text{ for all } z \in X.$$

Since $g(u) \in L^\sigma$, it follows that there exists a unique $\omega \in W^{2,\sigma}(\Omega) \cap W_0^{1,\sigma'}(\Omega)$ such that that $\Delta\omega = g(u)$. So we get from (4.2)

$$(4.3) \qquad \int \Delta u \Delta z = \int \Delta \omega z = \int \omega \Delta z \text{ for all } z \in X.$$

Now given $h \in L^{\sigma'}(\Omega)$ there exists a unique $z \in X$ such that $\Delta z = h$. Then (4.3) implies $\int (\Delta u - \omega) h = 0$ for all $h \in L^{\sigma'}(\Omega)$ which gives $\Delta u = \omega$.
So it is sufficient to prove the existence of a nontrivial critical point of I. We shall prove it by using the mountain pass Theorem which requires [6]

(i) There is $r > 0$ and $a > 0$ such that $I(u) \geq a$ for $\|u\|_H = r$.

(ii) There is $u_0 \in H$, $\|u_0\|_H > r$ such that $I(u_0) < 0$.

(iii) I satisfies (P.S.).

Proof of (i). It follows from $(g3)$ that there is $\mu < \lambda_1^2$ and $s_0 > 0$ such that $g(s)/s \leq \mu$ for $|s| \leq s_0$. So $(sgns)g(s) \leq \mu|s|$, for $|s| \leq s_0$. And we can choose $1 < p < \frac{N+4}{N-4}$ and a constant $c > 0$ such that $(sgns)g(s) \leq c|s|^p$. So

$$(sgns)g(s) \leq \mu|s| + c|s|^p \text{ for all } s \in \mathbb{R}.$$

This implies $G(s) \leq \frac{\mu}{2}s^2 + \frac{c}{p+1}|s|^{p+1}$. Then $G(u) \leq \frac{\mu}{2}\int u^2 + \frac{c}{p+1}\int |u|^{p+1} \leq \frac{\mu}{2}\frac{1}{\lambda_1^2}\|u\|_H^2 + \frac{c}{p+1}\|u\|_H^{p+1}$ and for $\|u\|_H = r$,

$$I(u) \geq \frac{1}{2}r^2 - \frac{\mu}{2}\frac{1}{\lambda_1^2}r^2 - \frac{c}{p+1}r^{p+1}. \tag{4.4}$$

We see that there is a value $r > 0$ such that the right hand side of (4.4) is positive.

Proof of (ii). Let $u = t\varphi_1$ where $t \in \mathbb{R}^+$. Then $I(t\varphi_1) = \frac{1}{2}\lambda_1 t^2 \int \varphi_1^2 - \int G(t\varphi_1)$. In view of $(g4)$, G is superquadratic and so $I(t\varphi_1) \to -\infty$ as $t \to \infty$.

Proof of (iii). Let $(u_n) \in H$ be such that

$$|I(u_n)| \leq c, |\langle I'(u_n), v\rangle| \leq \varepsilon_n \|v\|_H \text{ for all } v \in H \text{ with } \varepsilon_n \to 0. \tag{4.5}$$

Using the first inequality of (4.5) and the second one with $v = u_n$, we obtain from $(g4)$ that

$$(\frac{1}{2} - \theta)\|u_n\|_H^2 \leq c.$$

So there is a subsequence still denoted by (u_n) converging weakly to some $u_0 \in H$ and strongly in L^p for all $p < \frac{2N}{N-4}$. So $g(u_n) \to g(u_0)$ in L^q, for some $q > \frac{2N}{N+4}$, by using $(g2)$. Now using the second inequality in (4.5) with $v = u_n - u_0$ we obtain

$$|\int \Delta u_n \Delta(u_n - u_0) - \int g(u_n)(u_n - u_0)| \leq \varepsilon_n \|u_n - u_0\|_H$$

which implies

$$\int \Delta u_n \Delta(u_n - u_0) \to 0.$$

It follows that u_n converges strongly to u_0 in H, and I satisfies $(P.S.)$.

§5. BIFURCATION OF POSITIVE SOLUTIONS.

In this section we consider a nonlinear eigenvalue problem associated with the biharmonic equation subject to Navier boundary conditions.

$$(5.1) \qquad \Delta^2 u = \lambda u + h(u) \quad \text{in } \Omega, \quad u = \Delta u = 0 \quad \text{on } \partial\Omega.$$

Here Ω is a bounded domain of $\mathbb{R}^N$ with boundary C^5 and convex. We assume $h(0) = 0$, therefore $(\lambda, 0)$ is a solution for every $\lambda \in \mathbb{R}$. These solutions are called the trivial solutions. We are looking for a branch of positive solutions bifurcating from the line of trivial solutions and containing a solution of the form $(0, u)$, i.e. a solution of (4.1). We shall even obtain solutions of the form (λ, u) for some negative λ with positive u. Our assumptions are the following:

$(h1)$ $h \in C^2$, $h(0) = h'(0) = 0$.

$(h2)$ $h(s) > 0$, $h'(s) \geq 0$ for $s > 0$.

$(h3)$ there exist $\beta \in (0, \infty)$, $q \in [1, \infty)$ such that $\lim_{s \to \infty} \frac{h(s)}{s^q} = \beta$, with

$$q \in [1, \frac{N+4}{N-4}) \text{ when } N \geq 5.$$

$(h4)$ $\beta > \lambda_1^2$ if $q = 1$.

Theorem 5.1. Suppose $(h1)$, $(h2)$, $(h3)$ and $(h4)$ are satisfied and Ω is as above. Then there exists a set C^+ of nontrivial solutions of (5.1) connected in $\mathbb{R} \times C^4(\bar{\Omega})$ such that

(i) For every $(\lambda, u) \in C^+$, $u(x) > 0$ in Ω and $\frac{\partial u}{\partial n}(y) < 0$ on $\partial\Omega$.

(ii) There exists a sequence $(\lambda_n, u_n) \in C^+$ such that $\lim_{n \to \infty} (\lambda_n, u_n) = (\lambda_1^2, 0)$ in $\mathbb{R} \times C^4(\bar{\Omega})$.

(iii) There exists $\bar{\lambda} < 0$ such that

$$Proj_{\mathbb{R}} C^+ \supset [\bar{\lambda}, \lambda_1^2).$$

The proof will be done in several steps. We first prove a local bifurcation result.

Proposition 5.2. Assume $(h1)$. Then there is a neighbourhood U of $(\lambda_1^2, 0)$ in $\mathbb{R} \times C^1(\bar{\Omega})$, a C^1 map $s \longmapsto (\lambda(s), u(s)) \in \mathbb{R} \times C^1(\bar{\Omega})$ defined on $(-\varepsilon, \varepsilon)$ for some $\varepsilon > 0$, such that:

(i) $\lambda(0) = \lambda_1^2$, $u(0) = 0$.

(ii) For $s \in (-\varepsilon, \varepsilon)$, $u(s) \in C^{4,\delta}(\bar{\Omega})$ for every $\delta \in (0,1)$ and $(\lambda(s), (u(s))$ are solutions of (5.1).

(iii) The only solutions of (5.1) in U are either of the form $(\lambda(s), u(s))$ or the form $(\lambda, 0)$.

(iv) There exists a unique C^1 map $s \longmapsto v(s) \in C^1(\bar{\Omega})$ with $\int_\Omega v(s)\varphi_1(s) = 0$, $s \in (-\varepsilon, \varepsilon)$ such that

$$u(s) = s(\varphi_1 + v(s)), \quad s \in (-\varepsilon, \varepsilon),$$

and $v(0) = 0$.

(v) There exists $\varepsilon^+ \in (0, \varepsilon)$ such that

$$u(s) > 0 \text{ in } \Omega \text{ and } \frac{\partial u}{\partial n}(s) < 0 \text{ on } \partial\Omega \text{ for } s \in (0, \varepsilon^+).$$

Proof. We rewrite problem (5.1) as
(5.2) $u = G(\lambda, u)$ with $G : \mathbb{R} \times C^1(\bar{\Omega}) \to C^1(\bar{\Omega})$ defined by
(5.3) $G(\lambda, u) = K(\lambda u + h(u))$ where K denotes the solution operator of $\Delta^2 u = f$ in Ω, $u = \Delta u = 0$ on $\partial\Omega$ considered as an operator from $C(\bar{\Omega})$ into $C^1(\bar{\Omega})$. By using Schauder estimates for the Laplacian and assumption $(h1)$, one verifies that a solution (λ, u) of (5.2) in $\mathbb{R} \times C^1(\bar{\Omega})$ is a solution of (5.1) in $\mathbb{R} \times C^{4,\delta}(\bar{\Omega})$ for every $\delta \in (0,1)$ and conversely. By using $(h1)$ again, one verifies that

$G \in C^2(\mathbb{R} \times C^1(\bar{\Omega}), C^1(\bar{\Omega}))$ and that $(\lambda, 0)$ is a solution of (5.2) for all $\lambda \in \mathbb{R}$.

(Note that the substitution operator $h(u)$ belongs to $C^2(C^1(\bar{\Omega}), C(\bar{\Omega}))$. Next we apply Crandall-Rabinowitz Theorem on bifurcation from simple eigenvalues [5]. For convenience we define $F : \mathbb{R} \times C^1(\bar{\Omega}) \to C^1(\bar{\Omega})$ by $F(\lambda, u) = u - G(\lambda, u)$.
Then $F \in C^2(\mathbb{R} \times C^1(\bar{\Omega}), C^1(\bar{\Omega}))$, $F(\lambda, 0) = 0$ for all $\lambda \in \mathbb{R}$ and satisfies

$$\begin{aligned} N(F_u(\lambda_1^2, 0)) &= span\{\varphi_1\} \\ R(F_u(\lambda_1^2, 0)) &= \{v \in C^1(\bar{\Omega}); \textstyle\int_\Omega v\varphi_1 = 0\} \\ F_{\lambda u}(\lambda_1^2, 0)\varphi_1 &= -\lambda_1^2\varphi_1 \notin R(F_u(\lambda_1^2, 0)). \end{aligned}$$

It follows from [5] that (i) - (iv) hold.

Since $\lim_{s\downarrow 0}\|v(s)\|_{C^1(\bar\Omega)} = 0$ and $\varphi_1 \in C^1(\bar\Omega)$ satisfies $\varphi_1(x) > 0$ in Ω and $\frac{\partial\varphi}{\partial n}(x) < 0$ on $\partial\Omega$ there exists $\varepsilon^+ \in (0,\varepsilon]$ such that

$$\varphi_1 + v(s) > 0 \text{ in } \Omega \text{ and } \frac{\partial}{\partial n}(\varphi_1 + v(s)) < 0 \text{ on } \partial\Omega \text{ for } s \in (0,\varepsilon^+).$$

This completes the proof of the proposition.

Next we consider global bifurcation.

Lemma 5.3. Under the assumption of Proposition 5.3, there exists a subset $\mathcal{C}$ of nontrivial solutions of (5.1) connected in $\mathbb{R}\times C^4(\bar\Omega)$ and satisfying:

(i) $(\lambda(s),\, u(s)) \in \mathcal{C}$ for $s \in (0,\varepsilon)$ where $(\lambda(s),\, u(s))$ is defined in Proposition 5.2.

(ii) $\mathcal{C}$ is either unbounded in $\mathbb{R}\times C(\bar\Omega)$ or there exists a sequence $(\lambda_n, u_n) \in \mathcal{C}$ such that $\lim_{n\to\infty}(\lambda_n,u_n) = (\mu,0)$ in $\mathbb{R}\times C^4(\bar\Omega)$ where μ is an eigenvalue different from λ_1^2 of Δ^2 subject to Navier boundary conditions.

Proof. We study again equation (5.2) but with G considered as a map from $\mathbb{R}\times C(\bar\Omega)$ into $C(\bar\Omega)$. Then G is of class C^2, maps bounded subsets of $\mathbb{R}\times C(\bar\Omega)$ into relatively compact subsets of $C(\bar\Omega)$. We have $G(\lambda,0)=0$ for all $\lambda\in\mathbb{R}$. The partial Fréchet derivative of G with respect to u at $(\lambda,0)$, $G_u(\lambda,0)=\lambda K$. Since for $\lambda=\lambda_1^2$, one is an eigenvalue of λK with algebraic multiplicity one, Rabinowitz global bifurcation Theorem [14] can be applied. Let S denote the set of solutions of (5.2) in $\mathbb{R}\times C(\bar\Omega)$. Let $\mathcal{C}$ denote the component of $S\backslash\{\lambda\in\mathbb{R},\,(\lambda,0)\}$ containing the curve $\{(\lambda(s),\, u(s))\}_{s\in(0,\varepsilon)}$ defined in Proposition 5.2. By Rabinowitz Theorem, $\mathcal{C}$ is either unbounded in $\mathbb{R}\times C(\bar\Omega)$ or contains a sequence (λ_n,u_n) converging in $\mathbb{R}\times C(\bar\Omega)$ to $(\mu,0)$ where μ is an eigenvalue different from λ_1^2 of Δ^2 subject to Navier boundary conditions.
It follows from elliptic regularity that if $(\lambda,u)\in\mathcal{C}$, then $u\in C^{4,\delta}(\bar\Omega)$ for every $\delta\in(0,1)$ and (λ,u) satisfies (5.1). Similary if follows that $\mathcal{C}$ is not only connected in $\mathbb{R}\times C(\bar\Omega)$ but in $\mathbb{R}\times C^4(\bar\Omega)$ (even in $\mathbb{R}\times C^{4,\delta}(\bar\Omega)$, $\delta\in(0,1)$). For the same reason, the sequence (λ_n,u_n) converges in $\mathbb{R}\times C^4(\bar\Omega)$ and not only in $\mathbb{R}\times C(\bar\Omega)$. This completes the proof of the Lemma.

Proof of Theorem 5.1. It follows from [8] that there exists $\lambda^* < 0$ such that for every $\lambda^* < \lambda < \lambda_1^2$ the unique solution v to the problem $\Delta^2 v = \lambda v + f$ in Ω, $v = \Delta v = 0$ on $\partial\Omega$ satisfies $v(x) > 0$ in Ω and $\frac{\partial v}{\partial n}(x) < 0$ on $\partial\Omega$ whenever f is nonnegative and not identically zero. Choose $\tilde\lambda \in (\lambda^*, 0)$.

Clearly problem (5.1) can be rewritten as system (S) with $f(v) = v$ and $g_\lambda(u) = \lambda u + h(u)$. For $\lambda \in [0, \lambda_1^2]$, assumption $(f1)$ of Theorem 2.1 is satisfied. Assumption $(f3)$ is satisfied with $p = 1$ and q of $(h3)$. We have $\liminf_{s\to\infty} \frac{f(s)}{s} = 1$, hence $a = 1$ in $(f2)$, and $\liminf_{s\to\infty} \frac{g(s)}{s} = \infty$ if $q > 1$ and $\lambda + \beta$ if $q = 1$. Since $\beta > \lambda_1^2$ if $q = 1$, assumption $(f2)$ holds for all $\lambda \in [0, \lambda_1^2]$. Notice that if u is a positive solution of (5.1) then v is also positive since $-\Delta v = \lambda u + h(u)$ with $v = 0$ on $\partial\Omega$, and $\lambda u + h(u)$ is nonnegative. It follows from Theorem 2.1 that there is a constant $M > 0$ such that all solutions (λ, u) of (5.1) with $0 \le \lambda \le \lambda_1^2$ and $u \ge 0$ satisfy $u \le M$. Let $C_0^1(\bar\Omega) := \{v \in C^1(\bar\Omega)|\ v|_{\partial\Omega} = 0\}$.
We define $F = \{v \in C_0^1(\bar\Omega)| 0 \le v \le M + 1, \text{ in } \bar\Omega\}$. Clearly F is closed in $C_0^1(\bar\Omega)$ and $\overset{o}{F}$ the interior of F in $C_0^1(\bar\Omega)$ is given by:

$$\overset{o}{F} = \{v \in F | v > 0 \text{ in } \Omega, \frac{\partial v}{\partial n} < 0 \text{ on } \partial\Omega \text{ and } v < M + 1 \text{ in } \bar\Omega\}.$$

Finally let $A = [\tilde\lambda, \lambda_1^2] \times F$. The set A is closed in $\mathbb{R} \times C_0^1(\bar\Omega)$ and the interior of A is given by

$$\overset{o}{A} = (\tilde\lambda, \lambda_1^2) \times \overset{o}{F}.$$

Observe that if (λ, u) is a solution of (5.1) with u positive then $\lambda < \lambda_1^2$. Indeed multiplying (5.1) with φ_1 and integrating over Ω we obtain

$$(\lambda_1^2 - \lambda) \int u\varphi_1 = \int h(u)\varphi_1 > 0.$$

If follows that $\lambda(s) < \lambda_1^2$ for $0 < s < \epsilon^+$. Since $\lim_{s\to 0} \lambda(s) = \lambda_1^2$, we can choose $\bar s \in (0, \epsilon^+)$ such that $\lambda(s) \in (0, \lambda_1^2)$ for every $s \in (0, \bar s]$. Then $(\lambda(\bar s), u(\bar s)) \in \overset{o}{A}$. Denote by $\mathcal{C}^+$ the component of $(\lambda(\bar s), u(\bar s))$ in $\mathcal{C} \cap A$ endowed with the $\mathbb{R} \times C^1(\bar\Omega)$ topology. We claim that $\mathcal{C}^+$ satisfies the requirements of Theorem 5.1. Indeed the elements of $\mathcal{C}^+$ as elements of $\mathcal{C}$ are nontrivial solutions of (5.1) and $\mathcal{C}^+$ is also connected in $\mathbb{R} \times C^4(\bar\Omega)$ (even in $\mathbb{R} \times C^{4,\delta}(\bar\Omega)$ for every $\delta \in (0, 1)$). Let $(\lambda, u) \in \mathcal{C}^+$. Since u is nonnegative and $\lambda \in (\lambda^*, \lambda_1^2)$ it follows that $u(x) > 0$ in Ω and $\frac{\partial u}{\partial n}(y) < 0$ on $\partial\Omega$. Clearly $(\lambda(s), u(s)) \in \mathcal{C}^+$ for $s \in (0, \bar s]$, hence conclusion (ii) is satisfied. In order to prove (iii), due to the connectedness of $\mathcal{C}^+$, it is enough to prove the existence of $(\bar\lambda, u) \in \mathcal{C}^+$ with $\bar\lambda < 0$. For this purpose it is sufficient to show that $\mathcal{C}^+ \cap \partial A \neq \phi$, where ∂A denotes the boundary of A in $\mathbb{R} \times C_0^1(\bar\Omega)$. Indeed ∂A is the disjoint union of $\{\lambda_1^2\} \times F$, $(\tilde\lambda, \lambda_1^2) \times (F \setminus \overset{o}{F})$, and $\{\tilde\lambda\} \times F$. Let $(\lambda, u) \in \mathcal{C}^+$. Since u is positive, we have $\lambda < \lambda_1^2$ and for $\lambda \in [0, \lambda_1^2)$ from the a priori estimates and the strong maximum principle we have $u \in \overset{o}{F}$. Consequently if $\mathcal{C}^+ \cap \partial A \neq \phi$, then there are $\bar{\lambda} \in [\tilde{\lambda}, 0)$ and $u \in F$ such that $(\bar{\lambda}, u) \in \mathcal{C}^+$. Finally we prove that $\mathcal{C}^+ \cap \partial A \neq \phi$. Since, by Lemma 5.3, $\mathcal{C}$ is either unbounded in $\mathbb{R} \times C(\bar{\Omega})$ or contains a sequence (λ_n, u_n) converging to some $(\mu, 0)$ with $\mu > \lambda_1^2$, $\mathcal{C}$ must contain an element $(\bar{\lambda}, \bar{u})$ in the complement of A in $\mathbb{R} \times C(\bar{\Omega})$. Since $\bar{u} \in C^4(\bar{\Omega})$, $(\bar{\lambda}, \bar{u})$ belongs also to the complement of A in $\mathbb{R} \times C_0^1(\bar{\Omega})$. Due to the connectedness of $\mathcal{C}$ in $\mathbb{R} \times C_0^1(\bar{\Omega})$, we must have $\mathcal{C} \cap \partial A \neq \phi$ where ∂A denotes as above the boundary of A in $\mathbb{R} \times C_0^1(\bar{\Omega})$. Observe that $\mathcal{C} \cap A$ is compact in $\mathbb{R} \times C_0^1(\bar{\Omega})$ as well as $\mathcal{C}^+$ and $\mathcal{C} \cap \partial A$ as closed subsets of $\mathcal{C} \cap A$. It follows from a result of topology [17] that if $\mathcal{C}^+$ and $\mathcal{C} \cap \partial A$ are disjoint, then either there exists a connected subset c of $\mathcal{C} \cap A$ meeting $\mathcal{C}^+$ and $\mathcal{C} \cap \partial A$, or there exist two disjoint closed subsets of $\mathcal{C} \cap A$, denoted by c_1 and c_2 such that $c_1 \cup c_2 = \mathcal{C} \cap A$, $\mathcal{C}^+ \subset c_1$ and $\mathcal{C} \cap \partial A \subset c_2$. In the first case, c is contained in $\mathcal{C}^+$ due to the maximality of $\mathcal{C}^+$ and therefore $\mathcal{C}^+$ and $\mathcal{C} \cap \partial A$ are not disjoint. In the second case, $\mathcal{C}$ cannot be connected. Indeed $\mathcal{C}$ is then the union of the sets c_1 and $\mathcal{C} \backslash c_1$. Obviously c_1 is closed. Moreover we have

$$\mathcal{C} \backslash c_1 = c_2 \cup (\mathcal{C} \cap \overset{\circ}{A}{}^c) \text{ since } \mathcal{C} \cap \partial A \subset c_2.$$

Hence $\mathcal{C} \backslash c_1$ is also closed contradicting the connectedness of $\mathcal{C}$. Therefore $\mathcal{C}^+$ and $\mathcal{C} \cap \partial A$ are not disjoint and this completes the proof of Theorem 5.1. □

ACKNOWLEDGEMENTS

The authors would like to thank the referee for several helpful comments. The work of E. Mitidieri was supported by C.N.R. (Fondi 60%).

References

[1] H. Brezis, R.E.L. Turner, On a Class of Superlinear Elliptic Problems, Comm. in P.D.E., 2, 1977, 601-614.

[2] C. Cosner, Positive Solutions for Superlinear Elliptic Systems without Variational Structure, Nonlinear Analysis TMA, 8, 1984, 1427-1436.

[3] G. Caristi, E. Mitidieri, Further Results on Maximum Principles for Noncooperative Elliptic Systems, Nonlinear Analysis TMA, 17, 1991, 547-558.

[4] G. Caristi, E. Mitidieri, Maximum Principles for a Class of Noncooperative Elliptic Systems, Delft Progr. Rep., 14, 1990, 33-56.

[5] M.G. Crandall, P.H. Rabinowitz, Bifurcation from Simple Eigenvalue, J. Functional Analysis, 8, 1971, 321-340.

[6] D.G. de Figueiredo, *Lectures on the Ekeland Variational Principle with Applications and Détours*, Tata Lectures on Mathematics and Physics, vol. 81, 1989.

[7] D.G. de Figueiredo, P.L. Lions, R.D. Nussbaum, A Priori Estimates and Existence of Positive Solutions of Semilinear Elliptic Equations, J. Math. pures et appl., 61, 1982, 41-63.

[8] D.G. de Figueiredo, E. Mitidieri, Maximum Principles for Linear Elliptic Systems, Quaderni Matematici, Univ. di Trieste, 177, 1988.

[9] B. Gidas, W.-M.Ni., L. Nirenberg, Symmetry and Related Properties via the Maximum Principle, Comm. Math. Phys., 68, 1979, 209-243.

[10] B. Gidas, J. Spruck, A Priori Bounds for Positive Solutions of Nonlinear Elliptic Equations, Comm. in P.D.E., 6, (8), 1981, 883-901.

[11] P.L. Lions, On the Existence of Positive Solutions of Semilinear Elliptic Equations, SIAM Review, 24, 1982, 441-467.

[12] E. Mitidieri, A Rellich Type Identity and Applications, to appear.

[13] R.D. Nussbaum, Periodic Solutions of Some Nonlinear Autonomous Functional Differential Equations, II, J. Differential Equations, 14, 1973, 368-394.

[14] P.H. Rabinowitz, Some Global Results for Nonlinear Eigenvalue Problems, J. Functional Analysis, 7, 1971, 487-513.

[15] G. Sweers, A Strong Maximum Principle for a Noncooperative Elliptic System, SIAM J. Math. Anal. 20, 1989, 367-371.

[16] W.C. Troy, Symmetry Properties in Systems of Semilinear Elliptic Equations, J. Differential Equations, 42, 1981, 400-413.

[17] G.T. Whyburn, Topological Analysis, Princeton University Press, Princeton 1958.

ADDED IN PROOF

The assumption $(f1)$ can be relaxed if one assumes that Ω is a ball. This was done recently in:

L.A. Peletier and R.C.A.M. van der Vorst, Existence and Non-Existence of Positive Solutions of Non-Linear Elliptic Systems and the Bi-Harmonic Equation, to appear in Diff. and Int. Eq.

Received March 1991

Revised February 1992

TRANSACTIONS OF THE
AMERICAN MATHEMATICAL SOCIETY
Volume 343, Number 1, May 1994

ON SUPERQUADRATIC ELLIPTIC SYSTEMS

DJAIRO G. DE FIGUEIREDO AND PATRICIO L. FELMER

ABSTRACT. In this article we study the existence of solutions for the elliptic system

$$-\Delta u = \frac{\partial H}{\partial v}(u, v, x) \quad \text{in } \Omega,$$
$$-\Delta v = \frac{\partial H}{\partial u}(u, v, x) \quad \text{in } \Omega,$$
$$u = 0, \quad v = 0 \quad \text{on } \partial\Omega.$$

where Ω is a bounded open subset of $\mathbb{R}^N$ with smooth boundary $\partial\Omega$, and the function $H : \mathbb{R}^2 \times \overline{\Omega} \to \mathbb{R}$, is of class C^1. We assume the function H has a superquadratic behavior that includes a Hamiltonian of the form

$$H(u, v) = |u|^\alpha + |v|^\beta \quad \text{where } 1 - \frac{2}{N} < \frac{1}{\alpha} + \frac{1}{\beta} < 1 \text{ with } \alpha > 1, \ \beta > 1.$$

We obtain existence of nontrivial solutions using a variational approach through a version of the Generalized Mountain Pass Theorem. Existence of positive solutions is also discussed.

0. INTRODUCTION

This paper is devoted to the study of existence of solutions for certain superquadratic elliptic systems of the form

$$-\Delta u = \frac{\partial H}{\partial v}(u, v, x) \quad \text{in } \Omega, \tag{0.1}$$

$$-\Delta v = \frac{\partial H}{\partial u}(u, v, x) \quad \text{in } \Omega, \tag{0.2}$$

$$u = 0, \quad v = 0 \quad \text{on } \partial\Omega, \tag{0.3}$$

where Ω is a bounded open subset of $\mathbb{R}^N$, with smooth boundary $\partial\Omega$, and the function $H : \mathbb{R}^2 \times \overline{\Omega} \to \mathbb{R}$, which we call the Hamiltonian, is of class C^1.

For easy reference later on we call the above problem (ES). The term "superquadratic" used here comes from hypothesis (H2) on the Hamiltonian (see it in the sequel, as well as Remark 0.5). Such a condition actually takes into account the coupling of the system. It does not imply that both equations in (ES)

Received by the editors April 10, 1992.
1980 *Mathematics Subject Classification* (1985 *Revision*). Primary 35J50, 58E99.
Key words and phrases. Elliptic systems, positive solutions, variational method.
The first author was partially supported by CNPq.
The second author was supported by FONDECYT under grant No 1212-91.

0002-9947/94 \$1.00 + \$.25 per page

Originally published in: Trans. Amer. Math. Soc. 343 (1994), 99-116,

Reprint by Springer-Verlag Berlin Heidelberg 2013 with kind permission,
DOI: 10.1007/978-3-319-02856-9_26

are superquadratic, but it is implied by that, which follows from a condition like

$$\text{(0.4)} \qquad \text{(S)} \quad \frac{\partial H}{\partial u}(u, v, x) \cdot u + \frac{\partial H}{\partial v}(u, v, x) \cdot v \geq \mu H(u, v, x) > 0$$

for all $(u, v) \in \mathbb{R}^2 \setminus \{(0, 0)\}$, $x \in \overline{\Omega}$, and $\mu > 2$. Observe that (S) is a special case of (H2). We should mention that Benci and Rabinowitz [2] have already considered a special case of (ES) when both equations are superlinear, namely

$$-\Delta w = (w^2 + z^2)^{(s-1)/2} w, \quad -\Delta z = -(w^2 + z^2)^{(s-1)/2} z,$$

where $s > 1$. Such a system is of the form

$$\text{(0.5)} \qquad -\Delta w = \frac{\partial H}{\partial w}(w, z, x),$$

$$\text{(0.6)} \qquad -\Delta z = -\frac{\partial H}{\partial z}(w, z, x)$$

which is equivalent to our system (ES). In studying (0.5) and (0.6) Benci and Rabinowitz used the Generalized Mountain Pass Theorem in its infinite dimensional setting.

Recently in [4] Clement, de Figueiredo and Mitidieri studied a case of (ES) including Hamiltonians satisfying

$$\text{(0.7)} \qquad H(u, v) = |u|^\alpha + |v|^\beta$$

where $\frac{1}{\alpha} + \frac{1}{\beta} < 1$, $\alpha > 1$ and $\beta > 1$, but where the case $\alpha < 2$ is allowed. Thus not satisfying (S). In [4] a priori estimates for positive solutions were obtained and then degree theory arguments were used to prove the existence of positive solutions for (ES). In another work, Clement, de Figueiredo and Mitidieri [5] have considered some classes of superlinear elliptic systems with growth that allows the use of inequalities of the Hardy-Sobolev type, and have obtained a priori bounds for positive solutions.

Some other class of superlinear elliptic systems were also considered by Souto [13]. He used the techniques introduced by Gidas and Spruck [8] in order to obtain a priori bounds for positive solutions, and by degree theory existence of positive solutions was established. Sublinear systems with a Hamiltonian form were discussed in Costa and Magalhães [3].

In this paper we consider superquadratic Hamiltonians using a variational approach. This allows us to extend the results in [2 and 4]. This kind of Hamiltonian was studied recently by Felmer [6] in the context of Hamiltonian systems.

Next we describe our results in a more precise way. On the Hamiltonian H we will consider the following hypotheses:

(H0) $H: \mathbb{R}^2 \times \overline{\Omega} \to \mathbb{R}$ is of class C^1.
(H1) $H(u, v, x) \geq 0$ for all $(u, v, x) \in \mathbb{R}^2 \times \overline{\Omega}$.

Let us consider real constants $p \geq \alpha > p - 1 > 0$ and $q \geq \beta > q - 1 > 0$

such that

(i) $$\frac{1}{\alpha}+\frac{1}{\beta}<1,$$

(ii) $$\left\{2-\left(\frac{1}{p}+\frac{1}{q}\right)\right\}\max\left\{\frac{p}{\alpha},\frac{q}{\beta}\right\}<1+\frac{2}{N},$$

(iii) $$\frac{p-1}{p}\frac{q}{\beta}<1 \quad\text{and}\quad \frac{q-1}{q}\frac{p}{\alpha}<1.$$

In this paper we will always assume $N\geq 3$. If $N=2$ or $N=1$ less restrictive assumptions can be made.

Furthermore, in case $N\geq 5$ we also impose

(iv)
$$\left(1-\frac{1}{p}\right)\max\left\{\frac{p}{\alpha},\frac{q}{\beta}\right\}<\frac{N+4}{2N} \quad\text{and}\quad \left(1-\frac{1}{q}\right)\max\left\{\frac{p}{\alpha},\frac{q}{\beta}\right\}<\frac{N+4}{2N}.$$

With these constants α, β, p, q satisfying the above conditions (i)–(iv) we now state the further hypotheses on the Hamiltonian H:

(H2) There exists $R>0$ such that
$$\frac{1}{\alpha}\frac{\partial H}{\partial u}(u,v,x)\cdot u+\frac{1}{\beta}\frac{\partial H}{\partial v}(u,v,x)\cdot v\geq H(u,v,x)>0$$
for all $(u,v)\in\mathbb{R}^2$, $|(u,v)|\geq R$ and $x\in\overline{\Omega}$.

(H3) There exists $r>0$ and $a_1>0$ such that
$$H(u,v,x)\leq a_1(|u|^\alpha+|v|^\beta)\quad \text{if } |(u,v)|\leq r$$
and

(H4) There exists $a_2>0$ such that
$$\left|\frac{\partial H}{\partial u}(u,v,x)\right|\leq a_2(|u|^{p-1}+|v|^{(p-1)q/p}+1),$$
$$\left|\frac{\partial H}{\partial v}(u,v,x)\right|\leq a_2(|v|^{q-1}+|u|^{(q-1)p/q}+1).$$

Our first existence results consider the concept of strong solution. In our situation we have

Definition 0.1. We say that (u,v) is a strong solution of (ES) if
$$u\in W^{2,p/(p-1)}(\Omega)\cap W_0^{1,p/(p-1)}(\Omega),\quad v\in W^{2,q/(q-1)}(\Omega)\cap W_0^{1,q/(q-1)}(\Omega)$$
and (u,v) satisfies (0.1) and (0.2) a.e. in Ω.

We will prove the following results.

Theorem 0.1. *If H satisfies* (H0)–(H4) *then system* (ES) *possesses at least one strong solution.*

Theorem 0.2. *If H satisfies* (H0)–(H4) *and H is independent of $x\in\overline{\Omega}$ then* (ES) *possesses at least one nontrivial strong solution.*

Remark 0.1. When we assume more regularity on the Hamiltonian H, using standard arguments, it can be proved that strong solutions are indeed classical solutions. See [9]. In this direction a hypothesis to consider is

(H0′) $H:\mathbb{R}^2\times\overline{\Omega}\to\mathbb{R}$ is of class $C^{1,\varepsilon}$.

When some more assumptions are considered in H we also obtain a result on existence of positive classical solutions.

Theorem 0.3. *If* H *satisfies* (H0′), (H1)–(H4) *and also*

(H5) $\partial H(u,v,x)/\partial u\geq 0$, $\partial H(u,v,x)/\partial v\geq 0$ *for all* $(u,v)\in\mathbb{R}^2$, $u\geq 0$, $v\geq 0$, $x\in\overline{\Omega}$,

(H6) $\partial H(u,v,x)/\partial u=0$, $\partial H(u,v,x)/\partial v=0$ *if* $u=0$ *or* $v=0$.

Then (ES) *possesses at least one positive solution* (u,v) *with* $u(x)>0$, $v(x)>0$ *if* $x\in\Omega$.

Next we make some remarks about the hypotheses we considered.

Remark 0.2. Since $\frac{p}{\alpha}\geq 1$ and $\frac{q}{\beta}\geq 1$ we have from (ii) that

$$1>\frac{1}{\alpha}+\frac{1}{\beta}\geq\frac{1}{p}+\frac{1}{q}>1-\frac{2}{N}.$$

The first inequality above expresses the superquadratic character of the Hamiltonian H, as said before. The last inequality expresses the subcritical character of the system, as already pointed out in [4]. It seems however that these minimal assumptions do not suffice. Some further control on the constants α, β, p, q as expressed in (ii) and (iii) seems to be necessary.

Remark 0.3. Besides the critical curve $\frac{1}{p}+\frac{1}{q}>1-\frac{2}{N}$, in case $N\geq 5$ we have two extra critical lines as given by (iv).

Remark 0.4. If $\alpha=p$, $\beta=q$ then (i) and (ii) simply express

$$1>\frac{1}{\alpha}+\frac{1}{\beta}>1-\frac{2}{N}$$

and (iii) trivializes. Condition (iv) becomes

$$\frac{1}{p}>\frac{N-4}{2N},\quad\frac{1}{q}>\frac{N-4}{2N},$$

see [4].

Remark 0.5. It follows from (H2) that there are constants $c_1>0$ and $c_2>0$ such that

$$H(u,v,x)\geq c_1(|u|^{\alpha}+|v|^{\beta})-c_2. \tag{0.8}$$

For a proof see [6].

Our Theorem 0.3 generalizes Theorem 3.1 in [4]. Let $f,g:\mathbb{R}\to\mathbb{R}$ functions satisfying the hypothesis of Theorem 3.1 in [4]. Then we can define a Hamiltonian

$$H(u,v)=G(u)+F(v) \tag{0.9}$$

with $G(u) = \int_0^u g(s)\,ds$ and $F(v) = \int_0^v f(s)\,ds$. This Hamiltonian satisfies our hypotheses. In particular it satisfies (H5) and (H6).

1. The variational formulation of (ES)

In this section we set up the functional analytic framework needed to study problem (ES) from the variational point of view. We also give the variational formulation of (ES).

We shall work with spaces E^s, which are obtained as the domains of fractional powers of the operator

$$-\Delta : H^1(\Omega) \cap H_0^1(\Omega) \subset L^2(\Omega) \to L^2(\Omega)$$

where Δ denotes the Laplacian and $H^2(\Omega)$, $H_0^1(\Omega)$ are the usual Sobolev spaces. Namely $E^s = D((-\Delta)^{s/2})$ for $0 \le s \le 2$, and the corresponding operator is denoted by A^s

$$A^s : E^s \to L^2(\Omega).$$

The spaces E^s are Hilbert spaces with inner product

$$(u, v)_{E^s} = \int_\Omega A^s u A^s v\,dx \tag{1.1}$$

and associated norm

$$\|u\|_{E^s}^2 = \int_\Omega |A^s u|^2\,dx. \tag{1.2}$$

It is known that these spaces are indeed Sobolev spaces of fractional power as defined by interpolation, see Lions and Magenes [10]. In particular we have

$$E^s = H^s(\Omega) \quad \text{if } 0 \le s < \tfrac{1}{2}, \qquad E^{1/2} \subset H^{1/2}(\Omega),$$
$$E^s = \{u \in H^s(\Omega) \mid u = 0 \text{ on } \partial\Omega\} \quad \text{if } \tfrac{1}{2} < s \le 2,\ s \ne \tfrac{3}{2}, \quad \text{and}$$
$$E^{3/2} \subset \{u \in H^{3/2}(\Omega) \mid u = 0 \text{ on } \partial\Omega\}.$$

See also the results in Fujiwara [7].

If we consider a basis of $L^2(\Omega)$ constituted by eigenfunctions $\{\phi_n\}$ of

$$-\Delta\phi = \lambda\phi \quad \text{in } \Omega, \qquad \phi = 0 \quad \text{on } \partial\Omega$$

with associated eigenvalues $\{\lambda_n\}$, then we have the following characterization of E^s and A^s. If $u \in L^2(\Omega)$ we write $u = \sum_{n=1}^\infty a_n\phi_n$ its Fourier series with respect to the basis $\{\phi_n\}$. Then

$$E^s = \left\{ u \in L^2(\Omega) \,\middle|\, \sum_{n=1}^\infty \lambda_n^s |a_n|^2 < \infty \right\} \tag{1.3}$$

and

$$A^s u = \sum_{n=1}^\infty \lambda_n^{s/2} a_n \phi_n \tag{1.4}$$

for all $u \in E^s$. Now it is easy to see Poincaré's inequality for the operator A^s

$$\|A^s u\|_{L^2(\Omega)} \ge \lambda_1^{s/2} \|u\|_{L^2(\Omega)} \quad \forall u \in E^s \tag{1.5}$$

where λ_1 is the first eigenvalue of $-\Delta$.

The fractional order spaces have the following important embedding property.

Theorem 1.1. *Given $s > 0$ and $\sigma \geq 1$ so that $\frac{1}{\sigma} \geq \frac{1}{2} - \frac{2}{N}$ then the inclusion map $i : E^s \to L^\sigma(\Omega)$ is well defined and bounded. If above we have strict inequality then the inclusion is compact.*

The proof of this theorem is based on interpolation and the Kondrachov Theorem. See the article of Persson [11] for a proof.

With these preliminaries about the Laplacian and the spaces we can now define the functional associated to (ES). Let us consider first the quadratic part. For numbers $s > 0$ and $t > 0$ with $s + t = 2$ we define the Hilbert space $E = E^s \times E^t$ and the bilinear form $B : E \times E \to \mathbb{R}$ by the formula

$$B((u, v), (\phi, \psi)) = \int_\Omega A^s u A^t \psi + A^s \phi A^t v \, dx. \tag{1.6}$$

Using the Cauchy Schwarz inequality and (1.2) it is easy to see that B is continuous. We also see that B is symmetric. Then B induces a selfadjoint bounded linear operator $L : E \to E$ so that

$$B(z, \eta) = (Lz, \eta)_E \tag{1.7}$$

for all $z, \eta \in E$. Here and in what follows $(\cdot, \cdot)_E$ denotes the inner product in E induced by $(\cdot, \cdot)_{E^s}$ and $(\cdot, \cdot)_{E^t}$ on the product space E in the usual way. We can also define the quadratic form $Q : E \to \mathbb{R}$ associated to B and L as

$$Q(z) = \frac{1}{2}(Lz, z)_E = \int_\Omega A^s u A^t v \, dx \tag{1.8}$$

for all $z = (u, v) \in E$. A more explicit formula for L will be needed in the future.

Proposition 1.1. *The operator L defined above can be written as*

$$L(u, v) = ((A^s)^{-1} A^t v, (A^t)^{-1} A^s u). \tag{1.9}$$

Proof. If $z = (u, v)$, $\eta = (\phi, \psi) \in E$ and we write $L(z) = (w, y)$ then we have

$$(L(z), \eta)_E = ((w, y), (\phi, \psi))_E = \int_\Omega A^s w A^s \phi + A^t y A^t \psi \, dx. \tag{1.10}$$

On the other hand, from (1.6) we have

$$(L(z), \eta)_E = \int_\Omega A^s u A^t \psi + A^s \phi A^t v \, dx. \tag{1.11}$$

Taking $\psi = 0$ in (1.10) and (1.11) we obtain

$$\int_\Omega A^s \phi (A^t v - A^s w) \, dx = 0 \quad \forall \phi \in E^s. \tag{1.12}$$

Using that A^s is an isomorphism onto $L^2(\Omega)$ we conclude the equality $A^t v - A^s w = 0$ from where $w = (A^s)^{-1} A^t v$.

If we take $\phi = 0$ in (1.10) and (1.11) we obtain in a similar way $y = (A^t)^{-1} A^s u$. $\square$

In what follows we write $A^{-s} = (A^s)^{-1}$. Next we consider the eigenvalue problem

$$Lz = \lambda z \quad \text{in } E. \tag{1.13}$$

Using Proposition 1.1 we can write (1.13) in equivalent form as

$$A^{-s}A^{t}v = \lambda u, \tag{1.14}$$

$$A^{-t}A^{s}u = \lambda v, \tag{1.15}$$

where $z = (u, v)$. Since the operators A^s and A^t are isomorphisms λ cannot be zero. Then isolating u in (1.14) and substituting in (1.15) yields

$$v = \lambda^2 v \tag{1.16}$$

whence it follows that $\lambda = 1$ or $\lambda = -1$. The associated eigenvectors are

$$\text{for } \lambda = 1, \quad (u, A^{-t}A^{s}u) \quad \forall u \in E^s \tag{1.17}$$

and

$$\text{for } \lambda = -1, \quad (u, -A^{-t}A^{s}u) \quad \forall u \in E^s. \tag{1.18}$$

We can define the eigenspaces

$$E^+ = \{(u, A^{-t}A^{s}u) \mid u \in E^s\}$$

and

$$E^- = \{(u, -A^{-t}A^{s}u) \mid u \in E^s\}$$

which give a natural splitting $E = E^+ \oplus E^-$. The spaces E^+ and E^- are orthogonal with respect to the bilinear form B, that is

$$B(z^+, z^-) = 0 \quad \forall z^+ \in E^+, \ z^- \in E^-. \tag{1.19}$$

We also find that

$$\tfrac{1}{2}\|z\|_E^2 = Q(z^+) - Q(z^-) \tag{1.20}$$

where $z = z^+ + z^-$, $z_\pm \in E^\pm$.

Next we define the functional associated to the Hamiltonian. Using the growth hypothesis (H4) and integrating we obtain

$$H(u, v, x) \le |H(0, v, x)| + a_2(|u|^{p-1} + |v|^{(p-1)q/p} + 1)|u| \tag{1.21}$$

and

$$|H(0, v, x)| \le |H(0, 0, x)| + a_2(|v|^q + |v|). \tag{1.22}$$

Young's inequality implies

$$|v|^{(p-1)q/p}|u| \le \frac{|u|^p}{p} + \frac{(p-1)}{p}|v|^q. \tag{1.23}$$

From (1.21)–(1.23) we obtain

$$|H(u, v, x)| \le c_2(|u|^p + |v|^q) + c_3 \tag{1.24}$$

for certain constants c_2 and c_3. Now we will choose the numbers s and t defining the orders of the Sobolev spaces involved. From inequality (ii) in the Introduction we see the existence of $s, t \in \mathbb{R}$, $s + t = 2$ such that

$$\left(1 - \frac{1}{p}\right) \max\left\{\frac{p}{\alpha}, \frac{q}{\beta}\right\} < \frac{1}{2} + \frac{s}{N} \tag{1.25}$$

and

$$\left(1-\frac{1}{q}\right)\max\left\{\frac{p}{\alpha},\frac{q}{\beta}\right\}<\frac{1}{2}+\frac{t}{N}. \tag{1.26}$$

Using (iii) and (iv) if $N \geq 5$, we can choose $s > 0$ and $t > 0$. From the fact that $\frac{p}{\alpha} \geq 1$ and $\frac{q}{\beta} \geq 1$ we find, following from (1.25) and (1.26) that

$$\frac{1}{p} > \frac{1}{2} - \frac{s}{N} \quad \text{and} \quad \frac{1}{q} > \frac{1}{2} - \frac{t}{N}. \tag{1.27}$$

These last inequalities and Theorem 1.1 give the compact inclusions $E^s \to L^p(\Omega)$, $E^t \to L^q(\Omega)$. Then using (1.24) we can define the functional $\mathcal{H} : E \to \mathbb{R}$ as

$$\mathcal{H}(u, v) = \int_\Omega H(u(x), v(x), x)\, dx. \tag{1.28}$$

Proposition 1.2. *The functional* $\mathcal{H}$ *defined above is of class* C^1 *and its derivative is given by*

$$\mathcal{H}'(u, v)(\phi, \psi) = \int_\Omega \frac{\partial H}{\partial u}(u, v, x)\phi + \frac{\partial H}{\partial v}(u, v, x)\psi\, dx \tag{1.29}$$

for all $(u, v), (\phi, \psi) \in E$. *Moreover* $\mathcal{H}' : E \to E$ *is a compact operator.*

Proof. The expression given at the right-hand side of (1.29) is well defined. In fact, from the growth hypothesis (H4) we have

$$\int_\Omega \left|\frac{\partial H}{\partial u}(u, v, x)\phi\right| dx \leq a_2 \int_\Omega (|u|^{p-1} + |v|^{(p-1)q/p} + 1)|\phi|\, dx. \tag{1.30}$$

Using Hölder inequality and embeddings

$$\int_\Omega \left|\frac{\partial H}{\partial u}(u, v, x)\phi\right| dx \leq a_2(\|u\|_{E^s}^{p-1} + \|v\|_{E^t}^{(p-1)q/p} + 1)\|\phi\|_{E^s}. \tag{1.31}$$

In a similar way we obtain an inequality for the derivative with respect to v. Thus $\mathcal{H}'(u, v)$ is well defined and bounded in E.

Next, usual arguments give that $\mathcal{H}$ is Fréchet differentiable, $\mathcal{H}'$ is continuous and, as a consequence of the Sobolev embeddings, $\mathcal{H}'$ is also compact. See [12] for example. □

Now we can define a functional $\Phi : E \to \mathbb{R}$ as

$$\Phi(z) = Q(z) - \mathcal{H}(z) \quad \forall z \in E. \tag{1.32}$$

Φ is a functional of class C^1 and by previous considerations it has the structure needed to apply minimax techniques. See (3.6).

Definition 1.1. We say that $z = (u, v) \in E = E^s \times E^t$ is an (s, t)-weak solution of (ES) if z is a critical point of Φ. In other words, if for all $\eta = (\phi, \psi) \in E^s \times E^t$ we have

$$\int_\Omega A^s u A^t \psi + A^s \phi A^t v - \frac{\partial H}{\partial u}(u, v, x)\phi - \frac{\partial H}{\partial v}(u, v, x)\psi\, dx = 0. \tag{1.33}$$

The next theorem gives a regularity result for (s, t)-weak solutions.

Theorem 1.2. *If* $(u, v) \in E^s \times E^t$ *is an* (s, t)*-weak solution of* (ES) *then* $u \in W^{2,p/(p-1)}(\Omega) \cap W_0^{1,p/(p-1)}(\Omega)$, $v \in W^{2,q/(q-1)}(\Omega) \cap W_0^{1,q/(q-1)}(\Omega)$ *and*

$$-\Delta u = \frac{\partial H}{\partial v}(u, v, x), \tag{1.34}$$

$$-\Delta v = \frac{\partial H}{\partial u}(u, v, x) \tag{1.35}$$

a.e. in Ω. *In other words* (u, v) *is a strong solution of* (ES).

Proof. Let us consider $\psi = 0$ in (1.33), then

$$\int_\Omega A^t v A^s \phi\, dx = \int_\Omega \frac{\partial H}{\partial u}(u, v, x)\phi\, dx \tag{1.36}$$

for all $\phi \in E^s$. If $\phi \in H^2(\Omega) \cap H_0^1(\Omega)$ then we have

$$\int_\Omega A^t v A^s \phi\, dx = \int_\Omega v A^2 \phi\, dx = -\int_\Omega v \Delta\phi\, dx. \tag{1.37}$$

On the other hand, using estimates following from (H4) we find

$$\frac{\partial H}{\partial u}(u(x), v(x), x) \in L^{p/(p-1)}(\Omega),$$

and then from basic elliptic theory there exists one function $w \in W^{2,p/(p-1)}(\Omega) \cap W_0^{1,p/(p-1)}(\Omega)$ such that

$$-\Delta w = \frac{\partial H}{\partial u}(u(x), v(x), x), \tag{1.38}$$

see Gilbarg and Trudinger [9]. Following from (1.27)

$$\frac{1}{2} > \frac{p-1}{p} - \frac{s}{N} > \frac{p-1}{p} - \frac{2}{N}$$

so that from the embedding theorem of Sobolev $w \in L^2(\Omega)$ (see [1]). Thus we have

$$\int_\Omega \frac{\partial H}{\partial u}(u, v, x)\phi\, dx = -\int_\Omega \Delta w \phi\, dx = -\int_\Omega w \Delta\phi\, dx \tag{1.39}$$

for all $\phi \in H^2(\Omega) \cap H_0^1(\Omega)$. Next, from (1.37) and (1.39) we obtain

$$\int_\Omega (v - w)\Delta\phi\, dx \quad \forall \phi \in H^2(\Omega) \cap H_0^1(\Omega) \tag{1.40}$$

from where it follows that $v = w$. We have obtained that $v \in W^{2,p/(p-1)}(\Omega) \cap W_0^{1,p/(p-1)}(\Omega)$. Finally, since $v = w$ we conclude that v satisfies (1.35).

We can do the same reasoning for u. □

2. Palais Smale and geometric conditions

In this section we further study the functional Φ. We prove the Palais Smale condition for Φ and then obtain the geometric situation of the Generalized Mountain Pass Theorem.

Proposition 2.1. *Φ satisfies the Palais Smale condition.*

Proof. Let $\{z_n\}$ be a sequence in E so that

$$|\Phi(z_n)| \leq c \text{ and } \Phi'(z_n) \to 0, \quad \text{as } n \to \infty. \tag{2.1}$$

We prove first that (2.1) implies that $\{z_n\}$ is bounded. From (2.1) there is a sequence $\{\varepsilon_n\}$ converging to 0 so that

$$|\Phi'(z_n)\eta| \leq \varepsilon_n \|\eta\|_E \quad \forall \eta \in E. \tag{2.2}$$

Taking

$$\eta_n = \frac{\alpha\beta}{\alpha+\beta}\left(\frac{1}{\alpha}u_n, \frac{1}{\beta}v_n\right), \quad \text{where } z_n = (u_n, v_n),$$

and using (2.1) we find

$$\begin{aligned} c + \varepsilon_n\|\eta_n\|_E \geq \Phi(z_n) - \Phi'(z_n)\eta_n = \frac{\alpha\beta}{\alpha+\beta}\int_\Omega \frac{1}{\alpha}\frac{\partial H}{\partial u}(u_n, v_n, x)u_n \\ + \frac{1}{\beta}\frac{\partial H}{\partial v}(u_n, v_n, x)v_n - H(u_n, v_n, x)\,dx \\ + \left(\frac{\alpha\beta}{\alpha+\beta} - 1\right)\int_\Omega H(u_n, v_n, x)\,dx. \end{aligned} \tag{2.3}$$

Using (H2) we find a constant c_1 so that

$$c_1(1 + \|z_n\|_E) \geq \int_\Omega H(u_n, v_n, x)\,dx. \tag{2.4}$$

and then, using (0.8) we obtain a constant c_3 such that

$$\int_\Omega |u_n|^\alpha + |v_n|^\beta\,dx \leq c_2(1 + \|u_n\|_{E^s} + \|v_n\|_{E^t}). \tag{2.5}$$

Next, let us consider $\eta = (\phi, 0)$ with $\phi \in E^s$. Then from (2.2)

$$\left|\int_\Omega A^s\phi A^t v_n\,dx\right| \leq \int_\Omega \left|\frac{\partial H}{\partial u}(u_n, v_n, x)\phi\right|\,dx + \varepsilon_n\|\phi\|_{E^s}. \tag{2.6}$$

We estimate these terms next. From (1.25) and Theorem 1.1 we find

$$\|\phi\|_{L^{\alpha/(\alpha-p+1)}(\Omega)} \leq c_1\|\phi\|_{E^s} \tag{2.7}$$

then, using Hölder inequality with $a = \alpha/(p-1)$, $b = \alpha/(\alpha-p+1)$ we obtain

$$\int_\Omega |u_n|^{p-1}|\phi|\,dx \leq c_1\|u_n\|_{L^\alpha(\Omega)}^{p-1}\|\phi\|_{E^s}. \tag{2.8}$$

From (iii) in the Introduction and (1.25) we have

$$1 - \frac{p-1}{p}\frac{q}{\beta} > \frac{1}{2} - \frac{s}{N}$$

and then Theorem 1.1 implies the embedding of E^s into $L^{\beta p/(\beta p-(p-1)q)}(\Omega)$. This fact together with Hölder inequality with $a = \beta p/(p-1)q$ and $b = \beta p/(\beta p - (p-1)q)$ yields

$$\int_\Omega |v_n|^{(p-1)q/p}|\phi|\,dx \leq c_2\|v_n\|_{L^\beta(\Omega)}^{(p-1)q/p}\|\phi\|_{E^s}. \tag{2.9}$$

Finally, again using Theorem 1.1

$$\int_\Omega |\phi|\, dx \le c_3 \|\phi\|_{E^s}. \tag{2.10}$$

Putting together (H4), (2.6), (2.8)–(2.10) we obtain

$$\left|\int_\Omega A^s\phi A^t v_n\, dx\right| \le c_4(\|u_n\|_{L^\alpha(\Omega)}^{p-1} + \|v_n\|_{L^\beta(\Omega)}^{(p-1)q/p} + 1)\|\phi\|_{E^s} \tag{2.11}$$

for all $\phi \in E^s$, from where it follows

$$\|v_n\|_{E^t} \le c_4(\|u_n\|_{L^\alpha(\Omega)}^{p-1} + \|v_n\|_{L^\beta(\Omega)}^{(p-1)q/p} + 1). \tag{2.12}$$

By an analogous reasoning

$$\|u_n\|_{E^s} \le c_5(\|v_n\|_{L^\beta(\Omega)}^{q-1} + \|u_n\|_{L^\alpha(\Omega)}^{(q-1)p/q} + 1). \tag{2.13}$$

Replacing (2.12) and (2.13) into (2.5) we obtain

(2.14)

$$\begin{aligned}\|u_n\|_{E^s} &+ \|v_n\|_{E^t}\\ &\le c_6(\|u_n\|_{E^s}^{(p-1)/\alpha} + \|v_n\|_{E^t}^{(p-1)q/p\beta} + \|u_n\|_{E^s}^{(q-1)p/q\alpha} + \|v_n\|_{E^t}^{(q-1)/\beta} + 1).\end{aligned}$$

Since the exponents in the right-hand side of (2.14) are all less that 1, by the basic assumptions we made on α, β, p and q, we find that the sequence $\{z_n\}$ is bounded in E.

From here on a usual argument based on the compactness of $\mathscr{K}'$ and the invertibility of L gives the existence of a subsequence of $\{z_n\}$ that converges in E. □

We will consider now the study of the geometric characteristics of Φ leading to the Generalized Mountain Pass Theorem.

We will define subsets S and Q so that

(IS) There exists $\delta > 0$ such that

$$\Phi(z) \ge \delta \quad \forall z \in S. \tag{2.15}$$

(IQ) $\Phi(z) \le 0 \quad \forall z \in \partial Q$.

Where ∂Q denotes the boundary of Q relative to a certain subspace of E. It is shown also that S and ∂Q link in the sense of Benci and Rabinowitz.

For later reference we state a preliminary lemma giving the expressions of the projections over $E^\pm$.

Lemma 2.1. *The orthogonal projections $P^\pm : E \to E^\pm$ are given by the formula*

$$P^\pm(u, v) = \tfrac{1}{2}(u \pm A^{-s}A^t v, v \pm A^{-t}A^s u). \tag{2.16}$$

Proof. Direct from definitions. □

Now we choose numbers $\mu > 1$, $\nu > 1$ such that

$$\frac{1}{\alpha} < \frac{\mu}{\mu+\nu} \quad\text{and}\quad \frac{1}{\beta} < \frac{\nu}{\mu+\nu}. \tag{2.17}$$

Proposition 2.2. *There exist $\rho > 0$ and $\delta > 0$ such that if we define*

$$S = \{(\rho^{\mu-1}u, \rho^{\nu-1}v) \mid \|(u, v)\| = \rho, (u, v) \in E^+\}$$

then (IS) *is satisfied.*

Proof. Let $\tilde{z} = (u, v) \in E^+$ and put $z = (\rho^{\mu-1}u, \rho^{\nu-1}v)$. Recall that $v = A^{-t}A^s u$ and equivalently $u = A^{-s}A^t v$. We have

$$Q(z) = \int_\Omega \rho^{\mu-1}A^s u \rho^{\nu-1}A^t v\, dx = \rho^{\mu+\nu-2}\int_\Omega A^s u A^t v\, dx. \tag{2.18}$$

Using (1.20) and (2.18) we get

$$Q(z) = \tfrac{1}{2}\rho^{\mu+\nu-2}\|\tilde{z}\|_E^2. \tag{2.19}$$

On the other hand, from (H3) and (H4) we have

$$H(u, v, x) \le a_1(|u|^\alpha + |v|^\beta) + b_1(|u|^p + |v|^q) \tag{2.20}$$

for some constants a_1, b_1, then

$$\begin{aligned}\mathscr{H}(z) \le{}& a_1\left(\rho^{(\mu-1)\alpha}\int_\Omega |u|^\alpha dx + \rho^{(\nu-1)\beta}\int_\Omega |v|^\beta dx\right)\\ &+ b_1\left(\rho^{(\mu-1)p}\int_\Omega |u|^p dx + \rho^{(\nu-1)q}\int_\Omega |v|^q dx\right).\end{aligned} \tag{2.21}$$

Since $\alpha \le p$, $\beta \le q$, using Hölder inequality and Theorem 1.1 we find from (2.21) that

$$\mathscr{H}(z) \le b_2(\rho^{(\mu-1)\alpha}\|\tilde{z}\|_E^\alpha + \rho^{(\mu-1)p}\|\tilde{z}\|_E^p + \rho^{(\nu-1)\beta}\|\tilde{z}\|_E^\beta + \rho^{(\nu-1)q}\|\tilde{z}\|_E^q). \tag{2.22}$$

Putting (2.19) and (2.22) together and considering $\rho = \|\tilde{z}\|_E$ we have

$$\Phi(z) \ge \tfrac{1}{2}\rho^{\mu+\nu} - b_2(\rho^{\mu\alpha} + \rho^{\mu p} + \rho^{\nu\beta} + \rho^{\nu q}). \tag{2.23}$$

Since $\alpha \le p$, $\beta \le q$, for ρ small we obtain
(2.24)

$$\Phi(z) \ge \frac{1}{2}\rho^{\mu+\nu} - 2b_1(\rho^{\mu\alpha} + \rho^{\nu\beta}) = \left(\frac{\rho^{\mu+\nu}}{4} - 2b_1\rho^{\mu\alpha}\right) + \left(\frac{\rho^{\mu+\nu}}{4} - 2b_1\rho^{\nu\beta}\right).$$

From our choice of μ and ν in (2.17) we see that if ρ is small enough, there exists $\delta > 0$ such that $\Phi(z) \ge \delta$ if $\|\tilde{z}\|_E = \rho$. □

We next define the set Q. For some constants $\sigma > 0$ and $M > 0$ to be precised later on we define

$$Q = \{\tau(\sigma^{\mu-1}u_+, \sigma^{\nu-1}v_+) + (\sigma^{\mu-1}u, \sigma^{\nu-1}v) \mid 0 \le \tau \le \sigma, 0 \le \|(u, v)\|_E \le M, (u, v) \in E^-\}$$

where $z_+ = (u_+, v_+) \in E^+$ with u_+ some fixed eigenvector of $-\Delta$. We note that z_+ is an eigenvector of L associated to a positive eigenvalue (i.e. to 1). We assume $\|z_+\|_E = 1$. We denote by ∂Q the boundary of Q relative to the subspace

$$\{\tau(\sigma^{\mu-1}u_+, \sigma^{\nu-1}v_+) + (\sigma^{\mu-1}u, \sigma^{\nu-1}v) \mid \tau \in \mathbb{R}, (u, v) \in E^-\}.$$

Proposition 2.3. *There are constants $\sigma > 0$ and $M > 0$ such that for Q defined as above we have $\Phi(z) \le 0$ $\forall z \in \partial Q$.*

Proof. For $\tau \in \mathbb{R}_+$, $(u, v) \in E^-$ we set

$$z = \tau(\sigma^{\mu-1}u_+, \sigma^{\nu-1}v_+) + (\sigma^{\mu-1}u, \sigma^{\nu-1}v).$$

By the definitions of E^+ and E^- we have

$$v_+ = A^{-t}A^s u_+ \quad \text{and} \quad v = -A^{-t}A^s u, \tag{2.25}$$

and then evaluating Q in z we obtain

$$\begin{aligned} Q(z) &= \int_\Omega (\tau\sigma^{\mu-1}A^s u_+ + \sigma^{\mu-1}A^s u)(\tau\sigma^{\nu-1}A^s u_+ - \sigma^{\nu-1}A^s u)\,dx \\ &= \frac{1}{2}\sigma^{\mu+\nu-2}(\tau^2 - \|(u, v)\|_E^2). \end{aligned} \tag{2.26}$$

On the other hand, we have from Remark 0.5 that

$$\int_\Omega H(z, x)\,dx \geq c_1 \int_\Omega (\sigma^{\alpha(\mu-1)}|\tau u_+ + u|^\alpha + \sigma^{\beta(\nu-1)}|\tau v_+ + v|^\beta)\,dx - |\Omega| c_2. \tag{2.27}$$

Each u can be written as $u = \gamma u_+ + \hat{u}$, where $\hat{u}$ is orthogonal to u_+ in the $L^2(\Omega)$ sense, and γ is some real number. Using Hölder's inequality we obtain

$$\begin{aligned} (\tau + \gamma) \int_\Omega |u^+|^2\,dx &= \int_\Omega (\tau u^+ + u)u^+\,dx \\ &\leq \|\tau u^+ + u\|_{L^\alpha(\Omega)} \|u^+\|_{L^{\alpha'}(\Omega)} \end{aligned} \tag{2.28}$$

or

$$\tau + \gamma \leq c\|\tau u^+ + u\|_{L^\alpha(\Omega)}. \tag{2.29}$$

Similarly, observing that $A^{-t}A^s u^+ = \lambda_k^{-t+s}u^+$, where λ_k is the eigenvalue of $(-\Delta, H_0^1(\Omega))$ whose eigenfunction is u^+, we obtain

$$\begin{aligned} \lambda_k^{-t+s}(\tau - \gamma) \int_\Omega |u^+|^2\,dx &= \int_\Omega (\tau v^+ + v)u^+\,dx \\ &\leq \|\tau v^+ + v\|_{L^\beta(\Omega)} \|u^+\|_{L^{\beta'}(\Omega)} \end{aligned} \tag{2.30}$$

or

$$\tau - \gamma \leq c\|\tau v^+ + v\|_{L^\beta(\Omega)}. \tag{2.31}$$

If $\gamma \geq 0$ it follows from (2.26), (2.27) and (2.29) that

$$\Phi(z) \leq \frac{\sigma^{\mu+\nu-2}}{2}\tau^2 - c_6\tau^\alpha\sigma^{\alpha(\mu-1)} + |\Omega|c_2. \tag{2.32}$$

And if $\gamma \leq 0$ it follows from (2.26), (2.27) and (2.31) that

$$\Phi(z) \leq \frac{\sigma^{\mu+\nu-2}}{2}\tau^2 - c_6\tau^\beta\sigma^{\beta(\nu-1)} + |\Omega|c_2. \tag{2.33}$$

By the choice of μ and ν, taking $\tau = \sigma$ large we see in (2.32) and (2.33) that $\Phi(z) \leq 0$.

Now we choose M. If $\tau \in [0, \sigma]$, from (2.26) and (2.27) we have

$$\Phi(z) \leq \tfrac{1}{2}\sigma^{\mu+\nu} - \tfrac{1}{2}\sigma^{\mu+\nu-2}\|(u, v)\|_E^2 + |\Omega|c_2 \tag{2.34}$$

then taking $\|(u, v)\| = M$ and M large enough we find $\Phi(z) \leq 0$. We finish the proof if we show that when $\tau = 0$ then also $\Phi(z) \leq 0$. By hypothesis (H1) this is direct after the definition of Φ and (2.26). □

3. The minimax theorem

In this section we formulate a minimax theorem which is a version of Theorem 5.9 of [12] and it was proved in [6]. We describe this result and then we show how to use it in our situation. We consider a Hilbert space E with inner product $\langle\ ,\ \rangle$ and norm $\|\cdot\|$. We assume that E has a splitting $E = X \oplus Y$, where the subspace X and Y are not necessarily orthogonal and both of them can be infinite dimensional. Let $\Phi : E \to \mathbb{R}$ be a functional having the following structure

$$\Phi(z) = \tfrac{1}{2}\langle Lz, z\rangle + \mathscr{H}(z) \tag{3.1}$$

with

(I1) $L : E \to E$ is a linear, bounded, selfadjoint operator,

(I2) $\mathscr{H}'$ is compact.

There are two linear bounded, invertible operators $B_1, B_2 : E \to E$ satisfying

(I3) If $\omega \in \mathbb{R}_0^+$ then the linear operator

$$\widehat{B}(\omega) = P_X B_1^{-1} \exp(\omega L) B_2 : X \to X \tag{3.2}$$

is invertible.

Here P_X denotes the projection of E onto X induced by the splitting $E = X \oplus Y$. Let $\rho > 0$ and define

$$S = \{B_1 z \mid \|z\| = \rho,\ z \in Y\}. \tag{3.3}$$

For $z_+ \in Y$, $z_+ \neq 0$, $\sigma > \rho/\|B_1^{-1}B_2 z_+\|$ and $M > \rho$, we define

$$Q = \{B_2(\tau z_+ + z) \mid 0 \le \tau \le \sigma,\ \|z\| \le M,\ z \in X\}. \tag{3.4}$$

Where ∂Q denotes the boundary of Q relative to the subspace

$$\{B_2(\tau z_+ + z) \mid \tau \in \mathbb{R},\ z \in X\}.$$

Then we have the following theorem on existence of critical points of Φ.

Theorem 3.1. *Let $\Phi : E \to R$ be a C^1 functional satisfying the Palais Smale condition, and* (I1), (I2) *and* (I3). *Further assume there is a constant $\delta > 0$ such that*

(IS) $\Phi(z) \ge \delta\ \forall z \in S$,

(IQ) $\Phi(z) \le 0\ \forall z \in \partial Q$.

Then Φ possesses a critical point with critical value $C \ge \delta$.

The reader is referred to [6] for a proof of this theorem. The critical point given by Theorem 3.1 has a variational characterization we describe next. Let us consider the class of functions

$$\Gamma = \{h \in C(E \times [0, 1], E) \mid h \text{ satisfies } \Gamma_1, \Gamma_2 \text{ and } \Gamma_3\} \tag{3.5}$$

where

(Γ_1) h is given by $h(z, t) = \exp(\omega(z, t)L)z + K(z, t)$ where $\omega : E \times [0, 1] \to \mathbb{R}_0^+$ is continuous and transforms bounded sets into bounded sets, and $K : E \times [0, 1] \to E$ is compact.

(Γ_2) $h(z, t) = z$
$\forall z \in \partial Q$, $\forall t \in [0, 1]$.

(Γ_3) $h(z, 0) = z$
$\forall z \in Q$.

Then the minimax value

$$C = \inf_{h\in\Gamma} \sup_{z\in Q} \Phi(h(z, 1)) \tag{3.6}$$

is the critical value given in Theorem 3.1.

In order to use Theorem 3.1 to find critical points of our functional Φ we see that it is only left to be proved hypothesis (I3). In fact, the structure of our functional is that given by (I1) and (I2) as we saw in §1, and the P.S. condition together with the geometric conditions (IS) and (IQ) were proved in Propositions 2.1, 2.2 and 2.3 respectively.

Let us prove (I3). First we define the decomposition of E by taking $X = E^-$ and $Y = E^+$. The operators B_1 and B_2 are defined as $B_1, B_2 : E \to E$

$$B_1(u, v) = (\rho^{\mu-1}u, \rho^{\nu-1}v) \tag{3.7}$$

and

$$B_2(u, v) = (\sigma^{\mu-1}u, \sigma^{\nu-1}v), \tag{3.8}$$

certainly B_1 and B_2 are linear, bounded, invertible operators. Here the constants ρ and σ are those of Propositions 2.2 and 2.3 respectively. To show that $\widehat{B}(\omega)$ is invertible we first give a formula for $\exp(\omega L)$.

Lemma 3.1. *If* $\omega \in \mathbb{R}$ *then the operator* $\exp(\omega L) : E \to E$ *is given by*

$$\exp(\omega L)(u, v) = \cosh(\omega)(u, v) + \sinh(\omega)(A^{-s} \circ A^t v, A^{-t} \circ A^s u). \tag{3.9}$$

Proof. We recall from Proposition 1.1 that

$$L(u, v) = (A^{-s} \circ A^t v, A^{-t} \circ A^s u). \tag{3.10}$$

Then $L^2 = \mathrm{id}_E$. By writing explicitly the exponential operator as a series, using (3.10) and reordering the terms we obtain (3.9). □

Proposition 3.1. *The operator* $\widehat{B}(\omega) : E^- \to E^-$ *is invertible.*

Proof. Given $z \in E^-$ we have $z = (u, -A^{-t} \circ A^s u)$ with $u \in E^s$. By definition of B_2 in (3.8) we have

$$B_2 z = (\sigma^{\mu-1}u, -\sigma^{\nu-1}A^{-t} \circ A^s u), \tag{3.11}$$

then using Lemma 3.1 if we write $\exp(\omega L)B_2 z = (x, y)$ we have

$$x = (\cosh(\omega)\sigma^{\mu-1} - \sinh(\omega)\sigma^{\nu-1})u, \tag{3.12}$$

$$y = (-\cosh(\omega)\sigma^{\nu-1} + \sinh(\omega)\sigma^{\mu-1})A^{-t} \circ A^s u. \tag{3.13}$$

By definition of B_1 given in (3.7) and from (3.12) and (3.13), if we write $B_1 \exp(\omega L)B_2 z = (\bar{x}, \bar{y})$ we have

$$\bar{x} = \frac{\cosh(\omega)\sigma^{\mu-1} - \sinh(\omega)\sigma^{\nu-1}}{\rho^{\mu-1}}u, \tag{3.14}$$

$$\bar{y} = \frac{-\cosh(\omega)\sigma^{\nu-1} + \sinh(\omega)\sigma^{\mu-1}}{\rho^{\nu-1}}A^{-t} \circ A^s u. \tag{3.15}$$

Finally we project back into E^-. If we put $\widehat{B}(\omega)z = (\phi, \psi)$ then using the projection formula given in Lemma 2.1 and (3.14) and (3.15), after some calculations we obtain

$$\phi = \left\{\frac{1}{2}\left(\frac{\sigma^{\mu-1}}{\rho^{\mu-1}} + \frac{\sigma^{\nu-1}}{\rho^{\nu-1}}\right)\cosh(\omega) - \frac{1}{2}\left(\frac{\sigma^{\nu-1}}{\rho^{\mu-1}} + \frac{\sigma^{\mu-1}}{\rho^{\nu-1}}\right)\sinh(\omega)\right\}u. \tag{3.16}$$

If we put m as the coefficient of u in (3.16) we also find that

$$\psi = -mA^{-t} \circ A^s u. \tag{3.17}$$

In other words we have that $\widehat{B}(\omega)z = mz$. This constant m is positive if we assume, without loss of generality that $\sigma > 1$ and $\rho < 1$. In fact

$$\left(\frac{\sigma^{\nu-1}}{\rho^{\nu-1}} + \frac{\sigma^{\mu-1}}{\rho^{\mu-1}}\right) - \left(\frac{\sigma^{\mu-1}}{\rho^{\nu-1}} + \frac{\sigma^{\nu-1}}{\rho^{\mu-1}}\right) = \frac{(\rho^{\mu-1} - \rho^{\nu-1})(\sigma^{\nu-1} - \sigma^{\mu-1})}{\rho^{\mu+\nu-2}} \tag{3.18}$$

is positive so that $m > 0$ independently of the value of $\omega \in \mathbb{R}$. This in turn implies that $\widehat{B}(\omega)$ is invertible. □

4. Proof of the theorems

Now we are in a position of giving a proof of the theorems announced in the Introduction. The application of Theorem 3.1 will give existence of a critical point of the functional Φ and then by applying regularity results we obtain a solution of (ES).

Proof of Theorems 0.1 *and* 0.2. We apply minimax Theorem 3.1 to the functional Φ. Hypothesis (I1) and (I2) are satisfied by the considerations made in §1. Definitions (3.7) and (3.8) and Proposition 3.1 give (I3). The Palais Smale condition is proved in Proposition 2.1 and the geometric conditions leading to (IS) and (IQ) are proved in Propositions 2.2 and 2.3 respectively.

Thus, there exists $z \in E$ such that $\Phi'(z) = 0$ i.e. z is an (s, t)-weak solution of (ES). Next, Theorem 2.2 gives that $z = (u, v)$ is such that $u \in W^{2,p/(p-1)}(\Omega) \cap W_0^{1,p/(p-1)}(\Omega)$ and $v \in W^{2,q/(q-1)}(\Omega) \cap W_0^{1,q/(q-1)}(\Omega)$. That is, (u, v) is a strong solution of (ES).

When the Hamiltonian is independent of x we have that $(0, 0)$ is a solution of (ES). Since $\Phi(z) \geq \delta > 0$ and $\Phi((0, 0)) = 0$ we find that z is not trivial. □

In order to prove Theorem 0.3 we need to redefine the Hamiltonian. Let us define $\widetilde{H} : \mathbb{R}^2 \times \overline{\Omega} \to \mathbb{R}$ so that

$$\widetilde{H}(u, v, x) = \begin{cases} H(u, v, x) & \text{if } u \geq 0,\ v \geq 0, \\ H(0, v, x) & \text{if } u \leq 0,\ v \geq 0, \\ H(u, 0, x) & \text{if } u \geq 0,\ v \leq 0, \\ 0 & \text{if } u \leq 0,\ v \leq 0. \end{cases} \tag{4.1}$$

Since we are assuming (H6) the new Hamiltonian $\widetilde{H}$ is of class $C^{1,\varepsilon}$ as H. It also satisfies (H1), (H3) and (H4). However (H2) is satisfied in a restricted form. Next we modify the arguments given before to include this new situation.

Proof of Theorem 0.3. First we define the functional $\widetilde{\Phi} : E \to \mathbb{R}$ as

$$\widetilde{\Phi}(z) = Q(z) - \int_\Omega \widetilde{H}(z, x)\, dx. \tag{4.2}$$

It is clear that $\widetilde{\Phi}$ is of class C^1 in E. Assume there exists $z = (u, v)$ a critical point of $\widetilde{\Phi}$. Then by the argument given before we have a classical solution of

$$\begin{aligned} -\Delta u &= \frac{\partial \widetilde{H}}{\partial v}(u, v, x) \quad \text{in } \Omega, \\ -\Delta v &= \frac{\partial \widetilde{H}}{\partial u}(u, v, x) \quad \text{in } \Omega, \\ u = 0, &\quad v = 0 \quad \text{on } \partial\Omega. \end{aligned} \tag{4.3}$$

Since $\partial \widetilde{H}/\partial u \geq 0$ and $\partial \widetilde{H}/\partial v \geq 0$ by the maximum principle we find that $u > 0$ and $v > 0$ in Ω. We only need to prove that hypotheses of Theorem 3.1 still hold. Hypothesis (H2) was used only in proving the Palais Smale condition and the geometric condition (IQ).

Let us see first how we prove the Palais condition for $\widetilde{\Phi}$. Let $\{z_n\}$ be a sequence in E so that

$$|\widetilde{\Phi}(z_n)| \leq c \text{ and } \widetilde{\Phi}'(z_n) \to 0, \quad \text{as } n \to \infty. \tag{4.4}$$

As we did in Proposition 2.1 we have only to prove that (4.4) implies that $\{z_n\}$ is bounded. Proceeding as in Proposition 2.1 we find

$$c_1(1 + \|z_n\|_E) \geq \int_\Omega \widetilde{H}(u_n, v_n, x)\, dx, \tag{4.5}$$

where we used the hypothesis (H2) restricted to $u \geq 0$ and $v \geq 0$. Then, after Remark 0.5 we find c_2 so that

$$\int_\Omega |u_n^+|^\alpha + |v_n^+|^\beta\, dx \leq c_2(1 + \|u_n\|_{E^s} + \|v_n\|_{E^t}). \tag{4.6}$$

Where we denote by u_n^+ and v_n^+ the positive part of u_n and v_n respectively. Next, following the proof of Proposition 2.1 we obtain

$$\|v_n\|_{E^t} \leq c_4(\|u_n^+\|_{L^\alpha(\Omega)}^{p-1} + \|v_n^+\|_{L^\beta(\Omega)}^{(p-1)q/p} + 1). \tag{4.7}$$

By an analogous reasoning

$$\|u_n\|_{E^s} \leq c_5(\|v_n^+\|_{L^\beta(\Omega)}^{q-1} + \|u_n^+\|_{L^\alpha(\Omega)}^{(q-1)p/q} + 1). \tag{4.8}$$

Substituting (4.7) and (4.8) into (4.6) we obtain

$$\begin{aligned} &\|u_n^+\|_{L^\alpha(\Omega)}^\alpha + \|v_n^+\|_{L^\beta(\Omega)}^\beta \\ &\quad \leq c_6(\|u_n^+\|_{L^\alpha(\Omega)}^{p-1} + \|v_n^+\|_{L^\beta(\Omega)}^{(p-1)q/p} + \|u_n^+\|_{L^\alpha(\Omega)}^{(q-1)p/q} + \|v_n^+\|_{L^\beta(\Omega)}^{q-1} + 1). \end{aligned} \tag{4.9}$$

By our assumption (i) and (iii) on the exponents α, β, p and q we find that $\|u_n^+\|_{L^\alpha(\Omega)}$ and $\|v_n^+\|_{L^\beta(\Omega)}$ are bounded. Next we can apply (4.7) and (4.8) again to obtain that $\{z_n\}$ is bounded in E. This ends the proof of P.S. condition.

Now we consider (IQ). Keeping the definitions of Q given in §2 we take the function $z_+ = (u_+, v_+) \in E^+$ so that $u_+ = \phi_1$ and $v_+ = A^{-t}A^s\phi_1 = \lambda_1^{-t+s}\phi_1$, where ϕ_1 is the first eigenfunction of $-\Delta$. In particular $\phi_1 > 0$ in Ω. We need to prove the existence of the constants σ and M used in the definition of Q.

Since we have

$$(\tau\phi_1 + u)^+ = ((\tau + \gamma)\phi_1 + \hat{u})^+ \geq (\tau + \gamma)\phi_1 + \hat{u}, \tag{4.10}$$

we proceed as in the estimate (2.28) and get

$$\tau + \gamma \leq c\|(\tau u_+ + u)^+\|_{L^\alpha(\Omega)}. \tag{4.11}$$

Similarly we obtain

$$\tau - \gamma \leq c\|(\tau v_+ + v)^+\|_{L^\beta(\Omega)}. \tag{4.12}$$

From here on we proceed as in Proposition 2.3. □

Note added in proof. After this paper was finished we learned from Hulshof and van der Vorst that they have obtained similar results in their paper *Differential systems with strongly indefinite variational structure*, J. Funct. Anal. **114** (1993), 32–58.

References

1. R. A. Adams, *Sobolev spaces*, Academic Press, 1975.
2. V. Benci and P. Rabinowitz, *Critical point theorems for indefinite functionals*, Invent. Math. **52** (1979), 241–273.
3. D. G. Costa and C. A. Magalhães, *A variational approach to subquadratic perturbations of elliptic systems*, Preprint.
4. Ph. Clément, D. G. de Figueiredo and E. Mitidieri, *Positive solutions of semilinear elliptic systems*, Comm. Partial Differential Equations **17** (1992), 923–940.
5. ——, *Estimates of positive solutions of systems via Hardy-Sobolev inequalities*, Preprint.
6. P. Felmer, *Periodic solutions of 'superquadratic' Hamiltonian systems*, J. Differential Equations **102** (1993), 188–207.
7. D. Fujiwara, *Concrete characterization of the domains of fractional powers of some elliptic differential operators of the second order*, Proc. Japan Acad. **43** (1967).
8. B. Gidas and J. Spruck, *A priori bounds for positive solutions of nonlinear elliptic equations*, Comm. Partial Differential Equations **6** (1981), 883–901.
9. D. Gilbarg and N. S. Trudinger, *Elliptic partial differential equations of second order*, Springer-Verlag, Berlin, 1983.
10. J. L. Lions and E. Magenes, *Problèmes aux limites non homogènes et applications*, Vol. I, Dunod, Paris, 1968.
11. A. Persson, *Compact linear mappings between interpolation spaces*, Ark. Mat. **5** (1964), 215–219.
12. P. Rabinowitz, *Minimax methods in critical point theory with applications to differential equations*, CBMS Regional Conf. Ser. in Math., no. 65, Amer. Math. Soc., Providence, R.I., 1986.
13. M. A. Souto, Ph.D. Thesis, UNICAMP, Brasil, 1992.

IMECC-UNICAMP, Caixa Postal 6065, 13081 Campinas S. P. Brazil
E-mail address: djairo@ime.unicamp.br

Departamento de Ingeniería Matemática, F.C.F.M., Universidad de Chile, Casilla 170 Correo 3, Santiago, Chile
E-mail address: pfelmer@uchcecvm.bitnet

A Liouville-type Theorem for Elliptic Systems

D.G. DE FIGUEIREDO* - P.L. FELMER*

0. - Introduction

A priori estimates for solutions of superlinear elliptic problems can be established by a blow up technique. Such a method has been used by Gidas-Spruck [GS1] for the case of a single equation. Similar arguments can also be used in the case of systems. We refer to the work of Jie Qing [J] and M.A. Souto [S]. As in the scalar case, the treatment of systems poses the question of the validity of a result which is referred as a Liouville-type theorem for solutions of systems of elliptic equations in $\mathbb{R}^N$. Let us make precise this question.

We consider the elliptic system

$$\left\{\begin{array}{l} -\Delta u = v^{\alpha} \\ -\Delta v = u^{\beta} \end{array}\right. \tag{0.1}$$

in the whole of $\mathbb{R}^N$, $N \geq 3$. The question is to determine for which values of the exponents α and β the only non-negative solution (u, v) of (0.1) is $(u, v) = (0, 0)$. The notion of solution here is taken in the classical sense, i.e., $u, v \in C^2(\mathbb{R}^N)$. In the case of a single equation

$$\Delta u + u^p = 0, \quad u \geq 0 \quad \text{in} \quad \mathbb{R}^N \tag{0.2}$$

it has been proved in [GS2] that the only solution of (0.2) is $u = 0$ when

$$1 \leq p < \frac{N+2}{N-2}, \quad N \geq 3. \tag{0.3}$$

* The first author was partially supported by CNPq and the second by Fondecyt and D.T.I. University of Chile.

Pervenuto alla Redazione il 26 Marzo 1993.

Originally published in: Ann. Scuola Norm. Sup. Pisa XXI (1994), 387-397,
DOI: 10.1007/978-3-319-02856-9_27

In dimension $N = 2$, a similar conclusion holds for $0 \leq p < \infty$. This is a special case of the result that asserts that any superharmonic function bounded below in the whole plane $\mathbb{R}^2$ is necessarily constant, see [PW, Theorem 29, p. 130]. It is also well known that in the critical case, $p = (N+2)/(N-2)$, problem (0.2) has a two-parameter family of solutions given by:

$$u(x) = \frac{[N(N-2)\lambda^2]^{\frac{N-2}{4}}}{[\lambda^2 + |x - x_0|^2]^{\frac{N-2}{2}}}. \tag{0.4}$$

We see then that the critical exponent, in the scalar case, plays an important role in the validity of the Liouville-type theorem for equation (0.2). It is natural to conjecture that in the case of system (0.1) the condition (0.3) is replaced by

$$1 \geq \frac{1}{\alpha+1} + \frac{1}{\beta+1} > \frac{N-2}{N}. \tag{0.5}$$

The basis of this conjecture lies on the fact that the existence of positive solutions for the Dirichlet problem for system (0.1) in a bounded domain holds true if condition (0.5) is satisfied, see [CFM, FF, HV]. If this conjecture were true we would have a complete analogy with the scalar case. A further argument in favor of this conjecture is a theorem of Mitidieri [M, Theorem 3.2] which states that (0.1) has no nontrivial *radial* positive solutions of class $C^2(\mathbb{R}^N)$, $N \geq 3$, provided $1 < \alpha \leq \beta$ and (0.5) holds. The result proved here is the following:

THEOREM. A) If $\alpha > 0$ and $\beta > 0$ are such that

$$\alpha, \beta \leq \frac{N+2}{N-2}, \quad \text{but not both are equal to} \quad \frac{N+2}{N-2}, \tag{0.6}$$

then the only non-negative C^2 solution of (0.1) in the whole of $\mathbb{R}^N$ is the trivial one: $u = 0$, $v = 0$.

B) If $\alpha = \beta = \frac{N+2}{N-2}$, then u and v are radially symmetric with respect to some point of $\mathbb{R}^N$.

REMARK. In [S] it was proved that (0.1) has no nontrivial non-negative C^2 solutions in the whole of $\mathbb{R}^N$ provided

$$\alpha, \beta > 0 \quad \text{and} \quad \frac{1}{\alpha+1} + \frac{1}{\beta+1} \geq \frac{N-2}{N-1}. \tag{0.7}$$

We see then that this establishes the conjecture in a hyperbolic region of the plane (α, β) which is smaller than the region determined by (0.5). However (and this is important to stress) such a region contains points which are not included in the region defined in (0.6). Souto's argument in [S] is based on the non-existence of positive solutions in the whole of $\mathbb{R}^N$ of the inequality

$$\Delta u + u^{\sigma} \leq 0 \tag{0.8}$$

when $\sigma \leq N/(N-2)$. This fact was proved by Gidas [G]. The conjecture would be proved if we had this non-existence result (for inequality (0.8)) up to $(N+2)/(N-2)$. We do not know if such a result is true. In [J] there are some Liouville-type theorems; though Jie's results apply to more general nonlinearities, they do not cover our result.

An important feature of the present paper is to show how the Method of Moving Planes, as developed by Gidas-Ni-Nirenberg [GNN], and more recently by Berestycki-Nirenberg [BN], can be used to produce rather simple and elegant proofs of Liouville-type theorems for systems. The main difficulty stems from the fact that the domain where the solutions are considered is $\mathbb{R}^N \backslash \{0\}$. In such a case it is not clear where we can start the procedure, see details in Section 2. The present work has been motivated by recent papers by Caffarelli-Gidas-Spruck [CGS] and Chen-Li [CL], where Liouville-type theorems for a single equation (results formerly proved by Gidas-Spruck [GS2]) have received a treatment by the Method of Moving Planes.

The authors would like to thank Manuel del Pino for useful conversations.

1. - Some general facts about superharmonic (subharmonic) functions

Let us recall the so called Hadamard Three Spheres Theorem, see [PW, p. 131]:

"Let Ω be an open set containing the set

$$\{x \in \mathbb{R}^N : r_1 \leq |x| \leq r_2\}, \quad N \geq 3$$

and $u \in C^2(\Omega)$ with $\Delta u \geq 0$. For $r_1 \leq r \leq r_2$, let

$$M(r) = \max\{u(x) : |x| = r\}.$$

Then

$$M(r) \leq \frac{M(r_1)(r^{2-N} - r_2^{2-N}) + M(r_2)(r_1^{2-N} - r^{2-N})}{r_1^{2-N} - r_2^{2-N}}, \tag{1.1}$$

for any r in $[r_1, r_2]$."

The proof of this result is very simple. Let $\varphi(r) = a + br^{2-N}$. Choose a and b such that $\varphi(r_1) = M(r_1)$ and $\varphi(r_2) = M(r_2)$. Let $v(x) = u(x) - \varphi(|x|)$, and use the maximum principle for subharmonic functions.

LEMMA 1.1. Let $u \in C^2(\mathbb{R}^N \setminus \{0\})$ be such that $u < 0$ and $\Delta u \geq 0$. Then, for each $\varepsilon > 0$, one has

$$u(x) \leq M(\varepsilon), \quad 0 < |x| \leq \varepsilon \tag{1.2}$$

$$u(x) \leq \frac{M(\varepsilon)\varepsilon^{N-2}}{|x|^{N-2}}, \quad |x| \geq \varepsilon. \tag{1.3}$$

PROOF. Letting $r_1 \to 0$ in (1.1) we obtain $M(r) \leq M(r_2)$ for all $0 < r < r_2$, which implies (1.2) by taking $r_2 = \varepsilon$. Letting $r_2 \to +\infty$ in (1.1) we get $M(r)r^{N-2} \leq M(r_1)r_1^{N-2}$ for all $r \geq r_1$, which gives (1.3) by taking $r_1 = \varepsilon$. □

COROLLARY 1.1. Let $u \in C^2(R^N)$ be such that $u > 0$ and $\Delta u \leq 0$ in the whole of $\mathbb{R}^N$. Let

$$w(x) = \frac{1}{|x|^{N-2}} u\left(\frac{x}{|x|^2}\right)$$

be its Kelvin transform. Then, for each $\varepsilon > 0$, there are positive constants b_ε and c_ε such that

$$w(x) \geq c_\varepsilon, \quad 0 < |x| \leq \varepsilon \tag{1.4}$$

$$\frac{\varepsilon^{N-2}c_\varepsilon}{|x|^{N-2}} \leq w(x) \leq \frac{b_\varepsilon}{|x|^{N-2}}, \quad |x| \geq \varepsilon. \tag{1.5}$$

PROOF. The function $-w$ satisfies the hypothesis of Lemma 1.1. Choosing $c_\varepsilon = \min\{w(x) : |x| = \varepsilon\}$ we obtain (1.4) and the lower bound in (1.5). Choosing $b_\varepsilon = \max\{u(y) : |y| \leq \varepsilon^{-1}\}$ we obtain the upper bound in (1.5). □

2. - Some auxiliary facts

Let $u > 0$ and $v > 0$ be C^2 solutions of system (0.1) with both α and $\beta \leq (N+2)/(N-2)$. Let us introduce their Kelvin transforms

$$w(x) = \frac{1}{|x|^{N-2}} u\left(\frac{x}{|x|^2}\right), \quad z(x) = \frac{1}{|x|^{N-2}} v\left(\frac{x}{|x|^2}\right)$$

which are defined for $x \neq 0$. One verifies that w and z satisfy the system

$$\Delta w + \frac{1}{|x|^{N+2-\alpha(N-2)}} z^\alpha = 0 \tag{2.1}$$

$$\Delta z + \frac{1}{|x|^{N+2-\beta(N-2)}} w^\beta = 0. \tag{2.2}$$

We shall use the method of moving planes. Let us start by considering planes parallel to $x_1 = 0$, coming from $-\infty$. For each real λ let us define

$$\Sigma_\lambda = \{x = (x_1, \dots, x_N) \in \mathbb{R}^N : x_1 < \lambda\}, \quad T_\lambda = \partial\Sigma_\lambda$$

and x^λ as the reflection of x about the plane T_λ. Let $e_\lambda = (2\lambda, 0, \dots, 0)$. In what follows we consider $\lambda \le 0$ and set $\tilde{\Sigma}_\lambda = \Sigma_\lambda \backslash \{e_\lambda\}$. In $\tilde{\Sigma}_\lambda$ we define the following functions

$$w_\lambda(x) = w(x^\lambda), \qquad z_\lambda(x) = z(x^\lambda),$$
$$W_\lambda(x) = w_\lambda(x) - w(x), \qquad Z_\lambda(x) = z_\lambda(x) - z(x).$$

The first step in the use of the method of moving planes is to show that we can start the process. That is, there exists a $\lambda < 0$ such that $W_\lambda(x) \ge 0$ and $Z_\lambda(x) \ge 0$ for all $x \in \tilde{\Sigma}_\lambda$. Unfortunately, this cannot be shown as we shall see shortly. However this can be achieved with a slight modification of the functions. Let us explain how.

For $x \in \tilde{\Sigma}_\lambda$ we have

$$\Delta w(x) + \frac{1}{|x|^{N+2-\alpha(N-2)}} z^\alpha(x) = 0 \tag{2.3}$$

$$\Delta w(x^\lambda) + \frac{1}{|x^\lambda|^{N+2-\alpha(N-2)}} z^\alpha(x^\lambda) = 0. \tag{2.4}$$

Using the invariance of the Laplacian under a reflection and the fact that $|x^\lambda| \le |x|$, it follows from (2.4)

$$\Delta w_\lambda(x) + \frac{1}{|x|^{N+2-\alpha(N-2)}} z_\lambda^\alpha(x) \le 0. \tag{2.5}$$

Hence we get from (2.3) and (2.5)

$$\Delta W_\lambda(x) + \frac{1}{|x|^{N+2-\alpha(N-2)}} (z_\lambda^\alpha(x) - z^\alpha(x)) \le 0$$

which, using the mean value theorem, can be written as

$$\Delta W_\lambda(x) + c(x;\lambda) Z_\lambda(x) \le 0 \tag{2.6}$$

where

$$c(x;\lambda) = \frac{\alpha}{|x|^{N+2-\alpha(N-2)}} (\psi(x;\lambda))^{\alpha-1}$$

and $\psi(x;\lambda)$ is a real number between $z_\lambda(x)$ and $z(x)$. Similarly, we obtain

$$\Delta Z_\lambda(x) + \hat{c}(x;\lambda) W_\lambda(x) \le 0 \tag{2.7}$$

where

$$\hat{c}(x;\lambda) = \frac{\beta}{|x|^{N+2-\beta(N-2)}} (\hat{\psi}(x;\lambda))^{\beta-1}$$

and $\hat{\psi}(x;\lambda)$ is a real number between $w_\lambda(x)$ and $w(x)$. We would like to infer that $W_\lambda(x) \geq 0$ and $Z_\lambda(x) \geq 0$, for some $\lambda < 0$ and all $x \in \Sigma_\lambda$ but we observe that system (2.6)-(2.7) is not amenable for application of maximum principles. In order to obtain a system satisfying the hypothesis of maximum principles (see, for instance, [PW] or [FM]) we introduce the function g and show that the functions

$$\overline{W}_\lambda(x) = \frac{W_\lambda(x)}{g(x)}, \quad \overline{Z}(x) = \frac{Z_\lambda(x)}{g(x)},$$

where $g(x) = \sqrt{1 - x_1}$ for all x such that $x_1 < 0$, satisfy the following system of inequalities

$$\Delta\overline{W}_\lambda + 2\frac{\nabla g}{g} \cdot \nabla\overline{W}_\lambda + \frac{\Delta g}{g}\overline{W}_\lambda + c(x;\lambda)\overline{Z}_\lambda \leq 0 \tag{2.8}$$

$$\Delta\overline{Z}_\lambda + 2\frac{\nabla g}{g} \cdot \nabla\overline{Z}_\lambda + \frac{\Delta g}{g}\overline{Z}_\lambda + \hat{c}(x;\lambda)\overline{W}_\lambda \leq 0. \tag{2.9}$$

The latter system is cooperative, since both $c(x;\lambda)$ and $\hat{c}(x;\lambda)$ are positive. Thus we can apply the maximum principle (Thm. 1.1 in [FM]), provided

$$\frac{\Delta g(x)}{g(x)} + c(x;\lambda) \leq 0, \qquad \frac{\Delta g(x)}{g(x)} + \hat{c}(x;\lambda) \leq 0. \tag{2.10}$$

Our next purpose is to establish.

PROPOSITION 2.1. There exists $\lambda^* \leq 0$ such that for all $\lambda \leq \lambda^*$, $\overline{W}_\lambda(x) \geq 0$ and $\overline{Z}_\lambda(x) \geq 0$ for all $x \in \tilde{\Sigma}_\lambda$.

We shall use in the proof of the above proposition some properties of $\overline{W}_\lambda$ (and a similar set of properties of $\overline{Z}_\lambda$), which are collected in the lemmas below. We consider $\lambda < 0$.

LEMMA 2.1. (i) If $\overline{W}_\lambda(x) > 0$ for all x in some punctured ball $\dot{B}_\varepsilon(e_\lambda) \subset \tilde{\Sigma}_\lambda$, and if

$$c_\lambda =: \inf\{\overline{W}_\lambda(x) : x \in \tilde{\Sigma}_\lambda\} < 0$$

then $\overline{W}_\lambda$ attains its infimum in $\tilde{\Sigma}_\lambda$.

(ii) There exists $\overline{\lambda} < 0$, such that if $\lambda < \overline{\lambda}$ and $c(\lambda) < 0$, then $\overline{W}_\lambda$ attains its infimum in $\tilde{\Sigma}_\lambda$.

PROOF. (i) Since $\overline{W}_\lambda(x) \to 0$ as $|x| \to \infty$, we can find $r_1 > 0$ such that $W_\lambda(x) \geq \frac{1}{2}c(\lambda)$ for $|x| > r_1$. Thus W_λ attains its infimum on the compact set $\overline{B}_{r_1}(0) \cap \overline{(\tilde{\Sigma}_\lambda \backslash B_\varepsilon(e_\lambda))}$. Since W_λ vanishes on T_λ, the result follows.

(ii) Using Corollary 1.1 we find positive numbers c_1 and r such that $w(x) \geq c_1$, for $0 < |x| \leq 1$, and $w(x) \leq \frac{c_1}{2}$ for $|x| \geq r$. Choose $\overline{\lambda} = \min\{-r, -1\}$. Then for $\lambda \leq \overline{\lambda}$.

$$\overline{W}_\lambda(x) \geq \frac{1}{g(x)}\left(c_1 - \frac{c_1}{2}\right) > 0 \quad \text{if} \quad x \in B_1(e_\lambda). \qquad \square$$

LEMMA 2.2. (i) There is a positive constant d_λ, depending only on w and λ, such that

$$\hat{c}(x;\lambda) \le \frac{\beta d_\lambda^{\beta-1}}{|x|^4} \tag{2.11}$$

for all $x \in \tilde{\Sigma}_\lambda - B_1(e_\lambda)$, where $\overline{W}_\lambda(x) < 0$.

(ii) There is an $R_\lambda > 0$, depending only on λ and w, such that

$$\frac{\Delta g(x)}{g(x)} + \hat{c}(x;\lambda) < 0 \tag{2.12}$$

for all $x \in \tilde{\Sigma}_\lambda \backslash B_{R_\lambda}(0)$, where $\overline{W}_\lambda(x) < 0$.

PROOF. (i) If $\beta \ge 1$, it follows from $\overline{W}_\lambda(x) < 0$ that $w_\lambda(x) \le \hat{\psi}(x;\lambda) \le w(x)$. We then estimate $w(x)$ using inequality (1.5) of Corollary 1.1. We see that d_λ can be taken as $\max\{u(y) : |y| \le |\lambda|^{-1}\}$. Then $\hat{\psi}(x;\lambda) \le \frac{d_\lambda}{|x|^{N-2}}$. Using the definition of $\hat{c}(x;\lambda)$ we obtain (2.11) readily. Actually, inequality (2.11) holds for all $x \in \tilde{\Sigma}_\lambda$.

If $0 < \beta < 1$ then we estimate $w_\lambda(x) = w(x^\lambda)$ using the other part of inequality (1.5) to obtain

$$w_\lambda(x) \ge \frac{d_\lambda}{|x_\lambda|^{N-2}} \ge \frac{d_\lambda}{|x|^{N-2}}$$

where $d_\lambda = \min\{w(x) : |x| = 1\}$. From here we obtain (2.11) since $\beta - 1 < 0$.

(ii) Observe that

$$\frac{\Delta g(x)}{g(x)} = -\frac{1}{4(1-x_1)^2} \le -\frac{1}{8(1+x_1^2)}.$$

So (2.12) follows immediately from the above inequality together with (2.11). □

LEMMA 2.3. There exists $\lambda^* \le \bar{\lambda}$ such that, for all $\lambda \le \lambda^*$, inequality (2.12) holds for all $x \in \tilde{\Sigma}_\lambda$, where $\overline{W}_\lambda(x) < 0$.

PROOF. We first observe that $d_\lambda \le d_\mu$ if $\lambda \le \mu \le 0$. So for all $\lambda \le \bar{\lambda}$ we estimate

$$\frac{\Delta g(x)}{g(x)} + \hat{c}(x;\lambda) \le -\frac{1}{8(1+x_1^2)} + \frac{\beta d_{\bar{\lambda}}^{\beta-1}}{|x|^4} \tag{2.13}$$

for all $x \in \tilde{\Sigma}_\lambda$, where $\overline{W}_\lambda(x) < 0$. Hence the right side of (2.13) is negative if $|x| > R_{\bar{\lambda}}$ (as in Lemma 2.2(ii)). Now take λ^* such that $\tilde{\Sigma}_{\lambda^*} \subset \{x \in R^N : |x| > R_{\bar{\lambda}}\}$. □

LEMMA 2.4. Assume that there is $x_0 \in \Sigma_\lambda$ such that

$$\overline{W}_\lambda(x_0) = \inf\{\overline{W}_\lambda(x) : x \in \tilde{\Sigma}_\lambda\} < 0.$$

Then $\overline{Z}_\lambda(x_0) < 0$. If moreover $|x_0| > R_\lambda$ (as in the version of Lemma 2.2(ii) for $\overline{Z}_\lambda$), then $\overline{Z}_\lambda(x_0) < \overline{W}_\lambda(x_0)$.

PROOF. Since $\nabla\overline{W}_\lambda(x_0) = 0$ and $\Delta\overline{W}_\lambda(x_0) \geq 0$, it follows from (2.8) that

$$\frac{\Delta g(x_0)}{g(x_0)}\overline{W}_\lambda(x_0) + c(x_0;\lambda)\overline{Z}_\lambda(x_0) \leq 0, \tag{2.14}$$

which implies that $\overline{Z}_\lambda(x_0) < 0$. Next we can write (2.14) as

$$\left[\frac{\Delta g(x_0)}{g(x_0)} + c(x_0;\lambda)\right]\overline{W}_\lambda(x_0) + c(x_0;\lambda)\left[\overline{Z}_\lambda(x_0) - \overline{W}_\lambda(x_0)\right] \leq 0. \tag{2.15}$$

As we said before an analogue to Lemma 2.2 holds for $\overline{Z}_\lambda$. So the first bracket is negative, and the conclusion follows. □

PROOF OF PROPOSITION 2.1. Let us assume, by contradiction, that for some $\lambda \leq \lambda^*$ (λ^* = least of the two λ^*'s, the one defined in Lemma 2.3 and the one from the version of Lemma 2.3 for $\overline{Z}_\lambda$), we have

$$\overline{W}_\lambda(x) < 0, \quad \text{for some} \quad x \in \tilde{\Sigma}_\lambda \quad \textit{or} \quad \overline{Z}_\lambda(x) < 0, \quad \text{for some} \quad x \in \tilde{\Sigma}_\lambda. \tag{2.16}$$

Without loss of generality, we may assume that $\overline{W}_\lambda(x) < 0$ for some $x \in \tilde{\Sigma}_\lambda$. The argument below shows that the conjunction *or* in (2.16) can be replaced by *and*. Indeed, since $\lambda^* \leq \bar{\lambda}$, the assumption $\overline{W}_\lambda(x) < 0$, for some $x \in \tilde{\Sigma}_\lambda$, implies via Lemma 2.1(ii) that the infimum of $\overline{W}_\lambda$ is attained at a point x_0 in $\overline{W}_\lambda$; and Lemma 2.4 states that $\overline{Z}_\lambda(x_0) < 0$.

We next use Lemma 2.3 to see that the first bracket in (2.15) is negative, and then

$$\overline{Z}_\lambda(x_0) < \overline{W}_\lambda(x_0). \tag{2.17}$$

Now we do a similar argument with $\overline{Z}_\lambda$ and conclude that

$$\overline{W}_\lambda(x_1) < \overline{Z}_\lambda(x_1) \tag{2.18}$$

where $x_1 \in \tilde{\Sigma}_\lambda$ is the point where $\overline{Z}_\lambda$ attains its infimum. Finally we use (2.17), (2.18) and the fact that $\overline{Z}_\lambda(x_1) \leq \overline{Z}_\lambda(x_0)$ and $\overline{W}_\lambda(x_0) \leq \overline{W}_\lambda(x_1)$ to come to a contradiction. □

We next define

$$\lambda_0 = \sup\{\lambda < 0 : \overline{W}_\lambda(x) \geq 0 \quad \text{and} \quad \overline{Z}_\lambda(x) \geq 0, \quad \forall x \in \tilde{\Sigma}_\lambda\}.$$

PROPOSITION 2.2. If $\lambda_0 < 0$, then $\overline{W}_{\lambda_0}(x) \equiv 0$ and $\overline{Z}_{\lambda_0}(x) \equiv 0$ for all $x \in \tilde{\Sigma}_{\lambda_0}$.

REMARK. It follows readily from Proposition 2.2 that both w and z are symmetric with respect to the hyperplane T_{λ_0}.

PROOF OF PROPOSITION 2.2. By continuity we see that $\overline{W}_{\lambda_0}(x) \geq 0$ and $\overline{Z}_{\lambda_0}(x) \geq 0$ for $x \in \tilde{\Sigma}_{\lambda_0}$. Now observe that if $\overline{Z}_{\lambda_0}(x) \equiv 0$, it follows from (2.9) that $\overline{W}_{\lambda_0}(x) \leq 0$, which will imply $\overline{W}_{\lambda_0}(x) \equiv 0$. So if we assume, by contradiction, that the conclusion of Proposition 2.2 is not true, we conclude that *both* $\overline{W}_{\lambda_0}(x)$ and $\overline{Z}_{\lambda_0}(x)$ are not identically zero. It follows from (2.8) that

$$\Delta \overline{W}_{\lambda_0}(x) + 2\,\frac{\nabla g(x)}{g(x)} \cdot \nabla \overline{W}_{\lambda_0}(x) + \frac{\Delta g(x)}{g(x)}\,\overline{W}_{\lambda_0}(x) \leq 0$$

for all $x \in \tilde{\Sigma}_{\lambda_0}$. So by the maximum principle (recall that $\overline{W}_{\lambda_0}(x) \geq 0$) we see that $\overline{W}_{\lambda_0}(x) > 0$ for $x \in \tilde{\Sigma}_{\lambda_0}$. Similarly $\overline{Z}_{\lambda_0}(x) > 0$ for $x \in \tilde{\Sigma}_{\lambda_0}$.

Next using the Hopf maximum principle we obtain that

$$\frac{\partial \overline{W}_{\lambda_0}}{\partial \nu} < 0 \quad \text{and} \quad \frac{\partial \overline{Z}_{\lambda_0}}{\partial \nu} < 0 \tag{2.19}$$

on the boundary T_{λ_0}. (Here $\partial/\partial\nu$ is $\partial/\partial x_1$). We shall see now that this is impossible. From the definition of λ_0, we conclude that there exist a sequence of real numbers $\lambda_k \to \lambda_0$, with $\lambda_k > \lambda_0$, and a sequence of points in $\tilde{\Sigma}_{\lambda_k}$, where $\overline{W}_{\lambda_k}$ or $\overline{Z}_{\lambda_k}$ is negative. Since $W_{\lambda_0}(x)$ satisfies (2.6) and $Z_{\lambda_0}(x) > 0$ for $x \in \tilde{\Sigma}_{\lambda_0}$, we conclude that $W_{\lambda_0}(x)$ is superharmonic in $\tilde{\Sigma}_{\lambda_0}$. In particular W_{λ_0} is superharmonic in $B_{2R}(e_{\lambda_0})\backslash\{e_{\lambda_0}\}$, where $2R = |\lambda_0|$. The argument in the proof of Lemma 1.1 shows that there exists $c_0 > 0$ such that

$$W_{\lambda_0}(x) \geq c_0, \quad \text{if} \quad x \in B_R(e_{\lambda_0})\backslash\{c_{\lambda_0}\}.$$

A similar argument shows that

$$Z_{\lambda_0}(x) \geq c_0, \quad \text{if} \quad x \in B_R(e_{\lambda_0})\backslash\{e_{\lambda_0}\}.$$

(We may choose the same c, or decrease it). So

$$\overline{W}_{\lambda_0}(x) \geq \tilde{c}_0 > 0, \quad \overline{Z}_{\lambda_0}(x) \geq \tilde{c}_0 > 0, \quad \text{if} \quad x \in B_R(e_{\lambda_0})\backslash\{e_{\lambda_0}\}.$$

By continuity we have

$$\overline{W}_{\lambda_k}(x) \geq \frac{\tilde{c}_0}{2}, \quad \overline{Z}_{\lambda_k}(x) \geq \frac{\tilde{c}_0}{2}, \quad \text{if} \quad x \in B_{R/2}(e_{\lambda_k})\backslash\{e_{\lambda_k}\}, \tag{2.20}$$

for k sufficiently large.

It follows from Lemma 2.1(i) (using (2.20)) and Lemma 2.4 that for k sufficiently large both $\overline{W}_{\lambda_k}$ and $\overline{Z}_{\lambda_k}$ attain their negative infima in $\tilde{\Sigma}_{\lambda_k}$. [Recall that we are using the contradiction hypothesis that $\overline{W}_{\lambda_k}$ or $\overline{Z}_{\lambda_k}$ is negative somewhere in $\tilde{\Sigma}_{\lambda_k}]$. Let us denote by x_k and y_k, respectively, the points of minima of $\overline{W}_{\lambda_k}$ and $\overline{Z}_{\lambda_k}$. If follows from Lemma 2.4 that at least one of the sequences, $\{x_k\}$ or $\{y_k\}$ is bounded. Assume that $\{x_k\}$ is bounded and passing to a subsequence assume that $x_k \to \overline{x}$. By continuity we have that $\nabla\overline{W}_{\lambda_0}(\overline{x}) = 0$ and $\overline{W}_{\lambda_0}(\overline{x}) \leq 0$. Since $\overline{x} \neq e_{\lambda_0}$ and $\overline{W}_{\lambda_0}(x) > 0$ in $\tilde{\Sigma}_{\lambda_0}$, we conclude that $\overline{x} \in T_{\lambda_0}$, which contradicts the fact observed before that $\dfrac{\partial \overline{W}_{\lambda_0}}{\partial \nu} < 0$ on T_{λ_0}.

A similar argument applies if we assume that $\{y_k\}$ is bounded. □

3. - Proof of the theorem

We first prove **Part A**. Performing the moving plane procedure we have two possibilities:

(i) If $\lambda_0 < 0$, it follows from Proposition 2.2 that w and z are symmetric will respect to the plane T_{λ_0}. But looking at equations (2.1) and (2.2) we realize that this is impossible.

(ii) If $\lambda_0 = 0$, we conclude that $w_0(x) \geq w(x)$ and $z_0(x) \geq z(x)$ for all $x \in \Sigma_0$. We can perform the procedure from the right and we will reach a $\lambda_0^r = 0$, (λ_0^r cannot be positive, in virtue of an argument as in (i) above), from which we get $w(x) \geq w_0(x)$ and $z(x) \geq z_0(x)$ for all $x \in \Sigma_0$. So w and z are symmetric with respect to the plane $x_1 = 0$.

(iii) This reasoning can be made from any direction. And so the only possibility would be that both w and z are radially symmetric with respect to the origin 0. But 0 was chosen arbitrarily when we perform the Kelvin transform. Thus u and v are radially symmetric with respect to any point. Then they would be constant and from the equations we finally obtain $u = v = 0$.

Part B. We show that w and z are symmetric with respect to some plane parallel to $x_1 = 0$. Indeed, if $\lambda_0 < 0$, this follows from Proposition 2.2, and in this case the plane is T_{λ_0}. If $\lambda_0 = 0$, we perform the moving plane procedure from the right and find a corresponding $\lambda_0^r \geq 0$. If $\lambda_0^r > 0$, an analogue to Proposition 2.2 shows that w and z are symmetric with respect to $T_{\lambda_0^r}$. If $\lambda_0^r = 0$ we proceed as in (ii) above to conclude that both w and z are symmetric with respect to $x_1 = 0$. We perform this moving plane procedure taking planes perpendicular to any direction, and for each direction $\gamma \in R^N$, $|\gamma| = 1$ we find a plane T_γ with the property that both w and z are symmetric with respect to T_γ. A simple argument shows that all these planes intersect at a single point, or $w = z = 0$. □

REFERENCES

[BN] H. Berestycki - L. Nirenberg, *On the method of moving planes and the sliding method.* Bol. Soc. Brasil. Mat. **22** (1991), 1-22.

[CFM] Ph. Clément - D.G. de Figueiredo - E. Mitidieri, *Positive solutions of semilinear elliptic systems.* Comm. Partial Differential Equations **17** (1992), 923-940.

[CGS] L. Caffarelli - B. Gidas - J. Spruck, *Asymptotic Symmetry and local behavior of Semilinear Elliptic Equations with Critical Sobolev Growth.* Comm. Pure App. Math., **XLII** (1989), 271-297.

[CL] W. Chen - C. Li, *Classification of solutions of some nonlinear elliptic equations.* Duke Math. J. **63** (1991), 615-622.

[FF] D.G. de Figueiredo - P.L. Felmer, *On Superquadratic Elliptic Systems.* To appear in Trans. Amer. Math. Soc.

[FM] D.G. de Figueiredo - E. Mitidieri, *Maximum Principles for Linear Elliptic Systems.* Rend. Ist. Mat. Univ. Trieste **XXII** (1990), 36-66.

[HV] J. Hulshof - R. van der Vorst, *Differential Systems with Strongly Indefinite Variational Structure.* J. Fatl. Anal, vol. **114** n° 1 (1993), 32-58.

[G] B. Gidas, *Symmetry properties and isolated singularities of positive solutions of nonlinear elliptic equations.* In: Nonlinear Partial Differential Equations in Engineering and Applied Sciences. Editors R. Sternberg, A. Kalinovski and J. Papadakis. Marcel Dekker Inc., 1980.

[GNN] B. Gidas - W.M. Ni - L. Nirenberg, *Symmetry and related properties via the maximum principle.* Comm. Math. Phys. **68** (1979), 209-243.

[GS1] B. Gidas - J. Spruck, *A priori bounds for positive solutions of nonlinear elliptic equations*, Comm. Partial Differential Equations **6** (1981), 883-901.

[GS2] B. Gidas - J. Spruck, *Global and local behavior of positive solutions of nonlinear elliptic equations.* Comm. Pure App. Math. **34** (1981), 525-598.

[J] Qing Jie, *A priori estimates for positive solutions of semilinear elliptic systems*, J. Partial Differential Equations **1** (1988), 61-70.

[M] E. Mitidieri, *A Rellich type identity and applications.* To appear in Comm. Partial Differential Equations.

[PW] M.H. Protter - H.F. Weinberger, *Maximum principles in differential equations*, Prentice Hall (1967).

[S] M.A. Souto, *Sobre a existência de soluções positivas para sistemas cooperativos não lineares.* PhD thesis, Unicamp (1992).

IMECC - UNICAMP
Caixa Postal 6065
13081 Campinas, S.P.
Brasil

Depto. Ing. Matemática, FCFM
Universidad de Chile
Casilla 170, Correo 3
Santiago - Chile

NoDEA 1 (1994) 119-123

1021-9722/94/020119-5 $ 1.50+0.20

Monotonicity and symmetry of solutions of elliptic systems in general domains *

Djairo G. DE FIGUEIREDO
IMECC - UNICAMP
Caixa Postal 6065
13081 Campinas, S.P., Brasil

1 Introduction

A new Maximum Principle for elliptic equations has appeared recently in works of Berestycki, Nirenberg and Varadhan, see for instance [1]. The aim of this note is to extend this Maximum Principle to cooperative elliptic systems and then to apply it to monotonicity and symmetry properties of nonlinear elliptic systems. In this way we get more general results than the ones in [2] and [3], with even simpler proofs. The interest in these results comes from their use in obtaining a priori bounds for positive solutions of semilinear elliptic systems [6].

Let $\Omega \subset I\!R^N$ be an arbitrary domain and consider the differential operators

$$L_k(D) = \sum b_{ij}^k(x) D_i D_j + \sum b_i^k(x) D_i\,, \quad 1 \le k \le n$$

which are assumed to be uniformly elliptic with L^∞ coefficients. We may assume without loss of generality that for a suitable constant $c_0 > 0$ we have

$$c_0^{-1}|\xi|^2 \le \sum b_{ij}^k(x)\xi_i\xi_j \le c_0|\xi|^2\,, \qquad \forall x \in \Omega$$
$$\forall k \qquad \forall \xi \in I\!R^N$$

and

$$|b_i^k(x)| \le c_0\,, \qquad \forall\, i, k\,.$$

We consider these operators acting on functions which are in $E =: W_{\text{loc}}^{2,N}(\Omega) \cap C^0(\overline{\Omega})$. So the matricial operator

$$\mathcal{L} = \begin{bmatrix} L_1 & & \\ & \ddots & \\ & & L_n \end{bmatrix}$$

acts on $E \times \overset{n}{\ldots} \times E$.

*This work was done when the author visited Rutgers University under partial support of CNPq.

Originally published in: NoDEA Nonlinear Differential Equations and Appl. 1 (1994), 119-123, © 1994 Birkhäuser Verlag, Basel
Reprint by Springer-Verlag Berlin Heidelberg 2013,
DOI: 10.1007/978-3-319-02856-9_28

Let $A = (a_{ij}(x))$ be a cooperative $n \times n$ matrix with entries $a_{ij} \in L^\infty$. Cooperativeness means $a_{ij}(x) \geq 0$ for $i \neq j$. We use capital letters to denote vector functions $U(x) = (u_1(x), \ldots, u_n(x))$.

We write $U(x) \geq 0$ to mean $u_j(x) \geq 0, \quad \forall j = 1, \ldots, n$.

Following [1] we introduce the

Definition 1.1 *A Maximum Principle holds for $\mathcal{L} + A$ in Ω if*

$$(\mathcal{L} + A)U \geq 0 \quad \text{in} \quad \Omega \tag{1.1}$$

$$\limsup_{x \to \partial\Omega} U(x) \leq 0 \tag{1.2}$$

implies $U \leq 0$ in Ω.

We start by proving the Maximum Principle for bounded domains Ω.

Proposition 1.1 *Assume that $\operatorname{diam} \Omega \leq d$. Then there exists $\delta > 0$ depending on N, n, d and c_0, such that the maximum principle holds if the Lebesgue measure $|\Omega|$ of Ω is less than δ.*

Proof. Consider the kth equation

$$L_k u_k + \sum a_{kj}(x) u_j \geq 0$$

and rewrite it in the form

$$L_k u_k - a_{kk}^- u_k \geq -a_{kk}^+ u_k - \sum_{j \neq k} a_{kj} u_j .$$

Since A is cooperative we get

$$L_k u_k - a_{kk}^- u_k \geq -a_{kk}^+ u_k^+ - \sum_{j \neq k} a_{kj} u_j^+ .$$

Applying the Aleksandrov maximum principle, see for instance [4] for statement and proof, we get

$$\sup u_k \leq C \| a_{kk}^+ u_k^+ + \sum_{j \neq k} a_{kj}^+ u_j^+ \|_{L^N} ,$$

where C depends on N, d and c_0. Replacing $\sup u_k$ by $\sup u_k^+$ and adding up these n inequalities we get

$$\sum \sup u_k^+ \leq C n c_0 |\Omega|^{1/N} \sum \sup u_k^+ .$$

So $\sum \sup u_k^+ \leq 0$ if we choose $\delta = (Cnc_0)^{-N}$. ∎

2 Semilinear elliptic systems

Now we consider the system

$$\mathcal{L}U + F(U) = 0 \quad \text{in} \quad \Omega \tag{2.1}$$

where $\mathcal{L} = \begin{pmatrix} \Delta & & \\ & \ddots & \\ & & \Delta \end{pmatrix}$, and $F(U) = (f_1(u_1, \ldots, u_n), \ldots, f_n(u_1, \ldots, u_n))$.

We assume that $F : R^n \to R^n$ is C^1 and its Jacobian has all off-diagonal entries ≥ 0, i.e.

$$\frac{\partial f_i}{\partial \xi_j}(\xi) \geq 0 \qquad \forall \xi \in R^n, \ i \neq j\,.$$

We consider nonnegative solutions of system (1.2) which vanish on $\partial\Omega$. We remark that Ω is not necessarily bounded.

Fix a direction γ, i.e., $\gamma \in R^N$, $|\gamma| = 1$. We suppose that there is $a > -\infty$ such that

$$\gamma \cdot x > a \qquad \forall x \in \Omega\,.$$

This means that Ω is at one side of some hyperplane normal to γ. For simplicity, assume $\gamma = (1, 0, \ldots, 0)$. We recall the following standard notation, [5]. For $\lambda \in R$ and writing $x = (x_1, y)$:

$$\begin{aligned} T_\lambda &= \{x \in R^N : \quad \gamma \cdot x = \lambda\} \\ \Sigma(\lambda) &= \{x \in \Omega : \quad \gamma \cdot x < \lambda\} \\ \Sigma'(\lambda) &= \{x \in R^N : \quad x^\lambda \in \Sigma(\lambda)\} \end{aligned}$$

where x^λ is the reflection of x with respect to T_λ. I.e., if $x = (x_1, y)$ then $x^\lambda = (2\lambda - x_1, y)$.

Let $\lambda_0 = \sup\{\lambda : T_\lambda \cap \Omega = \emptyset\}$. Now we make the following assumption on the domain

$$(D) \qquad \exists \varepsilon > 0 \text{ s.t. for } \lambda_0 < \lambda < \lambda_0 + \varepsilon\,,\ \Sigma'(\lambda) \subset \Omega\,, \text{ and } \textstyle\sum(\lambda) \text{ is bounded.}$$

This assumption is satisfied in the case considered in [2]: Ω bounded, $\partial\Omega$ of class C^2. (D) is also satisfied on the assumptions in [3].

Now define

$$\overline{\lambda} = \sup\{\lambda : \Sigma'(\mu) \subset \Omega\,,\ \Sigma(\mu) \text{ bounded}, \quad \forall \mu < \lambda\}\,.$$

Theorem 2.1 *Let $U \geq 0$ be a solution of (2.1) in $W^{2,N}_{\text{loc}}(\Omega) \cap C^0(\overline{\Omega})$ with $U = 0$ on $\partial\Omega$.*

Then, each $u_k(x)$ is monotonically increasing with respect to x_1 for $x \in \Sigma(\overline{\lambda})$.

Note 2.1 *Suppose that $U = (u_1, \dots, u_n) \in E$ satisfies (1.1) with $U = 0$ on $\partial\Omega$ and $U \geq 0$ in Ω. If A is cooperative, it follows by the maximum principle that either $u_k \equiv 0$ or $u_k > 0$ in Ω.*

Note 2.2 *Without loss of generality we assume that $u_k \not\equiv 0$ for all k. Indeed if for one such a k, u_k where $\equiv 0$, we reduce the system to the remaining $k-1$ functions and equations.*

Proof. The proof follows the usual ideas of the moving planes method, as used by [5], [1] and later papers. For each λ, with $\lambda_0 < \lambda < \overline{\lambda}$ we define

$$v_k(x;\lambda) = v_k(x_1, y; \lambda) =: u_k(2\lambda - x_1, y; \lambda)\,, \qquad \forall x \in \Sigma(\lambda)$$

and $w_k(x;\lambda) = v_k(x;\lambda) - u_k(x), W(x,\lambda) = (w_1(x;\lambda), \dots, w_n(x;\lambda))$. Using the invariance of the Laplacian we get

$$\Delta w_k + f_k(v_1, \dots, v_n) - f_k(u_1, \dots, u_n) = 0$$

or

$$\Delta w_k + \sum \frac{\partial f_k}{\partial \xi_j}(\theta_1^k(x), \dots, \theta_n^k(x)) w_j = 0 \tag{2.2}$$

where $\theta_i^k(x)$ is between $u_i(x)$ and $v_i(x)$. So the coefficients of (2.2) are bounded, and a little argument shows that they are measurable. Now if $\lambda - \lambda_0$ is small enough the Lebesgue measure of $\sum(\lambda)$ is small and we are in condition to apply Proposition 1.1 to (2.2) and conclude that $W \geq 0$ in $\sum(\lambda)$. Now let

$$\lambda^* = \sup\{\lambda < \overline{\lambda} : W(x;\lambda) \geq 0 \quad \text{in} \quad \Sigma(\lambda)\}.$$

If $\lambda^* = \overline{\lambda}$, the proof is complete. Suppose by contradiction that $\lambda^* < \overline{\lambda}$. By continuity $W(x;\lambda^*) \geq 0$ and using Note 2.1 above we see that $W(x,\lambda^*) > 0$ in $\sum(\lambda^*)$. Let now d be the diameter of $\sum(\lambda^*)$. By Proposition 1.1 there is a $\delta > 0$ such that a Maximum Principle holds for system (2.2) in domains of measure less than δ. Take a compact set $K \subset \sum(\lambda^*)$ s.t. $|\sum(\lambda^*) \setminus K| < \delta/2$. Let $a > 0$ be such that $w_k(x;\lambda^*) \geq a$ for $x \in K$ and $k = 1, \dots, n$. By continuity we can find λ, with $\lambda^* < \lambda < \overline{\lambda}$, such that $|\sum(\lambda) \setminus K| < \delta$ and $w_k(x;\lambda) \geq a/2$ for $x \in K$ and $k = 1, \dots, n$. So applying Proposition 1.1 we see that $W(x;\lambda) \geq 0$ in $\sum(\lambda) \setminus K$. So $W(x;\lambda) \geq 0$ in $\sum(\lambda)$, contradicting the definition of λ^* ■

The next two results are immediate consequences of the above theorem. The first one is an extension of Theorem 1.3 [1]. In particular, it gives the radial symmetry of positive solutions of (2.1) subject to Dirichlet boundary conditions in the case that Ω is a ball. The second one is proved in [3]. It gives monotonicity of solutions of (2.1) in certain directions, for domains like cones, paraboloids, cylinders.

Corollary 2.1 *Let Ω be an arbitrary bounded domain in R^n which is convex on the x_1 direction and symmetric with respect to the plane $x_1 = 0$. Let $U \geq 0$ be a solution of the system (2.1) subject to Dirichlet boundary conditions. F satisfies the conditions put in the beginning of Section 2. Then U is symmetric with respect to x_1 and $u_{x_1} < 0$ for $x_1 < 0$ in Ω.*

Corollary 2.2 *Suppose Ω is an unbounded domain satisfying the following property: there are a $\gamma \in I\!R^n$, $|\gamma| = 1$ and a number $a > -\infty$ such that*

i) $x \cdot \gamma > a\,,\ \forall x \in \Omega,$

ii) $\forall x \in \Omega \Rightarrow x + t\gamma \in \Omega\,,\ \forall t \geq 0$

iii) $\forall \lambda$ *the set* $\{x \in \Omega : x \cdot \gamma < \lambda\}$ *is bounded.*

Let U be a positive solution of (2.1) under Dirichlet boundary conditions. Then U is (strictly) increasing in the direction γ.

References

[1] H. BERESTYCKI, L. NIRENBERG, On the Method of Moving Planes and the Sliding Method, *Bol. Soc. Bras. Mat.* **22**, 1-22 (1991)

[2] W. C. TROY, Symmetry Properties in Systems of Semilinear Elliptic Equations, *J. Diff. Eq.* **42**, 400-413 (1981)

[3] M. BADIALE, Monotonicity of Solutions for Elliptic Systems on Unbounded Domains, *Boll. U.M.I.* **6A**, 59-69 (1992)

[4] D. GILBARG, N. S. TRUDINGER, *Elliptic partial differential equations of second order*, Springer-Verlag, Berlin and New York (1977)

[5] B. GIDAS, W. -M. NI, L. NIRENBERG, Symmetry and related properties via the maximum principle, *Comm. Math. Phys.* **68**, 209-243 (1979)

[6] Ph. CLÉMENT, D. G. DE FIGUEIREDO, E. MITIDIERI, Positive Solutions of Semilinear Elliptic Systems, *Comm. P.D.E* **17**, 923-940 (1992)

Received December 15, 1992

Calc. Var. 3, 139–153 (1995)

Elliptic equations in R^2 with nonlinearities in the critical growth range

D.G. de Figueiredo[1,*], O.H. Miyagaki[2,*], B. Ruf[3]

[1] IMECC-UNICAMP, Caixa Postal 6065, 13081 Campinas, S.P., Brasile
[2] Departamento de Matemática, Universidade Federal de Uberlândia, 38400 Uberlândia, M.G.
[3] Dipartimento di Matematica, Università degli Studi, I-20133 Milano, Italy

Received June 10, 1993 / Accepted January 3, 1994

1. Introduction

In this paper we study the solvability of problems of the type

$$(1.1)\qquad -\Delta u = f(x,u) \text{ in } \Omega \;,\quad u=0 \text{ on } \partial\Omega,$$

where Ω is some bounded domain in R^2, and the function $f(x,s)$ has the maximal growth on s which allows to treat problem (1.1) variationally in $H_0^1(\Omega)$. More precisely, we treat the so-called subcritical case and also the critical case, which we define next.

We say that f has *subcritical growth* at $+\infty$ if for all $\alpha > 0$

$$(1.2)\qquad \lim_{t\to+\infty} \frac{|f(x,t)|}{e^{\alpha t^2}} = 0$$

and f has *critical growth* at $+\infty$ if there exists $\alpha_0 > 0$ such that

$$(1.3)\quad \lim_{t\to+\infty} \frac{|f(x,t)|}{e^{\alpha t^2}} = 0 \;,\; \forall\alpha > \alpha_0 \;;\; \lim_{t\to+\infty} \frac{|f(x,t)|}{e^{\alpha t^2}} = +\infty \;,\; \forall\, \alpha < \alpha_0.$$

Similarly we define subcritical and critical growth at $-\infty$. This notion of criticality is motivated by the so-called Trudinger-Moser inequality [12,9] which says that if u is a $H_0^1(\Omega)$ function then the integral $\int e^{u^2}$ is finite.

Problems of the above type have been studied recently by several authors, Atkinson-Peletier [5], Carleson-Chang [7] , Adimurthi et al. [1], [2], [3], [4]. In this paper we improve the existence conditions in [2], and extend the results to the nonsymmetrical case and to more general nonlinearities. Also we propose a

* These authors have been partially supported by CNPq

Originally published in: Calc. Var. and Partial Differential Equations 3 (1995), 139-153, © Springer-Verlag 1995
Reprint by Springer-Verlag Berlin Heidelberg 2013,
DOI: 10.1007/978-3-319-02856-9_29

unified approach by putting all questions in the framework of the by now classical Critical Point Theory as first developed in papers by Ambrosetti-Rabinowitz and Rabinowitz. For a complete reference of the results used here, see Rabinowitz [11; Theorems 2.2, 5.3 and 9.12]. As usual in the applications the hard points are the verification of conditions which allow the use of this Critical Point Theory, in particular the Palais-Smale condition.

In the results for nonlinearities with *critical growth* (Theorems 1.3 and 1.4 below) we give sufficient conditions for the existence of solutions. The proofs of these results follow the ideas introduced by H. Brezis and L. Nirenberg [6] in their pioneering work on the solvability of equations with critical growth in dimensions larger than 2. In fact, one observes that (as in their case) the functional under consideration satisfies the Palais-Smale condition only at certain levels. In order to assure that the constructed minimax levels are inside the Palais-Smale region we use test functions connected with the optimal Trudinger-Moser inequality (while Brezis- Nirenberg used test functions related to the optimal Sobolev imbedding).

For easy reference we state now conditions on f that will be assumed in all theorems below.

(H1) $$f : \overline{\Omega} \times \mathbb{R} \to \mathbb{R} \text{ is continuous, } f(x,0) = 0.$$

(H2) $\exists\ t_0 > 0\ ,\ \exists M > 0$ such that
$$0 < F(x,t) = \int_0^t f(x,s)ds \le M\,|f(x,t)|\ ,\ \ \forall |t| \ge t_0, \forall x \in \Omega.$$

(H3) $$0 < F(x,t) \le \frac{1}{2}\, f(x,t)t\ ,\ \ \forall\ t \in \mathbb{R} - \{0\}\ ,\ \ \forall x \in \Omega.$$

Now we state the results which will be proved here. We denote by $0 < \lambda_1 < \lambda_2 \le \lambda_3 \le \ldots$ the eigenvalues of $(-\Delta, H_0^1(\Omega))$. By "solution" in the theorems below we mean weak solution $u \in H_0^1(\Omega)$.

Theorem 1.1. (The subcritical case, local minimum at 0). *Assume (H1), (H2), (H3) and that f has subcritical growth at both $+\infty$ and $-\infty$. Furthermore suppose that*

(H4) $$\limsup_{t\to 0} \frac{2\,F(x,t)}{t^2} < \lambda_1\ ,\ \text{uniformly in } (x,t).$$

Then, problem (1.1) has a nontrivial solution. Moreover if $f(x,t)$ is an odd function in t, then (1.1) has infinitely many solutions.

Theorem 1.2. (The subcritical case, saddle at 0). *Assume (H1), (H2), (H3) and that f has subcritical growth at both $+\infty$ and $-\infty$. Furthermore suppose that*

(H5) $$\exists \delta > 0\ ,\ \exists\ \lambda_k \le \mu < \lambda_{k+1} \text{ such that } F(x,t) \le \frac{1}{2}\ \mu t^2\ ,\quad \forall x \in \Omega, \forall |t| \le \delta.$$

(H6) $$F(x,t) \geq \frac{1}{2}\lambda_k t^2 \qquad \forall x \in \Omega\ ,\ \forall t \in \mathbb{R}.$$

Then, problem (1.1) has a nontrivial solution. Moreover, if instead of (H6) we assume that $f(x,t)$ is an odd function in t, then (1.1) has infinitely many solutions.

For the problem with critical growth, Adimurthi [2] showed that (1.1) has a solution provided f satisfies (among other conditions) the asymptotic hypothesis $\lim_{t\to\infty} f(x,t)te^{-\alpha_0 t^2} = +\infty$. The next two theorems improve and generalize this result. In order to state these theorems we introduce the following notations: for $0 \leq \varepsilon < 1$

$$M_\varepsilon = \lim_{n\to\infty} \int_0^1 ne^{n[(1-\varepsilon)^2 t^2 - t]} dt$$

$$\widehat{M} = \lim_{\varepsilon\to 0} M_\varepsilon$$

Numerical calculations indicate that $M_0 = 2$ and $\widehat{M} = 1$. We recall the concept of inner radius of a set Ω:

$$d =: \text{radius of the largest open ball} \subset \Omega.$$

Theorem 1.3. (The critical case, local minimum at 0). *Assume $(H1), (H2), (H3)$ and that f has critical growth at both $+\infty$ and $-\infty$. Furthermore assume $(H4)$ and*

(H7) $$\lim_{t\to+\infty} f(x,t)e^{-\alpha_0 t^2} t \geq \beta\ ,\quad \beta > \frac{4}{(1+M_0)\alpha_0 d^2}$$

Then, problem (1.1) has a nontrivial solution.

Theorem 1.4. (The critical case, saddle point at 0). *Assume $(H1), (H2), (H3)$ and that f has critical growth at both $+\infty$ and $-\infty$. Furthermore assume $(H5), (H6)$ and*

(H8) $$\lim f(x,t)e^{-\alpha_0 t^2} t \geq \beta\ ,\quad \beta > \frac{4}{\widehat{M}\alpha_0 d^2}\ ,$$

(H9) $\exists \delta > 0$ and $C > 0$ such that $|f(x,t)| \leq C|t|, \forall x \in \Omega, \forall |t| \leq \delta$.

Then problem (1.1) has a nontrivial solution.

Remarks on the conditions above. Conditions (H1) and (H2) imply an exponential growth in t for both $f(x,t)$ and $F(x,t)$, see relation (2.2) later on.
Condition (H3) is satisfied if we assume as in [2] that f is C^1 and that $f'(x,t) \geq f(x,t)t^{-1}$, for all $t \neq 0$. Indeed, such a condition implies that $f(x,t)t^{-1}$ is a nondecreasing function of t, and from this fact (H3) follows readily. Condition (H5) is satisfied if one assumes as in [2] that f is C^1 and

$$\lambda_k \leq \inf_{x\in\overline{\Omega}} f'(x,0) \leq \sup_{x\in\overline{\Omega}} f'(x,0) < \lambda_{k+1}$$

In order to see this one uses the mean value theorem and the continuity of f' at $s = 0$. If one assumes in addition, that $f'(x,t) \geq f(x,t)t^{-1}$, for all $t \neq 0$, then condition (H6) follows. Finally condition (H9) is satisfied if f is C^1.

2. The variational formulation

We assume (H1), (H2) and the existence of positive constants C and β such that

(2.1) $$|f(x,t)| \leq C\, e^{\beta t^2} \qquad \forall\, x \in \Omega\ ,\ \forall\, t \in \mathbb{R}.$$

(In particular, this is the case if f has subcritical or critical growth). Then

$$\Phi(u) = \frac{1}{2}\int |\nabla u|^2 - \int F(x,u)$$

is a C^1 functional $\Phi : H_0^1(\Omega) \to \mathbb{R}$, and

$$< \Phi'(u), v >= \int \nabla u \nabla v - \int f(x,u)v\ ,\quad \forall v \in H_0^1(\Omega),$$

where $< .,. >$ denotes the inner product in H_0^1 (we also write $\|.\|$ for the corresponding H_0^1-norm). These statements follow from the fact that e^{v^2} is $L^1(\Omega)$ for all $v \in H_0^1(\Omega)$, see [12,9]. So in view of (2.1) we conclude that $f(x,u(x))$ is in $L^q(\Omega)$, for all $q > 1$, when $u \in H_0^1(\Omega)$.

It follows easily from (H1) and (H2) that
(i) there is a constant $C > 0$ such that

(2.2) $$F(x,t) \geq Ce^{\frac{1}{M}|t|} \qquad \forall\ |t| \geq t_0;$$

(ii) given $\varepsilon > 0$ there is $t_\varepsilon > 0$ such that

(2.3) $$F(x,t) \leq \varepsilon f(x,t)t \quad \forall x \in \Omega,\quad \forall |t| \geq t_\varepsilon.$$

Proposition 2.1. *Assume (H1), (H2), (H3) and (2.1). Then the functional Φ satisfies $(PS)_c$ for all $c < \frac{2\pi}{\beta}$.*

Corollary 2.1. *Assume (H1), (H2) and (H3). If f has subcritical growth at both $+\infty$ and $-\infty$ then Φ satisfies $(PS)_c$ for all $c \in \mathbb{R}$. If f has critical growth at both $+\infty$ and $-\infty$ with the same α_0, then Φ satisfies $(PS)_c$ for all $c \in (-\infty, \frac{2\pi}{\alpha_0})$.*

Proof of Proposition 2.1. Let $(u_n) \subset H_0^1(\Omega)$ be a Palais-Smale sequence, i.e.

(2.4) $$\frac{1}{2}\int |\nabla u_n|^2 - \int F(x,u_n) \to c$$

(2.5) $$\left|\int \nabla u_n \nabla v - f(x,u_n)v\right| \leq \varepsilon_n \|v\|, \qquad \forall v \in H_0^1(\Omega),$$

where $\varepsilon_n \to 0$. It follows from (2.4) using (2.3) that, for any $\varepsilon > 0$,

$$\frac{1}{2}\|u_n\|^2 \ \leq\ C + \int F(x,u_n) \leq C_\varepsilon + \varepsilon \int f(x,u_n)u_n,$$

and using (2.5) we obtain

$$\frac{1}{2}\|u_n\|^2 \leq C_\varepsilon + \varepsilon\|u_n\|^2 + \varepsilon_n\|u_n\|,$$

which implies that

$$\text{(2.6)} \qquad \|u_n\| \leq C \ , \quad \int f(x,u_n)u_n \leq C \qquad \int F(x,u_n) \leq C.$$

Observe that from (H3) the two integrals above are nonnegative. Now we take a subsequence denoted again by (u_n) such that, for some $u \in H_0^1$, we have

$$u_n \rightharpoonup u \quad \text{in } H_0^1 \ ; \ u_n \to u \ \text{ in } L^q(\Omega), \ \forall\, q \geq 1 \ ; \ u_n(x) \to u(x) \ \text{ a.e. in } \Omega.$$

Next we assume the following result, which will be proved later.

Lemma 2.1. *$f(x,u_n) \to f(x,u)$ in $L^1(\Omega)$.*

It follows from (H2) and Lemma 2.1, using the generalized Lebesgue dominated convergence theorem, that $F(x,u_n) \to F(x,u)$ in $L^1(\Omega)$. From (2.4) and (2.5) we obtain

$$\text{(2.7)} \ \lim\|u_n\|^2 = 2(c + \int F(x,u)) \ ; \ \lim \int f(x,u_n)u_n = 2(c + \int F(x,u)).$$

Using (H3) and (2.7) we conclude that $c \geq 0$. So any Palais-Smale sequence approaches a nonnegative level. It follows from Lemma 2.1 and relation (2.5) that

$$\int \nabla u \nabla \psi = \int f(x,u)\psi \qquad \forall\ \psi \in D(\Omega).$$

Since $f(x,u) \in L^2(\Omega)$ we conclude that $u \in H^2(\Omega)$ and $-\Delta u = f(x,u)$ in the strong sense. Hence

$$\int |\nabla u|^2 = \int f(x,u)u \geq 2\int F(x,u).$$

So $\Phi(u) \geq 0$. Now we separate the proof into three cases.

Case 1. $c = 0$. If this is the case we have using (2.7)

$$0 \leq \Phi(u) \leq \liminf \Phi(u_n) = \int F(x,u) - \int F(x,u) = 0.$$

So $\|u_n\| \to \|u\|$ and then $u_n \to u$ in H_0^1. The proof is finished in this case.

Case 2. $c \neq 0$, $u = 0$. We show that this cannot happen for a Palais-Smale sequence. First we claim that, for some $q > 1$, we have

$$\text{(2.8)} \qquad \int |f(x,u_n)|^q \leq \ \text{const.}$$

Assume that (2.8) has been proved. Then using (2.5) with $v = u_n$ we have

$$(2.9)\qquad \left| \int |\nabla u_n|^2 - \int f(x,u_n)u_n \right| \le \varepsilon_n \|u_n\| \le C\varepsilon_n .$$

We estimate the second integral above using Hölder's inequality. And then from (2.8) and the fact that $u_n \to 0$ in $L^{q'}$, we conclude that $\|u_n\| \to 0$. This contradicts (2.7), which says, in this case , that $\|u_n\|^2 \to 2c \neq 0$. So it remains to prove (2.8). Since $u = 0$, it follows form (2.7) that, given $\varepsilon > 0$, $\|u_n\|^2 \le 2c+\varepsilon$, for large n. So we estimate the integral in (2.8) using (2.1)

$$(2.10)\qquad \int |f(x,u_n)|^q \le C \int e^{q\beta u_n^2} = C \int e^{q\beta\|u_n\|^2(\frac{u_n}{\|u_n\|})^2}.$$

By the sharpened form of Trudinger's inequality proved by Moser [9], the integral in (2.10) is bounded, independently of n, if $q\beta\|u_n\|^2 < 4\pi$. But this will be indeed the case for $c < \frac{2\pi}{\beta}$, if we choose $q > 1$ sufficiently close to 1 and ε sufficiently small.

Case 3. $c \neq 0$, $u \neq 0$. In this case we claim that

$$(2.11)\qquad \Phi(u) = c.$$

If this is the case, it follows from (2.7) that $\|u_n\| \to \|u\|$ and the proof is also finished in this case. So it remains to prove (2.11). Assume by contradiction that $\Phi(u) < c$.
Then

$$(2.12)\qquad \|u\|^2 < 2(c + \int F(x,u))$$

Let $v_n = u_n/\|u_n\|$ and $v = u/\sqrt{2(c + \int F(x,u))}$. Since $v_n \rightharpoonup v$, $\|v_n\| = 1$, and $\|v\| < 1$, it follows by a result of Lions [8] that

$$(2.13)\qquad \sup \int e^{4\pi p v_n^2} < \infty \quad , \quad \forall p < \frac{1}{1 - \|v\|^2}.$$

Now we estimate the L^q-norm of $f(x,u_n)$ using (2.1)

$$\int |f(x,u_n)|^q \le C \int e^{q\beta\|u_n\|^2 v_n^2}$$

and this will be bounded if q and p can be chosen such that

$$q\beta\|u_n\|^2 \le 4\pi p < 4\pi\frac{c + \int F(x,u)}{c - \Phi(u)} = 4\pi\frac{1}{1 - \|v\|^2}.$$

This will be the case for large n if

$$\frac{\beta}{2\pi} < \frac{1}{c - \Phi(u)}$$

which is actually so, since $\Phi(u) \ge 0$ and $c < \frac{2\pi}{\beta}$.

Finally, using the fact just proved that $f(x,u_n)$ is bounded in some L^q, we see as in case 2 that $u_n \to u$ in $H_0^1(\Omega)$. This is impossible in view of (2.7) and (2.12). Thus, the proof of Proposition 2.1 is complete. □

Now we give the complete statement of lemma 2.1 and its respective proof.

Lemma 2.1. *Let (u_n) be a sequence of functions in $L^1(\Omega)$ converging to u in $L^1(\Omega)$. Assume that $f(x, u_n(x))$ and $f(x, u(x))$ are also L^1 functions. If*

$$\int |f(x, u_n(x))u_n(x)| \le C_1$$

then $f(x, u_n)$ converges in L^1 to $f(x, u)$.

Proof. It suffices to prove $\int |f(x, u_n)| \to \int |f(x, u)|$, cf. [10, p. 89]. Since $f(x, u(x)) \in L^1(\Omega)$ it follows that given $\varepsilon > 0$ there is a $\delta > 0$ such that

$$\int_A |f(x, u(x)| \le \varepsilon \qquad \text{if} \qquad |A| \le \delta \tag{2.14}$$

for all measurable subsets A of Ω. We use $|\cdot|$ to denote the Lebesgue measure. Next using the fact that $u \in L^1(\Omega)$ we find $M_1 > 0$ such that

$$|\{x \in \Omega : |u(x)| \ge M_1\}| \le \delta. \tag{2.15}$$

Let $M = \max\{M_1, C_1/\varepsilon\}$. We write

$$\left| \int |f(x, u_n)| - \int |f(x, u)| \right| \le I_1 + I_2 + I_3$$

and estimate each integral separately.

$$I_1 \equiv \int_{|u_n| \ge M} |f(x, u_n)| = \int_{|u_n| \ge M} \frac{f(x, u_n)u_n}{|u_n|} \le \frac{C_1}{M} \le \varepsilon.$$

By the choices made above we have

$$I_3 \equiv \int_{|u| \ge M} |f(x, u)| \le \varepsilon$$

Next we claim that

$$I_2 \equiv \int_{|u_n| < M} |f(x, u_n)| - \int_{|u| < M} |f(x, u)| \longrightarrow 0 \text{ as } n \to \infty$$

Indeed, $g_n(x) \equiv |f(x, u_n(x))|\chi_{|u_n| < M} - |f(x, u(x))|\chi_{|u| < M}$ tends to 0 a.e. in Ω. Moreover $|g_n(x)| \le |f(x, u(x))|$, if $|u_n| \ge M$ and $|g_n(x)| \le C + |f(x, u(x))|$, if $|u_n(x)| < M$, where $C = \sup\{|f(x, t)| : x \in \overline{\Omega}, |t| < M\}$. So, the claim follows from the Lebesgue dominated convergence theorem. □

The next proposition concerns the behaviour of Φ at ∞.

Proposition 2.2. *Assume (H1) and (H2). Let Z be a finite dimensional subspace of $H_0^1(\Omega)$ spanned by L^∞ functions. Then Φ is bounded above in Z, and moreover, given $M > 0$ there is an $R > 0$ such that*

$$\Phi(u) \le -M, \qquad \forall \|u\| \ge R \qquad u \in Z.$$

Proof. Given $u_0 \in Z$ with $\|u_0\|_{L^\infty} = 1$, let us define

$$\xi(t) = \Phi(tu_0) = \frac{t^2}{2}\int |\nabla u_0|^2 - \int F(x, tu_0) \qquad \forall\ t \in R.$$

It follows from (2.2) that, for $p > 2$, there is a constant $C > 0$ such that $F(x,t) \geq C|t|^p - C$ for all x and t. So

$$\xi(t) \leq \frac{t^2}{2}\|u_0\|^2 - C|t|^p \int |u_0|^p - C|\Omega|.$$

By the equivalence of norms in Z, we obtain

$$\xi(t) \leq \frac{t^2}{2}\|u_0\|^2 - C|t|^p\|u_0\|^p - C$$

which implies that $\xi(t) \to -\infty$ as $t \to \infty$. The result follows by compactness. □

Next we study the behavior of the functional Φ near $u = 0$.

Proposition 2.3. *Assume (H1), (H2), (H4) and condition (2.1). Then there exist* $a > 0$ *and* $\rho > 0$ *such that*

$$\Phi(u) \geq a \qquad \text{if} \qquad \|u\| = \rho.$$

Proof. From (H4) we have that there exist $\mu < \lambda_1$ and $\delta > 0$ such that

$$F(x,t) \leq \frac{1}{2}\ \mu\, t^2 \quad ,\text{if} \quad |t| \leq \delta.$$

On the other hand, from (2.1) we obtain for $q > 2$

$$F(x,t) \leq Ce^{\beta t^2}|t|^q \ , \quad \text{if} \quad |t| > \delta,$$

Putting these two estimates together we obtain

$$F(x,t) \leq \frac{1}{2}\ \mu\, t^2 + Ce^{\beta t^2}|t|^q \qquad \forall\ \ t \in \mathbb{R},$$

which implies

$$\begin{aligned}\Phi(u) \ &\geq\ \frac{1}{2}\int |\nabla u|^2 \ -\frac{1}{2}\mu \int u^2 - C\int e^{\beta u^2}|u|^q \\ &\geq\ \frac{1}{2}(1 - \frac{\mu}{\lambda_1})\|u\|^2 - C\left(\int e^{\beta p u^2}\right)^{1/p}\left(\int |u|^{qp'}\right)^{1/p'}\end{aligned}$$

Next we observe that

$$\int e^{\beta p u^2} = \int e^{\beta p \|u\|^2 (\frac{u}{\|u\|})^2} < \quad \text{const}$$

if $\|u\| \leq \delta$, where $\beta p \delta < 4\pi$. So

$$\Phi(u) \geq \frac{1}{2}(1 - \frac{\mu}{\lambda_1})\|u\|^2 - C\|u\|^q$$

Now choose $\rho > 0$ as the point where the function $g(s) = \frac{1}{2}(1 - \frac{\mu}{\lambda_1})s^2 - Cs^q$ assumes its maximum. Take $a = g(\rho)$. □

The above proposition will be used in the proofs of Theorems 1.1 and 1.3. For the proofs of the other theorems we shall need the next result. Let us denote by V the subspace of $H_0^1(\Omega)$ generated by the eigenfunctions φ_j of $(-\Delta,\ H_0^1(\Omega))$ corresponding to the eigenvalues λ_j for $j = 1, \ldots, k$. Let $W = V^\perp$.

Proposition 2.4. *Assume* $(H1)$, $(H2)$, $(H5)$ *and condition (2.1). Then, there exist* $a > 0$ *and* $\rho > 0$ *such that*

$$\Phi(u) \geq a \quad \text{if} \quad \|u\| = \rho \quad \text{and} \quad u \in W$$

Proof. The proof is completely analogous to the previous one; here we use the variational characterization of λ_{k+1}.

3. Proof of Theorems 1.1 and 1.2.

It follows from the hypotheses in both theorems that Φ satisfies $(PS)_c$ for all $c \in \mathbb{R}$, cf. Corollary 2.1. To finish the proof of Theorem 1.1 we use Propositions 2.2 and 2.3, and apply the Mountain Pass Theorem. To prove theorem 1.2 we apply the Generalized Mountain Pass Theorem. Proposition 2.4 and the following argument are used. Let $R > \rho$ and such that $\Phi(u) \leq 0$ for $\|u\| \geq R$ and $u \in V \oplus \mathbb{R}\varphi_{k+1}$, see Proposition 2.2. Let $Q = \{v + s\varphi_{k+1} : \|v\| \leq R,\ 0 \leq s \leq R\}$ and ∂Q its relative boundary in $V \oplus \mathbb{R}\varphi_{k+1}$. Clearly $\Phi(u) \leq 0$ in $\partial Q \cap \{u : \|u\| \geq R\}$. For $u \in \partial Q \cap V$, we use $(H6)$ to see that $\Phi(u) \leq 0$. If the function $f(x,t)$ is odd in t, then we apply Theorem 9.12 of [11] in order to conclude the existence of infinitely many solutions. □

4. Proof of Theorem 1.3.

It follows from the assumptions that Φ satisfies $(PS)_c$ for all $c < \frac{2\pi}{\alpha_0}$, see Corollary 2.1. From $(H4)$ we see that Φ has a local minimum at 0, see Proposition 2.3. To conclude via the Mountain Pass Theorem it suffices to show that there is a $\omega \in H_0^1$, $\|\omega\| = 1$, such $\max\{\Phi(t\omega) : t \geq 0\} < \dfrac{2\pi}{\alpha_0}$. For that matter we start by introducing the following functions

$$\overline{\omega}_n(x) = \frac{1}{\sqrt{2\pi}} \begin{cases} (\log n)^{1/2}, & 0 \leq |x| \leq \dfrac{1}{n} \\ \dfrac{\log \frac{1}{|x|}}{(\log n)^{1/2}}, & \dfrac{1}{n} \leq |x| \leq 1 \\ 0 & |x| \geq 1 \end{cases}$$

We see that $\overline{\omega}_n \in H_0^1(B_1(0))$ and $\|\overline{\omega}_n\| = 1$ for all $n = 1, 2, \ldots$ Here $B_1(0)$ denotes the ball of radius 1 centered at the origin of R^2. Next let d be the inner radius of Ω and $x_0 \in \Omega$ such that $B_d(x_0) \subset \Omega$. We then define the functions

$$\omega_n(x) = \overline{\omega}_n\left(\frac{x - x_0}{d}\right),$$

which are in $H_0^1(\Omega)$, $||\omega_n|| = 1$ and $\text{supp}\omega_n \subset B_d(x_0)$.

We claim that there exists n such that

$$\max\{\Phi(t\omega_n) : t \geq 0\} < \frac{2\pi}{\alpha_0}.$$

Suppose by contradiction that this is not the case. So, for all n, this maximum (it is indeed a maximum, in view of Proposition 2.2) is larger or equal to $\dfrac{2\pi}{\alpha_0}$. Let $t_n > 0$ be such that

$$\max\{\Phi(t\omega_n) : t \geq 0\} = \Phi(t_n\omega_n) \geq \frac{2\pi}{\alpha_0}. \tag{4.1}$$

It follows readily from (4.1) and $(H3)$ that

$$t_n^2 \geq \frac{4\pi}{\alpha_0}. \tag{4.2}$$

Also at $t = t_n$, we have $\dfrac{d}{dt}\Phi(t\omega_n) = 0$, i.e.

$$t_n - \int f(x, t_n\omega_n)\omega_n = 0 \ ,$$

which implies that

$$t_n^2 \geq \int_{B_d(x_0)} f(x, t_n\omega_n)t_n\omega_n. \tag{4.3}$$

Now, it follows from $(H7)$ that given $\varepsilon > 0$ there exists $s_\varepsilon > 0$ such that

$$f(x, s)s \geq (\beta - \varepsilon)e^{\alpha_0 s^2}, \ \ \forall s \geq s_\varepsilon.$$

So from (4.3) we obtain, for large n

$$\begin{aligned} t_n^2 \ &\geq \ (\beta - \varepsilon)\pi\frac{d^2}{n^2}e^{\alpha_0 t_n^2 \frac{1}{2\pi}\log n} = \\ &= \ (\beta - \varepsilon)\pi d^2 e^{2\log n(\frac{\alpha_0 t_n^2}{4\pi} - 1)} \end{aligned} \tag{4.4}$$

which implies readily that t_n is bounded. And moreover (4.4) together with (4.2) gives that $t_n^2 \to 4\pi/\alpha_0$.

Next let us estimate (4.3) more precisely. Let

$$A_n = \{x \in B_d(x_0) : t_n\omega_n(x) \geq t_\varepsilon\} \ \ , \ \ \ B_n = B_d(x_0)\backslash A_n \ ,$$

and break the integral in (4.3) into a sum of the integrals over A_n and B_n. Using (H7) we estimate (4.3):

$$t_n^2 \geq (\beta - \varepsilon)\int_{B_d(x_0)} e^{\alpha_0 t_n^2\omega_n^2} + \int_{B_n} f(x, t_n\omega_n)t_n\omega_n - (\beta - \varepsilon)\int_{B_n} e^{\alpha_0 t_n^2\omega_n^2} \tag{4.5}$$

Since $t_n\omega_n(x) < t_\varepsilon$ for $x \in B_n$, we see that the characteristic functions $\chi_{B_n} \to 0$ a.e. in $B_d(x_0)$ as $n \to \infty$. Hence, the two last integrals in (4.5) go to 0 as $n \to \infty$, in view of the Lebesgue Dominated Convergence Theorem. Passing to the limit in (4.5) we obtain

$$(4.6) \qquad \frac{4\pi}{\alpha_0} \geq (\beta - \varepsilon) \lim \int_{B_d(x_0)} e^{\alpha_0 t_n^2 \omega_n^2} \geq (\beta - \varepsilon) \lim \int_{B_d(x_0)} e^{4\pi\omega_n^2}$$

The last integral in (4.6), denote it by I_n, is evaluated as follows:

$$(4.7) \qquad I_n = d^2 \int_{B_1(0)} e^{4\pi\overline{\omega}_n^2} = d^2 \left\{ \frac{\pi}{n^2} e^{4\pi\frac{1}{2\pi}\log n} + 2\pi \int_{1/n}^{1} e^{4\pi\frac{1}{2\pi}\frac{(\log\frac{1}{r})^2}{\log n}} rdr \right\}$$

Changing variables in the integral in (4.7), $s = \log\left(\frac{1}{r}\right) / \log n$, we obtain

$$I_n = d^2 \left\{ \pi + 2\pi \log n \int_0^1 e^{2s^2 \log n - 2s \log n} ds \right\}$$

So finally from (4.6) we get

$$\frac{4\pi}{\alpha_0} \geq (\beta - \varepsilon) d^2 \pi (1 + M_0) \quad , \quad \forall \varepsilon > 0,$$

which implies $\beta \leq 4/\alpha_0 d^2(1 + M_0)$, a contradiction to (H7). □

5. Proof of Theorem 1.4.

It follows from the assumptions that Φ satisfies $(PS)_c$ for all $c < \frac{2\pi}{\alpha_0}$, see Corollary 2.1. The proof is accomplished by the use of the Generalized Mountain Pass Theorem. For that matter we have to select a $\omega \in W$, $||\omega|| = 1$, such that for some $R > \rho$, [the ρ of Proposition 2.4], one has the set

$$Q = \{v + s\omega : ||v|| \leq R,\ v \in V,\ 0 \leq s \leq R\}$$

with the properties: (p_1) $\Phi|_{\partial Q} \leq 0$ and (p_2) $\Phi(u) < \frac{2\pi}{\alpha_0}$, $\forall u \in Q$. Property (p_1) follows easily from $(H6)$ and Proposition 2.2; in fact this part is true no matter which ω we choose in W. However, in order to get (p_2) we choose a sequence ω_n like in the proof of Theorem 1.3. We then need to show that there is $n > \rho$ such that

$$\max\{\Phi(v + s\omega_n) : ||v|| \leq n,\ 0 \leq s \leq n\} < \frac{2\pi}{\alpha_0}.$$

Assume, by contradiction, that this is not the case. So for all n this maximum is $\geq \frac{2\pi}{\alpha_0}$. Let $u_n = v_n + t_n\omega_n$ be the point where this maximum is achieved. So

$$(5.1) \qquad \Phi(u_n) \geq \frac{2\pi}{\alpha_0}$$

$$\|u_n\|^2 - \int f(x, u_n)u_n = 0 \tag{5.2}$$

where we have used the fact that the derivative of Φ, restricted to $V \oplus \mathbb{R}\omega_n$, is zero at u_n. A contradiction will be obtained after the proof of the following assertions.

Assertion 1. (v_n), (t_n) *are bounded sequences. See proof later.*

So we may assume that $v_n \to v_0$ and $t_n \to t_0$.

Assertion 2. $v_0 = 0$ *and* $t_0^2 = \frac{4\pi}{\alpha_0}$. *See proof later.*

In view of Assertion 2, the part of the integral in (5.2) over $\Omega \setminus B_d(x_0)$ goes to 0.

So

$$\lim \|u_n\|^2 = \lim \int_{B_d(x_0)} f(x, u_n)u_n \tag{5.3}$$

The integral in (5.3), denote it by I_n, can be estimated as in the previous theorem

$$I_n \geq (\beta - \varepsilon) \int_{B_d(x_0)} e^{\alpha_0 u_n^2} + \int_{B_n} f(x, u_n)u_n - (\beta - \varepsilon) \int_{B_n} e^{\alpha_0 u_n^2} \tag{5.4}$$

Again the characteristic functions of B_n go to zero a.e, and consequently the last two integrals in (5.4) go to 0. Now given $\varepsilon > 0$, there is $c(\varepsilon)$ such that $u_n^2 \geq (1 - \varepsilon)t_n^2\omega_n^2 - c(\varepsilon)v_n^2$. Since $t_n^2 \to 4\pi/\alpha_0$, we get for large n

$$u_n^2 \geq (1 - \varepsilon)^2 \frac{4\pi}{\alpha_0}\omega_n^2 - c(\varepsilon)\|v_n\|_{L^\infty}^2 .$$

So the first integral in (5.4) is estimated from below by

$$(\beta - \varepsilon)e^{-c(\varepsilon)\|v_n\|_{L^\infty}^2} \int_{B_d(x_0)} e^{(1-\varepsilon)^2 4\pi\omega_n^2} \tag{5.5}$$

The integral in (5.5), denote it by J_n, can be evaluated as follows

$$J_n = d^2 \int_{B_1(0)} e^{(1-\varepsilon)^2 4\pi\overline{\omega}_n^2} = d^2 \left\{ \frac{\pi}{n^2} e^{(1-\varepsilon)^2 \log n} + 2\pi \int_{1/n}^{1} e^{(1-\varepsilon)^2 2(\log \frac{1}{r})^2 (\log n)^{-1}} r\, dr \right. \tag{5.6}$$

Using a change of variables as before we prove that the second integral in (5.6) goes to $\pi\widehat{M}$. So $J_n \to \pi\widehat{M}$. Hence

$$\lim J_n \geq (\beta - \varepsilon)d^2\pi\widehat{M} \tag{5.7}$$

On the other hand, for any ε,

$$\lim \|u_n\|^2 \leq (1 + \varepsilon) \lim t_n^2 = (1 + \varepsilon)\frac{4\pi}{\alpha_0}$$

So it follows from (5.3) and (5.7) that

$$(1+\varepsilon)\frac{4\pi}{\alpha_0} \geq (\beta - \varepsilon)d^2\pi\widehat{M} \quad ,\forall\varepsilon > 0$$

which implies $\beta \leq 4/\alpha_0 d^2\widehat{M}$, a contradiction to (H8). □

Proof of Assertion 1. Given (t_n) and (v_n) as above, one of the following two possibilities has to hold:
(i) either there exists a constant $C_0 > 0$ such that $t_n/||v_n|| \geq C_0$, or
(ii) there are subsequences, denoted again by (t_n) and (v_n), such that $t_n/||v_n|| \to 0$.

First, let us assume that (i) holds. Then there is a constant $C > 0$ such that

$$||u_n|| \leq ||v_n|| + t_n \leq Ct_n \quad ,$$

which applied to (5.2) gives

$$C^2t_n^2 \geq ||u_n||^2 \geq \int_{B_{d/n}(x_0)} f(x,u_n)u_n \geq (\beta - \varepsilon)\int_{B_{d/n}(x_0)} e^{\alpha_0 u_n^2} \quad , \tag{5.8}$$

where (H8) was used. Let $m_n = \frac{1}{\sqrt{2\pi}}(\log n)^{1/2}$. So

$$u_n(x) = t_n m_n\left(\frac{v_n(x)}{t_n}\frac{1}{m_n} + 1\right) \quad , \forall x \in B_{d/n}(x_0).$$

Hence, given $\varepsilon \in (0,1)$, we have $u_n(x) \geq (1-\varepsilon)t_n m_n$ for large n and $x \in B_{d/n}(x_0)$. Going on with the estimate in (5.8) we obtain

$$C^2t_n^2 \geq (\beta - \varepsilon)\frac{\pi}{n^2}e^{\alpha_0(1-\varepsilon)^2 t_n^2 m_n^2},$$

which can be written as

$$C^2t_n^2 \geq (\beta - \varepsilon)\pi d^2 e^{2\log n(\alpha_0(1-\varepsilon)^2 t_n^2 \frac{1}{4\pi} - 1)}. \tag{5.9}$$

It follows readily that t_n is bounded. Consequently $||v_n||$ is also bounded in case (i).

Next, we assume that (ii) holds. Then, for large n, we have $t_n \leq ||v_n||$, which implies $||u_n|| \leq 2||v_n||$. Suppose by contradiction that $||v_n|| \to \infty$. As before, let $s_\varepsilon > 0$ be such that

$$f(x,s)s \geq (\beta - \varepsilon)e^{\alpha_0 s^2} \quad ,\forall s \geq s_\varepsilon.$$

It follows then from (5.2)

$$1 \geq \int_{u_n \geq s_\varepsilon} \frac{f(x,u_n)u_n}{||u_n||^2} \geq \frac{\beta - \varepsilon}{4}\int_{u_n \geq s_\varepsilon} \frac{e^{\alpha_0 u_n^2}}{||v_n||^2} \quad . \tag{5.10}$$

Now in order to estimate the exponent in the last integral above we observe that

$$\frac{u_n}{||v_n||}\chi_{u_n \geq t_\varepsilon} = \frac{v_n}{||v_n||} + \frac{t_n}{||v_n||}\omega_n - \frac{u_n}{||v_n||}\chi_{u_n < t_\varepsilon}$$

converges a.e. to $\widehat{v} \in V$ where $\widehat{v}$ is the limit of $v_n/||v_n||$ in H_0^1, and $||\widehat{v}|| = 1$. So using Fatou's lemma in (5.10) and recalling that we have assumed $||v_n|| \to \infty$, we come to a contradiction. So $||v_n|| \leq$ const. and, consequently, also t_n is bounded.

Before proving Assertion 2 we show the folllowing auxiliary result.

Lemma 5.1. $t_0^2 \geq 4\pi/\alpha_0$.

Proof. First observe that ω_n converges weakly to zero in H_0^1. So $||u_n||^2 \to t_0^2 + ||v_0||^2$. On the other hand, we observe that $u_n \to v_0$ in $L^1(\Omega)$ and the other hypotheses of Lemma 2.1 are satisfied. So $\int f(x, u_n) \to \int f(x, v_0)$, and consequently $\int F(x, u_n) \to \int F(x, v_0)$. Using these informations in (5.1) we obtain

$$\frac{2\pi}{\alpha_0} \leq \frac{1}{2}(t_0^2 + ||v_0||^2) - \int F(x, v_0)$$

Using (H6) we get

$$\frac{2\pi}{\alpha_0} \leq \frac{1}{2}(t_0^2 + ||v_0||^2) - \frac{1}{2}\lambda_k \int v_0^2 \leq \frac{1}{2}t_0^2$$

where we have used the variational characterization of λ_k.

Remark 5.1. We may assume that in every neighbourhood of $t = 0$, $F(x, t) \not\equiv \frac{1}{2}\lambda_k t^2$. Since otherwise $f(x, t) \equiv \lambda_k t$ in some neighbourhood of $t = 0$, and this would imply that $\varepsilon\varphi_k$ would be a solution of (1.1) for small $\varepsilon > 0$ and φ_k an eigenfunction of $-\Delta$ corresponding to the eigenvalue λ_k. Hence if $v_0 \not\equiv 0$ then we actually would have $t_0^2 > 4\pi/\alpha_0$.

Proof of Assertion 2. We work again with the alternative set in the Proof of Assertion 1. We first observe that in view of lemma 5.1 the alternative (ii) cannot hold. So let us assume (i). We conclude readily from (5.9) that

$$t_0^2 \leq \frac{4\pi}{\alpha_0(1-\varepsilon)^2} \Rightarrow t_0^2 \leq \frac{4\pi}{\alpha_0}.$$

This, in conjunction with Lemma 5.1, implies $t_0^2 = 4\pi/\alpha_0$.Finally, as a consequence of Remark 5.1 we conclude that $v_0 = 0$. □

References

1. Adimurthi: Existence of Positive Solutions of the Semilinear Dirichlet Problem with critical Growth for the n-Laplacian. Ann. Sc. Norm. Sup. Pisa **XVII** (1990) 393–413
2. Adimurthi: Existence Results for the Semilinear Dirichlet Problem with Critical Growth for the n-Laplacian (preprint)
3. Adimurthi, S.L. Yadava: Multiplicity Results for Semilinear Elliptic Equations in a Bounded Domain of R^2 involving critical exponent. Ann. Sc. Norm. Sup. Pisa **XVII** (1990) 481–504

4. Adimurthi, P.N. Srikanth, S.L. Yadava: Phenomena of Critical Exponent in R^2. Proc. Royal Soc. Edinb. **119A** (1991) 19–25
5. F.V. Atkinson, L.A. Peletier: Elliptic equations with critical growth. Math. Inst. Univ. Leiden, Rep 21 (1986)
6. H. Brezis, L. Nirenberg: Positive solutions of nonlinear elliptic equations involving critical sobolev exponents. Comm. Pure Appl. Math. **36** (1983) 437–477
7. L. Carleson, A. Chang: On the existence of an extremal function for an inequality of J. Moser. Bull. Sc. Math. **110** (1986) 113–127
8. P.-L. Lions: The Concentration-compactness principle in the Calculus of Variations. Part I. Revista Iberoamericana **11** (1985) 185–201
9. J. Moser: A sharp form of an inequality by N. Trudinger. Ind. Univ. Math. J. **20** (1971) 1077–1092
10. H.L. Royden: Real Analysis. New York (1963)
11. P.H. Rabinowitz: Minimax methods in critical point theory with applications to differential equations. CBMS, no. 65 AMS (1986) 100
12. N.S. Trudinger: On imbedding into Orlicz spaces and some applications. J. Math. Mech. **17** (1967), 473–484

[illegible] equations [illegible]

4. A. [illegible], [illegible] phenomena of [illegible] [illegible] (1990) [illegible]
5. [illegible] A. [illegible], [illegible] equations with critical growth, [illegible] Inst. Henri [illegible] (1984)
6. H. Brezis, L. Nirenberg, Positive solutions of nonlinear elliptic equations involving critical [illegible] exponents, Comm. Pure Appl. Math. 36 (1983) [illegible]
7. [illegible], A. [illegible], [illegible] Bull. Sci. Math. 110 (1986) [illegible]
8. [illegible] The [illegible] from the Calculus of Variations [illegible] 17 (2003) 165–[illegible]
9. [illegible] Math. [illegible] (1972) [illegible]
10. R. [illegible], Real Analysis, [illegible] York (1963)
11. [illegible] methods in critical point theory with applications to differential equations, CBMS [illegible] AMS (1986) [illegible]
12. [illegible] [illegible] and [illegible] applications, [illegible] Math. [illegible] (1976) [illegible]

Pergamon

Nonlinear Analysis, Theory, Methods & Applications, Vol. 33, No. 3, pp. 211-234, 1998

Printed in Great Britain
0362-546X/98 \$19.00+0.00

PII: S0362-546X(97)00548-8

DECAY, SYMMETRY AND EXISTENCE OF SOLUTIONS OF SEMILINEAR ELLIPTIC SYSTEMS

DJAIRO G. DE FIGUEIREDO†‡ and YANG JIANFU§¶

†IMECC-UNICAMP, 13083-970 Campinas, S.P., Brazil and
§Department of Mathematics, Namchang University, Nanchang, Jiangxi 330047, People's Republic of China

(Received 8 November 1996; received for publication 21 May 1997)

Key words and phrases: Critical point theorems, symmetry of solutions, elliptic systems.

1. INTRODUCTION

In this paper we consider systems of semilinear elliptic equations of the form

$$-\Delta u + u = g(x, v) \tag{1.1}$$

$$-\Delta v + v = f(x, u) \tag{1.2}$$

in the whose of R^N. All along this paper we suppose $N \geq 3$. The basic assumptions on the functions f and g are listed below. Similar assumptions have been introduced in [1, 2] for the case of bounded domains, instead of R^N. Further assumptions and restrictions will be made, in each section, in order to get finer results. Our utmost aim was originally to establish the existence of solutions of system (1.1)–(1.2). However, in this effort we got sidetracked to consider other questions. In this way we come to interesting results on asymptotic behavior of solutions, on symmetry properties of such solutions, and the existence of ground states.

The difficulties in treating system (1.1)–(1.2) originate in at least three facts. First, there is a lack of compactness due to the fact that we are working in R^N. Second, due to the type of growth of the functions f and g, we cannot work with the usual $H^1(R^N)$, and then we need fractional Sobolev spaces. Third, although we have a variational problem, the functional associated to it is strongly indefinite, and so we have to recourse to the critical point theorems due to [3] and [4]. As to the type of lack of compactness showing up here, some authors have been able to overcome it due to the fact that, in their problems, the term u in the first equation comes with a coefficient $a(x) > 0$, $a(x) \to +\infty$ as $|x| \to \infty$; similarly v in the second equation with a coefficient $b(x) > 0$, $b(x) \to +\infty$, cf. [5–7]. With such assumptions, or similar ones, there is essentially no loss of compactness, since it is well known that the operator $-\Delta + a(x)$ in R^N behaves very much like $-\Delta$ in a bounded domain; that is the spectral theory is the same, cf. [8]. Another way of regaining some sort of compactness is working with radial symmetric functions. This was first observed in [9, 10], and used by several authors since then, [11], ... This is the approach we use in Section 5. Although the compactness in this case is retained for the space of radial symmetric functions, there is no compactness in the linear differential operator ($-\Delta + id$ in R^N). This contrasts with the case of $-\Delta$ in a bounded domain. In Section 4 we develop the spectral theory needed.

‡ With partial support of CNPq.
¶ Supported by FAPESP.

Originally published in: Nonlinear Anal. - TMA 33 (1998), 211-234,

Reprint by Springer-Verlag Berlin Heidelberg 2013 with kind permission,
DOI: 10.1007/978-3-319-02856-9_30

Next we list the basic assumptions on the functions f and g.

(H1) $f, g: R^N \times \mathbb{R} \to \mathbb{R}$ are continuous functions, with the property that there is an $\varepsilon > 0$ such that

$$tf(x, t) \geq 0 \qquad tg(x, t) \geq 0, \qquad \text{for all } |t| < \varepsilon.$$

(H2) There is a constant $c_1 > 0$ such that

$$|f(x, t)| \leq c_1(|t|^p + 1), \qquad |g(x, t)| \leq c_1(|t|^q + 1), \qquad \text{for all } t,$$

where $p, q > 1$, and

$$\frac{1}{p+1} + \frac{1}{q+1} > 1 - \frac{2}{N}, \qquad N \geq 3, \tag{1.3}$$

and

$$p, q \leq (N+4)/(N-4) \qquad \text{if } N \geq 5, \qquad \text{(this in addition to (1.3))} \tag{1.4}$$

(H3) There are constants $\alpha, \beta > 2$ and $c_0 > 0$ such that

$$ct^\alpha \leq \alpha F(x, t) \leq tf(x, t), \qquad ct^\beta \leq \beta G(x, t) \leq tg(x, t), \qquad \text{for all } t. \tag{1.5}$$

(H4) There are real numbers $a, b \geq 1$ and positive constants c_2 and r such that

$$|f(x, t)| \leq c_2|t|^a, \qquad |g(x, t)| \leq c_2|t|^b, \qquad \text{for } |t| \leq r.$$

Remarks on the above hypotheses. (1) Condition (H3) implies that both functions f and g are superlinear. Indeed, from the inequalities in (H3) we get

$$F(x, t) \geq c|t|^\alpha, \qquad G(x, t) \geq c|t|^\beta \tag{1.6}$$

$$|f(x, t)| \geq c|t|^{\alpha-1}, \qquad |g(x, t)| \geq c|t|^{\beta-1}, \tag{1.7}$$

for all t.

(2) Inequality (1.3) says that our problem is subcritical, since the pair (p, q) is below the so-called "critical hyperbola"

$$\frac{1}{p+1} + \frac{1}{q+1} = 1 - \frac{2}{N}, \qquad n \geq 3. \tag{1.8}$$

(3) The typical example of functions satisfying (H1)–(H4) is $f(t) = (t^+)^p$ and $g(t) = (t^+)^q$, with p and q satisfying (1.3). In this case, $\alpha = p + 1$, $\beta = q + 1$, $a = p$, $b = q$.

In Section 2, we prove a result on the decay of solutions of (1.1)–(1.2) as $|x| \to \infty$, and, in fact, exponential decay. For radial symmetric solutions we also prove the exponential decay of the derivatives up to order 2. The results of Section 2 are the following. We say that the pair (u, v) is a strong solution of (1.1)–(1.2) if $u \in W^{2,(p+1)/p}$ and $v \in W^{2,(q+1)/q}$.

THEOREM 2.1. Assume (H1), (H4) with $a = b = 1$, and (H2) with p, q satisfying

$$p, q < (N+2)/(N-2). \tag{1.9}$$

Then, the strong solutions u, v of (1.1)–(1.2) are such that

$$\lim_{|x| \to +\infty} u(x) = 0, \qquad \lim_{|x| \to \infty} v(x) = 0 \tag{1.10}$$

$$\lim_{|x| \to +\infty} |\nabla u(x)| = 0, \qquad \lim_{|x| \to \infty} |\nabla v(x)| = 0.$$

THEOREM 2.2. Assume (H1), (H2) with p, q satisfying (1.9), and (H4) with $a, b > 1$. Then, the strong positive solutions of (1.1)–(1.2) have the following asymptotic behavior

$$\lim_{|x| \to +\infty} u(x)\, e^{\theta |x|} = 0, \qquad \lim_{|x| \to \infty} v(x)\, e^{\theta |x|} = 0, \tag{1.12}$$

for $0 < \theta < 1$.

Remark. In the previous theorem, we do not need really that a and b are strictly greater than 1. All that is required in our proof is that $f(u) = o(u)$ and $g(v) = o(v)$ for u and v near zero.

For the case of positive radial symmetric solutions $u(r) = u(|x|)$ of (1.1)–(1.2), which are nonincreasing as a function of r, we have the next theorem. With further hypotheses on f and g, one can prove that, in fact, the positive radial solutions of (1.1)–(1.2) are nonincreasing, see Theorem 3.1.

THEOREM 2.3. Assume that f and g do not depend on x, and satisfy hypotheses (H1), (H2) with p, q fulfilling (1.9), and (H4) with $a, b > 1$. Suppose also that the strong positive radial solutions $u(r)$ and $v(r)$, $r = |x|$, of (1.1) are nonincreasing in $r \in [0, \infty)$. Then

$$u_r(r) = 0(e^{-\theta_1 r}), \qquad v_r(r) = 0(e^{-\theta_1 r}), \qquad u_{rr}(r) = 0(e^{-\theta_2 r}), \qquad v_{rr}(r) = 0(e^{-\theta_2 r}), \tag{1.13}$$

where $0 < \theta_1 \times, \theta_2 < 1$.

In Section 3, we consider autonomous systems (1.1)–(1.2) and discuss the radial symmetry property of their positive solutions. For that matter, we have to strengthen the basic hypotheses. Let us introduce a further set of assumptions.

(H5) $f, g\colon \mathbb{R} \to \mathbb{R}$ are C^1-functions, with $f(t) = g(t) = 0$ for $t \le 0$,

$$f(t) > 0 \quad \text{and} \quad g(t) > 0 \quad \text{for } t > 0.$$

(H6) $f'(t)$, $g'(t)$ are nondecreasing for $t \ge 0$.

(H7) There is a constant $c_1 > 0$, such that

$$f'(t) \le c_1(t^{p-1} + 1), \qquad g'(t) \le c_1(t^{q-1} + 1), \qquad t \ge 0, \tag{1.14}$$

with p, q satisfying (1.9).

(H8) There are constants $c_2 > 0$ and $r > 0$ such that

$$f'(t) \le ct^{a-1}, \qquad g'(t) \le ct^{b+1} \qquad 0 \le t \le r, \tag{1.15}$$

where $a, b > 1$.

THEOREM 3.1. Assume (H5)–(H8). Then all strong positive solutions of (1.1)–(1.2) are radially symmetric with respect to some point $x_0 \in R^N$. Moreover, if we translate the solutions, taking x_0 to the origin, we have

$$u'(r),\ v'(r) < 0, \qquad \text{for } r > 0.$$

In Section 5, we prove the existence of radial symmetric solutions of (1.1)–(1.2), assuming that f, g depend explicitly on $r = |x|$, instead of x. Namely the following theorem.

THEOREM 5.1. Assume (H1), (H2), (H3), (H4) with $a, b > 1$. Furthermore assume the following relationships involving the parameters p, q, α, β

$$\left\{2 - \left(\frac{1}{p+1} + \frac{1}{q+1}\right)\right\} \max\left\{\frac{p+1}{\alpha}, \frac{q+1}{\beta}\right\} < 1 + \frac{2}{N},$$

and $p < \alpha \leq p + 1$, $q < \beta \leq q + 1$.

Then (1.1)-(1.2) possesses at least one strong solution (u, v).

Finally in Section 6, we prove the existence of a ground state for the system

$$-\Delta u + u = v^q$$
$$-\Delta v + v = u^p.$$

2. DECAY OF SOLUTIONS AT INFINITE

In this section we prove Theorems 2.1, 2.2 and 2.3 stated in the Introduction. The restriction put by (1.9) could be eventually avoided. We conjecture that the result is true for any p, q below the critical hyperbola (1.8). We recall, however, that similar restriction is also found in [12].

In order to prove Theorem 2.1 we shall need the following lemma.

LEMMA 2.1. Let u and v be as in Theorem 2.1. Then, they belong to L^γ for $\gamma \in [2, \infty)$.

Proof. From the Sobolev imbedding theorem we have that

$$u \in L^\gamma(R^N) \qquad \text{for } \frac{p+1}{p} \leq \gamma \leq \frac{N(p+1)}{Np - 2(p+1)}$$

$$v \in L^\gamma(R^N) \qquad \text{for } \frac{q+1}{q} \leq \gamma \leq \frac{N(q+1)}{Nq - 2(q+1)}.$$

It follows from (1.9) that

$$u, v \in L^{2^*}(R^N), \qquad 2^* = 2N/(N-2).$$

A bootstrap argument as in [1] shows that u and v are continuous functions.

For each $k > 0$, define the open set

$$\Omega_k = \{x \in R^N : |u(x)| + |v(x)| < k\}.$$

Now given $x_0 \in R^N$, there exist $k_0 > 0$ and $r > 0$ such that the open ball $B_r(x_0) \subset \Omega_k$, for all $k \geq k_0$. Let $R(k) = \sup\{r > 0 : B_r(x_0) \subset \Omega_k\}$. Clearly $R(k) \to +\infty$ as $k \to +\infty$. Let $\varphi \in C_0^\infty(R^N)$ be such that

$$\varphi(x) = 1, \quad \text{for } x \in B_{1/2}(0); \qquad \varphi(x) = 0, \quad \text{for } x \in R^N \backslash B_1(0);$$

$$0 \leq \varphi(x) \leq 1 \qquad \text{and} \qquad |\nabla\varphi(x)| \leq \text{const}, \qquad \text{for all } x \in R^N.$$

Define $\varphi_R(x) = \varphi((x - x_0)/R)$ for $R := R(k)$. Multiplying equation (1.1) by $\varphi_R^2|u|^{s-1}u$, with $s > 1$, and integrating by parts, we get

$$\int_{R^N} \nabla u \nabla(\varphi_R^2 u|u|^{s-1}) + \int_{R^N} \varphi_R^2 |u|^{s+1} = \int_{R^N} \varphi_R^2 |u|^{s-1} u g(x, v).$$

For easy reference, let us denote the three above integrals by I_1, I_2 and I_3, in the order they appear. We estimate I_1:

$$I_1 = s \int \varphi_R^2 |u|^{s-1} |\nabla u|^2 + 2 \int \varphi_R u |u|^{s-1} \nabla u \cdot \nabla \varphi_R =: sI_4 + 2I_5.$$

Given $\varepsilon > 0$, we estimate I_5:

$$I_5 \leq \frac{\varepsilon}{2} \int \varphi_R^2 |u|^{s-1} |\nabla u|^2 + C_\varepsilon \int |u|^{s+1} |\nabla \varphi_R|^2$$

and so

$$I_1 \geq (s - \varepsilon) I_4 - 2C_\varepsilon \int |u|^{s+1} |\nabla \varphi_R|^2 =: (s - \varepsilon) I_4 - 2C_\varepsilon I_6.$$

On the other hand, we have

$$\nabla(\varphi_R u |u|^{(s-1)/2}) = (\nabla \varphi_R) u |u|^{(s-1)/2} + \frac{s+1}{2} \varphi_R |u|^{(s-1)/2} \nabla u$$

which gives

$$\int |\nabla(\varphi_R u|u|^{(s-1)/2})|^2 \leq \frac{1}{2} \int |u|^{s+1} |\nabla \varphi_R|^2 + \frac{1}{2}\left(\frac{s+1}{2}\right)^2 \int \varphi_R^2 |u|^{s-1} |\nabla u|^2$$

$$= \frac{1}{2} I_6 + \frac{(s+1)^2}{8} I_4.$$

Putting these estimates together we get

$$\int |\nabla(\varphi_R u|u|^{(s-1)/2})|^2 \leq \frac{(s+1)^2}{8} \frac{1}{s - \varepsilon} \{I_1 + 2C_\varepsilon I_6\} + \frac{1}{2} I_6.$$

Since $I_1 \leq I_3$ we get

$$\int |\nabla(\varphi_R u|u|^{(s-1)/2})|^2 \leq \frac{(s+1)^2}{8(s - \varepsilon)} I_3 + C(\varepsilon) I_6 \tag{2.1}$$

where $C(\varepsilon)$ is a constant which depends on ε. We next use Sobolev imbedding theorem to estimate the left side of (2.1) from below:

$$\left(\int (\varphi_R^2 |u|^{s+1})^{N/(N-2)}\right)^{(N-2)/N} \leq \frac{(s+1)^2}{8(s - \varepsilon)} I_3 + C(\varepsilon) I_6. \tag{2.2}$$

Now we estimate I_3 viewing to get integrals like the one in the left side of (2.2). Given $\varepsilon > 0$, it follows from assumptions (H2), (H4) and (1.9) that there is a constant $C(\varepsilon)$ such that

$$|g(x, v)| \le \varepsilon |v|^{(N+2)/(N-2)} + C(\varepsilon)|v|,$$

which we use to estimate I_3:

$$I_3 \le \varepsilon \int \varphi_R^2 |u|^s |v|^{(N+2)/(N-2)} + C(\varepsilon) \int \varphi_R^2 |u|^s v =: \varepsilon I_7 + C(\varepsilon) I_8 \tag{2.3}$$

We write the integrand of I_7 as $(\varphi_R^{2s/(s+1)}|u|^s)(\varphi_R^{2/(s+1)}|v|)|v|^{4/(N-2)}$ and use Hölder's inequality with exponents $p_1 = ((s+1)/s)(N/(N-2))$, $p_2 = (s+1)(N/(N-2))$ and $p_3 = N/2$. So

$$I_7 \le \left(\int (\varphi_R^2 |u|^{s+1})^{N/(N-2)} \right)^{(s/(s+1))(N/(N-2))}$$

$$\times \left(\int (\varphi_R^2 |v|^{s+1})^{N/(N-2)} \right)^{(1/(s+1))((N-2)/N)} \left(\int |v|^{2N/(N-2)} \right)^{2/N}. \tag{2.4}$$

Again by Hölder's and Young's inequalities we obtain

$$I_8 \le \left(\int \varphi_R^2 |u|^{s+1} \right)^{s/(s+1)} \left(\int \varphi_R^2 |v|^{s+1} \right)^{1/(s+1)} \le C \int \varphi_R^2 (|u|^{s+1} + |v|^{s+1}). \tag{2.5}$$

Let $A = (\int (\varphi_R^2 |u|^{s+1})^{N/(N-2)})^{(N-2)/N}$ and $B = (\int (\varphi_R^2 |v|^{s+1})^{N/(N-2)})^{(N-2)/N}$. It follows from (2.3) to (2.5) that

$$I_3 \le \varepsilon C A^{s/(s+1)} B^{1/(s+1)} + C(\varepsilon) \left(\int \varphi_R^2 |u|^{s+1} + \int \varphi_R^2 |v|^{s+1} \right). \tag{2.6}$$

So, from (2.2) and (2.6)

$$A \le \varepsilon C A^{s/(s+1)} B^{1/(s+1)} + C(\varepsilon) \int \varphi_R^2 (|u|^{s+1} + |v|^{s+1}) + C(\varepsilon) \int |u|^{s+1} |\nabla \varphi_R|^2. \tag{2.7}$$

A similar expression can be obtained with the roles of A and B interchanged:

$$B \le \varepsilon C A^{1/(s+1)} B^{s/(s+1)} + C(\varepsilon) \int \varphi_R^2 (|u|^{s+1} + |v|^{s+1}) + C(\varepsilon) \int |v|^{s+1} |\nabla \varphi_R|^2. \tag{2.8}$$

Assuming that $\int |u|^{s+1} < \infty$ and $\int |v|^{s+1} < \infty$ we obtain from (2.7) and (2.8)

$$A \le \varepsilon C A^{s/(s+1)} B^{1/(s+1)} + C(\varepsilon) \tag{2.9}$$

$$B \le \varepsilon C A^{1/(s+1)} B^{s/(s+1)} + C(\varepsilon). \tag{2.10}$$

Multiplying (2.9) by (2.10) we obtain

$$AB \le \varepsilon^2 CAB + \varepsilon C(\varepsilon) A^{s/(s+1)} B^{1/(s+1)} + \varepsilon C(\varepsilon) A^{1/(s+1)} B^{s/(s+1)} + C(\varepsilon).$$

For small fixed $\varepsilon > 0$, we get

$$AB \le \varepsilon C(\varepsilon) A^{s/(s+1)} B^{1/(s+1)} + \varepsilon C(\varepsilon) A^{1/(s+1)} B^{s/(s+1)} + C(\varepsilon). \tag{2.11}$$

Now letting $k \to \infty$, we have $R \to \infty$ and it follows from (2.11) that

$$\int |u|^{(s+1)(N/(N-2))} < \infty \quad \text{and} \quad \int |v|^{(s+1)(N/(N-2))} < \infty.$$

Repeating this procedure we see that $u, v \in L^\gamma$ for $\gamma = (s+1)(N/(N-2))^2$. So we may start with $s = (N+2)/(N-2)$ and obtain $u, v \in L^\gamma$ for all $\gamma = (2N/(N-2)(N(N-2))^n$, $n = 1, 2, \dots$. Using the Riesz–Thorin interpolation theorem, [13], we conclude that $u, v \in L^\gamma$ for all $\gamma \ge 2N/(N-2)$. And since $(p+1)/p$ and $(q+1)/q$ are <2, the proof of the lemma is complete. ■

Proof of Theorem 2.1. Let $B_2 = B_{2R}(x_0)$ be a ball of radius $2R$ centered at x_0. Since, from (H2) and (H4) we have

$$|g(x, v)| \le c(|v|^p + |v|),$$

it follows that

$$\|g(x, v)\|_{L^\gamma(B_2)} \le C(\|v\|^p_{L^{\gamma p}(B_2)} + \|v\|_{L^\gamma(B_2)}). \tag{2.12}$$

Using equation (1.1) and Lemma 2.1 we conclude that $\Delta u \in L^\gamma(B_2)$, for all $\gamma \ge 2$. By the Calderon–Zygmund inequality [14], Theorem 9.9, we conclude that $u \in W^{2,\gamma}(B_2)$. Next, by the interior L^p-estimates we have

$$\|u\|_{W^{2,\gamma}(B_1)} \le C(\|u\|_{L^\gamma(B_2)} + \|g(x, v)\|_{L^\gamma(B_2)}) \tag{2.13}$$

where $B_1 = B_R(x_0)$. From (2.13) and (2.12) we get

$$\|u\|_{W^{2,\gamma}(B_1)} \le C(\|u\|_{L^\gamma(B_2)} + \|v\|^p_{L^{\gamma p}(B_2)} + \|v\|_{L^\gamma(B_2)}).$$

If $\gamma > N$, we have by Sobolev imbedding

$$\|u\|_{C^{1,\lambda}(\bar{B}_1)} \le C(\|u\|_{L^\gamma(B_2)} + \|v\|^p_{L^\gamma(B_2)} + \|v\|_{L^\gamma(B_2)}). \tag{2.14}$$

Letting $|x_0| \to \infty$, we conclude that $\|u\|_{C^{1,\lambda}(B_R(x_0))} \to 0$, and similarly $\|v\|_{C^{1,\lambda}(B_R(x_0))} \to 0$, and so the proof of Theorem 2.1 is complete. ■

Next we prove that this decay is in fact exponential, by asking a little more on the functions f and g. And this is obtained by a strengthening of (H4).

Proof of Theorem 2.2. Fix $\theta \in (0, 1)$ and let $\delta = 1 - \theta^2$. Since from Theorem 2.1, $u(x), v(x) \to 0$ as $|x| \to \infty$, we conclude that there is $R > 0$, such that, if $|x| \ge R$ then

$$\frac{f(x, u(x))}{u(x)} \le \delta \qquad \frac{g(x, v(x))}{v(x)} \le \delta. \tag{2.15}$$

Let $w(x) = M e^{-\theta(|x|-R)}$ where $M = \max\{|u(x)| : |x| = R\} + \max\{|v(x)| : |x| = R\}$. Now let $L > R$ and let us define the set

$$\Omega_L = \{x \in R^N : R < |x| < L,\, u(x) + v(x) > w(x),\, u(x) > 0,\, v(x) > 0\}.$$

We claim that Ω_L is empty! Suppose, by contradiction, that this is not so. So Ω_L is a nonempty open set and in it we have

$$\Delta(w - u - v) = \left[\theta^2 - \frac{\theta(N-1)}{|x|}\right] w(x) + [f(x, u) - u] + [g(x, v) - v],$$

which can be estimated using (2.15), and by the choice of δ, by

$$\Delta(w - u - v) \leq \theta^2(w - u - v). \tag{2.16}$$

From the definition of Ω_L, we have from (2.16) that $\Delta(w - u - v) < 0$ in Ω_L. Using the maximum principle, we conclude that

$$w(x) - u(x) - v(x) \geq \min_{\partial\Omega_L}(w - u - v).$$

Since $|x| = R$ does not belong to the boundary of Ω_L, we have

$$w(x) - u(x) - v(x) \geq \min\left\{0, \min_{|x|=L}(w - u - v)\right\}.$$

Now, letting $L \to \infty$, and using the fact that u and v decay to 0 at ∞, we conclude that, for each fixed $|x| > R$, $w(x) - u(x) - v(x) \geq 0$, contradicting the very definition of Ω_L. So, for $|x| > R$ such that $u(x) > 0$ and $v(x) > 0$ we have $u(x) + v(x) \leq w(x)$. Consequently

$$u(x) + v(x) \leq w(x) \qquad \text{for all } |x| \geq R.$$

In a similar way, working with

$$\Omega_L = \{x \in R^N : R < |x| < L,\ u(x) + v(x) < -w(x),\ u(x) < 0,\ v(x) < 0\},$$

we come to $u(x) + v(x) \geq -w(x)$ for all $|x| \geq R$.

So, it follows that

$$|u| + |v| \leq w,$$

which proves Theorem 2.2. ∎

The next result proves the exponential decay of derivatives of radially symmetric positive solutions of system (1.1)-(1.2) in the autonomous case. In this case the system may be written as

$$-(r^{N-1}u_r)_r + r^{N-1}u = r^{N-1}g(v) \tag{2.17}$$

$$-(r^{N-1}v_r)_r + r^{N-1}v = r^{N-1}f(u). \tag{2.18}$$

THEOREM 2.3. Assume that f and g satisfy hypotheses (H1), (H2), (H3) with $p, q < (N + 2)/(N - 2)$, and (H4) with $a, b > 1$. Then the derivatives of the solutions of (2.17)-(2.18) decay exponentially at ∞. That is, there are θ_1, θ_2 such that

$$u_r(r) = 0(e^{-\theta_1 r}), \quad v_r(r) = 0(e^{-\theta_1 r}), \quad u_{rr}(r) = 0(e^{-\theta_2 r}), \quad v_{rr}(r) = 0(e^{-\theta_2 r}),$$

for some $\theta_1, \theta_2 \in (0, 1)$.

Proof. It follows from Theorem 2.1 that u, v decay exponentially. So, there are positive constants c_1, c_2 and r_0 such that

$$c_1 u \le u - f(u) \le u \qquad \text{for } r \ge r_0 \tag{2.19}$$

$$c_2 v \le v - g(v) \le v \qquad \text{for } r \ge r_0. \tag{2.20}$$

From equations (2.17) to (2.18), we obtain

$$-[r^{N-1}(u_r + v_r)]_r = [(f(u) - u) + (g(v) - v)]r^{N-1}, \tag{2.21}$$

which by integration from r to R leads to

$$[R^{N-1}(u_r(R) + v_r((R))] - [r^{N-1}(u_r(r) + v_r(r))]$$
$$= -\int_r^R s^{N-1}[(f(u(s)) - u(s)) + (g(v(s)) - v(s))]\,ds. \tag{2.22}$$

Using (2.19)–(2.20) and the exponential decay of u and v we conclude from (2.22) that $\lim_{r\to\infty} r^{N-1}(u_r(r) + v_r(r))$ exists. Call it σ, and let us prove that $\sigma \neq 0$. Using again (2.22) and the exponential decay of u and v, we get

$$u_r(r) + v_r(r) - \frac{\sigma}{r^{N-1}} = 0(e^{-\theta_1 r}), \qquad r \to \infty. \tag{2.23}$$

Hence, by integration of (2.23) we have

$$[u(R) + v(R)] - [u(r) + v(r)] + (N - 2)\sigma\left\{\frac{1}{R^{N-2}} - \frac{1}{r^{N-2}}\right\} = 0(e^{-\theta_1 r}),$$

and this is impossible if $\sigma \neq 0$, since u decays exponentially. Integrating (2.21) from r to ∞ and using that $\sigma = 0$ we get

$$r^{N-1}(u_r(r) + v_r(r)) = 0(e^{-\theta_1 r}), \qquad \text{as } r \to \infty \tag{2.24}$$

with $0 < \theta_1 < 1$.

Since $u_r < 0$ and $v_r < 0$, we conclude from (2.24) that

$$u_r(r) = 0(e^{-\theta_1 r}) \qquad \text{and} \qquad v_r(r) = 0(e^{-\theta_1 r}).$$

Using this information in (2.17)–(2.18) we obtain that

$$u_{rr}(r) = 0(e^{-\theta_2 r}) \qquad \text{and} \qquad v_{rr}(r) = 0(e^{-\theta_2 r}),$$

completing the proof of Theorem 2.3. ■

3. RADIAL SYMMETRY OF POSITIVE SOLUTIONS

In this section, we study the radial symmetry of positive solutions of the system

$$-\Delta u + u = g(v) \tag{3.1}$$

$$-\Delta v + v = f(u) \tag{3.2}$$

in the whole of R^N. The way of proving this symmetry is by using the method of moving parallel planes, as originally introduced by Alexandroff (see Hopf [15, Chap. 7]), later used by Serrin [16], and extensively used in recent times, after the work of Gidas–Ni–Nirenberg [17]. The scalar case on unbounded domains was studied in [18]. Although one can treat problems with the nonlinearities depending explicitly on x, in some appropriate way, we consider here the autonomous case as written in (3.1) and (3.2).

Remark 1. The restriction $p, q < (N+2)/(N-2)$ is used below only to obtain a decay of the positive solutions of the system (3.1)–(3.2). Such a decay was proved in the previous section in a more general situation. Typical example of nonlinearities satisfying the hypotheses of Theorem 3.1 is $f(u) = (u^+)^p$ and $g(v) = (v^+)^q$.

Remark 2. As in the previous section a strong solution of (3.1)–(3.2), that is $u \in W^{2,(p+1)/p}(R^N)$, $v \in W^{2,(q+1)/q}(R^N)$, is $C^{1,\alpha}_{\text{loc}}$. It follows from (H5) and Schauder interior estimates that u and v are in fact C^2.

The proof of Theorem 3.1 uses the method of moving planes. We consider, initially, planes parallel to the plane $x_1 = 0$, coming from $-\infty$. For each real λ, we define

$$\Sigma_\lambda = \{x = (x_1, \dots, x_N) \in R^N : x_1 < \lambda\}, \qquad T_\lambda = \partial\Sigma_\lambda,$$

$$x^\lambda = (2\lambda - x_1, x_2, \dots, x_n) = \text{the reflection of } x \text{ on the plane } T_\lambda.$$

In the sequel, we take $\lambda \leq 0$. Let us define in $\bar{\Sigma}_\lambda$ the following functions

$$u_\lambda(x) = u(x^\lambda), \qquad v_\lambda(x) = v(x^\lambda)$$

$$V_\lambda(x) = u_\lambda(x) - u(x), \qquad V_\lambda(x) = v_\lambda(x) - v(x).$$

Then

$$-\Delta U_\lambda(x) = -U_\lambda(x) + g(v(x^\lambda)) - g(v(x)).$$

By the mean value theorem

$$g(v(x^\lambda)) - g(v(x)) = g'(\psi(x;\lambda))V_\lambda(x),$$

where $\psi(x;\lambda)$ is a real number between $v(x^\lambda)$ and $v(x)$. Let us denote $g'(\psi(x;\lambda)) = c(x;\lambda)$. So

$$-\Delta U_\lambda(x) + U_\lambda(x) = c(x;\lambda)V_\lambda(x), \qquad \text{in } \Sigma_\lambda. \tag{3.3}$$

where $d(x;\lambda) = f'(\varphi(x;\lambda))$, and $\varphi(x;\lambda)$ is some value between $u(x^\lambda)$ and $u(x)$.

PROPOSITION 3.2. There exists $\lambda^* \leq 0$ such that for all $\lambda \leq \lambda^*$, $U_\lambda(x) \geq 0$ and $V_\lambda(x) \geq 0$, for $x \in \Sigma_\lambda$.

The proof of Proposition 3.2 relies on some properties of U_λ and V_λ, which are collected in the next lemmas. Since $u(x)$ and $v(x)$ have exponential decay as $|x| \to \infty$, it follows that both U_λ and V_λ are bounded below. The next lemma says more about their infima.

LEMMA 3.3. (i) If $U_\lambda(x) > 0$, for all $x \in B_\varepsilon(e_\lambda) \subset \Sigma_\lambda$, and if

$$c_\lambda = \inf\{U_\lambda(x) : x \in \Sigma_\lambda\} < 0,$$

Then U_λ attains its infimum in Σ_λ. (Here $B_\varepsilon(e_\lambda)$ denotes the open ball of radius $\varepsilon > 0$ centered at the point $e_\lambda = (2\lambda, 0, \ldots, 0)$).

(ii) There exists $\bar{\lambda} < 0$ such that if $\lambda < \bar{\lambda}$, and $c_\lambda < 0$, then U_λ attains its infimum.

Proof. (i) Since $U_\lambda(x) \to 0$ as $|x| \to \infty$, we can find $r > 0$, such that

$$|U_\lambda(x)| \le \frac{|c_\lambda|}{2}, \qquad \text{if } |x| \ge r,$$

which says, in particular, that

$$U_\lambda(x) \ge \frac{c_\lambda}{2}, \qquad \text{if } |x| \ge r.$$

This implies that U_λ attains its infimum in the compact set $\overline{B_r(0)} \cap \overline{\Sigma_\lambda \backslash B_\varepsilon(e_\lambda)}$. Since $U_\lambda = 0$ on T_λ, it follows that this infimum is achieved inside Σ_λ.

(ii) We may assume that $u(0) > 0$. Then, the continuity of u implies that there exists $\varepsilon > 0$ and $\sigma > 0$ such that

$$u(x) \ge \sigma, \qquad \text{for all } |x| \le \varepsilon.$$

By the fact that $u(x)$ decays to zero as $|x| \to \infty$, we see that there exists $r > 0$ (take $r > \varepsilon$) such that

$$u(x) \le \frac{1}{2}\sigma, \qquad \text{for } |x| \ge r.$$

Choosing $\bar{\lambda} = -r$, it follows that for all $x \in B_\varepsilon(e_\lambda)$, with $\lambda < \bar{\lambda}$, we have

$$U_\lambda(x) \ge \sigma - \frac{1}{2}\sigma > 0. \quad \blacksquare$$

LEMMA 3.4. There exist positive constants d and ξ, independent of λ and u, and a $\tilde{\lambda} < 0$, such that

$$c(x; \lambda) \le d\, e^{-\xi|r|}$$

for all $\lambda \le \tilde{\lambda}$, and $x \in \Sigma_\lambda$, where $U_\lambda(x) < 0$.

Proof. It follows from hypotheses (H7) and (H8) that there is a constant $d > 0$ such that

$$g'(t) \le d(t^{p-1} + t^{\alpha-1}). \tag{3.5}$$

From the assumption that $U_\lambda(x) < 0$, it follows that $u_\lambda(x) < u(x)$. Consequently

$$u_\lambda(x) \le \psi(x; \lambda) \le u(x).$$

Since g' is monotone nondecreasing, we obtain

$$c(x;\lambda) \le g'(u(x)) \le d(|u(x)|^{p-1} + |u(x)|^{\alpha-1}), \tag{3.6}$$

where we have used (3.5). Now, it follows from Theorem 2.2, that exists a $\tilde{\lambda} < 0$ such that

$$u(x) \le e^{-\theta|x|}, \qquad v(x) \le e^{-\theta|x|} \tag{3.7}$$

for $x \in \Sigma_\lambda$, and $\lambda \le \tilde{\lambda}$. Using (3.6) and (3.7) we obtain the inequality in the lemma with $\xi = \theta(\alpha - 1)$ and d given in (3.5). ■

Remark. It follows from (H5) and (H6) that $c(x;\lambda) \ge 0$.

Lemma 3.5. There exists $\hat{\lambda} \le \tilde{\lambda}$ such that for all $\lambda \le \hat{\lambda}$, we have

$$c(x;\lambda) < 1$$

for all $x \in \{x \in \Sigma_\lambda : U_\lambda(x) < 0\}$.

Proof. Let $r > 0$ be such that $d\,e^{-\xi r} < 1$, and take $\hat{\lambda} = \min\{-r, \tilde{\lambda}, \bar{\lambda}\}$. ■

Lemma 3.6. Let $\lambda < \bar{\lambda}$ and assume that $c_\lambda < 0$. Let $x_0 \in \Sigma_\lambda$ (it exists in view of Lemma 3.3) such that

$$U_\lambda(x_0) = \inf\{U_\lambda(x) : x \in \Sigma_\lambda\}.$$

Then $V_\lambda(x_0) < 0$. Moreover, if $\lambda < \hat{\lambda}$, then

$$V_\lambda(x_0) < U_\lambda(x_0). \tag{3.8}$$

Proof. Since x_0 is a point of minimum inside Σ_λ, we have that $-\Delta U_\lambda(x_0) \le 0$. It follows, then, from (3.3) that

$$c(x_0;\lambda)V_\lambda(x_0) \le U_\lambda(x_0).$$

Since $c(x_0;\lambda) \ge 0$, we obtain that $V_\lambda(x_0) < 0$. If $\lambda < \hat{\lambda}$ we use Lemma 3.5, and get

$$V_\lambda(x_0) < c(x_0;\lambda)V_\lambda(x_0) \le U_\lambda(x_0). \ ■$$

Proof of Proposition 3.2. Suppose, by contradiction, that, for all $\lambda < 0$, we have that $U_\lambda(x) < 0$ for some $x \in \Sigma_\lambda$, or $V_\lambda(x) < 0$, for some $x \in \Sigma_\lambda$. Take $\lambda \le \min\{\hat{\lambda}_u, \hat{\lambda}_v\}$, where $\hat{\lambda}_u$ is the $\hat{\lambda}$ obtained in Lemma 3.5. Analogous definition for $\hat{\lambda}_v$. It follows from Lemma 3.3 that c_λ is attained at some point $x_0 \in \Sigma_\lambda$, and by Lemma 3.6, that $V_\lambda(x_0) < U_\lambda(x_0)$. The same argument applies to V_λ, and we conclude that V_λ assumes its (negative) infimum at $x_1 \in \Sigma_\lambda$ and again by Lemma 3.6 (for the function V_λ) $U_\lambda(x_1) < V_\lambda(x_1)$. Since $V_\lambda(x_1) \le V_\lambda(x_0)$, we come to a contradiction. ■

Now Proposition 3.2 allows us to define

$$\lambda_0 = \sup\{\lambda < 0 : U_\lambda(x) \ge 0,\ V_\lambda(x) \ge 0, \text{ for all } x \in \Sigma_\lambda\}. \tag{3.9}$$

PROPOSITION 3.7. If $\lambda_0 < 0$, then $U_{\lambda_0}(x) = 0$ and $V_{\lambda_0}(x) = 0$, for all $x \in \Sigma_{\lambda_0}$.

Proof. By continuity, we see that

$$U_{\lambda_0}(x) \geq 0 \quad \text{and} \quad V_{\lambda_0}(x) \geq 0 \quad \text{for all } x \in \Sigma_{\lambda_0}.$$

It follows from (3.3) and (3.4) that $U_{\lambda_0} \equiv 0$ in Σ_{λ_0} if and only if $V_{\lambda_0} \equiv 0$. So, assume by contradiction, that $U_{\lambda_0} \not\equiv 0$, and consequently also $V_{\lambda_0} \not\equiv 0$. It follows from (3.3) and (3.4)

$$-\Delta U_{\lambda_0} + U_{\lambda_0} \geq 0, \qquad -\Delta V_{\lambda_0} + V_{\lambda_0} \geq 0 \qquad \text{in } \Sigma_{\lambda_0}.$$

So, by the maximum principle $U_{\lambda_0}(x)$, $V_{\lambda_0}(x) > 0$ in Σ_{λ_0} and $\partial U_{\lambda_0}/\partial x_1$, $\partial V_{\lambda_0}/\partial x_1 < 0$ in T_{λ_0}.

Now take a sequence $\lambda_k \to \lambda_0$ with $\lambda_k > \lambda_0$. By the definition of λ_0, for each λ_k there is a point $x_k \in \Sigma_{\lambda_0}$, where U_{λ_k} or V_{λ_k} is negative. Now, since U_{λ_0} and V_{λ_0} are positive in Σ_{λ_0}, it follows that there exist a $c_0 > 0$ and $R > 0$ such that

$$U_{\lambda_0}(x) \geq c_0 \quad \text{and} \quad V_{\lambda_0}(x) \geq c_0, \quad \text{for } x \in B_R(e_{\lambda_0}).$$

By continuity, we have that, for large k,

$$U_{\lambda_k}(x) \geq \frac{c_0}{2} \quad \text{and} \quad V_{\lambda_k}(x) \geq \frac{c_0}{2}, \quad \text{for } x \in B_{R/2}(e_{\lambda_k}).$$

So by Lemma 3.3, U_{λ_k} attains its infimum at a point x_k and V_{λ_k} at a point y_k. The sequence (x_k) and (y_k) are bounded, since U_{λ_k} and V_{λ_k} decay to zero at infinity. So, we may assume that $x_k \to \bar{x}$, with $\bar{x} \in \Sigma_{\lambda_0}$ and, by continuity, we have

$$\nabla U_{\lambda_0}(\bar{x}) = 0 \quad \text{and} \quad U_{\lambda_0}(\bar{x}) \leq 0,$$

which is impossible. ■

PROPOSITION 3.8. There exists $\lambda \in \mathbb{R}$ such that $x_1 = \lambda$ is a hyperplane of symmetry for the solutions u and v.

Proof. If the λ_0 defined in (3.9) above is less than zero, then it follows from Proposition 3.7 that $x_1 = \lambda_0$ is a hyperplane of symmetry. If that $\lambda_0 = 0$, then we employ the same procedure we used to define λ_0, but now taking parallel planes moving from $+\infty$. So in a similar way we define a $\tilde{\lambda}_0$ which is larger or equal to 0. If $\tilde{\lambda}_0$ is larger than zero, then as in the case of $\lambda_0 < 0$, the hyperplane $\lambda = \tilde{\lambda}_0$ is of symmetry. In $\tilde{\lambda}_0 = 0$, then this together with $\lambda_0 = 0$, implies that $x_1 = 0$ is a hyperplane of symmetry.

Proof of Theorem 3.1. Using moving planes parallel $x_j = \lambda$ for $j = 1, \ldots, N$ we determine N hyperplanes of symmetry, whose intersection is some $x_0 \in R^N$. Then, we proceed as in [18]. Taking x_0 as the new origin, we now take any other direction and conclude that there is a hyperplane of symmetry perpendicular to that direction and passing to x_0.

4. SOME ABSTRACT RESULTS AND FRACTIONAL SOBOLEV SPACES

System (1.1)–(1.2) is variational, and it is natural to consider the functional

$$\Psi(u, v) = \int_{R^N} (\nabla u \cdot \nabla v + uv) - \int_{R^N} F(x, u) - \int_{R^N} G(x, v),$$

since *formally* the Euler–Lagrange equations associated to Ψ are precisely the equations in the system. As usual we have the notation:

$$F(x, s) = \int_0^s f(x, t)\, \mathrm{d}t, \qquad G(x, s) = \int_0^s g(x, t)\, \mathrm{d}t.$$

Observe that the quadratic part of Ψ is well-defined if $u, v \in H^1(R^N)$. But, if we take such u and v, the nonlinear part is well defined if the growths of f and g at ∞, respectively p and q, are both less or equal to $(N + 2)/(N - 2)$, for $N \geq 3$. However, we would like to consider pairs (p, q) that do not satisfy this restriction. The basic requirement would be that (p, q) is below the critical hyperbola, one of them could be larger than $(N+2)/(N-2)$. For that matter, the quadratic part of the functional has to be redefined, and we then need fractional Sobolev spaces.

In this section, we first develop an abstract framework, which is suitable for other problems also. And then we apply this to the study of radial solutions of (1.1)–(1.2).

Let H be a separable real Hilbert space with scalar product denoted by $\langle\,,\rangle$ and corresponding norm by $\|\cdot\|$. Let $T\colon D(T) \subset H \to H$ be a self-adjoint linear operator semibounded from below. That is, there is a constant δ such that

$$\langle Tu, u\rangle \geq \delta\|u\|^2, \qquad \text{for } u \in D(T). \tag{4.1}$$

For simplicity, we take $\delta = 1$. So $0 \notin \sigma(T)$, where $\sigma(T)$ denotes the spectrum of T. Let $\{E(\lambda) : \lambda \in \mathbb{R}\}$ denote the unique right continuous spectral family associated with T. In view of (4.1) we have that $E(\lambda) = 0$, for $\lambda < 1$.

It is well known, cf. for instance [19–21], that

$$D(T) = \left\{u \in H : \int_1^\infty \lambda^2\, \mathrm{d}\langle E(\lambda)u, u\rangle < \infty\right\}, \tag{4.2}$$

$$\langle Tu, v\rangle = \int_1^\infty \lambda\, \mathrm{d}\langle E(\lambda)u, v\rangle, \qquad \text{for } u \in D(T),\ v \in H. \tag{4.3}$$

$$\langle Tu, u\rangle \leq \lambda\|u\|^2, \qquad \text{for } u \in E(\lambda)H. \tag{4.4}$$

$$\lambda\|u\|^2 \leq \langle Tu, u\rangle \leq \mu\|u\|^2, \qquad \text{for } u \in E(\mu)H \ominus E(\lambda)H. \tag{4.5}$$

$$\langle Tu, u\rangle \geq \mu\|u\|^2, \qquad \text{for } u \in [E(\mu)H]^\perp \cap D(T). \tag{4.6}$$

Since T is a positive operator, it has a square root

$$T^{1/2} = \int_1^\infty \lambda^{1/2}\, \mathrm{d}E(\lambda), \qquad T^{1/2}\colon D(T^{1/2}) \to H,$$

with

$$D(T^{1/2}) = \left\{u \in H : \int_1^\infty \lambda\, \mathrm{d}\langle E(\lambda)u, u\rangle < \infty\right\}.$$

We know that $T^{1/2}$ is self-adjoint, and, from (4.1) we have

$$\langle T^{1/2}u, u\rangle \geq \|u\|^2, \qquad \text{for } u \in D(T^{1/2}).$$

For each positive real s, we can define

$$T^{s/2} = \int_1^\infty \lambda^{s/2}\,\mathrm{d}E(\lambda).$$

Let us use the notation $A = T^{1/2}$, $A^s = T^{s/2}$. Let us define the spaces E^s as

$$E^s := D(A^s) = \left\{u \in H \;\middle|\; \int_1^\infty \lambda^s\,\mathrm{d}\langle E(\lambda)u, u\rangle < \infty\right\}. \tag{4.7}$$

Each E^s is a Hilbert space endowed with the graph norm

$$\langle u, v\rangle_{E^s} = \langle u, v\rangle + \langle A^s u, A^s v\rangle.$$

It follows from (4.1) that

$$\|A^s u\| \geq \|u\|, \qquad \text{for all } u \in E^s, \tag{4.8}$$

and as a consequence $\|u\|_{E^s}$ and $\|A^s u\|$ are equivalent norms in E^s. So, in E^s, we write from now on

$$\langle u,v\rangle_{E^s} = \langle A^s u, A^s v\rangle \qquad \text{and} \qquad \|u\|_{E^s} = \|A^s u\|. \tag{4.9}$$

In view of (4.8), $A^s\colon E^s \to H$ is an isomorphism. Let us denote by A^{-s} the inverse of A^s.

Now let $s, t > 0$ with $s + t = 2$. We define the Hilbert space $E = E^s \times E^t$, with inner product $\langle\cdot,\cdot\rangle$ induced by inner products $\langle\cdot,\ \rangle_{E^s}$, $\langle\cdot,\cdot\rangle_{E^t}$ in the usual way. Next we define a bilinear form $B\colon E \times E \to \mathbb{R}$ by

$$B[(u, v), (\phi, \psi)] = \langle A^s u, A^t\psi\rangle + \langle A^s\phi, A^t v\rangle.$$

Using the Cauchy–Schwarz inequality and (4.9) we obtain

$$|B[(u, v), (\phi, \psi)]| \leq \|u\|_{E^s}\|\psi\|_{E^t} + \|\phi\|_{E^s}\|v\|_{E^t},$$

which shows that B is continuous. Clearly B is symmetric. Hence B induces a self-adjoint bounded linear operator $L\colon E \to E$ such that

$$B[z, \eta] = \langle Lz, \eta\rangle_E, \qquad \text{for } z, \eta \in E.$$

It is easy to see that

$$Lz = (A^{-s}A^t v, A^{-t}A^s u), \qquad \text{for } z = (u, v) \in E.$$

We can then prove that L has two eigenvalues -1 and 1, whose corresponding eigenspaces are

$$E^- = \{(u, -A^{-t}A^s u) : u \in E^s\}, \qquad \text{for } \lambda = -1 \tag{4.10}$$

$$E^+ = \{(u, A^{-t}A^s u) : u \in E^s\}, \qquad \text{for } \lambda = 1. \tag{4.11}$$

We also have that

$$E \in E^+ \oplus E^- \tag{4.12}$$

and E^+ and E^- are mutually orthogonal with respect to the bilinear form, that is

$$B[z^+, z^-] = 0, \qquad \text{for } z^+ \in E^+ \text{ and } z^- \in E^-.$$

Let us consider the quadratic form associated with the bilinear form B, namely

$$Q(z) = \frac{1}{2} B[z, z] = \langle A^s u, A^t v\rangle \tag{4.13}$$

for $z = (u,v) \in E$. It follows then that

$$\frac{1}{2} \|z\|_E^2 = Q(z^+) - Q(z^-)$$

where $z = z^+ + z^-$, $z^+ \in E^+$, $z^- \in E^-$. We also have that there is a constant $c_0 > 0$ such that

$$Q(z) \geq c_0 \|z\|_E^2 \quad \text{for } z \in E^+ \qquad \text{and} \qquad Q(z) \leq -c_0\|z\|_E^2 \quad \text{for } z \in E^-. \tag{4.14}$$

We now apply the foregoing abstract results to the situation of our problem (1.1)-(1.2). As announced, we shall need fractional Sobolev spaces in order to formulate correctly the quadratic part of the functional Φ. We also work with radial symmetric functions. To put things in the above framework, let $H = L_r^2(R^N)$ be the space of L^2-functions in R^N which are radially symmetric. Let $T = -\Delta + id$, with domain $D(T) = H_r^2(R^N)$, which is the space of radial symmetric functions that are in L^2 and have second derivatives in L^2. For $0 \leq s \leq 2$, the spaced E^s, which is the domain $D(T^{s/2})$, is precisely the space obtained by interpolation between $H_r^2(R^N)$ and $L_r^2(R^N)$, namely

$$[H_r^2(R^N), L_r^2(R^N)]_{1-s/2}.$$

(See [22, p. 12]). In this case, the space E^s is the usual Sobolev space $H_r^s(R^N)$. So, denoting by $A = (-\Delta + id)^{1/2}$, we have for all $0 \leq s \leq 2$

$$D(A^s) = H_r^s(R^N) = [H_r^2(R^N), L_r^2(R^N)]_{1-s/2}.$$

We have the following imbedding theorem, see [23]. The case $s = 1$, was proved by Strauss [9].

THEOREM 4.1. Let $s > 0$. Then, the restriction to $H_r^s(R^N)$ of the Sobolev imbedding of $W^{s,2}(R^N)$ into $L^\gamma(R^N)$ is continuous if $2 \leq \gamma \leq 2N/(N-2s)$, and it is compact if $2 < \gamma < 2N/(N-2s)$.

So, if we choose $s, t > 0$, $s + t = 2$, such that

$$\begin{aligned} &\left(1 - \frac{1}{p+1}\right) \max\left(\frac{p+1}{\alpha}, \frac{q+1}{\beta}\right) < \frac{1}{2} + \frac{s}{N}, \\ &\left(1 - \frac{1}{q+1}\right) \max\left(\frac{p+1}{\alpha}, \frac{q+1}{\beta}\right) < \frac{1}{2} + \frac{t}{N}, \end{aligned} \tag{4.15}$$

then the inclusions $H_r^s(R^N) \hookrightarrow L^{p+1}(R^N)$ and $H_r^t(R^N) \hookrightarrow L^{q+1}(R^N)$ are compact.

Let $E = H_r^s(R^N) \times H_r^t(R^N)$, and the bilinear form $B: E \times E \to R$ is defined by

$$B[(u, v), (\phi, \psi)] = \int_{R^N} A^s u A^t \psi + A^s \phi A^t v,$$

and the corresponding quadratic form by

$$Q(z) = \int_{R^N} A^s u A^t v, \qquad z = (u, v) \in E.$$

This quadratic form will replace

$$\int \nabla u \cdot \nabla v + uv$$

in the functional Ψ, allowing in that way to consider different growths in f and g. This is done in the next section.

5. EXISTENCE OF RADIAL SOLUTIONS OF (1.1)–(1.2)

In this section we prove Theorem 5.1 on the existence of radial solutions of the system (1.1)–(1.2), with f and g depending explicitly on $r = |x|$ instead of x:

$$-\Delta u + u = g(r, v) \tag{5.1}$$

$$-\Delta v + v = f(r, u). \tag{5.2}$$

In view of hypothesis (H2) we choose s and t as in (4.15), and work with the Sobolev spaces $H_r^s(R^N)$ for u and $H_r^t(R^N)$ for v. So we consider the functional $\Phi: E \to R$, where $E = H_r^s(R^N) \times H_r^t(R^N)$, defined by

$$\Phi(z) = \int_{R^N} A^s u A^t v - \int_{R^N} F(r, u) - \int_{R^N} G(r, v), \tag{5.3}$$

for $z = (u, v) \in E$. Φ is a C^1 functional and

$$\langle \Phi'(z), \eta \rangle_E = \int_{R^N} A^s u A^t \psi + A^s \phi A^t v - \int_{R^N} f(r, u)\phi - \int_{R^N} g(r, v)\psi,$$

For $z = (u, v) \in E$ and $\eta = (\phi, \psi) \in E$. So, the critical points of Φ satisfy the equations

$$\int_{R^N} A^s u A^t \psi - \int_{R^N} g(r, v)\psi = 0, \qquad \text{for all } \psi \in H_r^t(R^N) \tag{5.4}$$

$$\int_{R^N} A^s \phi A^t v - \int_{R^N} f(r, u)\phi = 0, \qquad \text{for all } \phi \in H_r^s(R^N). \tag{5.5}$$

Equations (5.4)–(5.5) are the weak formulation of problem (5.1)–(5.2). We shall use a linking theorem due orginally to Liu and Li [24] and in the version needed here given in Li and Willem [22]. We could also use the generalized Mountain Pass Theorem of Benci–Rabinowitz [3] in a version due to Felmer [4], as in [25].

We shall need some finite dimensional subspaces of E, which we now introduce. Let (e_j), $j = 1, 2, \dots$, be a complete orthogonal system in $H_r^s(R^N)$. Let H_n denote the finite dimensional subspaces of $H_r^s(R^N)$ generated by (e_j), $j = 1, \dots, n$. Since $A^s: H_r^s(R^N) \to L_r^2(R^N)$ and $A^t: H_r^t(R^N) \to L_r^2(R^N)$ are isomorphisms, we see that $(\hat{e}_j)$, $j = 1, 2, \dots$, where $\hat{e}_j = A^{-t}A^s e_j$ is a complete orthogonal system in $H_r^t(R^N)$. Let $\hat{H}_n$ denote the finite dimensional subspace of $H_r^t(R^N)$ generated by $(\hat{e}_j)$, $j = 1, \dots, n$. For each $n \in \mathbb{N}$, we introduce the following subspaces of E^+ and E^-:

$$E_n^+ = \text{subspace of } E^+ \text{ generated by } (e_j, \hat{e}_j), \qquad j = 1, \dots, n.$$

$$E_n^- = \text{subspace of } E^- \text{ generated by } (e_j, -\hat{e}_j), \qquad j = 1, \dots, n.$$

Let $E_n = E_n^+ \oplus E_n^-$. Following [26], we say that Φ satisfies the (PS)* condition if all sequences (z_n), $z_n \in E_n$, such that

$$|\Phi(z_n)| \le \text{const}$$

$$|\langle \nabla_n \Phi(z_n), \eta \rangle| \le \varepsilon_n \|\eta\|_E, \qquad \varepsilon_n \to 0,\ \eta \in E_n,$$

possesses a subsequence converging to a critical point of Φ. Here we are using the symbol ∇_n to denote the derivative of the restriction of Φ to E_n.

We say that Φ has a *local linking* at 0 if, for some $r > 0$, one has

$$\Phi(z) \ge 0, \qquad \text{for } z \in E^+,\ \|z\| \le r,$$

and

$$\Phi(z) \le 0, \qquad \text{for } z \in E^-,\ \|z\| \le r.$$

For sake of reference we state the result from [26] used here.

THEOREM [26]. Suppose that $\Phi \in C^1(E, R)$ satisfies the following assumptions:
(B1) Φ has a local linking at 0,
(B2) Φ satisfies (PS)*,
(B3) Φ maps bounded sets into bounded sets,
(B4) for every $n \in \mathbb{N}$, $\Phi(z) \to -\infty$ as $|z| \to \infty$ and $z \in E_n^+ \oplus E^-$.
Then Φ has a nontrivial critical point.

Hence the proof of Theorem 5.1 follows readily from Lemmas 5.1 and 5.2 stated and proved in the sequel.

LEMMA 5.1. Assume the hypotheses of Theorem 5.1. Then, the functional defined in (5.3) satisfies the (PS)* condition.

Proof. Let $(z_n) = (u_n, v_n) \in E_n$ be a sequence such that

$$|\Phi(z_n)| \le c = \text{const}. \tag{5.6}$$

$$|\langle \nabla_n \Phi(z_n), \eta \rangle| \le \varepsilon_n \|\eta\|_E, \qquad \text{with } \varepsilon_n \to 0 \text{ and } \eta \in E_n. \tag{5.7}$$

Taking $\eta = z_n$ in (5.7), it follows from (5.6) and (5.7) that

$$c + \varepsilon_n \|z_n\|_E \ge -\int F(r, u_n) - \int G(r, v_n) + \frac{1}{2}\int f(r, u_n)u_n + \frac{1}{2}\int g(r, v_n)v_n. \tag{5.8}$$

Now, using hypothesis (H3) we obtain

$$C + \varepsilon_n \|z_n\|_E \geq \left(\frac{\alpha}{2} - 1\right) \int F(r, u_n) + \left(\frac{\beta}{2} - 1\right) \int G(r, v_n),$$

and then, in view of (1.6),

$$C + \varepsilon_n \|z_n\|_E \geq c \int |u_n|^\alpha + |v_n|^\beta. \tag{5.9}$$

Next, we estimate $\|u_n\|_{H_r^s}$ and $\|v_n\|_{H_r^t}$. It follows from hypotheses (H2) and (H4) that, given $\varepsilon > 0$, there is $c_\varepsilon > 0$

$$|f(r, u)| \leq \varepsilon |u| + c_\varepsilon |u|^p \qquad \text{for all } u. \tag{5.10}$$

From (5.7) with $\psi = 0$ we have

$$\left| \int A^s \phi A^t v_n \right| \leq \left| \int f(r, u_n) \phi \right| + \varepsilon_n \|\phi\|_{H_r^s}, \qquad \text{for all } \phi \in H_n.$$

Using (5.10) and Hölder's inequality, we obtain

$$\left| \int A^s \phi A^t v_n \right| \leq \varepsilon \|u_n\|_{L^2} \|\phi\|_{L^2} + c_\varepsilon \|u_n\|_{L^\alpha}^p \|\phi\|_{L^{\alpha/(\alpha - p)}} + \varepsilon_n \|\phi\|_{H_r^s}. \tag{5.11}$$

Since $2 \leq \alpha/(\alpha - p) \leq 2N/(N - 2s)$, we see from (5.11) that

$$\left| \int A^s \phi A^t v_n \right| \leq (\varepsilon \|u_n\|_{H_r^s} + c_\varepsilon \|u_n\|_{L^\alpha}^p) \|\phi\|_{H_r^s}, \qquad \forall \phi \in H_n,$$

which implies that

$$\|v_n\|_{H_r^t} \leq \varepsilon \|u_n\|_{H_r^s} + C_\varepsilon \|u_n\|_{L^\alpha}^p. \tag{5.12}$$

Similarly, we prove that

$$\|u_n\|_{H_r^s} \leq \varepsilon \|v_n\|_{H_r^t} + C_\varepsilon \|v_n\|_{L^\beta}^q. \tag{5.13}$$

Adding (5.12) and (5.13) we conclude that

$$\|u_n\|_{H_r^s} + \|v_n\|_{H_r^t} \leq C(\|u_n\|_{L^\alpha}^p + \|v_n\|_{L^\beta}^q). \tag{5.14}$$

Using this estimate in (5.9) we get

$$\|u_n\|_{L^\alpha}^\alpha + \|v_n\|_{L^\beta}^\beta \leq C(\|u_n\|_{L^\alpha}^p + \|v_n\|_{L^\beta}^q) + C.$$

Since $\alpha > p$ and $\beta > q$ we conclude that both $\|u_n\|_{L^\alpha}$ and $\|v_n\|_{L^\beta}$ are bounded, and consequently $\|u_n\|_{H_r^s}$, $\|v_n\|_{H_r^t}$ are also bounded in view of (5.14).

Next, we show that (z_n) contains a strongly convergent subsequence. And then using the continuity of Φ', we can easily conclude the proof of the (PS)* condition. It follows from Theorem 4.1 that (z_n) contains a subsequence, denoted again by $(z_n) = ((u_n, v_n))$ such that

$$u_n \rightharpoonup u \quad \text{in } H_r^s, \qquad v_n \rightharpoonup v \quad \text{in } H_r^t, \tag{5.15}$$

$$u_n \to u \quad \text{in } L^\gamma,\ 2 < \gamma < 2N/(N - 2s)$$

$$v_n \to v \quad \text{in } L^\gamma,\ 2 < \gamma < 2N/(N - 2t).$$

Given $\psi \in \hat{H}_n$, there is a $\tilde{\phi} \in H_r^s$ such that $A^t\psi = A^s\tilde{\phi}$. So

$$\int A^s u_n A^t \psi = \int A^s u_n A^s \tilde{\phi} \to \int A^s u A^s \tilde{\phi} = \int A^s u A^t \psi,$$

and a similar statement with v_n and ϕ.

It follows then from (5.7) that

$$\int [A^s u A^t \psi + A^s \phi A^t v] = \lim \int [\phi f(r, u_n) + \psi g(r, v_n)], \qquad \text{for } (\phi, \psi) \in E_n. \tag{5.16}$$

Now we claim that

$$\lim \int \phi f(r, u_n) = \int \phi f(r, u) \qquad \text{and} \qquad \lim \int \psi g(r, v_n) = \int \psi g(r, v). \tag{5.17}$$

Let $R > 0$, and let us estimate separately the integral of $\phi[f(r, u_n) - f(r, u)]$ over B_R, and over $R^N \backslash B_R$.

$$I_1 = \int_{B_R} |\phi[f(r, u_n) - f(r, u)]| \leq \|\phi\|_{L^{\theta_1'}(B_R)} \|f(r, u_n) - f(r, u)\|_{L^{\theta_1}(B_R)} \tag{5.18}$$

where $1/\theta_1 + 1/\theta_1' = 1$, $\theta_1 > 1$, and

$$2 \leq \theta_1' \leq \frac{2N}{N - 2s}. \tag{5.19}$$

On the other hand, it follows from (5.15) that $u_n \to u$ in $L^\gamma(B_R)$ for all $1 \leq \gamma < 2N/(N - 2s)$. In view of hypothesis (H2), the Nemytskii mapping $u \mapsto f(r, u)$ maps continuously $L^\gamma(B_R)$ into $L^{\gamma/p}(B_R)$, cf. [27]. So if $\theta_1 \leq \gamma/p$, then

$$\|f(r, u_n) - f(r, u)\|_{L^{\theta_1}(B_R)} \to 0. \tag{5.20}$$

It is easy to verify that θ_1 can be chosen in such a way that (5.19) holds and $p\theta_1 < 2N/(N - 2s)$. So, for each $R > 0$

$$I_1 \leq \|\phi\|_{H_r^s} \|f(r, u_n) - f(r, u)\|_{L^{\theta_1}(B_R)}. \tag{5.21}$$

Next we estimate the integral over $R^N \backslash B_R =: B_R^c$. It follows from (H2) and (H4) that

$$|f(r, u)| \leq c(|u|^a + c|u|^p). \tag{5.22}$$

So

$$I_2 := \left| \int_{B_R^c} \phi[f(r, u_n) - f(r, u)] \right| \leq c \int_{B_R^c} |\phi| [|u_n|^a + |u|^a + |u_n|^p + |u|^p].$$

Using Hölder's inequality

$$I_2 \leq c\|\phi\|_{L^{\theta_2'}(B_R^c)} \{\|u_n\|_{L^{a\theta_2}(B_R^c)}^a + \|u\|_{L^{a\theta_2}(B_R^c)}^a\}$$
$$+ C\|\phi\|_{L^{\theta_3'}(B_R^c)} \{\|u_n\|_{L^{p\theta_3}(B_R^c)}^p + \|u\|_{L^{p\theta_3}(B_R^c)}^p\}.$$

It is easy to see that one can choose θ_2, θ_3 in such a way that $2 \le \theta_2'$, $\theta_3' \le 2N/(N-2s)$ and $2 < a\theta_2$; $p\theta_3 > 2N/(N-2s)$. Then

$$I_2 \le c\|\phi\|_{H_r^s(R^N)}\{\|u_n - u\|^a_{L^{a\theta_2}(R^N)} + \|u\|^a_{L^{a\theta_2}(B_R^c)}$$

$$+ \|u_n - u\|^p_{L^{p\theta_3}(R^N)} + \|u\|^p_{L^{p\theta_3}(B_R^c)}\}. \tag{5.23}$$

So, from (5.18) and (5.23), we conclude (5.17).

On the other hand, from (5.7) we obtain

$$\left|\int A^s\phi A^t v_n - \int \phi f(r, u_n)\right| \le \varepsilon_n \|\phi\|_{H_r^s}, \qquad \phi \in H_n. \tag{5.24}$$

So, using in order (5.24), (5.16), (5.17), (5.21) and (5.23) we see that

$$\limsup \|v_n\|_{H_r^t} \le \|\sigma\|_{H_r^t} \tag{5.25}$$

and so we conclude that $v_n \to v$ strongly. In a similar way, we prove that $u_n \to u$ strongly. Thus the proof of Lemma 5.1 is complete. ■

Remark. It follows from the above lemma that Φ satisfies the usual (PS) condition.

We now study the geometry of the functional Φ.

LEMMA 5.2. Assume the Hypotheses of Theorem 5.1. Then, the functional Φ defined in (5.3) satisfies the assumptions (B1), (B3) and (B4) of Theorem [26].

Proof. Condition (B3) is an immediate consequence of the continuity of the bilinear form B and the growth restrictions of f and g. Next, to see that Φ has a local linking at zero we use (4.14) and conditions (H2) and (H4). For $z \in E^+$ we have

$$\Phi(z) \ge c_0\|z\|_E^2 - c\int |u|^{p+1} - c\int |u|^{a+1} - c\int |v|^{q+1} - c\int |v|^{q+1} - c\int |v|^{b+1}$$

(In this proof c denotes distinct constants.) Using Theorem 4.1 we get

$$\Phi(z) \ge c_0\|z\|_E^2 - c\|z\|^{a+1} - c\|z\|^{b+1}$$

for small $\|z\|$. And since $a, b > 1$ we conclude that $\Phi(z) \ge 0$ for $z \in E^+$ with $\|z\|$ small. A similar argument proves that $\Phi(z) \le 0$ for $z \in E^-$ with $\|z\|$ small.

Finally, let us prove condition (B4). Let $n \in \mathbb{N}$ be fixed and let $z \in E_n^+ \oplus E^-$, with $\|z\| \to \infty$.

We claim that $\Phi(z) \to -\infty$. Indeed, with $z = (u, v)$ and $z = z^+ + z^-$ we have

$$\Phi(z) = Q(z^+) + Q(z^-) - \int F(u) - \int G(v)$$
$$\leq -c_0\|z^-\|_E^2 + c\|z^+\|_E^2 - c\int |u|^\alpha - c\int |v|^\beta. \tag{5.26}$$

Letting $z^+ = (u^+, v^+)$ and $z^- = (u^-, v^-)$ we have $u^- = \gamma u^+ + \hat{u}$, where $\hat{u}$ is orthogonal to u^+ in L^2. Also $v^- = \tau v^+ + \hat{v}$, where $\hat{v}$ is orthogonal to v^- in L^2. It is easy to see that either γ or τ is positive.

Suppose $\gamma > 0$. Then we have

$$(1 + \gamma)\int |u^+|^2 = \int [(1 + \gamma)u^+ + \hat{u}]u^+ \leq \|u\|_{L^\alpha}\|u^+\|_{L^{\alpha'}}.$$

Using the fact that the norms in E_n^+ are equivalent we obtain

$$(1 + \gamma)\|u^+\|_{L^\alpha} \leq c\|u\|_{L^\alpha}$$

with constant c independent of u.

So from (5.26) we obtain

$$\Phi(z) \leq -c_0\|z^-\|_E^2 + c\|z^+\|_E^2 - c\|u^+\|_{L^\alpha}^\alpha$$

which implies that $\Phi(z) \to -\infty$ as $\|z\| \to \infty$.

6. EXISTENCE OF A GROUND STATE

In this section we consider the existence of a ground state for the problem

$$-\Delta u + u = v^q, \tag{6.1}$$
$$-\Delta v + v = u^p, \qquad \text{in } R^N \tag{6.2}$$

where $1 < p,\ q < (N + 2)/(N - 2)$. By a ground state we mean a solution which possesses the least energy among all solutions. That is, a ground state is a minimizer of the variational problem

$$\Phi^\infty = \inf\{\Phi(u, v) : (u, v) \text{ is a solution of (6.1)–(6.2)},\ (u, v) \neq (0, 0)\}.$$

From the previous sections, we know that all positive solutions of (6.1)–(6.2) are radially symmetric and decay exponentially to zero at infinite.

First we show that the set of nontrivial solutions is bounded below. In fact, from (6.1) we have

$$\|u\|_{H^1}^2 = \int_{R^N} v^q u \leq \left(\int_{R^N} v^{q+1}\right)^{q/(q+1)}\left(\int u^{q+1}\right)^{1/(q+1)} \leq c\|v\|_{L^{q+1}}^q\|u\|_{H^1}.$$

Since $u \neq 0$, we obtain then that

$$\|u\|_{H^1} \leq c\|v\|_{L^{q+1}}^q. \tag{6.3}$$

Now, using the two equations, we obtain

$$\int v^{q+1} = \int [\nabla u \, \nabla v + uv] = \int u^{p+1}. \tag{6.4}$$

From (6.3) and (6.4), it follows that

$$\|u\|_{H^1} \leq c\|u\|_{L^{p+1}}^{q(p+1)/(q+1)}. \tag{6.5}$$

We estimate the left side of (6.5) by Sobolev imbedding and get

$$\|u\|_{L^{p+1}} \leq c\|u\|_{L^{p+1}}^{q(p+1)/(q+1)},$$

which implies $\|u\|_{L^{p+1}} \geq c$, since $q(p\ 1)/(q+1) > 1$. Similarly we prove that $\|v\|_{L^{q+1}} \geq c$. Finally, we observe that Φ^∞ is attained. Indeed, let (u_n, v_n) be a sequence of solutions such that $\Phi(u_n, v_n) \to \Phi^\infty$. So (u_n, v_n) is a Palais–Smale sequence, and by Lemma (5.1), (u_n, v_n) converges strongly to (u_0, v_0), which is then a solution of (6.1)–(6.2).

REFERENCES

1. Clément, Ph., de Figueiredo, D. G. and Mitidieri, E., Positive solutions of semilinear elliptic systems. *Comm. P.D.E.*, 1992, **17**, 923–940.
2. Peletier, L. A. and van der Vorst, R. C. A. M., Existence and non-existence of positive solutions of non-linear elliptic systems and the bi-harmonic equation. *Diff. Int. Eq.*, 1992, **5**, 747–767.
3. Benci, V. and Rabinowitz, P., Critical point theorems for indefinite functionals. *Invent. Math.*, 1979, **52**, 241–273.
4. Felmer, P. L., Periodic solutions of "superquadratric" Hamiltonian systems. *J. Diff. Equat.*, 1993, **102**, 188–207.
5. Bartsch, T. and Wang, Z. Q., Existence and multiplicity results for some superlinear elliptic problems on R^N. *Comm. PDE*, 1995, **20**, 1725–1741.
6. Costa, D. G., On a class of elliptic systems in R^N. *Elect. J.D.E.*, 1994, **1994**(7), 1–14.
7. Yanheng, Ding and Shujie, Li, Existence of solutions for some superlinear or sublinear elliptic systems on R^N. Preprint.
8. Reed, M. and Simon, B., *Methods of Modern Mathematical Physics*, Vol. IV. Academic Press, 1978.
9. Strauss, W., Existence of solitary waves in higher dimensions. *Commun. Math. Phys.*, 1977, **55**, 149–162.
10. Coleman, S., Glazer, V. and Martin, A., Action minima among solutions to a class of Euclidean scalar field equations. *Commun. Math. Phys.*, 1978, **58**, 211–221.
11. Berestycki, H. and Lions, P. L., Nonlinear scalar field equations—I and II. *Arch. Rational Mech. Anal.*, 1983, **82**, 313–376.
12. de Figueiredo, D. G. and Felmer, P. L., A Liouville-type theorem for elliptic systems. *Ann. Scuola Norm. Sup. Pisa*, 1994, **21**(3), 387–397.
13. Bergh, J. and Löfström, J., *Interpolation Spaces—An Introduction*. Springer-Verlag, 1976.
14. Gilbarg, D. and Trudinger, N. S., *Elliptic Partial Differential Equations of Second Order*. Springer-Verlag, 1983.
15. Hopf, E., *Lectures on Differential Geometry in the Large*. Stanford University, 1956.
16. Serrin, J., A symmetry problem in potential theory. *Arch. Rat. Mech. Anal.*, 1971, **43**, 304–318.
17. Gidas, B., Ni, W. M. and Nirenberg, L., Symmetry and related properties via the maximum principle. *Comm. Math. Phys.*, 1979, **68**, 209–243.
18. Gidas, B., Ni, W. M. and Nirenberg, L., Symmetry of positive solutions of nonlinear elliptic equations in R^n. *Advances in Math. Suplemntary Studies*, 1981, **7**, 369–402.
19. Weidmann, J., *Linear Operators in Hilbert Spaces*, GMT 68. Springer-Verlag, Berlin, 1980.
20. Kato, T., *Perturbation Theory for Linear Operators*. Springer-Verlag, Berlin, 1966.
21. Yosida, K. , *Functional Analysis*. Springer-Verlag, Berlin, 1974.
22. Lions, J. L. and Magenes, E., *Non-homogeneous Boundary Value Problems and Applications, I*. Springer-Verlag, Berlin, 1972.

23. Lions, P. L., Symétrie et compacité dans les espaces de Sobolev. *J. Functl. Anal.*, 1982, **49**, 315–334.
24. Liu, J. Q. and Li, S., Some existence theorems on multiple critical points and their applications. *Kexue Tongbao*, 1984, **17**.
25. de Figueiredo, D. G. and Felmer, P. L., On superquadratic elliptic systems. *Trans. Amer. Math. Soc.*, 1994, **343**(1), 99–116.
26. Li, S. and Willem, M., Applications of local linking to critical point theory. *J. Math. Anal. Appl.*, 1995, **189**, 6–32.
27. de Figueiredo, D. G., *The Ekeland Variational Principle with Applications and Detours*. Springer-Verlag, Berlin, 1989.

INFINITELY MANY SOLUTIONS OF NONLINEAR ELLIPTIC SYSTEMS

THOMAS BARTSCH*
Mathematisches Institut
Universität Giessen
Arndtstr. 2
35392 Giessen
Germany

DJAIRO G. DE FIGUEIREDO
IMECC
UNICAMP
Caixa Postal 6065
13081-970 Campinas SP
Brazil

PNDE 35 1999 p.51-68

1 Introduction

In this paper we study elliptic systems of the form

$$\begin{cases} -\Delta u = H_v(x,u,v) & \text{in } \Omega \\ -\Delta v = H_u(x,u,v) & \text{in } \Omega \end{cases} \tag{1.1}$$

where $\Omega \subset \mathbb{R}^N$, $N \geq 3$, is a smooth bounded domain and $H: \overline{\Omega} \times \mathbb{R} \times \mathbb{R} \to \mathbb{R}$ is a C^1-function. We shall also consider the case when $\Omega = \mathbb{R}^N$, and in this case the system takes the form

$$\begin{cases} -\Delta u + u = H_v(x,u,v) & \text{in } \mathbb{R}^N \\ -\Delta v + v = H_u(x,u,v) & \text{in } \mathbb{R}^N \end{cases} \tag{1.2}$$

In the bounded case, we look for solutions of (1.1) subject to Dirichlet boundary conditions $u = v = 0$ on $\partial\Omega$. In the case when $\Omega = \mathbb{R}^N$ we assume that some symmetry with respect to x holds; for instance, that the x-dependence of H is radial, or that H is invariant with respect to certain subgroups of $O(N)$ acting on R^N. We shall obtain both radial and non-radial solutions in the radial symmetric case, thus observing a symmetry breaking effect.

*Research supported by GMD–CNPq. T.B. thanks the members of the Department of Mathematics at UNICAMP for their invitation and hospitality.

Originally published in: Prog. Nonlinear Differential Equations Appl. 35 (The Herbert Amann Anniversary Volume) (1999), 51-68, © 1999 Birkhäuser Verlag Basel, Switzerland Reprint by Springer-Verlag Berlin Heidelberg 2013,
DOI: 10.1007/978-3-319-02856-9_31

In order to illustrate the kind of results obtained here, let us state two theorems. We first consider the case when Ω is bounded. In such a case, the following set of hypotheses is assumed. First, the regularity of the Hamiltonian:

(H_1) $H : \overline{\Omega} \times \mathbb{R} \times \mathbb{R} \to \mathbb{R}$ is C^1 and $H \geq 0$.

Next we assume conditions related to the growth of the right side of (1.1).

(H_2) There exist constants $p, q > 1$ and $c_1 > 0$ with

$$1 > \frac{1}{p} + \frac{1}{q} > 1 - \frac{2}{N} \tag{1.3}$$

such that

$$|H_u(x,u,v)| \leq c_1(|u|^{p-1} + |v|^{(p-1)q/p} + 1) \tag{1.4}$$

and

$$|H_v(x,u,v)| \leq c_1(|v|^{q-1} + |u|^{(q-1)p/q} + 1) \tag{1.5}$$

for all $(x,u,v) \in \overline{\Omega} \times \mathbb{R} \times \mathbb{R}$.

The next condition is a "non-quadraticity" condition at infinity introduced by Costa-Magalhães [5]. It is related to the so-called Ambrosetti-Rabinowitz condition and it is devised to get some sort of Palais-Smale condition for the functionals involved.

(H_3) There exist $1 < \alpha < p$ and $1 < \beta < q$ with

$$\frac{1}{\alpha} + \frac{1}{\beta} = 1, \tag{1.6}$$

and such that

$$\frac{1}{\alpha} H_u(x,u,v)u + \frac{1}{\beta} H_v(x,u,v)v - H(x,u,v) \geq a(|u|^\mu + |v|^\nu - 1) \tag{1.7}$$

for all $(x,u,v) \in \overline{\Omega} \times \mathbb{R} \times \mathbb{R}$. Here a, μ, ν are positive constants satisfying

$$\mu > \frac{pN}{2} \max\left\{\frac{1}{2} - \frac{1}{p},\ 1 - \frac{1}{p} - \frac{1}{q}\right\}$$

and

$$\nu > \frac{qN}{2} \max\left\{\frac{1}{2} - \frac{1}{q},\ 1 - \frac{1}{p} - \frac{1}{q}\right\}$$

Remark 1.1. *Observe that* $1 - \frac{1}{p} - \frac{1}{q}$ *is always positive in view of* (H_2). *It follows from* (H_3) *that*

$$H(x,u,v) \geq c(|u|^\alpha + |v|^\beta - 1). \tag{1.8}$$

In fact, (1.8) *follows from a condition weaker than* (H_3), *namely*

$$\frac{1}{\alpha} H_u(x,u,v)u + \frac{1}{\beta} H_v(x,u,v)v \geq H(x,u,v)$$

for all $x \in \overline{\Omega}$ *and* $|(u,v)| \geq R$; *see Felmer* [10].

Remark 1.2. *Suppose* H *satisfies the following condition of Ambrosetti-Rabinowitz type: there is* $R > 0$ *and* $1 < \alpha' < p$ *and* $1 < \beta' < q$ *with* $\frac{1}{\alpha'} + \frac{1}{\beta'} < 1$ *and such that*

$$\frac{1}{\alpha'} H_u(x,u,v)u + \frac{1}{\beta'} H_v(x,u,v)v \geq H(x,u,v)$$

for $x \in \overline{\Omega}$ *and* $|(u,v)| \geq R$. *Then condition* (H_3) *holds. In this case, it follows that* H *is superquadratic, in the sense that*

$$H(x,u,v) \geq c_1(|u|^{\alpha'} + |v|^{\beta'}) - c_2.$$

The next condition provides the symmetry we assume here.

(H_4) $H(x,-u,-v) = H(x,u,v)$ for all $(x,u,v) \in \overline{\Omega} \times \mathbb{R} \times \mathbb{R}$.

Now we are prepared to state the result in the case of Ω bounded. For that matter we introduce a non-increasing sequence of constants δ_n, $n \in \mathbb{N}$, with $\delta_n \to 0$, which will be defined in Section 3, and which depend only on p, q, α and β.

Theorem 1.1. *Suppose that* (H_1)-(H_4) *hold. Then there is a* $k_0 \in \mathbb{N}$ *such that, if*

$$\liminf_{|(u,v)| \to \infty} \frac{2H(x,u,v)}{|u|^\alpha + |v|^\beta} > \frac{1}{\delta_K} \tag{1.9}$$

holds for $K \geq k_0$, *system* (1.1), *subject to Dirichlet boundary conditions, has* $K - k_0 + 1$ *pairs of nontrivial solutions.*
Moreover, if

$$\lim_{|(u,v)| \to \infty} \frac{H(x,u,v)}{|u|^\alpha + |v|^\beta} = +\infty,$$

(in particular, if H *is superquadratic) then system (1.1), subject to Dirichlet boundary conditions has infinitely many solutions.*

The solutions obtained in Theorem 1.1 are strong solutions in the sense that $u \in W^{2,p/(p-1)}(\Omega) \cap W_0^{1,p/(p-1)}(\Omega)$ and $v \in W^{2,q/(q-1)}(\Omega) \cap W_0^{1,q/(q-1)}(\Omega)$. The existence of at least one solution for the system (1.1), without symmetry assumptions, has been considered before. See the survey paper [6] for a list of references.

As a corollary of Theorem 1.1 we obtain solutions to two nonlinear eigenvalue problems. We consider first

$$\begin{cases} -\Delta u &= \delta u + \lambda |v|^{\beta-2} v + H_v(x,u,v), \\ -\Delta v &= \mu |u|^{\alpha-2} u + \delta v + H_u(x,u,v) \end{cases} \tag{1.10}$$

in a bounded domain $\Omega \subset \mathbb{R}^N$ subject to Dirichlet boundary conditions $u = v = 0$ on $\partial\Omega$. The constants α, β are those from (H_3).

Corollary 1.1. *If H satisfies* (H_1)-(H_4), *then for each $k \in \mathbb{N}$, there exists $\Lambda_k > 0$, such that (1.10) has k pairs of non-trivial solutions provided λ, $\mu > \Lambda_k$.*

Next we consider the eigenvalue problem

$$\begin{cases} -\Delta u &= \lambda H_v(x,u,v) \\ -\Delta v &= \lambda H_u(x,u,v) \end{cases} \tag{1.11}$$

in a bounded domain $\Omega \subset \mathbb{R}^N$ subject to Dirichlet boundary conditions $u = v = 0$ on $\partial\Omega$.

Corollary 1.2. *Suppose that H satisfies* (H_1)-(H_4), *and*

$$\liminf_{|(u,v)| \to \infty} \frac{H(x,u,v)}{|u|^\alpha + |v|^\beta} > 0.$$

Then for each $k \in \mathbb{N}$, there exists $\Lambda_k > 0$, such that (1.11) has k pairs of non-trivial solutions provided $\lambda > \Lambda_k$.

Let us now state a result for the case when system (1.2) is considered in the whole of $\mathbb{R}^N$. We need a distinct, but similar, set of hypotheses.

(H_1') $H : \mathbb{R}^N \times \mathbb{R} \times \mathbb{R} \to \mathbb{R}$ is C^1, $H \geq 0$, $H(x,u,v) > 0$ for $|(u,v)| > 0$ and H is radial in the variable x.

(H_2') There exist positive constants p, q, a, b and c_1 with

$$p, q > 2, \quad \frac{1}{p} + \frac{1}{q} > 1 - \frac{2}{N}, \quad 1 < a < p-1, \quad 1 < b < q-1, \tag{1.12}$$

such that

$$|H_u(x,u,v)| \leq c_1 (|u|^{p-1} + |v|^{(p-1)q/p} + |u|^a) \tag{1.13}$$

and

$$|H_v(x,u,v)| \le c_1(|v|^{q-1} + |u|^{(q-1)p/q} + |v|^b) \tag{1.14}$$

for all $(x,u,v) \in \mathbb{R}^N \times \mathbb{R} \times \mathbb{R}$.

(H_3') There exist $1 < \alpha < p$ and $1 < \beta < q$ with $\alpha^{-1} + \beta^{-1} < 1$ and such that

$$\frac{1}{\alpha} H_u(x,u,v)u + \frac{1}{\beta} H_v(x,u,v)v \ge H(x,u,v)$$

for all $(x,u,v) \in \mathbb{R}^N \times \mathbb{R} \times \mathbb{R}$.

(H_4') There are positive constants c and r such that

$$H(x,u,v) \ge c(|u|^p + |v|^q) \quad \text{for } x \in \mathbb{R}^N \text{ and } |(u,v)| \le r.$$

(H_5') $H(x,u,v) = H(x,-u,-v)$ for all $(x,u,v) \in \mathbb{R}^N \times \mathbb{R} \times \mathbb{R}$.

Remark 1.3. *It follows from* (H_3') *that there are positive constants c and R such that*

$$H(x,u,v) \ge c(|u|^p + |v|^q) \quad \textit{for } |(u,v)| \ge R. \tag{1.15}$$

Then (1.14) *and assumption* (H_4') *imply that*

$$H(x,u,v) \ge c(|u|^p + |v|^q) \quad \textit{for all } (x,u,v) \in \mathbb{R}^N \times \mathbb{R} \times \mathbb{R}. \tag{1.16}$$

Theorem 1.2. *Assume that the Hamiltonian H satisfies the hypotheses* (H_1')-(H_5'). *Then system* (1.2) *has infinitely many radial solutions.*

The solutions obtained in Theorem 1.2 are strong solutions in the sense that they satisfy $u \in W^{2,p/(p-1)}_{loc}(\mathbb{R}^N)$ and $v \in W^{2,q/(q-1)}_{loc}(\mathbb{R}^N)$. They also satisfy $u \in H^s(\mathbb{R}^N) \subset L^\gamma(\mathbb{R}^N)$ and $v \in H^t(\mathbb{R}^N) \subset L^\delta(\mathbb{R}^N)$ for some $s,t > 0$ with $s+t = 2$ and $2 < \gamma, 2N/(N-2s)$, $2 < \delta < 2N/(N-2t)$. The existence of at least one solution has been obtained before for special cases of system (1.2) in [9] and recently in [14].

The next result exhibits the breaking of symmetry in certain dimensions. The result extends to the type of systems we have here a result that Bartsch-Willem [3] proved in the scalar case.

Theorem 1.3. *Suppose that* (H_1')-(H_5') *holds. If $N = 4$ or $N \ge 6$ then system* (1.2) *has infinitely many non-radial solutions.*

2 Some Abstract Critical Point Theory

We consider a Hilbert space E and a functional $\Phi \in C^1(E,\mathbb{R})$. Given a sequence $\mathcal{F} = (X_n)$ of finite dimensional subspaces $X_n \subset X_{n+1}$, with $\overline{\bigcup X_n} = E$, we say that Φ satisfies $(PS)_c^{\mathcal{F}}$, at level $c \in \mathbb{R}$, if every sequence z_j, $j \in \mathbb{N}$, with $z_j \in X_{n_j}$, $n_j \to \infty$, such that

$$\Phi(z_j) \to c \quad \text{and} \quad (1 + \|z_j\|)(\Phi|_{X_{n_j}})'(z_j) \to 0 \tag{2.1}$$

has a subsequence which converges to a critical point of Φ. In the case when $X_n = E$ for all $n \in \mathbb{N}$ this form of the Palais-Smale condition is due to Cerami [4]. It is closely related to the standard Palais-Smale condition and to the $(PS)^*$ condition of [1] and [11]. It also yields a deformation lemma. In the present form $(PS)_c^{\mathcal{F}}$ was introduced in Bartsch-Clapp [2].

Remark 2.1. *If Φ has the form*

$$\Phi(z) = \frac{1}{2}\langle Lz, z\rangle - \Psi(z)$$

with $L : E \to E$ a linear Fredholm operator of index zero and $\nabla\Psi : E \to E$ completely continuous, then a bounded $(PS)_c^{\mathcal{F}}$ sequence (z_j) has a convergent subsequence. By a $(PS)_c^{\mathcal{F}}$ sequence we mean a sequence as in (2.1). *Let us prove the above statement. First select a subsequence, denoted again by (z_j) such that $z_j \rightharpoonup z$, weakly in E. Then $\nabla\Psi(z_j) \to \nabla\Psi(z)$, strongly in E. Let $P_n : E \to X_n$ denote the orthogonal projection onto X_n. We have that the sequence*

$$P_{n_j}\nabla\Phi(z_j) = P_{n_j}Lz_j + P_{n_j}\nabla\Psi(z_j)$$

converges to zero in view of (2.1). *So*

$$P_{n_j}Lz_j \to -\nabla\Psi(z) = Lz.$$

Hence $Lz_j \to Lz$. And as a consequence, $z_j \to z$, because $\ker L$ *is finite dimensional.*

Now suppose that E splits as a direct sum $E = E^+ \oplus E^-$. Let $E_1^\pm \subset E_2^\pm \subset \ldots$ be a strictly increasing sequence of finite dimensional subspaces of $E^\pm$ such that $\overline{\bigcup_{n=1}^\infty E_n^\pm} = E^\pm$. Setting $E_n = E_n^+ \oplus E_n^-$ we can formulate the hypotheses on Φ which are needed for our first abstract theorem.

(Φ_1) $\Phi \in C^1(E,\mathbb{R})$ and satisfies $(PS)_c^{\mathcal{F}}$ for $\mathcal{F} = (E_n)_{n\in\mathbb{N}}$ and $c > 0$.

(Φ_2) For some $k \geq 2$ and some $r > 0$ one has

$$b_k := \inf\{\Phi(z) : z \in E^+, z \perp E_{k-1}, \|z\| = r\} > 0. \tag{2.2}$$

(Φ_3) There exists an isomorphism $T : E \to E$ with $T(E_n) = E_n$, for all $n \in \mathbb{N}$, and there exist $K \geq k$ and $R > 0$ such that

$$\text{for } z = z^+ + z^- \in E_K^+ \oplus E^- \text{ with } \max\{\|z^+\|, \|z^-\|\} = R$$

one has

$$\|Tz\| > r \quad \text{and} \quad \Phi(Tz) \leq 0,$$

where k and r are the constants introduced in (Φ_2).

(Φ_4) $d_K := \sup\{\Phi \circ T(z^+ + z^-) : z^+ \in E_K^+, z^- \in E^-, \|z^+\|, \|z^-\| \leq R\} < \infty$.

(Φ_5) Φ is even, i.e. $\Phi(-z) = \Phi(z)$.

A stronger condition that implies (Φ_4) and holds in our application is:

(Φ_6) Φ maps bounded sets to bounded sets.

Theorem 2.1. *Assume* $(\Phi_1) - (\Phi_5)$. *Then, for every* $b < b_k$, Φ *has at least* $K - k + 1$ *pairs* $\pm z_i$ *of critical points with critical values in* $[b, d_K]$.

Proof: We need to recall the equivariant limit category defined in [2], specialized to our situation. We set $G = \mathbb{Z}/2$ which acts on E via the antipodal map. Given invariant subsets $Z \subset Y \subset X$ of E, we define the G–$\text{cat}_X(Y, Z)$ to be the least integer m such that there exists a covering $Y \subset U_0 \cup \ldots \cup U_m$ of Y with invariant open subsets $U_0, \ldots, U_m$ of X with the properties:

(i) $Z \subset U_0$ and there exists a continuous family $h^t : U_0 \to X$, $0 \leq t \leq 1$, of odd maps satisfying $h^0(z) = z$ and $h^1(z) \in Z$ for every $z \in U_0$, and $h^t(z) = z$ for every $z \in Z$ and every $t \in [0, 1]$.

(ii) For $i = 1, \ldots, m$ there exists a continuous family $h_i^t : U_i \to X$, $0 \leq t \leq 1$, of odd maps satisfying $h_i^0(z) = z$ for every $z \in U_i$ and such that $h_i^1(U_i) = \{\pm z_i\}$, for some $z_i \in X \setminus \{0\}$.

Now we define the equivariant limit category for G-invariant sets $Z \subset Y \subset E$ by

$$G\text{–cat}_E^{\mathcal{F}}(Y, Z) := \limsup_{n \to \infty} G\text{–cat}_{E_n}(Y \cap E_n, Z \cap E_n)\,.$$

Given $d > b > 0$ Theorem 2.8 of [2] says that Φ has at least $G\text{–cat}_E^{\mathcal{F}}(\Phi^d, \Phi^b)$ pairs of critical points with critical values in $[b, d]$. Therefore it suffices to prove that $G\text{–cat}_E^{\mathcal{F}}(\Phi^{d_K}, \Phi^b) \geq K - k + 1$ for $0 < b < b_k$. This follows from the next lemma.

Lemma 2.1. *Fix* $0 \leq b < b_k$ *and* $n \geq K$. *Then*

$$\gamma := G\text{–cat}_{E_n}(\Phi^{d_K} \cap E_n, \Phi^b \cap E_n) \geq K - k + 1.$$

Proof: For simplicity we set $d := d_K$, and $B := B_R E_K^+ \times B_R E_n^-$ with $R > 0$ from (Φ_3). We also write $S_r E_n$ for the sphere of radius r in E_n. Let

$$\Phi_n^d := \Phi^d \cap E_n \subset U_0 \cup \ldots \cup U_\gamma$$

be a covering as in the definition of $G\text{-cat}_{E_n}(\Phi_n^d, \Phi_n^b)$. There are odd mappings $h^1 : U_0 \to \Phi_n^b$ and $h_i^1 : U_i \to \{\pm z_i\}$. Making U_0 smaller if necessary we may assume that h^1 extends continuously to $\overline{U_0}$. Then we can extend h^1 to an odd mapping $h^1 : E_n \to E_n$ by using Tietze's extension theorem. Now we set

$$\mathcal{O} := \{z \in B : \|h^1(Tz)\| < r\}.$$

For $z \in \partial B$ we have $\|Tz\| > r$ and $\Phi(Tz) \leq 0$ by (Φ_3). Thus $Tz \in \Phi_n^0 \subset \Phi_n^b$ and $h^1(Tz) = Tz$, and hence $\|h^1(Tz)\| = \|Tz\| > r$. This implies that $\mathcal{O}$ is an open subset of B with $\overline{\mathcal{O}} \subset \operatorname{int} B$. Clearly $\mathcal{O}$ is an invariant neighborhood of 0 in $E_K^+ \oplus E_n^-$.

For $z \in T^{-1}(U_0)$ we have that $h^1(Tz) \in \Phi_n^b \subset E_n \backslash S_r(E_{k-1}^\perp \cap E_n^+)$, in virtue of (Φ_2). For $z \in \partial\mathcal{O}$, we have that $\|h^1(Tz)\| = r$. This implies that

$$h^1(\partial\mathcal{O} \cap T^{-1}(U_0)) \subset S_r E_n \setminus S_r(E_{k-1}^\perp \cap E_n^+).$$

The latter space has the sphere $S_r(E_{k-1} \oplus E_n^-)$ as a strong deformation retract. In particular, there exists an odd mapping

$$S_r E_n \setminus S_r(E_{k-1}^\perp \cap E_n^+) \longrightarrow S_r(E_{k-1} \oplus E_n^-).$$

Observe that $S_r(E_{k-1} \oplus E_n^-) \cong S^{k+n-2}$. Putting these mappings together we obtain an odd mapping

$$g_0 : \partial\mathcal{O} \cap T^{-1}(U_0) \longrightarrow S^{k+n-2} \subset \mathbb{R}^{k+n-1}.$$

The mappings $h_i^1 : U_i \to \{\pm z_i\}$ yield odd maps

$$g_i : \partial\mathcal{O} \cap T^{-1}(U_i) \longrightarrow S^0 = \{\pm 1\}.$$

By (Φ_4) we have that $T(B) \subset \Phi_n^d$. Therefore $T^{-1}(U_0), \ldots, T^{-1}(U_\gamma)$ cover B. Setting $V_i := \partial\mathcal{O} \cap T^{-1}(U_i)$, we obtain an open invariant covering of $\partial\mathcal{O}$. Choose then a partition of the unity $\pi_i : \partial\mathcal{O} \to [0,1]$, $i = 0, \ldots, \gamma$, subordinated to the covering $V_0, \ldots, V_\gamma$ of $\partial\mathcal{O}$. Since the V_i's are invariant we may assume that the π_i's are even. Now we define the mapping

$$g : \partial\mathcal{O} \longrightarrow \mathbb{R}^{k+n-1} \times \mathbb{R}^\gamma, \quad g(z) := (\pi_0(z)g_0(z), \ldots, \pi_\gamma(z)g_\gamma(z)).$$

First, observe that g is well defined. Namely, if $\pi_i(z) \neq 0$, then $z \in V_i$ and so $g_i(z)$ is defined. Obviously, g is odd, since the g_i are odd and the π_i are even. Also, g is continuous. In addition $g(z) \neq 0$ for every $z \in \partial\mathcal{O}$ because there exists $i \in \{0, \ldots, \gamma\}$, with $\pi_i(z) \neq 0$, and hence $z \in V_i$ and $|g_i(z)| = 1$.

Thus we have a continuous odd mapping $g : \partial\mathcal{O} \to \mathbb{R}^{k+n-1+\gamma} \setminus \{0\}$, where $\mathcal{O}$ is an invariant bounded open neighborhood of 0 in $E_K^+ \oplus E_n^-$. Now Borsuk's theorem implies that $k+n-1+\gamma \geq \dim E_K^+ \oplus E_n^- = K+n$. This shows that $\gamma \geq K-k+1$ as required. □

As an immediate corollary of Theorem 2.1, we obtain the Fountain Theorem, which we state below. First we introduce the following set of conditions.

(Φ_2') There exists a sequence $r_k > 0$, $k \in \mathbb{N}$, such that $b_k \to +\infty$ as $k \to \infty$. (Here b_k is defined as in (Φ_2) with r_k instead of r.)

(Φ_3') There exists a sequence of isomorphisms $T_k : E \to E$, $k \in \mathbb{N}$, with $T_k(E_n) = E_n$ for all k and n, and there exists a sequence $R_k > 0$, $k \in \mathbb{N}$, such that, for $z = z^+ + z^- \in E_k^+ \oplus E^-$ with $\max\{\|z^+\|, \|z^-\|\} = R_k$, one has

$$\|T_k z\| > r_k \quad \text{and} \quad \Phi(T_k z) < 0$$

where r_k is given in (Φ_2').

(Φ_4') $d_k := \sup\{\Phi\big(T_k(z^+ + z^-)\big) : z^+ \in E_k^+, z^- \in E^-, \|z^+\|, \|z^-\| \leq R_k\} < \infty$.

Theorem 2.2. (Fountain Theorem) *Suppose that (Φ_1), (Φ_2') – (Φ_4'), (Φ_5) hold. Then Φ has an unbounded sequence of critical values.*

Hypothesis (Φ_2') will be checked in the applications later on using the contents of the next remark.

Remark 2.2. *Let E be a Hilbert space and $E_1 \subset E_2 \subset \ldots$ be finite dimensional subspaces such that $\overline{\bigcup_{n=1}^{\infty} E_n} = E$. Let $\Phi \in C^1(E,\mathbb{R})$ be of the form $\Phi = P - \Psi$ such that*

$$P(z) \geq \alpha\|z\|^p \quad \text{for all } z \in E$$

and

$$|\Psi(z)| \leq \beta(1 + \|z\|_X^q) \quad \text{for all } z \in E$$

Here X is a Banach space such that $E \subset X$ compactly, and $q > p$, α, β are positive constants.

First we prove that

$$\mu_k := \sup\{\|z\|_X : z \in E, z \perp E_{k-1}, \|z\| = 1\} \to 0 \quad \text{as } k \to \infty.$$

Indeed, suppose by contradiction that this is not the case. So, there is $\varepsilon > 0$ and a sequence (z_j) in E with $z_j \perp E_{k_j-1}$, $\|z_j\| = 1$, $\|z_j\|_X \geq \varepsilon$ and $k_j \to \infty$ as $j \to \infty$. This implies that $z_j \rightharpoonup 0$, weakly in E hence $z_j \to 0$, strongly in X, which contradicting $\|z_j\|_X \geq \varepsilon$.

Next we prove that there are $r_k > 0$, $k \in \mathbb{N}$, so that

$$b_k := \inf\{\Phi(z) : z \in E, z \perp E_{k-1}, \|z\| = r_k\} \to \infty.$$

Indeed, for $z \in E$, $z \perp E_{k-1}$, we have

$$\begin{aligned} \Phi(z) &= P(z) - \Psi(z) \\ &\geq \alpha\|z\|^p - \beta(1 + \|z\|_X^q) \\ &\geq \alpha\|z\|^p - \beta - \beta\mu_k^q\|z\|^q. \end{aligned} \tag{2.3}$$

Taking $\|z\| = r_k$ with $r_k := (p\alpha/q\beta\mu_k^q)^{1/(q-p)}$, we obtain

$$\Phi(z) \geq c\mu_k^{p^2/(p-q)} \to +\infty,$$

where c depends only on p, q, α, β.

Although the Fountain Theorem is an immediate consequence of Theorem 2.1, we choose to give a direct proof of it which does not employ the equivariant limit category, since this is a result with many applications.

Proof of the Fountain Theorem: By (Φ_2') it suffices to show that Φ has a critical value in $[b_k, d_k]$, for every k with $b_k > 0$. Fix such a k and suppose that $[b_k, d_k]$ contains only regular values. By Proposition 2.6 in [2], for n large, there exists a continuous deformation $h_n^t : \Phi_n^{d_k} \to E_n$, $t \in [0,1]$, such that h_n^t is odd and $h_n^1(\Phi_n^{d_k}) \subset \Phi_n^{b_k - \varepsilon}$, for some $\varepsilon > 0$. Moreover $h_n^t(z) = z$ if $\Phi(z) \leq 0$. As above we set $B := B_{R_k}E_k^+ \times B_{R_k}E_n^-$, for $n \geq k$. Consider the set

$$\mathcal{O} = \{z \in B : \|h_1(T_k z)\| < r_k\}.$$

As in the proof of Theorem 2.1 one checks easily that $\mathcal{O}$ is an open invariant neighborhood of 0 in $E_k^+ \oplus E_n^-$, and that $\overline{\mathcal{O}} \subset \operatorname{int} B$. Now we set

$$g := P \circ h_1 \circ T_k : \partial\mathcal{O} \longrightarrow E_{k-1}^+ \oplus E_n^-,\ g(z) := P\big(h_1(T_k z)\big)$$

where $P : E_n \to E_{k-1}^+ \oplus E_n^-$ is the orthogonal projection. Since $\dim(E_k^+ \oplus E_n^-) > \dim(E_{k-1}^+ \oplus E_n^-)$, Borsuk's theorem tells us that g must have a zero. Now $z \in \partial\mathcal{O}$ implies that $\|h_1(Tz)\| = r_k$, and $g(z) = 0$ implies that $h_1(T_k z) \in E_n^+$, $h_1(T_k z) \perp E_{k-1}^+$. It follows from (Φ_2') that $\Phi(h_1(T_k z)) \geq b_k$. This contradicts the fact that $T_k z \in \Phi_n^{d_k}$ by (Φ_3') and $h_1(\Phi_n^{d_k}) \subset \Phi_n^{b_k - \varepsilon}$. □

3 The Variational Setting

3.1 The spaces in the case of a bounded domain in $\mathbb{R}^n$. Let φ_n, $n \in \mathbb{N}$, be an orthonormal basis of $L^2(\Omega)$ made up of eigenfunctions of the eigenvalue problem

$$-\Delta u = \lambda u \quad \text{in } \Omega, \qquad u = 0 \quad \text{on } \partial\Omega.$$

Let λ_n be the corresponding eigenvalues. For all real numbers $s > 0$ we define, for $u = \sum_{j=1}^{\infty} a_j\varphi_j$, $v = \sum_{j=1}^{\infty} b_j\varphi_j$:

$$H^s(\Omega) := \left\{u \in L^2(\Omega) : \sum_{j=1}^{\infty} \lambda_j^s |a_j|^2 < \infty\right\}.$$

This is a Hilbert space with respect to the inner product $\langle u, v\rangle_s := \sum_{j=1}^{\infty} \lambda_j^s a_j b_j$. Clearly, the operator

$$A^s : H^s(\Omega) \longrightarrow L^2(\Omega), \quad u \longmapsto \sum_{j=1}^{\infty} \lambda_j^{s/2} a_j\varphi_j$$

is an isometric isomorphism. It is easy to see that

$$\int_\Omega A^s u\, \phi = \int_\Omega u\, A^s\varphi \qquad \text{for all } u, \phi \in H^s(\Omega)$$

which is used to prove the regularity of weak solutions. One has also the Sobolev imbeddings

$$H^s(\Omega) \subset L^p(\Omega)$$

continuously if $1 \leq p \leq \frac{2N}{N-2s}$ and compactly if $1 \leq p < \frac{2N}{N-2s}$.

3.2 The spaces in the case $\Omega = \mathbb{R}^n$. In this case, the space $H^s(\mathbb{R}^n)$ is defined by

$$H^s(\mathbb{R}^N) := \{u \in L^2(\mathbb{R}^N) : \textstyle\int_{\mathbb{R}^N} (1 + |\xi|^s)^2 \cdot |\widehat{u}(\xi)|^2 d\xi < \infty\}$$

where $\widehat{u}$ denotes the Fourier transform of u. This is a Hilbert space with respect to

$$\langle u, v\rangle := \frac{1}{(2\pi)^n} \int_{\mathbb{R}^n} (1 + |\xi|^s)^2 \cdot \widehat{u}(\xi) \cdot \widehat{v}(\xi)\, d\xi.$$

The operator

$$A^s : H^s(\mathbb{R}^n) \longrightarrow L^2(\mathbb{R}^n), \quad u \longmapsto ((1 + |\xi|^s)\widehat{u})^\vee,$$

(where $\omega^\vee$ denotes the inverse Fourier transform of ω) is usually written as $A^s = (1+|D|^s)$. It is readily seen that it is an isometric isomorphism. It is also easy to see that

$$\int_{\mathbb{R}^n} A^s u\, \phi = \int_{\mathbb{R}^n} u\, A^s\phi \qquad \text{for all } u, \phi \in H^s(\mathbb{R}^N).$$

We observe that $A^2 = u - \Delta u$ since

$$A^2 u = ((1 + |\xi|^2)\widehat{u})^\vee \quad \text{for } u \in H^2(\mathbb{R}^N).$$

This explains the form of the system (1.2) in the case of $\Omega = \mathbb{R}^n$.

If G is a subgroup of $O(N)$, then we set

$$L_G^2(\mathbb{R}^n) := \{u \in L^2(\mathbb{R}^n) : u(gx) = u(x) \text{ for } g \in G, \text{ and } x \in \mathbb{R}^n \text{ a.e.}\}$$

and $H_G^s := H^s \cap L_G^2$. Clearly we have that

$$A^s(H_G^s(\mathbb{R}^n)) = L_G^2(\mathbb{R}^n)$$

In the case of $\mathbb{R}^n$ there is a loss of the compact imbeddings. However, depending on the group G acting on $\mathbb{R}^n$, we can still recover them. We mention the following result due to P.-L. Lions [12]:

Proposition 3.1. *If*

$$G = O(N_1) \times ... \times O(N_k)$$

with $N_i \geq 2$ *and* $\sum_{i=1}^k N_i = N$, *then the imbedding*

$$H_G^s(\mathbb{R}^n) \subset L^\gamma(\mathbb{R}^n) \quad \text{for } 2 < \gamma < \frac{2N}{N-2s}$$

is compact.

The case when $G = O(N)$ was first proved by Strauss [15].

3.3 The "quadratic" forms and the functionals. In the sequel we write E^s to denote both $H^s(\Omega)$ in the case of a bounded domain Ω, and $H^s(\mathbb{R}^N)$. Let us consider the Cartesian product $E := E^s \times E^t$ with $s, t \geq 0$, which is also a Hilbert space endowed with the inner product

$$\langle z, \eta \rangle := \langle u, \phi \rangle_s + \langle v, \psi \rangle_t, \quad \text{for } z = (u, v),\ \eta = (\phi, \psi) \in E.$$

We consider the bilinear form

$$B : E \times E \longrightarrow \mathbb{R}, \ B[z, \eta] := \int (A^s u\, A^t \psi + A^s \phi\, A^t v),$$

where $\int$ denotes the integral in both cases, over Ω or over $\mathbb{R}^n$. Associated to B, we have the quadratic form

$$Q(z) := \frac{1}{2} B[z, z] = \int A^s u\, A^t v\,.$$

It is easy to see (cf. [7]) that the bounded self-adjoint operator $L : E \to E$ defined by $\langle Lz, \eta \rangle := B[z, \eta]$ has exactly two eigenvalues $+1$ and -1, and that the corresponding eigenspaces E^+ and E^- are given by

$$E^+ = \{(u, A^{-t} A^s u) : u \in E^s\} \quad \text{and} \quad E^- = \{(u, -A^{-t} A^s u) : u \in E^s\}$$

where we are using the notation $A^{-t} = (A^t)^{-1}$.

Now consider the Hamiltonian $H : \overline{\Omega} \times \mathbb{R} \times \mathbb{R} \to \mathbb{R}$ from Section 1. We consider the functional

$$\Phi(z) := Q(z) - \int H(x, u, v)\, dx \tag{3.1}$$

where $z = (u, v)$. By (H_2) or (H_2'), there exist $s, t > 0$, with $s + t = 2$ and

$$\frac{1}{p} > \frac{1}{2} - \frac{s}{n} \quad \text{and} \quad \frac{1}{q} > \frac{1}{2} - \frac{t}{n}. \tag{3.2}$$

This implies that we have continuous imbeddings $E^s \subset L^p$ and $E^t \subset L^q$. We fix s and t with this property so that Φ is well defined in E by (H_2) or (H_2'). Moreover, (H_2) or (H_2') imply that $\Phi \in C^1(E, \mathbb{R})$ with

$$\langle \Phi'(z), \eta \rangle = B[z, \eta] - \int \Big(H_u(x, u, v)\phi \, + \, H_v(x, u, v)\psi \Big)\, dx$$

for $z, \eta \in E$. From this one deduces that a critical point $z = (u, v)$ of Φ corresponds to a weak solution of (1.1) or (1.2). Namely

$$\int A^s u\, A^t \psi = \int H_v(x, u, v)\, \psi \qquad \text{for all } \psi \in E^t$$

and

$$\int A^s \phi\, A^t v = \int H_u(x, u, v)\, \phi \qquad \text{for all } \phi \in E^s.$$

As shown in [7] for the case of Ω bounded these solutions are strong in the sense that

$$u \in W^{2,p/(p-1)}(\Omega) \cap W_0^{1,p/(p-1)}(\Omega) \text{ and } v \in W^{2,q/(q-1)}(\Omega) \cap W_0^{1,q/(q-1)}(\Omega),$$

and they satisfy (1.1). In the case of $\mathbb{R}^n$ we conclude that $u \in W_{loc}^{2,p/(p-1)}(\mathbb{R}^N)$ and $v \in W_{loc}^{2,q/(q-1)}(\mathbb{R}^N)$, and that they satisfy (1.2).

In order to apply Theorem 2.1 in the next sections, let us introduce the following notations. Let $(e_j)_{j\in\mathbb{N}}$ be an orthonormal basis of E^s. Clearly the $f_j := A^{-t}A^s e_j$, $j \in \mathbb{N}$, constitute an orthonormal basis of E^t. We set

$$E_n^s := \operatorname{span}\{e_j : j = 1, ..., n\} \quad \text{and} \quad E_n^t := \operatorname{span}\{f_j : j = 1, ..., n\}.$$

The following result can be readily seen.

Lemma 3.1. *With the above notations, we have*

$$E^{\pm} = \overline{\textstyle\bigcup_{n=1}^{\infty} E_n^{\pm}}, \quad E = E^+ \oplus E^-, \quad E_n := E_n^+ \oplus E_n^- = E_n^s \times E_n^t$$

□

Next we check that the functional Φ defined in (3.1) satisfies the condition (Φ_2') in both cases, Ω bounded or $\mathbb{R}^N$.

Lemma 3.2. *Assume* (H_1)–(H_2) *or* (H_1')-(H_2'). *Then there exists a sequence of positive real numbers* r_k, $k \geq 1$, *such that*

$$b_k := \inf\{\Phi(z) : z \in E^+, z \perp E_{k-1}^+, \|z\| = r_k\} \to +\infty.$$

Proof: We use Remark 2.2 applied to the Hilbert space E^+ with $P(z) := Q(z)$, and

$$\Psi(z) := \int H(x,u,v)$$

where $z = (u,v)$. We know that $Q(z) = \frac{1}{2}\|z\|^2$ for $z \in E^+$, and from (H_2) we obtain

$$\begin{aligned}\left|\int H(x,u,v)\right| &\leq C\left(\int |u|^p + \int |v|^q + 1\right)\\ &\leq C'(\|z\|_X^r + 1)\end{aligned}$$

where $X = L^p(\Omega) \times L^q(\Omega)$ and $r = \max\{p,q\} > 2$. Also, assuming (H_2') we obtain:

$$\begin{aligned}\left|\int H(x,u,v)\right| &\leq C\left(\int |u|^p + \int |v|^q + \int |u|^{a+1} + \int |v|^{b+1}\right)\\ &\leq C'\|z\|_X^r\end{aligned}$$

for some $r > 2$. □

4 The case Ω bounded.

In this section we prove Theorem 1.1. With the notation from Section 3, we want to apply Theorem 2.1 to the functional

$$\Phi : E \to \mathbb{R}, \quad \Phi(z) = Q(z) - \int H(x,u,v)\,dx \tag{4.1}$$

where $z = (u,v) \in E = E^s \times E^t$. First we show that Φ satisfies a Palais-Smale condition.

Lemma 4.1. *Assume* (H_1), (H_2) *and* (H_3). *Then* Φ *as defined in* (4.1) *satisfies* $(PS)_c^{\mathcal{F}}$ *for every* $c \in \mathbb{R}$ *and* $\mathcal{F} = (E_n)_{n\in\mathbb{N}}$.

Proof: In view of Remark 2.1, it suffices to prove that a $(PS)_c^{\mathcal{F}}$ sequence is bounded in E. This follows as in Section 3 of [8], up to the point where we are to get some bounds for $\|u_n\|_{E^s}$ and $\|v_n\|_{E^t}$. At this point we then use the fact that $\int A^s u_n\, A^t \psi = 0$, for all $\psi \in (E_n^t)^\perp$. □

Next we check the other assumptions of Theorem 2.1. It has already been proved that Φ satisfies condition (Φ_2'). In particular, condition (Φ_2) is satisfied.

For each $\lambda > 0$, let us define the isomorphism $T_\lambda : E \to E$ by

$$T_\lambda(u, v) = (\lambda^{\beta-1}u, \lambda^{\alpha-1}v).$$

Clearly $T_\lambda E_n = E_n$ for all $\in \mathbb{N}$. Observe however that $E_n^+ \cap T_\lambda E_n^+ = \{0\}$ and $E_n^- \cap T_\lambda E_n^- = \{0\}$ if $\alpha \neq \beta$.

For each $k \in \mathbb{N}$ consider the finite dimensional subspace E_k^s. Since all norms are equivalent in finite dimensional spaces, we have positive constants σ_k, σ_k', τ_k and τ_k' such that

$$\|u\|_{L^2} \geq \sigma_k \|u\|_{E^s} \quad \text{and} \quad \|u\|_{E^s} \geq \sigma_k' \|u\|_{L^\beta} \quad \text{for all } u \in E_k^s$$

and

$$\|v\|_{L^2} \geq \tau_k \|v\|_{E^t} \quad \text{and} \quad \|v\|_{E^t} \geq \tau_k' \|v\|_{L^\alpha} \quad \text{for all } v \in E_k^t.$$

These constants are going to compose the constants δ_k announced in the introduction.

Lemma 4.2. *Assume that there are constants c_2 and c_3, such that*

$$H(x, u, v) \geq \frac{1}{2} c_2(|u|^\alpha + |v|^\beta) - c_3 \tag{4.2}$$

for all $(x, u, v) \in \overline{\Omega} \times \mathbb{R} \times \mathbb{R}$. Then, for each $\lambda > 0$ and each $k \in \mathbb{N}$, one has

$$\Phi(T_\lambda z) \leq \lambda^{\alpha+\beta}(1 - c_2\delta_k) + c_3|\Omega|$$

for all $z = z^+ + z^- \in E_k^+ \oplus E^-$, with $\lambda := \|z^+\|$, where

$$\delta_k := \min\left\{ \left(\frac{\sigma_k^2 \sigma_k'}{2}\right)^\alpha, \left(\frac{\tau_k^2 \tau_k'}{2}\right)^\beta \right\}$$

Proof: For $z = z^+ + z^- \in E_k^+ \oplus E^-$ we write $z^- = z_1^- + z_2^-$ where $z_1^- \in E_k^-$, $z_2^- \perp E_k^-$. We also write $\overline{z} := z^+ + z_1^-$ and $z = (u, v)$, $z^\pm = (u^\pm, v^\pm), \ldots$ Using (4.2) we have

$$\int_\Omega H(x, T_\lambda z) \geq \frac{1}{2} c_2 \big(\lambda^{(\beta-1)\alpha} \|u\|_{L^\alpha}^\alpha + \lambda^{(\alpha-1)\beta} \|v\|_{L^\beta}^\beta\big) - c_3|\Omega|. \tag{4.3}$$

Since α and β are conjugate exponents and $u_2^- \perp \overline{u}$ in L^2, we obtain

$$\begin{aligned} \|u\|_{L^\alpha} &\geq |\langle u, \overline{u}\rangle_{L^2}| \cdot \|\overline{u}\|_{L^\beta}^{-1} \\ &= \|\overline{u}\|_{L^2}^2 \|\overline{u}\|_{L^\beta}^{-1} \\ &\geq \sigma_k^2 \sigma_k' \|\overline{u}\|_{E^s} \\ &=: \widetilde{\sigma}_k \|\overline{u}\|_{E^s}. \end{aligned} \tag{4.4}$$

Similarly

$$\|v\|_{L^\beta} \geq \tau_k^2\, \tau_k' \|\overline{v}\|_{E^t} =: \widetilde{\tau}_k \|\overline{v}\|_{E^t}. \tag{4.5}$$

Next observe that $\overline{u} = u^+ + u_1^-$ and $\overline{v} = A^{-t} A^s (u^+ - u_1^-)$. So

$$\|\overline{u}\|_{E^s} = \|u^+ + u_1^-\|_{E^s} \quad \text{and} \quad \|\overline{v}\|_{E^t} = \|u^+ - u_1^-\|_{E^s}. \tag{4.6}$$

Using (4.3)-(4.6) we obtain the following estimate on the Hamiltonian:

$$\int_\Omega H(x, T_\lambda z)$$
$$\geq \frac{1}{2} c_2 \left(\lambda^{(\beta-1)\alpha} \widetilde{\sigma}_K^\alpha \|u^+ + u_1^-\|_{E^s}^\alpha + \lambda^{(\alpha-1)\beta} \widetilde{\tau}_K^\beta \|u^+ - u_1^-\|_{E^s}^\beta \right) - c_3|\Omega|$$

Since $\|u^+\|_{E^s} = \|z^+\|/2 = \lambda/2$, we have that either $\|u^+ + u_1^-\|_{E^s} \geq \lambda/2$ or $\|u^+ - u_1^-\|_{E^s} \geq \lambda/2$. In either case

$$\int_\Omega H(x, T_\lambda z) \geq \frac{1}{2} c_2\, \delta_k\, \lambda^{\alpha+\beta} - c_3|\Omega|. \tag{4.7}$$

On the other hand we have that

$$Q(T_\lambda z) = \lambda^{\alpha+\beta-2} Q(z) = \frac{1}{2} \lambda^{\alpha+\beta-2} (\|z^+\|^2 - \|z^-\|^2)$$

and so for $\|z^+\| = \lambda$ we obtain

$$Q(T_\lambda z) \leq \frac{1}{2} \lambda^{\alpha+\beta}. \tag{4.8}$$

Thus it follows from (4.7) and (4.8) that

$$\Phi(T_\lambda z) \leq \frac{1}{2}(1 - c_2 \delta_k) \lambda^{\alpha+\beta} + c_3|\Omega| \quad \text{for all } \lambda = \|z^+\| \leq \|z^-\|. \tag{4.9}$$

□

An immediate consequence of Lemma 4.2 is the next result which establishes (Φ_3).

Lemma 4.3. *Suppose that* (4.2) *holds and there is* $K \in \mathbb{N}$ *such that*

$$1 - c_2\, \delta_K < 0 \tag{4.10}$$

Then, fixing $r > 0$ *there is a* $\lambda_0 > 0$ *such that, for all* $\lambda \geq \lambda_0$ *one has, for* $z = z^+ + z^- \in E_K^+ \oplus E^-$*:*

$$\Phi(T_\lambda z) < 0 \quad \text{if } \|z^+\| = \lambda, \quad \text{and } \|T_\lambda z\| > r \quad \text{if } \max\{\|z^+\|, \|z^-\|\} = \lambda.$$

□

Proof of Theorem 1.1: As mentioned before we apply Theorem 2.1. First we observe that Lemma 4.1 gives condition (Φ_1). Lemma 3.2 implies that there exists a $k_0 \in \mathbb{N}$ such that $b_{k_0} > 0$, which then gives (Φ_2) and (Φ_2'). Now suppose that (1.9) holds for some $K \geq k_0$. Then we can apply Lemmas 4.2 and 4.3 and conclude that (Φ_3) holds. (Φ_4) is implied by the fact that Φ maps bounded sets of E into bounded sets of $\mathbb{R}$. Finally, condition (Φ_5) is a consequence of (H_4). □

5 The case $\Omega = \mathbb{R}^n$

With the notations introduced in Section 3 the weak solutions of (1.2) are the critical points of the functional

$$\Phi(z) = \int_{\mathbb{R}^N} A^s u\, A^t v \; - \; \int_{\mathbb{R}^N} H(x,u,v) \tag{5.1}$$

acting in $E = E^s \times E^t := H^s(\mathbb{R}^N) \times H^t(\mathbb{R}^N)$, where s and t satisfy (3.2). We shall consider the functional Φ restricted to certain subspaces of E where we have compact imbeddings due to symmetry properties. Let us start with the group $G = O(N)$ acting in $\mathbb{R}^n$, and let us look for critical points of Φ in the subspace X of E given by $X = H^s_{O(N)}(\mathbb{R}^n) \times H^t_{O(N)}(\mathbb{R}^n)$. All subspaces introduced in Section 3 are now restricted to spherically symmetric functions. Observe that

$$X := \mathrm{Fix}(G) = \{(u,v) \in E : gu = u,\ gv = v, \ \text{ for all } g \in O(N)\}$$

where gu means $(gu)(x) = u(gx)$, for all $x \in \mathbb{R}^n$. We see also that Φ is invariant with respect to G, i.e. $\Phi(gu, gv) = \Phi(u,v)$. Hence, it follows from the Palais Principle of Symmetric Criticality, see [13] or [16], that the critical points of Φ restricted to X are critical points of Φ considered in the whole space E.

In order to prove Theorem 2.1, we have to check that $\Phi\big|_X$ satisfies the assumptions of the Fountain Theorem.

For each $\lambda > 0$, let us define the isomorphism $T_\lambda : E \to E$, by

$$T_\lambda(u,v) = (\lambda^\mu u,\ \lambda^\nu v),$$

where $\mu = \frac{m-p}{p}$ and $\nu = \frac{m-q}{q}$, and $m > \max(p,q)$. Clearly $T_\lambda E_n = E_n$.

Lemma 5.1. *Assume conditions* (H_1')-(H_4'). *Then, there is a sequence* $\lambda_k > 0$, $k \in \mathbb{N}$, *such that* (Φ_3') *holds with* $T_k := T_{\lambda_k}$ *and* $R_k := \lambda_k$.

Proof: Let us use the notation introduced in the proof of Lemma 4.2. It follows from (1.16) that for any $\lambda > 0$ we have

$$\int_{\mathbb{R}^N} H(x, T_\lambda z) \geq c\left(\lambda^{\mu p} \int_{\mathbb{R}^N} |u|^p \, + \, \lambda^{\nu q} \int_{\mathbb{R}^N} |v|^q\right). \tag{5.2}$$

Using Hölder's inequality, we obtain

$$\|u\|_{L^p}\,\|\overline{u}\|_{L^{p'}} \geq |\langle u,\overline{u}\rangle_{L^2}| = \|\overline{u}\|_{L^2}^2 \quad \text{where } \frac{1}{p}+\frac{1}{p'}=1,$$

Next, from the finite dimensionality of E_k^s we have that there is a positive constant γ_k such that

$$\|u\|_{L^p} \geq \gamma_k\|\overline{u}\|_{E^s} \quad \text{for } u\in E_k^s.$$

Similarly there is $\widetilde{\gamma}>0$ with

$$\|v\|_{L^q} \geq \widetilde{\gamma}_k\|\overline{v}\|_{E^t} \quad \text{for } v\in E_k^t.$$

Thus it follows from (5.2) that

$$\int_{\mathbb{R}^N} H(x,T_\lambda z) \geq c\big(\lambda^{\mu p}\gamma_k^p\|\overline{u}\|_{E^s}^p + \lambda^{\nu q}\widetilde{\gamma}_k^q\|\overline{v}\|_{E^t}^q\big).$$

As in the proof of Lemma 4.2 we obtain

$$\int_{\mathbb{R}^N} H(x,T_\lambda z) \geq c\cdot\min\left\{\frac{1}{2^p}\lambda^{\mu p}\gamma_k^p\lambda^p,\ \frac{1}{2^q}\lambda^{\nu q}\widetilde{\gamma}_k^q\lambda^q\right\} \geq \sigma_k\lambda^m$$

provided $\|z^+\|=\lambda$.

On the other hand,

$$Q(T_\lambda z) = \lambda^{\mu+\nu}(\|z^+\|^2-\|z^-\|^2) \leq \lambda^{\mu+\nu+2}$$

for $\|z^+\|=\lambda$. Consequently we have

$$\Phi(T_\lambda z) \leq \lambda^{\mu+\nu+2} - \sigma_k\lambda^m.$$

Since $m>\mu+\nu+2$, it follows that there is a $\lambda_0(k)>0$ such that $\Phi(T_{\lambda_k}z)<0$ if $\lambda_k>\lambda_0(k)$. Also

$$\|T_\lambda z\| \geq \lambda^{\min\{\mu,\nu\}}\|z\|^2,$$

which implies that

$$\|T_{\lambda_k}z\| \geq \lambda_k^{\min\{\mu,\nu\}+2} \qquad \text{for } \max\{\|z^+\|,\|z^-\|\}=\lambda_k.$$

Therefore, we can choose λ_k such that

$$\Phi(T_{\lambda_k}z)<0 \quad \text{and} \quad \|T_{\lambda_k}z\|\geq r_k$$

for any given r_k. □

Proof of Theorem 1.2: First we observe that hypotheses (H_1') and (H_2') imply that Φ is C^1 in E. And using (H_3') we prove easily that a $(PS)_c^{\mathcal{F}}$ sequence is bounded in X. So it follows from Remark 2.1 and Proposition 3.1 that $\Phi|_X$ satisfies the $(PS)_c^{\mathcal{F}}$ condition. Hence (Φ_1) holds. Condition (Φ_2') has already been checked in Lemma 3.2. Condition (Φ_3') is proved in Lemma 5.1. Condition (Φ_4') is trivially verified, and finally (Φ_5) is a consequence of (H_5). So we apply the Fountain Theorem and conclude. □

We omit the **Proof of Theorem 1.3** since it parallels a similar result of Bartsch-Willem [3] for the scalar case. The result in our case follows from an application of the Fountain Theorem, using Proposition 3.1. □

References

[1] A. Bahri and H. Berestycki, Existence of forced oscillations for some nonlinear differential equations, Comm. Pure Appl. Math. 37 (1984), 403-442.

[2] T. Bartsch and M. Clapp, Critical point theory for indefinite functionals with symmetries, J. Fctl. Anal. 138 (1996), 107-136.

[3] T. Bartsch and M. Willem, Infinitely many nonradial solutions of a Euclidean scalar field equation, J. Fctl. Anal. 117 (1993), 447-460.

[4] G. Cerami, Un criterio di esistenza per i punti critici su varietà illimitate, Rec. Ist. Lomb. Sci. Lett. 112 (1978), 332-336.

[5] D.G. Costa and C.A. Magalhães, A unified approach to a class of strongly indefinite functionals, J. Diff. Eq. 122 (1996), 521-547.

[6] D.G. de Figueiredo, Semilinear elliptic systems, Proceedings of the Second School on Nonlinear Functional Analysis and Applications to Differential Equations, ICTP Trieste 1997, World Scientific Publishing Company (to appear).

[7] D.G. de Figueiredo and P. Felmer, On superquadratic elliptic systems, Trans. AMS 343 (1994), 99-116.

[8] D.G. de Figueiredo and C.A. Magalhães, On nonquadratic Hamiltonian elliptic systems, Advances in Diff. Eq. 1 (1996), 881-898.

[9] D.G. de Figueiredo and J. Yang, Decay, Symmetry and existence of solutions of semilinear elliptic systems, to appear in Nonlinear Analysis, TMA.

[10] P. Felmer, Periodic solutions of "superquadratic" Hamiltonian systems, J. Diff. Eq. 102 (1993), 188-207.

[11] S. Li and J.Q. Liu, Some existence theorems on multiple critical points and their applications, Kexue Tongbao 17 (1984) [in chinese].

[12] P.-L. Lions, Symétrie et compacité dans les espaces de Sobolev, J. Funct. Anal. 49 (1982) 315-334.

[13] R.S. Palais, The principle of symmetric criticality, Comm. Math. Phys. 69(1979), 19-30.

[14] B. Sirakov, On the existence of solutions of elliptic systems in $\mathbb{R}^N$, Preprint.

[15] W.A. Strauss, Existence of solitary waves in higher dimensions, Comm. Math. Phys. (1977), 149-162.

[16] M. Willem, Minimax Theorems, Progress in Nonlinear Differential Equations and their Applications, vol. 24, Birkhäuser 1996.

COMMUN. IN PARTIAL DIFFERENTIAL EQUATIONS, 26(11&12), 2305–2321 (2001)

A PRIORI BOUNDS FOR POSITIVE SOLUTIONS OF A NON-VARIATIONAL ELLIPTIC SYSTEM

Djairo Guedes de Figueiredo and Jianfu Yang

Dept. de Matemática IMECC, UNICAMP Cx. P. 6065
13083–970, Campinas, Brazil

ABSTRACT

A proof is given of an a priori bound for positive solutions of semilinear elliptic systems with nonlinearities depending on the gradients.

AMS Classification: 35B45; 35J55

1. INTRODUCTION

The question of the existence of solutions of semilinear elliptic systems of the form

$$-\Delta u_1 = f(x, u_1, u_2), \quad -\Delta u_2 = g(x, u_1, u_2) \text{ in } \Omega \subset \mathbb{R}^N, \tag{1.1}$$

subject to zero Dirichlet boundary conditions has been the object of intensive research recently. In this paper we discuss only the case when Ω is some bounded domain in $\mathbb{R}^N$, $N \geq 3$. Variational methods have been frequently used, since there is by now a remarkable collection of abstract results on the existence of critical points. A large body of results in the Theory of Critical

Originally published in: Comm. Partial Differential Equations 26 (2001), No. 11-12, 2305-2321, © 2001 Marcel Dekker, Inc.
Reprint by Springer-Verlag Berlin Heidelberg 2013 with kind permission,
DOI: 10.1007/978-3-319-02856-9_32

Points have been developed and applied to semilinear elliptic equations (th scalar case) and to Hamiltonian systems. In addition to the classical result on minimization of functionals, some new results on critical points fo functionals unbounded from below have been established since 1970. Th pioneering work of Ambrosetti-Rabinowitz [1] where the so-calle Mountain-Pass Theorem was proved, has been followed by many othe results like the Saddle Point Theorem and several Linking Theorems Most of these results are already collected in several monographs, fo instance [13], [15], [18], [21]. The development mentioned above refers to a Theory of Critical Points for functionals without symmetry. Functional with some symmetry have been studied since 1930, cf the results o Ljusternik-Schnirelmann. In this direction, there has been a great deal o further work, see the bibliographies in [2] and [18].

Of course, variational methods are of some use when the differentia problem (equation or system) is the Euler-Lagrange equation(s) of some functional. In the case of a scalar equation $-\Delta u = f(x, u)$, this is always the case, provided some growth condition on f with respect to u is imposed. Such a growth is required in order to have the corresponding functional well defined in some appropriate Sobolev space. For the case of a system like (1.1), it is not always the case that there is a natural functional associated to it. However, two cases have been studied: gradient systems and Hamiltonian systems, for which there are functionals whose corresponding Euler-Lagrange equations yield (1.1). We refer to the survey paper [7].

Apart from these two cases, there is apparently no possibility of using variational methods; instead, one has to use topological methods. Existence of positive solutions for systems like (1.1) with superlinear nonlinearities were already discussed in [6], even for systems with more unknowns. The main difficulty when using topological methods lies in the need of obtaining a priori bounds. In [6] the author used a method developed in [9] for the case of one equation. Another method, the application of Hardy-Sobolev inequalities as introduced in [3] was used in [5]. The blow-up method, as introduced in [11] for the scalar case, was employed in [19], [16]. The most general results obtained by this method are in [16]. See also the survey paper [7].

In this paper we treat a more general class of systems, namely

$$-\Delta u_1 = f(x, u_1, u_2, \nabla u_1, \nabla u_2), \quad -\Delta u_2 = g(x, u_1, u_2, \nabla u_1, \nabla u_2) \text{ in } \Omega, \tag{1.2}$$

subject to zero Dirichlet boundary conditions on $\partial\Omega$, where Ω is some smooth bounded domain in $\mathbb{R}^N$. The corresponding scalar case has been treated in [22]. If we separate the leading part in (1.2), it becomes a system of the form

$$\begin{cases} -\Delta u_1 = a(x)u_1^{\alpha_{11}} + b(x)u_2^{\alpha_{12}} + h_1(x, u_1, u_2, \nabla u_1, \nabla u_2) \\ -\Delta u_2 = c(x)u_1^{\alpha_{21}} + d(x)u_2^{\alpha_{22}} + h_2(x, u_1, u_2, \nabla u_1, \nabla u_2) \end{cases} \tag{1.3}$$

System (1.3) is not variational, and consequently we will use topological methods to prove the existence of positive solutions. As we said before, the main point in such a method is obtaining a priori bounds for the eventual solutions. Here in contrast with system (1.1), there are difficulties due to the presence of gradients. In [5] the authors have been able to treat problems of this type, but rather strong restrictions were necessary. Namely in [5] a bounded dependence on the gradients was assumed. Moreover the superlinearity (i.e., the polynomial dependence of f and g with respect to u and v) is also restricted to a growth much lower than what it is to be expected from a subcritical problem. However such restrictions came naturally from the method used in [5], namely Hardy-Sobolev inequalities.

Here the a priori bounds for the solutions of (1.3) are obtained via the so-called blow-up method. We obtain results much more general than the ones in [5] as far as handling better superlinearities and unbounded dependence on the gradients. A serious difficulty comes when one proceeds to estimate the gradients of sequences of solutions that appear in the blow-up method. To handle it, we have to use some norm with weights depending on the distance to the boundary of the domains involved.

Assume the following conditions:

(A1) The coefficients $a, b, c, d : \overline{\Omega} \to [0, \infty)$ are continuous functions.
(A2) The exponents $\alpha_{ij} \geq 0 \quad i, j = 1, 2$.
(A3) There exist positive constants c_1 and c_2 such that

$$|h_1(x, s, t, \xi, \eta)| \leq c_1(1 + |s|^{\beta_{11}} + |t|^{\beta_{12}} + |\xi|^{\gamma_{11}} + |\eta|^{\gamma_{12}})$$
$$|h_2(x, s, t, \xi, \eta)| \leq c_2(1 + |s|^{\beta_{21}} + |t|^{\beta_{22}} + |\xi|^{\gamma_{21}} + |\eta|^{\gamma_{22}})$$

where

$$0 \leq \beta_{ij} < \alpha_{ij} \quad i, j = 1, 2.$$

We say that system (1.3) is *weakly coupled* if $\alpha_{11} > 1, \alpha_{22} > 1$ and

$$\alpha_{12} < \frac{\alpha_{22} - 1}{\alpha_{11} - 1}\alpha_{11}, \qquad \alpha_{21} < \frac{\alpha_{11} - 1}{\alpha_{22} - 1}\alpha_{22}$$

and (1.3) is *strongly coupled* if $\alpha_{12}\alpha_{21} > 0$ and

$$\alpha_{11} < \frac{\alpha_{21}+1}{\alpha_{12}+1}\alpha_{12}, \qquad \alpha_{22} < \frac{\alpha_{12}+1}{\alpha_{21}+1}\alpha_{21}$$

In order to take care of the gradients, we assume further that

(A4) If (1.3) is weakly coupled, $\gamma_{ij}, i,j = 1,2$ satisfy

$$\gamma_{11} < \frac{2\alpha_{11}}{\alpha_{11}+1}, \qquad \gamma_{22} < \frac{2\alpha_{22}}{\alpha_{22}+1}$$

$$\gamma_{12} < \frac{2\alpha_{11}(\alpha_{22}-1)}{(\alpha_{11}-1)(\alpha_{22}+1)}, \qquad \gamma_{21} < \frac{2\alpha_{22}(\alpha_{11}-1)}{(\alpha_{22}-1)(\alpha_{11}+1)};$$

(A5) If (1.3) is strongly coupled, $\gamma_{ij}, i,j = 1,2$ satisfy

$$\gamma_{11} < \frac{2\alpha_{12}(\alpha_{21}+1)}{2\alpha_{12}+\alpha_{12}\alpha_{21}+1}, \qquad \gamma_{22} < \frac{2\alpha_{21}(\alpha_{12}+1)}{2\alpha_{21}+\alpha_{12}\alpha_{21}+1}$$

$$\gamma_{12} < \frac{2\alpha_{12}(\alpha_{21}+1)}{2\alpha_{21}+\alpha_{12}\alpha_{21}+1}, \qquad \gamma_{21} < \frac{2\alpha_{21}(\alpha_{12}+1)}{2\alpha_{12}+\alpha_{12}\alpha_{21}+1}.$$

We will prove:

Theorem 1.1. (*The weakly coupled case*). *Let* (1.3) *be a weakly coupled system satisfying conditions (A1)–(A4) and* $a(x), d(x) \geq c_0 > 0$ *for* $x \in \bar{\Omega}$. *Assume also that*

$$0 < \alpha_{11}, \alpha_{22} < \frac{N+2}{N-2}. \tag{1.4}$$

Then there is a constant $C > 0$ *such that*

$$\|u_1\|_{L^\infty}, \|u_2\|_{L^\infty} \leq C$$

for all positive solutions $u_1, u_2 \in C^2(\Omega) \cap C^0(\bar{\Omega})$ *of system* (1.3).

The terminology used here, namely weakly coupled, comes from the fact that after the use of the blow-up method, system (1.3) "decouples at infinity", i.e. we obtain a system in $\mathbb{R}^N$ of the form

$$-\Delta v_1 = v_1^{\alpha_{11}}, \quad -\Delta v_2 = v_2^{\alpha_{22}}, \; v_1, v_2 \geq 0, \; v_1, v_2 \in C^2(\mathbb{R}^N) \tag{1.5}$$

or in $\mathbb{R}^N_+$

$$-\Delta v_1 = v_1^{\alpha_{11}}, \quad -\Delta v_2 = v_2^{\alpha_{22}}, \; v_1, , v_2 \geq 0, \; v_1, v_2 \in C^2(\mathbb{R}^N_+) \tag{1.6}$$

$$v_1(x', 0) = 0, \; v_2(x', 0) = 0, \tag{1.7}$$

where $x' = (x_1, \ldots, x_{N-1})$. So, hypothesis (1.4) in the above theorem implies that the only solution of problem (1.5), or problem (1.6), is $v_1 = v_2 = 0$. (See [12].)

In order to state the second result, we need two further conditions.

(A6) The problem below

$$- \Delta v_1 = v_2^{\alpha_{12}}, \quad - \Delta v_2 = v_1^{\alpha_{21}} \quad v_1, \quad v_2 \geq 0 \quad \text{in } \mathbb{R}^N, v_1, v_2 \in C^2(\mathbb{R}^N) \tag{1.8}$$

has only the trivial solution $v_1 = v_2 = 0$.

(A7) The problem below

$$- \Delta v_1 = v_2^{\alpha_{12}}, \quad - \Delta v_2 = v_1^{\alpha_{21}}, \quad v_1, \quad v_2 \geq 0, \quad \text{in } \mathbb{R}_+^N, \quad v_1, v_2 \in C^2(\mathbb{R}_+^N) \tag{1.9}$$

$$v_1(x', 0) = 0, \ v_2(x', 0) = 0 \tag{1.10}$$

has only the trivial solution $v_1 = v_2 = 0$

It is conjectured that (A6) and (A7) hold if

$$\frac{1}{\alpha_{12} + 1} + \frac{1}{\alpha_{21} + 1} > 1 - \frac{2}{N}. \tag{1.11}$$

Some partial results on this conjecture have been proved in [8], [19], [20]. Precisely, it was proved in [8] that the conjecture is true for $1 < \alpha_{12}, \alpha_{21} < (N+2)/(N-2)$ and for α_{12}, α_{21} in the full range by Serrin-Zou [20] in the class of radial functions, whereas in [19] it is required that $(1/(\alpha_{12}+1)) + (1/(\alpha_{21}+1)) \geq (N-2)/(N-1)$. Several other partial results have been established. A more complete discussion of this conjecture can be seen in [7].

Theorem 1.2. (*The strongly coupled case*). *Let* (1.3) *be a strongly coupled system satisfying conditions* (A1)–(A3) *and* (A5)–(A7), $b(x), c(x) \geq c_0 > 0$ *for* $x \in \bar{\Omega}$. *Then there is a constant* $C > 0$ *such that*

$$\|u_1\|_{L^\infty}, \|u_2\|_{L^\infty} \leq C$$

for all positive solutions $u_1, u_2 \in C^2(\Omega) \cap C^0(\bar{\Omega})$ *of system* (1.3).

2. SOME USEFUL INEQUALITIES

In this section we make use of general interpolation inequalities, known as the Gagliardo-Nirenberg inequalities. See [17] or [10]. For easy reference, we state here the inequality which gives as particular case the one used in our work.

"Let Ω be a bounded domain of class C^m in $\mathbb{R}^N$ and let $u \in W^{m,r}(\Omega) \cap L^q(\Omega)$, with $1 \le r, q \le \infty$. For any integer $j \in [0, m)$ and for any number $\alpha \in [j/m, 1]$, let

$$\frac{1}{p} = \frac{j}{N} + \alpha\left(\frac{1}{r} - \frac{m}{N}\right) + (1 - \alpha)\frac{1}{q}.$$

If $m - j - (N/r)$ is not a nonnegative integer, then

$$\| D^j u \|_{L^p} \le C \| u \|_{W^{m,r}}^{\alpha} \| u \|_{L^q}^{1-\alpha},$$

where C is a constant depending only on $\Omega, r, q, m, j, \alpha$. In case $m - j - (N/r)$ is a nonnegative integer then the above inequality holds with $\alpha = (j/m)$."

It follows from the above inequality that

$$\|\nabla u\|_{L^p(\Omega)} \le C\|u\|_{W^{2,r}(\Omega)}^{1/2} \|u\|_{L^q(\Omega)}^{1/2} \tag{2.1}$$

where $1/p = 1/2r + 1/2q$ and $N < r, 1 \le r, q \le \infty$ and Ω is a bounded smooth domain in $\mathbb{R}^N$. In particular,

$$\|\nabla u\|_{L^{2r}(\Omega)} \le C\|u\|_{W^{2,r}(\Omega)}^{1/2} \|u\|_{L^\infty(\Omega)}^{1/2} \tag{2.2}$$

Some weighted norms will play an essential role in the proof of a priori bounds for system (1.3) via the blow-up method. This has already been observed in [22] for the scalar case. Let us introduce them next. Denote by

$$\|u\|_{k,q,\Omega}^{(l)} = \sup_{0<\delta<1} \delta^{k+l} \|u\|_{W^{k,q}(\Omega_\delta)}$$

where

$$\Omega_\delta = \{x \in \Omega, \operatorname{dist}(x, \partial\Omega) > \delta\}, \ k, l \ge 0, \ q \in (1, \infty).$$

Lemma 2.1. *Suppose that* $u \in W^{2,q}(\Omega) \cap L^\infty(\Omega)$ *satisfies* $\Delta u = f(x)$ *in* Ω *with zero Dirichlet boundary condition. Then there is a constant* $C = C(q, \Omega)$ *such that*

$$\|u\|_{2,q,\Omega}^{(0)} \le C(\|u\|_{L^q(\Omega)} + \|f\|_{0,q,\Omega}^{(2)}). \tag{2.3}$$

Proof. It follows from the definition of $\|u\|_{2,q,\Omega}^{(0)}$ that there exists $0 < \delta < \frac{1}{2}$ such that

$$\|u\|_{2,q,\Omega}^{(0)} \le 8\delta^2 \|u\|_{2,q,\Omega_{2\delta}}. \tag{2.4}$$

Let $\varphi \in C_0^\infty(\Omega)$ be a cut-off function with $\varphi \equiv 1$ in $\Omega_{2\delta}$, $\varphi \equiv 0$ outside Ω_δ, and $|\nabla\varphi| \le C\delta^{-1}$. Then $w = \varphi u$ satisfies

$$\Delta w = \varphi f + u\Delta\varphi + 2\nabla u.\nabla\varphi. \tag{2.5}$$

By the elliptic L^p- estimates we obtain

$$\|w\|_{W^{2,q}(\Omega_\delta)} \le C\Big(\|w\|_{L^q(\Omega)} + \|\varphi f + u\Delta\varphi + 2\nabla u.\nabla\varphi\|_{L^q(\Omega)}\Big),$$

which implies

$$\|w\|_{W^{2,q}(\Omega_\delta)} \le C\Big(\|w\|_{L^q(\Omega)} + \|f\|_{L^q(\Omega_\delta)} + 2\|\nabla u\|_{L^q(\Omega_\delta)}\|\nabla\varphi\|_{L^\infty}\Big) \tag{2.6}$$

Next we estimate the right hand side of (2.4):

$$8\delta^2\|u\|_{W^{2,q}(\Omega_{2\delta})} = 8\delta^2\|w\|_{W^{2,q}(\Omega_{2\delta})} \le 8\delta^2\|w\|_{W^{2,q}(\Omega_\delta)}, \tag{2.7}$$

and using (2.6) we obtain

$$8\delta^2\|u\|_{W^{2,q}(\Omega_{2\delta})} \le C(\|u\|_{L^q(\Omega)} + \|f\|^{(2)}_{0,q,\Omega} + \|u\|^{(0)}_{1,q,\Omega}). \tag{2.8}$$

The interpolation inequality (2.1) yields

$$\delta\|\nabla u\|_{L^q(\Omega_\delta)} \le C\delta\|u\|^{1/2}_{W^{2,q}(\Omega_\delta)}\|u\|^{1/2}_{L^q(\Omega_\delta)}.$$

Then

$$\|u\|^{(0)}_{1,q,\Omega} \le C(\|u\|^{(0)}_{2,q,\Omega})^{1/2}\|u\|^{1/2}_{L^q(\Omega)}.$$

Consequently,

$$\|u\|^{(0)}_{2,q,\Omega} \le C\Big(\|u\|_{L^q(\Omega)} + \|f\|^{(2)}_{0,q,\Omega} + (\|u\|^{(0)}_{2,q,\Omega})^{1/2}\|u\|^{1/2}_{L^q(\Omega)}\Big).$$

The proof is completed by using Young's inequality.

3. A PRIORI ESTIMATES

The Proof of Theorem 1.1. We use blow up arguments. Suppose, by contradiction, that there exists a sequence of solutions $\{(u_{1,n}, u_{2,n})\}$ of (1.3) such that

$$\max\{\|u_{1,n}\|_{L^\infty(\Omega)}, \|u_{2,n}\|_{L^\infty(\Omega)}\} \to \infty$$

as $n \to \infty$. We may assume without loss of generality that $\|u_{1,n}\|_{L^\infty} \to +\infty$ and that

$$\|u_{1,n}\|_{L^\infty}^{1/\beta_1} \geq \|u_{2,n}\|_{L^\infty}^{1/\beta_2},$$

for some constants $\beta_1, \beta_2 > 0$ which are to be determined later.

Let $x_n \in \Omega$ be the point where $u_{1,n}$ assumes its maximum. Let $\lambda_n > 0$ be such that $\lambda_n^{\beta_1}\|u_{1,n}\|_{L^\infty} = 1$. Clearly, $\lambda_n \to 0$ as $n \to \infty$. Let $\Omega_n = \{y \in \mathbb{R}^N : \lambda_n y + x_n \in \Omega\}$ and define $v_{i,n}\colon \Omega_n \to \mathbb{R}$ by

$$v_{i,n}(y) := \lambda_n^{\beta_i} u_{i,n}(\lambda_n y + x_n), \quad i = 1, 2.$$

We observe that $v_{1,n}(0) = 1$ and $0 \leq v_{i,n}(y) \leq 1$ for all $y \in \Omega_n$ and $i = 1, 2$.

Due to the compactness of $\overline{\Omega}$ we may assume that $x_n \to x_0 \in \overline{\Omega}$. So there are two cases to consider, $x_0 \in \Omega$ and $x_0 \in \partial\Omega$.

(i) First case, $x_0 \in \Omega$. Then given $R > 0$ there is an $n \in \mathbb{N}$ such that $\overline{B}_R(0) \subset \Omega_n$. Since

$$\Delta_y v_{i,n} = \lambda_n^{\beta_i}\Delta_y u_{i,n}(\lambda_n y + x_n) = \lambda_n^{\beta_i+2}\Delta_x u_{i,n}(\lambda_n y + x_n),$$

$$\nabla_y v_{i,n} = \lambda_n^{\beta_i+1}\nabla_x u_{i,n}(\lambda_n y + x_n), \; i = 1, 2,$$

we see that

$$\begin{cases} -\Delta v_{1,n} = \lambda_n^{\beta_1+2-\beta_1\alpha_{11}} a(z) v_{1,n}^{\alpha_{11}}(y) + \lambda_n^{\beta_1+2-\beta_2\alpha_{12}} b(z) v_{2,n}^{\alpha_{12}} \\ +\lambda_n^{\beta_1+2} h_1(z, \lambda_n^{-\beta_1} v_{1,n}, \lambda_n^{-\beta_2} v_{2,n}, \lambda_n^{-(\beta_1+1)}\nabla v_{1,n}, \lambda_n^{-(\beta_2+1)}\nabla v_{2,n}) \\ -\Delta v_{2,n} = \lambda_n^{\beta_2+2-\beta_1\alpha_{21}} c(z) v_{1,n}^{\alpha_{21}}(y) + \lambda_n^{\beta_2+2-\beta_1\alpha_{22}} d(z) v_{2,n}^{\alpha_{22}} \\ +\lambda_n^{\beta_2+2} h_2(z, \lambda_n^{-\beta_1} v_{1,n}, \lambda_n^{-\beta_2} v_{2,n}, \lambda_n^{-(\beta_1+1)}\nabla v_{1,n}, \lambda_n^{-(\beta_2+1)}\nabla v_{2,n}) \end{cases} \tag{3.1}$$

where $z = \lambda_n y + x_n$. By assumption (A3) we have

$$\begin{aligned} &|\lambda_n^{\beta_1+2} h_1(z, \lambda_n^{-\beta_1} v_{1,n}, \lambda_n^{-\beta_2} v_{2,n}, \lambda_n^{-(\beta_1+1)}\nabla v_{1,n}, \lambda_n^{-(\beta_2+1)}\nabla v_{2,n})| \\ &\quad \leq C(\lambda_n^{\beta_1+2} + \lambda_n^{\beta_1+2-\beta_1\beta_{11}}|v_{1,n}|^{\beta_{11}} + \lambda_n^{\beta_1+2-\beta_2\beta_{12}}|v_{2,n}|^{\beta_{12}} \\ &\quad +\lambda_n^{\beta_1+2-(\beta_1+1)\gamma_{11}}|\nabla v_{1,n}|^{\gamma_{11}} + \lambda_n^{\beta_1+2-(\beta_2+1)\gamma_{12}}|\nabla v_{2,n}|^{\gamma_{12}}), \end{aligned} \tag{3.2}$$

and a similar estimate for $\lambda_n^{\beta_2+2}h_2$.

Choosing $\beta_1 = 2/(\alpha_{11} - 1)$, $\beta_2 = 2/(\alpha_{22} - 1)$, since (1.3) is a weakly coupled system, we obtain

$$\begin{aligned} &\beta_1 + 2 - \beta_1\alpha_{11} = 0, \quad \beta_1 + 2 - \beta_2\alpha_{12} > 0 \\ &\beta_2 + 2 - \beta_1\alpha_{21} > 0, \quad \beta_2 + 2 - \beta_2\alpha_{22} = 0. \end{aligned} \tag{3.3}$$

Since it is assumed that $\beta_{ij} < \alpha_{ij}$, we obtain the inequalities

$$\beta_1 + 2 - \beta_1\beta_{11} > 0, \quad \beta_1 + 2 - \beta_2\beta_{12} > 0 \tag{3.4}$$

$$\beta_2 + 2 - \beta_1\beta_{21} > 0, \quad \beta_2 + 2 - \beta_2\beta_{22} > 0. \tag{3.5}$$

Using assumption (A4) we also obtain

$$\beta_1 + 2 - (\beta_1 + 1)\gamma_{11} > 0, \quad \beta_1 + 2 - (\beta_2 + 1)\gamma_{12} > 0 \tag{3.6}$$

$$\beta_2 + 2 - (\beta_1 + 1)\gamma_{21} > 0, \quad \beta_2 + 2 - (\beta_2 + 1)\gamma_{22} > 0. \tag{3.7}$$

So by Lemma 2.1 we obtain from (3.1) using (3.2):

$$\begin{aligned} \|v_{1,n}\|^{(0)}_{2,q,B_R} &\le C\Big[\|v_{1,n}\|_{L^q(B_R)} + \|1 + |\nabla v_{1,n}|^{\gamma_{11}} + |\nabla v_{2,n}|^{\gamma_{12}}\|^{(2)}_{0,q,B_R}\Big] \\ &\le C\Big[1 + (\|\nabla v_{1,n}\|^{(2/\gamma_{11})}_{0,\gamma_{11}q,B_R})^{\gamma_{11}} + (\|\nabla v_{2,n}\|^{(2/\gamma_{12})}_{0,\gamma_{12}q,B_R})^{\gamma_{12}}\Big] \end{aligned} \tag{3.8}$$

An application of the interpolation inequality (2.2) gives

$$\begin{aligned} \delta^{2/\gamma_{11}}\|\nabla v_{1,n}\|_{L^{\gamma_{11}q}(B^\delta_R(0))} &\le \delta^{2/\gamma_{11}} C\|v_{1,n}\|^{1/2}_{W^{2,\gamma_{11}q/2}(B^\delta_R(0))}\|v_{1,n}\|^{1/2}_{L^\infty(B_R(0))} \\ &\le C[\delta^2\|v_{1,n}\|_{W^{2,q}(B^\delta_R(0))}]^{1/2}\|v_{1,n}\|^{1/2}_{L^\infty(B_R(0))} \end{aligned} \tag{3.9}$$

since $\gamma_{11} < 2$. Here $B^\delta_R(0) = \{x \in B_R(0)\colon \operatorname{dist}(x, \partial B_R) > \delta\}$. Therefore

$$\|\nabla v_{1,n}\|^{(2/\gamma_{11})}_{0,q,B_R(0)} \le C(\|v_{1,n}\|^{(0)}_{2,q,B_R(0)})^{1/2}\|v_{1,n}\|^{1/2}_{L^\infty(B_R(0))}$$

and then

$$(\|\nabla v_{1,n}\|^{(2/\gamma_{11})}_{0,q,B_R(0)})^{\gamma_{11}} \le C\epsilon(\|v_{1,n}\|^{(0)}_{2,q,B_R(0)})^{s\gamma_{11}/2} + C(\epsilon)\|v_{1,n}\|^{s'/2}_{L^\infty(B_R(0))},$$

where $1/s + 1/s' = 1$. Taking $(1/2)s\gamma_{11} = 1$, $s' = 2/(2 - \gamma_{11})$, we get

$$(\|\nabla v_{1,n}\|^{(2/\gamma_{11})}_{0,q,B_R(0)})^{\gamma_{11}} \le C\epsilon\|v_{1,n}\|^{(0)}_{2,q,B_R(0)} + C(\epsilon).$$

In the same way we infer that

$$(\|\nabla v_{2,n}\|^{(2/\gamma_{11})}_{\gamma_{12}q,B_R(0)})^{\gamma_{12}} \le C\epsilon\|v_{2,n}\|^{(0)}_{2,q,B_R(0)} + C(\epsilon).$$

Combining the above two inequalities with (3.8), one obtains

$$\|v_{1,n}\|^{(0)}_{2,q,B_R(0)} + \|v_{2,n}\|^{(0)}_{2,q,B_R(0)} \le C.$$

Since we can take $q > k$, we conclude that the sequences $\{v_{i,n}\}, i = 1, 2$, are uniformly bounded in $C^{1,\alpha}(\overline{B}_R)$. Passing to a subsequence we may assume that

$$v_{i,n} \to v_i, \quad \text{in} \quad C^{1,\alpha}(\overline{B}_R(0)), i = 1, 2 \tag{3.10}$$

for $n \to \infty$ and $0 < \alpha < 1$; we also have that

$$v_{i,n} \to v_i, \quad \text{weakly in} \quad W^{2,q}(B_R(0)), i = 1, 2 \tag{3.11}$$

Let $\eta \in C_0^\infty(B_R(0))$. Multiplying (3.1) by η and passing to the limit as $n \to \infty$ we conclude that

$$-\int_{B_R} \eta \Delta v_1 \, dx = \int_{B_R} \eta a(x_0) v_1^{\alpha_{11}} \, dx, \; -\int_{B_R} \eta \Delta v_2 \, dx = \int_{B_R} \eta d(x_0) v_2^{\alpha_{22}} \, dx. \tag{3.12}$$

This implies that (v_1, v_2) is a solution of

$$-\Delta v_1 = a(x_0) v_1^{\alpha_{11}}, \quad -\Delta v_2 = d(x_0) v_2^{\alpha_{22}} \quad \text{in } B_R(0). \tag{3.13}$$

Choosing a sequence $\{R_j\}$ with $R_j \to +\infty$ as $j \to \infty$, the above process applied with $R = R_j$ in $B_{R_j}(0)$ produces a subsequence of $\{(v_{1,n}, v_{2,n})\}$ whose limits satisfy (3.13) in $B_{R_j}(0)$, $j = 1, 2, \ldots$. Using a diagonal processd we obtain a subsequence $\{(v_{1,n_j}, v_{2,n_j})\}$ satisfying

$$v_{i,n_j} \to v_I, \quad \text{in} \quad C^{1,\alpha}(\overline{B}_{R_k}(0)), \; i = 1, 2 \tag{3.14}$$

and

$$v_{i,n_j} \to v_i, \quad \text{weakly in } W^{2,q}(B_{R_k}(0)), i = 1, 2 \tag{3.15}$$

for any R_k fixed. Hence (v_1, v_2) is a solution of

$$-\Delta v_1 = a(x_0) v_1^{\alpha_{11}}, \quad -\Delta v_2 = d(x_0) v_2^{\alpha_{22}} \quad \text{in } \mathbb{R}^N. \tag{3.16}$$

and $v_1(0) = 1$. By the assumption that $a(x), d(x) \geq c_0 > 0$, it follows by scaling that there are functions w_1 and w_2 of class C^2 defined in the whole of $\mathbb{R}^N$ with $w_1 \not\equiv 0$, $w_1 \geq 0$, $w_2 \geq 0$ satisfying

$$-\Delta w_1 = w_1^{\alpha_{11}}, \quad -\Delta w_2 = w_2^{\alpha_{22}} \quad \text{in } \mathbb{R}^N, \tag{3.17}$$

which is not possible.

MARCEL DEKKER, INC
270 Madison Avenue, New York, New York 10016

(ii) Second case, $x_0 \in \partial\Omega$.

Through translation and rotation of the coordinate system we may assume that x_0 is the origin and the inner normal to $\partial\Omega$ at x_0 is pointing in the direction of the positive x_N–axis. Then there is a smooth function $\psi(x'), x' = (x_1, \ldots, x_{N-1})$, defined for $|x'| < \delta_0$ such that (a) $\psi(0) = 0$ and $\nabla\psi(0) = 0$; (b) $\partial\Omega \cap \mathcal{N} = \{(x', x_N): x_N = \psi(x')\}$ and $\Omega \cap \mathcal{N} = \{(x', x_N): x_N > \psi(x')\}$, where $\mathcal{N}$ is a neighborhood of x_0. For $y \in \mathbb{R}^N$ with $|y|$ sufficiently small, we define a mapping $x = \Phi(y)$ with $\Phi(y) = (\Phi_1(y), \ldots, \Phi_N(y))$ by

$$\Phi_j(y) = \begin{cases} y_j - y_N(\partial\psi/\partial x_j)(y') & \text{for } j = 1, 2, \ldots, N-1, \\ y_N + \psi(y') & \text{for } j = N. \end{cases} \tag{3.18}$$

Since $\nabla\psi(0) = 0$, the differential map $D\Phi$ of Φ satisfies $D\Phi(0) = I$, the identity map. Thus Φ can be inverted $y = \Phi^{-1}(x)$ in $|x| < \delta'$. We write $\Psi(x) = (\Psi_1(x), \ldots, \Psi_N(x)) := \Phi^{-1}(x)$

Let $x_n \in \Omega$ be the maximum point of $u_{1,n}$, and again we assume $\|u_{1,n}\|_{L^\infty}^{1/\beta_1} \geq \|u_{2,n}\|_{L^\infty}^{1/\beta_2}$ and $\lambda^{\beta_1}\|u_{1,n}\|_{L^\infty} = 1$. Denote $y_n = \Psi(x_n)$. We assume $y_n \in B_\delta^+$, and write $y_n = (y_n', \alpha_n)$, $\alpha_n \geq 0$. Let

$$v_{1,n}(y) = u_{1,n}(x) = u_{1,n}(\Phi(y)), \quad v_{2,n}(y) = u_{2,n}(x) = u_{2,n}(\Phi(y)),$$

Denote

$$a_{ij}(y) = \sum_{k=1}^{N} \frac{\partial \Psi_i}{\partial x_k}(\Phi(y)) \frac{\partial \Psi_j}{\partial x_k}(\Phi(y)), \quad 1 \leq i, j \leq N,$$

$$b_j(y) = (\Delta\Psi_j)(\Phi(y)), \quad j = 1, \ldots, N.$$

Then the new variables $(v_{1,n}, v_{2,n})$ satisfy a system of semilinear elliptic equations in $\{y_N > 0\} \cap B_{2\delta}$ with the Laplacian replaced by a more general elliptic operator. The equation for $v_{1,n}$ is as follows, and a similar one holds for $v_{2,n}$.

$$-\left(\sum_{i,j=1}^{N} a_{ij}(y) \frac{\partial^2 v_{1,n}}{\partial y_i \partial y_j} + \sum_{j=1}^{N} b_j(y) \frac{\partial v_{1,n}}{\partial y_j}\right) = a(\Phi(y)) v_{1,n}^{\alpha_{11}}(y) + b(\Phi(y)) v_{2,n}^{\alpha_{12}} + h_1(\Phi(y), v_{1,n}, v_{2,n}, \nabla_x v_{1,n}(\Psi(x)), \nabla_x v_{2,n}(\Psi(x))). \tag{3.19}$$

Also the boundary conditions below are satisfied.

$$v_{1,n} = 0, \quad v_{2,n} = 0 \quad \text{on } \{y_N = 0\} \cap B_{2\delta}. \tag{3.20}$$

Marcel Dekker, Inc
270 Madison Avenue, New York, New York 10016

Now the proof splits into two cases according to the behavior of α_n/λ_n as $n \to \infty$.

Case 1. $\{\alpha_n/\lambda_n\}$ remains bounded.
We may assume $\alpha_n/\lambda_n \to \alpha \geq 0$ as $n \to \infty$. Define

$$w_{1,n}(z) = \lambda_n^{\beta_1} v_{1,n}(\lambda_n z' + y_n', \lambda_n z_N), \quad w_{2,n}(z) = \lambda_n^{\beta_2} v_{2,n}(\lambda_n z' + y_n', \lambda_n z_N).$$

$(w_{1,n}, w_{2,n})$ is well-defined in B^+_{δ/λ_n} and $0 < w_{i,n}(z) \leq 1$ for all n and $i = 1, 2$. From (3.19), (3.20) we see that $w_{1,n}$ satisfies the equation below, and a similar equation holds for $w_{2,n}$.

$$\begin{aligned}
&-\left(\sum_{i,j=1}^{N} a_{ij}^n(z)\frac{\partial^2 w_{1,n}}{\partial z_i \partial z_j} + \lambda_n \sum_{j=1}^{N} b_j^n(z)\frac{\partial w_{1,n}}{\partial z_j}\right) \\
&= \lambda_n^{\beta_1+2-\alpha_{11}\beta_1} a^n(z) w_{1,n}^{\alpha_{11}} + \lambda_n^{\beta_1+2-\alpha_{12}\beta_2} b^n(z) w_{2,n}^{\alpha_{12}} \\
&+ \lambda_n^{\beta_1+2} h_1(z_n, \lambda_n^{-\beta_1} w_{1,n}, \lambda_n^{-\beta_2} w_{2,n}, \lambda_n^{-(\beta_1+1)} \\
&\times \nabla_x w_{1,n}(x), \lambda_n^{-(\beta_2+1)} \nabla_x w_{2,n}(x)). \qquad (3.21)
\end{aligned}$$

with $a_{ij}^n(z) = a_{ij}(\lambda_n z' + y_n', \lambda_n z_N), b_j^n(z) = b_j(\lambda_n z' + y_n', \lambda_n z_N)$. Again the boundary conditions are

$$w_{1,n} = 0, \quad w_{2,n} = 0 \quad \text{on } \{z_N = 0\} \cap B_{\delta/\lambda_n} \qquad (3.22)$$

Choosing a sequence$\{R_m\}$, $R_m \to +\infty$ as $m \to \infty$. For fixed m, $B^+_{4R_m} \subset B_{\delta/\lambda_n}$ for n large, a_{ij}^n, b_j^n are uniformly bounded in n with respect to the $C^2(B_{\delta/\lambda_n})$-norm. Thus by (A3) we obtain

$$\begin{aligned}
&|\lambda_n^{\beta_1+2} h_1(z^n, \lambda_n^{-\beta_1} w_{1,n}, \lambda_n^{-\beta_2} w_{2,n}, \lambda_n^{-(\beta_1+1)} \nabla w_{1,n}, \lambda_n^{-(\beta_2+1)} \nabla w_{2,n})| \\
&\quad \leq C(\lambda_n^{\beta_1+2} + \lambda_n^{\beta_1+2-\beta_1\beta_{11}} |w_{1,n}|^{\beta_{11}} + \lambda_n^{\beta_1+2-\beta_2\beta_{12}} |w_{2,n}|^{\beta_{12}} \\
&\quad + \lambda_n^{\beta_1+2-(\beta_1+1)\gamma_{11}} |\nabla w_{1,n}|^{\gamma_{11}} + \lambda_n^{\beta_1+2-(\beta_2+1)\gamma_{12}} |\nabla w_{2,n}|^{\gamma_{12}}) \qquad (3.23)
\end{aligned}$$

and a similar estimate for h_2.

Because the system is weakly coupled, by Lemma 2.1 and the interpolation inequality we obtain that $\{(w_{1,n}, w_{1,n})\}$ is uniformly bounded in $W^{2,q}(B^+_{2R_m})$ for $q > 1$. Choosing $q > N$, a uniform bound on the $C^{1,\alpha}(\overline{B^+_{2R_m}})$-norm of $(w_{1,n}, w_{2,n})$, $0 < \alpha < 1$, is obtained. Applying the Schauder estimates to the system in $B^+_{B_{R_m}}$, we see that on each open subset $D \subset\subset B^+_{R_m}$, $(w_{1,n}, w_{2,n})$ is uniformly bounded in $C^{2,\alpha}(\overline{D})$. By a diagonal process, we may assume $w_{1,n} \to w_1, w_{2,n} \to w_2$ in $C^{2,\alpha'}(\mathbb{R}^N_+) \cap C^{1,\alpha'}(\mathbb{R}^N_+)$, $0 < \alpha' < \alpha$ on any compact subset of $\mathbb{R}^N_+$. Since $a_{ij}^n(z) \to a_{ij}(0)$ as $n \to \infty$

Marcel Dekker, Inc
270 Madison Avenue, New York, New York 10016

and $a_{ij}(0) = \delta_{ij}$ because $D\Psi(0) = I$, we conclude that (w_1, w_2) is a non-negative solution of

$$\begin{cases} -\Delta w_1 = a(0)w_1^{\alpha_{11}}, \quad -\Delta w_2 = d(0)w_2^{\alpha_{22}} \quad \text{in } \mathbb{R}^N_+, \\ w_1 = 0, w_2 = 0 \quad \text{on } z_N = 0 \end{cases} \tag{3.24}$$

Because $w_{1,n}(0, \dots 0, \alpha_n/\lambda_n) = \lambda_n^{\beta_1} v_{1,n}(y'_n, \alpha_n) = \lambda_n^{\beta_1} u_{1,n}(x_n) = 1$, we have $w_1(0, \dots, \alpha_n/\lambda_n) = 1$. It is a contradiction to the fact that (1.6), (1.7) possesses only trivial solution.

Case $\alpha_n/\lambda_n \to +\infty$ as $n \to \infty$.

In this case we define

$$w_{1,n}(z) = \lambda_n^{\beta_1} v_{1,n}(\lambda_n z + y_n), \quad w_{2,n}(z) = \lambda_n^{\beta_2} v_{2,n}(\lambda_n z + y_n)$$

Then $w_{1,n}(0) = 1$ for all n, $(w_{2,n}, w_{1,n})$ satisfies (3.21) and (3.22) with $a_{ij}^n(z) = a_{ij}(\lambda_n z + y_n), b_j^n(z) = b_j(\lambda_n z + y_n)$, in $B_{\delta/\lambda_n} \cap \{z_N > -\alpha_n/\lambda_n\}$ and $w_{1,n} = 0, w_{2,n} = 0$ on $\{z_N = -\alpha_n/\lambda_n\} \cap B_{\delta/\lambda_n}$. For any $R > 0$, we have $\alpha_n/\lambda_n > R$ if n is sufficiently large, so $\overline{B}_R \subset B_{\delta/\lambda_n} \cap \{z_N > -\alpha_n/\lambda_n\}$. Repeating the compactness arguments as in first case, we obtain a subsequence of $(w_{1,n}, w_{2,n})$ such that it is uniformly convergent to $C^{2,\alpha}$ function (w_1, w_2) on any compact set of $\mathbb{R}^N$, and (w_1, w_2) is a nonnegative solution to

$$-\Delta w_1 = a(x_0)w_1^{\alpha_{11}}, \quad -\Delta w_2 = d(x_0)w_2^{\alpha_{22}} \quad \text{in } \mathbb{R}^N. \tag{3.25}$$

This is a contradiction to the fact that (1.5) has only trivial solution because $w_1(0) = \lim_{n\to\infty} w_{1,n}(0) = 1$.

Thus in all cases we obtain a contradiction. The proof is completed.

The Proof of Theorem 1.2. The proof follows similar lines as those in the proof of Theorem 1.1. So we give only a sketch of it. Hence as in the proof of Theorem 1.1, we distinguish two cases, $x_0 \in \Omega$ and $x_0 \in \partial\Omega$. We only consider the case $x_0 \in \Omega$, the other is similar.

Suppose $\{\|u_{1,n}\|_\infty, \|u_{2,n}\|_\infty\} \to \infty$. Proceeding as in the proof of Theorem 1.1, we have (3.1) and (3.2). We choose

$$\beta_1 = \frac{2(\alpha_{12}+1)}{\alpha_{12}\alpha_{21}-1}, \qquad \beta_2 = \frac{2(\alpha_{21}+1)}{\alpha_{12}\alpha_{21}-1}.$$

Since the system is strongly coupled, we obtain

$$\begin{aligned} &\beta_1 + 2 - \beta_1\alpha_{11} > 0, \quad \beta_1 + 2 - \beta_2\alpha_{12} = 0 \\ &\beta_2 + 2 - \beta_1\alpha_{21} = 0, \quad \beta_2 + 2 - \beta_2\alpha_{22} > 0 \end{aligned} \tag{3.26}$$

and (A5) implies

$$\begin{aligned} &\beta_1 + 2 - \beta_1\gamma_{11} > 0, \qquad \beta_1 + 2 - \beta_2\gamma_{12} > 0 \\ &\beta_2 + 2 - \beta_1\gamma_{21} > 0, \qquad \beta_2 + 2 - \beta_2\gamma_{22} > 0 \end{aligned} \tag{3.27}$$

Using Lemma 1.2 and the L^p–estimates and Schauder estimates, we come to a contradiction as before.

4. EXISTENCE OF POSITIVE SOLUTIONS

In this section we discuss the existence of positive solutions for system (1.3). It seems that the only available technique to treat such nonvariational systems is a topological one, that is, the Topological Degree of Leray-Schauder. Let us then set to solve the problem on the existence of **positive solutions** of the Dirichlet problem for system (1.3). We assume

(A8) $h_1, h_2 : \overline{\Omega} \times \mathbb{R}^+ \times \mathbb{R}^+ \times \mathbb{R}^N \times \mathbb{R}^N \to \mathbb{R}^+$ are C^1, $h_1(x, 0, 0, 0, 0) = h_2(x, 0, 0\ ,0,0) = 0$.

We shall work here with the Banach space $C_0^0(\overline{\Omega})$ endowed with the norm of the maximum. Our notation is

$$C_0^0(\overline{\Omega}) = \{u : \overline{\Omega} \to \mathbb{R}, \text{ continuous and } u = 0 \text{ on } \partial\Omega\}.$$

Let $E := C_0^0(\overline{\Omega}) \times C_0^0(\overline{\Omega})$, and $K := \{(u, v) \in E : u \geq 0, v \geq 0\}$

Condition (A8) implies that, if $u, v \in C_0^0(\overline{\Omega})$, and $u, v \geq 0$, then the system

$$\begin{cases} -\Delta\hat{u} = a(x)u^{\alpha_{11}} + b(x)v^{\alpha_{12}} + h_1(x, u, v, \nabla u, \nabla v), \\ -\Delta\hat{v} = c(x)u^{\alpha_{21}} + d(x)v^{\alpha_{22}} + h_2(x, u, v, \nabla u, \nabla v) \text{ in } \Omega, \end{cases} \tag{4.1}$$

has a **positive** solution in the space $C^1(\Omega) \cap C_0^0(\overline{\Omega})$. These assertions come from the maximum principle and the theory of solvability of elliptic equations. Let $f(x, u, v, \nabla u, \nabla v)$ and $g(x, u, v, \nabla u, \nabla v)$ be the right-hand side of first and second equation of (4.1) respectively. We may define a mapping $S: K \to K$ by $S(f(x, u, v, \nabla u, \nabla v), g(x, u, v, \nabla u, \nabla v)) = (\hat{u}, \hat{v})$, then $S(K) \subset K$. In this step, the solutions $\hat{u}, \hat{v}$ are not classical, but what matters is that the weak solvability of (4.1) defines a mapping $T: K \to K$ as follows $T(u, v) = S(f(x, u, v, \nabla u, \nabla v), g(x, u, v, \nabla u, \nabla v))$. So a fixed point of T is a solution of system (1.3). In this way, one obtains a weak solution in $C^1(\Omega) \cap C_0^0(\overline{\Omega})$, and after, by the regularity theory the solution is in fact in $C^{2,\alpha}$. The operator T is compact, in view of the compact imbedding of

$C_0^0(\overline{\Omega})$ into $C^1(\Omega) \cap C_0^0(\overline{\Omega})$. So we can use topological degree through an Index Point Theorem, namely

Theorem 4.1. [Krasnosel'skii [14]]. *Let $\mathcal{C}$ be a cone in Banach space X and $T : \mathcal{C} \to \mathcal{C}$ a compact map such that $T(0) = 0$. Suppose that there exists $0 < r < R$, $t > 0$ such that*

(*i*) $x \neq tTx$ *for* $0 \leq t \leq 1$, $x \in \mathcal{C}$, $\|x\| = r$,
(*ii*) $\exists$ *compact map* $H\colon \overline{B}_R \times [0, \infty) \to \mathcal{C}$ *such that*
  (*a*) $H(x, 0) = Tx$ *for* $\|x\| = R$,
  (*b*) $H(x, t) \neq x$ *for* $\|x\| = R$ *and* $t \geq 0$,
  (*c*) $H(x, t) = x$ *has no solution* $x \in \overline{B}_R$ *for* $t \geq t_0$.

Then

$$i_{\mathcal{C}}(T, B_r) = 1, \quad i_{\mathcal{C}}(T, B_R) = 0, \quad i_{\mathcal{C}}(T, U) = -1,$$

where $U = \{x \in \mathcal{C} : r < \|x\| < R\}$. *So T has a fixed point in U.*

Let us illustrate the use of the above Theorem to system (1.3) under some further conditions. So let us assume

(A9) $h_1(x, u, v, \xi, \eta) = o(|u| + |v|)$, $h_2(x, u, v, \xi, \eta) = o(|u| + |v|)$, as $|u| + |v| \to 0$, uniformly in $x \in \overline{\Omega}$

(A10) There are positive real numbers μ_1, μ_2 and a constant $C > 0$ such that $\mu_1\mu_2 > \lambda_1^2$, and

$$h_1(x, u, v, \xi, \eta) \geq \mu_1 u - C, \text{ uniformly in } x \in \overline{\Omega},\ v \in \mathbb{R}^+,\ \xi, \eta \in \mathbb{R}^N \text{and}$$

$$h_2(x, u, v, \xi, \eta) \geq \mu_2 v - C, \text{ uniformly in } x \in \overline{\Omega},\ u \in \mathbb{R}^+\ \xi, \eta \in \mathbb{R}^N.$$

We recall that λ_1 is the first eigenvalue of $(-\Delta, H_0^1(\Omega))$.

Theorem 4.2. *Assume (A8), (A9) and (A10). And suppose that there is an a priori bound for all eventual positive solutions of the one-parameter family of Dirichlet problems*

$$\begin{cases} -\Delta u = a(x)(u + t)_+^{\alpha_{11}} + b(x)(v + t)_+^{\alpha_{12}} + h_1(x, u + t, v + t, \nabla u, \nabla v), \\ -\Delta v = c(x)(u + t)_+^{\alpha_{11}} + d(x)(v + t)_+^{\alpha_{12}} + h_2(x, u + t, v + t, \nabla u, \nabla v) \\ \text{in } \Omega,\ t \geq 0. \end{cases} \tag{4.2}$$

Then system (1.3) has a nonnegative nontrivial solution (u, v).

In order to show that condition (i) in Theorem 4.1 is satisfied, assume by contradiction that for all $r > 0$ there is a $t \in [0, 1]$ such that

MARCEL DEKKER, INC
270 Madison Avenue, New York, New York 10016

$$-\Delta u = tf(x, u, v, \nabla u, \nabla v), \quad -\Delta v = tg(x, u, v, \nabla u, \nabla v).$$

Taking $\epsilon < \lambda_1/2$ and using (A9), we see that there is $r_0 > 0$ such that for all $0 < r \leq r_0$ one has $|f(x, u, v, \nabla u, \nabla v)| \leq \epsilon(|u| + |v|)$ and a similar estimate for g. Multiplying the equations by ϕ_1, integrating by parts and using these estimates we get $\lambda_1 \int u\phi_1 \leq t\epsilon(\int u\phi_1 + \int v\phi_1)$ and a similar one corresponding to the second equation. Using these inequalities we conclude that $\lambda_1 \leq 2\epsilon$, which is a contradiction.

In order to verify condition (ii) of Theorem 4.1, we note that system (4.2) enables us to construct the homotopy

$$H((u, v), t) = S(f(x, u + t, v + t, \nabla u, \nabla v), g(x, u + t, v + t, \nabla u, \nabla v))$$

required in Theorem 4.1. Condition (A10) gives us condition (ii)(c). And of course the assumption that the eventual solutions of system (4.2) are uniformly bounded with respect to the parameter, gives (ii)(b). So Theorem 4.2 is a simple application of Theorem 4.1.

REFERENCES

1. Ambrosetti, A.; Rabinowitz, P.H. Dual variational methods in critical point theory and applications. J. Fctl. Anal. **1973**, *14*, 349–381.
2. Bartsch, T. *Topological Methods for Variational Problems with Symmetries*. Lect. Notes in Math. vol 1560. Springer Verlag 1993.
3. Brézis, H.; Turner, R.E.L. On a Class of Superlinear Elliptic Problems, Comm. Part. Diff. Eq. **1977**, *2*, 601–614.
4. Clément, Ph.; deFigueiredo, D.G.; Mitidieri, E. *Positive Solutions of Semilinear Elliptic Systems*. Comm. Part. Diff. Eq. **1992**, *17*, 923–940.
5. Clément, Ph.; deFigueiredo, D.G.; Mitidieri, E. A priori estimates for positive solutions of semilinear elliptic systems via Hardy-Sobolev inequalities. Pitman Res. Notes in Math. 1996, 73–91.
6. Cosner, C. Positive solutions For superlinear elliptic systems without variational structure, Nonlinear Anal. TMA. **1984**, *8*, 1427–1436.
7. deFigueiredo, D.G. *Semilinear Elliptic Systems,Proceedings of the Second School on Nonlinear Functional Analysis and Applications to Differential Equations*, Ed. Ambrosetti, A.; Chang, K.-C.; Ekeland, I. World Scientific 1998, 122–154.
8. deFigueiredo, D.G.; Felmer, P.L. A Liouville-type Theorem for systems, Ann. Sc. Norm. Sup. Pisa, **1994**, *XXI*, 387–397.
9. deFigueiredo, D.G.; Lions, P.-L.; Nussbaum, R. A priori estimates and existence of positive solutions of semilinear elliptic of positive solutions of semilinear elliptic equations. J. Math. Pures et Appl. **1982**, *61*, 41–63.

10. Friedman, A. Partial Differential Equations, Holt, Rinehart and Winston, Inc., 1969.
11. Gidas, B.; Spruck, J. A priori bounds for positive solutions of non-linear elliptic equations. Comm. Part. Diff. Eq. **1981**, *6*, 883–901.
12. Gidas, B.; Spruck, J. Global and local behaviour of positive solutions of nonlinear elliptic equations. Comm. Pure Appl. Math. **1981**, *34*, 525–598.
13. Kavian, O. Introduction à le therórie des points critiques, Mathmatiques & Applications vol. 13. Springer Verlag: 1993.
14. Krasnosels'kii, M.A. "Positive Solutions of Operator Equations". P. Noordhoff Groningen, 1964.
15. Mawhin, J.; Willem, M. *Critical Point theory and Hamiltonian Systems*, Applied Mathematical Sciences. Vol 74. Sringer Verlag, 1989.
16. Montenegro, M.S. *Criticalidade, superlinearidade e sublinearidade para sistemas elíptico semilineares*, Tese de Doutoramento, Unicamp, 1997.
17. Nirenberg, L. On elliptic partial differential equations. Ann. Sc. Norm. Sup. Pisa **1959**, *13*, 1–48.
18. Rabinowitz, P.H. Minimax methods in critical point theory with applications to differential equations. CBMS, Vol. 65, Amer. Math. Soc. 1986.
19. Souto, M.A.S. Sobre a existôncia de soluções positivas para sistemas cooperativos não-lineares, Tese de doutorado 1992, UNICAMP.
20. Serrin, J.; Zou, H. Non-existence of positive solutions of Lane-Emden system. Diff. Int. Eq. **1996**, *9*, 635–653.
21. Willem, M. Minimax Theorems, Progress in Nonlinear Differential Equations and Their Applications, vol. 24 Birkhäuser (1996).
22. Wang, X.-J.; Deng, Y.-B. Existence of Multiple Solutions to Nonlinear Elliptic Equations of Nondivergence Form, preprint.

Received September 2000
Revised February 2001

NoDEA
Nonlinear differ. equ. appl. 9 (2002) 309–323
1021–9722/02/030309–15 $ 1.50+0.20/0

Nonlinear Differential Equations
and Applications NoDEA

Some remarks on a system of quasilinear elliptic equations

Lucio BOCCARDO
Università di Roma I
Dipartimento di Matematica
Piazza A. Moro 2
00185 Roma - Italia
e-mail: boccardo@mat.uniroma1.it

Djairo GUEDES DE FIGUEIREDO
IMECC-UNICAMP
Caixa Postal 6065
13081-970 - Campinas-SP
Brasil
e-mail: djairo@ime.unicamp.br

Abstract. We study the existence of critical points (minima and saddle points) of functionals of the type

$$\Phi(u,v) = \frac{1}{p}\int_\Omega |\nabla u|^p + \frac{1}{q}\int_\Omega |\nabla v|^q - \int_\Omega F(x,u,v),$$

where p and q are real numbers larger than 1.

2000 *Mathematics Subject Classification*: 35J60, 35J20, 58E30.
Key words: Systems, quasilinear elliptic equations, critical points.

1 Introduction

In this paper we study the functional

$$\Phi(u,v) = \frac{1}{p}\int_\Omega |\nabla u|^p + \frac{1}{q}\int_\Omega |\nabla v|^q - \int_\Omega F(x,u,v), \quad (1.1)$$

Originally published in: NoDEA Nonlinear Differential Equations and Appl. 9 (2002), 309-323, © Birkhäuser Verlag, Basel 2002
Reprint by Springer-Verlag Berlin Heidelberg 2013,
DOI: 10.1007/978-3-319-02856-9_33

where p and q are real numbers larger than $1, \Omega$ is some bounded domain in R^N, u and v are real-valued functions defined in $\overline{\Omega}$ and belonging to appropriate spaces of functions and F (sometimes refered as a *potential*) is a real-valued differentiable function with domain $\overline{\Omega} \times \mathbb{R} \times \mathbb{R}$. Our aim is to study the geometry of this functional viewing to determining its critical points. Such critical points are the solutions of associated Euler-Lagrange equations, which in the present case is the system of quasilinear elliptic equations below

$$\begin{array}{l} -\Delta_p u = F_u(x,u,v) \\ -\Delta_q v = F_v(x,u,v) \end{array} \tag{1.2}$$

where F_u designates the partial derivative of F with respect to u and Δ_p is the so-called p-Laplacian operator

$$\Delta_p u = div(|\nabla u|^{p-2}\nabla u).$$

The geometry of Φ is sort of similar to the one of the functional

$$\frac{1}{p}\int |\nabla u|^p - \int F(x,u),$$

which corresponds to a single quasilinear equation. However, some interesting features appear due to the coupling in the equations (1.2). Our theorems include and unify some previous results by Boccardo – Fleckinger de Thelin [BFT] and de Thélin-Vélin [VT].

Let us introduce the precise assumptions under which our problem is studied. Our functional Φ is to be defined in the Cartesian product of Sobolev spaces $W_0^{1,p}(\Omega) \times W_0^{1,q}(\Omega)$. For that matter, the following assumption on F has to be made, although stronger restriction will come timely:

$$\begin{aligned} &F : \overline{\Omega} \times \mathbb{R} \times \mathbb{R} \to \mathbb{R}\, is\ C^1\ and \\ &|F(x,u,v)| \le c(1 + |u|^{p^*} + |v|^{q^*}), \end{aligned} \tag{F$_1$}$$

where $p^* = pN/(N-p)$ and c is some positive constant. Similarly q^*.

In this work, we assume that both p and q are less than N. If this is so, we have the continuous imbeddings $W_0^{1,p}(\Omega) \subset L^{p^*}(\Omega)$ and $W_0^{1,q}(\Omega) \subset L^{q^*}(\Omega)$, which then tell us that Φ is well defined. In order to have it of class C^1, we require a stronger condition than (F_1), namely: F is C^1 and

$$\begin{aligned} |F_u(x,u,v)| &\le C(1 + |u|^{p^*-1} + |v|^{\frac{q^*(p^*-1)}{p^*}}) \\ |F_v(x,u,v)| &\le C(1 + |v|^{q^*-1} + |u|^{\frac{p^*(q^*-1)}{q^*}}). \end{aligned} \tag{F$_2$}$$

It is easy to prove that, if (F_2) is satisfied, then also is (F_1). Under (F_2), it follows that the critical points of Φ are the weak solutions of system (1.2), subject to

Dirichlet boundary conditions. For easy reference, we summarize the aforesaid as follows: • under hypothesis (F_2), with $E = W_0^{1,p}(\Omega) \times W_0^{1,q}(\Omega)$, we have

$$\Phi : E \to \mathbb{R} \quad \text{is a } C^1\text{-functional.}$$

The geometry of Φ depends strongly on the values of r and s in the estimate below

$$|F(x,u,v)| \leq c(1 + |u|^r + |v|^s), \tag{F$_3$}$$

where c is some positive constant, and $r \leq p^*, s \leq q^*$. We discuss three distinct cases

(I) $r < p$ and $s < q$, ("sublinear-like")

(II) $r > p$ or $s > q$, and $r < p^*, s < q^*$, ("superlinear-like")

(III) $r = p$ and $s = q$, ("of resonant-type").

The expressions in parenthesis are to remind us of similar terminology in the case $p = q = 2$. Of course, there are several other situations, which could be of interest to consider. Observe that we are considering only subcritical cases. The cases where either $r = p^*$ or $s = q^*$ or both equalities hold lead to a loss of compactness, and to problems which should be investigated.

Next we state the main results of this paper.

Theorem 1 (The coercive case). *Assume* (F_2) *and* (F_3) *with r and s as in* (I). *Then Φ achieves a global minimum at some $(u_0, v_0) \in E$, which is then a weak solution of system* (1.2).

If we are in the situation that

$$F(x,0,0) = F_u(x,0,0) = F_v(x,0,0), \text{ for all } x \in \overline{\Omega}, \tag{F$_4$}$$

then $u \equiv 0$ and $v \equiv 0$ are a trivial solution of system (1.2). In this case the relevant question is obtaining a non-trivial solution of (1.2). This will be possible under appropriate assumptions on the function F, as it is stated in the next results.

Theorem 2 (The coercive case, non-trivial solution). *Assume* (F_2), (F_4) *and* (F_3) *with r and s as in* (I). *Then Φ achieves a global minimum at a point $(u_0, v_0) \neq (0,0)$, provided there exist positive constant R and $\theta < 1$, and a continuous function $K : \overline{\Omega} \times \mathbb{R} \times \mathbb{R} \to \mathbb{R}$ such that*

$$F(x, t^{\frac{1}{p}}u, t^{\frac{1}{q}}v) \geq t^\theta K(x,u,v), \textit{ for } x \in \overline{\Omega}, |u|, |v| \leq R, \textit{ and small } t > 0. \tag{F$_5$}$$

In this case a Palais-Smale condition holds (see Lemma 4.1) if we assume that there are numbers $R > 0$, θ_p and θ_q with

$$\frac{1}{p^*} < \theta_p < \frac{1}{p} \quad \frac{1}{q^*} < \theta_q < \frac{1}{q}$$

such that

$$0 < F(x,u,v) \le \theta_p u F_u(x,u,v) + \theta_q v F_v(x,u,v) \tag{F$_6$}$$

for all $x \in \overline{\Omega}$ and $|u|, |v| \ge R$.

Theorem 3 *Assume* $(\mathrm{F}_2), (\mathrm{F}_4), (\mathrm{F}_6)$ *and* (F_3) *with* r *and* s *as in* (II). *Assume also that there are constants* $c > 0$ *and* $\varepsilon > 0$ *and numbers* $\overline{r} > p, \overline{s} > q$, *such that*

$$|F(x,u,v)| \le c(|u|^{\overline{r}} + |v|^{\overline{s}}), \text{ for } |u|, |v| \le \varepsilon, x \in \overline{\Omega}. \tag{F$_7$}$$

Then Φ *has a non-trivial critical point.*

Remark. Without loss of generality we may assume $\overline{r} < p^*$ and $\overline{s} < q^*$.

Next we study the situation when (F_3) holds with r and s as in (III), the case we called "of resonant type". In this case, it is quite adequate to assume a condition on F that implies that the functional Φ satisfies the so-called Cerami condition (see Section 4 for the definition). The assumption on F is: there are positive numbers c, R, μ and ν such that

$$\frac{1}{p} u F_u + \frac{1}{q} v F_v - F \ge c(|u|^\mu + |v|^\nu) \tag{F$_8$}$$

for $|u|, |v| > R$.

This type of condition has been introduced by Costa-Magalhães [CM1], [CM2]. In order to avoid resonance we shall assume a condition on F involving an eigenvalue problem, which we introduce next. Let $G : R^2 \to [0, \infty)$ be a C^1 even function such that

$$G(t^{\frac{1}{p}} u, t^{\frac{1}{q}} v) = t G(u,v) \tag{G$_1$}$$

$$G(u,v) \le k(|u|^p + |v|^q). \tag{G$_2$}$$

Examples of such functions are

(i) $G(u,v) = c_1 |u|^p + c_2 |v|^q$

(ii) $G(u,v) = c|u|^\beta |v|^\gamma$, with $\dfrac{\beta}{p} + \dfrac{\gamma}{q} = 1$,

where c_1, c_2 and c are positive constants.

We shall prove in Section 3 that the eigenvalue problem

$$-\Delta_p u - aG_u = \lambda|u|^{p-2}u$$

$$-\Delta_q v - aG_v = \lambda|v|^{q-2}v$$

subject to Dirichlet boundary conditions, with $a \in L^\infty(\Omega)$, has an eingenvalue $\lambda_1(a)$, characterized variationally by

$$\frac{1}{p}\int|\nabla u|^p + \frac{1}{q}\int|\nabla v|^q - \int aG(u,v) \geq \lambda_1(a)\left(\frac{1}{p}\int|u|^p + \frac{1}{q}\int|v|^q\right)$$

for all $(u,v) \in E$.

Now we introduce the following assumption:

$$\lambda_1(a) > 0, \text{ where } \limsup_{|u|,|v|\to\infty} \frac{F(x,u,v)}{G(u,v)} \leq a(x) \in L^\infty(\Omega) \tag{F$_9$}$$

and state the next results.

Theorem 4 *Assume* $(F_2), (F_8), (F_9)$ *and* (F_3) *with* r *and* s *as in* (III). *Then the functional* Φ *is bounded below and the infimum is achieved.*

Theorem 5 *Assume* $(F_2), (F_4), (F_8)$ *and* (F_3) *with* r *and* s *as in* (III). *Suppose also that there are positive numbers* R *and* ε, *and* $L^\infty(\Omega)$ *functions* $b(x)$ *and* $c(x)$ *such that*

$$\lambda_1(b) < 0, \; F(x,u,v) \geq b(x)G(u,v), \; |u|,|v| \geq R \tag{F$_{10}$}$$

$$\lambda_1(c) > 0, \; F(x,u,v) \leq c(x)\tilde{G}(u,v), \; |u|,|v| \leq \varepsilon, \tag{F$_{11}$}$$

where G *and* $\tilde{G}$ *are functions satisfying the conditions* (G_1) *and* (G_2). *Then, the functional* Φ *possesses a non-trivial critical point.*

2 Special classes of potentials F

(i) The following class of potentials F (and its perturbations) have been considered by [dT], [VT], [FMT]:

$$F(x,u,v) = c(x)|u|^\beta|v|^\gamma$$

where $c(x) \in L^\infty(\overline{\Omega})$ and $\beta, \gamma \geq 1$. Using Young's inequality

$$|u|^\beta |v|^\gamma \leq \frac{1}{m}|u|^{\beta m} + \frac{1}{n}|v|^{\gamma n}$$

where $1/m + 1/n = 1$. Let $r = \beta m$ and $s = \gamma n$. So $\dfrac{\beta}{r} + \dfrac{\gamma}{s} = 1$. Consequently, for this class the three cases are

$$\frac{\beta}{p} + \frac{\gamma}{q} < 1. \tag{I$_\mathrm{i}$}$$

$$\frac{\beta}{p} + \frac{\gamma}{q} > 1 \quad \text{and} \quad \frac{\beta}{p^*} + \frac{\gamma}{q^*} < 1. \tag{II$_\mathrm{i}$}$$

$$\frac{\beta}{p} + \frac{\gamma}{q} = 1. \tag{III$_\mathrm{i}$}$$

In this example, condition (F_5) is precisely the inequality in (I) above. Condition (F_6) holds if β and γ are such that

$$\theta_p \beta + \theta_q \gamma \geq 1. \tag{2.1}$$

Observe that, if (2.1) holds with θ_p and θ_q as in the Introduction, then we are necessarily in case (II). That is, the problem is "superlinear-like".

The theorems stated in the Introduction contain and extend some of the results of the above mentioned papers.

(ii) In [BFT] the following system was studied

$$\begin{aligned} -\Delta_p u &= a(x)|u|^{\alpha-2}u + b(x)|v|^{\beta-2}v + f \\ -\Delta_q v &= c(x)|u|^{\gamma-2}u + d(x)|v|^{\delta-2}v + g \end{aligned}$$

subject to Dirichlet boundary conditions. In this generality, the system is not variational. However, if $b(x) = c(x)$ and $\beta = \gamma = 2$, the above equations are the Euler Lagrange equations of a functional Φ as in (1.1) with

$$F(x,u,v) = a(x)|u|^\alpha + b(x)uv + d(x)|v|^\delta + fu + gv,$$

where we assume that $a, b, d \in L^\infty(\Omega), \alpha, \delta \geq 1$ and $f \in L^{(p^*)'}(\Omega), g \in L^{(q^*)'}(\Omega)$. Here $(p^*)' = \frac{pN}{p+N(p-1)}$ and a similar expression for $(q^*)'$. The fact that f and g are not necessarily in $L^\infty(\Omega)$ implies that the various pointwise estimates (F) cannot hold. However, since the terms where they appear are linear in u and v, their presence essentially do not change the proofs of the theorems. So, in this example, the three cases studied are:

$$\alpha < p, \ \delta < q, \ \frac{1}{p} + \frac{1}{q} < 1. \tag{I$_\mathrm{ii}$}$$

$$\alpha > p \text{ or } \delta > q \text{ or } \frac{1}{p} + \frac{1}{q} > 1. \tag{II$_{\text{ii}}$}$$

$$\alpha = p,\ \delta = q,\ \frac{1}{p} + \frac{1}{q} = 1. \tag{III$_{\text{ii}}$}$$

We remark that case $(\text{II})_{\text{ii}}$ was not considered in [BFT]. Our results for case $(\text{III})_{\text{ii}}$ extend the ones in [BFT].

Remark. For the special examples above, condition (F5) can be replaced by (F5)′ there are positive constants c and ε such that

$$F(x, t^q, t^p) > ct^{pq} \qquad \text{for all} \quad x \in \overline{\Omega},\ 0 < t < \varepsilon.$$

3 The eigenvalue problem

Let $G : R^2 \to [0, \infty)$ be a C^1 even function satisfying conditions (G_1) and (G_2) given in the Introduction.

Lemma 3.1 *Given $a \in L^\infty(\Omega)$, there are a real number $\lambda_1(a)$ and $(u_0, v_0) \in E$, such that*

$$\begin{cases} -\Delta_p u_0 - aG_u(u_0, v_0) &= \lambda_1(a) u_0 |u_0|^{p-2} \\ -\Delta_q v_0 - aG_v(u_0, v_0) &= \lambda_1(a) v_0 |v_0|^{q-2} \end{cases} \tag{3.1}$$

and

$$\frac{1}{2} \int |\nabla u|^p + \frac{1}{q} \int |\nabla v|^q - \int aG(u, v) \geq \lambda_1(a) \left[\frac{1}{p} \int |u|^p + \frac{1}{q} \int |v|^q \right] \tag{3.2}$$

for all $(u, v) \in E$, with equality for (u_0, v_0).

Proof. Choose $M > k\|a\|_{L^\infty}$, where k is the constant in (G_2). Then the functional

$$\begin{aligned} J(u, v) = &\frac{1}{p} \int |\nabla u|^p + \frac{1}{q} \int |\nabla v|^q - \int aG(u, v) \\ &+ M \left[\frac{1}{p} \int |u|^p + \frac{1}{q} \int |v|^q \right] \end{aligned} \tag{3.3}$$

is non-negative for $(u, v) \in E$. Let

$$S = \left\{ (u, v) \in E : \frac{1}{p} \int |u|^p + \frac{1}{q} \int |v|^q = 1 \right\} \tag{3.4}$$

and let us look for *Inf*$\{J(u, v) : (u, v) \in S\}$. Let us denote this infimum by μ, and let us take a minimizing sequence $(u_n, v_n) \in S$. It follows that $\|u_n\|_{W^{1,p}}$ and

$\|v_n\|_{W^{1,q}}$ are bounded. So we may choose subsequences (denoted again by (u_n) and (v_n)) such that (u_n) converges to u_0, weakly in $W_0^{1,p}$ and strongly in L^p. Similarly for (v_n). Passing to the limit

$$\frac{1}{p}\int |\nabla u_0|^p + \frac{1}{q}\int |\nabla v_0|^q - \int a + G(u_0, v_0)$$
$$+M\left[\frac{1}{p}\int |u_0|^p + \frac{1}{q}\int |v_0|^q\right] \leq \mu$$

which indeed is an equality because $(u_0, v_0) \in S$. So the above infimum is achieved. If follows then that

$$\begin{cases} -\Delta_p u_0 - aG_u(u_0, v_0) + Mu_0|u_0|^{p-1} = \mu_M u_0|u_0|^{p-1} \\ -\Delta_q v_0 - aG_v(u_0, v_0) + Mv_0|v_0|^{q-1} \ = \mu_M v_0|v_0|^{p-1} \end{cases} \tag{3.5}$$

where μ_M is the Lagrange multiplier. It follows from $(G1)$ that

$$G(u, v) = \frac{1}{p}uG_u(u, v) + \frac{1}{q}vG_v(u, v) \tag{3.6}$$

for all $(u, v) \in R^2$. From the minimization above we have:

$$\mu\left[\frac{1}{p}\int |u|^p + \frac{1}{q}\int |v|^q\right]$$
$$\leq \frac{1}{p}\int |\nabla u|^p + \frac{1}{q}\int |\nabla v|^q - \int aG(u, v)$$
$$+ M\left[\frac{1}{p}\int |u|^p + \frac{1}{q}\int |v|^q\right] \tag{3.7}$$

for all $(u, v) \in E$. It follows from (3.5), (3.6) and (3.7) that $\mu = \mu_M$. In this way we get (3.1) and (3.2) with $\lambda_1(a) = \mu - M$, and the eigenfunction pair (u_0, v_0) as obtained above. □

Remark. From the minimization argument above, it follows that both u_0 and v_0 can be taken ≥ 0 in Ω. As in [dT] it can be proved that u_0, v_0 are $C^1(\overline{\Omega})$ as consequence of regularity results of [T]. Then by the Vasquez Maximum Principle [V] for the p-Laplacian, we conclude that u_0 and v_0 are indeed > 0 in Ω.

Lemma 3.2 *$\lambda_1(a)$ is a continuous function of a in the L^∞ norm.*

Proof. Let us denote by J_a the functional defined in (3.3) and J_b the corresponding one with a replaced by b, where b is some other L^∞-function. Let $\lambda_1(b)$ be eigenvalue corresponding to b given in Lemma 3.1. We show that, given $\varepsilon > 0$ there is a $\delta > 0$ such that $|\lambda_1(a) - \lambda_1(b)| < \varepsilon$ provided $\|a - b\|_{L^\infty} \leq \delta$.

Indeed, given $\varepsilon > 0$, choose $(u_\varepsilon, v_\varepsilon) \in S_1$ such that

$$J_a(u_\varepsilon, v_\varepsilon) \leq \lambda_1(a) + \frac{\varepsilon}{2}. \tag{3.8}$$

Next (G_2) implies

$$G(u,v) \leq K\left(\frac{1}{p}|u|^p + \frac{1}{q}|v|^q\right) \tag{3.9}$$

where $K \geq \max\{kp, kq\}$. Using (3.9) we obtain

$$|J_b(u_\varepsilon, v_\varepsilon) - J_a(u_\varepsilon, v_\varepsilon)| \leq \|b-a\|_{L^\infty}.K. \tag{3.10}$$

So, from (3.10) and (3.8)

$$\lambda_1(b) \leq J_b(u_\varepsilon, v_\varepsilon) \leq J_a(u_\varepsilon, v_\varepsilon) + K\|b-a\|_{L^\infty} \leq \lambda_1(a) + \frac{\varepsilon}{2} + K\|b-a\|_{L^\infty}$$

Choosing $\delta = \dfrac{\varepsilon}{2k}$ we have $\lambda_1(b) \leq \lambda_1(a) + \varepsilon$.

Reversing the roles of a and b, the result follows. □

4 Compactness conditions

We say that $\Phi : E \to \mathbb{R}$ satisfies the *(PS) condition* if, all $(u_n, v_n) \in E$ such that

$$|\Phi(u_n, v_n)| \leq const \qquad \Phi'(u_n, v_n) \to 0 \tag{4.1}$$

contains a convergent subsequence in the norm of E.

Lemma 4.1 *Suppose that F satisfies* $(F_2), (F_6)$ *and* (F_3) *as in* (II). *Then the functional Φ defined in the Introduction satisfies the (PS) condition.*

Proof. It follows from (4.1) that

$$\left|\frac{1}{p}\int |\nabla u_n|^p + \frac{1}{q}\int |\nabla v_n|^q - \int F(u_n, v_n)\right| \leq const,$$

$$\left|\int |\nabla u_n|^p - \int F_u(u_n, v_n)u_n\right| \leq \varepsilon_n \|u_n\|_{W^{1,p}}$$

and a similar one for v_n, where $\varepsilon_n \to 0$. Using these expressions we get

$$\left(\frac{1}{p} - \theta_p\right)\int |\nabla u_n|^p + \left(\frac{1}{q} - \theta_q\right)$$

$$\int |\nabla v_n|^q - \int (F(u_n, v_n) - \theta_p u_n F_u - \theta_q v_n F_v)$$

$$\leq c + c(\|u_n\|_{W^{1,p}} + \|v_n\|_{W^{1,q}})$$

which implies, using (F_6) that both $\|u_n\|_{W^{1,p}}$ and $\|v_n\|_{W^{1,q}}$ are bounded. The existence of convergent subsequences follows in a standard way, since the growth of F is below the critical exponents p^* and q^*. □

We say that $\Phi : E \to \mathbb{R}$ satisfies the *Cerami condition* (Ce) for short, if all $(u_n, v_n) \in E$ such that

$$|\Phi(u_n, v_n)| \leq const \qquad (1 + \|u_n\|_{W^{1,p}} + \|v_n\|_{W^{1,q}})\Phi'(u_n, v_n) \to 0 \tag{4.2}$$

contains a convergent subsequence in the norm of E.

Lemma 4.2 *Suppose that F satisfies* (F_2), (F_8) *and* (F_3) *as in* (III). *Then the functional Φ satisfies the (Ce) condition.*

Proof. Take a sequence pair $(u_n, v_n) \in E$ satisfying (4.2). It suffices to prove that $\|u_n\|_{W^{1,p}}$ and $\|v_n\|_{W^{1,q}}$ are bounded, as remarked in the proof of the previous lemma. It follows readily (4.2) that

$$\begin{aligned} C &\geq \left\langle \Phi'(u_n, v_n), \left(\frac{1}{p}u_n, \frac{1}{q}v_n\right)\right\rangle - \Phi(u_n, v_n) \\ &= \int \left(\frac{1}{p}u_n F_u(u_n, v_n) + \frac{1}{q}v_n F_v(u_n, v_n) - F(u_n, v_n)\right). \end{aligned}$$

Then using (F_8) we obtain

$$\int |u_n|^\mu + \int |v_n|^\nu \leq const. \tag{4.3}$$

Next we use the following interpolation inequality, see [CM2]: let $0 < a < b < c$ and suppose that for some measurable function $u : \Omega \to R$ we have that

$$\int |u|^a < \infty \quad \text{and} \quad \int |u|^c < \infty$$

then

$$\int |u|^b \leq \left(\int |u|^a\right)^{\frac{c-b}{c-a}} \cdot \left(\int |u|^c\right)^{\frac{b-a}{c-a}} \tag{4.4}$$

We use (4.4) for $0 < \mu < p < p^*$ and $0 < \nu < q < q^*$. So using (4.3) and (4.4) we get

$$\int |u_n|^p \leq C\left(\int |u_n|^{p^*}\right)^{\frac{p-\mu}{p^*-\mu}}, \quad \int |v_n|^q \leq C\left(\int |v_n|^{q^*}\right)^{\frac{q-\nu}{q^*-\nu}},$$

which implies by Sobolev imbedding

$$\int |u_n|^p \le c\|u_n\|^{\tilde{p}}_{W^{1,p}}, \int |v_n|^q \le c\|v_n\|^{\tilde{q}}_{W^{1,q}}, \tag{4.5}$$

where

$$\tilde{p} = \frac{p-\mu}{p^* - \mu}p^*, \qquad \tilde{q} = \frac{q-\nu}{q^*-\nu}q^*$$

Using (F_3) as in (III) we get

$$\Phi(u_n, v_n) \ge \frac{1}{p}\int |\nabla u_n|^p + \frac{1}{q}\int |\nabla v_n|^q - c\int |u_n|^p - c\int |v_n|^q$$

Which estimate by (4.5) leads to

$$\Phi(u_n, v_n) \ge \frac{1}{p}\|u_n\|^p_{W^{1,p}} + \frac{1}{q}\|v_n\|^q_{W^{1,q}} - c\|u_n\|^{\tilde{p}}_{W^{1,p}} - c\|v_n\|^{\tilde{q}}_{W^{1,q}}$$

Since $\Phi(u_n, v_n)$ is bounded and $\tilde{p} < p$ and $\tilde{q} < q$, it follows the boundedness of $\|u_n\|_{W^{1,p}}$ and $\|v_n\|_{W^{1,q}}$. □

5 Proofs of Theorems completed

Theorem 1 *Condition* (F_3) *with r and s as in* (I) *implies that Φ is weakly lower semicontinuous and coercive in E. So, it assumes its infimum at a point $(u_0, v_0) \in E$. Condition* (F_2) *implies that this is a critical point Φ and consequently a weak solution of* (1.2).

Theorem 2 *As in Theorem* 1, *Φ assumes its infimum at a point $(u_0, v_0) \in E$. To prove that this is not $(0,0)$ it is enough to show that there is $(u_1, v_1) \in E$ such that $\Phi(u_1, v_1) < 0$.*

Let φ be a first eigenfunction for the p-Laplacian

$$\begin{cases} -\Delta_p \varphi = \lambda_1(p)|\varphi|^{p-2}\varphi, & \text{in } \Omega \\ \varphi = 0 & \text{on } \partial\Omega, \end{cases}$$

and ψ for the q-Laplacian. We know that $\varphi, \psi \in C^{1,\alpha}$, see [dB], [T]. *So we can take $\|\varphi\|_{L^\infty}, \|\psi\|_{L^\infty} \le R$, where R is the constant in* (F_5). *So*

$$\begin{aligned}\Phi(t^{\frac{1}{p}}\varphi, t^{\frac{1}{q}}\psi) \ \le\ & t\left\{\frac{1}{p}\lambda_1(p)\int |\varphi|^p + \frac{1}{q}\lambda_1(q)\int |\psi|^q\right\} \\ & - t^\theta \int K(x, \varphi, \psi),\end{aligned}$$

which is negative for $t > 0$ small.

Theorem 3 *It is easy to see that in this case* Φ *has the geometry of the Mountain Pass Theorem. Indeed, it follows from* (F_3) *and* (F_7) *that*

$$|F(x,u,v)| \le c(|u|^{\overline{r}} + |v|^{\overline{s}} + |u|^r + |v|^s)$$

for all $x \in \overline{\Omega}$ *and* $(u,v) \in \mathbb{R}^2$, *where* $p < r, \overline{r} < p^*$ *and* $q < s, \overline{s} < q^*$. *So by the Sobolev imbedding*

$$\int F(x,u,v) \le c(\|u\|^r_{W^{1,p}} + \|v\|^s_{W^{1,q}} + \|u\|^{\overline{r}}_{W^{1,p}} + \|v\|^{\overline{s}}_{W^{1,q}})$$

for all $(u,v) \in E$. *So, we can estimate* Φ *by*

$$\Phi(u,v) \ge \frac{1}{p}\|u\|^p_{W^{1,p}} + \frac{1}{q}\|v\|^q_{W^{1,q}}$$
$$-c(\|u\|^r_{W^{1,p}} + \|v\|^s_{W^{1,q}} + \|u\|^{\overline{r}}_{W^{1,p}} + \|v\|^{\overline{s}}_{W^{1,q}})$$

which implies that there is an $\varepsilon > 0$ *and* $\rho > 0$ *such that* $\Phi(u,v) \ge \varepsilon$ *for* $\|u\|_{W^{1,p}} + \|v\|_{W^{1,q}} = \rho$. *On the other hand, using* (F_6) *we have*

$$\frac{d}{dt}\{F(x,t^{\theta_p}u,t^{\theta_q}v)\} \ge \frac{1}{t}F(x,t^{\theta_p}u,t^{\theta_q}v)$$

which implies that $F(x,t^{\theta_p}u,t^{\theta_q}v) \ge tK(x,u,v)$ *for some function* K. *Since*

$$\frac{1}{p}\int |\nabla(t^{\theta_p}u)|^p + \frac{1}{q}\int |\nabla(t^{\theta_q}v)|^q = t^{\theta_p p}c_1 + t^{\theta_q q}c_2$$

and from the hypothesis (F_6), $\theta_p p < 1$ *and* $\theta_q q < 1$, *we conclude that for any fixed*

$$(u,v) \neq (0,0), \quad \Phi(t^{\theta_p}u, t^{\theta_q}v) \to -\infty \quad \text{as} \quad t \to +\infty.$$

Since Φ *satisfies the (PS) condition, as proved in Lemma 4.1, we may apply the Mountain Pass Theorem and conclude.*

Theorem 4 *It follows from* (F_9) *that given* $\varepsilon > 0$ *there is a constant* $C_\varepsilon > 0$ *such that*

$$F(x,u,v) \le (a(x)+\varepsilon)G(u,v) + C_\varepsilon, \quad all \quad x \in \overline{\Omega}, (u,v) \in R^2.$$

Hence

$$\Phi(u,v) \ge \frac{1}{p}\int |\nabla u|^p + \frac{1}{q}\int |\nabla v|^q - \int (a+\varepsilon)G(u,v) - C_\varepsilon|\Omega|. \tag{5.1}$$

Now let $0 < \delta < 1$ *be a real number to be chosen later, and write* (5.1) *as follows*

$$\Phi(u,v) \ge \delta\left\{\frac{1}{p}\int |\nabla u|^p + \frac{1}{q}\int |\nabla v|^q\right\}$$
$$+ (1-\delta)\left\{\frac{1}{p}\int |\nabla u|^p + \frac{1}{q}\int |\nabla v|^q - \int \frac{a+\varepsilon}{1-\delta}G(u,v)\right\} - c_\varepsilon|\Omega|$$

The expression in the second bracket is estimated from below by

$$\lambda_1\left(\frac{a+\varepsilon}{1-\delta}\right)\left(\frac{1}{p}|u|^p+\frac{1}{q}\int|v|^q\right).$$

Now choose ε *and* δ *such that* $\lambda_1(\frac{a+\varepsilon}{1-\delta})>0$*, which is possible in view of the continuity of* $\lambda_1(a)$ *with respect to* a *(Lemma 3.2), in the* L^∞*-norm. So*

$$\Phi(u,v)\geq\delta\left\{\frac{1}{p}\int\nabla u|^p+\frac{1}{q}\int|\nabla v|^q\right\}-C_\varepsilon|\Omega|$$

which shows that Φ *is coercive. Since* Φ *is weakly lower semicontinuous, the result follows.*

Theorem 5 *It suffices to show that* Φ *in this case has the geometry of the Mountain Pass Theorem. The compactness condition* (Ce) *has already been proved in Lemma 4.2. First we prove that* $(0,0)$ *is a local minimum. Indeed, it follows from* (F_{11}) *and* (F_3) *that there is a constant* $c>0$ *such that*

$$F(x,u,v)\leq c(x)\tilde{G}(u,v)+c(|u|^{\tilde{r}}+|v|^{\tilde{s}}) \tag{5.2}$$

for all $x\in\overline{\Omega},(u,v)\in R^2$*, and* $\tilde{r}>p,\tilde{s}>q$*. Next using (5.2) and a* $\delta\in(0,1)$ *to be chosen later we have*

$$\begin{aligned}\Phi(u,v)\ \geq\ &\delta\left(\frac{1}{p}\int|\nabla u|^p+\frac{1}{q}\int|\nabla v|^q\right)\\ &+(1-\delta)\left\{\frac{1}{p}\int|\nabla u|^p+\frac{1}{q}\int|\nabla u|^q-\int\frac{c(x)}{1-\delta}G(u,v)\right\}\\ &-c\|u\|^{\tilde{r}}_{W^{1,p}}-c\|v\|^{\tilde{s}}_{W^{1,q}}\end{aligned}$$

Choosing δ *such that the expression in the bracket is positive, as a consequence of* (F_{11})*, we get*

$$\Phi(u,v)>c_1\|u\|^p_{W^{1,p}}+c_2\|v\|^q_{W^{1,q}}-C\|u\|^{\tilde{r}}_{W^{1,p}}-c\|v\|^{\tilde{s}}_{W^{1,q}}$$

which shows that there are positive numbers ρ *and* ε *such that* $\Phi(u,v)>\varepsilon$ *for* $\|u\|_{W^{1,p}}+\|v\|_{W^{1,q}}=\rho$*. Finally we use* (F_{10}) *to estimate*

$$\Phi(u,v)\leq\frac{1}{p}\int|\nabla u|^p+\frac{1}{q}\int|\nabla v|^q-\int b(x)G(u,v)+C \tag{5.3}$$

Choosing $u=t^{1/p}u_0$ *and* $v=t^{1/q}v_0$*, where* (u_0,v_0) *is the eigenfunction pair corresponding to* $\lambda_1(b)$ *and using Lemma 3.1 we get from*

$$\begin{aligned}\Phi(t^{1/p}u_0,t^{1/q}v_0)\ &\leq\ \frac{t}{p}\int|\nabla u_0|^p+\frac{t}{q}\int|\nabla v_0|^q-t\int bG(u_0,v_0)+C\\ &=\ t\lambda_1(b)\left\{\frac{1}{p}\int|u_0|^p+\frac{1}{q}\int|v_0|^q\right\}+C,\end{aligned}$$

which goes to $-\infty$ *as* $t\to+\infty$*, in view of* (F_{10})*.*

6 Final comments

There are many open problems connected with the study of the functional Φ depending on other assumptions on the potential F.

(i) Suppose that the growth of F with respect to u is like $|u|^{p^*}$ as $|u| \to \infty$. A similar question with respect to v. The scalar case has been studied by several people, Guedda-Veron [GV], Egnell [E]

(ii) Problems with Ω replaced by R^N. For example, in the scalar case see [0] and references there in.

Acknowledgments This paper was presented by the second author at the Workshop in Partial Differential Equations and Numerical Analysis, at the University of Rio de Janeiro, in October 1996.

References

[CM1] D.G. COSTA, C.A. MAGALHÃES, Variational elliptic problems which are non-quadratic at infinity, *Nonl. Anal. TMA* **23** (1994), 1401–1412.

[FMT] P. FELMER, R.F. MANÁSEVICH, F. DE THÉLIN, Existence and uniqueness of positive solutions for certain quasilinear elliptic systems, *Comm. Part. Dif. Eq.* **17** (1992), 2013–2029.

[V] J.L. VASQUEZ, A strong maximum principle for some quasilinear elliptic equations, *Appl. Math and Optimization* **12** (1984), 191–202.

[dT] F. DE THÉLIN, Première valeur propre d'un système elliptique non linéaire, *C.R. Acad. Sci. paris* **311** (1990), 603–606.

[BFT] L. BOCCARDO, J. FLECKINGER, F. DE THÉLIN, Elliptic systems with various growths, preprint.

[VT] J. VÉLIN, F. DE THÉLIN, Existence and non existence of nontrivial solutions for some nonlinear elliptic systems, *Rev. Mat. Univ. Compl. Madrid* **6** (1993), 153–194.

[O] J.M.B. DO Ó, Nontrivial solutions for perturbations of p-Laplacian in $\mathbb{R}^N$ interacting with the first eigenvalue, preprint.

[T] P. TOLKSDORF, Regularity for a more general class of quasilinear elliptic equations *J. Dif. Eq.* **51** (1984), 126–150.

[CM2] D.G. COSTA, C.A. MAGALHÃES, A variational approach to non-cooperative elliptic systems, *Nonl. Anal. TMA* **25** (1995), 699–715.

[GV] M. GUEDDA, L. VÉRON, Quasilinear elliptic equations involving critical Sobolev exponents, *Nonl. Anal. TMA* **13** (1989), 879–902.

[E] H. EGNELL, Elliptic boundary value problems with singular coefficients and critical nonlinearities, *Indiana Univ. Math. J.* **38** (1989), 235–251.

DISCRETE AND CONTINUOUS
DYNAMICAL SYSTEMS
Volume **8**, Number **3**, July **2002** pp. 563–584

SOLUTIONS OF A NONLINEAR SCHRÖDINGER EQUATION

D.G. deFigueiredo† and Y.H. Ding‡

† Dipartmento de Matematica, IMECC-UNICAMP
13081-970 Campinas S.P., Brasil

‡ Institute of Mathematics, AMSS, Chinese Academy of Sciences
100080 Beijing, P.R. China

(Communicated by A. Ambrosetti)

Abstract. This paper deals with existence and multiplicity of solutions to the nonlinear Schrödinger equation of the type

$$-\Delta u + (\lambda a(x) + a_0(x))u = f(x,u), \qquad u \in H^1(\mathbb{R}^N).$$

We improve some previous results in two respects: we do not require a_0 to be positive on one hand, and allow $f(x,u)$ to be critical nonlinear on the other hand.

1. Introduction and main results

In this paper we study the existence and multiplicity of solutions of nonlinear Schödinger equations

$$-\Delta u + V_\lambda(x)u = g(x,u) \qquad \text{in} \quad \mathbb{R}^N \tag{1.1}$$

which satisfy the condition $u(x) \to 0$ as $|x| \to \infty$. In fact we obtain solutions $u \in H^1(\mathbb{R}^N)$. The potential is assumed to be of the form $V_\lambda(x) = \lambda a(x) + a_0(x)$, where $\lambda > 0$ is a real parameter, and $a(x)$ and $a_0(x)$ satisfy conditions which will be stated shortly. Our work has been motivated by Bartsch-Pankov-Wang [6]. Our results extend those in [6] in two ways. First, we study cases where, for any λ, $V_\lambda(x)$ may be negative somewhere. In [6], $V_\lambda(x) > 0$ for all $x \in \mathbb{R}^N$. In view of our assumptions, the operator $H_\lambda := -\Delta + V_\lambda$, which will be defined precisely in Section 2, has a spectrum with parts below and above 0, so the related functional is in general indefinite. The second relevant feature of our work is the treatment of the critical case. More precisely, the right side of (1.1) has the form $K(x)|u|^{2^*-2}u + g(x,u)$, $2^* := 2N/(N-2)$, with appropriate conditions on g, which then appears as a subcritical perturbation. We emphasize that contrary to [4], [9], [12], [15], our assumptions on the potential do not give that H_λ has a discrete spectrum. Also there is no periodicity on the potential, as in [10] or in [1], [3], [8], [17] and references therein. All through the paper we will be referring to the following basic conditions on the potential:

(V_1) $a_0 \in L^\infty(\mathbb{R}^N)$;
(V_2) $a \in L^\infty_{loc}(\mathbb{R}^N)$, and there is a closed subset $Z \subset \mathbb{R}^N$ with nonempty $\Omega = \text{int} Z$ such that $a(x) = 0$ for $x \in Z$ and $a(x) > 0$ a.e. in $\mathbb{R}^N \backslash Z$.

AMS 2000 Subject Classification: Primary 35Q55; Secondary 58E05.
Key words and phrases: Schrödinger equation, multiple solutions, variational method.
Ding was supported by the Special Funds for Major State Basic Research Projects and the State Education Commission of China under G20000773 and J9904, and the National Research Council of Brazil.

Originally published in: Discrete Cont. Dyn. Syst. 8 (2002), 563-584,

DOI: 10.1007/978-3-319-02856-9_34

(V_3) there is $M_0 > 0$ such that the Lebesgue measure mes $\left(\{x \in \mathbb{R}^N : a(x) \le M_0\}\right) < \infty$.

These conditions will determine completely the spectral properties of the operator H_λ. Those will be described in Section 2.

Let us now state our first result on the existence and multiplicity of solutions of (1.1) in the subcritical case, which we write as

$$\begin{cases} -\Delta u + (\lambda a(x) + a_0(x))u = g(x,u) \\ u \in H^1(\mathbb{R}^N). \end{cases} \tag{P$_\lambda$}$$

We assume the following conditions on the nonlinear term $g(x,u)$.

(g_1) $g \in \mathcal{C}(\mathbb{R}^N \times \mathbb{R}, \mathbb{R})$ and $g(x,u) = o(|u|)$ uniformly in x as $u \to 0$;
(g_2) there is $p \in (2, \ 2^*)$ and $c > 0$ such that $|g(x,u)| \le c\left(1 + |u|^{p-1}\right)$ on $\mathbb{R}^N \times \mathbb{R}$;
(g_3) there is $q > 2$ such that $0 < qG(x,u) \le g(x,u)u$ whenever $u \ne 0$, where $G(x,u) = \int_0^u g(x,s)ds$;
(g_4) $g(x,-u) = -g(x,u)$ for all (x,u).

Theorem 1.1. *Assume $(V_1) - (V_3)$ and $(g_1) - (g_3)$. Then there is $\Lambda > 0$ such that (P$_\lambda$) has at least one nontrivial solution denoted by $\mathcal{F}_\lambda := \{v^\lambda\}$ provided $\lambda \ge \Lambda$. If (g_4) also holds, then for each $k \in \mathbb{N}$ there is $\Lambda_k > 0$ such that (P$_\lambda$) has at least k distinct pairs of solutions denoted by $\mathcal{F}_\lambda := \{\pm v_1^\lambda, \cdots, \pm v_k^\lambda\}$ provided $\lambda \ge \Lambda_k$. In addition, any sequence (u_n), $u_n \in \mathcal{F}_{\lambda_n}$ with $\lambda_n \to \infty$, has a subsequence converging in $H^1(\mathbb{R}^N)$ to a nontrivial solution of the equation*

$$-\Delta u + a_0(x)u = g(x,u) \quad \text{in} \quad \Omega, \qquad u = 0 \quad \text{a.e. on} \quad \mathbb{R}^N \backslash Z. \tag{P}$$

The following results are easy consequences of Theorem 1.1.

Corollary 1.1. *Assume that $a \in \mathcal{C}(\mathbb{R}^N)$, $a(x) \ge 0$ and $\liminf_{|x|\to\infty} a(x) > 0$ such that $\Omega := \text{int}\, a^{-1}(0)$ is nonempty. Then for any $k \in \mathbb{N}$ there is $\lambda_k > 0$ such that, if $\lambda \ge \Lambda_k$, the equation*

$$-\Delta u + \lambda a(x)u = |u|^{p-2}u, \qquad u \in H^1(\mathbb{R}^N) \tag{1.2$_\lambda$}$$

has k distinct pairs of solutions $\mathcal{F}_\lambda := \{\pm v_1^\lambda, \cdots, \pm v_k^\lambda\}$. Moreover, if $\partial\Omega$ is locally Lipschitz, then any sequence (u_n), $u_n \in \mathcal{F}_{\lambda_n}$ with $\lambda_n \to \infty$, has a subsequence converging in $H^1(\mathbb{R}^N)$ to a nontrivial solution of the equation

$$-\Delta u = |u|^{p-2}u, \qquad u \in H_0^1(\Omega). \tag{1.3}$$

Setting $v(x) := \varepsilon^{2/(p-2)}u(x)$ with $\varepsilon^{-2} = \lambda$, (1.2)$_\lambda$ is transformed to

$$-\varepsilon^2\Delta v + a(x)v = |v|^{p-2}v, \qquad v \in H^1(\mathbb{R}^N). \tag{1.4$_\varepsilon$}$$

Such an equation has been studied in many papers, see, e.g., [2], [11], [13], [14], [15] and their references. The following new multiplicity result in this direction follows from Corollary 1.1.

Corollary 1.2. *Assume that $a \in \mathcal{C}(\mathbb{R}^N)$, $a(x) \ge 0$ and $\liminf_{|x|\to\infty} a(x) > 0$ such that $\Omega := \text{int}\, a^{-1}(0)$ is nonempty. Then for any $k \in \mathbb{N}$ there is $\bar{\varepsilon}_k > 0$ such that, if $\varepsilon \in (0, \ \bar{\varepsilon}_k]$, equation (1.4)$_\varepsilon$ has k distinct pairs of solutions.*

It should be pointed out that the solutions of $(1.4)_\varepsilon$ stated in Corollary 1.2 could be quite different from those found in the cited references. The solutions found in the latter papers have a concentration at certain critical point of the potential as $\varepsilon \to 0$. This does not happen, except 0 is a critical point of $a(x)$, because as $\varepsilon = \lambda^{-2} \to 0$, $v_\varepsilon = \varepsilon^{2/(p-2)}u_\lambda \to 0$.

Next we handle the Schrödinger equation with critical nonlinearity:

$$\begin{cases} -\Delta u + (\lambda a(x) + a_0(x))u = K(x)|u|^{2^*-2}u + g(x,u) \\ u \in H^1(\mathbb{R}^N). \end{cases} \tag{Q_λ}$$

In order to describe an additional condition on the potential, let A_0 denote the unique selfadjoint operator on Z associated to the quadratic form $A_0(u,u) = \int_{\mathbb{R}^N} \left(|\nabla u|^2 + a_0 u^2\right)$, with the form domain

$$H_D^1(Z) := \left\{u \in H^1(\mathbb{R}^N) : \quad u(x) = 0 \quad \text{for a.e.} \quad x \in \mathbb{R}^N \backslash Z\right\}.$$

We assume

(V_4) $a_0 \in \mathcal{C}(\mathbb{R}^N)$ and $0 \notin \sigma(A_0)$.

Note that when ∂Z is smooth $H_D^1(Z)$ coincides with $H_0^1(Z)$, and then $A_0 = -\Delta + a_0$ with Dirichlet 0-boundary condition on ∂Z. In particular, if Z is bounded, with $\partial\Omega$ smooth and $a_0(x) \equiv -\mu$, then (V_4) is satisfied provided that μ is not an eigenvalue of $-\Delta$ on Ω with 0-boundary condition. On the critical case, we can use a weak version of (g_3), that is,

$(g_3)'$ $0 \le 2G(x,u) \le g(x,u)u$ on $\mathbb{R}^N \times \mathbb{R}$.

Finally K satisfies the following condition

(K) $K \in \mathcal{C}(\mathbb{R}^N)$, $\inf K(\mathbb{R}^N) > 0$, and there is $x_0 \in \Omega$ with $a_0(x_0) < 0$ such that $K(x_0) = \max K(\mathbb{R}^N)$, $K(x) - K(x_0) = o(|x - x_0|^2)$ as $x \to x_0$.

We will prove the following result.

Theorem 1.2. *Let $N \ge 4$. Assume $(V_1) - (V_4), (g_1) - (g_2), (g_3)'$ and (K). Then there is $\Lambda > 0$ such that (Q_λ) has at least one nontrivial solution u_λ provided $\lambda \ge \Lambda$. Moreover, any sequence $u_n := u_{\lambda_n}$, $\lambda_n \to \infty$, has a subsequence converging in $H^1(\mathbb{R}^N)$ to a nontrivial solution of the equation*

$$-\Delta u + a_0(x)u = K(x)|u|^{2^*-2}u + g(x,u) \quad \text{in} \quad \Omega, \qquad u = 0 \quad \text{a.e. on} \quad \mathbb{R}^N \backslash Z. \tag{Q}$$

As in the subcritical case, we have the following typical example.

Corollary 1.3. *Let $N \ge 4$. Assume that $a \in \mathcal{C}(\mathbb{R}^N)$ satisfies that $a(x) \ge 0$, $\liminf_{|x|\to\infty} a(x) > 0$ and $\Omega := \text{int}\, a^{-1}(0)$ is a nonempty set with $\partial\Omega$ smooth. Then, if $\mu > 0$ is not an eigenvalue of $(-\Delta, H_0^1(\Omega))$, there is $\Lambda > 0$ such that the equation*

$$-\Delta u + \lambda a(x)u = \mu u + |u|^{2^*-2}u, \quad u \in H^1(\mathbb{R}^N)$$

has at least one nontrivial solution u_λ provided $\lambda \ge \Lambda$. Moreover any sequence $u_n := u_{\lambda_n}$, $\lambda_n \to \infty$, has a subsequence converging to a nontrivial solution of the following equation

$$-\Delta u = \mu u + |u|^{2^*-2}u, \quad u \in H_0^1(\Omega).$$

The paper is organized as follows. In Section 2 we study in detail the operator H_λ, its spectrum, and the spectral decompositions of the spaces where it acts. The proofs of the two theorems are variational, and use well known Linking Theorems in a formulation that does not assume a Palais-Smale condition. Theorem 1.1 is proved in Section 3 and Theorem 1.2 in Section 4.

2. Preliminaries. Let $K_R = \{x \in \mathbb{R}^N : |x_i| < R, i = 1, \cdots, N\}$ and $K_R^c = \mathbb{R}^N \backslash \overline{K}_R$. Define

$$\mu(\lambda, R) := \inf_{\psi \in H^1(K_R^c)\backslash\{0\}} \frac{\int_{K_R^c} (|\nabla\psi|^2 + V_\lambda \psi^2)}{|\psi|^2_{L^2(K_R^c)}}.$$

It is known from [6] that (V_3) implies

(V_5) there is $R_j \to \infty$ such that $\lim\limits_{\lambda\to\infty} \lim\limits_{j\to\infty} \mu(\lambda, R_j) = \infty$.

In the lemmas below, we do not use the full force of (V_3). Just (V_5) suffices.

Let

$$E = \left\{ u \in H^1(\mathbb{R}^N) : \int_{\mathbb{R}^N} a(x)u^2 < \infty \right\}$$

endowed with the norm

$$\|u\| = \left(\|u\|^2_{H^1} + \int_{\mathbb{R}^N} a(x)u^2 \right)^{1/2},$$

which is clearly equivalent to the norms

$$\|u\|_\lambda = \left(\|u\|^2_{H^1} + \lambda \int_{\mathbb{R}^N} a(x)u^2 \right)^{1/2}$$

for all $\lambda > 0$.

Let $H_\lambda := -\Delta + V_\lambda$ be the selfadjoint operator on $L^2(\mathbb{R}^N)$ with the form domain $Q(H_\lambda) = E$. Let $(\cdot,\cdot)$ denote the L^2-inner product and write $(H_\lambda u, v) = \int_{\mathbb{R}^N} (\nabla u \nabla v + V_\lambda uv)$ for $u, v \in E$. For $\psi_1, \cdots, \psi_m \in E$, define

$$\begin{aligned} &U_\lambda(\psi_1, \cdots, \psi_m) \\ &= \inf\{(H_\lambda\psi, \psi) : \psi \in E,\ |\psi|_2 = 1,\ (\psi, \psi_j) = 0 \text{ for } j = 1, \cdots, m\}. \end{aligned}$$

For each $k \in \mathbb{N}$, define the spectral values

$$\mu_k^\lambda = \sup_{\psi_1, \cdots, \psi_{k-1}} U_\lambda(\psi_1, \cdots, \psi_{k-1}).$$

Choose a domain $\Omega_0 \subset \Omega$ with $\text{mes}(\Omega_0) < \infty$ and let $L := -\Delta + a_0$ be the selfadjoint operator on $L^2(\Omega_0)$ with Dirichlet boundary condition. Then $\sigma(L)$ consists of eigenvalues $\mu_k(L)$ (all of finite multiplicity) arranged by $-\infty < \mu_1(L) \le \mu_2(L) \le \cdots \le \mu_k(L) \le \cdots \to \infty$ $(k \to \infty)$. Notice that $\mu_k^\lambda \le \mu_k(L)$ for all k. Since μ_k^λ is nondecreasing in λ, we define $\mu_k := \lim_{\lambda\to\infty} \mu_k^\lambda$. Then it follows that

$$\mu_1 \le \mu_2 \le \cdots \quad \text{and} \quad \mu_k \le \mu_k(L) \text{ for every } k.$$

Lemma 2.1. *Assume* $(V_1) - (V_2)$ *and* (V_5). *Then*

(a) $\mu_k \to \infty$ *as* $k \to \infty$;

(b) $\inf \sigma_{ess}(H_\lambda) \to \infty$ *as* $\lambda \to \infty$*;*

(c) for any $k \in \mathbb{N}$ *there is* $\lambda_k > 0$ *such that* H_λ *has at least* k *eigenvalues below the essential spectrum provided* $\lambda \geq \lambda_k$*.*

Proof. Since $a_0 \in L^\infty$, we take $\alpha_0 > 0$ such that $\tilde{H}_\lambda := H_\lambda + \alpha_0 \geq 1$. The conclusions with H_λ replaced by $\tilde{H}_\lambda$ hold by [6]. The lemma then follows readily, because $H_\lambda = \tilde{H}_\lambda - \alpha_0$.

Let k^- and $k^0 \in \mathbb{N}$ be such that $\mu_k < 0$ for all $k \leq k^-$, $\mu_k = 0$ for all $k^- < k \leq k^- + k^0$ and $\mu_k > 0$ for all $k > k^- + k^0$. Set $k^+ = k^- + k^0 + 1$ and fix a number ν with $0 < \nu < \min\{-\mu_{k^-}, \mu_{k^+}\}$. Such numbers exist by *(a)* of Lemma 2.1. We remark that in an extreme (and noninteresting) case one could have $\mu_k > 0$ for all $k \in \mathbb{N}$.

Lemma 2.2. *Assume* $(V_1) - (V_2)$ *and* (V_5)*. Then there is* $\overline{\lambda} > 0$ *such that, for all* $\lambda \geq \overline{\lambda}$, $\dim\ker(H_\lambda) = k^0$, $\mu_k^\lambda < -\nu$ *for all* $k \leq k^-$*, and* $\mu_k^\lambda > \nu$ *for all* $k \geq k^+$*.*

Proof. This follows directly from Lemma 2.1 since $\mu_k^\lambda \to \mu_k$ as $\lambda \to \infty$.

Let e_k^λ with $|e_k^\lambda|_2 = 1$ be the eigenfunction of H_λ associated to the eigenvalue μ_k^λ : $H_\lambda e_k^\lambda = \mu_k^\lambda e_k^\lambda$. There is a direct sum decomposition

$$E = E_\lambda^- \oplus E_\lambda^0 \oplus E_\lambda^+, \qquad u = u^- + u^0 + u^+$$

where $E_\lambda^- = \text{span}\{e_1^\lambda, \cdots, e_{k^-}^\lambda\}$, $E_\lambda^0 = \ker H_\lambda = \text{span}\{e_{k^-+1}^\lambda, \cdots, e_{k^-+k^0}^\lambda\}$. We will also consider the following inner product and norm on E:

$$(u,v)_\lambda = (|H_\lambda|^{1/2}u, |H_\lambda|^{1/2}v) + (u^0, v^0), \qquad |u|_\lambda^2 = (u,u)_\lambda = |u^-|_\lambda^2 + |u^0|_\lambda^2 + |u^+|_\lambda^2.$$

Clearly

$$(H_\lambda u, u) = |u^+|_\lambda^2 - |u^-|_\lambda^2.$$

Letting $\gamma_1 = 1/(1 + |a_0|_\infty)$, $\gamma_2 = (1 + |a_0|_\infty)(1 + 1/\nu)$, using Lemma 2.2, if $\lambda \geq \overline{\lambda}$ one can prove that

$$\gamma_1 |u|_\lambda^2 \leq \|u\|_\lambda^2 \leq \gamma_2 |u|_\lambda^2, \qquad \text{for all } u \in E. \tag{2.1}$$

Since $E \hookrightarrow H^1$ and $H^1 \hookrightarrow L^p$ for all $p \in [2, 2^*]$, it follows that there is $c_p > 0$ such that

$$|u|_p \leq c_p |u|_\lambda, \qquad \text{for all } u \in E, \tag{2.2}$$

provided $\lambda \geq \overline{\lambda}$.

Lemma 2.3. *Assume* $(V_1) - (V_2)$ *and* (V_5)*. For each* $k \in \mathbb{N}$*, there is* $d_k > 0$*, independent of* λ*, such that* $|e_k^\lambda|_\infty \leq d_k$ *for all* $\lambda \geq \overline{\lambda}$*.*

Proof. We adapt a bootstrap argument similar to the one in [6]. By the L^p elliptic regularity theory $e_k^\lambda \in L^\infty(\mathbb{R}^N)$. The issue here is proving that, for each fixed $k \in \mathbb{N}$, the e_k^λ are uniformly bounded in $L^\infty(\mathbb{R}^N)$, for all λ. Let $\alpha_0 = 1 - \inf a_0(\mathbb{R}^N)$ and $\tilde{H}_\lambda := H_\lambda + \alpha_0$. Then $\tilde{H}_\lambda \geq 1$ and $\tilde{H}_\lambda e_k^\lambda = (\mu_k^\lambda + \alpha_0)e_k^\lambda$. We only consider the case $N \geq 3$ (it is simpler for $N = 1, 2$). It is known that the operator $\tilde{H}_\lambda^{-1} : L^s(\mathbb{R}^N) \to L^r(\mathbb{R}^N)$ is bounded linear for all $1 \leq s \leq r \leq \infty$ with $\frac{1}{s} - \frac{1}{r} < \frac{2}{N}$, and that there is $C_{r,s} > 0$ such that $|\tilde{H}_\lambda^{-1}u|_r \leq C_{r,s}|u|_s$, for all λ and $u \in L^s(\mathbb{R}^N)$, cf. [16] or [6]. Let χ_0^λ be the characteristic function of the set

$\{x \in \mathbb{R}^N : |e_k^\lambda(x)| \le 1\}$ and set $u_0^\lambda = \chi_0^\lambda e_k^\lambda$, $v_0^\lambda = e_k^\lambda - u_k^\lambda$. Observe that $|u_0^\lambda|_\infty \le 1$ and $|v_0^\lambda|_{2^*} \le c\|v_0^\lambda\|_{H^1} \le c\gamma_2 |v_0^\lambda|_\lambda \le c\gamma_2(1+|\mu_k|)|v_0^\lambda|_2 = c\gamma_2(1+|\mu_k|)$. Let $p_0 = 2^*$ and $u_1^\lambda = (\mu_k^\lambda + \alpha_0)\tilde{H}_\lambda^{-1}u_0^\lambda$, $v_1^\lambda = (\mu_k^\lambda + \alpha_0)\tilde{H}_\lambda^{-1}v_0^\lambda$. Then $e_k^\lambda = u_1^\lambda + v_1^\lambda$ and there is $C_1 > 0$ depending on k, but independent of λ, such that $|u_1^\lambda|_\infty \le C_1$ and $|v_1^\lambda|_{p_1} \le C_1$ for all λ, where $p_1 = \infty$ if $p_0 > N/2$, and p_1 is arbitrarily close to and less than $p_0 N/(N-2p_0)$ if $p_0 \le N/2$. The lemma then follows if $p_0 > N/2$. If $p_0 \le N/2$, we iterate the previous argument. Observe that, in this case $p_1 > p_0$, and so we have improved the estimate of v_1^λ. Let χ_1^λ be the characteristic function of the set $A_\lambda := \{x \in \mathbb{R}^N : |v_1^\lambda(x)| \le C_1 + 1\}$, and define $\overline{u}_1^\lambda = \chi_1^\lambda e_k^\lambda$, $\overline{v}_1^\lambda = (1-\chi_1^\lambda)e_k^\lambda = (1-\chi_1^\lambda)(u_1^\lambda + v_1^\lambda)$. Clearly, $|\overline{u}_1^\lambda|_\infty$ is bounded uniformly in λ. Note that

$$\text{mes}(A_\lambda^c) \le \frac{1}{(C_1+1)^{p_1}} \int_{A_\lambda^c} |v_1^\lambda|^{p_1} \le \left(\frac{C_1}{C_1+1}\right)^{p_1},$$

$$|(1-\chi_1^\lambda)u_1^\lambda|_{p_1} \le |u_1^\lambda|_\infty \left(\text{mes}(A_\lambda^c)\right)^{1/p_1},$$

and $|(1-\chi_1^\lambda)v_1^\lambda|_{p_1} \le |v_1^\lambda|_{p_1}$. Hence $|\overline{v}_1^\lambda|_{p_1}$ is bounded uniformly in λ. Now letting $u_2^\lambda = (\mu_k^\lambda + \alpha_0)\tilde{H}_\lambda^{-1}\overline{u}_1^\lambda$, $v_2^\lambda = (\mu_k^\lambda + \alpha_0)\tilde{H}_\lambda^{-1}v_1^\lambda$, and repeating the above argument we obtain $C_2 > 0$ independent of λ such that

$$|u_2^\lambda|_\infty \le C_2 \qquad \text{and} \qquad |v_2^\lambda|_{p_2} \le C_2$$

where $p_2 = \infty$ if $p_1 > N/2$, and p_2 is arbitrarily close to and less than $p_1 N/(N-2p_1)$. After a finite number of such steps we obtain the desired conclusion.

For every $m \in \mathbb{N}$ set

$$E_m^\lambda := \text{span}\{e_1^\lambda, \cdots, e_m^\lambda\}.$$

Lemma 2.4. *Assume* $(V_1)-(V_2)$ *and* (V_5). *For each* m *there exist* $0 < a_m \le b_m$, *both independent of* λ, *such that* $a_m|u|_2^2 \le |u|_\lambda^2 \le b_m|u|_2^2$ *for all* $u \in E_m^\lambda$ *provided* $\lambda \ge \overline{\lambda}$. *Moreover, for each* $p \in [2, \infty]$ *there is* $0 < a_m(p) \le b_m(p)$ *such that* $a_m(p)|u|_2^2 \le |u|_p^2 \le b_m(p)|u|_2^2$ *for all* $u \in E_m^\lambda$ *provided* $\lambda \ge \overline{\lambda}$.

Proof. For each $u \in E_m^\lambda$, $u = \sum_{i=1}^m c_i e_i^\lambda$ and $|u|_2^2 = \sum_{i=1}^m c_i^2$. Write $u = u^- + u^0 + u^+$. Then

$$\begin{aligned} |u|_\lambda^2 &= -\sum_{i=1}^{k^-} \mu_i^\lambda c_i^2 + |u^0|_2^2 + \sum_{i=k^+}^{m} \mu_i^\lambda c_i^2 \\ &\ge \nu|u^-|_2^2 + |u^0|_2^2 + \mu_{k^+}^\lambda |u^+|_2^2 \\ &\ge \nu|u^-|_2^2 + |u^0|_2^2 + \nu|u^+|_2^2 \end{aligned}$$

and

$$\begin{aligned} |u|_\lambda^2 &\le |a_0|_\infty |u^-|_2^2 + |u^0|_2^2 + \mu_m^\lambda |u^+|_2^2 \\ &\le |a_0|_\infty |u^-|_2^2 + |u^0|_2^2 + \mu_m |u^+|_2^2. \end{aligned}$$

Clearly, the constants $a_m = \min\{\nu, 1\}$ and $b_m = \max\{|a_0|_\infty, 1, \mu_m\}$ are as required.

By Lemma 2.3 we have, with $\alpha_m := \max\{d_1, \cdots, d_m\}$:

$$|u|_\infty \le \sum_{i=1}^m |c_i||e_i^\lambda|_\infty \le |u|_2 \left(\sum_{i=1}^m |e_i^\lambda|_\infty^2\right)^{1/2} \le m^{1/2}\alpha_m |u|_2.$$

For $p \in (2, \infty)$,

$$\int_{\mathbb{R}^N} |u|^p = |u|_\infty^{p-2}|u|_2^2 \le \left(m\alpha_m^2\right)^{(p-2)/2} |u|_2^p.$$

Letting $b_m(p) := \left(m\alpha_m^2\right)^{(p-2)/p}$ we have $|u|_p^2 \le b_m(p)|u|_2^2$.

Recall the following estimate, valid for all $u \in E$:

$$\begin{aligned}\mu(\lambda, R_j)\int_{K_{R_j}^c} u^2 &\le \int_{K_{R_j}^c} \left(|\nabla u|^2 + \lambda a u^2 + a_0 u^2\right)\\ &= (H_\lambda u, u) - \int_{K_{R_j}} \left(|\nabla u|^2 + \lambda a u^2 + a_0 u^2\right)\\ &\le (H_\lambda u, u) + |a_0|_\infty \int_{K_{R_j}} u^2.\end{aligned} \tag{2.3}$$

Hence, letting $|F|$ =mes(F) and using (2.3), one obtains

$$\begin{aligned}|u|_p^2 &\ge \left(\int_{K_{R_j}} |u|^p\right)^{2/p} \ge |K_{R_j}|^{(2-p)/p}\int_{K_{R_j}} u^2\\ &\ge |K_{R_j}|^{(2-p)/p}\left(|u|_2^2 - \int_{K_{R_j}^c} u^2\right)\\ &\ge |K_{R_j}|^{(2-p)/p}\left(|u|_2^2 - \frac{(H_\lambda u, u)}{\mu(\lambda, R_j)} - \frac{|a_0|_\infty}{\mu(\lambda, R_j)}\int_{K_{R_j}} u^2\right)\\ &\ge \frac{1}{|K_{R_j}|^{(p-2)/p}}\left(\left(1 - \frac{|a_0|_\infty}{\mu(\lambda, R_j)}\right)|u|_2^2 - \frac{(H_\lambda u, u)}{\mu(\lambda, R_j)}\right).\end{aligned} \tag{2.4}$$

For $u \in E_m^\lambda$, since $|(H_\lambda u, u)| \le (|H_\lambda|u, u) = |u|_\lambda^2 \le b_m|u|_2^2$, (2.4) implies

$$|u|_p^2 \ge \frac{1}{|K_{R_j}|^{(p-2)/p}}\left(1 - \frac{|a_0|_\infty + b_m}{\mu(\lambda, R_j)}\right)|u|_2^2.$$

Now by (V_5) we can take $R = R_{j_0}$ large and $\overline{\lambda}$ also large such that

$$a_m(p) := \frac{1}{|K_R|^{(p-2)/p}}\left(1 - \frac{|a_0|_\infty + b_m}{\mu(\overline{\lambda}, R)}\right) > 0.$$

Since $\mu(\lambda, K_R)$ is nondecreasing in λ, for $\lambda \ge \overline{\lambda}$ there holds $a_m(p)|u|_2^2 \le |u|_p^2$ for all $u \in E_m^\lambda$. This finishes the proof.

The last lemma of this section shows that if (V_4) also holds, then $E_\lambda^0 = \{0\}$. This fact will be used for dealing with the critical nonlinearity case.

Lemma 2.5. *Assume* $(V_1)-(V_2)$ *and* $(V_4)-(V_5)$. *Then* $[-\nu, \nu]\cap\sigma(H_\lambda) = \emptyset$, *hence* $E_\lambda^0 = \{0\}$ *for all* $\lambda \ge \overline{\lambda}$.

Proof. Since $[-\nu, \nu] \cap \sigma(H_\lambda) \subset \{0\}$ by Lemma 2.2, it remains to show that $0 \notin \sigma(H_\lambda)$ for all $\lambda \ge \overline{\lambda}$.

Arguing indirectly, assume that there are $\lambda_n \to \infty$ and $(u_n) \subset E$ with $H_{\lambda_n} u_n = 0$ and $|u_n|_2 = 1$. It follows from (2.1) and Lemma 2.4 that $\|u_n\|_{\lambda_n}^2 \le \gamma_2 |u_n|_{\lambda_n}^2 = \gamma_2 |u_n|_2^2 = \gamma_2$. We can suppose $u_n \rightharpoonup \overline{u}$ in $H^1(\mathbb{R}^N)$. Set

$$C_k = \{x : \ |x| \le k, \ a(x) \ge 1/k\}, \qquad k \in \mathbb{N}.$$

Then

$$\int_{C_k} |u_n|^2 \le k \int_{C_k} a u_n^2 \le \frac{k\gamma_2}{\lambda_n} \to 0,$$

hence, $\int_{C_k} \overline{u}^2 = 0$ for every k. This implies $\overline{u}(x) = 0$ for a.e. $x \in R^N \backslash Z$, i.e., $\overline{u} \in H_D^1(Z)$. Since $a|_Z = 0$, $0 = (H_{\lambda_n} u_n, v) = \int_{\mathbb{R}^N} (\nabla u_n \nabla v + a_0 u_n v) \to A(\overline{u}, v)$, i.e., $A(\overline{u}, v) = 0$ for any $v \in H_D^1(Z)$. This implies that $\overline{u} = 0$ by (V_4).

It follows from (2.3) and $H_{\lambda_n} u_n = 0$ that

$$\mu(\lambda_n, R_j) \int_{K_{R_j}^c} u_n^2 \le |a_0|_\infty \int_{K_{R_j}} u_n^2 \le |a_0|_\infty.$$

Using (V_5) we can find j_0 such that $\lim_{n\to\infty} \int_{K_{R_{j_0}}^c} u_n^2 < \frac{1}{2}$. Since $H^1 \hookrightarrow L_{loc}^2$ compactly, $\int_{K_{R_{j_0}}} u_n^2 \to 0$, and one sees

$$1 = \lim_{n\to\infty} |u_n|_2^2 < \frac{1}{2},$$

which is a contradiction.

3. The subcritical case In this section we provr Theorem 1.1. In fact, we shall prove the following more general results.

Theorem 3.1. *Assume $(V_1) - (V_2), (V_5)$ and $(g_1) - (g_3)$. Then there exist $\Lambda > 0$ and $0 < \alpha_0 \le \beta_0$, independent of λ, such that (P_λ) has a solution $u^\lambda \in E$ satisfying $\alpha_0 \le \int_{\mathbb{R}^N} g(x, u^\lambda) u^\lambda \le \beta_0$ for all $\lambda \ge \Lambda$.*

Theorem 3.2. *Assume $(V_1)-(V_2), (V_5)$ and $(g_1)-(g_4)$. Then for every $k \in \mathbb{N}$ there is $\lambda_k > 0$ and $0 < \alpha_k \le \beta_k$, independent of λ, such that (P_λ) has at least k distinct pairs of solutions $\pm u_j^\lambda \in E$ satisfying $\alpha_k \le \int_{\mathbb{R}^N} g(x, u_j^\lambda) u_j^\lambda \le \beta_k$ for $j = 1, \cdots, k$ and all $\lambda \ge \lambda_k$.*

Theorem 3.3. *Assume $(V_1) - (V_2), (V_5)$ and $(g_1) - (g_3)$. Let u_n be a solution of (P_{λ_n}) with $\lambda_n \to \infty$ such that $\alpha \le \int_{\mathbb{R}^N} g(x, u_n) u_n \le \beta$ for some constants $0 < \alpha \le \beta$. Then along a subsequence $u_n \to u$ in $H^1(\mathbb{R}^N)$ with $u \ne 0$ and solving* (P).

Consider the functional

$$I_\lambda(u) = \frac{1}{2}(H_\lambda u, u) - \int_{\mathbb{R}^N} G(x, u) = \frac{1}{2}|u^+|_\lambda^2 - \frac{1}{2}|u^-|_\lambda^2 - \int_{\mathbb{R}^N} G(x, u).$$

Then $I_\lambda \in \mathcal{C}^1(E, \mathbb{R})$ and critical points of I_λ are solutions of (P_λ).

The next two lemmas establish the geometry of the Linking Theorem [18] for the functional I_λ.

Lemma 3.4. *Assume $(V_1)-(V_2)$, (V_5) and $(g_1)-(g_3)$. For any $m\in\mathbb{N}$ there are λ_m and R_m, $\beta_m>0$, independent of λ, such that $I_\lambda(u)\le 0$ for all $u\in E_m^\lambda\backslash B_{R_m}$ and $\max I_\lambda\left(E_m^\lambda\cap B_{R_m}\right)\le\beta_m$ provided $\lambda\ge\lambda_m$.*

Proof. In the following d or d_i will denote positive constants independent of λ, however, they may depend on m.

By $(g_1)-(g_3)$ for any $\varepsilon>0$ there is $C_\varepsilon>0$ such that

$$G(x,u)\ge C_\varepsilon|u|^q-\varepsilon u^2. \tag{3.1}$$

Thus,

$$I_\lambda(u)\le\frac{1}{2}|u|_\lambda^2+\varepsilon|u|_2^2-C_\varepsilon|u|_q^q$$

for all $u\in E$. Fixing ε, it follows by Lemma 2.4 that for $u\in E_m^\lambda$ one has

$$I_\lambda(u)\le d_1|u|_2^2-d_2|u|_2^q\le d_3|u|_\lambda^2-d_4|u|_\lambda^q$$

and then the lemma follows because $q>2$.

Fix $e^\lambda=e_{k^+}^\lambda$, $R=R_{k^+}$ and set

$$Q^\lambda:=\left\{u=u^-+u^0+se^\lambda:\ u^-+u^0\in E_\lambda^-\oplus E_\lambda^0,\ s\ge 0,\ |u|_\lambda\le R\right\}.$$

Let $\beta=\beta_{k^+}$ and $\Lambda=\lambda_{k^+}$. As a consequence of Lemma 3.4 we have

Lemma 3.5. *Under the assumptions of Lemma 3.4, $\max I_\lambda(Q^\lambda)\le\beta$ and $I_\lambda(u)\le 0$ for all $u\in\partial Q^\lambda$ provided $\lambda\ge\Lambda$.*

Remark that $|u|_2^2\le\nu^{-1}|u|_\lambda^2$ for all $u\in E_\lambda^+$ whenever $\lambda\ge\overline{\lambda}$. By $(g_1)-(g_3)$ there is $d>0$ such that

$$G(x,u)\le\frac{\nu}{4}u^2+d|u|^p\qquad\text{for all }(x,u).$$

Thus

$$\begin{aligned}I_\lambda(u)&=\frac{1}{2}|u|_\lambda^2-\int_{\mathbb{R}^N}G(x,u)\ge\frac{1}{2}|u|_\lambda^2-\frac{\nu}{4}|u|_2^2-d|u|_p^p\\&\ge\frac{1}{4}|u|_\lambda^2-d|u|_p^p\end{aligned}\tag{3.2}$$

for all $u\in E_\lambda^+$ whenever $\lambda\ge\overline{\lambda}$.

In the following we always assume that $m\ge k^+$. Set

$$F_m^\lambda=(E_{m-1}^\lambda)^\perp,$$

the orthogonal complement of E_{m-1}^λ in E with respect to $(\cdot,\cdot)_\lambda$. Consequently $F_m^\lambda\subset E_\lambda^+$, and we shall use (3.2) in order to estimate I_λ in a small sphere about the origin in F_m^λ, see Lemma 3.6 next. Indeed, by definition one has

$$\mu_m^\lambda|u|_2^2\le|u|_\lambda^2\qquad\text{for all }u\in F_m^\lambda.$$

This, together with (2.1) and the Gagliardo-Nirenberg inequality ($\theta=(2p+2N-Np)/2p$)

$$|u|_p\le c\|u\|_{H^1}^{1-\theta}|u|_2^\theta\qquad\text{for all }u\in H^1,$$

shows that there is $d_1 > 0$ such that

$$|u|_p \le d_1(\mu_m^\lambda)^{-\theta/2}|u|_\lambda \qquad \text{for all } u \in F_m^\lambda$$

whenever $\lambda \ge \overline{\lambda}$. Therefore by (3.2),

$$I_\lambda(u) \ge \frac{1}{4}|u|_\lambda^2 - \frac{dd_1^p}{|\mu_m^\lambda|^{p\theta/2}}|u|_\lambda^p$$

for all $u \in F_m^\lambda$. With $d_2 = (2pdd_1^p)^{-1/(p-2)}$ we set

$$r_m^\lambda = d_2|\mu_m^\lambda|^{p\theta/2(p-2)} \quad \text{and} \quad \alpha_m^\lambda = \left(\frac{1}{4} - \frac{1}{2p}\right) d_2^2|\mu_m^\lambda|^{p\theta/(p-2)}.$$

Remark that r_m^λ and α_m^λ are nondecreasing in λ and k. Moreover by (V_5) one has $\alpha_m^\lambda \to \infty$ as λ, $m \to \infty$. The above arguments yield the following

Lemma 3.6. *Under the assumptions of Lemma 3.4, if $\lambda \ge \overline{\lambda}$ then*

$$I_\lambda(u) \ge \alpha_m^\lambda \qquad \text{for all } u \in \partial B_{r_m^\lambda} \cap F_m^\lambda.$$

In particular, letting $\alpha = \alpha_{k+}^{\overline{\lambda}}$ and $\rho = r_{k+}^{\overline{\lambda}}$, we get

Lemma 3.7. *Under the assumptions of Lemma 3.4, for all $\lambda \ge \overline{\lambda}$,* $\inf I_\lambda(B_\rho \cap E_\lambda^+) \ge 0$, $\inf I_\lambda(\partial B_\rho \cap E_\lambda^+) \ge \alpha$.

We now turn to the analysis of Palais-Smale sequences. Recall that a sequence $(u_n) \subset E$ is called a $(\mathrm{PS})_c$ sequence for I_λ if $I_\lambda(u_n) \to c$ and $I'_\lambda(u_n) \to 0$ as $n \to \infty$.

Lemma 3.8. *Under the assumptions of Lemma 3.4, given $b > 0$ there exists $d = d(b) > 0$ such that if $\lambda \ge \overline{\lambda}$ and (u_n) is a $(\mathrm{PS})_c$ sequence for I_λ with $c \le b$, then (u_n) is bounded and*

$$\limsup_{n\to\infty} \|u_n\|_\lambda^2 \le d(b). \tag{3.3}$$

Moreover, $c \ge 0$ and $c = 0$ if and only if $u_n \to 0$.

Proof. Since $I_\lambda(u_n) \to c$ and $I'_\lambda(u_n) \to 0$, we have

$$\begin{aligned} &I_\lambda(u_n) + o(1)|u_n|_\lambda \\ \ge&\frac{q-2}{2q}\int_{\mathbb{R}^N} g(x,u_n)u_n \\ =&\frac{q-2}{2q}\left(\int_{|u_n(x)|\le 1} g(x,u_n)u_n + \int_{|u_n(x)|>1} g(x,u_n)u_n\right). \end{aligned} \tag{3.4}$$

Conditions $(g_1) - (g_3)$ imply that

$$|g(x,u)|^2 \le d_1 g(x,u)u \qquad \text{if } |u| \le 1,$$

and for $p' := p/(p-1)$

$$|g(x,u)|^{p'} \le d_1 g(x,u)u \qquad \text{if } |u| \ge 1.$$

Thus, using (3.4), for n large

$$\begin{aligned}|u_n^+|_\lambda^2 \le & o(1)|u_n^+|_\lambda + \left(\int_{|u_n(x)|\le 1} |g(x,u_n)|^2\right)^{1/2} |u_n^+|_2 \\ & + \left(\int_{|u_n(x)|>1} |g(x,u_n)|^{p'}\right)^{1/p'} |u_n^+|_p \\ \le & o(1)|u_n^+|_\lambda + \left(\int_{|u_n(x)|\le 1} d_1 g(x,u_n)u_n\right)^{1/2} |u_n^+|_2 \\ & + \left(\int_{|u_n(x)|>1} d_1 g(x,u_n)u_n\right)^{1/p'} |u_n^+|_p \\ \le & o(1)|u_n^+|_\lambda + d_2\left(b^{1/2} + b^{1/p'} + |u_n|_\lambda^{1/2} + |u_n|_\lambda^{1/p'}\right)|u_n^+|_\lambda,\end{aligned}$$

or

$$|u_n^+|_\lambda \le d_2\left(1 + b^{1/2} + b^{1/p'} + |u_n|_\lambda^{1/2} + |u_n|_\lambda^{1/p'}\right). \tag{3.5}$$

Similarly,

$$|u_n^-|_\lambda \le d_2\left(1 + b^{1/2} + b^{1/p'} + |u_n|_\lambda^{1/2} + |u_n|_\lambda^{1/p'}\right). \tag{3.6}$$

Using (3.1)

$$\begin{aligned}I_\lambda(u_n) + o(1)|u_n|_\lambda &\ge \frac{q-2}{2}\int_{\mathbb{R}^N} G(x,u_n) \\ &\ge \frac{q-2}{2}\left(C_\varepsilon |u_n|_q^q - \varepsilon|u_n|_2^2\right).\end{aligned}$$

Hence for n large

$$|u_n|_q^q \le d_3\left(1 + b + |u_n|_\lambda^2\right).$$

Since $|u_n^0|_q \le |u_n|_q + |u_n^- + u_n^+|_q$, we have

$$|u_n^0|_2 \le d_4\left(1 + b^{1/q} + |u_n|^{2/q}\right) + d_4|u_n^- + u_n^+|_\lambda,$$

which together with (3.5) and (3.6) shows that (u_n) is bounded, and so there is a constant $d(b) > 0$, independent of λ, such that (3.3) holds.

A passage to the limit in (3.4) shows that $c \ge 0$. It is apparent that if $u_n \to 0$ in E then $c = 0$. On the other hand, suppose $c = 0$. (3.4) implies that $\int_{\mathbb{R}^N} g(x,u_n)u_n \to 0$, hence $\int_{|u_n(x)|\le 1} g(x,u_n)^2 \to 0$, $\int_{|u_n(x)|>1} |g(x,u_n)|^{p'} \to 0$. We get

$$|u_n^\pm|_\lambda^2 = o(1) \pm \int_{\mathbb{R}^N} g(x,u_n)u_n^\pm \to 0.$$

Since $\dim E_\lambda^0 < \infty$ we have also $u_n^0 \to 0$. Thus $u_n \to 0$.

As a consequence, we see that for any $(\mathrm{PS})_c$ sequence (u_n) for I_λ $(\lambda \ge \overline{\lambda})$ the following relations hold

$$\limsup_{n\to\infty} \int_{R^N} g(x,u_n)u_n = \limsup_{n\to\infty} (H_\lambda u_n, u_n) \le \frac{2qc}{q-2}, \tag{3.7}$$

and

$$2c \le \liminf_{n\to\infty} \int_{R^N} g(x, u_n)u_n = \liminf_{n\to\infty} (H_\lambda u_n, u_n). \tag{3.8}$$

Lemma 3.9 *Under the assumptions of Lemma 3.4, given $b > 0$, for any $\varepsilon > 0$ there exist $\Lambda_\varepsilon, R_\varepsilon > 0$ such that if (u_n) is a $(\mathrm{PS})_c$ sequence for I_λ with $\lambda \ge \Lambda_\varepsilon$ and $c \le b$, then*

$$\limsup_{n\to\infty} \int_{K^c_{R_\varepsilon}} (|u_n|^2 + |u_n|^p) < \varepsilon.$$

Proof. Note that

$$\begin{aligned}\mu(\lambda, R_j) \int_{K^c_{R_j}} u_n^2 &\le (H_\lambda u_n, u_n) - \int_{K_{R_j}} (|\nabla u_n|^2 + \lambda a u_n^2 + a_0 u_n^2) \\ &\le (H_\lambda u_n, u_n) + |a_0|_\infty |u_n|_2^2\end{aligned}$$

or

$$\int_{K^c_{R_j}} u_n^2 \le \frac{1}{\mu(\lambda, R_j)} \left((H_\lambda u_n, u_n) + |a_0|_\infty \|u_n\|^2_\lambda\right).$$

Then

$$\limsup_{n\to\infty} \int_{K^c_{R_j}} u_n^2 \le \frac{1}{\mu(\lambda, R_j)} \left(\frac{2qb}{q-2} + |a_0|_\infty d(b)\right). \tag{3.9}$$

It follows from the Gagliardo-Nirenberg inequality that

$$\begin{aligned}\int_{K^c_{R_j}} |u_n|^p &\le d|\nabla u_n|^{p(1-\theta)}_{L^2(K^c_{R_j})} |u_n|^{p\theta}_{L^2(K^c_{R_j})} \\ &\le d|u_n|^{p\theta}_{L^2(K^c_{R_j})}\end{aligned} \tag{3.10}$$

where we have used the boundedness of the $(\mathrm{PS})_c$ sequence. The conclusion then follows from (3.9)-(3.10) and (V_5).

We are finally ready to prove Theorem 3.1.

Proof of Theorem 3.1. By virtue of the linking theorem, Lemmas 3.5 and 3.7 imply that for λ large, I_λ has a $(\mathrm{PS})_c$ sequence (u_n^λ) with $c \in [\alpha, \beta]$. By Lemma 3.8 we can assume that $u_n^\lambda \rightharpoonup u$ as $n \to \infty$. Clearly, u is a solution of (P_λ). All we have to do is to show that $u \ne 0$. Note that it follows from $(g_1) - (g_3)$ that there is $d > 0$ satisfying $g(x,u)u \le u^2 + d|u|^p$ on $\mathbb{R}^N \times \mathbb{R}$. By Lemma 3.9, for $\varepsilon = c/(1+d)$ there exist $\Lambda > 0$ and $R > 0$ such that for all $\lambda \ge \Lambda$

$$\limsup_{n\to\infty} \int_{K^c_R} g(x, u_n)u_n \le \limsup_{n\to\infty} \int_{K^c_R} (u_n^2 + d|u_n|^p) < c.$$

This, together with (3.7)-(3.8), implies that

$$\begin{aligned}2c &\le \liminf_{n\to\infty} \int_{\mathbb{R}^N} g(x, u_n)u_n \\ &= \liminf_{n\to\infty} \left(\int_{K_R} g(x, u_n)u_n + \int_{K^c_R} g(x, u_n)u_n\right) \\ &\le \liminf_{n\to\infty} \int_{K_R} g(x, u_n)u_n + c \\ &= \int_{K_R} g(x, u)u + c\end{aligned}$$

and hence

$$\alpha_0 := \alpha \le c \le \int_{K_R} g(x,u)u \le \int_{\mathbb{R}^N} g(x,u)u \le \beta_0 := \frac{2q\beta}{q-2}.$$

The proof is complete.

Next we turn to the proof of Theorem 3.2. Fix $m \ge k^+$ and set $B_m^\lambda = \{u \in E_m^\lambda : |u|_\lambda \le R_m\}$ where R_m is as in Lemma 3.4. Define

$$\Gamma_m^\lambda = \{\gamma \in \mathcal{C}(B_m^\lambda, E) : \ \gamma \text{ is odd and } \gamma(u) = u \text{ if } |u|_\lambda = R_m\},$$

and

$$c_m^\lambda = \inf_{\gamma \in \Gamma_m^\lambda} \max_{u \in B_m^\lambda} I_\lambda(\gamma(u)).$$

It follows from Lemma 3.4 that

$$c_m^\lambda \le \beta_m = \max I_\lambda(B_m^\lambda) \quad \text{for all } m \in \mathbb{N},\ \lambda \ge \lambda_m.$$

A standard deformation argument yields a sequence $(u_{m,n}^\lambda)_{n\in\mathbb{N}}$ satisfying

$$I_\lambda(u_{m,n}^\lambda) \to c_m^\lambda \quad \text{and} \quad I'_\lambda(u_{m,n}^\lambda) \to 0 \quad \text{uniformly in } m \text{ and } \lambda \ge \lambda_m \text{ as } n \to \infty.$$

Lemma 3.8 shows that $(u_{m,n}^\lambda)_{n\in\mathbb{N}}$ is bounded and we can suppose $u_{m,n}^\lambda \rightharpoonup u_m^\lambda$ in E and $u_{m,n}^\lambda \to u_m^\lambda$ in $L_{loc}^s(\mathbb{R}^N)$ as $n \to \infty$. Clearly, u_m^λ is a weak solution of (P_λ). And again we shall prove that $u_m^\lambda \ne 0$.

Lemma 3.10. *Assume* $(V_1) - (V_2)$, (V_5) *and* $(g_1) - (g_4)$. *If* $\lambda \ge \lambda_m$, *then for every* $\gamma \in \Gamma_m^\lambda$, $\gamma(B_m^\lambda) \cap \partial B_{r_m^\lambda} \cap F_m^\lambda \ne \emptyset$, *where* $B_{r_m^\lambda}$ *is as in Lemma 3.6 above. Consequently,* $c_m^\lambda \ge \alpha_m^\lambda$.

Proof. Let

$$\mathcal{O} = \{u \in B_m^\lambda : \ |\gamma(u)|_\lambda < r_m^\lambda\}$$

and let $P_m : E \to E_m^\lambda$ be the orthogonal projection with respect to $(\cdot,\cdot)_\lambda$. Since γ is odd, $0 \in \mathcal{O}$ and $\mathcal{O}$ is an open neighborhood of 0 in E_m^λ. Clearly $\overline{\mathcal{O}} \subset \mathrm{int}(B_m^\lambda)$. The Borsuk-Ulam theorem yields $u \in \partial\mathcal{O}$ such that $P_m\gamma(u) = 0$, hence, $\gamma(u) \in F_m^\lambda$ and $|\gamma(u)|_\lambda = r_m^\lambda$. The desired conclusion follows.

Now we choose a sequence of integers $k^+ \le m_1 < m_2 < \cdots \to \infty$ such that

$$\frac{2qc_{m_j}^\lambda}{q-2} \le \frac{c_{m_{j+1}}^\lambda}{2}.$$

This is possible by Lemma 3.10 and the fact that $\alpha_m^\lambda \to \infty$ as $\lambda, m \to \infty$.

Lemma 3.11 *Under the assumptions of Lemma 3.4, for any* $k \in \mathbb{N}$, *there are* λ_k, $R_k > 0$ *such that*

$$\limsup_{n\to\infty} \int_{K_{R_k}^c} g(x, u_{m_j,n}^\lambda)u_{m_j,n}^\lambda \le c_{m_j}^\lambda \qquad \text{for } j = 1, \cdots, k,\ \lambda \ge \lambda_k.$$

Proof. Similarly to Lemma 3.9, using (3.9) with $b = \beta_{m_k}$ and (3.10) we find λ_k and R_k such that

$$\limsup_{n\to\infty} \int_{K_{R_k}^c} |u_{m_j,n}^\lambda|^2 < c_{m_j}^\lambda \qquad \text{and} \qquad \limsup_{n\to\infty} \int_{K_{R_k}^c} |u_{m_j,n}^\lambda|^p < c_{m_j}^\lambda$$

for all $j \le k$ and $\lambda \ge \lambda_k$. Since $g(x,u)u \le u^2 + d|u|^p$ on $\mathbb{R}^N \times \mathbb{R}$, we obtain the desired conclusion by enlarging λ_k and R_k if necessary.

Proof of Theorem 3.2. By Lemma 3.11,

$$\begin{aligned}\int_{K_{R_k}} g(x,u_{m_j}^\lambda)u_{m_j}^\lambda &= \lim_{n\to\infty} \int_{K_{R_k}} g(x,u_{m_j,n}^\lambda)u_{m_j,n}^\lambda \\ &= \left(\int_{\mathbb{R}^N} g(x,u_{m_j,n}^\lambda)u_{m_j,n}^\lambda - \int_{K_{R_k}^c} g(x,u_{m_j,n}^\lambda)u_{m_j,n}^\lambda \right) \\ &\ge 2c_{m_j}^\lambda - c_{m_j}^\lambda = c_{m_j}^\lambda .\end{aligned}$$

Therefore, for $\lambda \ge \lambda_k$ and $j = 1, \cdots, k$, using Fatou's lemma,

$$\begin{aligned} c_{m_j}^\lambda &\le \int_{K_{R_k}} g(x,u_{m_j}^\lambda)u_{m_j}^\lambda \\ &\le \liminf_{n\to\infty} \int_{\mathbb{R}^N} g(x,u_{m_j,n}^\lambda)u_{m_j,n}^\lambda \\ &\le \frac{2qc_{m_j}^\lambda}{q-2} \le \frac{c_{m_{j+1}}^\lambda}{2} \\ &< c_{m_{j+1}}^\lambda \le \int_{K_{R_k}} g(x,u_{m_{j+1}}^\lambda)u_{m_{j+1}}^\lambda .\end{aligned}$$

The proof is complete.

Proof of Theorem 3.3. By assumption

$$\frac{\alpha(q-2)}{2q} \le I_{\lambda_n}(u_n) \le \frac{\beta}{2}.$$

An argument similar to that of Lemma 3.8 shows that $\|u_n\|_{\lambda_n}$ is bounded. We can assume that $u_n \rightharpoonup u$ in E, $u_n \to u$ in L^2_{loc} and $u_n(x) \to u(x)$ a.e. on $\mathbb{R}^N$. As in the proof of Lemma 2.5 it follows easily that $u(x) = 0$ for a.e. $x \in \mathbb{R}^N \backslash Z$. The Schauder elliptic regularity theory implies that $u_n \to u$ in $\mathcal{C}^2_{loc}(\mathbb{R}^N)$, and so u is a (weak) solution of (P). As before our task reduces to proving that $u \ne 0$. In this case, we prove that $u_n \to u$ strongly in E. Set $v_n = u_n - u$. By $(V5)$, for $\lambda_n \ge \lambda$

$$\begin{aligned}\mu(\lambda, R_j) \int_{K_{R_j}^c} v_n^2 &\le \int_{K_{R_j}^c} (|\nabla v_n|^2 + (\lambda_n a + a_0)v_n^2) - (\lambda_n - \lambda) \int_{K_{R_j}^c} v_n^2 \\ &\le (H_{\lambda_n} v_n,\ v_n) + |a_0|_\infty |v_n|_2^2 \\ &\le d.\end{aligned}$$

For any $\varepsilon > 0$ one can choose λ_ε and $R_\varepsilon = R_{j_\varepsilon}$ so that

$$\int_{K^c_{R_\varepsilon}} v_n^2 \le \frac{d}{\mu(\lambda, R_\varepsilon)} < \varepsilon$$

for all $\lambda \ge \lambda_\varepsilon$. On the other hand, one has $\int_{K_{R_\varepsilon}} v_n^2 \to 0$. This proves that $v_n \to 0$ in L^2. By interpolation,

$$|v_n|_p \le |v_n|_2^{1-\theta} |v_n|_{2^*}^{\theta} \to 0$$

where $\theta = 2^*(p-2)/p(2^*-2)$. Thus $u_n \to u$ in L^τ for all $\tau \in [2, 2^*)$. We obtain that

$$\int_{\mathbb{R}^N} g(x, v_n) v_n \to 0 \qquad \text{and} \qquad \int_{\mathbb{R}^N} g(x, u_n) u_n \to \int_{\mathbb{R}^N} g(x, u) u.$$

Consequently

$$\begin{aligned}(H_{\lambda_n} v_n, v_n) &= (H_{\lambda_n} u_n, u_n) - (H_{\lambda_n} u, u) + o(1) \\ &= \int_{\mathbb{R}^N} g(x, u_n) u_n - \int_{\mathbb{R}^N} g(x, u) u + o(1) \\ &= o(1).\end{aligned}$$

Therefore, $|v_n|_{\lambda_n} \to 0$. The proof is complete.

Proof of Theorem 1.1. Since $(V_1) - (V_3)$ imply $(V5)$, the first conclusion follows from Theorem 3.1, the second one from Theorem 3.2 and the last one from Theorem 3.3.

4. The critical case. Throughout this section we assume that all the assumptions of Theorem 1.2 are satisfied.

Consider the functional

$$\begin{aligned}I_\lambda(u) &= \frac{1}{2}(H_\lambda u, u) - \frac{1}{2^*}\int_{\mathbb{R}^N} K|u|^{2^*} - \int_{\mathbb{R}^N} G(x, u) \\ &= \frac{1}{2}|u^+|_\lambda^2 - \frac{1}{2}|u^-|_\lambda^2 - \frac{1}{2^*}\int_{\mathbb{R}^N} K|u|^{2^*} - \int_{\mathbb{R}^N} G(x, u) \\ &= J_\lambda(u) - \int_{\mathbb{R}^N} G(x, u).\end{aligned}$$

Then $I_\lambda \in \mathcal{C}^1(E, \mathbb{R})$ and critical points of I_λ are solutions of (Q_λ).

We always assume $\lambda \ge \overline{\lambda}$. Note that $E_\lambda^0 = \{0\}$ by Lemma 2.5.

The assumptions $(g_1), (g_2)$ and $(g_3)'$ imply that for any $\delta > 0$ there is $c_\delta > 0$ such that

$$|g(x, u)| \le \delta|u| + c_\delta |u|^{2^*-1} \qquad \text{and} \qquad G(x, u) \le \delta|u|^2 + c_\delta|u|^{2^*} \tag{4.1}$$

for all (x, u). These inequalities, jointly with (2.2), yield

Lemma 4.1. *There are α, $\rho > 0$, independent of λ, such that* $\inf I_\lambda(B_\rho \cap E_\lambda^+) = 0$ *and $I_\lambda(u) \ge \alpha$ for all $u \in E_\lambda^+$ with $|u|_\lambda = \rho$.*

For $z_0 \in E_\lambda^+ \backslash \{0\}$ and $R > 0$ let

$$M := \left\{ u = u^- + s z_0 : \ u^- \in E_\lambda^-, \ s \ge 0, \ |u|_\lambda \le R \right\}.$$

It follows from conditions $(g_3)'$ and (K) that $(2^*)^{-1}K(x)|u|^{2^*} + G(x,u) \geq c_0|u|^{2^*}$ for some $c_0 > 0$. Thus we readily obtain

Lemma 4.2. *There exists $R > \rho$ (depending on z_0) such that $I_\lambda(u) \leq 0$ for all $u \in E_\lambda^- \oplus \mathbb{R}z_0 \backslash M$.*

Now the standard deformation argument yields the following

Lemma 4.3. *There is a $(\mathrm{PS})_{c_\lambda}$ sequence for I_λ with $c_\lambda \in [\alpha, \max I_\lambda(M)]$*

Observe that if (u_n) is a $(\mathrm{PS})_c$ sequence for I_λ, then

$$I_\lambda(u_n) - \frac{1}{2}I'_\lambda(u_n)u_n \geq \frac{1}{N}\int_{\mathbb{R}^N} K|u_n|^{2^*}. \tag{4.2}$$

Using (4.1) and (4.2), similarly to that of Lemma 3.7, it is easy to verify that

Lemma 4.4. *Given $b > 0$ there exists $d(b) > 0$ such that if $\lambda \geq \bar{\lambda}$ and (u_n) is a $(\mathrm{PS})_c$ sequence for I_λ with $c \leq b$, then*

$$\limsup_{n\to\infty} \|u_n\|_\lambda^2 \leq d(b).$$

Moreover, $c \geq 0$ and $c = 0$ if and only if $u_n \to 0$.

In addition, if u is a critical point of I_λ, then by (4.1)

$$\begin{aligned} |u|_\lambda^2 &= \int_{\mathbb{R}^N} \Big(K(x)|u|^{2^*-2}u(u^+ - u^-) + g(x,u)(u^+ - u^-)\Big) \\ &\leq \delta|u|_2^2 + c_\delta|u|_{2^*}^{2^*}. \end{aligned}$$

Therefore, using (2.1) and (2.2), one can find $c > 0$ independently of λ such that $\|u\|^2 \leq \frac{1}{2}\|u\|^2 + c\|u\|^{2^*}$. Consequently we obtain that

Lemma 4.5. *There exists $\gamma > 0$, independent of λ, such that $\|u\| \geq \gamma$ for all nontrivial critical points u of I_λ.*

The combination of Lemmas 4.1-4.4 shows that I_λ has a bounded $(\mathrm{PS})_{c_\lambda}$ sequence (u_n) with $c_\lambda \geq \alpha$. We can assume that $u_n \rightharpoonup u$ in E. It is clear that u is a (weak) solution of (Q_λ). Unfortunately, we do not know if this u is nontrivial. In the following we will prove that with a special choice of z_0 the corresponding u is nontrivial.

Without loss of generality we assume that $x_0 = 0$, where x_0 is from (K), and there is $\bar{r} > 0$ such that $B_{\bar{r}}(0) \subset \Omega$ and $a_0(x) \leq -\beta$ with $\beta > 0$ for all $x \in B_{\bar{r}}(0)$ due to the continuity of a_0 (Only here the continuity of a_0 enters. Clearly it can be replaced by supposing there is $\bar{r} > 0$ such that $B_{\bar{r}}(x_0) \subset \Omega$). For any $0 < r \leq \bar{r}$ choose $\psi \in C_0^\infty(\mathbb{R}^N, [0,1])$ satisfying $\psi(x) = 1$ for $x \in B_{r/2}(0)$ and $\psi(x) = 0$ for $x \in B_r^c$. For $\varepsilon > 0$ define

$$\varphi_\varepsilon(x) = \frac{C_N\psi(x)\varepsilon^{(N-2)/2}}{(\varepsilon^2 + |x|^2)^{(N-2)/2}}$$

where $C_N = (N(N-2))^{(N-2)/4}$. Let

$$S := \inf_{u\in H^1\backslash\{0\}} \frac{|\nabla u|_2^2}{|u|_{2^*}^2}.$$

The following estimates are well known (see [7] or [18]):

$$|\nabla\varphi_\varepsilon|_2^2 = S^{N/2} + O(\varepsilon^{N-2}), \qquad |\nabla\varphi_\varepsilon|_1 = O(\varepsilon^{(N-2)/2}),$$

$$|\varphi_\varepsilon|_{2^*}^{2^*} = S^{N/2} + O(\varepsilon^{N}), \qquad |\varphi_\varepsilon|_{2^*-1}^{2^*-1} = O(\varepsilon^{(N-2)/2})$$

and

$$|\varphi_\varepsilon|_1 = O(\varepsilon^{(N-2)/2}), \qquad |\varphi_\varepsilon|_2^2 = \begin{cases} b\varepsilon^2|\ln\varepsilon| + O(\varepsilon^2) & \text{if } N=4,\\ b\varepsilon^2 + O(\varepsilon^{N-2}) & \text{if } N\geq 5\end{cases}$$

where b is a positive constant. Since $\dim(E_\lambda^-) < \infty$, as a consequence of the above estimates, $|\varphi_\varepsilon^-|_{2^*} \leq c_1|\varphi_\varepsilon^-|_2 \leq c_1|\varphi_\varepsilon|_2 \to 0$, so $|\varphi_\varepsilon^+|_{2^*}^{2^*} \to S^{N/2}$ as $\varepsilon\to 0$. This means in particular that $\varphi_\varepsilon^+ \neq 0$ for all $\varepsilon>0$ small. Therefore we can consider the M with $z_0 = \varphi_\varepsilon^+$, and then denote it by M_λ^ε. Consider also

$$F_\lambda^\varepsilon := E_\lambda^- + \mathbb{R}\varphi_\varepsilon = E_\lambda^- \oplus \mathbb{R}\varphi_\varepsilon^+.$$

Finally, set

$$c^* := \frac{S^{N/2}}{N|K|_\infty^{(N-2)/2}}.$$

Lemma 4.6. *For $\varepsilon>0$ small enough, there is $c_\varepsilon < c^*$, independent of λ, such that* $\sup I_\lambda(F_\lambda^\varepsilon) \leq c_\varepsilon$, *hence* $\sup I_\lambda(M_\lambda^\varepsilon) \leq c_\varepsilon$, *whenever $\lambda \geq \bar\lambda$.*

Proof. Since $G(x,u)\geq 0$, it suffices to find $c_\varepsilon < c^*$ satisfying $\sup J_\lambda(F_\lambda^\varepsilon) \leq c_\varepsilon$.

It is easy to see that, if $u\neq 0$, then

$$\max_{t\geq 0} J_\lambda(tu) = \frac{(H_\lambda u,u)^{N/2}}{N\left(\int_{\mathbb{R}^N} K|u|^{2^*}\right)^{(N-2)/2}}.$$

Using the notation $|u|_{2^*,K}^{2^*} = \int_{\mathbb{R}^N} K|u|^{2^*}$, we define

$$m_\lambda^\varepsilon = \sup_{u\in F_\lambda^\varepsilon,\, |u|_{2^*,K}=1} (H_\lambda u,u).$$

It is sufficient to prove that there exists $\alpha_\varepsilon > 0$, independent of λ, satisfying

$$m_\lambda^\varepsilon \leq \alpha_\varepsilon < \frac{S}{|K|_\infty^{(N-2)/N}}. \tag{4.3}$$

We remark that there is $d_1>0$, independent of ε and λ, such that if $u := u^- + s\varphi_\varepsilon \in F_\lambda^\varepsilon$ with $|u|_{2^*,K} = 1$, then

$$|u^-|_{2^*} \leq d_1 \qquad \text{and} \qquad s \leq d_1. \tag{4.4}$$

Indeed, setting $u = v + s\varphi_\varepsilon^+$, by Lemma 2.4,

$$\begin{aligned}|v|_2^2 \leq |u|_2^2 &= \int_{K_{R_j}} |u|^2 + \int_{K_{R_j}^c} |v|^2\\ &\leq |K_{R_j}|^{(2^*-2)/2^*}|u|_{2^*}^2 + \frac{1}{\mu(\lambda,R_j)}\left(|v|_\lambda^2 + |a_0|_\infty |v|_2^2\right)\\ &\leq c_1|K_{R_j}|^{(2^*-2)/2^*}\frac{c_2}{\mu(\lambda,R_j)}|v|_2^2.\end{aligned}$$

By (V_5) we choose $\overline{\lambda}$ and $R = R_j$ large so that $c_2/\mu(\overline{\lambda}, R) \le 1/2$. Then we obtain $|v|_2^2 \le 2c_1|K_R|^{(2^*-2)/2^*}$, and so (4.4) follows.

The convexity of $|\cdot|_{2^*,K}$ implies

$$\begin{aligned}1 = |u|_{2^*,K}^{2^*} &\ge |s\varphi_\varepsilon|_{2^*,K}^{2^*} + 2^* \int_{\mathbb{R}^N} (s\varphi_\varepsilon)^{2^*-1} u^- \\ &\ge |s\varphi_\varepsilon|_{2^*,K}^{2^*} - c_3|\varphi_\varepsilon|_{2^*-1}^{2^*-1}|u^-|_2\end{aligned}$$

that is,

$$\begin{aligned}|s\varphi_\varepsilon|_{2^*,K}^{2^*} &\le \left(1 + c_3|\varphi_\varepsilon|_{2^*-1}^{2^*-1}|u^-|_2\right) \\ &= \left(1 + O(\varepsilon^{(N-2)/2})|u^-|_2\right).\end{aligned}$$

Here and below $O(\varepsilon^\tau)$ (and $o(\varepsilon^\tau)$) is uniform in λ. Moreover, using Lemma 2.4 again

$$\begin{aligned}(H_\lambda u^-, \varphi_\varepsilon) &= \int_{\mathbb{R}^N} H_\lambda u^- \varphi_\varepsilon \le |H_\lambda u^-|_\infty |\varphi_\varepsilon|_1 \\ &\le c_4|\varphi_\varepsilon|_1|u^-|_2 \\ &= O(\varepsilon^{(N-2)/2})|u^-|_2.\end{aligned}$$

Since $a_0(x) \le -\beta$ for $x \in \operatorname{supp}\varphi_\varepsilon$ and $K(x) - K(0) = o(|x|^2)$ as $x \to 0$, we have

$$\int_{\mathbb{R}^N} a_0\varphi_\varepsilon^2 \le \begin{cases} -d\varepsilon^2 & \text{if } N \ge 5 \\ -d\varepsilon^2|\ln\varepsilon| & \text{if } N = 4 \end{cases}$$

for some $d > 0$ independent of λ, and

$$\begin{aligned}|\varphi_\varepsilon|_{2^*,K}^{2^*} &= |K|_\infty|\varphi_\varepsilon|_{2^*}^{2^*} + \int_{\mathbb{R}^N} (K(x) - K(0))|\varphi_\varepsilon|^{2^*} \\ &= |K|_\infty S^{N/2} + o(\varepsilon^2).\end{aligned} \tag{4.5}$$

In addition,

$$\begin{aligned}(H_\lambda\varphi_\varepsilon, \varphi_\varepsilon) &= \int_{\mathbb{R}^N} (|\nabla\varphi_\varepsilon|^2 + a_0\varphi_\varepsilon^2) \\ &\le S^{N/2} + \begin{cases} -d\varepsilon^2|\ln\varepsilon| + O(\varepsilon^{N-2}) & \text{if } N = 4, \\ -d\varepsilon^2 + O(\varepsilon^{N-2}) & \text{if } N \ge 5 \end{cases}.\end{aligned}$$

First consider $N \ge 5$. Remark that

$$-|u^-|_2^2 + O(\varepsilon^{(N-2)/2})|u^-|_2 \le O(\varepsilon^{N-2}).$$

We obtain

$$\begin{aligned}m_\lambda^\varepsilon \le& -|u^-|_\lambda^2 + \frac{(H_\lambda\varphi_\varepsilon, \varphi_\varepsilon)}{|\varphi_\varepsilon|_{2^*,K}^2}|s\varphi_\varepsilon|_{2^*,K}^2 + O(\varepsilon^{(N-2)/2})|u^-|_2 \\ \le& -c_5|u^-|_2^2 + \frac{(H_\lambda\varphi_\varepsilon, \varphi_\varepsilon)}{|K|_\infty^{(N-2)/N}S^{(N-2)/2} + o(\varepsilon^2)}\left(1 + O(\varepsilon^{(N-2)/2})|u^-|_2\right)^{2/2^*} \\ &+ O(\varepsilon^{(N-2)/2}) \\ \le& -c_5|u^-|_2^2 + \frac{S^{N/2} - d\varepsilon^2 + O(\varepsilon^{N-2})}{|K|_\infty^{(N-2)/N}S^{(N-2)/2} + o(\varepsilon^2)} + O(\varepsilon^{(N-2)/2})|u^-|_2 \\ \le& \frac{S}{|K|_\infty^{(N-2)/N}} - d_0\varepsilon^2 + o(\varepsilon^2)\end{aligned}$$

where $d_0 > 0$ is independent of λ. In a similar way we obtain for $N = 4$

$$m_\lambda^\varepsilon \le \frac{S}{|K|_\infty^{(N-2)/N}} - d_0\varepsilon^2|\ln\varepsilon| + O(\varepsilon^2).$$

Consequently, (4.3) holds for ε sufficiently small.

We note that if $K(x) - K(0) = O(|x|^2)$ as $x \to 0$, then (4.5) remains true with $O(\varepsilon^2)$ replacing $o(\varepsilon^2)$. This does not affect the above estimates for $N = 4$. Hence Theorem 1.2 is valid for $N = 4$ with such a replacement.

Lemma 4.7. *There is $\Lambda > 0$ such that if (u_n^λ) is a $(\mathrm{PS})_{c_\lambda^\varepsilon}$ sequence with $u_n^\lambda \rightharpoonup u_\lambda$ and $\alpha \le c_\lambda^\varepsilon < c^*$ then $u_\lambda \ne 0$ provided $\lambda \ge \Lambda$.*

Proof. It follows from Lemma 4.4 that $\limsup_{n\to\infty} \|u_n^\lambda\|_\lambda^2 \le d(c^*)$ with $d(c^*)$ independent of λ. Observe that

$$\frac{1}{N}\int_{\mathbb{R}^N} K|u_n^\lambda|^{2^*} = \left(\frac{1}{2} - \frac{1}{2^*}\right)\int_{\mathbb{R}^N} K|u_n^\lambda|^{2^*} \le I_\lambda(u_n^\lambda) - \frac{1}{2}I_\lambda'(u_n^\lambda)u_n^\lambda.$$

Thus up to a subsequence

$$\lim_{n\to\infty}\frac{1}{N}\int_{\mathbb{R}^N} K|u_n^\lambda|^{2^*} = b \le c_\lambda^\varepsilon.$$

Recall that for any $\delta > 0$ there is $c_\delta > 0$ such that

$$G(x,u) \le \delta|u|^2 + c_\delta|u|^p \qquad \text{and} \qquad g(x,u)u \le \delta|u|^2 + c_\delta|u|^p \tag{4.6}$$

on $\mathbb{R}^N \times \mathbb{R}$. If $b = 0$ then, by Hölder inequality

$$|u_n^\lambda|_p \le |u_n^\lambda|_2^{1-\theta}|u_n^\lambda|_{2^*}^\theta \to 0$$

where $\theta = 2^*(p-2)/p(2^*-2)$, and so using (4.6) we obtain

$$\int_{\mathbb{R}^N} G(x,u_n^\lambda) \to 0 \qquad \text{and} \qquad \int_{\mathbb{R}^N} g(x,u_n^\lambda)u_n^\lambda \to 0. \tag{4.7}$$

This implies $c_\lambda^\varepsilon = 0$ a contradiction. Therefore $b > 0$.

Arguing indirectly, suppose that $u_\lambda = 0$ along a subsequence as $\lambda \to \infty$. Notice that either $|u_n^\lambda|_2 \to 0$ along a subsequence or $|u_n^\lambda|_2 \ge \gamma > 0$ for all n.

Assume $|u_n^\lambda|_2 \to 0$. Then (4.7) remains true, and

$$\begin{aligned} S|K|_\infty^{-2/2^*}|u_n^\lambda|_{2^*,K}^2 &\le S|u_n^\lambda|_{2^*}^2 \le |\nabla u_n^\lambda|_2^2 \\ &\le (H_\lambda u_n^\lambda, u_n^\lambda) + o(1) \\ &= I_\lambda'(u_n^\lambda)u_n^\lambda + \int_{\mathbb{R}^N} K|u_n^\lambda|^{2^*} + \int_{\mathbb{R}^N} g(x,u_n^\lambda)u_n^\lambda + o(1) \\ &= |u_n^\lambda|_{2^*,K}^{2^*} + o(1). \end{aligned}$$

Passing to the limit we obtain

$$S|K|_\infty^{-2/2^*}(bN)^{2/2^*} \le bN.$$

Hence $c^* \le b \le c_\lambda^\varepsilon < c^*$, we come to a contradiction.

Assume $|u_n^\lambda|_2 \ge \gamma$. Set

$$F := \{x \in \mathbb{R}^N : \ a(x) \le M_0\}$$

where M_0 is from (V_3). We claim that

$$\int_F |u_n^\lambda|^2 \to 0 \qquad \text{and} \qquad \int_F g(x, u_n^\lambda)u_n^\lambda \to 0 \tag{4.8}$$

as $n \to \infty$. Choosing $r \in (1,\ N/(N-2))$ and $r' = r/(r-1)$, we have

$$\int_{B_R^c \cap F} |u_n^\lambda|^2 \le |u_n^\lambda|_r^2 \,(\mathrm{mes}(B_R^c \cap F))^{1/r'} \le c\|u_n^\lambda\|_{H^1}^2 \,(\mathrm{mes}(B_R^c \cap F))^{1/r'} \to 0$$

as $R \to \infty$ by (V_3), and

$$\int_{B_R} |u_n^\lambda|^2 \to 0$$

as $n \to \infty$ since $H^1(\mathbb{R}^N) \hookrightarrow L_{loc}^r(\mathbb{R}^N)$ compactly. Thus

$$\int_F |u_n^\lambda|^2 \to 0 \qquad \text{as } n \to \infty.$$

By Hölder inequality again,

$$|u_n^\lambda|_{L^p(F)} \le |u_n^\lambda|_{L^2(F)}^{1-\theta} |u_n^\lambda|_{L^{2^*}(F)}^\theta \le |u_n^\lambda|_{L^2(F)}^{1-\theta} |u_n^\lambda|_{2^*}^\theta \to 0$$

as $n \to \infty$. Therefore, $\int_F g(x, u_n^\lambda)u_n^\lambda \to 0$ and (4.8) is proved. Remark that

$$\begin{aligned}
&\int_{F^c} (\lambda a|u_n^\lambda|^2 + a_0|u_n^\lambda|^2 - g(x, u_n^\lambda)u_n^\lambda) \\
\ge &\int_{F^c} ((\lambda M_0 - |a_0|_\infty - 1)|u_n^\lambda|^2 - c|u_n^\lambda|^p) \\
\ge &(\lambda M_0 - |a_0|_\infty - 1)|u_n^\lambda|_{L^2(F^c)}^2 - c|u_n^\lambda|_{L^2(F^c)}^{p(1-\theta)} |u_n^\lambda|_{2^*}^{p\theta} \\
\ge &|u_n^\lambda|_{L^2(F^c)}^{p(1-\theta)} \left((\lambda M_0 - |a_0|_\infty - 1)|u_n^\lambda|_{L^2(F^c)}^{2-p(1-\theta)} - c|u_n^\lambda|_{2^*}^{p\theta} \right)
\end{aligned}$$

which, together with (4.8), implies that

$$\int_{F^c} (\lambda a|u_n^\lambda|^2 + a_0|u_n^\lambda|^2 - g(x, u_n^\lambda)u_n^\lambda) \ge 0$$

for all λ large and n large. Now

$$\begin{aligned}
S|K|_\infty^{-2/2^*} |u_n^\lambda|_{2^*,K}^2 &\le S|u_n^\lambda|_{2^*}^2 \le |\nabla u_n^\lambda|_2^2 \\
&\le \int_{\mathbb{R}^N} |\nabla u_n^\lambda|^2 + \int_{F^c} (\lambda a|u_n^\lambda|^2 + a_0|u_n^\lambda|^2 - g(x, u_n^\lambda)u_n^\lambda) \\
&\le \int_{\mathbb{R}^N} (|\nabla u_n^\lambda|^2 + \lambda a|u_n^\lambda|^2) + \int_{F^c} (a_0|u_n^\lambda|^2 - g(x, u_n^\lambda)u_n^\lambda) \\
&= (H_\lambda u_n^\lambda, u_n^\lambda) - \int_{\mathbb{R}^N} g(x, u_n^\lambda)u_n^\lambda - \int_F (a_0|u_n^\lambda|^2 - g(x, u_n^\lambda)u_n^\lambda) \\
&= |u_n^\lambda|_{2^*,K}^{2^*} + o(1).
\end{aligned}$$

Therefore, as above, we obtain that $S|K|_\infty^{-2/2^*}(bN)^{2/2^*} \le bN$, and so $c^* \le b \le c_\lambda^\varepsilon < c^*$, a contradiction.

Proof of Theorem 1.2. Lemmas 4.1-4.4 and 4.6-4.7 imply that there exits $\overline{\lambda} > 0$ such that, for all $\lambda \ge \overline{\lambda}$, I_λ has a nontrivial critical point u_λ with $I_\lambda(u_\lambda) = c_\lambda \le c_\varepsilon$, where $c_\varepsilon < c^*$ and is independent of λ. Clearly, u_λ is a nontrivial solution of (Q_λ). In order to show the concentration, let $\lambda_n \to \infty$, $I_n := I_{\lambda_n}$ and $u_n := u_{\lambda_n}$. Then

$$c_\varepsilon \ge I_n(u_n) \ge \frac{1}{N}|u_n|_{2^*,K}^{2^*}$$

which implies that $|u_n|_{2^*}$ is bounded. Using (4.1) and (2.2) we find $d_1 > 0$ independent of λ such that

$$\begin{aligned}|u_n|_{\lambda_n}^2 &= \int_{\mathbb{R}^N} (K(x)|u_n|^{2*-2}u_n(u_n^+ - u_n^-) + g(x,u_n)(u_n^+ - u_n^-)) \\ &\le \frac{1}{2}|u_n|_{\lambda_n}^2 + d_1|u_n|_{2^*}^{2^*}.\end{aligned}$$

This proves that (u_n) is bounded in E. Going to a subsequence we can suppose that $u_n \rightharpoonup u$ in E, $u_n \to u$ in L^2_{loc} and $u_n(x) \to u(x)$ a.e. in $\mathbb{R}^N$. As before it is easy to check that $u(x) = 0$ for a.e. $x \in \mathbb{R}^N \backslash Z$. The elliptic theory implies $u_n \to u$ in $C^2_{loc}(\mathbb{R}^N)$ and so u satisfies (Q).

Letting $v_n = u_n - u$, as (4.8) we have $\int_F |v_n|^2 \to 0$. Moreover,

$$\int_{F^c} v_n^2 \le \frac{1}{\lambda_n M_0}\int_{F^c} \lambda_n a(x)v_n^2 = \frac{1}{\lambda_n M_0}\int_{F^c} \lambda_n a(x)u_n^2 \to 0.$$

Thus $u_n \to u$ in L^2 and we obtain $\int_{\mathbb{R}^N} g(x,v_n)v_n \to 0$. The Brézis-Lieb lemma implies that

$$|u_n|_{2^*,K}^{2^*} = |u|_{2^*,K}^{2^*} + |v_n|_{2^*,K}^{2^*} + o(1).$$

Since u verifies (Q), it holds that

$$I_n'(v_n)v_n = \int_{\mathbb{R}^N} (g(x,u_n)u_n - g(x,u)u) + o(1) = o(1),$$

that is,

$$(H_{\lambda_n}v_n,\ v_n) = |v_n|_{2^*,K}^{2^*} + o(1).$$

We can suppose, up to a subsequence, that

$$\frac{1}{N}|v_n|_{2^*,K}^{2^*} \to b.$$

Clearly $0 \le b \le c_\varepsilon$. If $b = 0$, then we have $\|v_n\| \to 0$ and the proof is complete. Assume $b > 0$. Then

$$\begin{aligned}S|K|_\infty^{-2/2^*}|v_n|_{2^*,K}^2 \le S|v_n|_{2^*}^2 &\le |\nabla v_n|_2^2 \\ &\le (H_{\lambda_n}v_n,\ v_n) + o(1) \\ &= |v_n|_{2^*,K}^{2^*} + o(1).\end{aligned}$$

Passing to the limit one obtains $c^* \leq b \leq c_\varepsilon < c^*$, which is a contradiction.

Finally, Lemma 4.5 implies $u \neq 0$. The proof is complete.

REFERENCES

[1] S. Alama and Y. Y. Li, *"Multibump" bound states for certain semilinear elliptic equations*, Indiana J. Math., 41 (1992), 983–1026.

[2] A. Ambrosetti, M. Badiale and S. Cingolani, *Semiclassical states of nonlinear Schrödinger equations*, Arch. Rat. Mech. Anal., 140 (1997), 285–300.

[3] T. Bartsch and Y. H. Ding, *On a nonlinear Schrödinger equation with periodic potential*, Math. Ann., 313 (1999), 15–37.

[4] T. Bartsch and Z. Q. Wang, *Existence and multiplicity results for some superlinear elliptic problems on* $\mathbb{R}^n$, Comm. Part. Diff. Eqs., 20 (1995), 1725–1741.

[5] T. Bartsch and Z. Q. Wang, *Multiple positive solutions for a nonlinear Schrödinger equation*, ZAMP, 51 (2000), 366–384.

[6] T. Bartsch, A. Pankov and Z. Q. Wang, *Nonlinear Schrödinger equations with steep potential well*, preprint.

[7] H. Brézis and L. Nirenberg, *Positive solutions of nonlinear elliptic equations involving critical Sobolev exponents*, Comm. Pure Appl. Math., 36 (1983), 437–477.

[8] J. Chabrowski and A. Szulkin, *On a semilinear Schrödinger equation with critical Sobolev exponent*, preprint.

[9] D. G. Costa, *On a class of elliptic systems in* $\mathbb{R}^N$, Electr. J. of Diff. Eq., 7 (1994), 1–4.

[10] V. Coti-Zelati and P. Rabinowitz, *Homoclinic type solutions for a semilinear elliptic PDE on* $\mathbb{R}^N$, Comm. Pure Appl. Math., 45 (1992), 1217–1269.

[11] M. del Pino and P. Felmer, *Multi-peak bounded states for nonlinear Schödinger equations*, AIHP, Anal. Non Linéaire, 15 (1998), 127–149.

[12] Y. H. Ding and S. J. Li, *Existence of entire solutions for some elliptic systems*, Bull. Austral. Math. Soc., 50 (1994), 501–519.

[13] A. Floer and A. Weinstein, *Nonspreading wave packets for the cubic Schrödinger equation with a bounded potential*, J. Funct. Anal., 69 (1986), 397–408.

[14] Y. Oh, *Existence of semi-classical bound states of nonlinear Schrödinger equations with potentials of the class* (v_a), Comm. Part. Diff. Eqs., 13 (1988), 1499–1519.

[15] P. Rabinowitz, *On a class of nonlinear Schrödinger equations*, ZAMP, 43 (1992), 270–291.

[16] B. Simon, *Schrödinger semigroups*, Bull. Amer. Math. (N.S.), 7 (1982), 447–526.

[17] C. Troestler and M. Willem, *Nontrivial solution of a semilinear Schrödinger equation*, Comm. Part. Diff. Eqs., 21 (1996), 1413–1449.

[18] M. Willem, "Minimax Theorems", Birkhäuser, 1996.

E-mail addresses : djairo@ime.unicamp.br dingyh@math.ac.cn

Received February 2001; revised July 2001.

TRANSACTIONS OF THE
AMERICAN MATHEMATICAL SOCIETY
Volume 355, Number 7, Pages 2973–2989
S 0002-9947(03)03257-4
Article electronically published on March 14, 2003

STRONGLY INDEFINITE FUNCTIONALS AND MULTIPLE SOLUTIONS OF ELLIPTIC SYSTEMS

D. G. DE FIGUEIREDO AND Y. H. DING

ABSTRACT. We study existence and multiplicity of solutions of the elliptic system

$$\begin{cases} -\Delta u = H_u(x,u,v) & \text{in } \Omega, \\ -\Delta v = -H_v(x,u,v) & \text{in } \Omega, \quad u(x) = v(x) = 0 \quad \text{on } \partial\Omega, \end{cases}$$

where $\Omega \subset \mathbb{R}^N, N \geq 3$, is a smooth bounded domain and $H \in \mathcal{C}^1(\bar{\Omega} \times \mathbb{R}^2, \mathbb{R})$. We assume that the nonlinear term

$$H(x,u,v) \sim |u|^p + |v|^q + R(x,u,v) \quad \text{with} \quad \lim_{|(u,v)| \to \infty} \frac{R(x,u,v)}{|u|^p + |v|^q} = 0,$$

where $p \in (1,\ 2^*)$, $2^* := 2N/(N-2)$, and $q \in (1,\ \infty)$. So some supercritical systems are included. Nontrivial solutions are obtained. When $H(x,u,v)$ is even in (u,v), we show that the system possesses a sequence of solutions associated with a sequence of positive energies (resp. negative energies) going toward infinity (resp. zero) if $p > 2$ (resp. $p < 2$). All results are proved using variational methods. Some new critical point theorems for strongly indefinite functionals are proved.

1. INTRODUCTION AND MAIN RESULTS

Consider the following elliptic system:

$$\text{(E)} \qquad \begin{cases} -\Delta u = H_u(x,u,v) & \text{in } \Omega, \\ -\Delta v = -H_v(x,u,v) & \text{in } \Omega, \\ u(x) = v(x) = 0 & \text{on } \partial\Omega, \end{cases}$$

where $\Omega \subset \mathbb{R}^N, N \geq 3$, is a smooth bounded domain and $H : \bar{\Omega} \times \mathbb{R}^2 \to \mathbb{R}$ is a $\mathcal{C}^1$-function. Here H_u denotes the partial derivative of H with respect to the variable u. Writing $z := (u,v)$, we suppose $H(x,0) \equiv 0$ and $H_z(x,0) \equiv 0$. Then $z = 0$ is a trivial solution of the system. In this paper we discuss the existence of nontrivial solutions. Roughly speaking, we are mainly interested in the class of Hamiltonians H such that

$$H(x,u,v) \sim |u|^p + |v|^q + R(x,u,v) \quad \text{with} \quad \lim_{|z| \to \infty} \frac{R(x,u,v)}{|u|^p + |v|^q} = 0,$$

where $1 < p < 2^* := 2N/(N-2)$ and $q > 1$. The most interesting results obtained here refer to the case when $q \geq 2^*$, which correspond to critical and supercritical problems. The case when $q < 2^*$ has been studied by Costa and Magalhães [5],

Received by the editors June 18, 2001.
2000 *Mathematics Subject Classification.* Primary 35J50; Secondary 58E99.
Key words and phrases. Elliptic system, multiple solutions, critical point theory.

Originally published in: Trans. Amer. Math. Soc. 355 (2003), 2973-2989,

DOI: 10.1007/978-3-319-02856-9_35

[6] and Benci and Rabinowitz [3]. See also Bartsch and De Figueiredo [2], De Figueiredo and Magalhães [7], De Figueiredo and Felmer [8] and Hulshof and van der Vorst [11], where similar systems also leading to strongly indefinite functionals have been studied. However, only subcritical systems have been considered in those papers.

Letting $2_* = 2^*/(2^* - 1) = 2N/(N+2)$, we assume that $H(x, z)$ satisfies the following condition:

(H_0) there are $p \in (1, 2^*)$, $q \in (1, \infty)$ and $\tau \in (1,\ 1 + q/2_*)$ such that, for all (x, z),

$$|H_u(x, u, v)| \leq \gamma_0(1 + |u|^{p-1} + |v|^{\tau-1})$$

and

$$|H_v(x, u, v)| \leq \gamma_0(1 + |u|^{p-1} + |v|^{q-1}).$$

In all hypotheses on $H(x, z)$ the γ_i's denote positive constants independent of (x, z). We note that if $q < 2^*$, then $2_* < q/(q-1)$, i.e., $q - 1 < q/2_*$. Hence, it is possible that $q \leq \tau < 1 + q/2_*$. However, if $q \geq 2^*$, then $\tau < q$. Furthermore, we remark that τ can be very large, if q is sufficiently large.

In addition, we need distinct conditions on H corresponding to the cases when $p > 2$, $p < 2$ or $p = 2$.

First, consider the case when $p > 2$. In this case, we assume the following three conditions:

(H_1) there are $\mu > 2$, $\nu > 1$ and $R_1 \geq 0$ such that

$$\frac{1}{\mu} H_u(x, z)u + \frac{1}{\nu} H_v(x, z)v \geq H(x, z) \quad \text{whenever } |z| \geq R_1\,,$$

with the provision that $\nu = \mu$ if $q > 2$;

(H_2) there are $2_*(p-1) \leq \alpha \leq p$ and $2_*(\tau - 1) < \beta$ such that

$$H(x, z) \geq \gamma_1\left(|u|^\alpha + |v|^\beta\right) - \gamma_2 \quad \text{for all } (x, z),$$

and $\beta = q$ if $q > 2^*$;

(H_3) $H(x, 0, v) \geq 0$ and $H_u(x, u, 0) = o(|u|)$ as $u \to 0$ uniformly in x.

We prove the following results.

Theorem 1.1. *Let (H_0) be satisfied with $p > 2$. If (H_1) – (H_3) hold, then* (E) *has at least one nontrivial solution.*

In order to provide some more transparent hypotheses under which the above result holds, we next present some conditions on H that are sufficient for (H_0), (H_1) and (H_2) to hold:

(H_0') there are $p \in (1, 2^*)$ and $q \in (2, \infty)$ such that, for all (x, z),

$$|H_u(x, u, v)| \leq \gamma_0(1 + |u|^{p-1} + |v|^{\frac{q}{2}-1})$$

and

$$|H_v(x, u, v)| \leq \gamma_0(1 + |u|^{p-1} + |v|^{q-1});$$

(H_1') there are $\mu > 2$ and $R_1 \geq 0$ such that

$$H_u(x, z)u + H_v(x, z)v \geq \mu H(x, z) \quad \text{whenever } |z| \geq R_1\,;$$

(H_2') for p and q as above,

$$H(x, z) \geq \gamma_1\left(|u|^p + |v|^q\right) - \gamma_2 \quad \text{for all } (x, z).$$

Theorem 1.1′. *Let (H_0') be satisfied with $p > 2$. If $(H_1'), (H_2')$, and (H_3) hold, then* (E) *has at least one nontrivial solution.*

Theorem 1.2. *Let (H_0) be satisfied with $p > 2$. If $H(x,z)$ is even in z and satisfies (H_1) and (H_2), then* (E) *has a sequence (z_n) of solutions with energies $I(z_n) := \int_\Omega \left(\frac{1}{2}(|\nabla u_n|^2 - |\nabla v_n|^2) - H(x,z_n)\right)$, going to ∞ as $n \to \infty$.*

In order to describe the other results, let $\sigma(-\Delta)$ denote the set of all eigenvalues of $(-\Delta, H_0^1(\Omega))$: $\lambda_1 < \lambda_2 \le \lambda_3 \le \cdots$.

We now consider the case when $p < 2$. We make the following assumptions:

(H_4) there are $\mu \in (1,\ 2)$, $\nu \ge 2$ and $\gamma_3 \ge 0$ ($\gamma_3 = 0$, if $q > 2^*$) such that

$$H(x,u,v) \ge \frac{1}{\mu} H_u(x,u,v)u + \frac{1}{\nu} H_v(x,u,v)v - \gamma_3 \quad \text{for all } (x,z);$$

(H_5) there are $\alpha \in (1,\ 2)$ and $\delta \in (0,\ 1/2)$ such that $H(x,u,v) \ge \gamma_4 |u|^\alpha - \delta\lambda_1 v^2$ for all (x,z);

(H_6) if $q \ge 2^*$, then $H_v(x,z)v \ge \gamma_5 |v|^q - \gamma_6(|v| + u^2)$ for all (x,z).

With these assumptions we have the following three results, for the case when $p < 2$.

Theorem 1.3. *Suppose that (H_0) holds with $p < 2$ and $q \ge 2$. If $H(x,z)$ also satisfies $(H_4) - (H_6)$, then* (E) *has at least one nontrivial solution.*

Theorem 1.4. *Suppose that $H(x,z)$ is even in z and (H_0) holds with $p < 2$ and $q \ge 2$. If $H(x,z)$ also satisfies $(H_4)-(H_6)$, then* (E) *has a sequence (z_n) of solutions with negative energies $I(z_n)$ going to 0 as $n \to \infty$.*

Theorem 1.5. *Let (H_0), with $p, q \in (1,\ 2)$, and (H_5) be satisfied. Then* (E) *has at least one nontrivial solution. If, in addition, $H(x,z)$ is even in z, then* (E) *has a sequence (z_n) of solutions with negative energies $I(z_n)$ going to 0 as $n \to \infty$.*

Finally, we consider the case when $p = 2$, which presents some sort of resonance. Assume

(H_7) there exist $b_0 \le 0 < a_0$ such that $R_0(x,z) := H(x,z) - \frac{1}{2}(a_0 u^2 + b_0 v^2) = o(|z|^2)$ as $z \to 0$ uniformly in x;

(H_8) there exist $\sigma \in (1,\ 2)$, $a_\infty \in [a_0,\ \infty) \backslash \sigma(-\Delta)$, such that $R_\infty(x,z) := H(x,z) - \frac{1}{2} a_\infty u^2$ satisfies $|\partial_u R_\infty(x,z)| \le \gamma_7(1 + |u|^{\sigma-1} + |v|^{\tau-1})$ and $R_\infty(x,z) \ge \gamma_8 |v|^q - \gamma_9 (1 + |u|^\sigma)$.

The position of the numbers a_0, a_∞, b_0 with respect to the spectrum $\sigma(-\Delta)$ plays a very essential role in the next result. For that matter, let i, j, k be nonnegative integers such that $\lambda_i = \min\{\lambda \in \sigma(-\Delta) : \ \lambda > a_0\}$, $\lambda_j = \max\{\lambda \in \sigma(-\Delta) : \ \lambda < -b_0\}$, $\lambda_k = \max\{\lambda \in \sigma(-\Delta) : \ \lambda < a_\infty\}$, and set

$$\ell = \begin{cases} j & \text{if } a_\infty = a_0, \\ j + k - i + 1 & \text{if } a_\infty > a_0. \end{cases}$$

Now we can state our last result.

Theorem 1.6. *Let (H_0) be satisfied with $p = 2$ and $\tau < 1 + q/2$. Assume that $H(x,z)$ is even in z and satisfies (H_7) and (H_8). Then* (E) *has at least one pair of nontrivial solutions if $\ell = 1$, and infinitely many solutions if $\ell \ge 2$.*

The cases covered in Theorem 1.6 include some asymptotically linear systems. Such systems have been studied in [5], [6] and Silva [13]. However, their results are not comparable with the ones obtained here.

We organize the paper as follows. In order to establish multiplicity of solutions we need some new abstract propositions on critical point theory for strongly indefinite functionals, which will be provided in Section 2. These propositions are based on certain Galerkin approximations, and we emphasize that the functionals do not satisfy the usual Palais-Smale condition. In Section 3 we study systems that are superlinear in the variable u, and prove Theorems 1.1 and 1.2. In Section 4 we consider systems that are sublinear in the variable u, and prove Theorems 1.3, 1.4 and 1.5. In both Sections 3 and 4, the variable v can have subcritical growth as well as supercritical growth. Finally, in Section 5, we consider a special asymptotically linear system and prove existence of multiple solutions.

2. Critical points for strongly indefinite functionals

Let E be a Banach space with norm $\|\cdot\|$. Suppose that E has a direct sum decomposition $E = E^1 \oplus E^2$ with both E^1 and E^2 being infinite dimensional. Let P^1 denote the projection from E onto E^1. Assume (e_n^1) (resp. (e_n^2)) is a basis for E^1 (resp. E^2). Set

$$X_n := \operatorname{span}\{e_1^1, \cdots, e_n^1\} \oplus E^2, \quad X^m := E^1 \oplus \operatorname{span}\{e_1^2, \cdots, e_m^2\},$$

and let $(X^m)^\perp$ denote the complement of X^m in E. For a functional $I \in \mathcal{C}^1(E, \mathbb{R})$ we set $I_n := I|_{X_n}$, the restriction of I on X_n. Recall that a sequence $(z_j) \subset E$ is said to be a $(\mathrm{PS})_c^*$ sequence if $z_j \in X_{n_j}$, $n_j \to \infty$, $I(z_j) \to c$ and $I'_{n_j}(z_j) \to 0$ as $j \to \infty$. If any $(\mathrm{PS})_c^*$ sequence has a convergent subsequence, then we say that I satisfies the $(\mathrm{PS})_c^*$ condition.

Denote the upper and lower level sets, respectively, by $I_a = \{z \in E : I(z) \geq a\}$, $I^b = \{z \in E : I(z) \leq b\}$ and $I_a^b = I_a \cap I^b$ (denote similarly $(I_n)_a$, $(I_n)^b$ and $(I_n)_a^b$). We also set $\mathcal{K} = \{z \in E : I'(z) = 0\}$, $\mathcal{K}_c = \mathcal{K} \cap I_c$, $\mathcal{K}^c = \mathcal{K} \cap I^c$ and $\mathcal{K}_a^b = \mathcal{K}_a \cap \mathcal{K}^b$.

Proposition 2.1. *Let E be as above and let $I \in \mathcal{C}^1(E, \mathbb{R})$ be even with $I(0) = 0$. In addition, suppose that, for each $m \in \mathbb{N}$, the conditions below hold:*

- (I_1) *there is $R_m > 0$ such that $I(z) \leq 0$ for all $z \in X^m$ with $\|z\| \geq R_m$;*
- (I_2) *there are $r_m > 0$ and $a_m \to \infty$ such that $I(z) \geq a_m$ for all $z \in (X^{m-1})^\perp$ with $\|z\| = r_m$;*
- (I_3) *I is bounded from above on bounded sets of X^m;*
- (I_4) *if $c \geq 0$, any $(\mathrm{PS})_c^*$ sequence (z_n) has a subsequence along which $z_n \rightharpoonup z \in \mathcal{K}_c$.*

Then the functional I has a sequence (c_k) of critical values, with the property that $c_k \to \infty$.

Remark 2.1. This proposition is more or less known if the condition (I_4) is replaced by the $(\mathrm{PS})^*$ condition (cf. [1], [9]), or by the usual Palais-Smale condition, that is, *any sequence $(z_k) \subset E$ such that $|I(z_k)| \leq c$ and $I'(z_k) \to 0$ has a convergent subsequence* (cf. [3]).

Proposition 2.2. *Let E be as above and let $I \in \mathcal{C}^1(E, \mathbb{R})$ be even. Assume that $I(0) = 0$ and that, for each $m \in \mathbb{N}$, the two conditions below hold:*

- (I_5) *there are $r_m > 0$ and $a_m > 0$ such that $I(z) \geq a_m$ for all $z \in X^m$ with $\|z\| = r_m$;*
- (I_6) *there is $b_m > 0$ with $b_m \to 0$ such that $I(z) \leq b_m$ for all $z \in (X^{m-1})^\perp$.*

Moreover, suppose that either I satisfies the $(\mathrm{PS})_c^*$ *condition for all* $c > 0$, *or that the condition below holds:*

(I_7) $\inf I(\mathcal{K}) = 0$, *and, for all* $c \geq 0$, *any* $(\mathrm{PS})_c^*$ *sequence* (z_n) *has a subsequence along which* $z_n \rightharpoonup z \in \mathcal{K}^c$ *with* $z = 0$ *only if* $c = 0$.

Then I has a sequence (c_k) *of positive critical values satisfying* $c_k \to 0$.

Proof. Let Σ be the family of symmetric, closed subsets of $E \backslash \{0\}$, and let $\gamma : \Sigma \to \mathbb{N} \cup \{0, \infty\}$ denote the Krasnoselski genus map. Set

$$c_n^m := \sup_{A \in \Sigma_n^m} \inf_{z \in A} I(z),$$

where

$$\Sigma_n^m := \{A \in \Sigma : \ A \subset X_n \text{ and } \gamma(A) \geq n + m\}.$$

Fix $m \in \mathbb{N}$. The Borsuk-Ulam theorem implies that $A \cap (X^{m-1})^{\perp} \neq \emptyset$ for each $A \in \Sigma_n^m$. It follows from (I_6) that

$$\inf_{z \in A} I(z) \leq \sup_{z \in (X^{m-1})^{\perp}} I(z) \leq b_m.$$

On the other hand, since $\gamma(\partial B_{r_m} \cap X_n^m) = n + m$, one has $S_n^m := \partial B_{r_m} \cap X_n^m \in \Sigma_n^m$, and so, by ($I_5$), we obtain

$$\inf_{z \in S_n^m} I(z) \geq a_m.$$

Therefore,

$$a_m \leq c_n^m \leq b_m. \tag{2.1}$$

A standard deformation argument, using a positive pseudo-gradient flow, yields the existence of a sequence $(z_n^m)_{n=1}^{\infty}$, with $z_n^m \in X_n$ satisfying

$$|I(z_n^m) - c_n^m| \leq \frac{1}{n} \quad \text{and} \quad \|I_n'(z_n^m)\| \leq \frac{1}{n}.$$

We can assume that $I(z_n^m) \to c_m$ as $n \to \infty$. So, (z_n^m) is a $(\mathrm{PS})_{c_m}^*$ sequence with

$$a_m \leq c_m \leq b_m. \tag{2.2}$$

Now, if we assume that I satisfies the $(\mathrm{PS})_c^*$ condition for $c > 0$, then the conclusion follows. Next, suppose instead that (I_7) holds. Then, along a subsequence, $z_n^m \rightharpoonup z_m$ as $n \to \infty$ with $I'(z_m) = 0$ and $0 < I(z_m) \leq c_m$. Finally, by (2.2),

$$I(z_m) \leq b_m \to 0,$$

and the proof is complete. □

Proposition 2.3. *Let E be as above and let* $I \in \mathcal{C}^1(E, \mathbb{R})$ *be even with* $I(0) = 0$. *Suppose, in addition, that the three conditions below hold:*

(I_8) *there are* $\ell \in \mathbb{N}$ *and* r, $a > 0$ *such that* $I(z) \geq a$ *for all* $z \in X^{\ell}$ *with* $\|z\| = r$;
(I_9) *there is* $b > 0$ *such that* $\sup I(E^2) \leq b$;
(I_{10}) *any* $(\mathrm{PS})_c^*$, $c > 0$, *sequence* (z_n) *has a subsequence along which* $z_n \rightharpoonup z \in \mathcal{K}_c$ *and* $P^1 z_n \to P^1 z$.

Then I has at least one pair of nontrivial critical points if $\ell = 1$, *and infinitely many critical points if* $\ell > 1$, *with positive critical values.*

Proof. Let Σ, γ, Σ_n^m and c_n^m be as in the proof of Proposition 2.2. As before, by (I_8) and (I_9), we obtain

$$a \leq c_n^m \leq b \quad \text{for all } n \in \mathbb{N} \text{ and } m = 1, \cdots, \ell,$$

and we find sequences $z_n^m \in X_n$ such that, going to subsequences if necessary, $I(z_n^m) \to c_m$ and $I_n'(z_n^m) \to 0$ as $n \to \infty$ with

$$b \geq c_1 \geq c_2 \geq \cdots \geq c_\ell \geq a.$$

Using (I_{10}), we can assume furthermore that $z_n^m \rightharpoonup z_m \in \mathcal{K}_{c_m}$ for $m = 1, \cdots, \ell$, as $n \to \infty$. If $\ell = 1$ the proof is complete.

Consider $\ell > 1$. Let $F = \{z \in \mathcal{K} : \; I(z) > 0\}$. We are going to prove that F is an infinite set. Arguing by contradiction, we suppose that F is finite. Choose $0 < \mu < a \leq b < \nu$ satisfying

$$\mu < \inf I(F) \leq \sup I(F) < \nu.$$

Let $k \in \mathbb{N}$ be so large that $0 \notin A := Q^k F$, where $Q^k : E \to X^k$ denotes the projection. Then A is also finite, and $\gamma(A) = 1$. By the continuity of γ, for all $\delta > 0$ small, $\gamma(N_\delta^k(A)) = \gamma(A)$, where $N_\delta^k(A) = \{z \in X^k : \; \operatorname{dist}(z, A) \leq \delta\}$. Set $C_\delta = N_\delta^k(A) \oplus (X^k)^\perp$. Since $N_\delta^k(A) \subset C_\delta$ and $Q^k : C_\delta \to N_\delta^k(A)$, it follows from the properties of γ that $\gamma(C_\delta) = \gamma(N_\delta^k(A))$. We remark that $Q^k = P^1 + (Q^k - P^1)$ and that the range of $Q^k - P^1$ is k-dimensional. So by virtue of (I_{10}), we conclude that, for all $c \geq 0$, any $(\mathrm{PS})_c^*$ sequence (z_n) has a subsequence along which $z_n \rightharpoonup z \in \mathcal{K}_c$ and $Q^k z_n \to Q^k z$. Hence there are $n_0 \in \mathbb{N}$ and $\sigma > 0$ such that for all $n \geq n_0$,

$$\|I_n'(w)\| \geq \sigma \quad \text{for all } w \in (I_n)_\mu^\nu \setminus C_\delta^n,$$

where $C_\delta^n = C_\delta \cap X_n$. By a standard deformation argument, we can then construct a sequence of odd homeomorphisms $\eta_n : X_n \to X_n$ such that

$$\eta_n \left((I_n)_\mu \setminus C_\delta^n \right) \subset (I_n)_\nu$$

(cf. [12]). For n_0 sufficiently large, we can suppose that

$$\mu < c_n^\ell \leq c_n^{\ell-1} \leq \cdots \leq c_n^1 < \nu \quad \text{for all } n \geq n_0.$$

Let $G \in \Sigma_n^\ell$ be such that $\inf I(G) > (\mu + c_n^\ell)/2$. One then has

$$\eta_n \left(G \backslash C_\delta^n \right) \subset (I_n)_\nu$$

and

$$\begin{aligned} \gamma(\eta_n(G \backslash C_\delta^n)) &= \gamma(G \backslash C_\delta^n) \geq \gamma(G) - \gamma(C_\delta^n) \\ &\geq n + \ell - \gamma(C_\delta^n) \geq n + \ell - 1. \end{aligned}$$

Thus $\eta_n(G \backslash C_\delta^n) \in \Sigma_n^{\ell-1}$ and $\nu \leq \inf I(\eta_n(G \backslash C_\delta^n)) \leq c_n^{\ell-1}$. One finally comes to $\nu \leq c_n^{\ell-1} < \nu$, which is a contradiction. □

From now on we turn to the system (E). We denote by $|\cdot|_t$ the usual $L^t(\Omega)$ norm for all $t \in [1, \infty]$. For $q > 1$ let $V_q = H_0^1(\Omega)$ if $q \leq 2^*$ and $V_q = H_0^1(\Omega) \cap L^q(\Omega)$, the Banach space equipped with the norm $\|v\|_{V_q} = \left(|\nabla v|_2^2 + |v|_q^2 \right)^{1/2}$, if $q > 2^*$. Let E_q be the product space $H_0^1(\Omega) \times V_q$ with elements denoted by $z = (u, v)$. We denote the norm in E_q by $\|z\|_q = (|\nabla u|_2^2 + \|v\|_{V_q}^2)^{1/2}$. E_q has the direct sum decomposition

$$E_q = E_q^- \oplus E^+, \quad z = z^- + z^+$$

where

$$E_q^- = \{0\} \times V_q \quad \text{and} \quad E^+ = H_0^1(\Omega) \times \{0\}.$$

For convenience, we will write $z^+ = u$ and $z^- = v$. Recall that by $(\lambda_n)_{n\in\mathbb{N}}$ we denote the sequence of eigenvalues of $(-\Delta, H_0^1(\Omega))$. Let e_n, $|e_n|_2 = 1$, be the eigenfunction corresponding to λ_n for each $n \in \mathbb{N}$. Clearly, $e_n^+ := (e_n, 0)$, $n \in \mathbb{N}$, is a basis for E^+, and $e_n^- = (0, e_n), n \in \mathbb{N}$, is a basis for E_q^-.

Suppose that the assumption (H_0) holds. Then

$$H(x,z) \le c(1 + |u|^{2^*} + |v|^q) \quad \text{for all } (x,z). \tag{2.3}$$

So the functional

$$I(z) := \frac{1}{2}\int_\Omega (|\nabla u|^2 - |\nabla v|^2) - \int_\Omega H(x,z) \tag{2.4}$$

is well defined in E_q. Moreover, $I \in \mathcal{C}^1(E_q, \mathbb{R})$, and the critical points of I are the solutions of (E).

Lemma 2.1. *If (H_0) holds, then I' is weakly sequentially continuous, that is, $I'(z_n) \rightharpoonup I'(z)$ provided $z_n \rightharpoonup z$.*

Proof. If $q < 2^*$ this statement is well known. Assume now that $q \ge 2^*$. Let $z_n \rightharpoonup z$ in E_q. Clearly, for all $w = (\varphi, \psi) \in E_q$, we have

$$\int_\Omega (\nabla u_n \nabla\varphi - \nabla v_n \nabla\psi) \to \int_\Omega (\nabla u \nabla\varphi - \nabla v \nabla\psi).$$

So it remains to show that

$$\int_\Omega H_u(x,z_n)\varphi \to \int_\Omega H_u(x,z)\varphi \quad \text{for all } \varphi \in H_0^1(\Omega) \tag{2.5}$$

and

$$\int_\Omega H_v(x,z_n)\psi \to \int_\Omega H_v(x,z)\psi \quad \text{for all } \psi \in V_q. \tag{2.6}$$

By the Sobolev embedding theorem and using interpolation, we obtain that $u_n \to u$ in L^t for $t \in [1,\ 2^*)$ and $v_n \to v$ in L^t for $t \in [1,\ q)$. Noting that $|H_u(x,u,v)| \le \gamma_0(1 + |u|^{p-1} + |v|^{\tau-1})$ with $2_*(\tau - 1) < q$, (2.5) follows easily since $u_n \to u$ in L^p, $v_n \to v$ in $L^{2_*(\tau-1)}$ and $\varphi \in H_0^1(\Omega) \subset L^{2^*}$. Next we see that (2.6) is clearly true when $\psi \in L^\infty$. In general, for a $\psi \in V_q$ we proceed as follows. Let $\tilde\psi_m \in L^\infty$ with $\tilde\psi_m \to \psi$ in L^q as $m \to \infty$. So

$$\left|\int_\Omega (H_v(x,z_n) - H_v(x,z))\psi\right| = \left|\int_\Omega (H_v(x,z_n) - H_v(x,z))(\tilde\psi_m + (\psi - \tilde\psi_m))\right|,$$

and using (H_0) we see that this expression is less than the following sum:

$$\left|\int_\Omega (H_v(x,z_n) - H_v(x,z))\tilde\psi_m\right| + c_1\left(|\psi - \tilde\psi_m|_1 + |u_n|_p^{p-1}|\tilde\psi_m - \psi|_p + |v_n|_q^{q-1}|\tilde\psi_m - \psi|_q\right),$$

which by its turn is estimated by

$$\left|\int_\Omega (H_v(x,z_n) - H_v(x,z))\tilde\psi_m\right| + c_2\left(|\tilde\psi_m - \psi|_p + |\tilde\psi_m - \psi|_q\right),$$

since (z_n) is bounded in E_q and L^∞ is dense in L^q. So (2.6) is proved, and it follows that

$$I'(z_n)w \to I'(z)w \text{ for all } w \in E_q.$$

□

3. The case $p > 2$

Throughout this section let (H_0) be satisfied with $p > 2$, and assume that (H_1) and (H_2) hold. Observe that, by (H_2), there exists $R > 0$ such that $H(x, z) > 0$ whenever $|z| \geq R$. This, jointly with (H_1), implies

$$H(x,z) \geq c_1(|u|^\mu + |v|^\nu) - c_2 \quad \text{for all } (x,z) \tag{3.1}$$

(see [10]). This, together with (2.3) and (H_2), shows that

$$\nu \leq q \text{ and } \beta \leq q. \tag{3.2}$$

Moreover, by virtue of (3.1) and (H_2), we may assume, without loss of generality, that (since $\mu > 2$)

$$\alpha > 2\,. \tag{3.3}$$

Now we set $E^1 = E_q^-$, $E^2 = E^+$ and $e_n^1 = e_n^-$, $e_n^2 = e_n^+$ for all $n \in \mathbb{N}$. So $E_q = E_1 \oplus E_2$. Consider the functional defined by (2.4), which has the properties stated in Section 2.

Lemma 3.1. *Any* $(\mathrm{PS})_c^*$ *sequence is bounded.*

Proof. Let $z_n \in X_n$ be such that

$$I(z_n) \to c \quad \text{and} \quad I_n'(z_n) \to 0.$$

Case 1: $q \leq 2$. In this case $E_q = (H_0^1(\Omega))^2$. By (H_1), for $w_n := (\frac{1}{\mu}u_n, \frac{1}{\nu}v_n)$, we have

$$\begin{aligned}
&I(z_n) - I_n'(z_n)w_n \\
&= (\frac{1}{2} - \frac{1}{\mu})|\nabla u_n|_2^2 + (\frac{1}{\nu} - \frac{1}{2})|\nabla v_n|_2^2 \\
&\quad + \int_\Omega \left(\frac{1}{\mu}H_u(x,z_n)u_n + \frac{1}{\nu}H_v(x,z_n)v_n - H(x,z_n)\right) - c_1 \\
&\geq (\frac{1}{2} - \frac{1}{\mu})|\nabla u_n|_2^2 + (\frac{1}{\nu} - \frac{1}{2})|\nabla v_n|_2^2 - c_2.
\end{aligned} \tag{3.4}$$

If $q < 2$, then (3.2) shows that $\nu < 2$, and so $\|z_n\|_q^2 \leq c_3(1 + \|z_n\|_q)$, which implies that (z_n) is bounded in E_q. Assume $q = 2$. Invoking (3.2), we get $\nu \leq 2$, and so $|\nabla u_n|_2^2 \leq c(1 + \|z_n\|_q)$ by (3.4). Since $H(x,z) > 0$ for all $|z|$ large, and

$$\frac{1}{2}|\nabla v_n|_2^2 + \int_\Omega H(x,z_n) = -I(z_n) + \frac{1}{2}|\nabla u_n|_2^2 \leq c(1 + \|z_n\|_q),$$

one sees that $\|z_n\|_q^2 \leq c(1 + \|z_n\|_q)$. Hence, (z_n) is bounded.

Case 2: $q > 2$. Note that in this case $\nu = \mu > 2$ in (H_1). So

$$\begin{aligned}
I(z_n) - \frac{1}{2}I_n'(z_n)z_n &= \int_\Omega (\frac{1}{2}H_z(x,z_n)z_n - H(x,z_n)) \\
&\geq (\frac{\mu}{2} - 1)\int_\Omega H(x,z_n) - c,
\end{aligned} \tag{3.5}$$

which, together with (H_2), yields

$$|u_n|_\alpha^\alpha + |v_n|_\beta^\beta \leq c(1 + \|z_n\|_q). \tag{3.6}$$

Using (H_0), we get

$$
\begin{aligned}
|\nabla u_n|_2^2 &= I_n'(z_n)(u_n,0) + \int_\Omega H_u(x,z_n)u_n \\
&\le c_1\|z_n\|_q + c_2\int_\Omega (|u_n|^p + |v_n|^{\tau-1}|u_n|).
\end{aligned}
\tag{3.7}
$$

Next we estimate the integrals in the right side of (3.7). Since $2_*(p-1) \le \alpha \le p$, we have that $\theta := \alpha/(1+\alpha-p) \le 2^*$. Using the Hölder inequality, the Sobolev embedding theorem and (3.6), we obtain

$$\int_\Omega |u_n|^p \le |u_n|_\alpha^{p-1}|u_n|_\theta \le c_1 + c_2\|z_n\|_q^{1+(p-1)/\alpha}.$$

Similarly, since $\tau - 1 < \beta/2_*$, we have $1 < \omega := \beta/(1+\beta-\tau) < 2^*$, and hence

$$\int_\Omega |v_n|^{\tau-1}|u_n| \le |v_n|_\beta^{\tau-1}|u_n|_\omega \le c_1 + c_2\|z_n\|_q^{1+(\tau-1)/\beta}.$$

Therefore, using the estimate in (3.7), we obtain

$$|\nabla u_n|_2^2 \le c(1 + \|z_n\|_q^{1+(p-1)/\alpha} + \|z_n\|_q^{1+(\tau-1)/\beta}). \tag{3.8}$$

Since

$$|\nabla v_n|_2^2 = -I_n'(z_n)(0,v_n) - \int_\Omega H_z(x,z_n)z_n + \int_\Omega H_u(x,z_n)u_n,$$

and using (3.5) and the above arguments, we obtain

$$|\nabla v_n|_2^2 \le c(1 + \|z_n\|_q^{1+(p-1)/\alpha} + \|z_n\|_q^{1+(\tau-1)/\beta}). \tag{3.9}$$

Recall that, in view of our assumptions, $(p-1)/\alpha \le 1/2_*$, $(\tau-1)/\beta < 1/2_*$, and $\beta = q$ if $q > 2^*$. Hence, it follows from (3.6) and (3.8)-(3.9) that (z_n) is bounded in E_q. □

Lemma 3.2. *Let $z_n \in X_n$ be a* $(\mathrm{PS})_c^*$ *sequence. If $q \le 2^*$, then (z_n) contains a convergent subsequence. If $q > 2^*$, then there is a $z \in E_q$ such that, along a subsequence, $z_n \rightharpoonup z$ and $I'(z) = 0$ and $I(z) \ge c$.*

Proof. By Lemma 3.1, (z_n) is bounded. We can assume that $z_n \rightharpoonup z$ in E_q, $z_n \to z$ in $(L^s(\Omega))^2$ for all $1 \le s < 2^*$, and $z_n(x) \to z(x)$ a.e. on Ω. It follows from the weak sequential continuity of I' (see Lemma 2.1) that $I'(z) = 0$. Since $I_n'(z_n) \to 0$, we obtain

$$
\begin{aligned}
(\nabla u_n, \nabla u_n - \nabla u)_{L^2} &= I_n'(z_n)(u_n - u, 0) + \int_\Omega H_u(x,z_n)(u_n - u) \\
&= o(1) + \int_\Omega H_u(x,z_n)(u_n - u).
\end{aligned}
$$

Using (H_0) and the Hölder inequality, we obtain the estimate

$$
\begin{aligned}
&\left|\int_\Omega H_u(x,z_n)(u_n - u)\right| \\
&\le c\left(|u_n - u|_1 + |u_n|_p^{p-1}|u_n - u|_p + |v_n|_\beta^{\tau-1}|u_n - u|_\omega\right) = o(1),
\end{aligned}
$$

where ω is as in the proof of Lemma 3.1. Hence $|\nabla u_n|_2^2 \to |\nabla u|_2^2$, which implies $u_n \to u$ in $H_0^1(\Omega)$. Let $P_n : E_q \to X_n$ denote the projection. Observe that $P_n z \to z$

in E_q for all $z \in E_q$. Moreover, using again (H_0) and the Hölder inequality, we estimate

$$\left|\int_\Omega H_v(x, z_n)(v - P_n v)\right|$$
$$\leq c\left(|v - P_n v|_1 + |u_n|_p^{p-1}|v - P_n v|_p + |v_n|_q^{q-1}|v - P_n v|_q\right) \to 0.$$

On the other hand,

$$\begin{aligned}(\nabla v_n, \nabla v - \nabla v_n)_{L^2} &= o(1) + I_n'(z_n)(0, v_n - P_n v) + \int_\Omega H_v(x, z_n)(v_n - P_n v)\\ &= o(1) + \int_\Omega H_v(x, z_n)(v_n - v)\\ &= o(1) + \int_\Omega H_z(x, z_n)(z_n - z) - \int_\Omega H_u(x, z_n)(u_n - u)\\ &= o(1) + \int_\Omega H_z(x, z_n)z_n - \int_\Omega H_z(x, z_n)z.\end{aligned}$$

Lebesgue's theorem and the weak sequential continuity of $H_z(x, \cdot)$ (see the proof of Lemma 2.1) yield

$$|\nabla v|_2^2 - \limsup_{n\to\infty} |\nabla v_n|_2^2 = \liminf_{n\to\infty}\left(\int_\Omega H_z(x, z_n)z_n - \int_\Omega H_z(x, z_n)z\right) \geq 0,$$

i.e., $|\nabla v|_2^2 \geq \limsup_{n\to\infty} |\nabla v_n|_2^2$. This, together with the weak lower semicontinuity of norms, implies $|\nabla v_n|_2 \to |\nabla v|_2$. So $v_n \to v$ in $H_0^1(\Omega)$.

Therefore, if $q \leq 2^*$, we obtain that, along a subsequence, $z_n \to z$ in E_q and consequently $I(z) = c$. Next assume that $q > 2^*$. Observe that

$$\begin{aligned}I(z) - I(z_n) = \frac{1}{2}(|\nabla u|_2^2 - |\nabla u_n|_2^2) - \frac{1}{2}(|\nabla v|_2^2 - |\nabla v_n|_2^2)\\ + \int_\Omega H(x, z_n) - \int_\Omega H(x, z);\end{aligned}$$

hence,

$$I(z) - c = o(1) + \int_\Omega H(x, z_n) - \int_\Omega H(x, z).$$

Lebesgue's theorem then yields

$$I(z) - c = \liminf_{n\to\infty} \int_\Omega H(x, z_n) - \int_\Omega H(x, z) \geq 0,$$

that is, $I(z) \geq c$. □

Lemma 3.3. *If (H_3) also holds, there are $r, \rho > 0$ such that $\inf I(\partial B_r E^+) \geq \rho$.*

Proof. By (H_0) and (H_3), for any $\varepsilon > 0$, there is $c_\varepsilon > 0$ such that

$$H(x, u, 0) \leq \varepsilon|u|^2 + c_\varepsilon|u|^{2^*}.$$

Hence,

$$I(u) \geq \frac{1}{2}|\nabla u|_2^2 - \varepsilon|u|_2^2 - c_\varepsilon|u|_{2^*}^{2^*},$$

and the conclusion follows easily. □

Let $e \in E^+$ with $|\nabla e|_2^2 = 1$, and set

$$Q = \{(se, v) : \ 0 \leq s \leq r_1, \|v\|_q \leq r_2\}.$$

Lemma 3.4. *If (H_3) also holds, there are $r_1, r_2 > 0$, with $r_1 > r$, such that $I(z) \leq 0$ for all $z \in \partial Q$.*

Proof. By (H_3), $I(z) \leq 0$ for all $z \in E_q^-$. By (H_2),

$$I((se, v)) \leq \frac{s^2}{2} - \frac{1}{2}|\nabla v|_2^2 - c_1 \int_\Omega (|se|^\alpha + |v|^\beta) + c_2.$$

The conclusion follows since $\alpha > 2$. □

We are now in a position to prove Theorem 1.1.

Proof of Theorem 1.1. Lemmas 3.3 and 3.4 say that I has the linking geometry. Let $Q_n := Q \cap X_n$, and define

$$c_n := \inf_{\gamma \in \Gamma_n} \max I(\gamma(Q_n)),$$

where $\Gamma_n := \{\gamma \in \mathcal{C}(Q_n, X_n) : \gamma|_{\partial Q_n} = id\}$. Then $\rho \leq c_n \leq \kappa := \sup I(Q)$. A standard deformation argument shows that there is $z_n \in X_n$ such that $|I(z_n) - c_n| \leq 1/n$ and $\|I_n'(z_n)\| \leq 1/n$. So we obtain a $(\mathrm{PS})_c^*$ sequence (z_n) with $c \in [\rho, \kappa]$. Lemma 3.2 implies $z_n \rightharpoonup z$ with $I'(z) = 0$ and $I(z) \geq c$. The proof is complete. □

We now consider the multiplicity of solutions using Proposition 2.1.

Lemma 3.5. *I satisfies (I_1).*

Proof. Using (H_2), we obtain

$$I(z) \leq \frac{1}{2}|\nabla u|_2^2 - \frac{1}{2}|\nabla v|_2^2 - c_1 \int_\Omega (|u|^\alpha + |v|^\beta) + c_2.$$

Since all norms in $\mathrm{span}\{e_1, \cdots, e_m\}$ are equivalent, we obtain

$$I(z) \leq -\left(c_3|\nabla u|_2^{\alpha-2} - \frac{1}{2}\right)|\nabla u|_2^2 - \left(\frac{1}{2}|\nabla v|_2^2 + c_1|v|_\beta^\beta\right) + c_2,$$

for all $z = (u, v) \in X^m \simeq \mathrm{span}\{e_1, \cdots, e_m\} \times V_q$. So (I_1) follows easily. □

Lemma 3.6. *I satisfies (I_2).*

Proof. Since $(X^m)^\perp \subset H_0^1(\Omega)$ and $H_0^1(\Omega)$ embeds compactly in $L^p(\Omega)$, we have that $\eta_m > 0$ and $\eta_m \to 0$ as $m \to \infty$, where

$$\eta_m := \sup_{u \in (X^m)^\perp \setminus \{0\}} \frac{|u|_p}{|\nabla u|_2}; \tag{3.10}$$

see Lemma 3.8 in [14]. For $z = (u, 0) \in (X^m)^\perp$, it follows from (H_0) that

$$I(z) = \frac{1}{2}|\nabla u|_2^2 - \int_\Omega H(x, u, 0) \geq \frac{1}{2}|\nabla u|_2^2 - c_1|u|_p^p - c_2$$
$$\geq \frac{1}{2}|\nabla u|_2^2 - c_1\eta_m^p|\nabla u|_2^p - c_2.$$

Setting $r_m = (pc_1\eta_m^p)^{1/(2-p)}$ and $a_m = (p-2)r_m^2/2p - c_2$, we come to the desired conclusion. □

Proof of Theorem 1.2. Since $H(x, z)$ is even in z, I is even. Lemma 3.2 shows that I satisfies the assumption (I_4) of Proposition 2.1. Lemmas 3.5 and 3.6 show that (I_1) and (I_2) hold. Clearly (I_3) is also true. Therefore by Proposition 2.1, there is a sequence $(z_n) \subset E_q$ satisfying $I'(z_n) = 0$ and $I(z_n) \to \infty$. The proof is complete. □

4. The case $p < 2$

Throughout this section we assume that (H_0) is satisfied with $p \in (1, 2)$. We also suppose that $(H_4) - (H_6)$ hold.

Let $E_q = E^1 \oplus E^2$ be as in Section 3. Consider the functional

$$J(z) = -I(z) = \int_\Omega H(x, z) + \frac{1}{2}|\nabla v|_2^2 - \frac{1}{2}|\nabla u|_2^2.$$

Lemma 4.1. *Any* $(\mathrm{PS})_c^*$ *sequence* (z_n) *has a subsequence converging weakly to a critical point* z *of* J *with* $J(z) \le c$*, and* $z = 0$ *only if* $z_n \to 0$ *in* E_q*.*

Proof. The proof is divided into two parts.

Part I. The sequence (z_n) is bounded in E_q. By (H_4) it follows that

$$J(z_n) - J_n'(z_n)(\frac{1}{\mu}u_n, \frac{1}{\nu}v_n) \ge \Big(\frac{1}{2} - \frac{1}{\nu}\Big)|\nabla v_n|_2^2 + \Big(\frac{1}{\mu} - \frac{1}{2}\Big)|\nabla u_n|_2^2 - c.$$

Hence $|\nabla u_n|_2^2 \le c(1 + \|z_n\|_q)$. If $\nu > 2$, we also get $|\nabla v_n|_2^2 \le c(1 + \|z_n\|_q)$. If $\nu = 2$, we use (H_5) and the fact that $|\nabla v|_2^2 \ge \lambda_1 |v|_2^2$ in order to obtain

$$\Big(\frac{1}{2} - \delta\Big)|\nabla v_n|_2^2 \le \frac{1}{2}|\nabla v_n|_2^2 + \int_\Omega H(x, z_n) = J(z_n) + \frac{1}{2}|\nabla u_n|_2^2.$$

Hence, $|\nabla v_n|_2^2 \le c(1 + \|z_n\|_q)$, and we get

$$|\nabla u_n|_2^2 + |\nabla v_n|_2^2 \le c(1 + \|z_n\|_q).$$

Thus, if $q \le 2^*$, then (z_n) is bounded in E_q. Assume next that $q > 2^*$. It follows from (H_6) that

$$J_n'(z_n)(0, v_n) \ge c_1|v_n|_q^q + |\nabla v_n|_2^2 - c_2\left(|v_n|_1 + |u_n|_2^2\right). \tag{4.1}$$

Thus $|\nabla u_n|_2^2 + |\nabla v_n|_2^2 + |v_n|_q^q \le c(1 + \|z_n\|_q)$, which implies that (z_n) is bounded in E_q also in the case when $q > 2^*$.

Part II. We can now suppose that $z_n \rightharpoonup z$ in E_q, $z_n \to z$ in $(L^s(\Omega))^2$ for all $1 \le s < 2^*$, and $z_n(x) \to z(x)$ a.e. in $x \in \Omega$. It follows that z is a critical point of J. As in the proof of Lemma 3.2, using (H_0) and

$$J_n'(z_n)(u_n - u, 0) = \int_\Omega H_u(x, z_n)(u_n - u) - (\nabla u_n, \nabla(u_n - u))_{L^2},$$

we obtain that

$$\begin{aligned}&|(\nabla u_n, \nabla(u_n - u))_{L^2}| \\ &\le o(1) + c(|u_n - u|_1 + |u_n|_p^{p-1}|u_n - u|_p + |v_n|_\beta^{\tau-1}|u_n - u|_\omega) = o(1),\end{aligned}$$

and so $u_n \to u$ in $H_0^1(\Omega)$. Let $P_n : E_q \to X_n$ be the projection as in the proof of Lemma 3.2. So we obtain

$$\begin{aligned}(\nabla v_n, \nabla(v - v_n))_{L^2} &= o(1) + (\nabla v_n, \nabla(P_n v - v_n))_{L^2} \\ &= o(1) + \int_\Omega H_v(x, z_n)(v_n - P_n v) - J_n'(z_n)(0, v_n - P_n v) \\ &= o(1) + \int_\Omega H_v(x, z_n)(v_n - v) + \int_\Omega H_v(x, z_n)(v - P_n v).\end{aligned}$$

Using (H_0), we have

$$\left|\int_\Omega H_v(x, z_n)(v - P_n v)\right| \le c\left(1 + |u_n|_{2^*}^{2^*} + |v_n|_q^{q-1}\right)\|v - P_n v\|_q \to 0.$$

Consequently,

$$(\nabla v_n, \nabla(v - v_n))_{L^2} = \int_\Omega H_v(x, z_n)(v_n - v) + o(1). \tag{4.2}$$

Thus if $q < 2^*$, it follows from (4.2) that $|\nabla v_n|_2 \to |\nabla v|_2$, which implies $v_n \to v$, and so $z_n \to z$. This proves that J satisfies the $(\mathrm{PS})_c^*$ condition in this case, and that $J(z) = c$.

Consider next $q \geq 2^*$. The weak sequential continuity of $H_v(x, \cdot)$ (see the proof of Lemma 2.1) yields $\int_\Omega H_v(x, z_n)v \to \int_\Omega H_v(x, z)v$. By (H_6), $f_n(x) := H_v(x, z_n)v_n + \gamma_6(|v_n| + |u_n|^2) \geq 0$. Using the fact that $|v_n|_1 \to |v|_1$ and $|u_n|_2 \to |u|_2$, and applying Fatou's lemma to the sequence (f_n), we get

$$\liminf_{n\to\infty} \int_\Omega H_v(x, z_n)v_n \geq \int_\Omega H_v(x, z)v.$$

Using this estimate in (4.2), we obtain that $|\nabla v|_2^2 \geq \limsup_{n\to\infty} |\nabla v_n|_2^2$, which implies that $v_n \to v$ in $H_0^1(\Omega)$. In order to conclude that $J(z) \leq c$, we use the estimate

$$J(z_n) - J(z) = \int_\Omega (H(x, z_n) - H(x, z)) + o(1),$$

(H_4) and Fatou's lemma. Finally, if $z = 0$, then $z_n \to 0$ in $(H_0^1(\Omega))^2$. By (4.1),

$$|v_n|_q^q \leq o(1) + c\left(|v_n|_1 + |u_n|_2^2\right) \to 0,$$

and so $z_n \to 0$. □

Remark 4.1. In a similar way, using even simpler arguments, one checks that, if (H_0) holds with $p, q \in (1,\ 2)$, J satisfies the $(\mathrm{PS})_c^*$ condition for all c.

Remark 4.2. Let $\tilde{J}_m = J|_{X^m}$ denote the restriction of J on X^m. As in Lemma 4.1, it is not difficult to check that, if the sequence $(z_m) \subset E_q$, with $z_m \in X^m$, satisfies $J(z_m) \to c$ and $\tilde{J}_m'(z_m) \to 0$ as $m \to \infty$, then it possesses a subsequence converging weakly to a critical point z of J with $J(z) \leq c$, and $z = 0$ only if $z_n \to 0$ in E_q. We also have, as in Remark 4.1, that, if (H_0) holds with $p, q \in (1,\ 2)$, then any such sequence has a convergent subsequence.

Lemma 4.2. *There is an $R > 0$ such that $J(z) \leq 0$ for all $z = (u, 0)$ with $\|z\| \geq R$.*

Proof. By (H_0), we have $H(x, u, 0) \leq c(1 + |u|^p)$. Hence

$$\begin{aligned} J((u,0)) &= \int_\Omega H(x, u, 0) - \frac{1}{2}|\nabla u|_2^2 \leq c_1 + c_2|u|_p^p - \frac{1}{2}|\nabla u|_2^2 \\ &\leq c_1 - \Big(\frac{1}{2}|\nabla u|_2^{2-p} - c_3\Big)|\nabla u|_2^p, \end{aligned}$$

and the lemma follows, since $p < 2$. □

Lemma 4.3. *For $\varepsilon > 0$ small there is $\rho > 0$ such that $J((\varepsilon e_1, v)) \geq \rho$ for all $v \in V_q$, where e_1 is the eigenfunction corresponding to the first eigenvalue λ_1 of $(-\Delta,\ H_0^1(\Omega))$.*

Proof. By (H_5), for $\varepsilon > 0$ small, $H(x, \varepsilon e_1, v) \geq \gamma_4 \varepsilon^\alpha e_1^\alpha - \delta\lambda_1 v^2$; hence,

$$J((\varepsilon e_1, v)) = \int_\Omega H(x, \varepsilon e_1, v) + \frac{1}{2}|\nabla v|_2^2 - \frac{1}{2}\lambda_1\varepsilon^2 \geq (\gamma_4|e_1|_\alpha^\alpha - \frac{1}{2}\lambda_1\varepsilon^{2-\alpha})\varepsilon^\alpha.$$

The conclusion follows. □

We are now ready to prove Theorem 1.3.

Proof of Theorem 1.3. Recall that $X^m \simeq \text{span}\{e_1, \cdots, e_m\} \times V_q$, and consider the restrictions $\tilde{J}_m$ as defined in Remark 4.2. Set $D_R = B_R \cap E^2 = B_R \cap (H_0^1(\Omega) \times \{0\})$ and $D_m = D_R \cap X^m$, where $R > 0$ comes from Lemma 4.2. Define

$$c_m := \inf_{\gamma \in \Gamma_m} \max J(\gamma(D_m)),$$

where $\Gamma_m := \{\gamma \in C(D_m, X^m) : \gamma(z) = z \text{ for all } z \in \partial D_m\}$. It is well known that $\gamma(D_m) \cap W \neq \emptyset$ for all $\gamma \in \Gamma_m$, where $W = \{(\varepsilon e_1, 0)\} \times V_q$ with $\varepsilon > 0$ small. Invoking Lemma 4.3, we fix an $\varepsilon > 0$ so small that there is $\rho > 0$ satisfying $\inf J(W) \geq \rho$. Then we have

$$\rho \leq c_m \leq b := \max J(D_R).$$

The well-known saddle point theorem (cf. [12] or [4], [14]) implies that there is $z_m \in X^m$ satisfying $|J(z_m) - c_m| \leq 1/m$ and $\|\tilde{J}'_m(z_m)\| \leq 1/m$. Now by virtue of Remark 4.2, along a subsequence, $z_m \rightharpoonup z$ with $J'(z) = 0$ and $z \neq 0$, ending the proof. □

We now turn to the proof of Theorems 1.4 and 1.5.

Lemma 4.4. *If, in addition,* $\gamma_3 = 0$ *in* (H_4), *then* J *satisfies* (I_5).

Proof. It follows from (H_5) that

$$\begin{aligned} J(z) &\geq c_1 |u|_\alpha^\alpha + \Big(\frac{1}{2} - \delta\Big)|\nabla v|_2^2 - \frac{1}{2}|\nabla u|_2^2 \\ &\geq \Big(c_2 - \frac{1}{2}|\nabla u|^{2-\alpha}\Big)|\nabla u|_2^\alpha + \Big(\frac{1}{2} - \delta\Big)|\nabla v|_2^2. \end{aligned} \tag{4.3}$$

Since $\alpha < 2$, the result follows in the case when $q \leq 2^*$. Next consider $q > 2^*$. Suppose (I_5) does not hold. Then for any $r > 0$ there is a sequence $z_j \in X^m$ such that $\|z_j\| = r$ and $J(z_j) \to 0$. It follows from (4.3) with $z = z_j$, and for r small, that $|\nabla u_j|_2 \to 0$ and $|\nabla v_j|_2 \to 0$. All this implies that $\int_\Omega H(x, z_j) \to 0$. From assumption (H_0) and the fact that (u_j) lies in a finite-dimensional subspace, it follows that $\int_\Omega H_u(x, z_j) u_j \to 0$. Consequently, by (H_4) with $\gamma_3 = 0$, $\int_\Omega H_v(x, z_j) v_j \to 0$. This, jointly with (H_6), yields

$$|v_j|_q^q \leq c_1 \int_\Omega H_v(x, z_j) v_j + c_2(|v_j|_1 + |u_j|_2^2) \to 0.$$

Hence, $z_j \to 0$ in E_q, which is a contradiction. □

Lemma 4.5. J *satisfies* (I_6).

Proof. By (H_0), $H(x, u, 0) \leq c(|u| + |u|^p)$, and so, for $u \in (X^{m-1})^\perp$, one has

$$\begin{aligned} J((u, 0)) &\leq c_1 \left(|u|_p + |u|_p^p\right) - \frac{1}{2}|\nabla u|_2^2 \\ &\leq \Big(c_1 |u|_p - \frac{1}{4}|\nabla u|_2^2\Big) + \Big(c_1 |u|_p^p - \frac{1}{4}|\nabla u|_2^2\Big) \\ &\leq \Big(c_1 \eta_m - \frac{1}{4}|\nabla u|_2\Big)|\nabla u|_2 + \Big(c_1 \eta_m^p - \frac{1}{4}|\nabla u|_2^{2-p}\Big)|\nabla u|_2^p, \end{aligned}$$

where η_m was defined by (3.10). Let $b_m := (c_1 \eta_m)^2 + (1 - p/2) c_1 \eta_m^p (2 p c_1 \eta_m^p)^{p/(2-p)}$. Then $0 < b_m \to 0$ and $J((u, 0)) \leq b_m$ for all $(u, 0) \in (X^{m-1})^\perp$. □

Proof of Theorem 1.4. Since $H(x,z)$ is even in z, J is even. If $q \le 2^*$, then J satisfies the $(\mathrm{PS})_c^*$ condition for all c (see the proof of Lemma 4.1). If $q > 2^*$, then, using assumption (H_4) applied to a critical point z, we obtain

$$J(z) = J(z) - J_n'(z)\Big(\frac{1}{\mu}u, \frac{1}{\nu}v\Big) \ge \Big(\frac{1}{2} - \frac{1}{\nu}\Big)|\nabla v|_2^2 + \Big(\frac{1}{\mu} - \frac{1}{2}\Big)|\nabla u|_2^2 \ge 0.$$

This, jointly with Lemma 4.1, shows that (I_7) is satisfied. It follows from Lemmas 4.4 and 4.5 that J satisfies (I_5) and (I_6). Therefore, the desired conclusion follows. □

Finally, we prove Theorem 1.5.

Proof of Theorem 1.5. The proof of the existence of one nontrivial solution is similar to that of Theorem 1.3, using Remark 4.2 and Lemmas 4.2 and 4.3. The other conclusion can be obtained along the lines of the proof of Theorem 1.4, using Remark 4.1 and Lemmas 4.4 and 4.5. □

5. The case $p = 2$

In this section we always assume that (H_0) holds with $p = 2$ and $\tau < 1 + q/2$. We also suppose that (H_7) and (H_8) are satisfied. We will apply Proposition 2.3 in order to prove Theorem 1.6. Thus, set

$$E^2 = \mathrm{span}\{e_1^+, \cdots, e_k^+\} \oplus E_q^- \simeq \mathrm{span}\{e_1, \cdots, e_k\} \times V_q\,, \qquad E^1 = E_q \ominus E^2,$$

and

$$X^\ell = E^1 \oplus \mathrm{span}\{e_i^+, \cdots, e_k^+, e_1^-, \cdots, e_j^-\}.$$

One may arrange the bases as $e_n^1 = e_{k+n}^+$ for $n \in \mathbb{N}$, and $e_n^2 = e_{n+i-1}^+$ for $1 \le n \le \ell - j$, $e_n^2 = e_{n-\ell+j}^-$ for $\ell - j < n \le \ell$, $e_n^2 = e_{n-\ell}^+$ for $\ell < n \le \ell + i - 1$, and $e_n^2 = e_{n-k}^-$ for $n > \ell + i - 1$. Consider the functional I given by (2.4).

Lemma 5.1. *I satisfies (I_8); that is, there exist $r, a > 0$ such that $I(z) \ge a$ for all $z \in X^\ell$ with $\|z\|_q = r$.*

Proof. Let $z = (u,v) \in X^\ell$. Since $v \in \mathrm{span}\{e_1, \cdots, e_j\}$, we have $v \in L^\infty$. By (H_0) and (H_7), for any $\varepsilon > 0$, there exists $c_\varepsilon > 0$ such that

$$R_0(x,z) \le \varepsilon|z|^2 + c_\varepsilon(|u|^{2^*} + |v|^q).$$

Thus

$$\begin{aligned} I(z) &= \frac{1}{2}\left(|\nabla u|_2^2 - a_0|u|_2^2\right) - \frac{1}{2}\left(|\nabla v|_2^2 - b_0|v|_2^2\right) - \int_\Omega R_0(x,z) \\ &\ge \frac{1}{2}\Big(1 - \frac{a_0}{\lambda_i}\Big)|\nabla u|_2^2 + \frac{1}{2}\Big(\frac{-b_0}{\lambda_j} - 1\Big)|\nabla v|_2^2 - \varepsilon|z|_2^2 - c_\varepsilon\left(|u|_{2^*}^{2^*} + |v|_q^q\right). \end{aligned}$$

Now the conclusion follows easily. □

Lemma 5.2. *I satisfies (I_9); that is, $\sup I(E^2) < \infty$.*

Proof. For $z \in E^2$ we have, using (H_8), that

$$I(z) = \frac{1}{2}\left(|\nabla u|_2^2 - a_\infty |u|_2^2\right) - \frac{1}{2}|\nabla v|_2^2 - \int_\Omega R_\infty(x,z)$$
$$\leq -\frac{1}{2}\Big(\frac{a_\infty}{\lambda_k} - 1\Big)|\nabla u|_2^2 - \frac{1}{2}|\nabla v|_2^2 + \gamma_9 |u|_\sigma^\sigma - \gamma_8 |v|_q^q + \gamma_9 |\Omega|$$
$$\leq -\Big(\frac{1}{2}\Big(\frac{a_\infty}{\lambda_k} - 1\Big)|\nabla u|^{2-\sigma} - c_1\Big)|\nabla u|_2^\sigma - \Big(\frac{1}{2}|\nabla v|_2^2 + \gamma_8 |v|_q^q\Big) + c_2,$$

which implies that $I(z) \leq 0$ for all $z \in E^2$ with $\|z\|_q$ large. □

Lemma 5.3. *Let $c > 0$. Then any $(\mathrm{PS})_c$ sequence is bounded.*

Proof. We decompose $H_0^1(\Omega)$ as

$$H_0^1(\Omega) = U^- \oplus U^+, \quad u = u^- + u^+,$$

where $U^- = \operatorname{span}\{e_1, \cdots, e_k\}$ and U^+ is the orthogonal complement of U^- in $H_0^1(\Omega)$.

Let (z_n) be a $(\mathrm{PS})_c^*$ sequence. Using the expression of I_n':

$$I_n'(z_n)u_n^+ = |\nabla u_n^+|_2^2 - a_\infty |u_n^+|_2^2 - \int_\Omega \partial_u R_\infty(x, z_n) u_n^+ ,$$

plus (H_8) and the Hölder inequality, we obtain

$$\Big(1 - \frac{a_\infty}{\lambda_{k+1}}\Big)|\nabla u_n^+|_2^2 \leq c_1 |\nabla u_n^+|_2 + \gamma_7\Big(|u_n^+|_1 + |u_n|_\sigma^{\sigma-1}|u_n^+|_\sigma + |v_n|_q^{\tau-1}|u_n^+|_r\Big)$$

where $r = q/(1+q-\tau)$. By assumption, $1 < r < 2$. It then follows from the Sobolev embedding theorems that

$$\Big(1 - \frac{a_\infty}{\lambda_{k+1}}\Big)|\nabla u_n^+|_2^2 \leq c_2\Big(1 + |u_n|_\sigma^{\sigma-1} + |v_n|_q^{\tau-1}\Big)|\nabla u_n^+|_2.$$

Similarly, we deduce that

$$\Big(\frac{a_\infty}{\lambda_k} - 1\Big)|\nabla u_n^-|_2^2 \leq c_2\Big(1 + |u_n|_\sigma^{\sigma-1} + |v_n|_q^{\tau-1}\Big)|\nabla u_n^-|_2.$$

The two previous inequalities imply the estimate

$$|\nabla u_n|_2^2 \leq c_3\left(1 + |u_n|_\sigma^{2(\sigma-1)} + |v_n|_q^{2(\tau-1)}\right). \tag{5.1}$$

Using the expression of H given in (H_8), and recalling that $I(z_n) > 0$ for large n, we obtain

$$\frac{1}{2}|\nabla v_n|_2^2 + \int_\Omega R_\infty(x, z_n) = \frac{1}{2}|\nabla u_n|_2^2 - \frac{a_\infty}{2}|u_n|_2^2 - I(z_n) \leq \frac{1}{2}|\nabla u_n|_2^2. \tag{5.2}$$

Next using (5.2), assumption (H_8) and (5.1), we obtain

$$|\nabla v_n|_2^2 + c_4 |v_n|_q^q \leq c_5\left(1 + |u_n|_\sigma^\sigma + |u_n|_\sigma^{2(\sigma-1)} + |v_n|_q^{2(\tau-1)}\right). \tag{5.3}$$

The combination of (5.1) and (5.3) implies

$$|\nabla z_n|_2^2 + |v_n|_q^q \leq c_6\left(1 + |u_n|_2^\sigma + |v_n|_q^{2(\tau-1)}\right).$$

Since $\sigma < 2$ and $2(\tau - 1) < q$, we see that (z_n) is bounded. □

Lemma 5.4. *I satisfies (I_{10}).*

Proof. Let (z_n) be a $(\mathrm{PS})_c^*$ sequence with $c > 0$. Using Lemma 5.3, an argument similar to that of Lemma 3.2 shows that along a subsequence $z_n \rightharpoonup z \in \mathcal{K}_c$, we have $u_n \to u$ in $H_0^1(\Omega)$. Since $E^1 \subset H_0^1(\Omega)$, we have $P^1 z_n \to P^1 z$. □

Proof of Theorem 1.6. Since $H(x, z)$ is even in z, I is even. By assumption, $I(0) = 0$. Lemmas 5.1, 5.2 and 5.4 show that I satisfies $(I_8) - (I_{10})$. Now Proposition 2.3 applies, and the proof is complete. □

Acknowledgments

De Figueiredo was supported by CNPq-FAPESP-PRONEX. Ding was supported by the Special Funds for Major State Basic Research Projects of China, the funds of CAS/China R9902, 1001800, and the CNPq of Brazil.

References

1. T. Bartsch, *Infinitely many solutions of a symmetric Dirichlet problem*, Nonlinear Anal. TMA, **20** (1993), 1205–1216. MR **94g:**35093
2. T. Bartsch and D. G. De Figueiredo, *Infinitely many solutions of nonlinear elliptic systems*, Progress in Nonlinear Differential Equations and Their Applications, vol. 35, Birkhäuser, Basel/Switzerland, 1999, pp. 51–67. MR **2000j:**35072
3. V. Benci and P. Rabinowitz, *Critical point theorems for indefinite functionals*, Invent. Math. **52** (1979), 241–273. MR **80i:**58019
4. K. C. Chang, *Infinite-Dimensional Morse Theory and Multiple Solution Problems*, Birkhäuser, Boston, 1993. MR **94e:**58023
5. D. G. Costa and C. A. Magalhães, *A variational approach to noncooperative elliptic systems*, Nonlinear Anal. TMA **25** (1995), 699–715. MR **96g:**35070
6. D. G. Costa and C. A. Magalhães, *A unified approach to a class of strongly indefinite functionals*, J. Differential Equations **125** (1996), 521–547. MR **96m:**58061
7. D. G. De Figueiredo and C. A. Magalhães, *On nonquadratic Hamiltonian elliptic systems*, Adv. Differential Equations **1** (1996), 881–898. MR **97f:**35049
8. D. G. De Figueiredo and P. L. Felmer, *On superquadratic elliptic systems*, Trans. Amer. Math. Soc. **343** (1994), 97–116. MR **94g:**35072
9. Y. H. Ding, *Infinitely many entire solutions of an elliptic system with symmetry*, Topological Methods in Nonlinear Anal. **9** (1997), 313–323. MR **99a:**35062
10. P. L. Felmer, *Periodic solutions of "superquadratic" Hamiltonian systems*, J. Differential Equations **102** (1993), 188–207. MR **94c:**58160
11. J. Hulshof and R. van der Vorst, *Differential systems with strongly indefinite variational structure*, J. Funct. Anal. **114** (1993), 32-58. MR **94g:**35073
12. P. H. Rabinowitz, *Minimax Methods in Critical Point Theory with Applications to Differential Equations*, C. B. M. S. vol. 65, Amer. Math. Soc., Providence, RI, 1986. MR **87j:**58024
13. E. A. B. Silva, *Nontrivial solutions for noncooperative elliptic systems at resonance*, Electronic J. Differential Equations **6** (2001), 267–283. MR **2001j:**35097
14. M. Willem, *Minimax Theorems*, Progress in Nonlinear Differential Equations and their Applications, vol. 24, Birkhäuser, Boston, 1996. MR **97h:**58037

IMECC-UNICAMP, Caixa Postal 6065, 13083-970 Campinas S.P. Brazil
E-mail address: djairo@ime.unicamp.br

Institute of Mathematics, AMSS, Chinese Academy of Sciences, 100080 Beijing, People's Republic of China
E-mail address: dingyh@math03.math.ac.cn

Available online at www.sciencedirect.com
SCIENCE DIRECT®

ACADEMIC PRESS

Journal of Functional Analysis 199 (2003) 452–467

JOURNAL OF Functional Analysis

http://www.elsevier.com/locate/jfa

Local superlinearity and sublinearity for indefinite semilinear elliptic problems

Djairo G. De Figueiredo,[a,1] Jean-Pierre Gossez,[b,*,2] and Pedro Ubilla[c,3]

[a] *IMECC, UNICAMP, Caixa Postal 6065, 13081-970 Campinas, SP, Brazil*
[b] *Département de Mathématique, C.P. 214, Université Libre de Bruxelles, 1050 Bruxelles, Belgium*
[c] *Departamento de Matemáticas y C. C., Universidad de Santiago de Chile, Casilla 307, Correo 2, Santiago, Chile*

Received 17 January 2002; accepted 8 July 2002

Abstract

In this paper the usual notions of superlinearity and sublinearity for semilinear problems like $-\Delta u = f(x,u)$ are given a local form and extended to indefinite nonlinearities. Here $f(x,s)$ is allowed to change sign or to vanish for s near zero as well as for s near infinity. Some of the well-known results of Ambrosetti–Brézis–Cerami are partially extended to this context.

Keywords: Superlinearity; Sublinearity; Indefinite nonlinearity; Concave–convex nonlinearity; Semilinear elliptic problem

1. Introduction

This paper is concerned with the study of existence, nonexistence and multiplicity of solutions for the problem

$$\begin{cases} -\Delta u = f(x,u) & \text{in } \Omega, \\ u \geqslant 0, \ u \not\equiv 0 & \text{in } \Omega, \\ u = 0 & \text{on } \partial\Omega. \end{cases} \tag{1.1}$$

*Corresponding author.
E-mail addresses: djairo@ime.unicamp.br (D.G. De Figueiredo), gossez@ulb.ac.be (J.-P. Gossez), pubilla@lauca.usach.cl (P. Ubilla).
[1] Supported by CNPq-FNRS, PRONEX, FAPESP.
[2] Supported by CNPq-FNRS.
[3] Supported by grants DICYT-USACH and FONDECYT 1990183.

Originally published in: J. Funct. Anal. 199 (2003), 452-467,

Reprint by Springer-Verlag Berlin Heidelberg 2013 with kind permission,
DOI: 10.1007/978-3-319-02856-9_36

Here Ω is a bounded domain in $\mathbb{R}^N$ and $f:\Omega\times\mathbb{R}^+\to\mathbb{R}$ is a Carathéodory function.

There has been a great deal of interest on problem (1.1) since the pioneering work of Ambrosetti–Rabinowitz [2]. Our purpose here is to introduce some kind of local analogues to the classical conditions of sublinearity at 0 and of superlinearity at ∞.

Let us recall the definitions of these classical conditions. Problem (1.1) is said to be sublinear at 0 if there exist $\alpha > \lambda_1(\Omega)$ and $s_0 > 0$ such that

$$f(x,s)\geqslant \alpha s \quad \text{for a.e. } x\in\Omega \text{ and all } 0\leqslant s\leqslant s_0. \tag{1.2}$$

Here $\lambda_1(\Omega)$ denotes the principal eigenvalue of $-\Delta$ on $H_0^1(\Omega)$. Problem (1.1) is said to be superlinear at ∞ if there exist $\beta > \lambda_1(\Omega)$ and $s_1\geqslant 0$ such that

$$f(x,s)\geqslant \beta s \text{ for a.e } x\in\Omega \text{ and all } s\geqslant s_1. \tag{1.3}$$

Problem (1.1) under (1.2) and (1.3) was studied for instance in [1,8].

One feature of the hypothesis on the nonlinearity f we consider in the present paper is that some sort of sublinearity at 0 and some sort of superlinearity at ∞ will be required to hold only on open subsets Ω_1 and Ω_2 of Ω, subsets which may be small and possibly disjoint (cf. hypothesis (H_4) and (H_5) in Section 2). In particular, we will investigate in this context how much of the sublinearity at 0 and of the superlinearity at ∞ is needed to give the existence of multiple solutions. We will also show that for the class of sublinear problems considered here, the existence of at least one solution (when no multiplicity is aimed at) can be established with essentially no restrictions on the growth of the nonlinearity at infinity. We will also derive some nonexistence results.

Our approach to obtain multiple solutions in Theorem 2.1 is variational. In this respect another feature of the problems we consider here is the following: although the (PS) condition will hold, the classical superquadraticity condition of Ambrosetti–Rabinowitz [2] will generally not be satisfied (cf. Remark 2.3) and some substitute for it has to be introduced (cf. hypothesis (H_2) in Section 2). Our approach to derive the existence of at least one solution in Theorem 2.5 is based on the method of weak upper and lower solutions.

A typical example to which our results apply is

$$\begin{cases} -\Delta u = \lambda a(x)u^q + b(x)u^p & \text{in } \Omega,\\ u\geqslant 0,\ u\not\equiv 0 & \text{in } \Omega,\\ u=0 & \text{on } \partial\Omega, \end{cases} \tag{1.4}$$

where $\lambda > 0$ is a parameter and the exponents p and q satisfy $0\leqslant q<1<p$ with $p\leqslant 2^*-1$ if $N\geqslant 3$, $p<\infty$ if $N=1$ or 2. Here $2^* := 2N/(N-2)$. Problems of this form have drawn much attention after the work [1]. In there $a(x)\equiv 1$, $b(x)\equiv 1$, so that the corresponding $f(x,u)$ clearly satisfies (1.2) and (1.3). Here we will allow the coefficients $a(x)$, $b(x)$ in (1.4) to vanish on parts of Ω as well as to change sign in Ω. As we will see some of the results of [1] can be partially extended to this situation (cf. Remark 2.11). As a byproduct of our results, we also obtain an estimate from above and from below for the parameter value Λ separating in [1] existence from nonexistence (cf. Remark 2.12).

Several papers have been devoted in the last years to the study of nonlinearities with indefinite sign. Most of them however deal with problems which are not directly comparable to those considered here (cf. e.g. [3–6]). The present work is more related to results in [12], although we consider more general nonlinearities. In particular Theorem 2 of [12], when applied to example (1.4), requires $b(x)$ to be positive, which is not the case of our Corollary 2.2 below.

In Section 2 we state in detail our results relative to (1.1). We also derive there their consequences for (1.4). Proofs are given in Section 3.

2. Statement of results

In this section, we state our results relative to (1.1) and derive their consequences for (1.4). We will always assume $N \geqslant 3$ but our results can easily be adapted by standard modifications to the cases $N = 1$ or 2 for nonlinearities with power growth. Also by a solution we will always mean a weak solution in the $H_0^1(\Omega)$ sense. We will denote by σ' the Hölder conjugate of σ, and by $\| \|_\sigma$ the $L^\sigma(\Omega)$ norm.

Let $f(x,s)$ be a Carathéodory function from $\Omega \times \mathbb{R}^+$ into $\mathbb{R}$, where Ω is a bounded domain in $\mathbb{R}^N$ with $N \geqslant 3$. The following assumptions on f will be used below in our existence results:

(H_0) $f(x,0) \geqslant 0$ for a.e. $x \in \Omega$.

(H_1) There exist $1 \leqslant \sigma < 2^*$, $d_1 \in L^{\sigma'}(\Omega)$, $d_2 > 0$ such that

$$|f(x,s)| \leqslant d_1(x) + d_2|s|^{\sigma-1}$$

for a.e. $x \in \Omega$ and all $s \geqslant 0$.

(H_2) There exist $\Theta > 2$, $1 \leqslant r < 2$, $d \in L^{(\frac{2^*}{r})'}(\Omega)$, with $d \geqslant 0$ a.e. in Ω, $s_0 \geqslant 0$, such that

$$\Theta F(x,s) \leqslant s f(x,s) + d(x) s^r$$

for a.e. $x \in \Omega$ and all $s \geqslant s_0$, where F denotes the primitive of f:

$$F(x,s) := \int_0^s f(x,t)\, dt.$$

(H_3) There exist $0 \leqslant q < 1 < p < 2^* - 1$, $a_0 \in L^{\sigma_q}(\Omega)$, with $\sigma_q := (\frac{2^*}{q+1})'$ and $a_0 \geqslant 0$ a.e. in Ω, $b_0 \in L^{\sigma_p}(\Omega)$, with $\sigma_p := (\frac{2^*}{p+1})'$ and $b_0 \geqslant 0$ a.e. in Ω, such that

$$f(x,s) \leqslant a_0(x)s^q + b_0(x)s^p$$

for a.e. $x \in \Omega$ and all $s \geqslant 0$.

(H_4) There exist a nonempty subdomain $\Omega_1 \subset \Omega$, $\Theta_1 > \lambda_1(\Omega_1)$, $s_1 > 0$, such that

$$F(x,s) \geqslant \Theta_1 \frac{s^2}{2}$$

for a.e. $x \in \Omega_1$ and all $0 \leqslant s \leqslant s_1$.

(H_5) There exist a nonempty open subset $\Omega_2 \subset \Omega$, $\Theta_2 > 0$, $s_2 \geqslant 0$, such that

$$F(x,s) \geqslant \Theta_2 s^2$$

for a.e. $x \in \Omega_2$ and all $s \geqslant s_2$, with the additional requirement that the function $d(x)$ appearing in (H_2) is bounded on Ω_2.

Our first result concerns the existence of multiple solutions for a subcritical problem using a variational method.

Theorem 2.1. *Assume* (H_0), ..., (H_5). *Then one can find explicitly* $\eta = \eta(p,q,N) > 0$ *such that if*

$$||a_0||_{\sigma_q}^{p-1} ||b_0||_{\sigma_p}^{1-q} < \eta,$$

then problem (1.1) *has at least two solutions. One of them, call it* v, *satisfies* $\Phi(v) > 0$, *while the other, call it* w, *satisfies* $\Phi(w) < 0$, *where* Φ *denotes the associated energy*:

$$\Phi(u) := \frac{1}{2} \int_\Omega |\nabla u|^2 - \int_\Omega F(x,u).$$

Moreover, if f *varies in such a way that the coefficients in* (H_3) *fulfill*

$$a_0 \to 0 \text{ in } L^{\sigma_q}(\Omega) \text{ and } b_0 \text{ bounded in } L^{\sigma_p}(\Omega), \tag{2.1}$$

then the solution $w = w_f$ *can be constructed such that* $w_f \to 0$ *in* $H_0^1(\Omega)$.

Here are some comments on the hypotheses (H_0), ..., (H_5).

When assuming (H_0), we will always understand that $f(x,s)$ has been extended for $s<0$ by putting $f(x,s)=f(x,0)$ for a.e. $x\in\Omega$ and all $s<0$. Doing so, any solution $u\in H_0^1(\Omega)$ of

$$\begin{cases} -\Delta u = f(x,u) & \text{in } \Omega, \\ u=0 & \text{on } \partial\Omega \end{cases} \tag{2.2}$$

is automatically $\geqslant 0$ in Ω (indeed take $-u^-$ as testing function in (2.2)). Solving (1.1) consequently reduces to looking for nontrivial solutions of (2.2) where f has been extended as indicated.

Assumption (H_1) is the standard subcritical growth condition.

Assumption (H_2) is a weaker form of the classical superquadraticity condition of Ambrosetti–Rabinowitz [2] (the latter corresponds to $d(x)\equiv 0$, with the additional requirement that $F(x,s)$ is >0 for a.e. $x\in\Omega$ and all $s\geqslant s_0$). As it will be clear from Corollary 2.2, (H_2) allows $f(x,s)$ to change sign near $s=0$ and does not imply the positivity of $F(x,s)$ at infinity. Another generalization of this superquadraticity condition of Ambrosetti–Rabinowitz was introduced in [7], which was called there nonquadraticity condition. As we shall see later, it turns out that these conditions of [2,7] are generally not satisfied in the problems we consider (cf. Remark 2.3).

Assumption (H_3) is a bound from above for f of the same type as the one that appears in the right-hand side of example (1.4). Note that when $0<q<1$, (H_3) combined with (H_0) imply $f(x,0)\equiv 0$.

The inequality in (H_4) is clearly implied by the inequality $f(x,s)\geqslant\Theta_1 s$. The latter is itself implied by the following "local" sublinearity condition at 0:

$$\lim_{\substack{s\to 0\\ s>0}} \frac{f(x,s)}{s} = +\infty \tag{2.3}$$

uniformly for $x\in\Omega_1$.

Finally, the first part of (H_5) is clearly implied by the following "local" superlinearity condition at ∞:

$$\lim_{s\to\infty} \frac{f(x,s)}{s} = +\infty \tag{2.4}$$

uniformly for $x\in\Omega_2$, together with the requirement that, for some constants c and S,

$$F(x,s)\geqslant -c \text{ for a.e. } x\in\Omega_2 \text{ and all } s\geqslant S. \tag{2.5}$$

We now turn to example (1.4).

Corollary 2.2. *Assume in* (1.4) $\lambda>0$, $0\leqslant q<1<p<2^*-1$, $a\in L^{\tau_q}(\Omega)$ *with* $\tau_q>\sigma_q$, $b\in L^{\tau_p}(\Omega)$ *with* $\tau_p>\sigma_p$, *with in addition* $a(x)\geqslant 0$ *a.e. in* Ω *in case* $q=0$. *Suppose*

(i) *there exists a nonempty open subset* $\Omega_1\subset\Omega$ *such that, on* Ω_1, $a(x)\geqslant\varepsilon_1$ *for some* $\varepsilon_1>0$ *and* $b(x)$ *is bounded from below;*

(ii) *there exists a nonempty open subset* $\Omega_2 \subset \Omega$ *such that, on* Ω_2, $b(x) \geqslant \varepsilon_2$ *for some* $\varepsilon_2 > 0$ *and* $a(x)$ *is bounded from above and from below.*

Then one can find explicitly $\bar{\eta} = \bar{\eta}(p, q, N) > 0$ *such that if*

$$\lambda < \frac{\bar{\eta}}{\|a^+\|_{\sigma_q} \|b^+\|_{\sigma_p}^{(1-q)/(p-1)}},$$

then problem (1.4) *has at least two solutions* v *and* w *which satisfy* $\Psi(v) > 0$ *and* $\Psi(w) < 0$, *where* Ψ *denotes the energy functional associated to* (1.4). *Moreover, if* $\lambda \to 0$, *then the solution* $w = w_\lambda$ *can be constructed such that* $w_\lambda \to 0$ in $H_0^1(\Omega)$.

Proof. It suffices to verify that for each $\lambda > 0$, hypothesis $(H_0), \ldots, (H_5)$ of Theorem 2.1 hold. Assumptions (H_0) and (H_3) are clear. The verification of (H_1) follows easily from the strict inequalities $\tau_q > \sigma_q$ and $\tau_p > \sigma_p$, by using Young's inequality. In the verification of (H_2), one can take for example $\Theta = p + 1$, $r = q + 1$, $d(x) = \lambda(\frac{\Theta}{q+1} - 1)a^+(x)$ and $s_0 = 0$. Using hypothesis (i), one deduces that the sublinear condition at 0 (2.3) holds on Ω_1, which yields (H_4). Using hypothesis (ii), one deduces that the superlinear condition at ∞ (2.4), as well as (2.5), hold on Ω_2; since in addition $d(x)$ in (H_2) can be taken as $\lambda(\frac{\Theta}{q+1} - 1)a^+(x)$, which is bounded on Ω_2 by assumption (ii), (H_5) follows. □

Remark 2.3. The superquadraticity condition of [2] and the nonquadraticity condition of [7] are generally not satisfied in the framework of Corollary 2.2. This is the case for instance if $a(x) \equiv 0$ on Ω_2 and $b(x) \equiv 0$ on Ω_1.

Remark 2.4. Assume in Corollary 2.2 that Ω is of class $C^{1,1}$ and that a and b belong to $L^\infty(\Omega)$. Regularity theory then implies that w_λ remains bounded in every $W^{2,r}(\Omega)$ with $r < \infty$, and consequently $w_\lambda \to 0$ in $L^\infty(\Omega)$. Combining this $L^\infty(\Omega)$ convergence with Theorem 2.2 of [1], one deduces that for $a(x) \equiv b(x) \equiv 1$, the solution w_λ above coincides for λ sufficiently small with the minimal solution constructed in [1].

Our next result concerns the existence of at least one solution for a possibly supercritical problem using the upper-lower solutions method.

Theorem 2.5. *Assume* (H_0) *as well* $(H_3)'$ *and* $(H_4)'$ *below*:

$(H_3)'$ *There exist* $0 \leqslant q < 1 < p$ *and* $a_0, b_0 \geqslant 0$ *such that*

$$f(x, s) \leqslant a_0 s^q + b_0 s^p$$

for a.e. $x \in \Omega$ *and all* $s \geqslant 0$.

$(H_4)'$ *There exist a nonempty* $C^{1,1}$ *subdomain* $\Omega_1 \subset \Omega$, $s_1 > 0$, *such that*

$$f(x, s) \geqslant \lambda_1(\Omega_1)s$$

for a.e. $x \in \Omega$ *and all* $0 \leqslant s \leqslant s_1$.

Assume also Ω *of class* $C^{1,1}$. *Then there exists* $\delta = \delta(p, q, \Omega, b_0) > 0$ *such that if*

$$a_0 < \delta,$$

then problem (1.1) *has at least one solution.*

We finally turn to the statement of our nonexistence results. It will be convenient from now on to denote by $\{a > 0\}$ the set of all $x \in \Omega$ such that $a(x) > 0$, and similarly for $a \geqslant 0, \dots$. All the inclusions or equalities of sets in the statements below should be understood "up to a set of zero measure".

Let f be as before a Carathéodory function from $\Omega \times \mathbb{R}^+$ into $\mathbb{R}$, where Ω is a bounded domain in $\mathbb{R}^N$, $N \geqslant 3$. The following assumptions on f will now be used:

(H_6) There exist $d_1 \in L^{(2^*)'}(\Omega)$, $d_2 > 0$ such that

$$|f(x, s)| \leqslant d_1(x) + d_2 s^{2^*-1}$$

for a.e. $x \in \Omega$ and all $s \geqslant 0$.

(H_7) Ω can be partitioned into $\Omega = A_+ \cup A_- \cup A_0$ where

$$A_+ := \{x \in \Omega \colon f(x, \cdot) \geqslant 0 \text{ with } f(x, \cdot) \not\equiv 0\},$$

$$A_- := \{x \in \Omega \colon f(x, \cdot) \leqslant 0 \text{ with } f(x, \cdot) \not\equiv 0\},$$

$$A_0 := \{x \in \Omega \colon f(x, \cdot) \equiv 0\};$$

moreover, there exists a nonempty domain $\tilde{\Omega}$ such that $A_+ \subset \tilde{\Omega} \subset A_+ \cup A_0$, with in addition $\tilde{\Omega}$ of class $C^{1,1}$ in case $\tilde{\Omega} \neq \Omega$.

(H_8) There exist $0 \leqslant q < 1 < p$ and nonnegative functions $\tilde{a}$, $\tilde{b}$ on $\tilde{\Omega}$ such that

$$f(x, s) \geqslant \tilde{a}(x) s^q + \tilde{b}(x) s^p \quad \text{for a.e } x \in \tilde{\Omega} \text{ and all } s \geqslant 0,$$

$$\tilde{m} := (\tilde{a})^{\frac{p-1}{p-q}} (\tilde{b})^{\frac{1-q}{p-q}} \not\equiv 0 \quad \text{on } \tilde{\Omega},$$

$$\tilde{m} \in L^r(\tilde{\Omega}) \quad \text{for some } r > \frac{N}{2}.$$

Theorem 2.6. *Suppose* (H_6), (H_7), (H_8) *and denote by* $\lambda_1(\tilde{m}, \tilde{\Omega})$ *the principal eigenvalue of* $-\Delta$ *on* $H_0^1(\tilde{\Omega})$ *for the weight* $\tilde{m}$. *Then one can find explicitly* $c = c(p, q) > 0$ *such that problem* (1.1) *has no solution if*

$$\lambda_1(\tilde{m}, \tilde{\Omega}) < c(p, q).$$

Here are some comments on hypotheses (H_6)–(H_8).

Hypothesis (H_6) is analogous to (H_1).

Hypothesis (H_7) is a restriction on the possible changes of sign of f. The fact that $A_+ \cup A_- \cup A_0$ is the whole of Ω means that for any given x, $f(x,\cdot)$ does not change sign in $\mathbb{R}^+$. Note that (H_7) is automatically satisfied, with $\tilde{\Omega} = \Omega$, when $f \geqslant 0$ on $\Omega \times \mathbb{R}^+$.

Hypothesis (H_8) compared to (H_3) shows an estimate on $\tilde{\Omega}$ in the other direction.

Corollary 2.7. *Assume in* (1.4) $\lambda > 0$, $0 \leqslant q < 1 < p \leqslant 2^* - 1$, $a \in L^{\sigma_q}(\Omega)$ *and* $b \in L^{\sigma_p}(\Omega)$, *where the* σ*'s are defined in* (H_3). *Suppose that there exists a domain* $\tilde{\Omega}$ (*with* $\tilde{\Omega}$ *of class* $C^{1,1}$ *in case* $\tilde{\Omega} \neq \Omega$) *such that*

$$\{a > 0\} \cup \{b > 0\} \subset \tilde{\Omega} \subset \{a \geqslant 0\} \cap \{b \geqslant 0\}, \tag{2.6}$$

$$meas(\tilde{\Omega} \cap \{a > 0\} \cap \{b > 0\}) > 0, \tag{2.7}$$

$$a^{\frac{p-1}{p-q}} \cdot b^{\frac{1-q}{p-q}} \in L^r(\tilde{\Omega}) \text{ for some } r > \frac{N}{2}. \tag{2.8}$$

Then there exists $\bar{c} = \bar{c}(p, q, \tilde{\Omega}, a, b)$ *such that problem* (1.4) *has no solution if*

$$\lambda > \bar{c}.$$

Note that the assumptions $a \in L^{\sigma_q}(\Omega)$ and $b \in L^{\sigma_p}(\Omega)$ imply $a^{\frac{p-1}{p-q}} b^{\frac{1-q}{p-q}} \in L^{\frac{N}{2}}(\tilde{\Omega})$, which is almost (2.8).

Here are some situations where Corollary 2.7 applies. In the examples below we only consider the question of existence of $\tilde{\Omega}$ such that (2.6) and (2.7) hold.

Example 2.8. If $a \geqslant 0$ and $b \geqslant 0$ a.e. in Ω, with $meas(\{a > 0\} \cap \{b > 0\}) > 0$, then Corollary 2.7 applies. It suffices to take $\tilde{\Omega} = \Omega$.

Example 2.9. If $\{a > 0\} = \{b > 0\}$ and if this set is a nonempty $C^{1,1}$ domain, then Corollary 2.7 applies. It suffices to take this $C^{1,1}$ domain as $\tilde{\Omega}$.

Example 2.10. If $meas(\{a > 0\} \cap \{b > 0\}) > 0$ and if $\{a > 0\} \cup \{b > 0\}$ has a compact closure contained in a domain which itself is contained in $\{a \geqslant 0\} \cap \{b \geqslant 0\}$, then Corollary 2.7 applies.

Remark 2.11. One of the main results of [1] is the following. Suppose in (1.4) $q > 0$, $a(x) \equiv 1$, $b(x) \equiv 1$ and $p \leqslant 2^* - 1$; then there exists $0 < \Lambda < \infty$ such that (1.4) has at least two solutions if $0 < \lambda < \Lambda$, at least one solution if $\lambda = \Lambda$, and no solution if $\lambda > \Lambda$. Corollaries 2.2 and 2.7 show that part of these conclusions remain valid when variable coefficients are present in (1.4), which may vanish on parts of Ω and change sign in Ω.

Remark 2.12. Corollary 2.2 and Theorem 2.6 provide, when $0<q<1<p<2^*-1$, estimates from above and from below for the parameter value Λ considered in Remark 2.11:

$$\frac{\bar{\eta}(p,q,N)}{|\Omega|^{\frac{1}{\sigma_q}+\frac{1}{\sigma_p}\left(\frac{1-q}{p-1}\right)}}\leqslant\Lambda\leqslant\left(\frac{\lambda_1(\Omega)}{c(p,q)}\right)^{\frac{p-q}{p-1}}.$$

Note that the constant $\bar{\eta}(p,q,N)$ has an explicit form in terms of p,q and the best Sobolev constant S. In fact $\bar{\eta}(p,q,N)=\eta(p,q,N)^{1/(p-1)}$ where $\eta(p,q,N)$ is the constant of Theorem 2.1, and the expression of $\eta(p,q,N)$ in terms of p,q,S appears in the proof of Theorem 2.1 (cf. (3.3)). Similarly $c(p,q)$ has on explicit form in terms of p,q (cf. the proof of Lemma 3.6).

3. Proofs

This section is devoted to the proofs of Theorems 2.1, 2.5 and 2.6. We start with the former.

Proposition 3.1. *Assume* (H_0)–(H_2). *Then* Φ *satisfies the (PS) condition on* $H_0^1(\Omega)$.

Proof. As already observed in connection with (H_0), f is extended on all $\Omega\times\mathbb{R}$ by putting $f(x,s)=f(x,0)$ for a.e. $x\in\Omega$ and all $s<0$. It is then clear that (H_1) implies that Φ is a C^1 functional on $H_0^1(\Omega)$. Let u_n be a (PS) sequence, i.e. $\Phi(u_n)$ bounded and $\Phi'(u_n)\to 0$. So, for Θ as in (H_2) and for some $\varepsilon_n\to 0$ and some constant C,

$$\Theta\Phi(u_n)-\Phi'(u_n)u_n\leqslant C+\varepsilon_n||u_n||,$$

where $||u||$ denotes the $H_0^1(\Omega)$ norm $(\int_\Omega ||\nabla u||^2)^{1/2}$. This relation means that

$$\left(\frac{\Theta}{2}-1\right)||u_n||^2-\int_\Omega(\Theta F(x,u_n)-u_nf(x,u_n))\leqslant C+\varepsilon_n||u_n||$$

and consequently, by (H_2),

$$\left(\frac{\Theta}{2}-1\right)||u_n||^2\leqslant c'+\int_\Omega d(x)(u_n^+)^r+\varepsilon_n||u_n||$$

for another constant c'. Since $r<2$, one deduces that $||u_n||$ remains bounded. The existence of a subsequence of u_n which converges in $H_0^1(\Omega)$ then follows by standard arguments, using (H_1). $\square$

The following lemma will be used in the proof of Theorem 2.1.

Lemma 3.2. *Let $0 \leqslant q < 1 < p$, $A > 0$, $B > 0$, and consider the function*

$$\Psi_{A,B}(t) := t^2 - At^{q+1} - Bt^{p+1}$$

for $t \geqslant 0$. Then $\max\{\Psi_{A,B}(t) : t \geqslant 0\}$ is > 0 if and only if

$$A^{p-1}B^{1-q} < \frac{(p-1)^{p-1}(1-q)^{1-q}}{(p-q)^{p-q}} := \eta_1(p,q).$$

Moreover, for $t = t_B := [(1-q)/B(p-q)]^{1/(p-1)}$, one has

$$\Psi_{A,B}(t_B) = t_B^2\left[\frac{p-1}{p-q} - AB^{\frac{1-q}{p-1}}\left(\frac{p-q}{1-q}\right)^{\frac{1-q}{p-1}}\right]. \tag{3.1}$$

Proof. Since $\Psi_{A,B}(t) = t^{q+1}(t^{1-q} - A - Bt^{p-q})$, $\Psi_{A,B}(t)$ is > 0 if and only if $t^{1-q} - A - Bt^{p-q}$ is > 0. The derivative of $t^{1-q} - A - Bt^{p-q}$ vanishes exactly for $t = t_B$ and one readily computes $\Psi_{A,B}(t_B)$, as indicated in (3.1). The conclusion of the lemma then follows easily. □

Proof of Theorem 2.1. The first solution v will be constructed by using the classical mountain pass theorem of [2]. For that purpose we first observe that $\Phi(0) = 0$. We also have, by (H_3),

$$\begin{aligned}\Phi(u) &\geqslant \frac{1}{2}||u||^2 - \int_\Omega \left[\frac{a_0(u^+)^{q+1}}{q+1} + \frac{b_0(u^+)^{p+1}}{p+1}\right] \\ &\geqslant \frac{1}{2}||u||^2 - c_1||a_0||_{\sigma_q}||u||^{q+1} - c_2||b_0||_{\sigma_p}||u||^{p+1}\end{aligned} \tag{3.2}$$

for all $u \in H_0^1(\Omega)$, where the positive constants c_1, c_2 depend only on p, q and N. In fact $c_1 = (q+1)^{-1}S^{-\frac{q+1}{2}}$ and $c_2 = (p+1)^{-1}S^{-\frac{p+1}{2}}$, where

$$S := \inf\left\{\int_\Omega |\nabla u|^2 : u \in H_0^1(\Omega) \text{ and } \int_\Omega |u|^{2^*} = 1\right\}$$

is the best Sobolev constant, which is independent of Ω. We now apply to (3.2) Lemma 3.2 above with $A = 2c_1||a_0||_{\sigma_q}$ and $B = 2c_2||b_0||_{\sigma_p}$ (which are > 0 because (H_4) and (H_5) imply a_0 and $b_0 \not\equiv 0$). This gives that for all $u \in H_0^1(\Omega)$ with $||u|| = t_B$, one has

$$\Phi(u) \geqslant \tfrac{1}{2}\Psi_{A,B}(t_B) > 0$$

provided that $A^{p-1}B^{1-q}<\eta_1(p,q)$, i.e., provided that

$$||a_0||_{\sigma_q}^{p-1}||b_0||_{\sigma_p}^{1-q}<\frac{\eta_1(p,q)}{(2c_1)^{p-1}(2c_2)^{1-q}}:=\eta(p,q,N). \tag{3.3}$$

So we have obtained, under (3.3), a "range of mountains" around 0.

We will now look for some $u_2\in H_0^1(\Omega)$ such that $\Phi(tu_2)\to-\infty$ as $t\to+\infty$. Assumption (H_5) will be used here. Let us choose s_3 sufficiently large so that $s_3\geqslant s_0$ (from (H_2)) and for some $\Theta_3>0$, $F(x,s)\geqslant\Theta_3 s^2+1$ for a.e. $x\in\Omega_2$ and all $s\geqslant s_3$, which is clearly possible by (H_5). For $x\in\Omega_2$ and $s\geqslant s_3$, we then divide the inequality of (H_2) by $sF(x,s)$, integrate from s_3 to s and take the exponential to get

$$F(x,s)\geqslant F(x,s_3)\left(\frac{s}{s_3}\right)^{\Theta}\exp\left(-d(x)\int_{s_3}^{s}\frac{t^{r-1}}{F(x,t)}dt\right).$$

Consequently, using the second part of (H_5),

$$F(x,s)\geqslant cs^{\Theta} \tag{3.4}$$

for a.e. $x\in\Omega_2$ and all $s\geqslant s_3$, where $c>0$ is a constant. We now take a smooth function u_2 with support in Ω_2 and $u_2\geqslant 0$, $u_2\not\equiv 0$, and consider tu_2 with $t\geqslant t_2$ where t_2 is such that the measure of $\{x\in\Omega_2:\ tu_2(x)\geqslant s_3\}$ is >0. We have

$$\Phi(tu_2)=\frac{t^2}{2}||u_2||^2-\int_{\Omega_2}F(x,tu_2)$$

and consequently, splitting the integral over Ω_2 into an integral over $\{x\in\Omega_2:\ tu_2(x)<s_3\}$ and an integral over $\{x\in\Omega_2:\ :tu_2(x)\geqslant s_3\}$, and applying (3.4) to the latter, we get, for some constants c', c'' with $c''>0$,

$$\begin{aligned}\Phi(tu_2)&\leqslant\frac{t^2}{2}||u_2||^2+c'-ct^{\Theta}\int_{\{x\in\Omega_2:\ tu_2\geqslant s_3\}}(u_2)^{\Theta}\\&\leqslant\frac{t^2}{2}||u_2||^2+c'-c''\,t^{\Theta}.\end{aligned}$$

Since $\Theta>2$, this latter relation implies that $\Phi(tu_2)\to-\infty$ as $t\to+\infty$.

We are thus, under (3.3), in a position to apply the mountain pass theorem, which yields a critical point v of Φ with

$$\Phi(v)\geqslant\tfrac{1}{2}\Psi_{A,B}(t_B)>0. \tag{3.5}$$

This function v clearly is a nontrivial solution of (2.2) and so a solution of (1.1).

The second solution of (1.1) will be constructed through local minimization. We will first show that there exists $u_1\in H_0^1(\Omega)$ such that

$$\Phi(tu_1)<0 \tag{3.6}$$

for all $t > 0$ sufficiently small. For that purpose we use assumption (H_4). Take for u_1 the positive eigenfunction associated to the principal eigenvalue of $-\Delta$ on $H_0^1(\Omega_1)$. It is known that $u_1 \in L^\infty(\Omega_1)$. We have, for $t > 0$ sufficiently small (in fact $t \leqslant s_1/||u_1||_\infty$),

$$\Phi(tu_1) = \frac{t^2}{2}||u_1||^2 - \int_{\Omega_1} F(x, tu_1) \leqslant \frac{t^2}{2}\left(||u_1||^2 - \Theta_1 \int_{\Omega_1} u_1^2\right).$$

The expression into the parentheses is <0 since $\Theta_1 > \lambda_1(\Omega_1)$. Consequently (3.6) follows.

It follows from (3.6) that the minimum of the (weakly lower semicontinuous) functional Φ on any closed ball in $H_0^1(\Omega)$ with center 0 and radius R such that

$$\Phi(u) \geqslant 0 \text{ for all } u \text{ with } ||u|| = R \tag{3.7}$$

is achieved in the corresponding open ball and thus yields a nontrivial solution w of (2.2) (and consequently a solution of (1.1)), with

$$\Phi(w) < 0 \text{ and } ||w|| < R. \tag{3.8}$$

We will now show that if (3.3) holds, then a ball satisfying (3.7) can always be found. This is in fact clear since we have seen that under (3.3),

$$\Phi(u) \geqslant \tfrac{1}{2}\Psi_{A,B}(t_B) > 0$$

for all $u \in H_0^1(\Omega)$ with $||u|| = t_B$. This completes the proof of the existence of at least two solutions in Theorem 2.1.

We now turn to the study of the asymptotic behaviour of one of these two solutions. We will show that when f varies in such a way that $a_0 \to 0$ in $L^{\sigma_q}(\Omega)$ and b_0 remains bounded in $L^{\sigma_p}(\Omega)$, it is possible to construct balls verifying (3.7) with $R \to 0$. The corresponding solution w will then, by (3.8), converges to 0 in $H_0^1(\Omega)$. To construct such balls, we fix $\alpha \in]0, 1/(1-q)[$ and observe that for $R = ||a_0||_{\sigma_q}^\alpha$ and for all u with $||u|| = R$, one has, by (3.2),

$$\begin{aligned}\Phi(u) &\geqslant \tfrac{1}{2}||a_0||_{\sigma_q}^{2\alpha} - c_1||a_0||_{\sigma_q}^{1+\alpha(q+1)} - c_2||b_0||_{\sigma_p}||a_0||_{\sigma_q}^{\alpha(p+1)} \\ &= ||a_0||_{\sigma_q}^{2\alpha}\left(\tfrac{1}{2} - c_1||a_0||_{\sigma_q}^{1-\alpha(1-q)} - c_2||b_0||_{\sigma_p}||a_0||_{\sigma_q}^{\alpha(p-1)}\right).\end{aligned}$$

Since $1 - \alpha(1-q) > 0$ by the choice of α, the latter expression is > 0 for $||a_0||_{\sigma_q} > 0$ sufficiently small, and consequently (3.7) holds with $R = ||a_0||_{\sigma_q}^\alpha$. This completes the proof of Theorem 2.1. □

Remark 3.3. The combination on Ω_2 of assumptions (H_2) and (H_5) implies that F is superquadratic on Ω_2 (cf. (3.4) above).

Remark 3.4. Assumptions (H_0)–(H_5) suffice to construct a solution with positive energy. Assumptions (H_0)–(H_4) suffice to construct a solution with negative energy.

We now turn to the proof of Theorem 2.5. The following simple lemma will be needed, whose proof can be found for instance in [10].

Lemma 3.5. *Let $u \in H^2(\omega)$ and $v \in H^1(\omega)$ where ω is a bounded Lipschitz domain. Then*

$$\int_{\omega} \nabla u \nabla v = \int_{\partial\omega} \frac{\partial u}{\partial \nu} v - \int_{\omega} (\Delta u) v,$$

where ν denotes the unit exterior normal to ω and where $\frac{\partial u}{\partial \nu} = \nabla u \cdot \nu$ in the integral over $\partial\omega$ is understood in the sense of traces.

Proof of Theorem 2.5. We will use the method of upper–lower solutions as given in [11]. Our construction of the upper solution is similar to that carried out in [1] and we briefly describe it below for the sake of completeness. One starts with the solutions e of

$$\begin{cases} -\Delta e = 1 & \text{in } \Omega, \\ e = 0 & \text{on } \partial\Omega. \end{cases}$$

Clearly $e \in H^2(\Omega) \cap L^\infty(\Omega)$. Since $0 \leqslant q < 1 < p$, one can find $M = M(p, \Omega, b_0) > 0$ and $\delta = \delta(p, q, \Omega, b_0) > 0$ such that for each $a_0 \in [0, \delta]$,

$$M \geqslant a_0 M^q ||e||_\infty^q + b_0 M^p ||e||_\infty^p.$$

It then follows from $(H_3)'$ that the function Me satisfies

$$-\Delta(Me) = M \geqslant a_0 (Me)^q + b_0 (Me)^p \geqslant f(x, Me)$$

a.e. in Ω, and hence Me is an upper solution of (2.2). To construct a lower solution, we use the subdomain Ω_1 provided by $(H_4)'$. Denote by u_1 the positive normalized principal eigenfunction of $-\Delta$ on $H_0^1(\Omega_1)$, which clearly belongs to $H^2(\Omega_1) \cap C^1(\overline{\Omega_1})$. Extending u_1 by 0 in $\Omega \backslash \Omega_1$, one has that the extended function, still denoted by u_1, belongs to $H_0^1(\Omega) \cap L^\infty(\Omega)$. Call $u_\varepsilon = \varepsilon u_1$ for $\varepsilon > 0$. Using Lemma 3.5 above one has, for any $v \in C_c^\infty(\Omega)$ with $v \geqslant 0$,

$$\int_{\Omega} \nabla u_\varepsilon \nabla v = \int_{\partial\Omega_1} \frac{\partial u_\varepsilon}{\partial \nu} v - \int_{\Omega_1} (\Delta u_\varepsilon) v \leqslant \int_{\Omega_1} \lambda_1(\Omega_1) u_\varepsilon v,$$

where ν denotes the unit exterior normal to Ω_1. It now follows from $(H_4)'$ and (H_0) that for $\varepsilon\|u_1\|_\infty \leqslant s_1$,

$$\int_\Omega \nabla u_\varepsilon \nabla v \leqslant \int_{\Omega_1} f(x, u_\varepsilon)v \leqslant \int_\Omega f(x, u_\varepsilon)v,$$

which shows that εu_1 is a weak lower solution of (2.2). Taking $\varepsilon > 0$ smaller if necessary so that $\varepsilon u_1 \leqslant Me$ on Ω, one deduces from Theorem 2.4 of [11] the existence of a weak solution u to (2.2). □

Let us finally turn to the proof of the nonexistence result, Theorem 2.6. The following lemma will be used in its proof.

Lemma 3.6. *Let $A, B \geqslant 0$ and $0 \leqslant q < 1 < p$. Then there exists $c = c(p,q) > 0$ such that*

$$As^q + Bs^p \geqslant cA^{\frac{p-1}{p-q}} B^{\frac{1-q}{p-q}} s$$

for all $s \geqslant 0$.

Proof. Young's inequality implies that

$$\begin{aligned} A^{\frac{p-1}{p-q}} B^{\frac{1-q}{p-q}} s &\leqslant A\left(\frac{p-1}{p-q}\right)s^q + B\left(\frac{1-q}{p-q}\right)s^p \\ &\leqslant \max\left\{\frac{p-1}{p-q}, \frac{1-q}{p-q}\right\}(As^q + Bs^p) \end{aligned}$$

and the conclusion follows by taking for $c(p,q)$ the inverse of the above maximum. □

Proof of Theorem 2.6. Suppose that (1.1) admits a solution u. We first show that

$$u > 0 \text{ on } \tilde{\Omega}. \tag{3.9}$$

To prove (3.9) we start by observing that $u \not\equiv 0$ on A_+. Indeed, if $u \equiv 0$ on A_+, then, by (1.1) and (H_7),

$$\int_\Omega |\nabla u|^2 = \int_\Omega f(x,u)u = \int_{\Omega\setminus A_+} f(x,u)u \leqslant 0$$

and consequently $u \equiv 0$ on Ω, which is impossible. Since $A_+ \subset \tilde{\Omega}$, it follows that $u \not\equiv 0$ on $\tilde{\Omega}$. Now by (H_7) again, $\tilde{\Omega} \subset A_+ \cup A_0$, which gives $-\Delta u \geqslant 0$ on $\tilde{\Omega}$. The strong maximum principle can thus be applied (cf. e.g. [9, Theorem 8.19]), which yields (3.9).

Let now $v \in H_0^1(\tilde{\Omega})$ with $v \geqslant 0$. We have, by (H_8) and Lemma 3.5,

$$\int_{\Omega} \nabla u \nabla v \geqslant \int_{\tilde{\Omega}} (\tilde{a}u^q + \tilde{b}u^p)v \geqslant c(p,q) \int_{\tilde{\Omega}} \tilde{m}uv. \qquad (3.10)$$

We will take for v the positive eigenfunction φ associated to the principal Dirichlet eigenvalue $\lambda_1(\tilde{m}, \tilde{\Omega})$ of $-\Delta$ on $\tilde{\Omega}$ for the weight $\tilde{m}$. Note here that (H_8) guarantees that $\tilde{m} \geqslant 0$, $\tilde{m} \not\equiv 0$ on $\tilde{\Omega}$ and that $\tilde{m} \in L^r(\tilde{\Omega})$ for some $r > \frac{N}{2}$. Note also that φ is > 0 on $\tilde{\Omega}$. We distinguish now between the cases $\tilde{\Omega} \neq \Omega$ and $\tilde{\Omega} = \Omega$. Suppose first $\tilde{\Omega} \neq \Omega$. Since $\tilde{\Omega}$ is then of class $C^{1,1}$, the regularity theory applies and gives $\varphi \in C^1(\tilde{\Omega} \cup \partial\tilde{\Omega}) \cap H^2(\tilde{\Omega})$; moreover one clearly has $\partial\varphi/\partial v \leqslant 0$ on $\partial\tilde{\Omega}$, where v denotes the unit exterior normal to $\tilde{\Omega}$. By Lemma 3.5, we have

$$\int_{\tilde{\Omega}} \nabla u \nabla \varphi = \int_{\partial\tilde{\Omega}} u \frac{\partial \varphi}{\partial v} - \int_{\tilde{\Omega}} u \Delta \varphi. \qquad (3.11)$$

The trace function u on $\partial\tilde{\Omega}$ is clearly $\geqslant 0$ since $u \geqslant 0$ on $\tilde{\Omega}$. It follows that

$$\int_{\tilde{\Omega}} \nabla u \nabla \varphi \leqslant \lambda_1(\tilde{m}, \tilde{\Omega}) \int_{\tilde{\Omega}} \tilde{m}u\varphi. \qquad (3.12)$$

Now $\int_{\tilde{\Omega}} \tilde{m}u\varphi$ is > 0 since u and φ are > 0 on $\tilde{\Omega}$ and $\tilde{m} \geqslant 0$, $\tilde{m} \not\equiv 0$ on $\tilde{\Omega}$. Combining (3.10) and (3.12), we get

$$c(p,q) \leqslant \lambda_1(\tilde{m}, \tilde{\Omega}),$$

and the conclusion of Theorem 2.6 follows. Suppose now $\tilde{\Omega} = \Omega$. Here we obtain, instead of (3.11),

$$\int_{\Omega} \nabla u \nabla \varphi = \lambda_1(\tilde{m}, \tilde{\Omega}) \int_{\Omega} \tilde{m}u\varphi.$$

Combining with (3.10) leads as before to the conclusion of Theorem 2.6. □

References

[1] A. Ambrosetti, H. Brézis, G. Cerami, Combined effects of concave and convex nonlinearities in some elliptic problems, J. Funct. Anal. 122 (1994) 519–543.

[2] A. Ambrosetti, P.H. Rabinowitz, Dual variational methods in critical points theory and applications, J. Funct. Anal. 14 (1973) 349–381.

[3] S. Alama, M. Del Pino, Solutions of elliptic equations with indefinite nonlinearities via Morse theory and linking, Ann. Inst. H. Poincaré Anal. Non Lineaire 13 (1996) 95–115.

[4] S. Alama, G. Tarantello, On semilinear elliptic equations with indefintite nonlinearities, Calc. Var. 1 (1993) 439–475.

[5] S. Alama, G. Tarantello, Elliptic problems with nonlinearities indefinite in sign, J. Funct. Anal. 141 (1996) 159–215.

[6] H. Berestycki, I. Capuzzo Dolcetta, L. Nirenberg, Variational methods for indefinite superlinear homogeneous elliptic systems, NoDEA 2 (1995) 533–572.
[7] D.G. Costa, C.A. Magalhães, Variational elliptic problems which are nonquadratic at infinity, Nonlinear Anal. TMA 23 (1994) 1401–1412.
[8] D.G. de Figueiredo, P.-L. Lions, On pairs of positive solutions for class of semilinear elliptic problems, Indiana Univ. Math. J. 34 (1985) 591–606.
[9] D. Gilbarg, N.S. Trudinger, Elliptic Partial Differential Equations of Second Order, Springer, Berlin, 1983.
[10] P. Grisvard, Elliptic Problems in Nonsmooth Domains, Pitman, London, 1985.
[11] M. Struwe, Variational Methods, Springer, Berlin, 1990.
[12] H. Tehrani, On elliptic equations with nonlinearities that are sum of a sublinear and superlinear term, volume dedicated to Dr. S. Shahshahani, Iran (2002) 123–131.

Available online at www.sciencedirect.com

SCIENCE DIRECT

ELSEVIER

Journal of Functional Analysis 224 (2005) 471–496

JOURNAL OF Functional Analysis

www.elsevier.com/locate/jfa

An Orlicz-space approach to superlinear elliptic systems☆

Djairo G. de Figueiredo[a,*], João Marcos do Ó[b], Bernhard Ruf[c]

[a] *IMECC, Univ. Estadual di Campinas, 13081-970 Campinas, SP, Brazil*
[b] *Dep. de Matemática, Univ. Federal Paraíba, 58059-900 João Pessoa, PB, Brazil*
[c] *Dip. di Matematica, Università degli Studi, Via Saldini 50, 20133 Milano, Italy*

Received 27 July 2004; accepted 30 September 2004
Communicated by H. Brezis
Available online 8 December 2004

Abstract

In this paper we study superlinear elliptic systems in Hamiltonian form. Using an Orlicz-space setting, we extend the notion of critical growth to superlinear nonlinearities which do not have a polynomial growth. Existence of nontrivial solutions is proved for superlinear nonlinearities which are subcritical in this generalized sense.

MSC: 35J55; 35J20

Keywords: Elliptic system; Critical growth; Orlicz space; Variational methods

☆ Research partially supported by CNPq, PRONEX-MCT/Brazil and Millennium Institute for the Global Advancement of Brazilian Mathematics—IM-AGIMB.

* Corresponding author.

E-mail addresses: djairo@ime.unicamp.br (D.G. de Figueiredo), jmbo@mat.ufpb.br (J. Marcos do Ó), ruf@mat.unimi.it (B. Ruf).

Originally published in: J. Funct. Anal. 224 (2005), 471-496,

DOI: 10.1007/978-3-319-02856-9_37

1. Introduction

In this paper, we study nonlinear elliptic systems in Hamiltonian form

$$\begin{cases} -\Delta u = g(v) & \text{in } \Omega, \\ -\Delta v = f(u) & \text{in } \Omega, \\ v > 0, \ u > 0 & \text{in } \Omega, \\ u = 0, \ v = 0 & \text{on } \partial\Omega, \end{cases} \tag{1.1}$$

where Ω is a bounded open subset of $\mathbb{R}^N (N \geqslant 3)$, with smooth boundary $\partial\Omega$ and Δ is the Laplace operator.

For the scalar equation $-\Delta u = f(u)$ critical growth is given by $f(s) \sim s^{\frac{N+2}{N-2}}$. This is obtained by considering the related functional $\frac{1}{2}\int_\Omega |\nabla u|^2 - \int_\Omega F(u)$, where $F(s) = \int_0^s f(t)\,dt$. The natural space for the first term is $H_0^1(\Omega)$, and then the maximal growth allowed for $F(s)$ is given by the Sobolev embedding $H_0^1(\Omega) \subset L^{2^*}(\Omega)$. For the system (1.1) the associated functional is

$$\int_\Omega \nabla u \nabla v \, dx - \int_\Omega (F(u) + G(v)) \, dx. \tag{1.2}$$

If we consider this functional on $H_0^1(\Omega) \times H_0^1(\Omega)$, then we find again the maximal growths $F(s) \sim |s|^{2^*}$ and $G(s) \sim |s|^{2^*}$. However, in interesting papers by Hulshof–Van der Vorst [5] and Felmer–de Figueiredo [2] the use of Sobolev spaces of fractional order has been proposed. Roughly speaking, these spaces, denoted by $H^s(\Omega), s > 0$, consist of the functions whose derivative of order s is in $L^2(\Omega)$ (these spaces can be defined via interpolation or via Fourier expansion). Introducing suitable self-adjoint operators $A^s : H^s(\Omega) \to L^2(\Omega)$, the first term in the functional (1.2) can be substituted by

$$\int_\Omega A^s u A^t v \quad \text{with } s + t = 2.$$

The maximal growth condition on $F(s) \sim |s|^{p+1}$ and $G(s) \sim |s|^{q+1}$ is then given by the largest values p and q such that $H^s \subset L^{p+1}$ and $H^t \subset L^{q+1}$. This yields the so-called *critical hyperbola*

$$\frac{1}{p+1} + \frac{1}{q+1} = 1 - \frac{2}{N}. \tag{1.3}$$

One notes that now one of the nonlinearities may have a larger growth than $|s|^{2^*}$ provided the other nonlinearity has a suitably lower growth.

We propose here another approach: in order to have the term $\int_\Omega \nabla u \nabla v$ well-defined, we can use Hölder's inequality to estimate

$$\left|\int_\Omega \nabla u \nabla v\right| \leqslant \|u\|_{W_0^{1,\alpha}} \|v\|_{W_0^{1,\beta}} \quad \text{with } \frac{1}{\alpha} + \frac{1}{\beta} = 1 \tag{1.4}$$

and hence the first term in (1.2) can be considered on the space $W_0^{1,\alpha}(\Omega) \times W_0^{1,\beta}(\Omega)$. Again, the maximal growth for F and G is then given by the embeddings

$$W_0^{1,\alpha}(\Omega) \subset L^{p+1}(\Omega) \quad \text{and} \quad W_0^{1,\beta}(\Omega) \subset L^{q+1}(\Omega)$$

with

$$\frac{1}{p+1} = \frac{1}{\alpha} - \frac{1}{N} \quad \text{and} \quad \frac{1}{q+1} = \frac{1}{\beta} - \frac{1}{N}; \tag{1.5}$$

thus, this approach yields the same critical hyperbola.

Consequently, the same equations studied in [2,5] can be treated also by this approach. However, the second approach has the advantage that it can be generalized to more general situations by using an *Orlicz-space* setting. Namely, we can replace the spaces $W_0^{1,\alpha}$ and $W_0^{1,\beta}$ by Sobolev–Orlicz spaces $W_0^1 L_A$ and $W_0^1 L_{\tilde{A}}$; here $W_0^1 L_A$ is given by the functions u such that $\int_\Omega A(|\nabla u|) < \infty$, where A is a so-called N-function, and u vanishes in a certain sense at the boundary. The function $\tilde{A}$ is the Young-conjugate function to A (see below). Then we have again a Hölder-type inequality

$$\left|\int_\Omega \nabla u \nabla v\right| \leqslant 2\|u\|_{W_0^1 L_A} \|v\|_{W_0^1 L_{\tilde{A}}}.$$

The maximal growth for F and G are then determined by the Orlicz-space embeddings

$$W_0^1 L_A(\Omega) \subset L_\Phi(\Omega) \quad \text{and} \quad W_0^1 L_{\tilde{A}}(\Omega) \subset L_\Psi(\Omega), \tag{1.6}$$

where L_Φ and L_Ψ are Orlicz spaces with suitable N-functions Φ and Ψ. In this way, we can treat nonlinearities f and g in Eq. (1.1) that cannot be handled by the Hilbert space approach of Hulshof and van der Vorst [5] and de Figueiredo and Felmer [2].

Let us call two N-functions (Φ, Ψ) a *critical Orlicz pair*, if they have the maximal possible growth in the above embeddings (1.6), for suitable N-functions A and $\tilde{A}$. For example, $\Phi(s) = |s|^{p+1}$ and $\Psi(s) = |s|^{q+1}$ with $p, q \in (1, +\infty)$ satisfying (1.3) constitute a critical Orlicz pair, with $A(s) = |s|^\alpha$ and $\tilde{A}(s) = |s|^\beta$, where α and β are given by (1.5) (see Example 2.1).

We will prove

Theorem 1.1. *Let $\Omega \subset \mathbb{R}^N$ be a bounded, smooth domain. Let $\Phi \in C^1$ be a given N-function, and set $\varphi(t) = \Phi'(t)$. Assume that*

$$\lim_{s\to\infty} \frac{\varphi(s)\, s}{\Phi(s)} = \theta_\Phi > \frac{N}{N-2}. \tag{1.7}$$

Then there exists an associate N-function Ψ such that (Φ, Ψ) form a critical Orlicz pair. Furthermore, the limit

$$\lim_{s\to\infty} \frac{s\Psi'(s)}{\Psi(s)} = \theta_\Psi$$

exists, and

$$\frac{1}{\theta_\Phi} + \frac{1}{\theta_\Psi} = 1 - \frac{2}{N}. \tag{1.8}$$

We give some examples of critical Orlicz pairs:

Example 1.1. Let

$$\Phi(s) \sim |s|^{p+1}(\log|s|)^\alpha \quad \text{and} \quad \Psi(s) \sim |s|^{q+1}(\log|s|)^{-\alpha\frac{q+1}{p+1}}$$

with $\alpha > 0$ and $p, q \in (1, +\infty)$ satisfying (1.3). Then Φ and Ψ satisfy the two above limits, with $\theta_\Phi = p+1$ and $\theta_\Psi = q+1$, respectively, and (Φ, Ψ) form a critical Orlicz-pair (see the proof in *Section* 6).

Remark 1.2. The restriction $\theta_\Phi > N/(N-2)$ is necessary in order to obtain a Ψ which is θ-regular, in the sense of *Definition* 2.10. Also in the polynomial case such a restriction, which here is $p+1 > N/(N-2)$, is necessary in order to obtain $q > 1$.

Based on this characterization, we will prove an existence theorem for superlinear nonlinearities, which have subcritical growth with respect to the above specified Orlicz criticality.

We make the following hypotheses

(H1) let $f, g \in C(\mathbb{R})$ and let F, G denote their primitives;
(H2) there exist constants $\theta > 2$ and $t_0 > 0$ such that, for all $t \geqslant t_0$,

$$0 < \theta F(t) \leqslant t f(t) \quad \text{and} \quad 0 < \theta G(t) \leqslant t g(t).$$

(H3) F and G are uniformly superquadratic near zero (for the definition, see Definition 5.1 below).

Then we prove

Theorem 1.3. *Suppose that* (Φ, Ψ) *is a critical Orlicz pair. Suppose that* F *and* G *satisfy* (H1)–(H3), *and that* F *and* G *have an essentially slower growth than* Φ *and* Ψ, *respectively* (*see Section* 2 *below*). *Suppose also that*

$$\lim_{t\to 0} \frac{F(t)}{\Phi(t)} = C_F < \infty \quad \text{and} \quad \lim_{t\to 0} \frac{G(t)}{\Psi(t)} = C_G < \infty.$$

Then system (1.1) *has a nontrivial solution.*

Example 1.2. Let (Φ, Ψ) denote the critical Orlicz pair given in Example 1.1. Suppose that $F(s) \sim s^{p+1}(\log s)^{\beta}$ and $G(s) \sim s^{q+1}(\log s)^{-\gamma}$, for s positive and large with $\beta < \alpha$ and $\gamma > \alpha \frac{q+1}{p+1}$. Then F and G have essentially slower growth than Φ and Ψ, respectively.

2. Orlicz spaces

Here we recall some basic facts about Orlicz spaces, for more details see for instance [1,6,9]. Let M be a N-function, that is, $M : \mathbb{R} \to [0, +\infty)$ is continuous, convex, even, $M(t) = 0$ if and only if $t = 0$,

$$M(t)/t \to 0 \text{ as } t \to 0 \text{ and } M(t)/t \to +\infty \text{ as } t \to +\infty.$$

We say that an N-function A *dominates* an N-function B (near infinity) if, for some positive constant k, $B(x) \leqslant A(kx)$ (for $x \geqslant x_0$), and write $B \prec A$. A and B are *equivalent* if A dominates B and B dominates A; then we write $A \sim B$. Finally, we say that B *increases essentially more slowly* than A if $\lim_{t\to\infty} \frac{B(kt)}{A(t)} = 0$, for all $k > 0$; in this case we write $B \prec\prec A$.

Associated to the N-function M we have the following class of functions.

Definition 2.1 (*Orlicz class*). The Orlicz class is defined by

$$K_M(\Omega) := \{u : \Omega \to \mathbb{R} : u \text{ measurable and } \int_{\Omega} M(u(x))\, dx < \infty\}.$$

Orlicz classes are convex sets, but in general not linear spaces. One then defines

Definition 2.2 (*Orlicz space*). The vector space $L_M(\Omega)$ generated by $K_M(\Omega)$ is called Orlicz space.

Fact 1. The Orlicz class $K_M(\Omega)$ is a vector space, and hence equal to $L_M(\Omega)$ if and only if M satisfies the following

Definition 2.3 (*Δ_2-condition*). There exist numbers $k > 1$ and $t_0 \geqslant 0$ such that

$$M(2t) \leqslant kM(t) \quad \text{for } t \geqslant t_0.$$

Furthermore, we define

Definition 2.4 (*∇_2-condition*). There exist numbers $h > 1$ and $t_1 \geqslant 0$ such that

$$M(t) \leqslant \frac{1}{2h} M(ht) \quad \text{for } t \geqslant t_1.$$

We call a function satisfying the Δ_2- and the ∇_2-condition *Δ-regular*.

We remark that the Orlicz class depends only on the asymptotic growth of the function M; therefore, also the Δ_2-condition and the ∇_2-condition need to be satisfied only near infinity.

We define the following norm on $L_M(\Omega)$:

Definition 2.5 (*Luxemburg norm*).

$$\|u\|_{(M)} = \inf\{\ \lambda > 0 : \int_\Omega M(\frac{|u|}{\lambda}) \leqslant 1\}$$

Fact 2. $(L_M, \|\cdot\|_{(M)})$ is a Banach space.

Definition 2.6 (*Conjugate function*). *Let*

$$\widetilde{M}(x) = \sup_{y>0} \{xy - M(y)\}.$$

$\widetilde{M}$ is called the Young conjugate function of M.

It is clear that $\widetilde{\widetilde{M}} = M$, and M and $\widetilde{M}$ satisfy the Young inequality:

$$st \leqslant M(t) + \widetilde{M}(s) \quad \forall s, \quad t \in \mathbb{R},$$

with *equality* when $s = M'(t)$ or $t = \widetilde{M}'(s)$. In the spaces L_M and $L_{\widetilde{M}}$ the *Hölder inequality* holds

$$|\int_\Omega u(x)v(x)\,dx| \leqslant 2\|u\|_{(M)}\|v\|_{(\widetilde{M})}.$$

Hence, for every $\tilde{u} \in L_{\widetilde{M}}$ we can define a continuous linear functional $l_{\tilde{u}} v := \int_\Omega \tilde{u} v \, dx$ and $l_{\tilde{u}} \in (L_M)^*$. Then we can define

$$\|\tilde{u}\|_{\widetilde{M}} = \|l_{\tilde{u}}\| = \sup_{\|v\|_{(M)} \leqslant 1} \int_\Omega \tilde{u}(x) v(x) \, dx$$

Definition 2.7. $\|\tilde{u}\|_{\widetilde{M}}$ is called the *Orlicz norm* on the space $L_{\widetilde{M}}$, and analogously one defines the Orlicz norm $\|u\|_M$ on L_M.

Thus, we have two different norms on L_M, the Luxemburg (or gauge) norm $\|\cdot\|_{(M)}$ and the Orlicz norm $\|\cdot\|_M$; they are equivalent, and satisfy

$$\|u\|_{(M)} \leqslant \|u\|_M \leqslant 2\|u\|_{(M)}. \tag{2.1}$$

In order to be precise about which norm is considered in the spaces, we are going to use from now on the following notations:

$$\left(L_M, \|\cdot\|_M\right) := L_M \quad \text{and} \quad \left(L_M, \|\cdot\|_{(M)}\right) := L_{(M)}$$

and similarly for $\widetilde{M}$.

Fact 3. It follows from the definition of Orlicz norm that, if $u \in L_M$ and $\widetilde{w} \in L_{\widetilde{M}}$, then one has the following Hölder inequality:

$$\left| \int_\Omega u \widetilde{w} \, dx \right| \leqslant \|u\|_M \|\widetilde{w}\|_{(\widetilde{M})}. \tag{2.2}$$

Fact 4. L_M is *reflexive* if and only if M and $\widetilde{M}$ satisfy the Δ_2-condition, and then

$$(L_{(M)})^* = L_{\widetilde{M}} \quad \text{and} \quad (L_{(\widetilde{M})})^* = L_M$$

(see [4,9, p. 111]).

Fact 5. If Φ is Δ-regular, then there exists a $\Phi_1 \sim \Phi$ such that $L_\Phi = L_{\Phi_1}$ as sets, and their Luxemburg norms (respectively, Orlicz norms) are equivalent, with the following additional structure:

(a) L_Φ and L_{Φ_1} are isomorphic, and both are reflexive spaces,
(b) L_{Φ_1} is uniformly convex and uniformly smooth (see [9, Theorem 2, p. 297]).

Next, we define the *Orlicz–Sobolev* spaces: Let A be a N-function. Then set

Definition 2.8.

$$W^1L_A = \left\{u : \Omega \to \mathbb{R};\ \max_{|\alpha|\in\{0,1\}} \int_\Omega A(|D^\alpha u|) < +\infty\right\}$$

with Luxemburg norm

$$\|u\|_{W^1L_A} := \max\left\{\|D^\alpha u\|_{(M)} : |\alpha| \in \{0,1\}\right\}.$$

On the space $W_0^1L_{(A)}$, i.e. the space of functions in W^1L_A which vanish on the boundary, an equivalent Luxemburg norm is given by

$$\|u\|_{1,(A)} = \|\nabla u\|_{(A)} = \inf\left\{\lambda > 0 : \int_\Omega A\left(\frac{|\nabla u|}{\lambda}\right) \leqslant 1\right\}.$$

The equivalence of these two norms is a consequence of the Poincaré inequality,

$$\|u\|_{(M)} \leqslant C \sum_{i=1}^n \|D_i u\|_{(M)} \quad \forall u \in W_0^1L_M(\Omega),$$

(see [4]). In analogy with the above definition of the Orlicz norm in L_M we can define an Orlicz norm in $W_0^1L_{(A)}$ by

$$\|u\|_{1,A} := \sup\left\{\int_\Omega \nabla u \nabla \widetilde{w}\, dx : \widetilde{w} \in W_0^1L_{(\widetilde{A})},\ \|\widetilde{w}\|_{1,(\widetilde{A})} \leqslant 1\right\}.$$

The space $W_0^1L_{(A)}$ endowed with this new norm is denoted by $W_0^1L_A$.

Definition 2.9 (*Sobolev conjugate (Adams [1], p. 248)*). Suppose that $\int_1^\infty \frac{A^{-1}(t)}{t^{1+1/n}}\, dt = +\infty$. Then the Sobolev conjugate function $\Phi(t)$ is given by

$$\Phi^{-1}(t) = \int_0^t \frac{A^{-1}(\tau)}{\tau^{1+1/n}}\, d\tau, \quad t \geqslant 0.$$

Fact 6. Let Ω be bounded, and satisfying the cone property. Then

$$W^1L_A(\Omega) \hookrightarrow L_\Phi(\Omega) \text{ continuously}$$

and compactly into $L_G(\Omega)$, where G is any N-function increasing *essentially more slowly* than Φ, i.e. $\lim_{t\to\infty} \frac{G(kt)}{\Phi(t)} = 0$, for all $k > 0$.

Example 2.1. One easily checks that for $\Phi(s) = s^{p+1}$ the above formula yields $A(s) = cs^{\alpha}$, with α satisfying $\frac{1}{\alpha} = \frac{1}{p+1} + \frac{1}{N}$, i.e. we have the classical Sobolev imbedding theorem for $W^{1,\alpha}(\Omega) \hookrightarrow L^{p+1}(\Omega)$.

Next, we make the following:

Definition 2.10. Let $g \in C(\mathbb{R})$ be an N-function, and G its primitive. Then we say that G is θ-regular, if there exists a constant $\theta_G > 1$ such that

$$\lim_{s\to\infty} \frac{sg(s)}{G(s)} = \theta_G. \tag{2.3}$$

Let $F(t) = G^{-1}(t)$, and $f(t) = F'(t)$. Then the above condition is equivalent to

$$\lim_{t\to\infty} \frac{tf(t)}{F(t)} = \frac{1}{\theta_G}. \tag{2.4}$$

Indeed, we have $G(s) = t \Leftrightarrow F(t) = s$, and $f(t) = \frac{d}{dt}[G^{-1}(t)] = \frac{1}{g(s)}$.
Note that by Rao-Ren [9, p. 26] we have

Fact 7. If G is θ-regular, then G is Δ-regular, i.e. $G \in \Delta_2 \cap \nabla_2$.

3. Orlicz-space criticality

Definition 3.1 (*Critical Orlicz pair*). Let Φ and Ψ be Δ-regular N-functions. Then (Φ, Ψ) are a *critical Orlicz pair* if there exist Δ-regular and conjugate N-functions A and $\widetilde{A}$ such that L_Φ and L_Ψ are the smallest Orlicz spaces with

$$W^1L_A \hookrightarrow L_\Phi, \quad W^1L_{\widetilde{A}} \hookrightarrow L_\Psi.$$

Consider the following example:

Example 3.1. In Example 2.1 we saw that to $\Phi(s) = s^{p+1}$ corresponds the inverse Sobolev conjugate $A(s) = cs^{\alpha}$, with

$$\frac{1}{p+1} + \frac{1}{N} = \frac{1}{\alpha}.$$

The conjugate function $\widetilde{A}$ to A is given by $\widetilde{A}(s) = cs^{\beta}$, with $\frac{1}{\alpha} + \frac{1}{\beta} = 1$, which in turn has as Sobolev conjugate $\Psi(s) = s^{q+1}$, with

$$\frac{1}{q+1} + \frac{1}{N} = \frac{1}{\beta}.$$

Adding the two equation yields

$$\frac{1}{p+1} + \frac{1}{q+1} = 1 - \frac{2}{N}.$$

This is the critical hyperbola, see [2,5]. Thus, $(|s|^{p+1}, |s|^{q+1})$ are a critical Orlicz pair, and so the above theorem contains the critical hyperbola as a special case. We remark that the proof given here is also new in the polynomial case; in [2,5] fractional Sobolev spaces H^s were used in order to conserve the Hilbert space structure.

3.1. Proof of Theorem 1.1

(1) Hypothesis (1.7) expresses the fact that the function Φ is θ-regular with $\theta_\Phi > N/(N-2)$. Let A be the inverse Sobolev conjugate of Φ, see Definition 2.9. Note that $W^1 L_A$ is the largest Orlicz–Sobolev space that embeds into L_Φ.

Claim 1. *A is θ-regular, with* $\theta_A = \frac{N\theta_\Phi}{N+\theta_\Phi} > 1$.

Indeed, let $F(s) = \Phi^{-1}(s)$ and $B(t) = A^{-1}(t)$. Then $F(s) = \int_0^s \frac{B(t)}{t^{1+1/N}}\,dt$, and hence

$$f(s) = \frac{B(s)}{s^{1+1/N}}.$$

Then we have by (2.4)

$$\frac{1}{\theta_\Phi} = \lim_{s\to\infty} \frac{f(s)s}{F(s)} = \lim_{s\to\infty} \frac{B(s)\,s^{-1/N}}{F(s)}.$$

Then, by l'Hospital's rule

$$\lim_{s\to\infty} \frac{B(s)\,s^{-1/N}}{F(s)} = \lim_{s\to\infty} \frac{b(s)\,s^{-1/N} - \frac{1}{N}\,s^{-\frac{1}{N}-1}\,B(s)}{\frac{B(s)}{s^{1+1/N}}} = \lim_{s\to\infty} \frac{b(s)s}{B(s)} - \frac{1}{N}.$$

We conclude that

$$\frac{1}{\theta_\Phi} = \lim_{s\to\infty} \frac{b(s)s}{B(s)} - \frac{1}{N}$$

and thus

$$\lim_{s\to\infty} \frac{b(s)s}{B(s)} = \frac{1}{\theta_\Phi} + \frac{1}{N} < 1.$$

This implies that A is θ-regular, with $\theta_A = \frac{N\theta_\Phi}{N+\theta_\Phi} > 1$.

(2) Next, let $\widetilde{A}$ be the conjugate function of A, given by Definition 2.6. $\widetilde{A}$ is a N-function, and Δ-regular, see [9, Corollary 4, p. 26].

In the following, suppose that $s = A'(t)$ (iff $t = \widetilde{A}'(s)$); note that $t \to \infty$ iff $s \to \infty$. Then

$$\frac{1}{\theta_A} = \lim_{t\to\infty} \frac{A(t)}{tA'(t)} = \lim_{t\to\infty} \frac{A(t)}{ts} = \lim_{s\to\infty} \frac{st - \widetilde{A}(s)}{st} = 1 - \lim_{s\to\infty} \frac{\widetilde{A}(s)}{s\widetilde{A}'(s)} = 1 - \frac{1}{\theta_{\widetilde{A}}}.$$

Thus, $\widetilde{A}$ is θ-regular, with $\theta_{\widetilde{A}} > 1$.

We can now define the corresponding Orlicz–Sobolev space $W^1 L_{\widetilde{A}}$.

(3) Next, use Definition 2.9 again to define the function Ψ; by Adams [1, p. 248], Ψ is an N-function.

Claim 2. *Ψ is θ-regular, with* $\theta_\Psi = \frac{N\theta_{\widetilde{A}}}{N-\theta_{\widetilde{A}}}$.

This follows similarly as in Claim 1, reversing the direction in the arguments. Finally, L_Ψ is the smallest Orlicz space into which $W^1 L_{\widetilde{A}}$ imbeds continuously. Thus, we have shown that (Φ, Ψ) is a critical Orlicz pair.

Finally, we have

$$\frac{1}{\theta_\Phi} + \frac{1}{\theta_\Psi} = \frac{N-\theta_A}{N\theta_A} + \frac{N-\theta_{\widetilde{A}}}{N\widetilde{A}} = \frac{1}{\theta_A} - \frac{1}{N} + \frac{1}{\theta_{\widetilde{A}}} - \frac{1}{N} = 1 - \frac{2}{N}. \qquad \square$$

4. The tilde-map

In this section we define a map from $W_0^1 L_A$ to the space $W_0^1 L_{(\widetilde{A})}$, where $\widetilde{A}$ is the Young conjugate of A.

Theorem 4.1. *For each $u \in W_0^1 L_A$ consider*

$$S := \sup\left\{ \int_\Omega \nabla u(x) \nabla \widetilde{w}(x)\, dx : \widetilde{w} \in W_0^1 L_{(\widetilde{A})},\ \|\widetilde{w}\|_{1,(\widetilde{A})} = \|u\|_{1,A} \right\}. \tag{4.1}$$

Then there exists a unique $\widetilde{u} \in W_0^1 L_{(\widetilde{A})}$ such that

$$\|\widetilde{u}\|_{1,(\widetilde{A})} = \|u\|_{1,A} \quad \text{and} \quad S = \int_\Omega \nabla u(x) \nabla \widetilde{u}(x)\, dx = \|u\|_{1,A} \|\widetilde{u}\|_{1,(\widetilde{A})}$$

Furthermore, $\tilde{u}$ depends continuously (but nonlinearly) on u.

Proof. We first remark that by Fact 5 we may assume that all the spaces $W^1 L_A$, $W^1 L_{(A)}$, $W^1 L_{\widetilde{A}}$ and $W^1 L_{(\widetilde{A})}$ are reflexive and uniformly convex. Observe also that by (2.2)

$$S \leqslant \sup \|\widetilde{w}\|_{1,(\widetilde{A})} \|u\|_{1,A} = \|u\|_{1,A}^2.$$

• *Existence*: Let $\{\widetilde{v}_n\}$ be a maximizing sequence for (4.1). Since the sequence is bounded, we have by reflexivity that (for a subsequence) $\tilde{u} \rightharpoonup \tilde{v}$ weakly in $W_0^1 L_{(\tilde{A})}$. So $\int_\Omega \nabla u \nabla \widetilde{v}_n \to S$, and consequently $\int_\Omega \nabla u \nabla \widetilde{u} = S$, that is, the supremum is attained. It remains to prove that $\|\widetilde{u}\|_{1,(\tilde{A})} = \|u\|_{1,A}$. Suppose by contradiction that $\|\widetilde{u}\|_{1,(\tilde{A})} = k < \|u\|_{1,A}$. Take $\widetilde{w} = (\widetilde{u}/k)\|u\|_{1,A}$. Then for this $\widetilde{w}$ we have

$$S \geqslant \int_\Omega \nabla u \nabla \widetilde{w} = \int_\Omega \nabla u \nabla \widetilde{u} \frac{\|u\|_{1,A}}{k} > S,$$

which is impossible.

• *Uniqueness*: by the uniform convexity of $W_0^1 L_A$, $\tilde{u}$ is unique.

• *Continuity*: Let $u_n \to u \neq 0$ in $W_0^1 L_A$. By the above we have $\|\widetilde{u}_n\|_{1,(\tilde{A})} = \|u_n\|_{1,A}$ and $\|\widetilde{u}\|_{1,(\tilde{A})} = \|u\|_{1,A}$. Consequently $\|\widetilde{u}_n\|_{1,(\tilde{A})} \to \|\widetilde{u}\|_{1,A}$. So we have that, for some subsequence, $\widetilde{u}_n \rightharpoonup \widetilde{v}$ in $W_0^1 L_{(\tilde{A})}$. If we prove that $\widetilde{v} = \widetilde{u}$, then in fact we are concluding that the whole sequence $\widetilde{u}_n$ converges strongly to $\widetilde{u}$ in $W_0^1 L_{(\tilde{A})}$. To this end observe that

$$\int_\Omega \nabla u_n \nabla \widetilde{u}_n = \|\widetilde{u}_n\|_{1,(\tilde{A})} \|u_n\|_{1,A}$$

implies

$$\int_\Omega \nabla u \nabla \widetilde{v} = \|\widetilde{u}\|_{1,(\tilde{A})} \|u\|_{1,A}.$$

We claim that $\|\widetilde{v}\|_{1,(\tilde{A})} = \|u\|_{1,A}$, and then by the uniqueness of $\widetilde{u}$ it follows that $\widetilde{v} = \widetilde{u}$ and the proof is complete. The claim is proved by contradiction as we did in the existence above assuming that $\|\widetilde{v}\|_{1,(\tilde{A})} = k < \|u\|_{1,A}$. $\square$

Using the previous theorem we now define the "*tilde-map*"

$$\begin{aligned} \tilde{}: W_0^1 L_A &\longrightarrow W_0^1 L_{(\tilde{A})}, \\ u &\longmapsto \widetilde{u}, \end{aligned}$$

which is continuous.

Remark 4.2. It follows from the construction that the tilde-map is positively homogeneous, i.e.

$$\widetilde{\rho u} = \rho \widetilde{u} \quad \forall u \in W^1 L_A \quad \forall \rho \geqslant 0$$

With the help of the tilde-map, we define two continuous sub-manifolds of

$$E := W_0^1 L_A \times W_0^1 L_{(\tilde{A})}$$

by

$$E^+ = \{(u, \widetilde{u});\ u \in W_0^1 L_A\} \quad \text{and} \quad E^- = \{(u, -\widetilde{u});\ u \in W_0^1 L_A\}.$$

We remark that E^+ and E^- are nonlinear submanifolds of E when regarded with respect to the standard vector space structure of E. Surprisingly, E^+ and E^- turn out to be *linear* with respect to the following notion of *tilde-sum*:

Definition 4.3 (*Tilde sum*). Given elements $(u, \widetilde{v}) \in E$ and $(y, \widetilde{z}) \in E$, we set

$$(u, \widetilde{v}) \,\widetilde{+}\, (y, \widetilde{z}) := (u + y, \widetilde{v + z}).$$

Indeed, with this notion we can prove

Lemma 4.4. (1) *Let $(u, \widetilde{u}) \in E^+$ and $(v, \widetilde{v}) \in E^+$; then, for all $\alpha, \beta \in \mathbb{R}$*

$$\alpha(u, \widetilde{u}) \,\widetilde{+}\, \beta(v, \widetilde{v}) \in E^+ \quad \textit{and} \quad \alpha(u, \widetilde{u}) \,\widetilde{+}\, \beta(v, \widetilde{v}) = (\alpha u + \beta v, \widetilde{\alpha u + \beta v}).$$

(2) *For every $(y, \widetilde{z}) \in E$ there exist unique elements $(u, \widetilde{u}) \in E^+$ and $(v, -\widetilde{v}) \in E^-$ such that*

$$(y, \tilde{z}) = (u, \tilde{u}) \,\widetilde{+}\, (v, -\tilde{v}),$$

i.e. we can write

$$E = E^+ \,\widetilde{\oplus}\, E^-.$$

Proof. (1) We have

$$\begin{aligned}\alpha(u, \tilde{v}) \,\widetilde{+}\, \beta(y, \tilde{z}) &= (\alpha u, \alpha\tilde{v}) \,\widetilde{+}\, (\beta y, \beta\tilde{z}) \\ &= (\alpha u, \widetilde{\alpha v}) \,\widetilde{+}\, (\beta y, \widetilde{\beta z}) = (\alpha u + \beta y, \widetilde{\alpha v + \beta z}).\end{aligned}$$

(2) • *Uniqueness*: Suppose that $(u + v, \widetilde{u - v}) = (u_1 + v_1, \widetilde{u_1 - v_1})$; then $u + v = u_1 + v_1$ and $\widetilde{u - v} = \widetilde{u_1 - v_1}$, which implies $u = u_1$ and $v = v_1$.

• *Existence*: We look for elements u and v in $W_0^1 L_A$ such that

$$(y, \tilde{z}) = (u, \tilde{u}) \,\widetilde{+}\, (v, -\tilde{v}) = (u + v, \widetilde{u - v}).$$

That is, $y = u + v$ and $z = u - v$, and hence $u = \frac{y+z}{2}$ and $v = \frac{y-z}{2}$. □

5. An existence theorem

In this section we prove the existence of a nontrivial solution for system (1.1), in the case of superlinear nonlinearities which have subcritical growth with respect to a given critical Orlicz pair.

5.1. The functional

In this section, we define the framework for the functional I associated to problem (1.1) and given in (1.2).

We first give the precise definition of *uniformly superquadratic near zero.*

Definition 5.1. A continuous function $H : \mathbb{R} \to \mathbb{R}$ is *uniformly superquadratic near zero*, if there exist numbers $\sigma > 2$ and $c \geqslant 1$ such that

$$H(st) \leqslant cs^{\sigma}\, H(t) \quad \forall t > 0 \quad \forall s \in [0, 1]$$

Note that if $H(t) = t^p$ with $p > 2$, then H satisfies the definition with $\sigma = p$.

In Section 3 we have proved the existence of critical Orlicz pairs (Φ, Ψ). As specified in Theorem 1.3, we assume that the functions F and G grow essentially more slowly than Φ and Ψ, respectively. Since we are interested in positive solutions we redefine F and G to be zero on $(-\infty, 0]$.

Consider the functional $I : W_0^1 L_A(\Omega) \times W_0^1 L_{(\widetilde{A})}(\Omega) \to \mathbb{R}$ given by

$$I(u, \widetilde{v}) = \int_{\Omega} \nabla u \nabla \widetilde{v}\, dx - \int_{\Omega} [F(u) + G(\widetilde{v})]\, dx \tag{5.1}$$

Here $\tilde{v} \in W_0^1 L_{(\widetilde{A})}(\Omega)$ is an independent variable; we write $\tilde{v}$ to emphasize that $\tilde{v}$ belongs to the space $W_0^1 L_{(\widetilde{A})}(\Omega)$.

The functional I is well defined and belongs to the class C^1 with

$$I'(u, \widetilde{v})(\eta, \widetilde{\xi}) = \int_{\Omega} [\nabla u \nabla \widetilde{\xi} + \nabla \widetilde{v} \nabla \eta]\, dx - \int_{\Omega} [f(u)\eta + g(\widetilde{v})\widetilde{\xi}\ ]\, dx, \tag{5.2}$$

for all $(\eta, \widetilde{\xi}) \in W_0^1 L_A(\Omega) \times W_0^1 L_{(\widetilde{A})}(\Omega)$. Consequently, critical points of the functional I correspond to the weak solutions of (1.1).

5.2. The geometry of the linking theorem

We first prove that the functional I given in (5.1) has the geometry of the linking theorem.

Let E^+ and E^- be as in Section 4.

Lemma 5.2. *There exist* ρ_0, $\sigma_0 > 0$ *such that* $I(z) \geqslant \sigma_0$, *for all* $z \in \partial B_{\rho_0} \cap E^+$.

Proof.

$$I(u, \widetilde{u}) = \int_\Omega \nabla u \nabla \widetilde{u}\, dx - \int_\Omega F(u)\, dx - \int_\Omega G(\widetilde{u})\, dx.$$

Now, using that

$$\int_\Omega \nabla u \nabla \widetilde{u}\, dx = \|u\|_{1,A}\, \|\widetilde{u}\|_{1,(\widetilde{A})} = \|u\|^2_{1,A} = \|\widetilde{u}\|^2_{1,(\widetilde{A})}$$

and by (2.1)

$$\|u\|_{1,A} \geqslant \|u\|_{1,(A)}$$

we obtain that

$$I(u, \widetilde{u}) \geqslant \frac{1}{2}\, \|u\|^2_{1,(A)} - \int_\Omega F(u)\, dx + \frac{1}{2}\, \|\widetilde{u}\|^2_{1,(\widetilde{A})} - \int_\Omega G(\widetilde{u})\, dx.$$

Assume that $\rho \in (0, 1)$, $u \in W^1_0 L_A(\Omega)$ and $\|u\|_{1,(A)} = c_1^{-1}$, where $c_1 > 0$ is such that

$$\|u\|_{(\Phi)} \leqslant c_1 \|u\|_{1,(A)}.$$

Since

$$\|u\|_{(\Phi)} = \inf\{\ \lambda > 0 : \int_\Omega \Phi(\frac{|u|}{\lambda}) \leqslant 1\} \leqslant 1,$$

it follows that $\int_\Omega \Phi(|u|) \leqslant 1$, and thus $\int_\Omega F(|u|) \leqslant c$. By hypothesis (H3) we get for $0 \leqslant \rho \leqslant 1$

$$\int_\Omega F(\rho u)\, dx \leqslant c\rho^\sigma \int_\Omega F(u)\, dx \leqslant c\rho^\sigma.$$

Hence we obtain that

$$\frac{1}{2}\, \|\rho u\|^2_{1,(A)} - \int_\Omega F(\rho u)\, dx \geqslant \frac{1}{2}\, \rho^2\, \|u\|^2_{1,(A)} - c\rho^\sigma.$$

Arguing similarly for G and $\tilde{u}$ we get

$$\frac{1}{2}\,\|\rho\widetilde{u}\|^2_{1,(\tilde{A})} - \int_\Omega G(\rho\widetilde{u})\,dx \geqslant \frac{1}{2}\,\rho^2\,\|\tilde{u}\|^2_{1,(\tilde{A})} - c\rho^{\sigma_1}.$$

By joining the two estimates we can find a $\rho_0 > 0$ such that

$$I(u,\tilde{u}) \geqslant \sigma_0 > 0 \quad \text{for } \|(u,\tilde{u})\| = \rho_0 > 0$$

This concludes the proof. □

Let e_1 denote the first eigenfunction of the Laplacian, with $\|(e_1,\widetilde{e_1})\| = 1$ and set

$$Q = \{r(e_1,\widetilde{e_1}) \mathbin{\widetilde{+}} w : w \in E^-,\ \|w\| \leqslant R_0 \text{ and } 0 \leqslant r \leqslant R_1\}.$$

Lemma 5.3. *There exist positive constants R_0, R_1 such that $I(z) \leqslant 0$ for all $z \in \partial Q$.*

Proof. Notice that the boundary ∂Q of the set Q is taken in the set $\mathbb{R}(e_1,\widetilde{e_1}) \mathbin{\widetilde{+}} E^-$, and consists of three parts.

(i) If $z \in \partial Q \cap E^-$ we have $I(z) \leqslant 0$ because, for all $z = (\omega, -\widetilde{\omega}) \in E^-$,

$$I(z) = -\|\omega\|^2_{1,(A)} - \int_\Omega [F(\omega) + G(-\widetilde{\omega})]\,dx \leqslant 0.$$

(ii) If $z = r(e_1,\widetilde{e_1}) \mathbin{\widetilde{+}} (\omega,-\widetilde{\omega}) = (re_1+\omega, \widetilde{re_1-\omega}) \in \partial Q$ with $\|(\omega,-\widetilde{\omega})\| = R_0$ and $0 \leqslant r \leqslant R_1$, we proceed as follows:

First step: Assume that $R_1 = 1$:

$$\begin{aligned}
I(z) &\leqslant \int_\Omega \nabla(re_1+\omega)\nabla(\widetilde{re_1-\omega})\,dx \\
&= -\int_\Omega \nabla(\omega - re_1)\nabla(\widetilde{\omega - re_1})\,dx - 2r\int_\Omega \nabla e_1 \nabla(\widetilde{\omega - re_1}) \\
&\leqslant -\|\omega - re_1\|^2_{1,A} + 2\|re_1\|_{1,A}\|\widetilde{\omega - re_1}\|_{1,(\tilde{A})} \\
&\leqslant -\|\omega - re_1\|^2_{1,A} + 2\|re_1\|_{1,A}\|\omega - re_1\|_{1,A} \\
&\leqslant -\|\omega\|^2_{1,A} - \|re_1\|^2_{1,A} + 2\|re_1\|_{1,A}\|\omega\|_{1,A} \\
&\quad +2\|re_1\|_{1,A}\left(\|\omega\|_{1,A} + \|re_1\|_{1,A}\right) \\
&\leqslant -\|\omega\|^2_{1,A} + 4r\|\omega\|_{1,A} + r^2.
\end{aligned}$$

Since $2\|\omega\|^2_{1,A} \geqslant \|\omega\|^2_{1,A} + \|\widetilde{\omega}\|^2_{1,(\widetilde{A})} = R_0^2$, we conclude that the last expression is $\leqslant 0$, for R_0 sufficiently large.

Second step: Observe that using homogeneity this now holds for all $\rho \geqslant 1$ with $0 \leqslant r \leqslant \rho$ and $\|(\omega, -\widetilde{\omega})\| = \rho R_0$.

(iii) Finally, let $z = \rho(e_1, \widetilde{e}_1) \mp \rho(\omega, -\widetilde{\omega}) = (\rho e_1 + \rho\omega, \widetilde{\rho e_1 - \rho\omega}) \in \partial Q$ with $\|e_1\|_{1,(A)} = \frac{1}{2}$ and $\|(\omega, -\widetilde{\omega})\| \leqslant R_0$.

We show: there exists $R_1 > 0$ sufficiently large such that for all $\rho \geqslant R_1$, we have $I(z) \leqslant 0$. We use that $W_0^1 L_A(\Omega) \hookrightarrow L_\Phi(\Omega)$, $W_0^1 L_{\widetilde{A}}(\Omega) \hookrightarrow L_\Psi(\Omega)$, and that by assumption (H2): $F(s) \geqslant c|s|^\theta - c_1$ and $G(s) \geqslant c|s|^\theta - c_1$, for some $\theta > 2$; then

$$\begin{aligned} I(z) &= \int_\Omega \nabla(\rho e_1 + \rho\omega)\nabla(\widetilde{\rho e_1 - \rho\omega})\,dx - \int_\Omega [F(\rho e_1 + \rho\omega) + G(\widetilde{\rho e_1 - \rho\omega})]\,dx \\ &\leqslant \rho^2 \|e_1 + \omega\|_{1,A} \|\widetilde{e_1 - \omega}\|_{1,(\widetilde{A})} - c\int_\Omega |\rho e_1 + \rho\omega|^\theta\,dx + c_1 \\ &\quad - c\int_\Omega |\widetilde{\rho e_1 - \rho\omega}|^\theta\,dx + c_1 \\ &\leqslant \rho^2[\|e_1\|_{1,A} + \|\omega\|_{1,A}]^2 - c\rho^\theta\Big\{\int_\Omega |e_1 + \omega|^\theta\,dx + \int_\Omega |\widetilde{e_1 - \omega}|^\theta\,dx\Big\} + 2c_1. \end{aligned}$$

Thus

$$I(z) \leqslant \rho^2(1 + R_0)^2 - c\rho^\theta\delta_0 + 2c_1, \tag{5.3}$$

where

$$\delta_0 := \inf_{\|(\omega, -\widetilde{\omega})\| \leqslant R_0} \Big\{\int_\Omega |e_1 + \omega|^\theta\,dx + \int_\Omega |\widetilde{e_1 - \omega}|^\theta\,dx\Big\} > 0.$$

Indeed, suppose by contradiction that there exists a sequence $(\omega_n) \subset W_0^1 L_A(\Omega)$ such that $\|(\omega_n, -\widetilde{\omega}_n)\| \leqslant R_0$ and

$$\lim_{n\to\infty} \Big\{\int_\Omega |e_1 + \omega_n|^\theta\,dx + \int_\Omega |\widetilde{e_1 - \omega_n}|^\theta\,dx\Big\} = 0.$$

Taking a subsequence, we may assume that $\omega_n \to \omega \in L^\theta$ (since $W^1 L_A \subset\subset L_F \subset L^\theta$) which implies that $e_1 + \omega_n \to e_1 + \omega$ and $\widetilde{e_1 - \omega_n} \to \widetilde{e_1 - \omega}$ in L^θ, where we have used the continuity of the tilde mapping. Thus, taking the limit we see that

$$\int_\Omega |e_1 + \omega|^\theta\,dx + \int_\Omega |\widetilde{e_1 - \omega}|^\theta\,dx = 0$$

which implies that $e_1 + \omega = \widetilde{e_1 - \omega} = 0$. So, $e_1 = \omega = 0$, which is a contradiction.

Finally, using (5.3) we can find $R_1 > 0$ such that $I(z) \leqslant 0$ for all $\rho \geqslant R_1$, and hence, the geometry of the linking theorem holds. □

5.3. On Palais–Smale sequences

Proposition 5.4. *Let* $(u_m, \widetilde{v}_m) \in E$ *such that*

(I_1) $I(u_m, \widetilde{v}_m) = c + \delta_m$, *where* $\delta_m \to 0$ *as* $m \to +\infty$;

(I_2) $|I'(u_m, \widetilde{v}_m)(\eta, \xi)| \leqslant \varepsilon_m \|(\eta, \xi)\|$, *for* η, $\xi \in \{u_m, v_m\}$, *where* $\varepsilon_m \to 0$ *as* $m \to +\infty$,

then

$$\begin{array}{ll} \|u_m\|_{1,A} \leqslant C, & \|\widetilde{v}_m\|_{1,(\widetilde{A})} \leqslant C, \\ \int_\Omega f(u_m)u_m\,dx \leqslant C, & \int_\Omega g(\widetilde{v}_m)\widetilde{v}_m\,dx \leqslant C, \\ \int_\Omega F(u_m)\,dx \leqslant C, & \int_\Omega G(\widetilde{v}_m)\,dx \leqslant C. \end{array}$$

Proof. From (I_1) we have

$$\int_\Omega \nabla u_m \nabla \widetilde{v}_m\,dx - \int_\Omega F(u_m)\,dx - \int_\Omega G(\widetilde{v}_m)\,dx = c + \delta_m. \tag{5.4}$$

Taking $(\eta, \widetilde{\xi}) = (u_m, \widetilde{v}_m)$ in (I_2) we have

$$\left| 2\int_\Omega \nabla u_m \nabla \widetilde{v}_m\,dx - \int_\Omega f(u_m)u_m\,dx - \int_\Omega g(\widetilde{v}_m)\widetilde{v}_m\,dx \right| \leqslant \varepsilon_m \|(u_m, \widetilde{v}_m)\|, \tag{5.5}$$

which together with (I_1) and (H_3) implies that

$$(\theta - 2)\int_\Omega [F(u_m) + G(\widetilde{v}_m)]\,dx \leqslant 2c + 2\delta_m + \varepsilon_m \|(u_m, \widetilde{v}_m)\|.$$

Thus

$$\int_\Omega F(u_m)\,dx \leqslant c(1 + \delta_m + \varepsilon_m \|(u_m, v_m)\|),$$

$$\int_\Omega G(\widetilde{v}_m)\,dx \leqslant c(1 + \delta_m + \varepsilon_m \|(u_m, v_m)\|)$$

and then by (5.4)

$$\left| \int_\Omega \nabla u_m \nabla \widetilde{v}_m\,dx \right| \leqslant c(1 + \delta_m + \varepsilon_m \|(u_m, v_m)\|)$$

and finally by (5.5) also

$$\int_\Omega f(u_m)u_m\,dx \leqslant c(1+\delta_m+\varepsilon_m\|(u_m,v_m)\|),$$

$$\int_\Omega g(\widetilde{v}_m)\widetilde{v}_m\,dx \leqslant c(1+\delta_m+\varepsilon_m\|(u_m,v_m)\|).$$

Taking $(\eta,\widetilde{\xi})=(0,\widetilde{u}_m)$ in (I_2) we have

$$\left|\int_\Omega \nabla u_m\nabla\widetilde{u}_m\,dx-\int_\Omega g(\widetilde{v}_m)\widetilde{u}_m\,dx\right| \leqslant \varepsilon_m\|(0,\widetilde{u}_m)\| = \varepsilon_m\|\widetilde{u}_m\|_{1,(\widetilde{A})},$$

thus,

$$\|u_m\|_{1,A}^2-\int_\Omega g(\widetilde{v}_m)\widetilde{u}_m\,dx \leqslant \varepsilon_m\|\widetilde{u}_m\|_{1,(\widetilde{A})}. \tag{5.6}$$

Setting $\widetilde{U}_m=\widetilde{u}_m/C_0\|\widetilde{u}_m\|_{1,(\widetilde{A})}$ and $V_m=v_m/C_1\|v_m\|_{1,A}$ we have $\|\widetilde{U}_m\|_{1,(\widetilde{A})}=1/C_0$ and $\|\widetilde{U}_m\|_{(\Psi)}\leqslant C_0\|\widetilde{U}_m\|_{1,(\widetilde{A})}\leqslant 1$ and thus by (5.6)

$$\|u_m\|_{1,A}\leqslant C_0\int_\Omega g(\widetilde{v}_m)\widetilde{U}_m+\varepsilon_m. \tag{5.7}$$

Note also that

$$\frac{1}{C}\int_\Omega G(\widetilde{U}_m)\leqslant\int_\Omega \Psi(\widetilde{U}_m)\,dx\leqslant 1$$

since $\|\widetilde{U}_m\|_{(\Psi)}=\inf\{\lambda\colon \int_\Omega \Psi(\frac{\widetilde{U}_m}{\lambda})\leqslant 1\}$.

We now rely on the following elementary inequalities

$$xy\leqslant F(x)+f^{-1}(y)y \quad\text{and}\quad xy\leqslant G(x)+g^{-1}(y)y. \tag{5.8}$$

Applying (5.8) to the first term on the right-hand side in (5.7), with $y=g(\widetilde{v}_m)$ and $x=\widetilde{U}_m$ yields

$$\begin{aligned}\int_\Omega g(\widetilde{v}_m)\widetilde{U}_m\,dx &\leqslant \int_\Omega G(\widetilde{U}_m)+\int_\Omega g(\widetilde{v}_m)\widetilde{v}_m\,dx\\ &\leqslant C(1+\delta_m+\varepsilon_m\|(u_m,\widetilde{v}_m)\|).\end{aligned}$$

Now using (5.7), we get

$$\|u_m\|_{1,A} \leqslant \varepsilon_m + C(1 + \delta_m + \varepsilon_m \|(u_m, \widetilde{v}_m)\|).$$

Arguing similarly, choosing $(\eta, \tilde{\xi}) = (v_m, 0)$, yields

$$\|\widetilde{v}_m\|_{1,(\widetilde{A})} \leqslant \varepsilon_m + C(1 + \delta_m + \varepsilon_m \|(u_m, \widetilde{v}_m)\|).$$

Joining the two estimates yields the claim. □

5.4. *Approximation by finite-dimensional problem*

The functional I given by (1.2) is strongly indefinite near zero, since the first term is positive on the submanifold E^+, and negative on the submanifold E^-. Since both E^+ and E^- are infinite dimensional, the standard linking theorems do not apply. We overcome this difficulty by using a finite-dimensional approximation. Denoting by $e_1, e_2, \ldots$ an orthonormal basis of eigenfunctions associated to the eigenvalues $\lambda_1, \lambda_2, \ldots$ of the Laplacian (with Dirichlet boundary conditions), we set $E_n = \operatorname{span}\{e_1, e_2, \ldots, e_n\}$. Let

$$E_n^+ := \{(z, \tilde{z});\ z \in E_n\}, \quad E_n^- := \{(z, -\tilde{z});\ z \in E_n\}.$$

Setting $\widetilde{E}_n = \{\widetilde{v} \mid v \in E_n\}$, one shows exactly as in Lemma 4.4 that

$$E_n \times \widetilde{E}_n = E_n^+ \,\widetilde{\oplus}\, E_n^-.$$

We recall once more that E^+ and E^- are *linear* with respect to the "tilde-sum". Thus, we can define the following "projections":

$$P_n^- : E_n^+ \,\widetilde{\oplus}\, E_n^- \to E_n^-, \quad P_n^-((u, \widetilde{v})) = \left(\tfrac{u-v}{2}, -\widetilde{\tfrac{u-v}{2}}\right),$$
$$P_n^+ : E_n^+ \,\widetilde{\oplus}\, E_n^- \to E_n^+, \quad P_n^+((u, \widetilde{v})) = \left(\tfrac{u+v}{2}, \widetilde{\tfrac{u+v}{2}}\right),$$

which are clearly continuous mappings.

We now restrict the functional I to $E_n \times \widetilde{E}_n = E_n^+ \,\widetilde{\oplus}\, E_n^-$. Consider the set

$$Q_n := \{w \;\widetilde{+}\; r(e_1, \widetilde{e}_1);\ w \in E_n^-,\ \|w\| \leqslant R_0 \text{ and } 0 \leqslant r \leqslant R_1\} \subset E_n^+ \,\widetilde{\oplus}\, E_n^-,$$

where R_0 and R_1 are as in Lemma 5.3. Furthermore, define the class of mappings

$$H_n = \{h \in C(Q_n, E_n^+ \widetilde{\oplus} E_n^-);\ h(z) = z \text{ on } \partial Q_n\},$$

where ∂Q_n is the boundary of Q_n relative to $E_n^+ \tilde{\oplus} E_n^-$. Finally, set

$$c_n = \inf_{h\in H_n} \max_{z\in Q_n} I(h(z)).$$

We show

Lemma 5.5. *The sets Q_n and $\partial B_{\rho_0} \cap E_n^+$ link, i.e.*

$$h(Q_n) \cap (\partial B_{\rho_0} \cap E_n^+) \neq 0 \quad \forall h \in H_n. \tag{5.9}$$

Proof. The statement (5.9) is equivalent to saying that

$$\exists (u, \tilde{v}) \in Q_n \text{ such that } \|h(u, \tilde{v})\| = \rho_0 \quad \text{and} \quad P_n^- h(u, \tilde{v}) = 0. \tag{5.10}$$

Let $(u, \tilde{v}) = w \tilde{+} s(e_1, \tilde{e}_1) \in Q_n$. Define the continuous maps (here we use Remark 4.2)

$$\begin{aligned} &\psi_t : Q_n \to E_n^- \tilde{+} [(e_1, \tilde{e}_1)], \\ &\psi_t(w \tilde{+} s(e_1, \tilde{e}_1)) \\ &\quad = t P_n^- h((u, \tilde{v})) \tilde{+} (1-t) w \tilde{+} [t \| P_n^+ h((u, \tilde{v}))\| + (1-t)s - \rho_0](e_1, \tilde{e}_1). \end{aligned}$$

Note that for $(u, \tilde{v}) = w \tilde{+} t(e_1, \tilde{e}_1) \in \partial Q_n$, we have

$$\psi_t(w \tilde{+} s(e_1, \tilde{e}_1)) = w \tilde{+} (s - \rho_0)(e_1, \tilde{e}_1) \neq (0, 0) \quad \forall t \in [0, 1]$$

and hence

$$\psi_0(w \tilde{+} s(e_1, \tilde{e}_1)) = w \tilde{+} (s - \rho)(e_1, \tilde{e}_1)$$

is homotopic to

$$\psi_1(w \tilde{+} s(e_1, \tilde{e}_1)) = P_n^- h((u, \tilde{v})) \tilde{+} (\| P_n^+ h((u, \tilde{v}))\| - \rho_0)(e_1, \tilde{e}_1).$$

By the properties of the topological degree on oriented manifolds (see [7]) we have that the degree of the maps ψ_t with respect to Q_n and $(0, 0)$ is well defined, and that

$$deg(\psi_1, Q_n, (0, 0)) = deg(\psi_0, Q_n, (0, 0)) = 1.$$

Hence, there exists an element $(u, \widetilde{v}) \in Q_n$ such that $\psi_1(u, \widetilde{v}) = (0, 0)$, and hence satisfying (5.10). □

Choosing ρ_0 as in Lemma 5.2, we now conclude that

$$c_n \geqslant \sigma_0 > 0 \quad \text{for all } n \in \mathbb{N}.$$

Furthermore, since $id_{E_n^+ \widetilde{\oplus} E_n^-} \in \Gamma_n$, we have for $z = r(e_1, \widetilde{e}_1) \mp (u, -\tilde{u}) \in Q_n$

$$c_n \leqslant \max_{z \in Q_n} I(z) \leqslant R_1^2 \ \|e_1\|^2 \leqslant c\ R_1^2.$$

Thus, by the linking theorem (see [8]), we obtain a PS-sequence, which is bounded in view of Proposition 5.4. Since $E_n^+ \widetilde{\oplus} E_n^-$ is finite dimensional, we therefore get that c_n is a critical level of $I|_{E_n^+ \widetilde{\oplus} E_n^-}$, for each $n \in \mathbb{N}$, with a corresponding sequence of critical points $z_n \in E_n^+ \widetilde{\oplus} E_n^-$ with $\|z_n\| \leqslant c$, where c does not depend on n.

5.5. *Limit for $n \to \infty$*

By the last subsection we have a sequence $z_n = (u_n, \widetilde{v}_n) \in E_n \times E_n$ with

$$I(z_n) = c_n \in [\sigma_0, cR_1^2] \quad \text{and} \quad I'(z_n) = 0. \tag{5.11}$$

By Proposition 5.4 we have $\|z_n\| \leqslant c$ and hence, for a subsequence, $z_n = (u_n, \widetilde{v}_n) \rightharpoonup z = (u, \tilde{v})$ in $E = W_0^1 L_A \times W_0^1 L_{(\widetilde{A})}$. Again by Proposition 5.4 we have $\int_\Omega F(u_n)\,dx \leqslant c$, $\int_\Omega G(\widetilde{v}_n)\,dx \leqslant c$ and $\int_\Omega f(u_n)u_n\,dx \leqslant c$, $\int_\Omega g(u_n)u_n \leqslant c$. Using Lemma 2.1 in [3] we conclude that

$$f(u_n) \to f(u) \quad \text{and} \quad g(\widetilde{v}_n) \to g(\widetilde{v}) \text{ in } L^1(\Omega)$$

Taking arbitrary test functions $(0, \widetilde{\eta})$ and $(\zeta, 0)$ in $E_n \times \widetilde{E}_n$ we get

$$\int_\Omega \nabla u_n \nabla \widetilde{\eta}\,dx = \int_\Omega g(\widetilde{v}_n)\widetilde{\eta}\,dx, \quad \int_\Omega \nabla \widetilde{v}_n \nabla \zeta\,dx = \int_\Omega f(u_n)\zeta\,dx \quad \forall \eta, \quad \zeta \in E_n. \tag{5.12}$$

Using the fact that $\cup_{n \in \mathbb{N}}(E_n \times \widetilde{E}_n)$ is dense in E, we obtain by taking the limit $n \to \infty$,

$$\int_\Omega \nabla u \nabla \widetilde{\eta}\,dx = \int_\Omega g(\widetilde{v})\widetilde{\eta}\,dx \quad \forall \widetilde{\eta} \in W^1 L_{\widetilde{A}},$$

$$\int_\Omega \nabla \widetilde{v} \nabla \zeta\,dx = \int_\Omega f(u)\zeta\,dx \quad \forall \zeta \in W_0^1 L_A\ .$$

Thus, $(u, \widetilde{v}) \in W_0^1 L_A \times W_0^1 L_{(\widetilde{A})}$ is a weak solution of problem (1.1).

It remains to show that $(u, \widetilde{v})$ is nontrivial; assume by contradiction that $u = 0$, then by the Eq. (1.1) also $\widetilde{v} = 0$. Note that we can find a suitable Δ-regular N-function F_1 with $F_1 \prec\prec \Phi$ and the properties $F(x) \leqslant F_1(x)$, $f(x) \leqslant f_1(x) \quad \forall x \in \mathbb{R}^+$. Thus

$$\|u_n\|_{(F_1)} \to 0, \text{ i.e. } \inf\{\lambda > 0 \; ; \int_\Omega F_1(\frac{u_n}{\lambda}) \leqslant 1\} =: \lambda_n \to 0.$$

Since, for $\lambda_n < 1$ holds $\frac{1}{\lambda_n} \int_\Omega F_1(u_n) \leqslant \int_\Omega F_1(\frac{u_n}{\lambda_n}) \leqslant 1$, we conclude that

$$\int_\Omega F(u_n) \leqslant \int_\Omega F_1(u_n) \leqslant \lambda_n \to 0.$$

Since F_1 is Δ-regular, we have $x f_1(x) \leqslant c F_1(x)$, for some $c > 1$, and hence

$$0 \leqslant \int_\Omega f(u_n) u_n \leqslant \int_\Omega f_1(u_n) u_n \leqslant c \int_\Omega F_1(u_n)\, dx \to 0 \; . \tag{5.13}$$

This implies now by (5.12), choosing $\zeta = u_n$, that $\int_\Omega \nabla u_n \nabla \widetilde{v}_n \, dx \to 0$, and thus also $I(u_n, \widetilde{v}_n) \to 0$. But this contradicts that $I(u_n, \widetilde{v}_n) \geqslant \sigma_0 > 0$, for all $n \in \mathbb{N}$.

This concludes the proof of Theorem 1.3. □

6. Critical Orlicz pairs near the critical hyperbola

In this section, we consider N-functions of the (asymptotic) type

$$\Phi(s) \sim s^{p+1} (\log(1+s))^\alpha \tag{6.1}$$

with $p > 1$ and $\alpha > 0$.

It is natural to expect that the critical Orlicz associate Ψ (i.e. such that (Φ, Ψ) form a critical Orlicz pair) will be given by a N-function Ψ of the asymptotic form

$$\Psi(s) \sim s^{q+1} (\log(1+s))^{-\beta}, \tag{6.2}$$

where q satisfies $\frac{1}{p+1} + \frac{1}{q+1} = 1 - \frac{2}{N}$, and some relation between α and β. This is indeed so, and the relation between α and β will be given in Proposition 6.3 below.

We begin by showing that functions of type (6.1) and (6.2) satisfy the hypotheses of Theorem 1.3.

Lemma 6.1. *Suppose that $\Phi(t)$ is of the form*

$$\Phi(t) = t^{p+1} \, g(t),$$

where $p > 1$, and $g \in C(\mathbb{R}^+, \mathbb{R}^+)$ satisfying one of the following conditions:

(1) *$g(t)$ increasing,*
(2) *$g(t) \searrow 0$ and $g(t)\,t^{\varepsilon}$ increasing for large t, for any $\varepsilon > 0$.*

Then Φ is uniformly superquadratic near zero (see Definition 5.1).

Proof. Let $0 < s \leqslant 1$ and $t > 0$.

(1) We have, using that g is increasing,

$$\Phi(st) = (st)^{p+1} g(st) = s^{p+1} t^{p+1} g(t) \frac{g(st)}{g(t)} = s^{p+1} \frac{g(st)}{g(t)} \Phi(t) \leqslant s^{p+1} \Phi(t).$$

(2) We have, for some $0 < \delta < p - 1$

$$\Phi(st) = (st)^{p+1} g(st) = s^{p+1} \frac{g(st)}{g(t)} \Phi(t) = s^{2+\delta} \Phi(t)\, s^{p-1-\delta} \frac{g(st)}{g(t)} \leqslant s^{2+\delta} \Phi(t),$$

indeed, let $\varepsilon = p - 1 - \delta$, and suppose that $t^{\varepsilon} g(t)$ is increasing for $t \geqslant t_0$. Then we have, since $0 \leqslant s \leqslant 1$

$$s^{\varepsilon} \frac{g(st)}{g(t)} \leqslant \max_{0 \leqslant t \leqslant t_0} \frac{g(st)}{g(t)} + \max_{t \geqslant t_0} \frac{(st)^{\varepsilon} g(st)}{t^{\varepsilon} g(t)} \leqslant c\,. \qquad \square$$

Lemma 6.2. *Suppose that Φ is of class C^1, and (asymptotically) of the form*

$$\Phi(s) \sim c s^{p+1} g(s) \quad with \ p + 1 > \frac{N}{N-2}$$

and

$$\lim_{s \to \infty} \frac{g'(s)}{g(s)} = 0.$$

Then Φ is θ-regular, with $\theta = p + 1$ (see Definition 2.11).

Proof. Indeed, we have

$$\lim_{s \to \infty} \frac{s \varphi(s)}{\Phi(s)} = \lim_{s \to \infty} \frac{(p+1) s^{p+1} g(s) + s^{p+1} g'(s)}{s^{p+1} g(s)} = p + 1. \qquad \square$$

Proposition 6.3. *Suppose that Φ is (asymptotically) of the form*

$$\Phi(s) = cs^{p+1}(\log s)^{\alpha} \quad with \ p+1 > \frac{N}{N-2}.$$

Then the associate critical Orlicz function Ψ is (asymptotically) given by

$$\Psi(s) = ds^{q+1}(\log s)^{-\alpha\frac{q+1}{p+1}}$$

with

$$\frac{1}{p+1} + \frac{1}{q+1} = 1 - \frac{2}{N}. \tag{6.3}$$

Proof. It is easy to check that (asymptotically)

(1) $\Phi^{-1}(t) \sim c_1 t^{\frac{1}{p+1}}(\log t)^{-\frac{\alpha}{p+1}}$.
(2) $(\Phi^{-1})'(t) \sim c_2 t^{\frac{1}{p+1}-1}(\log t)^{-\frac{\alpha}{p+1}}$.
(3) Using Definition 2.9:

$$A^{-1}(t) \sim c_3 t^{\frac{N+(p+1)}{N(p+1)}}(\log t)^{-\frac{\alpha}{p+1}}.$$

(4) $A(s) \sim c_4 s^{\frac{N(p+1)}{N+(p+1)}}(\log s)^{\alpha\frac{N}{N+(p+1)}}$.
(5) $\widetilde{A}(s) \sim c_5 s^{\frac{N(p+1)}{Np-(p+1)}}(\log s)^{-\alpha\frac{N}{Np-(p+1)}}$.
(6) $\widetilde{A}^{-1}(t) \sim c_6 t^{\frac{Np-(p+1)}{N(p+1)}}(\log t)^{\frac{\alpha}{p+1}}$.
(7) $\frac{\widetilde{A}^{-1}(t)}{t^{1+\frac{1}{N}}} \sim c_6 t^{\frac{-2p-1}{Np}}(\log t)^{\frac{\alpha}{p+1}}$.
(8) Using again Definition 2.9:

$$\Psi^{-1}(t) \sim c_7 t^{\frac{(N-2)(p+1)-N}{N(p+1)}}(\log s)^{\frac{\alpha}{p+1}}.$$

(9) $\Psi(s) \sim c_8 s^{\frac{N(p+1)}{(N-2)(p+1)-N}}(\log s)^{-\frac{\alpha}{p+1}\frac{N(p+1)}{(N-2)(p+1)-N}}$. Setting $q+1 := \frac{N(p+1)}{(N-2)(p+1)-N}$, once checks that (6.3) holds, and thus finally
(10) $\Psi(s) \sim ds^{q+1}(\log s)^{-\alpha\frac{q+1}{p+1}}$. $\square$

We remark that M.A. Krasnoselskǐ and J.B. Rutickiĭ, in their book on Orlicz spaces [6, Chapter I, Section 7], consider the class of N-functions

$$\Phi(s) = cs^{p+1}(\log s)^{\alpha_1}(\log\log s)^{\alpha_2}\dots\ (\log\log\dots\log s)^{\alpha_k}, \quad \alpha_i \in \mathbb{R}.$$

Repeating the above calculations one shows that the critical Orlicz associates to these functions are given by

$$\Psi(s) = ds^{q+1}(\log s)^{-\beta_1}(\log\log s)^{-\beta_2}\ldots(\log\log\ldots\log s)^{-\beta_k}$$

with

$$\frac{1}{p+1} + \frac{1}{q+1} = 1 - \frac{2}{N} \quad \text{and} \quad \beta_i = \alpha_i \frac{q+1}{p+1}, \qquad i = 1, \ldots, k.$$

References

[1] R.A. Adams, Sobolev Spaces, Pure and Applied Mathematics, vol. 65, Academic Press, New York, London, 1975.

[2] D.G. de Figueiredo, P.L. Felmer, On superquadratic elliptic systems, Trans. Amer. Math. Soc. 343 (1994) 99–116.

[3] D.G. de Figueiredo, O. Miyagaki, B. Ruf, Elliptic equations in R^2 with nonlinearities in the critical growth range, Calculus Variations Partial Differential Equations 4 (1996).

[4] J.P. Gossez, Orlicz Spaces, Orlicz–Sobolev Spaces and Strongly Nonlinear Elliptic Problems, Trabalho de Matemática, No. 103, Departamento de Matematica, Universidade de Brasilia, 1976.

[5] J. Hulshof, R. van der Vorst, Differential systems with strongly indefinite variational structure, J. Funct. Anal. 114 (1993) 32–58.

[6] M.A. Krasnoselskiĭ, J.B. Rutickiĭ, Convex functions and Orlicz Spaces, Transl. first Russian edition by Leo F. Boron. P. Noordhoff Ltd., Groningen, 1961.

[7] L. Nirenberg, Topics in Nonlinear Functional Analysis, Courant Institute, New York University, 1974.

[8] P.H. Rabinowitz, Minimax Methods in Critical Point Theory with Applications to Differential Equations, CBMS Regional Conference Series in Mathematics, vol. 65, American Mathematical Society, Providence, RI, 1986.

[9] M.M. Rao, Z.D. Ren, Theory of Orlicz Spaces, Monographs and Textbooks in Pure and Applied Mathematics, vol. 146. Marcel Dekker, Inc., New York, 1991.

Math. Ann. 333, 231–260 (2005)

Mathematische Annalen

DOI: 10.1007/s00208-005-0639-1

Liouville type theorems, monotonicity results and a priori bounds for positive solutions of elliptic systems

Djairo G. de Figueiredo · Boyan Sirakov

Received: 14 November 2003 /
Published online: 16 July 2005 – © Springer-Verlag 2005

1. Introduction

This paper is a contribution to the study of boundary value problems for systems of elliptic partial differential equations of the form

$$\begin{cases} -\Delta u_1 = f(x, u_1, u_2) & \text{in } \Omega \\ -\Delta u_2 = g(x, u_1, u_2) & \text{in } \Omega \\ u_1 = u_2 = 0 & \text{on } \partial\Omega, \end{cases} \tag{1}$$

where u_1, u_2 are real-valued functions defined on a smooth bounded domain Ω in $\mathbb{R}^N$, $N \geq 3$, and f and g are Hölder continuous functions defined in $\overline{\Omega} \times \mathbb{R} \times \mathbb{R}$.

This type of systems has been extensively studied during the last two decades - see for example the survey paper [19] and the references therein. One of the important questions is the existence of *a priori* bounds for positive smooth solutions of these systems.

It is well known that the existence of a priori bounds depends on the growth of the functions f and g as u_1 and u_2 go to infinity. In view of what is known for scalar equations, one expects that some polynomial (subcritical) growth is to be required. In fact such a restriction comes from the Sobolev imbedding theorems in dimension $N \geq 3$. It is also known that a priori bounds are particularly interesting when superlinear equations are considered. In fact, it is classical (see [27], [4], [3], [34]) that establishing a priori bounds for a scalar equation permits, through use of Krasnoselskii's index theory, to obtain existence results for such an equation. Very recently it was shown that systems have an analogous property and definitions of superlinearity for nonvariational systems of two equations were given in [2] ; see also [19] and [40] for applications of index theory to some special types

D.G. De Figueiredo
IMECC-UNICAMP, Caixa Postal 6065, Campinas, S.P., Brazil

B. Sirakov
Laboratoire MODALX, UFR SEGMI, Université Paris 10, Bâtiment G, 92001 Nanterre Cedex, France

CAMS, EHESS, 54 bd Raspail, 75270 Paris Cedex 06, France

Originally published in: Math. Ann. 333 (2005), 231-260,

Reprint by Springer-Verlag Berlin Heidelberg 2013,
DOI: 10.1007/978-3-319-02856-9_38

of systems. More general results can be found in [37] where, in addition, systems of many equations are considered.

The simplest case of systems of type (1) — which is the only case in which a priori bounds have been studied up to now — is when the leading parts of f and g involve just pure powers of u_1 and u_2. More precisely, when f and g are such that (1) can be written in the form

$$\begin{cases} -\Delta u_1 = a(x)u_1^{\alpha_{11}} + b(x)u_2^{\alpha_{12}} + h_1(x, u_1, u_2) \\ -\Delta u_2 = c(x)u_1^{\alpha_{21}} + d(x)u_2^{\alpha_{22}} + h_2(x, u_1, u_2), \end{cases} \tag{2}$$

where the exponents α_{ij} are nonnegative real numbers, $a(x), b(x), c(x), d(x)$ are nonnegative continuous functions on $\overline{\Omega}$, and h_1, h_2 are locally bounded functions such that uniformly in $x \in \Omega$

$$\begin{cases} \lim\limits_{|(u_1,u_2)|\to\infty} \left(a(x)u_1^{\alpha_{11}} + b(x)u_2^{\alpha_{12}}\right)^{-1} |h_1(x, u_1, u_2)| = 0 \\ \lim\limits_{|(u_1,u_2)|\to\infty} \left(c(x)u_1^{\alpha_{21}} + d(x)u_2^{\alpha_{22}}\right)^{-1} |h_2(x, u_1, u_2)| = 0. \end{cases} \tag{3}$$

The method used here in order to obtain the a priori bounds, the so-called *blow-up method*, was introduced in [24] to treat the scalar case. The use of this method to treat systems like the one in (2) was first done in [38], and then in [33], [8], [19], [40]. Let us note that the blow-up method itself depends on results of nonexistence of positive solutions of equations and systems in the whole space or in a half-space. Such results are usually referred to as Liouville type theorems – see Section 2.

Our main result, Theorem 1.1, unifies and extends the previous results on a priori bounds for (2). In addition, it allows more general nonlinearities in systems of type (1), namely mixed powers of u_1 and u_2 in the principal part of the nonlinearities f and g.

Specifically, our results will concern the following system

$$\begin{cases} -\Delta u_1 = a(x)u_1^{\alpha_{11}} + b(x)u_2^{\alpha_{12}} + f_1(x)u_1^{\gamma_{11}}u_2^{\gamma_{12}} + h_1(x, u_1, u_2) \\ -\Delta u_2 = c(x)u_1^{\alpha_{21}} + d(x)u_2^{\alpha_{22}} + f_2(x)u_1^{\gamma_{21}}u_2^{\gamma_{22}} + h_2(x, u_1, u_2), \end{cases} \tag{4}$$

where we keep all the hypotheses made before for the system (2).

As for new hypotheses for system (4), we suppose that the continuous functions f_1 and f_2 are nonnegative in Ω, so (2) is a particular case of (4), and

(γ_1) we have $0 \leq \gamma_{ij} \leq \alpha_{ij}$, $i, j = 1, 2$,

$$\frac{\gamma_{11}}{\alpha_{11}} + \frac{\gamma_{12}}{\alpha_{12}} = 1 \qquad \text{and} \qquad \frac{\gamma_{21}}{\alpha_{21}} + \frac{\gamma_{22}}{\alpha_{22}} = 1. \tag{5}$$

Hypothesis (5) means that the powers γ_{ij} are such that the terms with coefficients f_i in (4) are exactly of the order of the principal parts in (2) and cannot be included in the functions h_1, h_2, via Young's inequality.

We make the following superlinearity assumptions on the exponents α_{ij} :

$$\text{either} \quad \alpha_{11} > 1, \quad \text{or} \quad \alpha_{22} > 1, \quad \text{or} \quad \alpha_{12}\alpha_{21} > 1. \tag{6}$$

We remark here that $\alpha_{ii} > 1$ means the Emden-Fowler equation $-\Delta u_i = u_i^{\alpha_{ii}}$ is superlinear, while the third inequality in (6) has long been used as a notion of superlinearity for the Lane-Emden system

$$\begin{cases} -\Delta u_1 = u_2^{\alpha_{12}} \\ -\Delta u_2 = u_1^{\alpha_{21}}. \end{cases} \tag{7}$$

Further, we have to specify the maximal growth of f and g in (1) that we can allow. This can best be done by using the following geometric construction. We denote $\vec{\beta} = (\beta_1, \beta_2) \in \mathbb{R}^2$, and introduce the following lines (see Figure 1 in Section 2)

$$l_1 = \left\{\vec{\beta} \mid \beta_1 + 2 - \beta_1\alpha_{11} = 0\right\}, \quad l_2 = \left\{\vec{\beta} \mid \beta_2 + 2 - \beta_2\alpha_{22} = 0\right\},$$

$$l_3 = \left\{\vec{\beta} \mid \beta_1 + 2 - \beta_2\alpha_{12} = 0\right\}, \quad l_4 = \left\{\vec{\beta} \mid \beta_2 + 2 - \beta_1\alpha_{21} = 0\right\}.$$

A thorough explanation of how these lines appear will be given in the next section, when we expound the blow-up method. We just note here that if we set $(\beta_1', \beta_2') = l_1 \cap l_2$ (in case $\alpha_{11}, \alpha_{22} > 1$) and $(\beta_1'', \beta_2'') = l_3 \cap l_4$ (in case $\alpha_{12}\alpha_{21} > 1$), then the Emden-Fowler equations $-\Delta u_i = u_i^{\alpha_{ii}}$ are subcritical provided $\beta_i' > \frac{N-2}{2}$, while system (7) is subcritical provided $\beta_1'' + \beta_2'' > N - 2$. Actually, the last condition is equivalent to saying that the exponents α_{12}, α_{21} in (7) are under the so-called "critical hyperbola" — a widely used notion of criticality for (7), introduced in [16] and [26].

We consider points $\vec{\beta} \geq 0$ which are to the left of or on l_1, below or on l_2 (note that both l_1 and l_2 can be empty, and then they introduce no restriction), below or on l_3, and above or on l_4. We call these points **admissible** (this notion will be completely understood in the next section). So given a system of type (2) the set of admissible points is automatically defined.

We divide the systems of type (2) into three classes, determined by the exponents α_{ij}. In each case we make a choice of (β_1, β_2) – which we shall use to state our theorem – and make some assumptions on the coefficients of the system.

Case A. *The intersection of l_1 and l_2 is admissible.* Then we set $(\beta_1, \beta_2) = l_1 \cap l_2$. In this case we shall assume that the functions $a(x)$ and $d(x)$ are bounded below on $\overline{\Omega}$ by a positive constant.

Case B. *The intersection of l_3 and l_4 is admissible.* For this type of systems we take $(\beta_1, \beta_2) = l_3 \cap l_4$. In this case we shall assume that the functions $b(x)$ and $c(x)$ are bounded below on $\overline{\Omega}$ by a positive constant. Further, in case B we have to assume in addition that $\alpha_{12} > 1$ and $\alpha_{21} > 1$ - see Remark 2 below.

Case C. *None of* $l_1 \cap l_2$ *and* $l_3 \cap l_4$ *is admissible.* Then either $l_1 \cap l_3$ or $l_2 \cap l_4$ is admissible and we take this intersection point to be our (β_1, β_2). In this case we shall assume that the function $b(x)$ (resp. $c(x)$) is bounded below on $\overline{\Omega}$ by a positive constant.

We stress that nothing prevents (β_1, β_2) from being the intersection of more than two lines. Note that a system can be simultaneously of type A and B, but type C is exclusive of the other two types.

In the special situation in Case B when β is the intersection of more than two lines we shall need in our arguments below a further technical assumption on the γ_{ij} (we believe this hypothesis can be removed) :

(γ_2) $\gamma_{11}, \gamma_{12} \geq 1$, if $\vec{\beta} = l_1 \cap l_3 \cap l_4$, ; $\gamma_{21}, \gamma_{22} \geq 1$, if $\vec{\beta} = l_2 \cap l_3 \cap l_4$.

Theorem 1.1. *Assume that system (4) satisfies the conditions stated above, and that the pair* (β_1, β_2) *which corresponds to the type of the system (A, B or C) satisfies the condition*

$$\min\{\beta_1, \beta_2\} > \frac{N-2}{2}. \tag{8}$$

Then system (4) admits a priori estimates, that is, each couple of positive classical solutions of (4) is bounded in the L^∞*-norm by a constant which depends only on* L^∞*-bounds for the coefficients of the system and on the domain.*

Remark 1. In Case C it is actually sufficient to suppose that $\beta_1 > \dfrac{N-2}{2}$ if $l_1 \cap l_3$ is admissible, and that $\beta_2 > \dfrac{N-2}{2}$ if $l_2 \cap l_4$ is admissible.

Remark 2. Note the additional assumption we have made in Case B - that both exponents α_{12} and α_{21} are greater than 1. This is due to the problem (7) in $\mathbb{R}^N_+$. We do not actually know of *any* Liouville-type result on this problem in a half-space, which does not require this hypothesis. See the detailed discussion at the end of the next section.

Remark 3. In [19] the first author presented the blow-up procedure for systems of type (2) and observed that there exist two special types of these systems (called weakly coupled and strongly coupled in [19]), contained in Cases A and B respectively, for which the blow-up procedure leads to Liouville type results for the equation $-\Delta u = u^p$ or for the Lane-Emden system (7).

In [40] Zou considered system (2) and showed that it admits a priori estimates under the supplementary assumptions that all exponents $\alpha_{ij} > 1$, that both coordinates of the point $l_1 \cap l_2$ are larger than $\dfrac{N-2}{2}$, that both coordinates of the point $l_3 \cap l_4$ are larger than $\dfrac{N-2}{2}$ (or one of them is larger than $N-2$), and that the point $l_3 \cap l_4$ lies neither on l_1 nor on l_2. Note that the hypotheses $\alpha_{12} > 1, \alpha_{21} > 1$ are not stated in [40] but they are actually used in the proofs of the results, because of the previous remark.

Remark 4. The question of non-existence of positive *bounded* solutions of (7) in $\mathbb{R}^N$ was completely solved when $N = 3$ by Serrin and Zou, [36]. So, when $N = 3$ one can get more precise results than Theorem 1.1 - see [40], Theorems 1.1, 2.1, and 3.1 for system (2).

As noted before, implementing the blow-up method depends on availability of Liouville type results in $\mathbb{R}^N$ or in a half-space. In a number of situations that we are led to consider the results available in the literature (see Section 2) are not sufficient, so we had to prove Liouville type theorems for systems. Let us stress that our paper differs in that respect from all previous works on a priori bounds for systems, where the conditions on the system were actually chosen so that the limit process in the blow-up method (see Section 2 for details) leads to known Liouville type theorems.

We prove nonexistence results in a half-space for limit systems of (4) by showing that whenever such a system does not admit bounded solutions in $\mathbb{R}^{N-1}$ then it does not possess bounded solutions in any half-space of $\mathbb{R}^N$ either (so actually it is enough to prove the required Liouville results in the whole space). We prove this fact by using a monotonicity result for autonomous systems in a half-space. Both these theorems – of clear independent interest – hold under some supplementary assumptions that we list next.

Suppose we have an autonomous system of the type

$$\begin{cases} \Delta u_1 + f_1(u_1, u_2) = 0 \\ \Delta u_2 + f_2(u_1, u_2) = 0, \end{cases} \tag{9}$$

where $f_i \in C^1(\mathbb{R}^2)$, $i = 1, 2$, and

$$\frac{\partial f_i}{\partial u_j}(\vec{u}) \geq 0 \qquad \text{for all } i \neq j,\ \vec{u} \in \mathbb{R}^2. \tag{10}$$

Systems satisfying the last property are usually referred to as cooperative (or quasi-monotone). We shall suppose in addition that we can write

$$f_1(u_1, u_2) = u_2^p + g_1(u_1, u_2)u_1, \qquad f_2(u_1, u_2) = u_1^q + g_2(u_1, u_2)u_2, \tag{11}$$

for some $p, q > 1$ and some continuous functions g_1, g_2, which have polynomial growth in u_1, u_2.

Here are the precise statements of the results.

Theorem 1.2. *Suppose we have a nontrivial nonnegative* bounded *classical solution (u_1, u_2) of system (9) in $\mathbb{R}^N_+ = \{x \in \mathbb{R}^N \mid x_N > 0\}$, such that $u_1 = u_2 = 0$ on $\partial\mathbb{R}^N_+$. Suppose (10) and (11) are satisfied. Then*

$$\frac{\partial u_i}{\partial x_N} > 0 \qquad in\ \mathbb{R}^N_+, \quad i = 1, 2. \tag{12}$$

Theorem 1.3. *Suppose we have a system of type (9) which satisfies (10) and (11). If problem (9) with Dirichlet boundary condition has a nontrivial nonnegative bounded solution in $\mathbb{R}^N_+$, then the same problem has a positive solution in $\mathbb{R}^{N-1}$ (the limit as $x_N \to \infty$ of the solution in $\mathbb{R}^N_+$).*

We note that Dancer [17] obtained property (12) for the scalar equation $-\Delta u = f(u)$ provided that either $f(0) > 0$ or both $f(0) = 0$ and $f'(0) \geq 0$. Dancer's result was proved to hold for unbounded solutions and globally Lipschitz continuous f with $f(0) \geq 0$ by Berestycki, Caffarelli and Nirenberg ([6]). In this paper the authors also showed that positive solutions of $-\Delta u = f(u)$ are functions of x_N only, provided $f(\sup u) \leq 0$. They obtained even stronger results in dimensions 2 and 3.

The proof of Theorem 1.2 is based on the moving planes method and has two main ingredients. First, we use a maximum principle for cooperative systems in unbounded narrow domains, which follows from a result on scalar equations by Cabre [13]. Second, we make use of a recent Harnack type inequality for nonlinear systems obtained in [12]. By using some ideas from [12] we prove the following Harnack inequality for system (9).

Theorem 1.4. *Let (u_1, u_2) be a positive solution of (9) in some domain G and suppose (10) and (11) hold. Suppose K is a compact set properly included in G and*

$$\max\left\{\inf_{x\in K} u_1,\ \inf_{x\in K} u_2\right\} \leq 1, \qquad \max\left\{\sup_{x\in G} u_1,\ \sup_{x\in G} u_2\right\} \leq M.$$

Then

$$\sup_{x\in K} \max\{u_1, u_2\} \leq C \min\left\{\left(\inf_{x\in K} u_1\right)^{\frac{1}{p}}, \left(\inf_{x\in K} u_2\right)^{\frac{1}{q}}\right\}.$$

where C depends only on N and M.

This inequality permits to us to use a technique inspired by the proof of a symmetry result for scalar equations in cylinders in [6]. However, contrary to [6], we avoid using boundary Harnack inequalities. A supplementary difficulty in the argument stems from the fact that we have to use a Harnack inequality on a sequence of systems in which the coupling degenerates.

Finally, we prove some Liouville type results for systems in the whole space by, on one hand, extending to general systems a recent monotonicity result by Busca and Manasevich [10], who obtained a Liouville type theorem for the Lane-Emden system (7), and, on the other hand, by noticing an identity between exponents appearing after a blow-up change of variables and after the passage to polar coordinates used in [10]. For instance, we obtain the following result.

Theorem 1.5. *The system*

$$\begin{cases} \Delta u_1 + u_1^{\alpha_1} + u_2^{\alpha_1 \frac{\alpha_2 - 1}{\alpha_1 - 1}} = 0 \\ \Delta u_2 + u_1^{\alpha_2 \frac{\alpha_1 - 1}{\alpha_2 - 1}} + u_2^{\alpha_2} = 0, \end{cases} \tag{13}$$

does not have bounded positive classical solutions in $\mathbb{R}^N$, *provided*

$$1 < \alpha_1, \alpha_2 < \frac{N+2}{N-2}.$$

The paper is organized as follows. In Section 2 we give some preliminary results and state the known Liouville-type results for scalar equations and Lane-Emden systems in $\mathbb{R}^N$ or $\mathbb{R}^N_+$. In Section 3 we prove Theorem 1.2 and Theorem 1.3. Finally, in Section 4 we prove Theorem 1.5 and Theorem 1.1.

2. Preliminaries

Let us first describe the blow-up procedure. We will only sketch this, a full presentation can be found in [19], where this procedure is explained for system (2) (for the scalar case see Gidas-Spruck [24]). Assume that positive solutions of (4) do not have an a priori bound, that is, there exists a sequence $(u_{1,n}, u_{2,n})$ of positive solutions of (4) such that at least one of the sequences $u_{1,n}$ and $u_{2,n}$ tends to infinity in the L^∞-norm. Let β_1, β_2 be fixed positive constants to be chosen later. We set

$$\lambda_n = ||u_{1,n}||^{1/\beta_1}_{L^\infty(\Omega)},$$

if $||u_{1,n}||^{\beta_2}_{L^\infty(\Omega)} \geq ||u_{2,n}||^{\beta_1}_{L^\infty(\Omega)}$ (up to a subsequence), and

$$\lambda_n = ||u_{2,n}||^{1/\beta_2}_{L^\infty(\Omega)}$$

otherwise. We shall suppose – without restricting the generality – that we are in the first of these two situations.

Note that we have $\lambda_n \to 0$ as $n \to \infty$. Let $x_n \in \Omega$ be a point where $u_{1,n}$ assumes its maximum. The functions

$$v_{i,n}(x) = \lambda_n^{\beta_i} u_{i,n}(\lambda_n x + x_n),$$

are such that $v_{1,n}(0) = 1$ and $0 \leq v_{i,n} \leq 1$ in Ω. One also verifies that the functions $v_{1,n}$ and $v_{2,n}$ satisfy

$$\begin{cases} -\Delta v_{1,n} = a(\cdot)\lambda_n^{\beta_1+2-\beta_1\alpha_{11}}\, v_{1,n}^{\alpha_{11}} + b(\cdot)\lambda_n^{\beta_1+2-\beta_2\alpha_{12}}\, v_{2,n}^{\alpha_{12}} \\ \qquad + f_1(\cdot)\lambda_n^{\beta_1+2-\beta_1\gamma_{11}-\beta_2\gamma_{12}}\, v_{1,n}^{\gamma_{11}} v_{2,n}^{\gamma_{12}} + \lambda_n^{\beta_1+2}\widetilde{h_{1,n}} \\ -\Delta v_{2,n} = c(\cdot)\lambda_n^{\beta_2+2-\beta_1\alpha_{21}}\, v_{1,n}^{\alpha_{21}} + d(\cdot)\lambda_n^{\beta_2+2-\beta_2\alpha_{22}}\, v_{2,n}^{\alpha_{22}} \\ \qquad + f_2(\cdot)\lambda_n^{\beta_2+2-\beta_1\gamma_{21}-\beta_2\gamma_{22}}\, v_{1,n}^{\gamma_{21}} v_{2,n}^{\gamma_{22}} + \lambda_n^{\beta_2+2}\widetilde{h_{2,n}} \end{cases} \tag{14}$$

in the domain $\Omega_n = \dfrac{1}{\lambda_n}(\Omega - x_n)$, where the dot stands for $\lambda_n x + x_n$, and $\widetilde{h_{i,n}} = h_i\left(\cdot, \lambda_n^{-\beta_1} v_{1,n}, \lambda_n^{-\beta_2} v_{2,n}\right)$. By compactness we can assume that $\{x_n\}$ tends to some point $x_0 \in \overline{\Omega}$.

The idea is to pass to the limit in (14) and obtain a system which can be proven to have only the trivial solution. This would then contradict the fact that the limit of $v_{1,n}$ has value 1 at the origin.

The next lemma deals with the passage to the limit.

Lemma 2.1. *The sequences $v_{1,n}, v_{2,n}$ converge in $W^{2,p}_{\text{loc}}$, $2 \le p < \infty$ to functions $v_1, v_2 \in C^2(G) \cap C^0(\overline{G})$, satisfying the limiting system of (14) in $G = \mathbb{R}^N$ or $G = \mathbb{R}^N_+$, provided all the powers of λ_n in (14) are non-negative. This limiting system is obtained by removing the terms in (14) where the powers of λ_n are strictly positive, the terms where the coefficient vanishes at x_0, and the terms containing $h_{i,n}, i = 1, 2$.*

Proof. The argument is standard. The proof of this lemma, except for the last part, can be found in [19]. Note that the passage to the limit in the terms containing products of $v_{1,n}$ and $v_{2,n}$ causes no problem, since $W^{2,p}_{\text{loc}} \cap L^\infty$ is an algebra. To prove that the terms in $h_{i,n}$ tend to zero we distinguish two cases : if the sequences $\lambda_n^{-\beta_1} v_{1,n}$ and $\lambda_n^{-\beta_2} v_{2,n}$ are both bounded this follows from the local boundedness of h_i, if one of these sequences is unbounded, it follows from hypothesis (3). □

Now we can explain the choice of the couple (β_1, β_2) which we made in the introduction (Cases A, B and C). As stated above, in order to be able to make a passage to the limit in (14) all the powers of λ_n in this system have to be nonnegative. Hence, geometrically, we have to pick up a point (β_1, β_2) which is to the left of or on l_1, below or on l_2, below or on l_3, and above or on l_4 (we forget for an instant the mixing terms with coefficients f_i). These are exactly the points which we called admissible - see Figure 1.

Further, it is important to observe that not all admissible points (β_1, β_2) would permit us to prove a priori bounds for positive solutions of (4), as we explain next. Indeed, we have necessarily to make the choice of β_1 and β_2 in such a way that at least two of the powers of λ_n in (14) are zero. Otherwise, after the passage to the limit in (14) we may end up with an uncoupled system in $\mathbb{R}^N$ in which at least one of the equations is the Laplace equation. Hence the corresponding one of the functions v_1 and v_2 can be identically equal to a positive constant, and in this way we do not come to a contradiction.

So (β_1, β_2) has to be chosen on the intersection of at least two of the lines l_1, l_2, l_3, l_4. It can be seen that this is actually possible provided the superlinearity conditions (6) hold. Indeed, the first inequality in (6) implies that the line l_1 is not empty, and similarly the second one implies that l_2 is not empty either. Note also

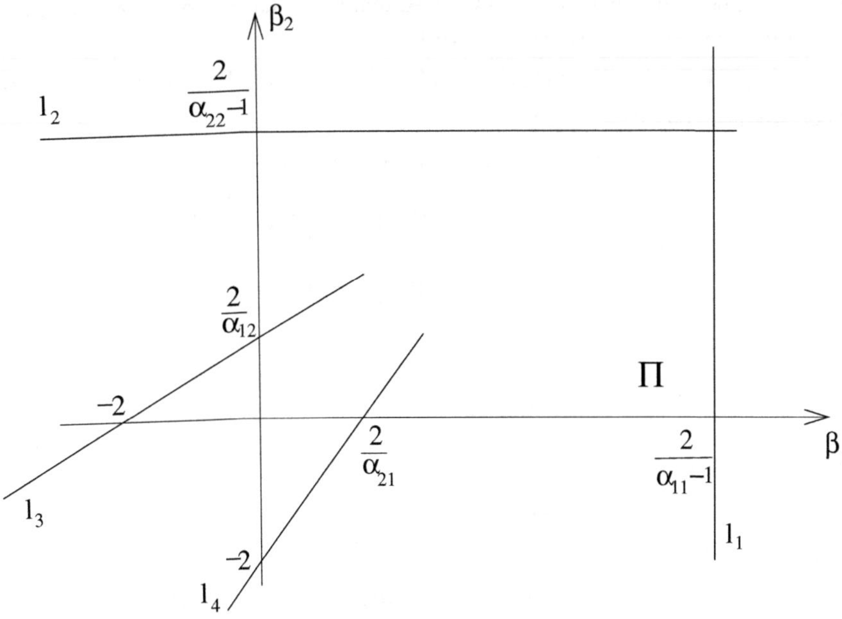

Fig. 1. Admissible couples (β_1, β_2) lie to the left of or on l_1, below or on l_2, below or on l_3, and above or on l_4

that the slope of l_3 is α_{12}^{-1} and the slope of l_4 is α_{21} ; consequently these two lines meet at a point with positive coordinates provided $\alpha_{12}\alpha_{21} > 1$.

Observe that the weakly coupled case of [19] corresponds to the situation when the intersection of l_1 and l_2 lies strictly below l_3 and strictly above l_4, while the strongly coupled case means l_3 and l_4 meet in Π, where Π denotes the rectangle enclosed by l_1, l_2, and the axes. We remark that l_1 and l_2 can be empty, in case $\alpha_{11} \le 1$ or $\alpha_{22} \le 1$; in this case Π would be a half strip or the whole positive quadrant of the (β_1, β_2)-plane.

The next lemma shows how the terms in (14) with coefficients f_i transform after the passage to the limit.

Lemma 2.2. *If hypothesis* (γ_1) *holds and* β_1, β_2 *are chosen as in the previous section then*

$$\beta_1 + 2 - \beta_1\gamma_{11} - \beta_2\gamma_{12} \ge 0 \quad \text{and} \quad \beta_2 + 2 - \beta_1\gamma_{21} - \beta_2\gamma_{22} \ge 0. \qquad (15)$$

In Case A the first (resp. the second) inequality in (15) is strict if and only if l_3 *(resp.* l_4*) does not pass through* $l_1 \cap l_2$*. In Case B the first (resp. the second) inequality is strict if and only if* l_1 *(resp.* l_2*) does not pass through* $l_3 \cap l_4$*. In Case C one of the inequalities is always strict and the other is an equality.*

Proof. Suppose first that we are in Case A, that is, we have chosen (β_1, β_2) to be the admissible point $l_1 \cap l_2$. This means that

$$\beta_1 + 2 = \alpha_{11}\beta_1 \geq \alpha_{12}\beta_2 \quad \text{and} \quad \beta_2 + 2 = \alpha_{22}\beta_2 \geq \alpha_{21}\beta_1. \tag{16}$$

Then we have, by (16) and (5),

$$\begin{aligned}\beta_1 + 2 - \beta_1\gamma_{11} - \beta_2\gamma_{12} &= (\alpha_{11} - \gamma_{11})\beta_1 - \gamma_{12}\beta_2 \\ &= \alpha_{11}\left(1 - \frac{\gamma_{11}}{\alpha_{11}}\right)\beta_1 - \gamma_{12}\beta_2 \\ &= \gamma_{12}\left(\frac{\alpha_{11}}{\alpha_{12}}\beta_1 - \beta_2\right) \\ &\geq 0\end{aligned}$$

and the inequality is strict if and only if $\beta_1\alpha_{11} = \beta_1 + 2 > \beta_2\alpha_{12}$, i.e. l_3 does not pass through $l_1 \cap l_2$. The second inequality in (15) is proved in a similar way.

Suppose now we are in Case B, that is, we have picked the admissible point $l_3 \cap l_4$ to be our (β_1, β_2). This means

$$\beta_1 + 2 = \alpha_{12}\beta_2 \geq \alpha_{11}\beta_1 \quad \text{and} \quad \beta_2 + 2 = \alpha_{21}\beta_1 \geq \alpha_{22}\beta_2. \tag{17}$$

Then, as before,

$$\begin{aligned}\beta_1 + 2 - \beta_1\gamma_{11} - \beta_2\gamma_{12} &= -\beta_1\gamma_{11} + (\alpha_{12} - \gamma_{12})\beta_2 \\ &= \gamma_{11}\left(-\beta_1 + \frac{\alpha_{12}}{\alpha_{11}}\beta_2\right) \\ &\geq 0,\end{aligned}$$

and a similar computation proves the second inequality in (15).

Finally, let us consider Case C. Suppose we have chosen $(\beta_1, \beta_2) = l_1 \cap l_3$ (a similar argument can be done when $(\beta_1, \beta_2) = l_2 \cap l_4$). Then

$$\beta_1 + 2 = \alpha_{11}\beta_1 = \alpha_{12}\beta_2 \quad \text{and} \quad \beta_2 + 2 > \max\{\alpha_{21}\beta_1, \alpha_{22}\beta_2\}, \tag{18}$$

so $\beta_1 + 2 - \beta_1\gamma_{11} - \beta_2\gamma_{12} = 0$ as in Case A. To prove the second inequality in (15) we distinguish two cases. First, if $\alpha_{21}\beta_1 \geq \alpha_{22}\beta_2$ we get

$$\begin{aligned}\beta_2 + 2 - \beta_1\gamma_{21} - \beta_2\gamma_{22} &> (\alpha_{21} - \gamma_{21})\beta_1 - \gamma_{22}\beta_2 \\ &= \gamma_{22}\left(\frac{\alpha_{21}}{\alpha_{22}}\beta_1 - \beta_2\right) \\ &\geq 0,\end{aligned}$$

while if $\alpha_{21}\beta_1 \leq \alpha_{22}\beta_2$ we have

$$\begin{aligned}\beta_2 + 2 - \beta_1\gamma_{21} - \beta_2\gamma_{22} &> -\beta_1\gamma_{21} + (\alpha_{22} - \gamma_{22})\beta_2 \\ &= \gamma_{21}\left(-\beta_1 + \frac{\alpha_{22}}{\alpha_{21}}\beta_2\right) \\ &\geq 0.\end{aligned}$$

This finishes the proof of the Lemma. □

We next list the known Liouville type theorems on the equation $-\Delta u = u^p$ and on the Lane-Emden system (7). We shall use some of these in the sequel.

Theorem 2.1. *The problem*

$$\begin{cases} -\Delta u = u^p \\ \quad u \geq 0 \end{cases} \quad \text{in} \quad \mathbb{R}^N \tag{19}$$

has no nontrivial classical solution provided $0 < p < \dfrac{N+2}{N-2}$ *(Gidas-Spruck [24], Chen-Li [15]). The same problem has no nontrivial classical supersolution provided* $1 < p \leq \dfrac{N}{N-2}$ *(Gidas [22], Souto [38], Mitidieri-Pohozaev [32]). The problem*

$$\begin{cases} -\Delta u = u^p & \text{in } \mathbb{R}^N_+ \\ \quad u \geq 0 & \text{in } \mathbb{R}^N_+ \\ \quad u = 0 & \text{on } \partial\mathbb{R}^N_+ \end{cases} \tag{20}$$

has no nontrivial classical solution provided $1 < p < \dfrac{N+1}{N-3}$ *(*$1 < p < \infty$ *if* $N = 3$*) - see Gidas-Spruck [24], Dancer [17]. The same problem has no classical supersolution provided* $1 < p \leq \frac{N+1}{N-1}$ *(Bandle-Essen [5], Laptev [29]).*

Theorem 2.2. *Consider the problem*

$$\begin{cases} -\Delta u = v^p \\ -\Delta v = u^q \\ \; u, v \geq 0 \end{cases} \quad \text{in} \quad \mathbb{R}^N \tag{21}$$

If $pq \leq 1$ *(Serrin-Zou [36]) or* $0 < p, q < \dfrac{N+2}{N-2}$ *(de Figueiredo-Felmer [20]) this problem has no nontrivial classical solutions. If* $pq > 1$ *we set*

$$\alpha = \frac{2(p+1)}{pq-1}, \qquad \beta = \frac{2(q+1)}{pq-1}.$$

Then problem (21) has no nontrivial classical solutions provided either $\max\{\alpha, \beta\} \geq N-2$ *(Mitidieri [30], Serrin-Zou [36]), or* $p, q > 1$ *and* $\min\{\alpha, \beta\} > \dfrac{N-2}{2}$ *(Busca-Manasevich [10]). The same problem has no nontrivial supersolutions provided* $pq > 1$ *and* $\max\{\alpha, \beta\} \geq N-2$ *(Mitidieri [31], Laptev [29]).*

The problem

$$\begin{cases} -\Delta u = v^p & \text{in } \mathbb{R}^N_+ \\ -\Delta v = u^q & \text{in } \mathbb{R}^N_+ \\ \; u, v \geq 0 & \text{in } \mathbb{R}^N_+ \\ \; u, v = 0 & \text{on } \partial\mathbb{R}^N_+ \end{cases} \tag{22}$$

has no nontrivial solutions provided $p, q > 1$ *and* $\max\{\alpha, \beta\} \geq N - 3$ *(Birindelli-Mitidieri [8]). The same problem does not have nontrivial classical supersolutions provided* $p, q > 1$ *and* $\max\{\alpha, \beta\} \geq N - 1$ *(Laptev [29]).*

The problem of non-existence of solutions of Lane-Emden systems in a half-space (system (22)) deserves some discussion. There are two types of results on this problem. On one hand, it is known that a Lane-Emden system in a half-space does not possess positive supersolutions when $\max\{\alpha, \beta\} \geq N - 1$. This is an exact result and a particular case of a more general theorem about existence of supersolutions in cones. However, it does not permit to obtain more precise results for *solutions*. On the other hand, in the framework of a scalar equation Dancer developed a technique, based on the moving planes method, which gives a monotonicity result for *bounded* solutions of the equation in a half-space. Then one gets as a corollary that the existence of a nontrivial bounded solution in $\mathbb{R}^N_+$ implies the existence of a non-trivial solution in $\mathbb{R}^{N-1}$. However, the moving planes method requires Lipschitz nonlinearities – that is why a result, where this method is directly employed has to require that all powers involved be greater or equal to one.

The next section contains the proofs of Theorems 1.2 and 1.3, as well as another monotonicity lemma, which will play an important role in the proof of Theorem 1.1.

3. Monotonicity results for systems

3.1. A maximum principle in narrow domains

In this section we extend to cooperative systems a maximum principle in narrow domains, proved by Cabre [13] in the case of a scalar equation.

We recall the following definition of [13]. For a given domain $\Omega \subset \mathbb{R}^N$, the quantity $R(\Omega)$ is defined to be the smallest positive constant R such that

$$\operatorname{meas}(B_R(x) \setminus \Omega) \geq \frac{1}{2} \operatorname{meas}(B_R(x)), \qquad \text{for all } x \in \Omega.$$

If no such radius R exists, we define $R(\Omega) = +\infty$.

It is easy to see that whenever the domain Ω is contained between two parallel hyperplanes at a distance d, we have

$$R(\Omega) \leq \frac{2^N d}{\omega_N}, \tag{23}$$

where ω_N is the volume of the unit ball in $\mathbb{R}^N$.

In the sequel we shall consider uniformly elliptic second-order operators in the form

$$L = \sum a_{ij}(x)\partial_{ij} + \sum b_i(x)\partial_i + c(x), \tag{24}$$

where $c_0 I \leq (a_{ij}) \leq C_0 I$, $\sup |b_i| \leq b$, for some positive constants c_0, C_0, and some $b \geq 0$.

In [13] (see also [14], Theorem 5.3) Cabre proved the following result.

Proposition 3.1. *Let Ω be a domain such that $R(\Omega) < \infty$ and let L be an operator in the form (24), such that $c \leq 0$ in Ω. Suppose $u \in W^{2,N}_{\text{loc}}(\Omega)$ and $f \in L^\infty(\Omega)$ satisfy $Lu \geq f$ in Ω, $\limsup_{x\to\partial\Omega} u(x) \leq 0$, and $\sup_\Omega u < \infty$. Then*

$$\sup_\Omega u \leq C R(\Omega)^2 ||f||_{L^\infty(\Omega)},$$

where C is a constant depending only on $N, c_0, C_0, bR(\Omega)$.

It is not difficult to deduce from Proposition 3.1 a maximum principle for systems in domains with small $R(\Omega)$.

Theorem 3.1. *Let L_k, $k = 1, \ldots, n$, be uniformly elliptic second-order operators with bounded coefficients and without zero-order term, that is, $L_k = \sum a^{(k)}_{ij}(x)\partial_{ij} + \sum b^{(k)}_i(x)\partial_i$, $c_0 I \leq (a^{(k)}_{ij}) \leq C_0 I$, $\sup |b^{(k)}_i| \leq b$. Let the functions $c_{ij} \in L^\infty(\Omega)$, $|c_{ij}| \leq b$, be such that $c_{ij} \geq 0$ for $i \neq j$. Then there exists a number $\bar{R}$ depending only on N, c_0, C_0 and b, such that $R(\Omega) \leq \bar{R}$ implies that each solution $u_i \in W^{2,N}_{\text{loc}}(\Omega) \cap C(\overline{\Omega})$ of*

$$\begin{cases} L_i u_i + \sum_{j=1}^n c_{ij} u_j \geq 0 \text{ in } \Omega, & i = 1, \ldots, n \\ u_i(x) \leq 0 \text{ on } \partial\Omega, & i = 1, \ldots, n, \end{cases} \tag{25}$$

satisfies $u_i \leq 0$ in Ω, $i = 1, \ldots, n$ (in this case we shall say, with obvious notation, that the matrix operator $\vec{L} + \mathcal{C}(x)$ satisfies the maximum principle in Ω).

Proof. For a function f we denote $f^+ = \max\{f, 0\}$, $f = f^+ - f^-$. Then (25) implies

$$L_i u_i - c^-_{ii} u_i \geq -c^+_{ii} u^+_i - \sum_{j\neq i} c_{ij} u^+_j, \quad i = 1 \ldots, n.$$

By applying Proposition 3.1 to these equations we obtain

$$\sup_\Omega u^+_i \leq C R(\Omega)^2 \sum_{j=1}^n \sup_\Omega |c_{ij}| \sup_\Omega u^+_j.$$

Hence, denoting $\vec{v} = \left(\sup_\Omega u^+_1, \ldots, \sup_\Omega u^+_n\right)^T \geq 0$ we have

$$\vec{v} \leq \bar{R}^2 B \vec{v}$$

where B is a constant matrix whose entries depend only on n, N and on bounds for the coefficients of the elliptic operators. By choosing $\bar{R}$ such that the matrix $I - \bar{R}^2 B$ is positive definite and by multiplying scalarly the last inequality by the nonnegative vector $\vec{v}$, we obtain $\vec{v} \equiv 0$. □

3.2. *Harnack type estimates for systems*

The moving planes argument in Section 3.3 will require some Harnack estimates which we state in this section. Such results were recently obtained in [12]. The first two theorems below are particular cases of Theorem 3.2 and Propositions 3.1 and 8.1 in [12]. We include them here for the reader's convenience.

In this section G denotes an arbitrary domain in $\mathbb{R}^N$ and Q_l $(l = 1, 2)$ are concentric cubes with side l, properly included in G.

Theorem 3.2 ([12]). *Assume $f_1(u_1, u_2)$, $f_2(u_1, u_2)$ are globally Lipschitz continuous functions, with Lipschitz constant A, which satisfy the cooperativeness assumption (10). Let (u_1, u_2) be a nonnegative solution of (9) in G. We suppose that the system is fully coupled, in the sense that $f_1(0, v) > 0$ for all $v > 0$, and $f_2(u, 0) > 0$ for $u > 0$. Then for any compact subset K of G there exists a function $\Phi(t)$ (depending on A, K and G), continuous on $[0, \infty)$, such that $\Phi(0) = 0$ and*

$$\sup_{x\in K} \max\{u_1, u_2\} \le \Phi(\inf_{x\in K} \min\{u_1, u_2\}).$$

In particular, if any of u_1, u_2 vanishes at one point in G then both u_1 and u_2 vanish identically in G.

Theorem 3.3 ([12]). *Assume $f_1(u_1, u_2)$, $f_2(u_1, u_2)$ are globally Lipschitz continuous functions, with Lipschitz constant A, which satisfy the cooperativeness assumption (10). Let (u_1, u_2) be a nonnegative subsolution of (9) in G. Then for each $p > 0$ there exists a constant C depending only on p, N, and A such that*

$$\sup_{x\in Q_1} \max\{u_1, u_2\} \le C\|\max\{u_1, u_2\}\|_{L^p(Q_2)}$$

The same result holds if f_1, f_2 depend also on x and the constant A is uniform in x.

In the sequel we shall need the following two classical results on scalar equations by Krylov and Safonov.

Theorem 3.4 ([25], Theorem 9.22). *Let L be a linear uniformly elliptic operator with bounded coefficients in G, in the form (24). Suppose c_0 is an ellipticity constant for L, and b is an upper bound for the L^∞-norms of the coefficients of L. Let $u \in W^{2,N}_{\mathrm{loc}}(G)$ be a positive function satisfying $Lu \le f$ a.e. in G, for some*

$f \in L^N_{loc}(G)$. *Then there exist positive constants* p_0 *and* C *depending on* c_0, b *and* N *such that*

$$\|u\|_{L^{p_0}(Q_2)} \leq C\left(\inf_{x\in Q_1} u + \|f\|_{L^N(Q_2)}\right).$$

Theorem 3.5 (Krylov). *Let L be a linear uniformly elliptic operator with bounded coefficients in G, in the form (24). Suppose* c_0 *is an ellipticity constant for L, and b is an upper bound for the* L^∞*-norms of the coefficients of L. Let* $u \in W^{2,N}_{loc}(G)$ *be a positive function satisfying* $Lu \leq 0$ *a.e. in G and* $Lu \leq -\rho$ *a.e. in a closed subset* $\omega \subset Q_2$*, for some* $\rho > 0$*. Then there exists a constant* $m > 0$*, depending only on* N, c_0, b*, and on a positive lower bound on* meas $(\omega) > 0$*, such that*

$$\inf_{Q_1} u \geq m\rho. \tag{26}$$

Theorem 3.5 is a consequence of Theorem 12 on p. 129 in [28] - we state it here in the form which was given in [7].

Next we state a partial Harnack inequality for a linear system, which will play a crucial role in the proof of the monotonicity result in Section 3.3.

Theorem 3.6. *Suppose* $a, b, c, d \in L^\infty(Q_2)$ *are such that* $|a|, |d| \leq A$*, and* $0 \leq b \leq A$, $0 \leq c \leq A$ *in* Q_2*. Suppose* (u_1, u_2) *is a positive solution of*

$$\begin{cases} \Delta u_1 + a(x)u_1 + b(x)u_2 = 0 \\ \Delta u_2 + c(x)u_1 + d(x)u_2 = 0 \end{cases}$$

in Q_2*. Assume in addition that* $b(x)$ *is bounded below by a positive constant on* Q_1*. Then*

$$\sup_{x\in Q_1} u_1 \leq C \inf_{x\in Q_1} u_1. \tag{27}$$

where the constant C depends on N, A, and on upper bound for $\dfrac{\sup_{Q_2} b}{\inf_{Q_1} b}$.

Proof. Note that inequality (27) was proved in [12] (Theorem 8.2 in that paper) with the constant C depending on N, A, and on constants $\rho > 0$, $\delta > 0$ such that $b(x) \geq \rho$ on a set with measure δ. Theorem 3.6 follows from this result applied to the system

$$\begin{cases} \Delta u_1 + a(x)u_1 + \widetilde{b}(x)\widetilde{u}_2 = 0 \\ \Delta \widetilde{u}_2 + \widetilde{c}(x)u_1 + \widetilde{d}(x)\widetilde{u}_2 = 0, \end{cases} \tag{28}$$

where $\widetilde{u}_2 = (\inf_{Q_1} b)u_2$, $\widetilde{c}(x) = (\inf_{Q_1} b)c(x)$, $\widetilde{d}(x) = (\inf_{Q_1} b)d(x)$, and $\widetilde{b}(x) = b(x)/(\inf_{Q_1} b)$, so that $\widetilde{b} \geq 1$ in Q_1.

We are going to give the argument of the proof of inequality (27), since we shall need it in the sequel. By Theorem 3.3 applied to (28) we have for each $p > 0$

$$\sup_{x\in Q_1} \max\{u_1, \widetilde{u}_2\} \le C\|\max\{u_1, \widetilde{u}_2\}\|_{L^p(Q_2)} \le C\left(\|u_1\|_{L^p(Q_2)} + \|\widetilde{u}_2\|_{L^p(Q_2)}\right).$$

In order to estimate the right-hand side of this inequality we apply Theorem 3.4 to the following two scalar inequalities

$$\begin{cases} \Delta u_1 + a(x)u_1 \le 0 \\ \Delta \widetilde{u}_2 + \widetilde{d}(x)\widetilde{u}_2 \le 0 \end{cases}$$

(these are a consequence of (28)) and obtain

$$\sup_{x\in Q_1} u_1 \le \sup_{x\in Q_1} \max\{u_1, \widetilde{u}_2\} \le C\left(\inf_{x\in Q_1} u_1 + \inf_{x\in Q_1} \widetilde{u}_2\right). \tag{29}$$

Finally, we note that the first equation in (28) implies

$$\Delta u_1 + a(x)u_1 \le 0 \quad \text{in } Q_2 \quad \text{and} \quad \Delta u_1 + a(x)u_1 \le -\inf_{x\in Q_1} \widetilde{u}_2 \quad \text{in } Q_1,$$

so Theorem 3.5 gives

$$\inf_{x\in Q_1} u_1 \ge m \inf_{x\in Q_1} \widetilde{u}_2.$$

We finish the proof by combining this inequality and (29). □

We finish this section by giving the proof of the strong Harnack inequality for fully coupled systems satisfying hypothesis (11) - Theorem 1.4.

Proof of Theorem 1.4. By (10) and (11) we have

$$\begin{cases} \Delta u_1 + a(x)u_1 + b(x)u_2 = 0 \\ \Delta u_2 + c(x)u_1 + d(x)u_2 = 0 \end{cases}$$

in Q_2, where the coefficients $a(x) = g_1(u_1(x), u_2(x))$, $b(x) = u_2^{p-1}(x)$, $c(x) = u_1^{q-1}(x)$, $d(x) = g_2(u_1(x), u_2(x))$ are bounded continuous functions such that $b(x), c(x) > 0$ in Q_2.

As in the proof of the previous theorem we have

$$\sup_{x\in Q_1} \max\{u_1, u_2\} \le C\left(\inf_{x\in Q_1} u_1 + \inf_{x\in Q_1} u_2\right). \tag{30}$$

By applying Theorem 3.5 to

$$\Delta u_1 + a(x)u_1 \le 0 \quad \text{in } Q_2 \quad \text{and} \quad \Delta u_1 + a(x)u_1 \le -\left(\inf_{x\in Q_1} u_2\right)^p \quad \text{in } Q_1,$$

we get

$$\inf_{x\in Q_1} u_1 \geq m \left(\inf_{x\in Q_1} u_2 \right)^p .$$

In the same way we obtain

$$\inf_{x\in Q_1} u_2 \geq m \left(\inf_{x\in Q_1} u_1 \right)^q .$$

We obtain the statement of Theorem 1.4 by combining (30) with the last two inequalities. □

3.3. Proof of Theorem 1.2

We use the moving planes method of Alexandrov [1], which was subsequently developed in the framework of partial differential equations by Serrin [36], Gidas-Ni-Nirenberg [23], Berestycki-Nirenberg [9].

We shall denote

$$M = \max \left\{ \sup_{\mathbb{R}^N_+} u_1,\ \sup_{\mathbb{R}^N_+} u_2 \right\}.$$

We can suppose that the functions f_1 and f_2 are globally Lipschitz continuous. Indeed, if they are not, we can replace them by $f_1\varphi$ and $f_2\varphi$, where φ is a cut-off function such that $\varphi = 1$ on the positive cube with side M, and $\varphi = 0$ outside a cube with side $M + 1$, containing properly the previous one.

Hence system (9) satisfies the hypotheses of Theorem 3.2, from which we deduce that either both functions u_1 and u_2 vanish identically on $\mathbb{R}^N_+$ or both u_1 and u_2 are strictly positive on $\mathbb{R}^N_+$. The first case is excluded by hypothesis. So, from now on we shall assume that u_1, u_2 are strictly positive in $\mathbb{R}^N_+$.

For each $\lambda > 0$ we denote

$$T_\lambda = \{x \in \mathbb{R}^N \mid x_N = \lambda\}, \qquad \Sigma_\lambda = \{x \in \mathbb{R}^N \mid 0 < x_N < \lambda\},$$

and introduce the functions

$$v_i^{(\lambda)}(x) = u_i(x', 2\lambda - x_N), \qquad w_i^{(\lambda)}(x) = v_i^{(\lambda)}(x) - u_i(x), \qquad i = 1, 2,$$

defined in Σ_λ. Since both (u_1, u_2) and $(v_1^{(\lambda)}, v_2^{(\lambda)})$ satisfy system (9) we obtain by subtracting the corresponding equations and by Taylor's expansion

$$\begin{cases} \Delta w_1^{(\lambda)} + c_{11}^{(\lambda)}(x)w_1^{(\lambda)} + c_{12}^{(\lambda)}(x)w_2^{(\lambda)} = 0 \\ \Delta w_1^{(\lambda)} + c_{21}^{(\lambda)}(x)w_1^{(\lambda)} + c_{22}^{(\lambda)}(x)w_2^{(\lambda)} = 0 \end{cases} \tag{31}$$

in Σ_λ, where $c_{ij}^\lambda(x)$ is the partial derivative of f_i with respect to u_j, evaluated at some point between $u_j(x)$ and $v_j^{(\lambda)}(x)$. Note that $c_{ij}^{(\lambda)}$ are bounded by a Lipschitz constant of $\vec{f} = (f_1, f_2)$ on $[0, M]^2$, and $c_{12}^{(\lambda)}, c_{21}^{(\lambda)} \geq 0$.

Obviously $\vec{w}^{(\lambda)} = (w_1^{(\lambda)}, w_2^{(\lambda)}) \equiv 0$ on T_λ and $\vec{w}^{(\lambda)} > 0$ on T_0 (recall that $u_i = 0$ on T_0 and $u_i > 0$ on T_λ, $\lambda > 0$). By Theorem 3.1, if λ is small enough, then $\vec{w}^{(\lambda)} \geq 0$ in Σ_λ. Hence

$$\lambda^* = \sup\{\lambda \mid \vec{w}^{(\mu)} \geq 0 \text{ in } \Sigma_\mu, \ \forall \mu < \lambda\} > 0.$$

We see that for each $0 < \lambda \leq \lambda^*$ the function $w_i^{(\lambda)} \geq 0$ satisfies the inequality $\Delta w_i^{(\lambda)} + c_{ii}^{(\lambda)} w_i^{(\lambda)} \leq 0$ in Σ_λ. Hence Hopf's lemma implies $w_i^{(\lambda)} > 0$ and

$$\frac{\partial u_i}{\partial x_N} = -\frac{1}{2}\frac{\partial w_i^{(\lambda)}}{\partial x_N} > 0 \qquad \text{on } T_\lambda.$$

Therefore, the theorem is proved if we show that $\lambda^* = +\infty$.

Suppose for contradiction that λ^* is finite.

By Theorem 3.1 we can fix $\varepsilon_0 > 0$ such that the matrix operator $\vec{\Delta} + C_\lambda(x)$ satisfies the maximum principle in the domain $\Sigma_{\lambda^*+\varepsilon_0} \setminus \Sigma_{\lambda^*-\varepsilon_0}$ (here $C_\lambda(x)$ denotes the matrix of the coefficients in (31)). For instance, we can take

$$\varepsilon_0 = \frac{\omega_N}{2^{N+1}}\overline{R}$$

where $\overline{R}$ is the number from Theorem 3.1 (see inequality (23)).

Lemma 3.1. *There exists $\delta_0 \in (0, \varepsilon_0]$, such that for each $\delta \in (0, \delta_0)$ we have*

$$w_i^{(\lambda^*+\delta)} \geq 0 \qquad \text{in } \Sigma_{\lambda^*-\varepsilon_0} \setminus \Sigma_{\varepsilon_0}, \qquad i = 1, 2.$$

Suppose this lemma is proved. Then we can apply Theorem 3.1 to (31) in $\Sigma_{\lambda^*+\delta} \setminus \Sigma_{\lambda^*-\varepsilon_0}$ and in Σ_{ε_0} (these domains are narrow enough) to conclude that $w_i^{(\lambda^*+\delta)} \geq 0$ in $\Sigma_{\lambda^*+\delta}$ for each $\delta \in (0, \delta_0)$. This contradicts the maximal choice of λ^* and proves Theorem 1.2. □

Proof of Lemma 3.1. We denote $y = (x_1, \ldots, x_{N-1})$. Suppose for contradiction that there exist sequences $\delta_m \to 0$ and $x^{(m)} = \left(y^{(m)}, x_N^{(m)}\right) \in \Sigma_{\lambda^*-\varepsilon_0} \setminus \Sigma_{\varepsilon_0}$ such that

$$w_1^{(\lambda^*+\delta_m)}\left(x^{(m)}\right) < 0. \tag{32}$$

We can suppose that $x_N^{(m)} \to x_N^0 \in [\varepsilon_0, \lambda^* - \varepsilon_0]$ as $m \to \infty$.

We define the functions

$$u_i^{(m)}(y, x_N) = u_i(y + y^{(m)}, x_N), \qquad i = 1, 2,$$

and, respectively,

$$w_{i,\lambda}^{(m)}(y, x_N) = u_i^{(m)}(y, 2\lambda - x_N) - u_i^{(m)}(y, x_N), \qquad i = 1, 2.$$

Note that system (9) is autonomous, so $\vec{u}^{(m)}$ satisfies the same system. Since $f_1(u_1^{(m)}, u_2^{(m)})$ and $f_2(u_1^{(m)}, u_2^{(m)})$ are uniformly bounded in m (by a Lipschitz constant of $\vec{f}$ on $[0, M]^2$), standard elliptic theory, applied to (9), then implies

$$||\vec{u}^{(m)}||_{W^{2,p}(K)} \le CM$$

for each compact set K in the closure of $\mathbb{R}^N_+$, where C depends on a Lipschitz constant of $\vec{f}$ on $[0, M]^2$. It follows, by imbedding theorems and elliptic theory, that $\vec{u}^{(m)}$ converges in $C^{1,\alpha}$ on compact sets to a classical solution $\widetilde{u} = (\widetilde{u}_1, \widetilde{u}_2)$ of system (9), satisfying the Dirichlet boundary condition on $\{x_N = 0\}$.

It follows from Theorem 3.2 that either both $\widetilde{u}_1$ and $\widetilde{u}_2$ are strictly positive on $\mathbb{R}^N_+$ or both these two functions vanish identically on $\mathbb{R}^N_+$. Suppose that at least one of the functions $\widetilde{u}_1, \widetilde{u}_2$ does not vanish identically in $\mathbb{R}^N_+$ so we are in the first of these two situations. By what we have already shown we know that $w_{i,\lambda}^{(m)}(y, x_N) = w_i^\lambda(y + y^{(m)}, x_N) > 0$ in Σ_λ for all $\lambda \le \lambda^*$. Hence the limit functions $\widetilde{w}_i^\lambda = \lim_{m\to\infty} w_{i,\lambda}^{(m)}$ are nonnegative in Σ_λ for all $\lambda \le \lambda^*$. However $\widetilde{w}_i^\lambda$ is to $\widetilde{u}$ what w_i^λ is to u, so by repeating the moving planes argument for $\widetilde{u}$ we see that $\widetilde{\lambda}^* \ge \lambda^*$ ($\widetilde{\lambda}^*$ is the critical value for $\widetilde{u}$), since we can write a system like (31) for $\widetilde{w}_i^\lambda$. Applying the strong maximum principle to this system we get, as before, that $\widetilde{w}_i^\lambda > 0$ in Σ_λ, for $\lambda \le \lambda^*$. On the other hand $\widetilde{w}_1^{\lambda^*}(0, x_N^0) = 0$ and $x_N^0 \in (0, \lambda^* - \varepsilon_0]$, a contradiction.

The argument is considerably more involved in case $\widetilde{u} \equiv 0$ in $\mathbb{R}^N_+$. We fix the rectangular domains

$$Q_0 = \{x \in \mathbb{R}^N_+ \mid -1 < x_1 < 1, \dots, -1 < x_{N-1} < 1,\ \varepsilon_0 < x_N < 2\lambda^* + 1\},$$

$$Q_1 = \{x \in \mathbb{R}^N_+ \mid -2 < x_1 < 2, \dots, -2 < x_{N-1} < 2,\ \frac{\varepsilon_0}{2} < x_N < 2\lambda^* + 2\},$$

$$Q_2 = \{x \in \mathbb{R}^N_+ \mid -3 < x_1 < 3, \dots, -3 < x_{N-1} < 3,\ \frac{\varepsilon_0}{4} < x_N < 2\lambda^* + 3\},$$

and note that all results from the previous section can trivially be applied in Q_1 and Q_2 (one simply has to cover Q_1 and Q_2 with a finite number of cubes and apply Harnack inequalities in these cubes).

Since $\vec{u}^{(m)}$ converges uniformly to zero in Q_2 we can suppose that $u_1^{(m)} \le 1$, $u_2^{(m)} \le 1$ in Q_2. We set

$$\alpha_m = u_1^{(m)}\left(0, x_N^{(m)}\right), \qquad \beta_m = u_2^{(m)}\left(0, x_N^{(m)}\right).$$

By Theorem 1.4 we have

$$\alpha_m \le C\beta_m^{\frac{1}{q}} \qquad \text{and} \qquad \beta_m \le C\alpha_m^{\frac{1}{p}}, \tag{33}$$

where C is independent of m. We also recall inequality (30)

$$\sup_{x\in Q_1} \max\{u_1^{(m)}, u_2^{(m)}\} \le C\left(\inf_{x\in Q_1} u_1^{(m)} + \inf_{x\in Q_1} u_2^{(m)}\right). \tag{34}$$

Next we introduce the functions

$$z_1^{(m)} = \frac{1}{\alpha_m} u_1^{(m)}, \quad z_2^{(m)} = \frac{1}{\beta_m} u_2^{(m)}, \quad \zeta_1^{(m)} = \frac{1}{\beta_m} u_1^{(m)}, \quad \zeta_2^{(m)} = \frac{1}{\alpha_m} u_2^{(m)}.$$

Note that $z_1^{(m)}(0, x_N^{(m)}) = z_2^{(m)}(0, x_N^{(m)}) = 1$.

We distinguish two cases : up to a subsequence

$$\inf_{Q_1} u_1^{(m)} \le \inf_{Q_1} u_2^{(m)} \quad \text{(Case 1)} \qquad \text{and} \qquad \inf_{Q_1} u_2^{(m)} \le \inf_{Q_1} u_1^{(m)} \quad \text{(Case 2)}.$$

Suppose we are in Case 1. Then, as in the previous section, $\left(z_1^{(m)}, \zeta_2^{(m)}\right)$ satisfies the linear system

$$\begin{cases} \Delta z_1^{(m)} + a_m(x) z_1^{(m)} + b_m(x)\zeta_2^{(m)} = 0 \\ \Delta \zeta_2^{(m)} + c_m(x) z_1^{(m)} + d_m(x)\zeta_2^{(m)} = 0 \end{cases} \tag{35}$$

in Q_2, where $a_m(x) = g_1(u_1^{(m)}(x), u_2^{(m)}(x))$, $b_m(x) = \left(u_2^{(m)}(x)\right)^{p-1} > 0$, $c_m(x) = \left(u_1^{(m)}(x)\right)^{q-1} > 0$, $d_m(x) = g_2(u_1^{(m)}(x), u_2^{(m)}(x))$ are uniformly bounded in m.

By combining inequality (34) with the hypothesis of Case 1 we obtain

$$\sup_{Q_1} u_2^{(m)} \le C \inf_{Q_1} u_2^{(m)},$$

which implies that the quantity

$$\frac{\sup_{x\in Q_1} b_m(x)}{\inf_{x\in Q_1} b_m(x)}$$

is bounded by a constant independent of m. Hence we can apply Theorem 3.6 to (35), and infer that

$$\sup_{Q_0} z_1^{(m)} \le C_1 \inf_{Q_0} z_1^{(m)} \le C_1,$$

where C_1 does not depend on m.

Next, recall that $w_1^{(\lambda^*)} \ge 0$ in Σ_{λ^*}, which implies

$$z_1^{(m)}(y, x_N) \le z_1^{(m)}(y, 2\lambda^* - x_N) \le C_1$$

for every (y, x_N) in the closure of Σ_{λ^*}. Hence

$$\|z_1^{(m)}\|_{L^\infty(Q)} \le C_1,$$

where

$$Q = \{x \in \mathbb{R}^N_+ \mid -1 < x_1 < 1, \dots, -1 < x_{N-1} < 1,\ 0 < x_N < 2\lambda^* + 1\}.$$

In Case 2 we write a linear system for $\left(\zeta_1^{(m)}, z_2^{(m)}\right)$ and use an analogous reasoning to infer that

$$\|z_2^{(m)}\|_{L^\infty(Q)} \le C_2,$$

where C_2 does not depend on m.

Next we prove that actually both $z_1^{(m)}$ and $z_2^{(m)}$ are bounded in $L^\infty(Q)$. To this end we remark that $\left(z_1^{(m)}, z_2^{(m)}\right)$ satisfies the system

$$\begin{cases} \Delta z_1^{(m)} + g_1(u_1^{(m)}, u_2^{(m)}) z_1^{(m)} + \frac{\beta_m^p}{\alpha_m}\left(z_2^{(m)}\right)^p = 0 \text{ in } \mathbb{R}^N_+ \\ \Delta z_2^{(m)} + \frac{\alpha_m^q}{\beta_m}\left(z_1^{(m)}\right)^q + g_1(u_1^{(m)}, u_2^{(m)}) z_2^{(m)} = 0 \text{ in } \mathbb{R}^N_+ \\ z_1^{(m)} = z_2^{(m)} = 0 \text{ on } \partial\mathbb{R}^N_+ \end{cases} \tag{36}$$

by (11). We recall that we have already proved that the sequences $\{\beta_m^p \alpha_m^{-1}\}$ and $\{\alpha_m^q \beta_m^{-1}\}$ are bounded (see (33)). It follows from the above argument that one of the sequences $\{z_1^{(m)}\}$, $\{z_2^{(m)}\}$ (say the first) is bounded in $L^\infty(Q)$. Then standard elliptic estimates applied to the second equation in (36), regarded as a scalar equation in $z_2^{(m)}$ with bounded coefficients and bounded right-hand side, imply that $\{z_2^{(m)}\}$ is also bounded in $L^\infty(Q)$ (if necessary, we restrict a little the upper and the lateral boundaries of Q).

Now, since both $\{z_1^{(m)}\}$, $\{z_2^{(m)}\}$ are bounded on Q, elliptic theory and (36) imply that (up to a subsequence) these two sequences converge uniformly to nonnegative functions $z_1, z_2 \in W^{2,p}_{\text{loc}}(Q) \cap C(\overline{Q})$, which satisfy the system

$$\begin{cases} \Delta z_1 + g_1(0,0) z_1 + \beta_0 (z_2)^p = 0 \text{ in } Q \\ \Delta z_2 + \alpha_0 (z_1)^q + g_1(0,0) z_2 = 0 \text{ in } Q \\ z_1 = z_2 = 0 \text{ on } \{x_N = 0\} \cap \partial Q, \end{cases} \tag{37}$$

for some constants $\alpha_0 \ge 0$, $\beta_0 \ge 0$. Since $z_1 \ge 0$ and $\Delta z_1 + g_1(0,0) z_1 \le 0$ in Q, the strong maximum principle implies that either z_1 vanishes identically in Q or $z_1 > 0$ in Q. The first possibility is excluded by $z_1(0, x_N^0) = 1$.

Introduce the comparison functions

$$\omega_i^{(\lambda)}(y, x_N) = z_i(y, 2\lambda - x_N) - z_i(y, x_N), \quad i = 1, 2,$$

defined in $\Sigma_\lambda \cap \overline{Q}$, for all $\lambda \leq \lambda^* + 1/2$. We have, by continuity,

$$\omega_i^{(\lambda^*)} \geq 0, \; i = 1, 2, \qquad \text{and} \qquad \omega_1^{(\lambda^*)}(0, x_N^0) = 0$$

(recall (32)). Since $\Delta\omega_1^{(\lambda^*)} + g_1(0,0)\omega_1^{(\lambda^*)} \leq 0$ the strong maximum principle implies $\omega_1^{(\lambda^*)} \equiv 0$ in $\Sigma_{\lambda^*} \cap \overline{Q}$. This contradicts the fact that $z_1 = 0$ on $\{x_N = 0\}$ and $z_1 > 0$ on $\{x_N = 2\lambda^*\}$.

The proof of Theorem 1.2 is finished. □

3.4. *Proof of Theorem 1.3*

Set

$$x' \in \mathbb{R}^{N-1}, \qquad u_i^{(t)}(x') = u_i(x', t), \quad i = 1, 2.$$

Since the sequence $\{u_i^{(t)}\}_t$ is uniformly bounded and pointwise increasing in t (by Theorem 1.2) the Lebesgue monotone convergence theorem implies that $u_i^{(t)}$ converges as $t \to \infty$ in $L^p_{\text{loc}}\left(\mathbb{R}^{N-1}\right)$, $p < \infty$, to a bounded function $\widetilde{u}_i$. Then the hypotheses we made on f_i imply

$$f_i(u_1^{(t)}, u_2^{(t)}) \longrightarrow f_i(\widetilde{u}_1, \widetilde{u}_2) \qquad \text{as } t \to \infty,$$

in $L^p_{\text{loc}}\left(\mathbb{R}^{N-1}\right)$, $p < \infty$, $i = 1, 2$. We shall prove that $(\widetilde{u}_1, \widetilde{u}_2)$ is a weak solution to the problem in $\mathbb{R}^{N-1}$ (then standard elliptic regularity theory implies that it is a classical solution). Let $\phi(x') \in C_c^\infty(\mathbb{R}^{N-1})$ be an arbitrary function and let

$$\psi(t) = \begin{cases} 0 & \text{if } t \leq 0, t \geq 2 \\ t & \text{if } 0 \leq t \leq 1 \\ 2 - t & \text{if } 1 \leq t \leq 2. \end{cases}$$

The function $\psi(t)$ is chosen so that $\psi \in C_c(\mathbb{R})$, $\operatorname{supp} \psi = [0, 2]$,

$$\int_0^1 \psi(t)\, dt = \int_1^2 \psi(t)\, dt = \frac{1}{2} \qquad \text{and} \qquad \psi''(t) = \delta(0) - 2\delta(1) + \delta(2),$$

where $\delta(t)$ is the Dirac mass at t. We set $\psi_m(t) = \psi(t - m)$.

We multiply each equation in (9) by $\phi(x')\psi_m(x_N)$ and integrate over $\mathbb{R}^N_+$. Integration by parts and Fubini's theorem then yield

$$\int_{\mathbb{R}^N_+} u_i \left(\Delta_{x'}\phi(x')\psi_m(x_N) + \phi(x')\psi_m''(x_N)\right) dx_N\, dx'$$
$$= \int_{\mathbb{R}^N_+} f_i(u_1, u_2)\phi(x')\psi_m(x_N)\, dx_N\, dx'$$

or

$$\int\limits_{\mathbb{R}^{N-1}} \left(\int\limits_0^2 u_i(x', x_N + m)\psi(x_N)\, dx_N \right) \Delta_{x'}\phi(x')\, dx'$$

$$+ \int\limits_{\mathbb{R}^{N-1}} \left(u_i^{(m)}(x') - 2u_i^{(m+1)}(x') + u_i^{(m+2)}(x') \right) \phi(x')\, dx'$$

$$= \int\limits_{\mathbb{R}^{N-1}} \left(\int\limits_0^2 f_i \left(u_1(x', x_N + m), u_2(x', x_N + m) \right) \psi(x_N)\, dx_N \right) \phi(x')\, dx'.$$

Since $u_i^{(m)}$ tends to $\widetilde{u}_i$ in any Lebesgue space, the following lemma finishes the proof of Theorem 1.3. Note that the second integral in the left-hand side of the last equality vanishes at the limit.

Lemma 3.2. *Under the hypotheses we made on* f_1, f_2

$$\int\limits_0^2 f_i \left(u_1(x', x_N + m), u_2(x', x_N + m) \right) \psi(x_N)\, dx_N \tag{38}$$

tends to $f_i\left(\widetilde{u_1}(x'), \widetilde{u_2}(x')\right)$ *in* $L^1_{\text{loc}}\left(\mathbb{R}^{N-1}\right)$ *as* $m \to \infty$.

Proof. We split the integral (38) in two and use the fact that ψ is continuous and monotonous in [0, 1] and in [1, 2]. By standard properties of the Riemann integral (38) is equal to

$$\frac{f\left(u_1(x', \xi_1 + m), u_2(x', \xi_1 + m)\right) + f\left(u_1(x', \xi_2 + m), u_2(x', \xi_2 + m)\right)}{2}$$

where $\xi_1 \in (0, 1)$, $\xi_2 \in (1, 2)$. Obviously $u_i(x', \xi_i + m)$ tends to $\widetilde{u}_i(x')$ in $L^p_{\text{loc}}\left(\mathbb{R}^{N-1}\right)$, $p < \infty$, so the lemma follows. □

3.5. A monotonicity lemma

In the proof of Theorem 1.1 we shall need the following monotonicity result, which is an extension of a recent result of Busca and Manasevich [10].

Lemma 3.3. *Let* $f_i \in C^1((\mathbb{R}_+)^2)$, $f_i(0,0) = 0$, $\nabla f_i(0,0) = 0$, $i = 1, 2$ *satisfy (10). Suppose that* $f_1(0, v) > 0$ *for all* $v > 0$, *and* $f_2(u, 0) > 0$ *for* $u > 0$. *Let* $u_i(t, \theta) \geq 0$, $i = 1, 2$ *be* C^2*-functions defined on* $\mathbb{R} \times S^{N-1}$ *and satisfying*

$$\begin{cases} \dfrac{\partial^2 u_i}{\partial t^2} + \Delta_\theta u_i - \delta_i \dfrac{\partial u_i}{\partial t} - \nu_i u_i + f_i(u_1, u_2) = 0 \quad \text{in} \quad \mathbb{R} \times S^{N-1} \\ \hfill u_i \to 0 \quad \text{as } t \to -\infty, \end{cases} \tag{39}$$

where $\delta_i \geq 0$, $\max\{\delta_1, \delta_2\} > 0$, $\nu_i > 0$, $i = 1, 2$ *are constants. Suppose also that there exists* $t_0 \in \mathbb{R}$ *such that* $\frac{\partial u_i}{\partial t} > 0$ *in* $(-\infty, t_0) \times S^{N-1}$, *for* $i = 1, 2$. *Then* $\frac{\partial u_i}{\partial t} > 0$ *in* $\mathbb{R} \times S^{N-1}$, *for* $i = 1, 2$.

In [10] the authors studied the case $n = 2$, $f_1(u_1, u_2) = u_2^p$, $f_2(u_1, u_2) = u_1^q$. Note that the result in [10] was stated without the hypothesis $p, q > 1$, but this hypothesis is actually used in their proof.

The proof of Lemma 3.3 is based on the moving planes method and uses both ideas from the proof in [10] and from the proof of the symmetry result for systems in $\mathbb{R}^N$ obtained in [11].

We set

$$h_i(u_1, u_2) = -\nu_i u_i + f_i(u_1, u_2)$$

and remark that all we shall need is the fact that the matrix of the partial derivatives of h_1 and h_2 is negative definite at $(0, 0)$. We use this fact as in the proof of Theorem 2 from [11] (see hypotheses (i)–(iii) in Section 2.1 of [11]).

For each $\lambda > 0$ we denote

$$T_\lambda = \{(t, \theta) \in \mathbb{R} \times S^{N-1} \mid t = \lambda\}, \qquad \Sigma_\lambda = \{(t, \theta) \in \mathbb{R} \times S^{N-1} \mid t < \lambda\},$$

and introduce the functions

$$w_i^{(\lambda)}(t, \theta) = u_i(2\lambda - t, \theta) - u_i(t, \theta), \qquad i = 1, \dots, n,$$

defined in Σ_λ. By Taylor's expansion $\vec{w}_\lambda$ satisfies the system

$$\frac{\partial^2 \vec{w}^{(\lambda)}}{\partial t^2} + \Delta_\theta \vec{w}^{(\lambda)} - D\frac{\partial \vec{w}^{(\lambda)}}{\partial t} + C_\lambda(x)\vec{w}^{(\lambda)} = -2D\frac{\partial \vec{u}}{\partial t} \quad \text{in } \Sigma_\lambda, \tag{40}$$

where

$$D = \operatorname{diag}(\delta_1, \delta_2), \quad C_\lambda(x) = \left(c_{ij}(x)\right)_{i,j=1}^2, \quad c_{ij}(x) = \frac{\partial h_i}{\partial u_j}(\xi_1, \xi_2),$$

and $\xi_{ij} = \xi_{ij}(t, \theta, \lambda)$,

$$\xi_{ij} \in \left[\min(u_j(t, \theta), u_j(2\lambda - t, \theta)), \max(u_j(t, \theta), u_j(2\lambda - t, \theta))\right].$$

System (40) is treated in basically the same way as inequality (3)–(4) from [11] - note that the right hand side of (40) is nonpositive in Σ_λ for any $\lambda \leq t_0$, by hypothesis. We shall only sketch the argument, since most of the details can be seen in [11] (note only that here the moving planes go "to the right", contrary to the choice made in [11]). First, there exists $\lambda^* < t_0$ such that $\vec{w}^{(\lambda)} \geq 0$ in Σ_λ, for any $\lambda \leq \lambda^*$. The proof of this fact goes like the proof of Step 1 in the proof of Theorem 2 in [11]. Second, we define

$$\lambda_0 = \sup\left\{\lambda \in \mathbb{R} \;:\; \vec{w}^{(\mu)} \geq 0 \quad \text{and} \quad \frac{\partial \vec{w}^{(\mu)}}{\partial t} > 0 \text{ in } \Sigma_\mu, \quad \text{for all } \mu < \lambda\right\}.$$

If $\lambda_0 = +\infty$ we are done (note that $\frac{\partial \vec{w}^{(\lambda)}}{\partial t} = 2\frac{\partial \vec{u}}{\partial t}$ on T_λ). If λ_0 is finite we reason like in Step 2 in the proof of Theorem 2 in [11], to infer that either $w_1^{(\lambda_0)}$ or $w_2^{(\lambda_0)}$ vanishes in Σ_{λ_0}. Since the system is fully coupled this implies that $\vec{w}^{(\lambda_0)} \equiv 0$ in Σ_{λ_0}, or, equivalently, $\frac{\partial \vec{u}}{\partial t}$ is odd with respect to T_{λ_0}. On the other hand, by (40) at least one of the derivatives with respect to t of u_1 and u_2 (the one which corresponds to a strictly positive δ_i) is even with respect to T_{λ_0}. Hence at least one of the functions u_i is equal to a constant, which contradicts the hypothesis that u_i is strictly increasing in t for $t \in (-\infty, t_0)$. □

In the next section we prove our main result.

4. Proof of Theorem 1.1

We divide the proof in three parts, according to the case we consider (A, B or C). We recall that the values of the parameters α_{ij} determine which case we are in.

In the sequel G will denote either $\mathbb{R}^N$ or $\mathbb{R}^N_+$.

Proof in Case A. In this case we choose $(\beta_1, \beta_2) = l_1 \cap l_2$, that is

$$\beta_1 = \frac{2}{\alpha_{11} - 1}, \qquad \beta_2 = \frac{2}{\alpha_{22} - 1}. \tag{41}$$

1. First, if none of the lines l_3 and l_4 passes through $l_1 \cap l_2$ we are precisely in the weakly coupled case considered in [19]. By using Lemmas 2.1 and 2.2, and by letting $n \to \infty$ in (14) we obtain (after scaling) the uncoupled system

$$\begin{cases} -\Delta v_1 = v_1^{\alpha_{11}} \\ -\Delta v_2 = v_2^{\alpha_{22}}, \end{cases} \tag{42}$$

in G, which has only the trivial solution because

$$\max\{\alpha_{11}, \alpha_{22}\} < \frac{N+2}{N-2}, \tag{43}$$

which is a consequence of (8) and (41).

2. Next, suppose exactly one of l_3 and l_4 (say l_3) passes through $l_1 \cap l_2$. Then again by Lemmas 2.1 and 2.2, after letting $n \to \infty$ in (14) we obtain the system

$$\begin{cases} -\Delta v_1 = a_0 v_1^{\alpha_{11}} + b_0 v_2^{\alpha_{12}} + c_1 v_1^{\gamma_{11}} v_2^{\gamma_{12}} \\ -\Delta v_2 = d_0 v_2^{\alpha_{22}} \\ u, v \geq 0 \end{cases} \quad \text{in } \; G, \tag{44}$$

where a_0, d_0, b_0, c_1 are constants such that $a_0, d_0 > 0, b_0, c_1 \geq 0$. Since (43) holds this system has no nontrivial solution. Indeed, the second equation in (44) implies $v_2 \equiv 0$. Then the first equation becomes a scalar equation, which again has no nontrivial solutions under (43).

3. Finally, suppose all four lines l_1, l_2, l_3, l_4 meet at one point (so we are simultaneously in Cases A and B). Letting $n \to \infty$ in (14), we come to the system

$$\begin{cases} -\Delta v_1 = a_0 v_1^{\alpha_{11}} + b_0 v_2^{\alpha_{12}} + c_1 v_1^{\gamma_{11}} v_2^{\gamma_{12}} \\ -\Delta v_2 = c_0 v_1^{\alpha_{21}} + d_0 v_2^{\alpha_{22}} + c_2 v_1^{\gamma_{21}} v_2^{\gamma_{22}} \quad \text{in} \quad G, \\ u, v \geq 0 \end{cases} \tag{45}$$

where $a_0, b_0, c_0, d_0 > 0, c_1, c_2 \geq 0$. Note that now

$$\alpha_{12} = \frac{\alpha_{11}(\alpha_{22}-1)}{\alpha_{11}-1}, \qquad \alpha_{21} = \frac{\alpha_{22}(\alpha_{11}-1)}{\alpha_{22}-1}.$$

Assume first $G = \mathbb{R}^N_+$. Then (45) satisfies the hypotheses of Theorem 1.2 and Theorem 1.3. Hence, by Theorem 1.3, if $(v_1, v_2) \not\equiv (0,0)$ then there exists a nontrivial solution of (45) in $\mathbb{R}^{N-1}$. So, if we manage to prove that (45) has only the trivial solution in $\mathbb{R}^N$ under (8) (note that a particular case of this result would be Theorem 1.5), then it has no nontrivial solutions in $\mathbb{R}^N_+$ under $\min\{\beta_1, \beta_2\} > \frac{N-3}{2}$ which is consequence of (8).

From now on we suppose $G = \mathbb{R}^N$ and distinguish two cases,

$$\max\{\beta_1, \beta_2\} \geq N-2 \text{ (Case 1)} \qquad \text{and} \qquad \max\{\beta_1, \beta_2\} < N-2 \text{ (Case 2)}.$$

In Case 1 (say $\beta_1 \geq N-2$) we have $\alpha_{11} \leq \frac{N}{N-2}$. But the first equality in (45) implies $-\Delta v_1 \geq a_0 v_1^{\alpha_{11}}$ in $\mathbb{R}^N$, so $v_1 \equiv 0$ in $\mathbb{R}^N$, by the results about non-existence of supersolutions (Theorem 2.1). Then the second equation in (45) becomes $-\Delta v_2 = d_0 v_2^{\alpha_{22}}$ in $\mathbb{R}^N$. So $v_2 \equiv 0$ in $\mathbb{R}^N$, because $\alpha_{22} < \frac{N+2}{N-2}$, which is a consequence of (8).

In Case 2 we write system (45) in polar coordinates $(r, \theta) \in \mathbb{R} \times S^{N-1}$ and make the change of variables, as in [10],

$$r = |x|, \qquad t = \ln|x| \in \mathbb{R}, \qquad \theta = \frac{x}{|x|} \in S^{N-1},$$

and set

$$w_i(t, \theta) = e^{\beta_i t} v_i(e^t, \theta).$$

Then system (45) transforms into

$$\begin{cases} -L_1 w_1 = a_0 e^{(\beta_1+2-\alpha_{11}\beta_1)t} w_1^{\alpha_{11}} + b_0 e^{(\beta_1+2-\alpha_{12}\beta_2)t} w_2^{\alpha_{12}} \\ \quad + c_1 e^{(\beta_1+2-\gamma_{11}\beta_1-\gamma_{12}\beta_2)t} w_1^{\gamma_{11}} w_2^{\gamma_{12}} \\ -L_2 w_2 = c_0 e^{(\beta_2+2-\alpha_{21}\beta_1)t} w_1^{\alpha_{21}} + d_0 e^{(\beta_2+2-\alpha_{22}\beta_2)t} w_2^{\alpha_{22}} \\ \quad + c_2 e^{(\beta_2+2-\gamma_{21}\beta_1-\gamma_{22}\beta_2)t} w_1^{\gamma_{21}} w_2^{\gamma_{22}} \end{cases} \tag{46}$$

in $\mathbb{R} \times S^{N-1}$, where

$$L_i = \frac{\partial^2}{\partial t^2} + \Delta_\theta - \delta_i \frac{\partial}{\partial t} - \nu_i, \quad i = 1, 2,$$

and

$$\delta_i = 2\beta_i - (N-2), \qquad \nu_i = \beta_i(N - 2 - \beta_i), \quad i = 1, 2.$$

It is remarkable that after this change of variables one obtains the same powers in the exponential functions in (46) as the powers of the parameter λ_n after the "blow-up" change of variables. So, in the case we consider all these powers are zero, and system (46) is *autonomous*.

Observe that condition (8) implies that $\delta_i > 0$, and that the condition of Case 2 gives $\nu_i > 0$. Further, by by applying Theorem 3.2 to system (45) we see that we can suppose that v_1 and v_2 are strictly positive in $\mathbb{R}^N$. This easily implies that the derivatives of w_1 and w_2 with respect to t are positive for large negative t. So we can now apply Lemma 3.3 to (46), and infer $\dfrac{\partial w_i}{\partial t} > 0$ in $\mathbb{R} \times S^{N-1}$, or

$$\beta_i v_i + r\frac{\partial v_i}{\partial r} > 0, \qquad i = 1, 2. \tag{47}$$

Then, using a reasoning from [10], we remark that all the above argument can be carried out for any translation of (v_1, v_2) (since (45) is autonomous), so (47) implies

$$\beta_i v_i(x) + \nabla v_i(x)(x - x_0) \geq 0,$$

for all $x, x_0 \in \mathbb{R}^N$. This easily implies $\nabla v_i(x) \equiv 0$ (write $x_0 = x - \tau e$, $\tau > 0$, $e \in S^{N-1}$, divide by τ, let $\tau \to \infty$ and observe that the resulting inequality holds for any $e \in S^{N-1}$), which contradicts (45).

Proof in Case B. We recall that in this case we choose $(\beta_1, \beta_2) = l_3 \cap l_4$, that is,

$$\beta_1 = \frac{2(1+\alpha_{12})}{\alpha_{12}\alpha_{21} - 1}, \qquad \beta_2 = \frac{2(1+\alpha_{21})}{\alpha_{12}\alpha_{21} - 1}.$$

1. First, suppose none of l_1 and l_2 passes through $l_3 \cap l_4$; this is precisely the strongly coupled case from [19]. Then by using Lemmas 2.1 and 2.2, after the passage to the limit in (14) we obtain (after scaling) the system

$$\begin{cases} -\Delta v_1 = v_2^{\alpha_{12}} \\ -\Delta v_2 = v_1^{\alpha_{21}} \\ \quad u, v \geq 0 \end{cases} \quad \text{in} \quad G, \tag{48}$$

Observe that it is known that under (8) system (48) in $\mathbb{R}^N$ has only the trivial solution, see Theorem 2.2. However, as in the beginning of point 3 of the proof of Case A above we see that we can restrict the analysis to $G = \mathbb{R}^N$. Indeed, system (48) satisfies the hypotheses of Theorem 1.2, since by assumption $b_0, c_0 > 0$, $\alpha_{12}, \alpha_{21} > 1$. Let us stress again that this argument requires Lipschitz nonlinearities.

2. Second, if one of l_1 or l_2 (say l_1) passes through $l_3 \cap l_4$, by letting $n \to \infty$ in (14) we obtain, again with the help of Lemmas 2.1 and 2.2, the following system

$$\begin{cases} -\Delta v_1 = a_0 v_1^{\alpha_{11}} + b_0 v_2^{\alpha_{12}} + c_1 v_1^{\gamma_{11}} v_2^{\gamma_{12}} \\ -\Delta v_2 = c_0 v_1^{\alpha_{21}} \\ u, v \geq 0 \end{cases} \quad \text{in} \quad \mathbb{R}^N, \tag{49}$$

where $b_0, c_0 > 0$, $a_0, c_1 \geq 0$. So (49) implies

$$\begin{cases} -\Delta v_1 \geq b_0 v_2^{\alpha_{12}} \\ -\Delta v_2 \geq c_0 v_1^{\alpha_{21}} \\ u, v \geq 0 \end{cases} \quad \text{in} \quad \mathbb{R}^N. \tag{50}$$

We split the argument in two cases once again.

- If $\max\{\beta_1, \beta_2\} \geq N - 2$, Theorem 2.2 applied to (50) gives $v_1 = v_2 \equiv 0$ in $\mathbb{R}^N$.
- Finally, if we have $\dfrac{N-2}{2} < \beta_1, \beta_2 < N - 2$, we apply Lemma 3.3 to (49), and conclude $v_1 = v_2 \equiv 0$ in $\mathbb{R}^N$, through the same argument as in the last part of the proof of Case A.

Proof in Case C. All that remains to consider are the cases when either l_3 meets l_1 at a point above l_4 or l_4 meets l_2 at a point below l_3. In both cases we take the couple (β_1, β_2) to be this intersection point, that is, for instance in the first of these two situations

$$\beta_1 = \frac{2}{\alpha_{11} - 1}, \qquad \beta_2 = \frac{2\alpha_{11}}{\alpha_{12}(\alpha_{11} - 1)}.$$

Then, after passing to the limit in (14) we obtain a nontrivial *bounded* solution to the system

$$\begin{cases} -\Delta v_1 = a_0 v_1^{\alpha_{11}} + b_0 v_2^{\alpha_{12}} + c_1 v_1^{\gamma_{11}} v_2^{\gamma_{12}} \\ -\Delta v_2 = 0 \end{cases} \tag{51}$$

in the whole space or in a half-space, with v_1 and v_2 vanishing on the boundary of the half space. Then Liouville's theorem applied to the second equation in (51) implies that v_2 is identically equal to a constant c. If $c = 0$ (this is the only case if (51) is in a half-space, because of the boundary condition) we replace in the first equation in (51) and obtain a nontrivial solution to the Emden-Fowler equation (19) with $p = \alpha_{11}$, which is known not to have solutions for

$$1 < \alpha_{11} < \frac{N+2}{N-2},$$

and this is a consequence of (8).

If $c > 0$ the first equation in (51) implies that the inequality

$$-\Delta v_1 \geq b_0\, c^{\alpha_{12}} > 0$$

has a bounded solution in $\mathbb{R}^N$, which is well-known to be impossible. □

References

1. Alexandrov, A.D.: A characteristic property of the spheres. Ann. Mat. Pura Appl. **58**, 303–354, (1962)
2. Alves, C., de Figueiredo, D.G.: Nonvariational elliptic systems. Discr. Cont. Dyn. Systems **8**(2), 289–302 (2002)
3. Amann, H.: Fixed point equation and nonlinear eigenvalue problems in ordered Banach spaces. SIAM Review **18**, 620–709 (1976)
4. Amann, H.: On the number of solutions of nonlinear equations in ordered Banach spaces. J. Funct. Anal. **14**, 349–381 (1973)
5. Bandle, C., Essen, M.: On positive solutions of Emden equations in cones. Arch. Rat. Mech. Anal. **112**(4), 319–338 (1990)
6. Berestycki, H., Caffarelli, L., Nirenberg, L.: Further qualitative properties for elliptic equations in unbounded domains. Dedicated to Ennio De Giorgi. Ann. Scuola Norm. Sup. Pisa Cl. Sci. (4) **25**(1–2), 69–94 (1997)
7. Berestycki, H., Nirenberg, L., Varadhan, S.R.S.: The principal eigenvalue and maximum principle for second order elliptic operators in general domains. Comm. Pure Appl. Math **47**(1), 47–92 (1994)
8. Birindelli, I., Mitidieri, E.: Liouville theorems for elliptic inequalities and inequations. Proc. Royal Soc. Edinburgh, **128A**, 1217–1247 (1998)
9. Berestycki, H., Nirenberg, L.: On the method of moving planes and the sliding method. Bull. Soc. Brazil Mat. Nova Ser **22**, 1–37 (1991)
10. Busca, J., Manasevich, R.: A Liouville type theorem for Lane-Emden systems. Indiana Univ. Math. J **51**(1), 37–51 (2002)
11. Busca, J., Sirakov, B.: Symmetry results for semilinear elliptic systems in the whole space. J. Diff. Eq **163**(1), 41–56 (2000)
12. Busca, J., Sirakov, B.: Harnack type estimates for nonlinear elliptic systems and applications. Ann. Inst. H. Poincare, Anal. Non Linéaire **21**, 543–590 (2005)
13. Cabre, X.: On the Alexandrov-Bakelman-Pucci estimate and the reversed Hölder inequality for solutions of elliptic and parabolic equations. Comm. Pure and Appl. Math **48**, 539–570 (1995)
14. Cabre, X.: Topics in regularity and qualitative properties of solutions of nonlinear elliptic equations. Discr. Cont. Dyn. Syst **8**(2), 331–359 (2002)
15. Chen, W.X., Li, C.: Classification of solutions of some nonlinear elliptic equations, Duke Math. J **63**, 615–623 (1991)
16. Clement, Ph., de Figueiredo, D.G., Mitidieri, E.: Positive solutions of semilinear elliptic systems. Comm. Part. Diff. Eq **17**, 923–940 (1992)

17. Dancer, E.N.: Some notes on the method of moving planes. Bull. Austral. Math. Soc **46**, 425–434 (1992)
18. de Figueiredo, D.G.: Monotonicity and symmetry of solutions of elliptic systems in general domains. NoDEA **1**, 119–123 (1994)
19. de Figueiredo, D.G.: Semilinear elliptic systems. Nonl. Funct. Anal. Appl. Diff. Eq. World Sci. Publishing, River Edge, 1998 pp. 122–152
20. de Figueiredo, D.G., Felmer, P.: A Liouville-type theorem for elliptic systems. Ann. Sc. Norm. Sup. Pisa **21**, 387–397 (1994)
21. de Figueiredo, D.G., Lions, P.-L., Nussbaum, R.: A priori estimates and existence of positive solutions of semilinear elliptic equations. J. Math. Pures Appl **61**, 41–63 (1982)
22. Gidas, B.: Symmetry properties and isolated singularities of positive solutions of nonlinear elliptic equations. Lect. Notes on Pure Appl. Math **54**, 255–273 (1980)
23. Gidas, B., Ni, W.M., Nirenberg, L.: Symmetry and related properties via the maximum principle. Comm. Math. Phys **6**, 883–901 (1981)
24. Gidas, B., Spruck, J.: A priori bounds for positive solutions of nonlinear elliptic equations. Comm. Part. Diff. Eq. **(6)**, 883–901 (1981)
25. Gilbarg, D., Trudinger, N.S.: Elliptic Partial Differential Equations of Second Order, 2nd edition, Revised Third Printing, Springer Berlin Heidelberg 1998
26. Hulshof, J., van der Vorst, R.C.A.M.: Differential systems with strongly indefinite variational structure. J. Funct. Anal **114**, 32–58 (1993)
27. Krasnoselskii, M.A.: Positive solutions of operator equations, P. Noordhoff, Groningen.
28. Krylov: Nonlinear elliptic and parabolic equations of second order. Coll. Math. and its Appl. (1987)
29. Laptev, G.G.: Absence of global positive solutions of systems of semilinear elliptic inequalities in cones. (Russian) Izv. Ross. Akad. Nauk Ser. Mat. **64**(6), 107–124, (2000); translation in Izv. Math. **64**(6), 1197–1215 (2000)
30. Mitidieri, E.: Non-existence of positive solutions of semilinear elliptic systems in $\mathbb{R}^N$. Quaderno Matematico **285**, (1992)
31. Mitidieri, E.: Non-existence of positive solutions of semilinear elliptic systems in $\mathbb{R}^N$. Diff. Int. Eq **9**(3), 465–479 (1996)
32. Mitidieri, E., Pohozaev, S.: A priori estimates and the absence of solutions of nonlinear partial differential equations and inequalities. Tr. Mat. Inst. Steklova **234**, 1–384 (2001)
33. Montenegro, M.S.: Criticalidade, superlinearidade e sublinearidade para sistemas elípticos semilineares, Tese de Doutoramento, Uniçamp (1997)
34. Nussbaum, R.: Positive solutions of nonlinear elliptic boundary value problems. J. Math. Anal. Appl **51**, 461–482 (1975)
35. Serrin, J.: A symmetry theorem in potential theory. Arch. Rat. Mech. Anal **43**, 304–318 (1971)
36. Serrin, J., Zou, H.: Non-existence of positive solutions of Lane-Emden systems. Diff. Int. Eq. **9**(4) (1996), 635–653.
37. Sirakov, B.: Notions of sublinearity and superlinearity for nonvariational elliptic systems. Discrete Cont. Dyn. Syst. **A13**, 163–174 (2005)
38. Souto, M.A.S.: A priori estimates and and existence of positive solutions of nonlinear cooperative elliptic systems. Diff. Int. Eq. **8**(5), 1245–1258 (1995)
39. Sweers, G.: Strong positivity in $C(\overline{\Omega})$ for elliptic systems. Math. Z. **209**, 251–271 (1992)
40. Zou, H.: A priori estimates for a semilinear elliptic system without variational structure and their application. Math. Ann **323**, 713–735 (2002)

Available online at www.sciencedirect.com
ScienceDirect
J. Differential Equations 240 (2007) 357–374
Journal of Differential Equations
www.elsevier.com/locate/jde

On the Ambrosetti–Prodi problem for non-variational elliptic systems

Djairo G. de Figueiredo [a], Boyan Sirakov [b,c,*]

[a] IMECC-UNICAMP, Caixa Postal 6065, 13083-970 Campinas, SP Brazil
[b] Université Paris 10, 92001 Nanterre cedex, France
[c] CAMS, EHESS, 54 bd Raspail, 75006 Paris, France

Received 9 June 2006; revised 31 May 2007
Available online 27 June 2007

Abstract

We study the Ambrosetti–Prodi problem for nonlinear elliptic equations and systems, with uniformly elliptic operators in non-divergence form and non-smooth coefficients, and with non-linearities with linear or power growth.

1. Introduction

In this paper we revisit an old and classical problem in the theory of elliptic partial differential equations, the so-called Ambrosetti–Prodi problem.

Although we deal with systems of equations, we start by some background on the scalar problem, for which we also have new results. Let $\Omega \subset \mathbb{R}^N$ be a smooth bounded domain, and $f(x,u)$, $h(x)$ be real-valued Hölder continuous in $x \in \overline{\Omega}$, with f locally Lipschitz continuous in $u \in \mathbb{R}$. The issue here is the existence of classical solutions to the problem

$$\begin{cases} -Lu = f(x,u) + t\varphi_1(x) + h(x) & \text{in } \Omega, \\ u = 0 & \text{on } \partial\Omega, \end{cases} \tag{1}$$

* Corresponding author. Fax: +33 149542109.
E-mail address: sirakov@ehess.fr (B. Sirakov).

Originally published in: J. Differential Equations 240 (2007), 357-374,

DOI: 10.1007/978-3-319-02856-9_39

where L is general second-order uniformly elliptic operator with Hölder continuous coefficients,

$$L=\sum_{i,j=1}^{N} a_{ij}(x)\frac{\partial^2}{\partial x_i \partial x_j}+\sum_{i=1}^{N} b_i(x)\frac{\partial}{\partial x_i}.$$

Here $t \in \mathbb{R}$ is a parameter and φ_1 is the first (positive) eigenfunction of L, i.e. $-L\varphi_1 = \lambda_1\varphi_1$ in Ω, with $\varphi_1 = 0$ on $\partial\Omega$; we refer to [6] for properties of $\lambda_1 > 0$ and φ_1.

We obtain Ambrosetti–Prodi type results (see below) for the solutions of (1), both in the case of a single equation and in the case of a system. Our main concern will be obtaining a priori bounds for the solutions. The need of these comes from the fact that topological methods (degree theory) have to be used in order to obtain multiplicity of solutions.

Problem (1) is said to be of Ambrosetti–Prodi type provided there exist constants a, b, C such that $b > \lambda_1 > a$, and for all $x \in \overline{\Omega}$,

$$f(x,s) \geqslant bs - C \quad \text{for } s \geqslant 0, \qquad f(x,s) \geqslant as - C \quad \text{for } s \leqslant 0. \tag{2}$$

This hypothesis is equivalent to the existence of constants a', b' such that $\limsup_{s\to-\infty} \frac{f(x,s)}{s} \leqslant a' < \lambda_1 < b' \leqslant \liminf_{s\to\infty} \frac{f(x,s)}{s}$.

A typical result in this setting states:

(AP) There exists $t_0 \in \mathbb{R}$ such that problem (1) has at least two solutions for $t < t_0$, at least one solution for $t = t_0$, and no solutions for $t > t_0$.

The first result in this line was obtained by Ambrosetti and Prodi in [2], and this fact originated the present terminology for this sort of problems. In [2], $L = \Delta$, $f(x,u) = f(u)$ was a convex function of class C^2 such that $0 < \lim_{s\to-\infty} f'(s) < \lambda_1 < \lim_{s\to+\infty} f'(s) < \lambda_2$. With $t\varphi_1(x) + h(x) = g(x)$ they proved, using results on differentiable mappings with singularities, that there is a closed connected C^1 manifold M of codimension 1 in the space $C^{0,\alpha}(\overline{\Omega})$ which splits the space into two connected components S_0, S_2 with the property that, if $g \in S_0$ then (1) has no solution, if $g \in M$ then (1) has exactly one solution, and if $g \in S_2$ then (1) has exactly two solutions.

The result in [2] received immediately attention by several authors trying to obtain similar conclusions and relaxing the original hypotheses. In [7] Berger and Podolak used the Liapunov–Schmidt method, so for them it was natural to use the decomposition of the function g as it appears in (1) above. The result obtained there and in most of the subsequent works is precisely the statement in (AP).

In [21] Kazdan and Warner consider more general functions f and smooth differential operators of second order. However only one solution is obtained. In [14], Dancer extended the result in [21], for differential operators in the divergence form, by getting a second solution. A result for a general L with smooth coefficients and a non-linearity f with linear growth is due to Hess [20]. In these problems, the existence of a second solution depends very heavily on the growth of the non-linearity at $+\infty$, that is: the existence of a $p \geqslant 1$ such that for all $x \in \overline{\Omega}, s \geqslant 0$, $|f(x,s)| \leqslant C(1+s^p)$. The method used in these papers is topological and the a priori bound for the solutions of (1) either depends on the linear growth of f or is obtained using the Hardy–Sobolev inequality (this method of obtaining a priori bounds is due to Brézis, Turner [8]). The use of Hardy–Sobolev inequalities requires divergence form operators and restricts the growth of the non-linearity at infinity to $p < (N+1)/(N-1)$.

Later variational methods were used for divergence form operators, and fairly general results as far as the growth of the non-linearity at $+\infty$ is concerned were obtained in [12,17]. In these papers the result holds for all subcritical problems, that is, $p < (N+2)/(N-2)$, when an additional condition of Ambrosetti–Rabinowitz (mountain-pass) type is assumed in order to get a Palais–Smale condition for the associated functional. The critical case $p = p_c$ was studied in [18] for dimensions $N > 6$ and the other dimensions in [11]. Lately the Ambrosetti–Prodi problem for operators of the type $\operatorname{div}(A(x,u)Du)$ and a non-linearity with pure power subcritical growth was considered in [3]. The problem for the m-Laplacian and f growing as $|u|^{m-2}u$ was studied in [4]. For further work on similar problems and for various multiplicity results see [5,9,23,27,29], as well as the references in these papers.

To our knowledge, there are no results on the Ambrosetti–Prodi problem with a superlinear subcritical non-linearity, when the operator is in non-divergence form, that is, the problem does not admit a weak formulation in terms of integrals. It is this situation that we want to study here. We will even not suppose that the adjoint operator of L has a principal eigenfunction. As we shall explain later, the methods from the papers quoted above do not apply in this case, and some new ideas are needed. An overview of the method we use, and of the novelties in the approach is given in the beginning of the next section.

We are going to show that (AP) holds for any operator L, provided f has a precise subcritical power growth at $+\infty$. The following theorem is a consequence of more general results for systems of equations, which are the main results of this paper.

Theorem 1. *Suppose that* (2) *holds and that there exists a bounded function $a(x)$, positive on $\overline{\Omega}$, such that for all $x \in \overline{\Omega}$*

$$\lim_{s\to\infty} \frac{f(x,s)}{s^p} = a(x), \quad \textit{for some } p \in \left(1, \frac{N+2}{N-2}\right). \tag{3}$$

Then (AP) *holds.*

Remark. We could weaken even further the regularity assumptions on the operator—instead of Hölder functions, we could consider operators with bounded (and continuous second-order) coefficients, and $h \in L^p(\Omega)$, $p \geqslant N$. Then the solutions we obtain belong to $W^{2,p}(\Omega)$.

Now we come to the discussion of Ambrosetti–Prodi results for systems of elliptic equations. Let us have a system of d equations, written in matrix format

$$(\mathcal{P}_t) \quad \begin{cases} -Lu = f(x,u) + t\varphi_1(x) + h(x) & \text{in } \Omega, \\ u = 0 & \text{on } \partial\Omega, \end{cases}$$

here $u = (u_1, \dots, u_d)^T$, $h = (h_1, \dots, h_d)^T$, $f = (f_1, \dots, f_d)^T$, $t = (t_1, \dots, t_d)^T$, $\varphi_1 = (\varphi_{1,1}, \dots, \varphi_{1,d})^T$, $t\varphi_1 = (t_1\varphi_{1,1}, \dots, t_d\varphi_{1,d})^T$, $L = \operatorname{diag}(L_1, \dots, L_d)$, with $L_k = \sum_{i,j=1}^N a_{ij}^{(k)}(x)\frac{\partial^2}{\partial x_i \partial x_j} + \sum_{i=1}^N b_i^{(k)}(x)\frac{\partial}{\partial x_i}$, where $a_{ij}^{(k)}(x)$, $b_i^{(k)}(x)$ are Hölder continuous (or, if one wants to have only $W^{2,p}$-solutions of the system we suppose that $a_{ij}^{(k)}(x)$ are continuous in $\overline{\Omega}$ and $b_i^{(k)}(x)$ are bounded), and $\varphi_{1,i}$ is the first eigenfunction of the operator L_i, normalized so that $\max_\Omega \varphi_{1,i} = 1$, see [6]. All (in)equalities between vectors will be understood to hold component-wise. Up to changing h we assume $f(x,0,\dots,0) = 0$. For any $u \in \mathbb{R}^d$, we denote $\|u\| = \max_{1\leqslant u\leqslant d} |u_i|$, and set u^+ (respectively u^-) to be the vector of the positive (respectively negative) parts of the components of u. So $u = u^+ - u^-$. We set $e = (1, \ldots, 1) \in \mathbb{R}^d$.

In order to state an Ambrosetti–Prodi problem for a system, one needs to define a first eigenvalue for a matrix operator of the type $L + A(x)$, which has the essential property to be a dividing value for the maximum principle to hold. This was recently done in [10], provided $A(x)$ is a bounded *cooperative* matrix, that is, all off-diagonal entries of A are nonnegative (and examples were given showing that for noncooperative matrices this may not be possible), and *fully coupled*. Note that any matrix A can be written in block-triangular form $A = (A_{ij})_{i,j=1}^m$, for some $m \in \{1, \ldots, d\}$, with A_{ii} fully coupled and $A_{ij} = 0$ for $i < j$ (see Section 2.1).

We shall suppose all along this paper that the map $f(x, \cdot) : \mathbb{R}^d \to \mathbb{R}^d$ is *quasi-monotone* for all $x \in \overline{\Omega}$, that is, $f_i(x, u)$ is nondecreasing in u_j, for any $i \neq j$ (this condition is of course void for a scalar equation). This is the usual condition to have a maximum principle for systems.

Condition (2) for systems will be written as follows: there exist bounded cooperative matrices $A_1(x)$, $A_2(x)$, and constants b_1, b_2, such that ($L_{1,2}^{(i)}$ will denote the minor $\operatorname{diag}(L_j)_{j \in J}$ where J contains the same indices as those in the fully coupled blocks $A_{1,ii}$, $A_{2,ii}$, for any $i \in \{1, \ldots, m\}$)

$$\lambda_1\big(L_1^{(i)} + A_{1,ii}\big) > 0 \quad \text{for all } i; \qquad \lambda_1\big(L_2^{(i)} + A_{2,ii}\big) < 0 \quad \text{for all } i; \tag{4}$$

$$f(x, s) \geqslant A_1(x)s - b_1 e \quad \text{in } \big\{s \in \mathbb{R}^d \colon s \leqslant 0\big\}; \tag{5}$$

$$f(x, s) \geqslant A_2(x)s - b_2 e \quad \text{in } \big\{s \in \mathbb{R}^d \colon s \geqslant 0\big\}. \tag{6}$$

We will also need the following (mild) assumption: for any sequence $\{s_n\} \subset \mathbb{R}^d$ such that $\{\|s_n^-\|\}$ is bounded and $\|s_n^+\| \to \infty$ as $n \to \infty$,

$$\liminf_{n \to \infty} \frac{f(x, s_n) - f(x, s_n^+)}{\|s_n^+\|} \geqslant 0. \tag{7}$$

Note that (7) is void when $d = 1$ and is trivial if f is globally Lipschitz in s.

An Ambrosetti–Prodi result for systems should state:

(APS) There exists a Lipschitz hypersurface $\Gamma \subset \mathbb{R}^d$ which divides $\mathbb{R}^d$ into two parts Y, N such that problem $(\mathcal{P}_t)$ has at least two solutions for $t \in Y$, at least one solution for $t \in \Gamma$, and no solutions for $t \in N$.

Theorem 2 *(Linear growth). Suppose that* (4)–(7) *hold and, in addition,*

$$f(x, s) \leqslant C\big(1 + \|s\|\big), \quad \textit{for all } x \in \Omega,\ s \in \mathbb{R}^d. \tag{8}$$

Then (APS) *holds.*

This result extends a previous one of K.C. Chang [13]. Note that, contrary to [13], we do not suppose that the matrices A_1, A_2 are fully coupled, nor that f is globally Lipschitz, nor that inequalities (5), (6) hold for all $s \in \mathbb{R}^d$. Note also that in [13] a different notion of first eigenvalue was used, namely concerning problems with weight. That eigenvalue exists under stronger hypotheses.

The result given by the particular case of Theorem 2, when we have only one equation, i.e. $d = 1$, has appeared in several papers (see the remarks above), under stronger restrictions on the differential operator as well as in the non-linearities involved. We state this result here, since we are unaware of a reference where it appears in the present generality.

Corollary 1.1. *Let f, h, L, φ_1 be scalar ($d = 1$). Suppose*

$$\limsup_{s\to-\infty} \frac{f(x,s)}{s} \leqslant a_1(x) \in L^\infty(\Omega), \qquad \liminf_{s\to\infty} \frac{f(x,s)}{s} \geqslant a_2(x) \in L^\infty(\Omega),$$

and $f(x,s) \leqslant C(1+|s|)$. Assume that the first eigenvalue of $L + a_1(x)$ is positive, and the first eigenvalue of $L + a_2(x)$ is negative. Then (AP) *holds.*

Next we turn to the more difficult case of superlinear systems. We will only consider systems of two equations (this is due to the necessity of using Liouville type results for positive solutions of such systems). We will also need the assumption that the second order coefficients of L_1 and L_2 coincide. So, let us have a system with two equations of the form

$$(\mathcal{P}_t) \quad \begin{cases} -L_1 u = f_1(x, u_1, u_2) + t_1\varphi_1(x) + h_1(x) & \text{in } \Omega, \\ -L_2 u = f_2(x, u_1, u_2) + t_2\varphi_2(x) + h_2(x) & \text{in } \Omega, \\ u_1 = u_2 = 0 & \text{on } \partial\Omega, \end{cases}$$

where $\varphi_i > 0$ is the first eigenfunction of L_i. We suppose that the functions f_i, $i = 1, 2$, in $(\mathcal{P}_t)$ satisfy

$$\lim_{s_j\to\infty} \frac{f_i(x, s_1, s_2)}{s_j^{\alpha_{ij}}} = a_{ij}(x), \quad i, j = 1, 2, \tag{9}$$

where the exponents $\alpha_{ij} > 1$, and $a_{ij}(x) \geqslant 0$, $a_{ij} \in C(\overline{\Omega})$.

We denote $\vec{\beta} = (\beta_1, \beta_2) \in \mathbb{R}^2$, and consider the lines

$$l_{11} = \{\vec{\beta} \mid \beta_1 + 2 - \beta_1\alpha_{11} = 0\}, \qquad l_{22} = \{\vec{\beta} \mid \beta_2 + 2 - \beta_2\alpha_{22} = 0\},$$
$$l_{12} = \{\vec{\beta} \mid \beta_1 + 2 - \beta_2\alpha_{12} = 0\}, \qquad l_{21} = \{\vec{\beta} \mid \beta_2 + 2 - \beta_1\alpha_{21} = 0\}.$$

The expressions of the lines above appear quite naturally when applying the blow-up method in order to obtain a priori bounds for the solutions of $(\mathcal{P}_t)$ in the case of systems, see [16]. We call a pair (β_1^0, β_2^0) *a blow-up pair* if it has the property to be in the intersection of two of those lines, and further $\vec{\beta}^0$ is to the left of or on l_{11}, below or on l_{22}, below or on l_{12}, and above or on l_{21}. Suppose the $a_{ij}(x)$ corresponding to these two (or more) lines are positive on $\overline{\Omega}$. We recall the following idea from [16]: if the exponents α_{ij} are such that one can choose a blow-up pair, then this will lead to statements of Liouville type, and consequently to a priori bounds for the solutions.

Theorem 3 *(Superlinear growth). Let the above hypotheses hold, and suppose that* (4)–(7) *are satisfied by the system of two equations* $(\mathcal{P}_t)$*, and, in addition,*

$$\min\{\beta_1^0, \beta_2^0\} > \frac{N-2}{2} \quad \text{or} \quad \max\{\beta_1^0, \beta_2^0\} > N-2. \tag{10}$$

Then (APS) *holds.*

We remark here that Theorem 1 is a particular case of Theorem 3 (with $a_{12} = a_{21} = 0$, $\alpha_{12} = \alpha_{21} = p$, $\beta_1^0 = \beta_2^0 = 2/(p-1)$), and that (10) with $\vec{\beta} = l_{12} \cap l_{21}$ is the best hypothesis under which the Lane–Emden system $-\Delta u_1 = u_2^{\alpha_{12}}$, $-\Delta u_2 = u_1^{\alpha_{21}}$ is known not to have classical positive solutions in $\mathbb{R}^N$.

Previous works on Ambrosetti–Prodi problems for subcritical superlinear systems are [24] and [25]. In [24] variational systems are considered and a variant of the result in [17] is obtained. The paper [25] is devoted to nonvariational systems. In that paper the restriction to the exponents was considerably stronger than (10) due to the use of Hardy–Sobolev inequalities; it was supposed that $L_1 = L_2 = \Delta$, and (4)–(6) were replaced by a stronger hypothesis concerning the first eigenvalue of Δ. Recently, [26] has used variational methods to obtain results for variational systems with non-linearities of the Ambrosetti–Prodi type, both critical and subcritical, extending results of [18] proved before for the scalar case.

The next section is devoted to the proof of the main theorems. We start by an overview, then in Section 2.1 we give some preliminaries and results on the applicability of Perron's method to our case. The heart of the paper is Section 2.2, where we prove a priori bounds for solutions of $(\mathcal{P}_t)$, as well as nonexistence of solutions for large t. The proof is concluded in Section 2.3.

2. Proofs

Here are the steps in the proof of Ambrosetti–Prodi type results:

1. Prove supersolutions exist for sufficiently small t, subsolutions of $(\mathcal{P}_t)$ exist for all t and can be chosen to be smaller than any solution of $(\mathcal{P}_t)$; deduce by Perron's method that solutions of $(\mathcal{P}_t)$ exist for $t \in (-\infty, t^*)$;
2. Prove an a priori bound on the negative part of u, for $t \geqslant -C$;
3. Prove an a priori upper bound on t, such that $(\mathcal{P}_t)$ has a solution;
4. Prove an a priori bound on u, for $t \geqslant -C$. There are two general ways to do this in the superlinear case:
   - use the Brezis–Turner technique; this restricts the growth of f to $(N+1)(N-1)$;
   - use the Gidas–Spruck blow-up technique; this requires exact power growth of f at ∞;
5. Use fixed point and degree theory to conclude.

This scheme is well known since the 1970s and has been used many times ever since, when Ambrosetti–Prodi results were to be established. We have followed this scheme too.

It is known how to prove step 1 when Perron's method is applicable and one has solvability of the Dirichlet problem and a maximum principle for L. Consequently, in Section 2.1, we recall some notations and results, essentially from [10], where these questions were studied for systems of equations, and prove some easy results on the application of the method of monotone iteration to our case.

The main difficulty is in proving steps 2, 3, and 4. In case the operator L is in divergence form, there are well-known techniques for proving steps 2 and 3, which consist in testing the equation with u^- and φ_1 respectively (this could easily be checked in the model case $L = \Delta$, $f(u) = (\lambda_1 - \varepsilon)u + (u^+)^p$). In particular, step 3 follows directly from testing with φ_1 and $\exists C$, $\delta > 0$: $f(u) \geqslant (\lambda_1 + \delta)u - C$, $\forall u \geqslant 0$. Then, once one has the uniform upper bound in t, one can prove step 4 supposing t is in a compact interval, that is, the $t\varphi_1$ term in the equation is a L^∞ right-hand side, and so trivially disappears after a blow-up.

None of the above can be done when the operator is in non-divergence form. We will now explain how we deal with the problem. First, the bound on u^- is obtained by showing that the (non-smooth) function u^- satisfies a linear inequality in the viscosity sense, and then by applying Caffarelli's ABP inequality for such solutions, and its extensions to systems, proved in [10]. Second, the proofs of steps 3 and 4 are carried out jointly. We perform a simultaneous blow-up argument in $\|u\|_{L^\infty(\Omega)}$ and t. This argument gives a bound neither on u nor on t, but rather leads to the inequality

$$\|u\|_{L^\infty(\Omega)} \leqslant Ct^{1/p}, \tag{11}$$

which is interesting in its own right. Then through a maximum principle argument we show that

$$t \leqslant C\big(1 + \|u\|_{L^\infty(\Omega)}\big).$$

These two inequalities together yield bounds both on u and t. To our knowledge, no similar approach has been used in other works.

We remark that the implementation of the blow-up method follows the standard blow-up procedure except that the term coming from $t\varphi_1$ does not necessarily disappear at the limit, since t is unbounded as well. Actually we find its disappearing is equivalent to the failure of (11), which permits to conclude the contradiction argument.

Finally, it is not difficult to prove step 5, when one has an uniform upper bound for t and $\|u\|$, such that u is a solution of $\mathcal{P}_t$.

2.1. *Preliminaries*

In this section we state some results, and consequences of results from [10] (see in particular Sections 8, 13 and 14 in that paper).

Let us consider d uniformly elliptic operators in the general non-divergence form

$$L_k = \sum_{i,j=1}^{N} a_{ij}^{(k)}(x)\frac{\partial^2}{\partial x_i \partial x_j} + \sum_{i=1}^{N} b_i^{(k)}(x)\frac{\partial}{\partial x_i},$$

where $a_{ij}^{(k)}(x)$ are continuous in $\overline{\Omega}$ and $b_i^{(k)}(x)$ are bounded. Let $c_{ij}(x)$ be bounded functions and set $C(x) = (c_{ij}(x))_{i,j=1}^d$. Let $f_i(x) \in L^N(\Omega)$. We will consider systems in the form

$$Lu + Cu = f, \tag{12}$$

where $L = \operatorname{diag}(L_1, \ldots, L_d)$, $C(x) = (c_{ij}(x))_{i,j=1}^d$, $u = (u_1, \ldots, u_d)^T$, and $f = (f_1, \ldots, f_d)^T$.

We shall need to consider solutions of this system in the viscosity sense, whose definition we recall next. A function $u \in C(\overline{\Omega}, \mathbb{R}^d)$ is called a *viscosity subsolution* of (12) provided for each i, each $x_0 \in \Omega$, and each $\varphi \in C^2(\Omega)$ such that $\varphi(x_0) = u_i(x_0)$, $\varphi \geqslant u_i$ in Ω, we have $L_i\varphi(x_0) \geqslant f_i(x_0) - \sum_k c_{ik}(x_0)u_k(x_0)$. A function is a *viscosity supersolution* if this definition with reverse inequalities holds, and u is a *viscosity solution* if it is both a viscosity subsolution and a viscosity supersolution. Note that this definition is valid if all functions in (12) are continuous in x; if this is not the case one needs to use the so-called L^N-viscosity solutions, see [10]. Any classical solution is of course a viscosity solution. It is very simple to check that, in the viscosity sense, the maximum of two subsolutions is a subsolution, and the minimum of two supersolutions is a supersolution (note that even if two functions are smooth their maximum is merely continuous, that is why we need this weaker notion). We shall work with solutions in the viscosity sense, without necessarily specifying each time.

Viewing to use Alexandrov–Bakelman–Pucci estimates, Harnack inequalities and maximum principles we consider cooperative systems. System (12) is called *cooperative* (or *quasi-monotone*) if $c_{ij} \geqslant 0$ for all $i \neq j$.

We recall that a system of this type is called *fully coupled* (and the matrix C is called *irreducible*) provided for any non-empty sets $I, J \subset \{1, \ldots, d\}$ such that $I \cap J = \emptyset$ and $I \cup J = \{1, \ldots, d\}$, there exist $i_0 \in I$ and $j_0 \in J$ for which

$$\operatorname{meas}\{x \in \Omega \mid c_{i_0 j_0}(x) > 0\} > 0. \tag{13}$$

For simplicity, when (13) holds we write $c_{i_0 j_0} \not\equiv 0$ in Ω. Simply speaking, a system is fully coupled provided it cannot be split into two subsystems, one of which does not depend on the other.

As explained in [10], any matrix can have its lines and columns renumbered in such a way that it is in block triangular form, with each block on the main diagonal being fully coupled. More precisely, $C = (C_{kl})_{k,l=1}^m$, where $1 \leqslant m \leqslant d$, C_{kl} are $t_k \times t_l$ matrices for some $t_k \leqslant d$ with $\sum_{k=1}^m t_k = d$, C_{kk} is an *irreducible* matrix for all $k = 1, \ldots, m$, and $C_{kl} \equiv 0$ in Ω, for all $k, l \in \{1, \ldots, m\}$ with $k < l$. Note that $m = 1$ means C itself is irreducible, while $m = d$ means C is in triangular form. We set $s_0 = 0$, $s_k = \sum_{j=1}^k t_j$, and $S_k = \{s_{k-1} + 1, \ldots, s_k\}$.

It was proved in [10, Theorem 13.1] that the matrix operator $L + C$ admits a principal eigenvalue with all the usual properties of the principal eigenvalue of a scalar operator (see [6]), provided C is cooperative and irreducible. We recall that the *principal eigenvalue* of $L + C$ is defined by

$$\begin{aligned}\lambda_1 &= \lambda_1(L + C)\\ &= \sup\{\lambda \in \mathbb{R}\colon \exists \psi \in W^{2,N}_{\mathrm{loc}}(\Omega, R^d),\ \text{s.t. } \psi > 0,\ (L + C + \lambda I)\psi \leqslant 0 \text{ in } \Omega\}.\end{aligned}$$

Hence, using the above explained block triangular representation of the cooperative matrix C, we can associate to C a set of eigenvalues $\lambda_1^{(1)}, \ldots, \lambda_1^{(m)}$, where $\lambda_1^{(k)}$ is the principal eigenvalue of $L^{(k)} + C_{kk}$. Here we have denoted $L^{(k)} = \operatorname{diag}(L_{s_{k-1}+1}, \ldots, L_{s_k})$ (see above for the notations).

By combining Theorems 8.1, 12.1, 13.1, 13.2, 14.1 and Lemma 14.1 in [10] we obtain the following result.

Theorem 4.

(i) *The following are equivalent*:
  (a) $\lambda_1^{(k)} > 0$ *for all* $k = 1, \ldots, m$;
  (b) *there exists a vector* $\psi(x) \in C^2(\Omega, \mathbb{R}^d)$ (*or* $W^{2,p}(\Omega, \mathbb{R}^d) \cap C(\overline{\Omega}, \mathbb{R}^d)$) *such that* $\psi \geqslant e$ *and* $L\psi + C\psi \leqslant 0$ *in* Ω;
  (c) *for any* $f \in (L^N(\Omega))^d$ *and any viscosity subsolution of* (12) *there holds*

$$\sup_{\Omega} \max\{u_1, \ldots, u_d\} \leqslant C\Big(\sup_{\partial\Omega} \max\{u_1, \ldots, u_d\} + \|f\|_{L^N(\Omega)}\Big),$$

where C depends only on Ω and on the coefficients of L and C. Respectively, for any supersolution we have

$$\inf_{\Omega} \min\{u_1, \ldots, u_d\} \geqslant C\Big(\inf_{\partial\Omega} \min\{u_1, \ldots, u_d\} - \|f\|_{L^N(\Omega)}\Big).$$

(d) *the operator $L + \mathcal{C}$ satisfies the maximum principle in Ω, that is, if $Lu + \mathcal{C}u \leqslant 0$ in Ω and $u \geqslant 0$ on $\partial\Omega$, then $u \geqslant 0$ in Ω.*

(ii) *If $\lambda_1^{(k)} > 0$ for all $k = 1, \ldots, m$, then for any $f \in C^\alpha(\Omega, \mathbb{R}^d)$ (or for any $f \in L^p(\Omega, \mathbb{R}^d)$, $p \geqslant N$) there exists a unique classical (respectively in $W^{2,p}(\Omega, \mathbb{R}^d) \cap C(\overline{\Omega}, \mathbb{R}^d)$) solution of* (12), *such that $u = 0$ on $\partial\Omega$.*

(iii) *Suppose $\psi \in C(\overline{\Omega}, \mathbb{R}^d)$ is such that $\psi \geqslant 0$ and $L\psi + \mathcal{C}\psi \leqslant 0$ in Ω. If $\psi_j \not\equiv 0$ in $\overline{\Omega}$ for some $j \in S_k$ and some $k \in \{1, \ldots, m\}$, then $\lambda_1^{(k)} \geqslant 0$.*

Proof. (i) Theorem 14.1 and Lemma 14.1 in [10] give (a) ⇔ (b) ⇔ (d). Theorem 8.1 in [10] gives (b) ⇒ (c), and (c) ⇒ (d) is obvious.

(ii) If $m = 1$ this is Theorem 13.2 in [10] (due to Sweers [28]). If $m > 1$ we apply this theorem m times: using the block-diagonal structure of $\mathcal{C}$, first we solve $(L^{(1)} + \mathcal{C}_{11})u^{(1)} = f^{(1)}$, then $(L^{(2)} + \mathcal{C}_2)u^{(2)} = f^{(2)} - \mathcal{C}_{21}u^{(1)}$, etc.

(iii) This follows from the cooperativeness of $\mathcal{C}$ and the definition of the first eigenvalue, together with Theorem 14.1 from [10]. □

Using this theorem, it is easy to prove the following two lemmas, classical in the Ambrosetti–Prodi setting. From now on, any time we write a norm of a function, we are going to mean the $L^\infty(\Omega)$-norm.

Lemma 2.1. *Under the hypotheses of either Theorem* 2 *or Theorem* 3, *for each $t \in \mathbb{R}^d$ there exists a classical subsolution $\underline{u} \leqslant 0$ of system $(\mathcal{P}_t)$.*

Proof. Set $K = 2\max_{i=1,\ldots,d}\{\|h_i\| + |t_i|\} + b_1$ (b_1 is the constant from hypothesis (5)). By the previous theorem and (4) we can find a solution of the system

$$Lu + A_1(x)u = Ke - h(x) - t\varphi_1(x)$$

with Dirichlet boundary condition. Clearly the solution of this problem is nonpositive (by the maximum principle, Theorem 4(d)) and can be taken as the subsolution we are searching for. □

Remark. We will show in the proof of Proposition 2.1 in the next section that $\underline{u}$ is smaller than any supersolution of $(\mathcal{P}_t)$.

Lemma 2.2. *Under the hypotheses of either Theorem* 2 *or Theorem* 3, *there exists $t_0 \in \mathbb{R}$ such that for each $t \leqslant t_0 e$ there exists a classical supersolution $\overline{u} \geqslant 0$ of system $(\mathcal{P}_t)$.*

Proof. By the hypotheses, there exist constants C_1 and $p_i \geqslant 1$ such that for all $u \geqslant 0$

$$f(x, u) \leqslant C_1\big(1 + u_1^{p_1} + \cdots + u_d^{p_d}\big)e.$$

Let $\bar{u}$ be the solution of the following d equations:

$$Lu + h^+ + C_1 e = 0$$

with $\bar{u} = 0$ on $\partial\Omega$. By the maximum principle $\bar{u} \geqslant 0$ in Ω. By the well-known properties of φ_1 and Hopf's lemma we can choose $t_0 \in \mathbb{R}$ such that

$$-t_0\varphi_1 \geqslant C_1\left(\bar{u}_1^{p_1} + \cdots + \bar{u}_d^{p_d}\right).$$

Using the two inequalities above in the equation for $\bar{u}$ we get the result. □

Once the results of the two previous lemmas are available, the method of monotone iteration (see [1]) can be applied to get a minimal solution of the problem for sufficiently small t, see Proposition 2.1 below. Observe that this method applies for cooperative elliptic systems, which is the case here, since the functions f are quasi-monotone. For the reader's convenience we state the following result, which will be sufficient for our purposes (see for example [22] for more general statement).

Theorem 5. *Suppose $f(x,u)$ is a quasi-monotone map, which is Hölder continuous in x and locally Lipschitz continuous in u. Suppose $\underline{u}, \bar{u} \in C(\overline{\Omega}, \mathbb{R}^d)$ are respectively a subsolution and a supersolution of the system*

$$Lu + f(x,u) = 0 \tag{14}$$

in Ω, such that $\underline{u} \leqslant \bar{u}$ in Ω, $\underline{u} \leqslant 0 \leqslant \bar{u}$ on $\partial\Omega$. Then there exists a classical solution u of (14) such that $\underline{u} \leqslant u \leqslant \bar{u}$ in Ω, $u = 0$ on $\partial\Omega$.

Sketch of the proof of Theorem 5. Set $u_0 = \underline{u}$, $m = \inf_\Omega \min_i \underline{u}_i$, $M = \sup_\Omega \max_i \bar{u}_i$ and let

$$k = \max_{1\leqslant i\leqslant n} \left\| \frac{\partial f_i}{\partial u_i} \right\|_{L^\infty(\Omega\times[m,M]^n)}.$$

By Theorem 4 we can solve the hierarchy of problems

$$-Lu^{(n+1)} + ku^{(n+1)} = f\left(x, u^{(n)}\right) + ku^{(n)} \quad \text{in } \Omega,$$
$$u^{(n+1)} = 0 \quad \text{on } \partial\Omega.$$

It is then easy to check, with the help of the maximum principle, that we have $\underline{u} \leqslant u^{(n)} \leqslant u^{(n+1)} \leqslant \bar{u}$ for all n, so $u^{(n)}$ converges to a solution of (14).

2.2. *A priori bounds*

We start with a lemma which shows that the negative parts of the solutions of $(\mathcal{P}_t)$ are uniformly bounded, provided t is above some fixed level. From now on, all constants we write may change from line to line and depend only on the data in $(\mathcal{P}_t)$—that is, on L, f, h, Ω (and on other quantities, if stated). We will also make the convention that any norm of a vector is the maximum of the corresponding norms of its components.

Lemma 2.3. *Under the hypotheses of either Theorem 2 or Theorem 3, for each $C_0 \in \mathbb{R}_+$ there exists a constant M such that for any $t \geqslant -C_0 e$ and any solution u of $(\mathcal{P}_t)$ with this t we have*

$$\|u^-\| \leqslant M.$$

Proof. Set $m = \max_i\{\|h_i\|_{L^\infty(\Omega)} + C_0\}$. So $(\mathcal{P}_t)$ yields

$$Lu + f(x,u) \leqslant me \quad \text{in } \Omega.$$

Since f is quasi-monotone this implies

$$L_i u_i + f_i\big(x, -u_1^-, \dots, -u_{i-1}^-, u_i, -u_{i+1}^-, \dots, -u_d^-\big) \leqslant m, \tag{15}$$

for each $i \in \{1, \dots, d\}$. On the other hand, again by the quasi-monotonicity of f,

$$f_i\big(x, -u_1^-, \dots, -u_{i-1}^-, 0, -u_{i+1}^-, \dots, -u_d^-\big) \leqslant f_i(x, 0, \dots, 0) = 0.$$

This means that $u_i \equiv 0$ is also a solution of (15), seen as a scalar equation in u_i. As explained in the previous section, the minimum of two supersolutions is a viscosity supersolution, hence (15) continues to hold if we replace u_i by $-u_i^-$, which gives,

$$L(-u^-) + f(x, -u^-) \leqslant m\,e,$$

in the viscosity sense. By hypothesis (5)

$$L(-u^-) + A_1(x)(-u^-) \leqslant (m + b_1)e.$$

Since (4) holds, Theorem 4(c) implies the lemma. □

Proposition 2.1. *Under the hypotheses of either Theorem 2 or Theorem 3, there exists $t_0 \in \mathbb{R}$ such that for each $t \leqslant t_0 e$ there exists a minimal solution of system $(\mathcal{P}_t)$.*

Proof. Using Theorem 5 and Lemmas 2.1 and 2.2 we show the existence of a minimal solution of $(\mathcal{P}_t)$, for any fixed $t \leqslant t_0 e$, where the t_0 is the one of Lemma 2.2. For that matter we first claim that any supersolution $\bar{u}$ of $(\mathcal{P}_t)$ satisfies $\bar{u} \geqslant \underline{u}$ in Ω, where $\underline{u}$ is the subsolution constructed in Lemma 2.1. Once this is done, we use Theorem 5 and finish. In order to prove the claim, we proceed as in the proof of Lemma 2.3 we have $L(-\bar{u}^-) + f(x, -\bar{u}^-) \leqslant me$ provided $\bar{u}$ is a supersolution of $(\mathcal{P}_t)$, so, by the way $\underline{u}$ is chosen,

$$L(-\bar{u}^- - \underline{u}) + f(x, -\bar{u}^-) - A_1(x)\underline{u} \leqslant 0.$$

Hence, by (5),

$$\big(L + A_1(x)\big)(-\bar{u}^- - \underline{u}) \leqslant 0,$$

and so the maximum principle implies $-\bar{u}^- - \underline{u} \geqslant 0$, which implies $\bar{u} \geqslant -\bar{u}^- \geqslant \underline{u}$ in Ω. □

Remark. The above proof shows that, if for some t problem $(\mathcal{P}_t)$ has a solution u, then $u \geqslant \underline{u}$, where $\underline{u}$ is the subsolution in Lemma 2.1. So as soon as $(\mathcal{P}_t)$ is solvable for some t, then a minimal solution exists.

Next, we show that for any unbounded sequence $\{t_n\}$, the growth of the corresponding solutions—if such solutions exist—controls the growth of $\{t_n\}$.

Lemma 2.4. *Under the hypotheses of either Theorem 2 or Theorem 3, for each $C_0 \in \mathbb{R}_+$ there exists a constant C_1 such that for any $t \geqslant -C_0 e$ and any solution u of $(\mathcal{P}_t)$ with this t we have*

$$t_i^+ \leqslant C_1\left(1 + \left\|u_i^+\right\|\right) \leqslant C_1\left(1 + \|u\|\right), \quad i = 1, \ldots, d.$$

Proof. By using successively the quasi-monotonicity of f, property (7) and finally property (6), we get

$$\begin{aligned} f_i(x,u) &\geqslant f_i(x, -M, \ldots, -M, u_i, -M, \ldots, -M) \\ &\geqslant f_i\left(x, 0, \ldots, 0, u_i^+, 0, \ldots, 0\right) - C\left|u_i^+\right| \\ &\geqslant -C\left(1 + u_i^+\right), \end{aligned} \tag{16}$$

for all $i \in \{1, \ldots, n\}$, $x \in \Omega$, $u \geqslant -Me$ (M is the constant from Lemma 2.3). Suppose now for some $t \in \mathbb{R}^d$ and some function u we have

$$Lu + f(x,u) + t\varphi_1 + h = 0,$$

and $u = 0$ on $\partial\Omega$. This implies, by (16), that we have the following d scalar inequalities:

$$L_i\left(u_i - t_i\lambda_{1,i}^{-1}\varphi_{1,i}\right) \leqslant C\left(1 + u_i^+\right),$$

and $u_i - t_i\lambda_{1,i}^{-1}\varphi_{1,i} = 0$ on $\partial\Omega$, where the constant C absorbs the norm of h. By the Alexandrov–Bakelman–Pucci inequality (Theorem 4(c) for $d = 1$) we get

$$u_i - t_i\lambda_{1,i}^{-1}\varphi_{1,i} \geqslant -C\left(1 + \left\|u_i^+\right\|\right), \quad i = 1, \ldots, d,$$

from which the result follows (we recall that the first eigenvectors $\varphi_{1,i}$ are normalized so that $\|\varphi_{1,i}\| = 1$). □

We can now deduce an a priori bound in the linear growth case.

Proposition 2.2. *Under the hypotheses of Theorem 2, for each $C_0 \in \mathbb{R}_+$ there exists a constant M such that for any $t \geqslant -C_0 e$ and any solution u of $(\mathcal{P}_t)$ with this t we have*

$$\|u\| \leqslant M.$$

Proof. Suppose for contradiction that there exist sequences $\{u_n\}$, $\{t_n\}$, such that $t_n \geqslant -C_0 e$, and

$$Lu_n + f(x, u_n) + t_n\varphi_1 + h = 0, \quad \|u_n\| \to \infty,$$

as $n \to \infty$. Using Lemma 2.3, we get $\lim \|u_n^+\| = \infty$, $\|u_n^+\| = \|u_n\|$ for n large. By dividing the equation by $\|u_n^+\|$ and by using $t_n \geqslant -C_0 e$ and (16), we see that $Lv_n \leqslant C$, where $v_n = u_n/\|u_n^+\|$. By using the linear growth (9) and Lemma 2.4 we have $Lv_n \geqslant -C$. Hence, by elliptic theory, v_n converges (up to a subsequence) to a function v in $W^{2,p}(\Omega)$. Note that $v \geqslant 0$, by Lemma 2.3. Of course $\|v_n\| = 1$ for n large, so $\|v\| = 1$.

By using the fact that (7) implies $f(x, u_n) \geqslant f(x, u_n^+) - o(1)\|u_n\|$, we obtain

$$Lv_n + \frac{f(x, u_n^+)}{\|u_n^+\|} \leqslant o(1),$$

Hence, by (6) and passage to the limit we see that $v \geqslant 0$ is a nontrivial solution to inequality $Lv + A_2(x)v \leqslant 0$, which contradicts (4) and Theorem 4(iii). □

We now turn to the superlinear case. The following bound plays an essential role.

Proposition 2.3. *Under the hypotheses of Theorem* 3, *there exists a constant C such that for every vector $t \geqslant e$ and every solution $u = (u_1, u_2)$ of $(\mathcal{P}_t)$ corresponding to this t, the following inequalities hold*:

$$\|u_1\|^{1+\frac{2}{\beta_1^0}} \leqslant Ct_1 \quad \textit{and} \quad \|u_2\|^{1+\frac{2}{\beta_2^0}} \leqslant Ct_2.$$

Here $(\beta_1^0, \beta_2^0) > 0$ is the vector which appears in Theorem 3.

For clarity, before proving Proposition 2.3, we state the particular case of it when only one equation is considered, that is, when we are in the framework of Theorem 1.

Proposition 2.4. *Under the hypotheses of Theorem* 1, *there exists a constant C such that for every number $t \geqslant 1$ and every solution u of $(\mathcal{P}_t)$, corresponding to this t, we have*

$$\|u\| \leqslant Ct^{\frac{1}{p}}.$$

More precisely, to get Proposition 2.4 from Proposition 2.3, we take $L_1 = L_2 = L$, $a_{11} = a_{22} = a$, $\alpha_{11} = \alpha_{22} = p$, $a_{12} = a_{21} = 0$, so $\beta_1^0 = \beta_2^0 = 2/(p-1)$.

Proof of Proposition 2.3. It follows from (9) that we can write, for $i = 1, 2$,

$$f_i(x, s_1, s_2) = a_{i1}(x)s_1^{\alpha_{i1}} + a_{i2}(x)s_2^{\alpha_{i2}} + g_i(x, s_1, s_2),$$

where

$$\lim_{|(s_1,s_2)|\to\infty} \left[a_{i1}s_1^{\alpha_{i1}} + a_{i2}s_2^{\alpha_{i2}}\right]^{-1} g_i(x, s_1, s_2) = 0.$$

Now suppose that the result of the proposition is false, that is, there exist sequences $\{u_n\}$, $\{t_n\}$, such that $t_n \geqslant e$,

$$-L_1 u_{1,n} = a_{11}(u_{1,n}^+)^{\alpha_{11}} + a_{12}(u_{2,n}^+)^{\alpha_{12}} + g_1(x, u_{1,n}, u_{2,n}) + t_{1,n}\varphi_{1,1} + \tilde{h}_1,$$
$$-L_1 u_{2,n} = a_{21}(u_{1,n}^+)^{\alpha_{21}} + a_{22}(u_{2,n}^+)^{\alpha_{22}} + g_2(x, u_{1,n}, u_{2,n}) + t_{2,n}\varphi_{1,2} + \tilde{h}_2$$

(here $\tilde{h}_i = h_i + d_i$, where d_i is some bounded function, that corresponds to the negative part of $u_{i,n}$, which is bounded by Lemma 2.3) and

$$\|u_{1,n}\|^{1+\frac{2}{\beta_1^0}} \geqslant n t_{1,n} \quad \text{or} \quad \|u_{2,n}\|^{1+\frac{2}{\beta_2^0}} \geqslant n t_{2,n}.$$

Taking a subsequence if necessary, we may suppose that one of these inequalities (say the first) holds for each n.

Assume first that

$$\lim_{n\to\infty} \frac{\|u_{1,n}\|^{1+\frac{2}{\beta_1^0}}}{t_{1,n}} = \lim_{n\to\infty} \frac{\|u_{2,n}\|^{1+\frac{2}{\beta_2^0}}}{t_{2,n}} = \infty. \tag{17}$$

We will use a blow-up type argument, originally due to Gidas and Spruck [19] in the case of a scalar equation, and developed for our type of systems in [16]—we refer to that paper for details. Set

$$\lambda_n = \|u_{1,n}\|^{-1/\beta_1^0} \quad \text{if } \|u_{1,n}\|^{\beta_2^0} \geqslant \|u_{2,n}\|^{\beta_1^0},$$

and $\lambda_n = \|u_{2,n}\|^{-1/\beta_2^0}$ otherwise (say we are in the first of these situations). Then $\lambda_n \to 0$ and the functions

$$v_{i,n} = \lambda_n^{\beta_i^0} u_{i,n}(\lambda_n x + x_n)$$

are such that $v_{1,n}(0) = 1$, $0 \leqslant v_{i,n} \leqslant 1$ in Ω (here x_n is the point in Ω where $u_{1,n}$ attains its maximum). Then

$$-L_{1,n} v_{1,n} = a_{11}(\cdot)\lambda_n^{\gamma_{11}}(v_{1,n}^+)^{\alpha_{11}} + a_{12}(\cdot)\lambda_n^{\gamma_{12}}(v_{2,n}^+)^{\alpha_{12}} + \lambda_n^{\beta_1^0+2} t_{1,n}\varphi_{1,1},$$
$$-L_{2,n} v_{2,n} = a_{21}(\cdot)\lambda_n^{\gamma_{21}}(v_{1,n}^+)^{\alpha_{21}} + a_{22}(\cdot)\lambda_n^{\gamma_{22}}(v_{2,n}^+)^{\alpha_{22}} + \lambda_n^{\beta_2^0+2} t_{2,n}\varphi_{1,2} \tag{18}$$

(we have omitted the terms coming from g_i, h_i, d_i, since they tend to zero as $n \to \infty$); here

$$\gamma_{ij} = \beta_i^0 + 2 - \beta_j^0 \alpha_{ij}$$

(recall the equations of the lines $l_{ij} = \{\vec{\beta} \mid \beta_i + 2 - \beta_j \alpha_{ij} = 0\}$), and

$$L_{k,n} = \sum_{i,j=1}^{N} a_{ij}(\cdot)\frac{\partial^2}{\partial x_i \partial x_j} + \lambda_n \sum_{i=1}^{N} b_i^{(k)}(\cdot)\frac{\partial}{\partial x_i},$$

the dot stands for $\lambda_n x + x_n$, and the equations are given in the domain $\lambda_n^{-1}(\Omega - x_n)$.

We can pass to the limit in (18) as in [16, Lemma 2.1], and conclude that $v_{i,n}$ converges (up to a subsequence) to a bounded function v_i (note that v_i are non-negative, since the negative parts of $u_{i,n}$ are bounded, by Lemma 2.3). The difference with [16] is in the last terms in the right-hand side of (18). However, these terms turn out to vanish as $n \to \infty$, under the hypothesis that Proposition 2.3 is false. Indeed, under (17),

$$\lambda_n^{\beta_1^0+2} t_{1,n} = \|u_{1,n}\|^{-\frac{\beta_1^0+2}{\beta_1^0}} t_{1,n} \to 0 \quad \text{as } n \to \infty,$$

while, by the choice of λ_n,

$$\lambda_n^{\beta_2^0+2} t_{2,n} \leqslant \|u_{2,n}\|^{-\frac{\beta_2^0+2}{\beta_2^0}} t_{2,n} \to 0 \quad \text{as } n \to \infty,$$

by (17).

Hence, after the passage to the limit, we obtain a system in $\mathbb{R}^N$ or in a half-space, which, by the results in [16] (see also the references there), has only the trivial solution, which contradicts $v_1(0) = 1$. Note that the differential operator in the limiting system has constant coefficients and can be transformed into the Laplacian through an orthogonal change of variables.

Next, suppose (17) does not hold, that is

$$\|u_{2,n}\|^{1+\frac{2}{\beta_2^0}} \leqslant K t_{2,n},$$

for some constant K. Combining this inequality with Lemma 2.4 (used twice), we see that both sequences $\{\|u_{2,n}\|\}$ and $\{t_{2,n}\}$ are bounded. Then we can repeat exactly the same argument as above to get a contradiction with $\|u_{1,n}\|^{1+\frac{2}{\beta_1^0}} \geqslant n t_{1,n} \geqslant n$—note that at the end $\lambda_n^{\beta_2^0+2} t_{2,n} \to 0$ trivially follows from $\lambda_n \to 0$. □

Actually, looking at the proof of Proposition 2.3, we see that it implies the following stronger statement.

Proposition 2.5. *Under the hypotheses of Theorem 3, for each $C_0 \in \mathbb{R}_+$ there exists a constant M such that for any $t \geqslant -C_0 e$ and any solution $u = (u_1, u_2)$ of $(\mathcal{P}_t)$ with this t we have*

$$\|u_1\|^{1+\frac{2}{\beta_1^0}} \leqslant M \max\{1, t_1\} \quad \text{and} \quad \|u_2\|^{1+\frac{2}{\beta_2^0}} \leqslant M \max\{1, t_2\}.$$

Proof. First, since $t\varphi_1 = t^+\varphi_1 - t^-\varphi_1$, we can think of the bounded term $t^-\varphi_1$ as being part of the function $h(x)$, and assume $t > 0$. Then we simply repeat the proof of Proposition 2.3, replacing t by $\max\{t, e\}$ in it. □

We can now conclude that solutions of our system admit a priori bounds and that the system does not have solutions if t is large.

Proposition 2.6. *Under the hypotheses of Theorem 2 or Theorem 3, there exists a constant M such that if for some $t \geqslant 0$ there exists a solution u of system $(\mathcal{P}_t)$, then*

$$t \leqslant Me \quad \textit{and} \quad \|u\| \leqslant M.$$

More generally, for each $t \in \mathbb{R}^d$ there exists a constant M with this property, and M depends only on t^-.

Proof. Combine Propositions 2.2 or 2.5 with Lemmas 2.3 and 2.4. □

2.3. Conclusion

In the previous sections we have established the following facts, which will now be used to complete the proofs of Theorems 2 or 3:

(i) if C is sufficiently large, $(\mathcal{P}_t)$ has a minimal solution for $t \leqslant -Ce$;
(ii) if C is sufficiently large, $(\mathcal{P}_t)$ does not have a solution for $\|t\| \geqslant C$;
(iii) a priori bound: given $t_0 \in \mathbb{R}^d$, the (eventual) solutions of $(\mathcal{P}_t)$ for all $t \geqslant t_0$ are bounded by the same constant.

Next, the surface Γ will be defined by parametrization with respect to the hyperplane $H = \{t \in \mathbb{R}^d \mid t_1 + \cdots + t_d = 0\}$. Let us define, for each $t_0 \in H$,

$$A(t_0) = \{k \in \mathbb{R}\colon\ (\mathcal{P}_{t_0+ke}) \text{ has a solution}\}.$$

By (i) above this set is not empty.

On the other hand we know ((ii) above) that for each $t_0 \in H$ there is a $k_0 \in \mathbb{R}$ such that problem $(\mathcal{P}_{t_0+ke})$ does not have a solution for all $t_0 + ke$ with $k \geqslant k_0$. So the function $K : H \to \mathbb{R}$, $K(t) = \sup A(t)$ is well defined. Further, if $k \in A(t)$ for some t then any $k' \leqslant k$ also belongs to $A(t)$. Indeed, a solution of $(\mathcal{P}_{t_0+ke})$ is a supersolution for $(\mathcal{P}_{t_0+k'e})$ and by Lemma 2.1 and Theorem 5 we have a solution of $(\mathcal{P}_{t_0+k'e})$. So $A(t)$ is an interval.

Next, the function $K(t) : H \to \mathbb{R}$ is Lipschitz continuous, with Lipschitz constant 1. Indeed, to show this one can use the following argument from [13]: given $t^1, t^2 \in H$ it follows from what we just saw that

$$t^1 + K(t^1)e \not< t^2 + K(t^2)e, \qquad t^2 + K(t^2)e \not< t^1 + K(t^1)e.$$

Hence there exist indices $i, j \in \{1, \ldots, d\}$ such that

$$t_i^2 + K(t^2) \leqslant t_i^1 + K(t^1), \qquad t_j^2 + K(t^2) \geqslant t_j^1 + K(t^1),$$

so

$$-|t^1 - t^2| \leqslant t_i^2 - t_i^1 \leqslant K(t^1) - K(t^2) \leqslant t_j^2 - t_j^1 \leqslant |t^1 - t^2|.$$

This yields that the hypersurface $\Gamma = \{t + K(t)e\colon\ t \in H\}$ is Lipschitz.

Next we prove that problem $(\mathcal{P}_{t_0+ke})$ has at least two solutions for $k < K(t_0)$. Viewing to use topological degree arguments, let us define the mapping $S_t : C^{1,\alpha}(\Omega)^d \to C^{1,\alpha}(\Omega)^d$ by $u = S_t v$, where

$$\begin{cases} -Lu = f(x,v) + t\varphi_1(x) + h(x) & \text{in } \Omega, \\ u = 0 & \text{on } \partial\Omega. \end{cases}$$

Finding a solution of $(\mathcal{P}_t)$ is equivalent to obtaining a fixed point of S_t. The search for fixed point will be done with the help of degree considerations. We will be brief here, we refer for example to [15], where a simple and thorough account of this type of argument is given.

Fix $t_0 \in H$ and $k_0 < K(t_0)$. By Proposition 2.1 problem $(\mathcal{P}_{t_0+k_0e})$ has a minimal solution. It is classical that the Leray–Schauder degree of this minimal solution is one, so there exists an open set $\mathcal{O}$ in $C^{1,\alpha}(\Omega)^d$ which contains the minimal solution and

$$\deg(I - S_{t_0+k_0e}, \mathcal{O}, 0) = 1.$$

On the other hand, by (ii) above there exists $\bar{k} \in \mathbb{R}$ such that problem $(\mathcal{P}_{t_0+ke})$ has no solution for $k \geqslant \bar{k}$. This implies

$$\deg(I - S_{t_0+\bar{k}e}, B_R, 0) = 0$$

for any ball $B_R \subset C^{1,\alpha}(\Omega)^d$.

However, the a priori bound (iii) implies that there exists R sufficiently large, such that

$$\deg(I - S_{t_0+ke}, B_R, 0)| \quad \text{is constant in } k \geqslant k_0.$$

This means that $\deg(I - S_{t_0+k_0e}, B_R, 0) = 0$, and by the excision property of the degree there exists a solution of $(\mathcal{P}_{t_0+k_0e})$ in $B_R \setminus \mathcal{O}$.

Finally, given $t \in H$ we take a sequence $k_n \nearrow K(t)$ and a sequence of solutions of $(\mathcal{P}_{t+k_ne})$. Thanks to the a priori bounds this sequence is bounded in $L^\infty(\Omega)$ and elliptic theory permits us to pass in the limit in $(\mathcal{P}_{t+k_ne})$, which gives one solution of $(\mathcal{P}_{t+K(t)e})$.

References

[1] H. Amann, Fixed point equation and nonlinear eigenvalue problems in ordered Banach spaces, SIAM Rev. 18 (1976) 620–709.

[2] A. Ambrosetti, G. Prodi, On the inversion of some differentiable mappings with singularities between Banach spaces, Ann. Mat. Pura Appl. (4) 93 (1972) 231–246.

[3] D. Arcoya, J. Carmona, Quasilinear elliptic problems interacting with its asymptotic spectrum, Nonlinear Anal. 52 (6) (2003) 1591–1616.

[4] D. Arcoya, D. Ruiz, The Ambrosetti–Prodi problem for the p-Laplacian, Comm. Partial Differential Equations 31 (2006) 841–865.

[5] H. Berestycki, Le nombre de solutions de certains problèmes semi-linéaires elliptiques, J. Funct. Anal. 40 (1981) 1–29.

[6] H. Berestycki, L. Nirenberg, S.R.S. Varadhan, The principal eigenvalue and maximum principle for second order elliptic operators in general domains, Comm. Pure Appl. Math. 47 (1) (1994) 47–92.

[7] M.S. Berger, E. Podolak, On the solutions of a nonlinear Dirichlet problem, Indiana Univ. Math. J. 24 (1974/1975) 837–846.

[8] H. Brézis, R.E.L. Turner, On a class of superlinear elliptic problems, Comm. Partial Differential Equations 2 (1977) 601–614.

[9] H. Bueno, A. Zumpano, Remarks about the geometry of the Ambrosetti–Prodi problem, Nonlinear Anal. 48 (5) (2002) 651–661.

[10] J. Busca, B. Sirakov, Harnack type estimates for nonlinear elliptic systems and applications, Ann. Inst. H. Poincaré Anal. Non Linéaire 21 (5) (2004) 543–590.

[11] M. Calanchi, B. Ruf, Elliptic equations with one-sided critical growth, Electron. J. Differential Equations 2002 (89) (2002) 1–21.

[12] K.C. Chang, Variational method and sub- and super-solutions, Sci. Sinica, Ser. A 26 (1983) 1256–1265.

[13] K.C. Chang, Ambrosetti–Prodi type results in elliptic systems, Nonlinear Anal. 51 (4) (2002) 553–566.

[14] E.N. Dancer, On the ranges of certain weakly nonlinear elliptic partial differential equations, J. Math. Pures Appl. (9) 57 (4) (1978) 351–366.

[15] D.G. de Figueiredo, Lectures on boundary value problems of the Ambrosetti–Prodi type, in: Atas do 12 Seminario Brasileiro de Analyse (12th Brazilian Analysis Seminar), Universidade de Sao Paulo, Sao Paulo, 1980, pp. 230–292.

[16] D.G. de Figueiredo, S. Sirakov, Liouville type theorems, monotonicity results and a priori bounds for positive solutions of elliptic systems, Math. Ann. 333 (2) (2005) 231–260.

[17] D.G. de Figueiredo, S. Solimini, A variational approach to superlinear elliptic problems, Comm. Partial Differential Equations 9 (7) (1984) 699–717.

[18] D.G. de Figueiredo, J. Yang, Critical superlinear Ambrosetti–Prodi problems, Topol. Methods Nonlinear Anal. 14 (1999) 59–80.

[19] B. Gidas, J. Spruck, A priori bounds for positive solutions of nonlinear elliptic equations, Comm. Partial Differential Equations (6) (1981) 883–901.

[20] P. Hess, On a nonlinear elliptic boundary value problem of the Ambrosetti–Prodi type, Boll. Unione Mat. Ital. Sez. A (5) 17 (1) (1980) 187–192.

[21] J. Kazdan, F.W. Warner, Remarks on some quasilinear elliptic equations, Comm. Pure Appl. Math. 28 (5) (1975) 567–597.

[22] P. Korman, A.W. Leung, A general monotone scheme for elliptic systems with applications to ecological models, Proc. Roy. Soc. Edinburgh Sect. A 102 (3–4) (1986) 315–325.

[23] A. Marino, A.M. Micheletti, A. Pistoia, A nonsymmetric asymptotically linear elliptic problem, Topol. Methods Nonlinear Anal. 4 (2) (1994) 289–339.

[24] D.C. de Morais Filho, A variational approach to an Ambrosetti–Prodi type problem for a system of elliptic equations, Nonlinear. Anal. 26 (10) (1996) 1655–1668.

[25] D.C. de Morais Filho, An Ambrosetti–Prodi-type problem for an elliptic system of equations via monotone iteration method and Leray–Schauder degree theory, Abstr. Appl. Anal. 1 (2) (1996) 137–152.

[26] F.R. Pereira, Problemas do tipo de Ambrosetti–Prodi para sistemas envolvendo expoentes subcrítico e crítico, Tese de Doutorado, UNICAMP, 2005, 95 p.

[27] B. Rynne, Half-eigenvalues of elliptic operators, Proc. Roy. Soc. Edinburgh Sect. A 132 (2002) 1439–1451.

[28] G. Sweers, Strong positivity in $C(\overline{\Omega})$ for elliptic systems, Math. Z. 209 (1992) 251–271.

[29] H.T. Tehrani, A multiplicity result for the jumping nonlinearity problem, J. Differential Equations 188 (1) (2003) 272–305.

Available online at www.sciencedirect.com
ScienceDirect
ELSEVIER
Ann. I. H. Poincaré – AN 25 (2008) 149–161
ANNALES DE L'INSTITUT HENRI POINCARÉ
ANALYSE NON LINÉAIRE
www.elsevier.com/locate/anihpc

Solitary waves for some nonlinear Schrödinger systems

Ondes solitaires pour certains systèmes d'équations de Schrödinger non linéaires

Djairo G. de Figueiredo, Orlando Lopes *

IMECC-UNICAMP, Caixa Postal 6065, Campinas, SP-13083-859, Brasil

Received 6 February 2006; received in revised form 10 April 2006; accepted 11 November 2006

Available online 28 December 2006

Abstract

In this paper we study the existence of radially symmetric positive solutions in $H^1_{\text{rad}}(\mathbb{R}^N) \times H^1_{\text{rad}}(\mathbb{R}^N)$ of the elliptic system:

$$-\Delta u + u - (\alpha u^2 + \beta v^2)u = 0,$$
$$-\Delta v + \omega^2 v - (\beta u^2 + \gamma v^2)v = 0,$$

$N = 1, 2, 3$, where α and γ are positive constants (β will be allowed to be negative). This system has trivial solutions of the form $(\phi, 0)$ and $(0, \psi)$ where ϕ and ψ are nontrivial solutions of scalar equations. The existence of nontrivial solutions for some values of the parameters $\alpha, \beta, \gamma, \omega$ has been studied recently by several authors [A. Ambrosetti, E. Colorado, Bound and ground states of coupled nonlinear Schrödinger equations, C. R. Acad. Sci. Paris, Ser. I 342 (2006) 453–458; T.C. Lin, J. Wei, Ground states of N coupled nonlinear Schrödinger equations in R^n, $n \leqslant 3$, Comm. Math. Phys. 255 (2005) 629–653; T.C. Lin, J. Wei, Ground states of N coupled nonlinear Schrödinger equations in R^n, $n \leqslant 3$, Comm. Math. Phys., Erratum, in press; L. Maia, E. Montefusco, B. Pellacci, Positive solutions for a weakly coupled nonlinear Schrödinger system, preprint; B. Sirakov, Least energy solitary waves for a system of nonlinear Schrödinger equations in R^N, preprint; J. Yang, Classification of the solitary waves in coupled nonlinear Schrödinger equations, Physica D 108 (1997) 92–112]. For $N = 2, 3$, perhaps the most general existence result has been proved in [A. Ambrosetti, E. Colorado, Bound and ground states of coupled nonlinear Schrödinger equations, C. R. Acad. Sci. Paris, Ser. I 342 (2006) 453–458] under conditions which are equivalent to ours. Motivated by some numerical computations, we return to this problem and, using our approach, we give a more detailed description of the regions of parameters for which existence can be proved. In particular, based also on numerical evidence, we show that the shape of the region of the parameters for which existence of solution can be proved, changes drastically when we pass from dimensions $N = 1, 2$ to dimension $N = 3$. Our approach differs from the ones used before. It relies heavily on the spectral theory for linear elliptic operators. Furthermore, we also consider the case $N = 1$ which has to be treated more extensively due to some lack of compactness for even functions. This case has not been treated before.

* Corresponding author.
E-mail addresses: djairo@ime.unicamp.br (D.G. de Figueiredo), lopes@ime.unicamp.br (O. Lopes).

Originally published in: Ann. Inst. H. Poincaré, Anal. Non Linéaire 25 (2008), 149-161, © 2007 Elsevier Masson SAS. All rights reserved
Reprint by Springer-Verlag Berlin Heidelberg 2013 with kind permission,
DOI: 10.1007/978-3-319-02856-9_40

Résumé

Dans cet article, on étudie l'existence des solutions positives radialement symétriques dans $H^1_{\text{rad}}(\mathbb{R}^N) \times H^1_{\text{rad}}(\mathbb{R}^N)$ du système elliptique

$$-\Delta u + u - (\alpha u^2 + \beta v^2)u = 0,$$
$$-\Delta v + \omega^2 v - (\beta u^2 + \gamma v^2)v = 0,$$

$N = 1, 2, 3$ où α et γ sont des constantes positives (il est permis que β soit négatif). Ce système a des solutions triviales de la forme $(\phi, 0)$ et $(0, \psi)$ où ϕ et ψ sont des solutions non triviales des équations scalaires. L'existence de solutions non triviales pour certaines valeurs des paramètres $\alpha, \beta, \gamma, \omega$ a été étudiée récemment par plusieurs auteurs. Pour $N = 2, 3$ peut-être le résultat le plus général d'existence a été prouvé dans [A. Ambrosetti, E. Colorado, Bound and ground states of coupled nonlinear Schrödinger equations, C. R. Acad. Sci. Paris, Ser. I 342 (2006) 453–458] sous des conditions qui sont équivalentes aux nôtres. Motivé par quelques calculs numériques on retourne à ce problème et en utilisant notre approche on donne une description plus détaillée des régions de l'espace des paramètres pour lesquels l'existence peut être prouvée. En particulier, en se basant sur des résultats numériques, on démontre que la forme de la région de l'espace des paramètres pour lesquels l'existence de solutions peut être prouvée, change drastiquement quand on passe des dimensions $N = 1, 2$ à la dimension $N = 3$. Notre approche diffère des précédentes. Elle repose fortement sur la théorie spectrale des opérateurs linéaires. De plus, on considère aussi les cas $N = 1$ qui nécessite un traitement plus détaillé dût au manque de compacité pour les fonctions paires. Ce cas n'a pas été traité avant.

Keywords: Nonlinear Schrödinger systems; Nehari manifold

1. Introduction and statement of the results

In this paper we study the existence of radially symmetric positive solutions in $H^1_{\text{rad}}(\mathbb{R}^N) \times H^1_{\text{rad}}(\mathbb{R}^N)$ of the system

$$\begin{aligned} &-\Delta u + u - (\alpha u^2 + \beta v^2)u = 0,\\ &-\Delta v + \omega^2 v - (\beta u^2 + \gamma v^2)v = 0, \end{aligned} \tag{1.1}$$

$N = 1, 2, 3$, where α and γ are positive constants (β will be allowed to be negative). System (1.1) has motivated a large amount of research, both theoretically and numerically, due to the fact that it gives solitary waves for Schrödinger systems that govern phenomena in many physical problems, specially nonlinear optics (see [3] and [8] and the references therein).

System (1.1) has unique trivial solutions of the form $(\phi, 0)$ and $(0, \psi)$ where ϕ and ψ are radially symmetric positive (nontrivial) functions satisfying

$$-\Delta\phi + \phi - \alpha\phi^3 = 0$$

and

$$-\Delta\psi + \omega^2\psi - \gamma\psi^3 = 0.$$

By a nontrivial solution of (1.1) we mean a pair (u, v) such that $u \not\equiv 0 \not\equiv v$.

To state our main existence results we need some preliminaries. We denote by ϕ_0 the unique radial positive function satisfying

$$-\Delta\phi_0 + \phi_0 - \phi_0^3 = 0 \tag{1.2}$$

and for $\eta > 0$ we define

$$-\lambda_1(\eta) = \text{principal eigenvalue of } M_0 k = -\Delta k - \eta\phi_0^2 k. \tag{1.3}$$

The behavior of the function $\lambda_1(\eta)$ for η small depends on the dimension N. In fact, as we will see later, if $N = 1, 2$ then, for $\eta > 0$, $\lambda_1(\eta)$ is positive, increasing and

$$\lim_{\eta\to 0} \lambda_1(\eta) = 0 \quad \text{and} \quad \lim_{\eta\to\infty} \lambda_1(\eta) = \infty. \tag{1.4}$$

However, if $N = 3$ then there is a $\eta_0 > 0$ such that for $0 \leqslant \eta \leqslant \eta_0$ the whole spectrum of M_0 is $[0, \infty)$ and for $\eta > \eta_0$ we have $\lambda_1(\eta) > 0$, increasing with η and

$$\lim_{\eta \to \eta_0} \lambda_1(\eta) = 0 \quad \text{and} \quad \lim_{\eta \to \infty} \lambda_1(\eta) = \infty. \tag{1.5}$$

For $N = 3$ and $0 \leqslant \eta \leqslant \eta_0$ we set $\lambda_1(\eta) = 0$. From these properties of the function $\lambda_1(\eta)$ we see that for each α and γ fixed and $N = 1, 2, 3$ there is a unique $\beta > 0$ such that $\lambda_1(\beta/\alpha) = 1/\lambda_1(\beta/\gamma)$. This value will be denoted by $\beta(\alpha, \gamma)$. Our main result is the following:

Theorem 1.1.

(i) *Suppose first $N = 1, 2$. Then*
- *if $0 \leqslant \beta < \beta(\alpha, \gamma)$ then system* (1.1) *has a nontrivial positive radially symmetric solution if*

$$\lambda_1(\beta/\alpha) < \omega^2 < \frac{1}{\lambda_1(\beta/\gamma)} \quad (\textit{region A in Fig. 1}); \tag{1.6}$$

- *if $\beta(\alpha, \gamma) < \beta$ then system* (1.1) *has a nontrivial positive radially symmetric solution if*

$$\frac{1}{\lambda_1(\beta/\gamma)} < \omega^2 < \lambda_1(\beta/\alpha) \quad (\textit{region B in Fig. 1}). \tag{1.7}$$

(ii) *Suppose now $N = 3$ and $\alpha < \gamma$ (for the case $\alpha \geqslant \gamma$ the conclusions are similar). Let $\eta_0 > 0$ be the number previously defined. Then* (1.1) *has a nontrivial positive radially symmetric solution if the parameters β and ω^2 belong either to the region $A = A_1 \cup A_2 \cup A_3$ or region B in Fig. 2 where $\beta_1 = \eta_0\alpha$, $\beta_2 = \eta_0\gamma$ and the regions A_1, A_2 and A_3 are defined by the following inequalities involving the parameters $\beta < \beta(\alpha, \gamma)$ and ω^2:*

$$A_1 = \{0 \leqslant \beta \leqslant \beta_1, \omega^2 > 0\}, \qquad A_2 = \left\{\beta_1 < \beta \leqslant \beta_2, 0 < \omega^2 < \frac{1}{\lambda_1(\beta/\gamma)}\right\},$$

$$A_3 = \left\{\beta_2 < \beta < \beta(\alpha, \gamma), \lambda_1(\beta/\alpha) < \omega^2 < \frac{1}{\lambda_1(\beta/\gamma)}\right\}$$

and for $\beta > \beta(\alpha, \gamma)$ the region B is described by (1.7).

It is worthwhile to remark that the numerical experiments performed by [9] for system (1.1) in the case $N = 1$ and $\alpha = \gamma = 1$ have not detected existence of positive solutions outside the regions A and B. This may be an indication that, for $N = 1, 2, 3$, the regions A and B for which our existence results hold may be optimal in the sense that, outside them, positive solutions of (1.1) with finite energy do not exist. If this is indeed the case, then Figs. 1 and 2 show how the shape of the region of existence changes when we pass from dimensions $N = 1, 2$ to dimension $N = 3$.

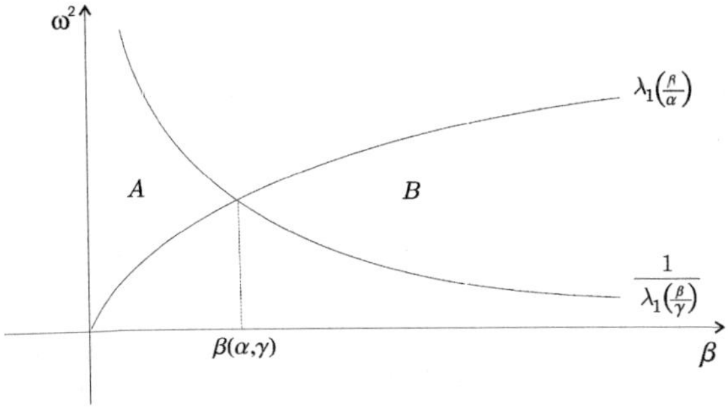

Fig. 1.

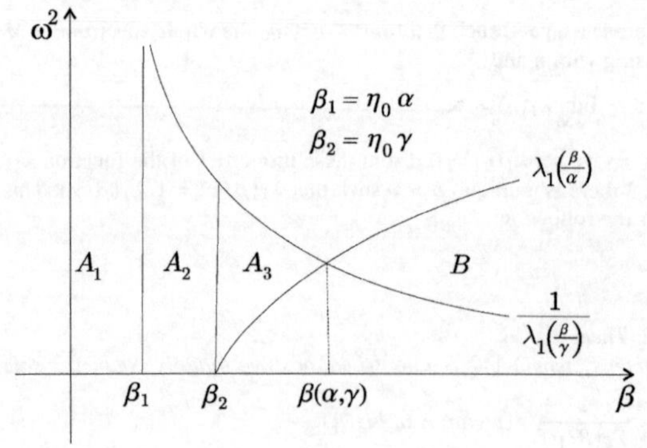

Fig. 2.

Next we give sufficient conditions for having the hypotheses of Theorem 1.1 satisfied in dimensions $N = 2, 3$ (the case $N = 1$ will be treated separately).

Theorem 1.2. *If*

$$\frac{\beta}{\alpha} < \min\{1, \omega^2\} \quad \text{and} \quad \frac{\beta}{\gamma} < \min\left\{1, \frac{1}{\omega^2}\right\} \tag{1.8}$$

then we are in region A of Figs. 1 *or* 2. *Hence, system* (1.1) *has a nontrivial positive solution.*

Theorem 1.3. *If*

$$\frac{\beta}{\alpha} > \max\{1, \omega^2\} \quad \text{and} \quad \frac{\beta}{\gamma} > \max\left\{1, \frac{1}{\omega^2}\right\} \tag{1.9}$$

then we are in region B of Figs. 1 *or* 2. *Hence, system* (1.1) *has a nontrivial positive solution.*

In the case $N = 1$, we have $\phi_0(x) = \sqrt{2}\,\mathrm{sech}(x)$ and the function $\lambda_1(\eta)$ is known explicitly and we have the following:

Theorem 1.4. *Suppose $N = 1$ and let $\beta(\alpha, \gamma) > 0$ be the unique solution of*

$$\frac{1}{2}\left(\sqrt{1 + \frac{8\beta}{\alpha}} - 1\right) = \frac{2}{\sqrt{1 + 8\beta/\gamma} - 1}. \tag{1.10}$$

Then system (1.1) *has a nontrivial positive solution if either $0 < \beta < \beta(\alpha, \gamma)$ and*

$$\frac{1}{2}\left(\sqrt{1 + \frac{8\beta}{\alpha}} - 1\right) < \omega < \frac{2}{\sqrt{1 + 8\beta/\gamma} - 1} \quad (\text{region } A); \tag{1.11}$$

or $\beta > \beta(\alpha, \gamma)$ and

$$\frac{2}{\sqrt{1 + 8\beta/\gamma} - 1} < \omega < \frac{1}{2}\left(\sqrt{1 + \frac{8\beta}{\alpha}} - 1\right) \quad (\text{region } B). \tag{1.12}$$

In the particular case $\alpha = \gamma = 1$, the existence region of Theorem 1.4 is precisely the region for which existence of positive solutions has been verified numerically in [9]. In that case, the curves intersect at $\beta = 1$ and $\omega = 1$.

Notice that if $\gamma < \alpha$ then, as we will see later, $\beta(\alpha, \gamma) > \sqrt{\alpha\gamma}$. In this case, our existence theorem applies to a region of the parameters that lies to the right of $\sqrt{\alpha\gamma}$ (see (1.20)).

Next we state a theorem which is a more geometric version of Theorem 1.1. This version applies to region A in both Figs. 1 and 2. First we recall that ϕ and ψ, respectively, are the unique positive radially symmetric decreasing functions tending to zero exponentially at infinity and satisfying the equations

$$-\Delta\phi + \phi - \alpha\phi^3 = 0 \tag{1.13}$$

and

$$-\Delta\psi + \omega^2\psi - \gamma\psi^3 = 0. \tag{1.14}$$

We define the selfadjoint operators

$$Lh = -\Delta h + \left(1 - \beta\psi^2\right)h \tag{1.15}$$

and

$$Mk = -\Delta k + \left(\omega^2 - \beta\phi^2\right)k \tag{1.16}$$

acting on functions belonging to $H^2_{\mathrm{rad}}(\mathbb{R}^N)$. Then we can state the following:

Theorem 1.5. *If $\beta > 0$ and the operators L and M are positive definite then system* (1.1) *has a nontrivial positive radially symmetric decreasing solution.*

As we will see, in the region A the trivial solutions $(\phi, 0)$ and $(0, \psi)$ have Morse index equal to one and in region B their Morse index is at least two. The case of Morse index larger than one has been considered in [5] for more general nonlinearities. Therefore, we focus our attention on region A.

The solutions of (1.1) are critical points of the functional:

$$E(u, v) = \int\limits_{\mathbb{R}^N} \left(\frac{1}{2}|\operatorname{grad} u|^2 + \frac{1}{2}|\operatorname{grad} v|^2 + \frac{u^2}{2} + \frac{\omega^2 v^2}{2} - \frac{\alpha u^4}{4} - \frac{\beta u^2 v^2}{2} - \frac{\gamma v^4}{4}\right) dx, \tag{1.17}$$

which is well defined and smooth in $H^1_{\mathrm{rad}}(\mathbb{R}^N) \times H^1_{\mathrm{rad}}(\mathbb{R}^N)$ for $N = 1, 2, 3$.

The first attempt to find nontrivial critical points of $E(u, v)$ is to minimize $E(u, v)$ on the Nehari manifold

$$\mathcal{N} = \Bigg\{(u, v) \in H^1_{\mathrm{rad}}(\mathbb{R}^N) \times H^1_{\mathrm{rad}}(\mathbb{R}^N), (u, v) \neq (0, 0): \\ I(u, v) \widehat{=} \int\limits_{\mathbb{R}^N} \left(|\operatorname{grad} u|^2 + |\operatorname{grad} v|^2 + u^2 + \omega^2 v^2 - \alpha u^4 - 2\beta u^2 v^2 - \gamma v^4\right) dx = 0\Bigg\}. \tag{1.18}$$

As we explain next, such a procedure does not give nontrivial critical points when the parameters lie in region A in Figs. 1 and 2.

First notice that if $(u, v) \in \mathcal{N}$ is a critical point of E restricted to $\mathcal{N}$, then (u, v) is a critical point of E because the Lagrange multiplier is zero. Moreover, the Morse index of a minimizer of $E(u, v)$ on $\mathcal{N}$ is at most one because it is a minimizer of a functional under one constraint (actually, in the present case, the Morse index of the minimizer is exactly one because if (u, v) solves (1.1) then $E''(u, v)((u, v), (u, v)) < 0$). For values of the parameters in the region B in Fig. 1 and in Fig. 2, the Morse index of each trivial solution $(\phi, 0)$ and $(0, \psi)$ is at least two (see [1]). Therefore, in this case, the minimizer of $E(u, v)$ on $\mathcal{N}$ gives indeed a nontrivial solution of (1.1).

However, in region A the Morse index of each trivial solutions $(\phi, 0)$ and $(0, \psi)$ is exactly one (see [1]). In this case, it is not clear that the minimization procedure produces a nontrivial solution of (1.1). In fact, for values of the parameters in region A, the level of the nontrivial solution is higher than the level of the trivial solutions $(\phi, 0)$ and $(0, \psi)$. Hence, such solution cannot be obtained by the minimization of $E(u, v)$ on $\mathcal{N}$.

In particular, in the narrower region $0 < \beta < \sqrt{\alpha\gamma}$, it is not difficult to show that the Morse index of any eventual nontrivial solution of (1.1) is at least two. Therefore, any method used to prove the existence of nontrivial solutions has to take this fact in account.

The existence of positive solutions of (1.1) has been also considered in [3] and [4] for $N = 2, 3$. Existence and nonexistence results are also proved in [8]. In both papers, the method of proving consists of minimizing the energy $E(u, v)$ on the double Nehari manifold given by the elements (u, v), $u \not\equiv 0$, $u \not\equiv 0$ satisfying:

$$\int_{\mathbb{R}^3} \left(|\operatorname{grad} u|^2 + u^2 - \alpha u^4 - \beta u^2 v^2\right) dx = 0,$$

$$\int_{\mathbb{R}^3} \left(|\operatorname{grad} v|^2 + \omega^2 u^2 - \beta u^2 v^2 - \gamma v^4\right) dx = 0. \tag{1.19}$$

In order to show that the minimizer is indeed a solution of (1.1) the assumption

$$\beta^2 < \alpha\gamma \tag{1.20}$$

is used in those papers. Here in this paper (and in [1] as well) such a condition is not needed. In fact, at least in the case $N = 1$, we can show existence of solution for values of β that are to the right of $\sqrt{\alpha\gamma}$.

As we have said before, our approach was motivated by the numerical experiments of [9] for $N = 1$ and $\alpha = \gamma = 1$. According to them, positive nontrivial solutions exist if either (1.11) or (1.12) with $\alpha = \gamma = 1$ hold. Since we are focusing our attention to region A, the first thing is to give a geometric meaning for condition (1.11). As we will see, it turns out that if (1.11) holds then the trivial solutions $(\phi, 0)$ and $(0, \psi)$ are local minimizers of the functional $E(u, v)$ restricted to the manifold $\mathcal{N}$. In other words, if we consider $E(u, v)$ as a map from $\mathcal{N}$ into $\mathbb{R}$, then under the condition (1.9), the functional $E(u, v)$ has a mountain pass geometry on the manifold $\mathcal{N}$.

In the general case, the region A in Theorem 1.1 is precisely the region where the functional $E(u, v)$ has the mountain pass geometry. Under certain conditions which are equivalent to ours, such mountain pass geometry has already been observed in [1] and, for $N = 2, 3$, the existence of solutions is proved there.

The case $N = 1$ requires a different approach because certain compactness argument in the space of radial functions fails in the one-dimensional case. Our approach consists in showing that the problem in a finite interval $(-n, n)$ with Dirichlet boundary conditions has a nontrivial solution if n is large and we pass to the limit as n tends to infinity. Careful estimates are needed to show that the limit is not trivial.

Remark 1.1. If $N = 2, 3$ and $-\sqrt{\alpha\gamma} < \beta < 0$ then we can prove the existence of positive nontrivial solutions but we do not know if they are decreasing. Due to this technical difficulty, our proof for the case $N = 1$ does not work for $\beta < 0$ in that range.

This paper is organized as follows:

- Section 2: proof of Theorem 1.5 for $N = 2, 3$;
- Section 3: proof of Theorem 1.5 in the case $N = 1$;
- Section 4: proof of Theorem 1.1;
- Section 5: proofs of Theorems 1.2, 1.3 and 1.4.

2. Proof of Theorem 1.5 for $N = 2, 3$

In this section we prove Theorem 1.5 in the case $N = 2, 3$. Under the assumptions of Theorem 1.5, it is easy to show that the trivial solutions $(\phi, 0)$ and $(0, \psi)$ are strict local minimizers of the functional $E(u, v)$ restricted to the manifold $\mathcal{N}$. Therefore a mountain pass argument can be used. If we define the quantities:

$$\gamma_1^2 = \inf \frac{\int_{\mathbb{R}^N} (|\operatorname{grad} k|^2 + \omega^2 k^2)\, dx}{\int_{\mathbb{R}^N} \phi^2 k^2\, dx}, \qquad \gamma_2^2 = \inf \frac{\int_{\mathbb{R}^N} (|\operatorname{grad} h|^2 + \omega^2 h^2)\, dx}{\int_{\mathbb{R}^N} \psi^2 h^2\, dx}$$

where ϕ and ψ are defined by (1.13) and (1.14) respectively, then it is also easy to see that L and M to be positive definite is equivalent to say that $\beta < \min(\gamma_1^2, \gamma_2^2)$. But, under these conditions, the mountain pass argument has been carried out in [1] and we omit it. Therefore, Theorem 1.5 is proved for $N = 2, 3$.

3. Proof of Theorem 1.5 in the case $N = 1$

We consider the ODE system:

$$-u'' + u - (\alpha u^2 + \beta v^2)u = 0,$$
$$-v'' + \omega^2 v - (\beta u^2 + \gamma v^2)v = 0 \tag{3.1}$$

and, as before, we denote by ϕ and ψ the unique nontrivial positive symmetric functions tending to zero exponentially at infinity and satisfying

$$-\phi'' + \phi - \alpha\phi^3 = 0. \tag{3.2}$$

and

$$-\psi'' + \omega^2\psi - \gamma\psi^3 = 0, \tag{3.3}$$

respectively. We also define the selfadjoint operators

$$Lh = -h'' + (1 - \beta\psi^2)h \tag{3.4}$$

and

$$Mk = -k'' + (\omega^2 - \beta\phi^2)k. \tag{3.5}$$

The proof of Theorem 1.1 for the ODE case will be done by approximation; namely, we first show that for n large the system

$$-u'' + u - (\alpha u^2 + \beta v^2)u = 0,$$
$$-v'' + \omega^2 v - (\beta u^2 + \gamma v^2)v = 0, \tag{3.6}$$
$$x \in [-n, n], \quad u(-n) = v(-n) = 0 = u(n) = v(n)$$

has a unique nontrivial positive solution (u_n, v_n) and that its limit as n tends to infinity converges to a nontrivial solution of (3.1).

Since ϕ solves (3.2) we have

$$-\frac{\phi'(x)^2}{2} + \frac{\phi^2}{2} - \frac{\alpha\phi(x)^4}{4} = 0, \tag{3.7}$$

and this implies

$$\phi(0) = \sqrt{\frac{2}{\alpha}}. \tag{3.8}$$

We denote by ϕ_n the approximate positive symmetric (unique) solution of the problem

$$-\phi_n'' + \phi_n - \alpha\phi_n^3 = 0, \qquad \phi(-n) = 0 = \phi(n). \tag{3.9}$$

If we define $a_n = \phi_n(0)$, it follows that $a_n > \phi(0) = \sqrt{2/\alpha}$. We also define $\tilde{\phi}_n$ as the extension of ϕ_n to the entire line setting it equal to zero outside the interval $[-n, n]$. $\tilde{\psi}_n$ is defined similarly.

Lemma 3.1. *As a_n converges to $\phi(0)$, the approximate solution $\tilde{\phi}_n$ converges to ϕ in the $H^1(\mathbb{R})$ norm. In particular, $\tilde{\phi}_n$ converges to ϕ uniformly on $(-\infty, \infty)$. A similar statement holds for $\tilde{\psi}_n$.*

Proof. Using ODE methods, we obtain some estimates for ϕ_n on the interval $[0, +\infty)$. By symmetry, the same estimates hold for $(-\infty, 0]$. From (3.9) we get:

$$\phi_n'^2(x) = \phi_n^2(x) - 2\alpha\phi_n^4(x) + c_n$$

where $c_n = \phi_n'^2(-n) > 0$. Since ϕ_n converges to ϕ uniformly on compact sets and $\phi(x)$ tends to zero as x tends to $+\infty$, there is are $x_1 > 0$ and n_0 such that for $x > x_1$ and $n > n_0$ we have

$$\phi_n^2(x) - 2\alpha\phi_n^4(x) + c_n \geqslant \frac{1}{4}(\phi_n^2(x) + c_n).$$

Moreover, $\phi_n'(x) < 0$, for $x > x_1$ and then

$$\frac{d\phi}{\sqrt{\phi_n^2(x) + c_n}} \leqslant -\frac{dx}{2}.$$

Integrating this last inequality we get

$$\log \frac{\phi_n(x) + \sqrt{\phi_n^2(x) + c_n}}{\phi_n(x_1) + \sqrt{\phi_n^2(x_1) + c_n}} \leqslant -\frac{1}{2}(x - x_1)$$

and then

$$\frac{\phi_n(x) + \sqrt{\phi_n^2(x) + c_n}}{\phi_n(x_1) + \sqrt{\phi_n^2(x_1) + c_n}} \leqslant \mathrm{e}^{-\frac{1}{2}(x-x_1)}.$$

From this last inequality we conclude that for $x_1 < x < n$ and some constant K we have

$$2\phi_n(x) \leqslant \phi_n(x) + \sqrt{\phi_n^2(x) + c_n} \leqslant K\mathrm{e}^{-\frac{1}{2}x}. \tag{3.10}$$

Clearly (3.10) implies that $\tilde{\phi}_n$ converges to ϕ uniformly on $[0, +\infty)$ and that the norm $L^p([0, +\infty))$ of $\tilde{\phi}_n$ is bounded for $1 \leqslant p \leqslant \infty$. Next, if we subtract (3.2) from (3.9), multiply the result by $\phi_n(x) - \phi(x)$ and integrate we get:

$$\int_{-n}^{n} \left[\left(\phi_n'(x) - \phi'(x)\right)^2 + \left(\phi_n(x) - \phi(x)\right)^2 \right] \mathrm{d}x = 2\alpha \int_{-n}^{n} \left(\phi_n(x) - \phi(x)\right)^2 \left(\phi_n^2(x) + \phi_n(x)\phi(x) + \phi^2(x)\right) \mathrm{d}x.$$

Since this last integral converges to zero (because $\tilde{\phi}_n$ converges to ϕ uniformly on $(-\infty, \infty)$), the lemma is proved. □

We will also need the following

Lemma 3.2. *The solution ϕ_n of* (3.9) *is unique and zero is not an eigenvalue of the linearized operator*

$$T_n h = -h'' + \left(1 - 3\phi_n^2\right)h$$

with domain $H^2(-n, n) \cap H_0^1(-n, n)$.

Proof. The uniqueness follows from the fact, that for the nonlinearity we are treating, the period of a solution of (3.9) is a decreasing function of the amplitude. There are several general results about that subject (see for instance [2] and the references therein) but in our case we can give a simple direct proof. In fact, changing the notation, if we have the equation $u'' + f(u) = 0$ and $F'(u) = f(u)$ and $u(x)$ is a solution with $u'(0) = 0$ and $u(0) = u_0$ then the "time" the solution takes to vanish is given by

$$X(u_0) = 2 \int_0^{u_0} \frac{\mathrm{d}u}{\sqrt{F(u_0) - F(u)}}.$$

In our case, $F(u) = -u^2/2 - \alpha u^4/4$ and $u_0 > \sqrt{2/\alpha}$; then

$$X(u_0) = 4 \int_0^{u_0} \frac{\mathrm{d}u}{\sqrt{(u_0^2 - u^2)(\alpha(u_0^2 + u^2) - 2)}}.$$

If we make the change of variable $u = tu_0$ we get

$$X(u_0) = 4 \int_0^{1} \frac{\mathrm{d}t}{\sqrt{(1 - t^2)(\alpha(1 + t^2)u_0^2 - 2))}}$$

and this shows that $X(u_0)$ is a decreasing function of u_0 and the uniqueness is proved.

To show the second part of the lemma, we multiply (3.9) by ϕ_n and integrating it we see that

$$\langle T_n\phi_n, \phi_n\rangle = -2\alpha \int_{-n}^{n} \phi_n^4 \,\mathrm{d}x < 0.$$

Therefore the principal eigenvalue of T_n is negative. Now suppose zero is an eigenvalue of T_n with eigenfunction $p(x)$. Then, besides vanishing at $-n$ and n, p has to vanish at an interior point of $(-n, n)$ because it is not the principal eigenfunction. Moreover, ϕ_n' also satisfies the linearized equation and then, by the Sturm oscillation theorem, ϕ_n' has at least two zeroes on $(-n, n)$ and this is a contradiction (because $\phi_n'(x)$ vanishes at $x = 0$ only) and the lemma is proved. □

Next we show the existence of a nontrivial even solution of (3.6) if n is large. As in the three-dimensional case, this will be done by a mountain pass argument on a manifold. We work in the space $H^1_{\mathrm{ev}}(\mathbb{R}) \times H^1_{\mathrm{ev}}(\mathbb{R})$, where the subscript ev stands for even, and we define the truncated energy

$$E_n(u, v) = \int_{-n}^{n} \left(\frac{1}{2}u'^2 + \frac{1}{2}v'^2 + \frac{u^2}{2} + \frac{\omega^2 v^2}{2} - \frac{\alpha u^4}{4} - \frac{\beta u^2 v^2}{2} - \frac{\gamma v^4}{4}\right) \mathrm{d}x \tag{3.11}$$

on $H^1_{\mathrm{ev},0}$ and the Nehari manifold

$$\mathcal{N}_n = \left\{(u, v) \in H^1_{\mathrm{ev},0}(\mathbb{R}) \times H^1_{\mathrm{ev},0}(\mathbb{R}), (u, v) \neq (0, 0): \int_{-n}^{n} \left(u'^2 + v'^2 + u^2 + \omega v^2 - \alpha u^4 - 2\beta u^2 v^2 - \gamma v^4\right) \mathrm{d}x = 0\right\}. \tag{3.12}$$

We also define the selfadjoint operators:

$$L_n h = -h'' + \left(1 - \beta\tilde{\psi}_n^2\right)h \tag{3.13}$$

and

$$M_n k = -k'' + \left(\omega^2 - \beta\tilde{\phi}_n^2\right)k \tag{3.14}$$

with domain $H^2_{\mathrm{ev}}(-n, n) \cap H^1_{\mathrm{ev},0}(-n, n)$. The functions $\tilde{\phi}_n$ and $\tilde{\psi}_n$ have been defined right before Lemma 3.1. Since by assumption the operators L and M are positive definite, using Lemma 3.1 we see that there are n_0 and a constant m such that for $n \geqslant n_0$ we have

$$\langle L_n h, h\rangle \geqslant m\|h\|^2, \qquad \langle M_n k, k\rangle \geqslant m\|k\|^2 \tag{3.15}$$

for any $h, k \in H^2_{\mathrm{ev}}(-n, n) \cap H^1_{\mathrm{ev},0}(-n, n)$. Arguing exactly as in the proof of Theorem 1.5 for the three-dimensional case (the verification of Palais–Smale condition is trivial for the case of bounded interval) we construct a sequence of nontrivial symmetric, positive and decreasing functions (u_n, v_n) satisfying

$$\begin{aligned} &-u_n'' + u_n - \left(\alpha u_n^2 + \beta v_n^2\right)u_n = 0, \\ &-v_n'' + \omega^2 v_n - \left(\beta u_n^2 + \gamma v_n^2\right)v_n = 0, \\ &x \in [-n, n], \quad u_n(-n) = v_n(-n) = 0 = u_n(n) = v_n(n). \end{aligned} \tag{3.16}$$

Moreover, if we consider the path $(t\phi_n, (1 - t)\psi_n)$ and we lift it to the manifold $\mathcal{N}_n$ defined by (3.12), on this lifted path the functional E_n is uniformly bounded because, according to Lemma 3.1, $\tilde{\phi}_n$ and $\tilde{\psi}_n$ tend to ϕ and ψ, respectively, in $H^1(\mathbb{R})$. Therefore, $E(u_n, v_n)$ is bounded by a constant independent of n and since (u_n, v_n) belongs to $\mathcal{N}_n$, we conclude that the $H^1(-n, n)$ norm of u_n and v_n are also bounded by a constant independent of n. In particular, the sequences $u_n(0)$ and $v_n(0)$ are bounded. Passing to a subsequence we can assume that $u_n(0)$ and $v_n(0)$ converge and our next proposition is:

Lemma 3.3. *$u_n(0)$ and $v_n(0)$ converge to a and b, respectively, where $a \neq 0 \neq b$.*

Proof. First we notice that both $u_n(0)$ and $v_n(0)$ cannot converge to zero because, otherwise, the $L^\infty(-n, n)$ norms of u_n and v_n would converge to zero (because u_n and v_n are decreasing) and this contradicts the definition (3.12) of $\mathcal{N}_n$. Therefore we have to exclude the possibility, say, $a \neq 0$ and $b = 0$. Suppose $a \neq 0$ and $b = 0$. Then $\|v_n\|_{L_\infty(-n,n)}$

tends to zero (because v_n is decreasing) and, multiplying the second equation of (3.16) by v_n and integrating, we see that the $H^1(-n,n)$ norm of v_n tends to zero. In particular, the extended functions $\widetilde{u_n}, \widetilde{v_n}$ converge to $(\phi, 0)$ uniformly on compact sets. If we multiply the first equation of (3.16) by u_n and define the function

$$f_n(x) = -\frac{u_n'^2}{2} + \frac{u_n^2}{2} - \frac{\alpha u_n^4}{4} - \frac{1}{2}\beta u_n^2 v_n^2$$

we see that $f_n'(x) = -2\beta u_n^2 v_n' > 0$ on $[0,n]$. Therefore for $x \in [0,n]$ we have $f_n(x) \leqslant f_n(n) = --u_n'^2(n)/2 = -c_n$ where c_n is a positive constant. Then for $x \in [0,n]$ we have

$$-\frac{u_n'^2}{2} + \frac{u_n^2}{2} - \frac{\alpha u_n^4}{4} - \frac{1}{2}\beta u_n^2 v_n^2 \leqslant -c_n$$

and

$$u_n'^2 \geqslant \frac{u_n^2}{2} - \frac{\alpha u_n^4}{4} - \frac{1}{2}\beta u_n^2 v_n^2 + c_n.$$

If we choose $x_2 > 0$ such that

$$\frac{u_n^2}{2} - \frac{\alpha u_n^4}{4} - \frac{1}{2}\beta u_n^2 v_n^2 + c_n \geqslant \frac{1}{4}\left(u_n^2 + c_n\right)$$

for $x > x_2$ then arguing exactly as in the proof of Lemma 3.1, we conclude that $\widetilde{u_n}$ converges to ϕ uniformly on $(-\infty, \infty)$. Then the is n_0 and a constant $m_1 > 0$ such that for $n > n_0$ we have $\langle M_n k, k\rangle \geqslant m_1\|k\|^2$ for any $k \in H^1(-\infty,\infty)$. However, if we multiply the second equation of (3.16) by v_n and integrate we get

$$\int_{-\infty}^{\infty} \left(\widetilde{v_n}'^2 + \left(\omega^2 - \beta\widetilde{u_n}^2\right)\widetilde{v_n}^2 - \gamma\widetilde{v_n}^4\right) dx = 0$$

and then

$$m_1\|\widetilde{v_n}\|^2 \leqslant \int_{-\infty}^{\infty} \left(\widetilde{v_n}'^2 + \left(\omega^2 - \beta\widetilde{u_n}^2\right)\widetilde{v_n}^2\right) dx \leqslant \gamma\|\widetilde{v_n}\|_{L_\infty}^2 \|\widetilde{v_n}\|^2$$

and then $v_n \equiv 0$ for n large and this is a contradiction and the lemma is proved. □

End of proof of Theorem 1.5 in the case $N = 1$. Since $\widetilde{u_n}$ and $\widetilde{v_n}$ are bounded in $H^1(-\infty,\infty)$, passing to a subsequence we can assume that $\widetilde{u_n}$ and $\widetilde{v_n}$ converge weakly in $H^1(-\infty,\infty)$ to functions u and v. Clearly (u,v) solves system (3.1) and, due to Lemma 3.3, none of them are trivial. So, Theorem 1.5 is also proved in the case $N = 1$.

4. Proof of Theorem 1.1

To begin the proof of Theorem 1.1, we recall that $-\lambda_1(\eta)$ is the principal eigenvalue of the operator

$$M_0 k = -\Delta k - \eta\phi_0^2 k \tag{4.1}$$

and we need the following proposition:

Lemma 4.1.

(i) *If $N = 1, 2$ then for $\eta > 0$ $\lambda_1(\eta)$ is positive, increasing and*

$$\lim_{\eta\to 0}\lambda_1(\eta) = 0 \quad \textit{and} \quad \lim_{\eta\to\infty}\lambda_1(\eta) = \infty. \tag{4.2}$$

(ii) *If $N = 3$ then there is a $\eta_0 > 0$ such that for $0 \leqslant \eta \leqslant \eta_0$ the whole spectrum of M_0 is $[0,\infty)$ and for $\eta > \eta_0$ we have $\lambda_1(\eta) > 0$, increasing with η and*

$$\lim_{\eta\to\eta_0}\lambda_1(\eta) = 0 \quad \textit{and} \quad \lim_{\eta\to\infty}\lambda_1(\eta) = \infty. \tag{4.3}$$

Proof. For $N = 1, 2$, the fact that $\lambda_1(\eta)$ is positive for $\eta > 0$ follows from Theorem XIII.11 (volume 4) in [6]. The other properties of the function $\lambda_1(\eta)$ are trivially verified. For $N = 3$ using Theorem XIII.9 (volume 4) in [6] we conclude that for $\eta > 0$ and small, the spectrum of M_0 is $[0, \infty)$ (recall that $\phi_0^2 \in L^{3/2}(\mathbb{R}^3)$ and then it is a Rollnik potential according to [7], Section X). Since $\lambda_1(\eta)$ is positive for $\eta > 0$ and large, the existence of η_0 is clear as well as the other properties of $\lambda_1(\eta)$. So the lemma is proved. □

Proof of Theorem 1.1. First of all we notice that, in terms of ϕ_0 defined by (1.2), the solutions ϕ and ψ of (1.13) and (1.14) are $\phi(x) = (1/\sqrt{\alpha})\phi_0(x)$ and $\psi(x) = (\omega/\sqrt{\gamma})\phi_0(\omega x)$, respectively. We define the operators

$$L_1 h = -\Delta h - \beta\psi^2 h = -\Delta h - \beta\frac{\omega^2}{\gamma}\phi_0^2(\omega x)h \tag{4.4}$$

and

$$M_1 k = -\Delta k - \beta\phi^2(x)k = -\Delta k - \frac{\beta}{\alpha}\phi_0^2(x)k. \tag{4.5}$$

Therefore

$$Lh = L_1 h + h \quad \text{and} \quad Mk = M_1 k + \omega^2 k. \tag{4.6}$$

Concerning the operator L_1, using a scaling argument we see that μ is an eigenvalue of L_1 if and only if μ/ω^2 is an eigenvalue of

$$L_2 h = -\Delta h - \frac{\beta}{\gamma}\phi_0^2(x)h. \tag{4.7}$$

Therefore, the principal eigenvalues of L_1 and M_1 are $-\lambda_2$ and $-\lambda_3$, respectively where λ_2 and λ_3 are given in terms of the function $\lambda_1(\eta)$ by

$$\lambda_2 = \omega^2\lambda_1\left(\frac{\beta}{\gamma}\right) \quad \text{and} \quad \lambda_3 = \lambda_1\left(\frac{\beta}{\alpha}\right), \tag{4.8}$$

respectively.

Consider first the case $N = 1, 2$. From the considerations above we can draw the following conclusions:

1. L is positive definite iff $1 > \omega^2\lambda_1(\beta/\alpha)$; in particular, if $1 < \omega^2\lambda_1(\beta/\alpha)$ then L has at least one negative eigenvalue.
2. M is positive definite iff $\omega^2 > \lambda_1(\beta/\gamma)$; in particular, if $\omega^2 < \lambda_1(\beta/\gamma)$ then M has at least one negative eigenvalue.

With those facts at hand, if we look at Fig. 1 we see that in region A the operators L and M are positive definite. Hence, the existence of solution follows from Theorem 1.5. Moreover, as we have pointed out in the introduction, in region B the Morse index of the trivial solutions $(\phi, 0)$ and $(0, \psi)$ is at least two. Hence, as we have also mentioned in the introduction, the existence of solutions follows from the minimization of $E(u, v)$ on $\mathcal{N}$. An alternative is to use the usual mountain pass theorem on the entire space $H^1_{rad}(\mathbb{R}^3) \times H^1_{rad}(\mathbb{R}^3)$ at the solution $(0, 0)$.

Suppose now $N = 3$ and $\alpha < \gamma$ and let us define $\beta_1 = \alpha\eta_0$ and $\beta_2 = \gamma\eta_0$. If $\beta \leqslant \beta_1 = \alpha\eta_0$, then $\beta/\alpha \leqslant \eta_0$ and $\beta/\gamma < \eta_0$. According to Lemma 4.1, this implies that the spectra of both L_1 and M_1 is $[0, \infty)$ and then L and M are positive definite. Therefore, in region A_1 Theorem 1.5 applies. If $\beta_1 \leqslant \beta \leqslant \beta_2$ then $\beta/\gamma \leqslant \eta_0$ and then the spectrum of L_1 is $[0, \infty)$ and this implies that L is positive definite. Moreover, for $\beta_1 \leqslant \beta \leqslant \beta_2$ the operator M is positive definite if $\omega^2 < 1/\lambda_1(\beta/\gamma)$. All this mean that in region A_2 Theorem 1.5 also applies. The analysis for the regions A_3 and B follows exactly as in the case $N = 1, 2$ and Theorem 1.1 is proved.

5. Proofs of Theorems 1.2, 1.3 and 1.4

Proof of Theorem 1.2. We define $f(x) = \phi(\lambda x)$ where $0 < \lambda < 1$ will be chosen later. Since $\phi(x)$ is decreasing, we have $f(x) > \phi(x)$ for $x \neq 0$. Moreover, f satisfies

$$-\Delta f + \lambda^2 f - \lambda^2\alpha f^3 = 0. \tag{5.1}$$

To impose $Mf > 0$ we use (1.20) and calculate

$$Mf = -\Delta f + (\omega^2 - \beta\phi^2)f = f(\omega^2 - \lambda^2 + \lambda^2\alpha f^2 - \beta\phi^2) > f(\omega^2 - \lambda^2 + \lambda^2\alpha\phi^2 - \beta\phi^2).$$

Therefore, $Mf > 0$ if there is a $\lambda < 1$ such that $\lambda^2 < \omega^2$ and $\beta < \lambda^2\alpha$. Clearly under the first assumption of Theorem 1.2 such λ exists. Also clearly, $f > 0$ and $Mf > 0$ implies that M is positive definite. The positivity of L is verified in a similar way and Theorem 1.2 is proved as a consequence of Theorem 1.5. □

Proof of Theorem 1.3. We define $f(x) = \phi(\lambda x)$ with $\lambda > 1$ and arguing as in the proof of Theorem 1.2 we can show that, under the assumptions of Theorem 1.3, it is possible to choose $\lambda > 1$ such that $Mf < 0$. This implies that M has at least one negative eigenvalue. According to Remark 2.1, the Morse index of the trivial solution $(0, \psi)$ is at least two. Similarly we show that the Morse index of the trivial solution $(\phi, 0)$ is at least two. Therefore, as a consequence of Theorem 1.1, Theorem 1.3 is proved. □

For the proof of Theorem 1.4 we need the following proposition taken from [9]:

Lemma 5.1. *If we define the operator*

$$P_1 h = -h'' + (\eta^2 - 2\beta\,\mathrm{sech}^2(x))h \tag{5.2}$$

where $\beta > 0$, then the eigenvalues λ of P_1 are given by $\lambda = \eta^2 - (s-n)^2$ where

$$s = \frac{1}{2}(\sqrt{1+8\beta} - 1), \tag{5.3}$$

n is a nonnegative integer and $s - n > 0$. In particular, P_1 is positive definite if $s < \eta$.

The proof of next lemma follows immediately from the preceding lemma and a scaling argument.

Lemma 5.2. *If*

$$P_2 h = -h'' + (\eta^2 - 2\beta c^2\,\mathrm{sech}^2(cx))h \tag{5.4}$$

then P_2 is positive definite if $s < \eta/c$ where s is define by (5.3).

Proof of Theorem 1.4. The solution of $-v'' + \omega^2 v - \gamma v^3 = 0$ is $\psi(x) = \sqrt{2/\gamma}\,\omega\,\mathrm{sech}(\omega x)$ and then $-\beta\psi^2(x) = -2(\beta/\gamma)\omega^2\,\mathrm{sech}^2(\omega x)$. Therefore, according to Lemma 5.2, the operator $Lh = -h'' + (1 - \beta\psi^2(x))h$ is positive definite if

$$\frac{1}{2}\left(\sqrt{1+\frac{8\beta}{\gamma}} - 1\right) < \frac{1}{\omega}.$$

Similarly, M is positive definite if

$$\frac{1}{2}\left(\sqrt{1+\frac{8\beta}{\alpha}} - 1\right) < \omega$$

and Theorem 1.4 is proved. □

Remark 5.1. An elementary calculation shows that

$$\frac{\sqrt{1+8y} - 1}{2} < \frac{2}{\sqrt{1+8/y} - 1} \quad \text{for } 0 < y < 1.$$

Therefore, if $\beta^2 = \alpha\gamma$ and we define $y = \beta/\alpha = \gamma/\beta$ and $y < 1$ (that is, $\gamma < \beta < \alpha$) we see that $\beta(\alpha, \gamma) > \sqrt{\alpha\gamma}$, where $\beta(\alpha, \gamma)$ is defined in the statement of Theorem 1.4. This means that part of the region of the parameters for which Theorem 1.1 applies lies to the right of $\sqrt{\alpha\gamma}$.

References

[1] A. Ambrosetti, E. Colorado, Bound and ground states of coupled nonlinear Schrödinger equations, C. R. Acad. Sci. Paris, Ser. I 342 (2006) 453–458.
[2] L. Bonorino, E. Brietzke, J.P. Lukaszczyk, C. Taschetto, Properties of the period function for some hamiltonian systems and homogeneous solutions of a semilinear elliptic equation, J. Differential Equations 214 (2005) 156–175.
[3] T.C. Lin, J. Wei, Ground states of N coupled nonlinear Schrödinger equations in R^n, $n \leqslant 3$, Comm. Math. Phys. 255 (2005) 629–653.
[4] T.C. Lin, J. Wei, Ground states of N coupled nonlinear Schrödinger equations in R^n, $n \leqslant 3$, Comm. Math. Phys., Erratum, in press.
[5] L. Maia, E. Montefusco, B. Pellacci, Positive solutions for a weakly coupled nonlinear Schrödinger system, preprint.
[6] M. Reed, B. Simon, Methods of Modern Mathematical Physics, IV, Analysis of Operators, Academic Press, New York, 1972.
[7] M. Reed, B. Simon, Methods of Modern Mathematical Physics, II, Fourier Analysis, Academic Press, New York, 1972.
[8] B. Sirakov, Least energy solitary waves for a system of nonlinear Schrödinger equations in R^N, preprint.
[9] J. Yang, Classification of the solitary waves in coupled nonlinear Schrödinger equations, Physica D 108 (1997) 92–112.

J. fixed point theory appl. *Online First*
© 2008 Birkhäuser Verlag Basel/Switzerland
DOI 10.1007/s11784-008-0069-2

Journal of Fixed Point Theory and Applications

Non-variational elliptic systems in dimension two: a priori bounds and existence of positive solutions

Djairo G. de Figueiredo, João Marcos do Ó and Bernhard Ruf

Dedicated to Felix Browder on the occasion of his 80th birthday

Abstract. We establish a priori bounds for positive solutions of semilinear elliptic systems of the form

$$\begin{cases} -\Delta u = g(x,u,v) & \text{in } \Omega, \\ -\Delta v = f(x,u,v) & \text{in } \Omega, \\ u > 0, \quad v > 0 & \text{in } \Omega, \\ u = v = 0 & \text{on } \partial\Omega, \end{cases}$$

where Ω is a bounded and smooth domain in $\mathbb{R}^2$. We obtain results concerning such bounds when f and g depend exponentially on u and v. Based on these bounds, existence of positive solutions is proved.

Mathematics Subject Classification (2000). 35J60, 35B45.

Keywords. Elliptic systems, a priori estimates.

1. Introduction

In this paper we establish the existence of positive solutions for semilinear elliptic systems of the form

$$\begin{cases} -\Delta u = g(x,u,v) & \text{in } \Omega, \\ -\Delta v = f(x,u,v) & \text{in } \Omega, \\ u > 0, \quad v > 0 & \text{in } \Omega, \\ u = v = 0 & \text{on } \partial\Omega, \end{cases} \tag{1.1}$$

where Ω is a bounded domain in $\mathbb{R}^2$ with smooth boundary $\partial\Omega$, Δ is the Laplace operator, and $f, g : \Omega \times \mathbb{R}^2 \to \mathbb{R}$ are continuous functions.

Originally published in: J. Fixed Point Theory Appl. 4 (2008), 77-96,

Reprint by Springer-Verlag Berlin Heidelberg 2013,
DOI: 10.1007/978-3-319-02856-9_41

This class of systems has been extensively studied when Ω is a bounded domain in $\mathbb{R}^N$ with $N \geq 3$ (see the recent survey paper [6]).

The case of dimension 2 presents some different features than the case of higher dimensions, and has been less studied. Some material in the case of dimension two has been collected in the recent survey paper [17]. Using variational methods, systems of the form

$$\begin{cases} -\Delta u = g(x, v) & \text{in } \Omega, \\ -\Delta v = f(x, u) & \text{in } \Omega, \\ u > 0, \quad v > 0 & \text{in } \Omega, \\ u = v = 0 & \text{on } \partial\Omega, \end{cases} \tag{1.2}$$

have been treated in [10], [16]. Existence of solutions to (1.2) has been obtained for nonlinearities both subcritical and critical in the sense of the Trudinger–Moser exponential growth [14].

Later, in [12], we investigated the question of a priori bounds for positive solutions of (1.2). Such bounds were obtained under much more restrictive hypotheses on the growth of the nonlinearities as compared with the above growth used in the variational approach.

This fact is in agreement with a similar phenomenon already observed in the scalar case by the authors of [3] and [5]. We recall that existence of solutions for the scalar problem

$$\begin{cases} -\Delta u = f(x, u) & \text{in } \Omega, \\ u > 0 & \text{in } \Omega, \\ u = 0 & \text{on } \partial\Omega, \end{cases} \tag{1.3}$$

can be proved for cases where the non-linearity $f(x, u)$ behaves at ∞ as $e^{\alpha|u|^2}$ (plus some additional conditions needed to use variational methods). However, the a priori bounds for positive solutions of (1.3) have been obtained in [3] and [5] under more restrictive conditions, namely f behaving as e^u at ∞. And in [3] it is even shown that in general one cannot expect to get the bounds if f has higher growth in the non-homogeneous case.

It is apparent that one is in the presence of phenomena similar to those observed in [18] in dimension $N \geq 3$. These phenomena show that a priori bounds in these dimensions, for general non-homogeneous nonlinearities, can be obtained only if $f(x, u)$ has polynomial growth at ∞ like u^p with $p \leq (N+1)/(N-1)$; this is the so-called Brezis–Turner exponent, which is the best exponent to get a priori bounds for positive solutions of (1.3) using the Hardy inequality (see [4]).

The purpose of the present paper is to study systems of the form (1.1) which are not necessarily variational. We will derive a priori estimates for positive solutions for non-linearities whose growth will be restricted by exponentials whose exponents are related and compensate each other. Then we use topological methods to prove the existence of positive solutions for these systems.

We assume throughout the paper that the non-linearities satisfy the hypothesis:

(H_0) $f, g : \Omega \times \mathbb{R} \times \mathbb{R} \to \mathbb{R}^+$ *are continuous.*

To start with, we consider *very weak solutions* of system (1.1), that is, *solutions in the sense of distributions*. More precisely, we assume that u and v are L^1-functions such that

$$f(x,u,v), g(x,u,v) \in L^1(\Omega)$$

and

$$\begin{aligned} -\int_\Omega u\Delta\varphi\, dx &= \int_\Omega g(x,u,v)\varphi\, dx, \\ -\int_\Omega v\Delta\varphi\, dx &= \int_\Omega f(x,u,v)\varphi\, dx, \end{aligned} \qquad \forall\varphi \in E = \{u \in C^2(\overline{\Omega}) : u|_{\partial\Omega} = 0\}.$$

Before stating our main results on a priori bounds, we present a result on the regularity of the distribution solutions of (1.1). For that purpose, a growth assumption on only one of the non-linearities suffices, namely,

(H_1) $f(x,t,s) \le ce^{pt+qs}$ *and* $g(x,t,s) \le e^{p't}h(s)$

or

(H_1') $g(x,t,s) \le ce^{pt+qs}$ *and* $f(x,t,s) \le e^{q's}k(t)$,

where c, p, q, p' and q' are some positive constants, and $h(s)$ and $k(t)$ are positive, continuous functions.

Theorem 1.1 (Regularity of distribution solutions). *Assume* (H_1) *or* (H_1'). *Then the distribution solutions of system* (1.1) *are in fact in* $L^\infty(\Omega)$.

It follows from Theorem 1.1 that any solution (u,v) of system (1.1) satisfies $\int_\Omega g(x,u,v)\,dx < \infty$ and $\int_\Omega f(x,u,v)\,dx < \infty$. Our next result states that under some additional hypotheses there is a *uniform bound* for these integrals.

Let $c \in C(\overline{\Omega})$ be a positive function and denote by $\lambda_1(c)$ the first eigenvalue of the following eigenvalue problem:

$$-\Delta\varphi = \lambda c(x)\varphi \quad \text{in } \Omega, \qquad \varphi = 0 \quad \text{on } \partial\Omega.$$

We now state the "superlinearity" assumptions we need.

(H_2) *There exist positive functions* $a, b \in C(\overline{\Omega})$ *and constants* a_1, b_1 *with*

$$f(x,t,s) \ge a(x)t - a_1 \quad \text{and} \quad g(x,t,s) \ge b(x)s - b_1$$

and such that

$$a(x)\lambda_1(a)^2 < b(x), \qquad x \in \overline{\Omega}. \tag{1.4}$$

(H_2') *There exist positive functions* $a, b \in C(\overline{\Omega})$ *and constants* a_1, b_1 *with*

$$f(x,t,s) \ge a(x)t - a_1 \quad \text{and} \quad g(x,t,s) \ge b(x)s - b_1$$

and such that

$$b(x)\lambda_1(b)^2 < a(x), \qquad x \in \overline{\Omega}. \tag{1.5}$$

The next assumption concerns the behaviour of the nonlinearities near the boundary:

(H_3) *For some* $r > 0$,

$$\frac{\partial}{\partial t} f(x,t,s) \geq 0 \quad \text{and} \quad \frac{\partial}{\partial s} g(x,t,s) \geq 0 \quad \text{in } \Omega_r \times \mathbb{R} \times \mathbb{R},$$

where $\Omega_r := \{x \in \overline{\Omega} : \operatorname{dist}(x, \partial\overline{\Omega}) \leq r\}$.

We note that condition (H_3) will allow the use of the maximum principle for cooperative systems; this is a basic tool in order to apply the moving planes technique (cf. [1], [8], [19]).

Due to the fact that we are considering non-autonomous problems, we also need geometric assumptions concerning the behavior of f and g with respect to x near the boundary.

(H_4) (Ω a convex domain) *There exist* $r, \delta > 0$ *such that* $g(\cdot,t,s), f(\cdot,t,s) \in C^1(\Omega_r)$, *and*

$$\nabla_x g(x,t,s) \cdot \theta \leq 0 \quad \text{and} \quad \nabla_x f(x,t,s) \cdot \theta \leq 0$$

for all $x \in \Omega_r$, $t, s \geq 0$, *and unit vectors* θ *such that* $|\theta - \nu(\bar{x})| < \delta$, *where* $\bar{x}$ *is the closest point to* x *in* $\partial\Omega$ *and* $\nu(\bar{x})$ *denotes the unit outward normal to* $\partial\Omega$ *at the point* $\bar{x}$.

If Ω is not convex, we use the Kelvin transform as in [5] to reduce the problem to the convex case. So we follow [5] and assume

(H_5) (Ω a general domain) *There exist* $r, C > 0$ *such that* $g(\cdot,t,s), f(\cdot,t,s) \in C^1(\Omega_r)$ *and*

$$|\nabla_x g(x,t,s)| \leq C g(x,t,s) \quad \text{and} \quad |\nabla_x f(x,t)| \leq C f(x,t,s)$$

for all $x \in \Omega_r$ *and* $t, s \geq 0$.

Theorem 1.2 (Uniform estimates). *Assume* (H_1) (*or* (H_1')), (H_2) (*or* (H_2')), (H_3), (H_4) *and* Ω *convex* (*or* (H_5)). *Then there exists a positive constant* C, *depending only on* f, g *and* Ω, *such that*

$$\int_\Omega g(x,u,v)\,dx \leq C \quad \text{and} \quad \int_\Omega f(x,u,v)\,dx \leq C$$

for all solutions (u,v) *of system* (1.1).

In order to obtain a priori bounds for the solutions of system (1.1) we have to assume further conditions regarding the growth at infinity of the non-linearities f and g:

(H_6) *There exists a constant* c *such that*

$$g(x,t,s) \leq c e^{t^\alpha + s^\beta} \quad \text{and} \quad f(x,t,s) \leq c e^{t^{\beta'} + s^{\alpha'}},$$

where

(a) $0 \leq \alpha \leq \beta$ *and* $0 \leq \alpha' \leq \beta'$,

(b) $\alpha + \beta < 2$ *and* $\alpha' + \beta' < 2$,
(c) $\beta + \beta' < 2$.

Remark 1.1. From our main growth assumption (H_6) it follows that both (H_1) and (H_1') are satisfied.

Example 1.1. The nonlinearities below satisfy condition (H_6):

$$f(x,t,s) = e^{t+s^{1/2}}, \qquad g(x,t,s) = e^{t^{1/2}+s^{3/4}}$$

Remark 1.2. If $f(x,t,s) \le c' e^{pt^\alpha + qt^\beta}$ for some $p, q > 1$, then condition (H_6) will be satisfied for suitable choices of c and $\alpha < \tilde{\alpha} < 1$, $\beta < \tilde{\beta} < 1$, and similarly for $g(x,t,s)$. So these conditions are not more general than (H_6).

The next theorem gives a priori estimates for positive solutions of system (1.1) under the growth restrictions (H_6):

Theorem 1.3. *Assume (H_2) (or (H_2')), (H_3), (H_4) and Ω convex (or (H_5)), and (H_6). Then there exists a constant $C > 0$ such that*

$$\|u\|_{L^\infty} \le C \quad \text{and} \quad \|v\|_{L^\infty} \le C$$

for all solutions (u,v) of system (1.1).

Remark 1.3. 1) Condition (H_6)(a) says that the system is *strongly coupled.*

2) If $\alpha = 0 = \alpha'$, then the system reduces to

$$-\Delta u = g(x,v), \qquad -\Delta v = f(x,u) \tag{1.6}$$

which was considered in [11]; indeed, in this case Theorem 1.3 contains and generalizes Theorems 1.4 and 1.5 from that article, which covers the case $\beta + \beta' < 2$.

3) The general case $\beta + \beta' = 2$ (and $\alpha = \alpha' = 0$) was left open in [11], except for the particular case $\beta = \beta' = 1$, for which it was possible to obtain a priori bounds following the techniques developed by Brezis–Merle [3].

Here we come back to this situation, which we call "critical growth", and generalize it to the case where both non linearities depend on the variables u and v, both with exponential growth, i.e. with $\beta = \beta' = 1$ together with $\alpha = \alpha' = 1$; see the next result.

Theorem 1.4. *Assume (H_2) (or (H_2')), (H_3), (H_4) and Ω convex (or (H_5)), and assume that $f(x,t,s)$ and $g(x,t,s)$ satisfy:*

(H_7) *There exist constants $c, d > 0$ such that*

$$ce^{t+s} \le g(x,t,s) \le de^{t+s} \quad \text{and} \quad 0 \le f(x,t,s) \le de^{t+s}.$$

Then all positive solutions (u,v) of system (1.1) are uniformly bounded in L^∞.

Example 1.2. The nonlinearities below satisfy condition (H_7):

$$g(x,t,s) = be^{t+s}, \qquad f(x,t,s) = \sin^2(x)e^{t+s},$$

where $b > \sin^2(x) \cdot \lambda_1^2(\sin^2(x))$ (see (H_2)).

Finally, we give an existence theorem. Denote by

$$(-\Delta)^{-1} : C^0(\overline{\Omega}) \to C^0(\overline{\Omega}), \qquad y \mapsto u \text{ with } -\Delta u = y,\ u|_{\partial\Omega} = 0,$$

the inverse Laplace operator, with M such that

$$\|(-\Delta)^{-1} y\|_{C^0} \le M\|y\|_{C^0}, \qquad \forall y \in C^0(\overline{\Omega}). \tag{1.7}$$

Theorem 1.5. *Assume that f and g satisfy the hypotheses of Theorem 1.3 or 1.4, and in addition:*

(H_8) *There exists $\varepsilon > 0$ and constants c_1, c_2 with*

$$f(x,t,s) \le c_1(|t| + |s|) \quad \text{and} \quad g(x,t,s) \le c_2(|t| + |s|) \quad \text{for } |t| + |s| \le \varepsilon,$$

and with $M(c_1 + c_2) < 1$.

Then system (1.1) has a nontrivial solution (u, v).

2. Regularity of distribution solutions

For easy reference, we state a result due to Brezis–Merle [3] which will be used to prove Theorem 1.1 above, and also in the proofs of Theorems 1.3 and 1.4.

Proposition 2.1 (Brezis–Merle). *Let u be a distribution solution of the linear equation*

$$\begin{cases} -\Delta u = h(x) & \text{in } \Omega, \\ u = 0 & \text{on } \partial\Omega, \end{cases} \tag{2.1}$$

where Ω is a bounded domain in $\mathbb{R}^2$, and $h \in L^1(\Omega)$. Then

(i) *for every $\delta \in (0, 4\pi)$, we have*

$$\int_\Omega \exp\left(\frac{(4\pi - \delta)|u(x)|}{\|h\|_{L^1}}\right) dx \le \frac{4\pi^2}{\delta}(\operatorname{diam}\Omega)^2,$$

(ii) *for every $k > 0$, $e^{ku} \in L^1(\Omega)$.*

Proof of Theorem 1.1. Let (u, v) be a distribution solution of system (1.1). By Proposition 2.1(ii) it follows that e^u, e^v are in L^p for every $p \ge 1$. Consequently, if we assume (H_1), i.e. $f(x,u,v) \le ce^{pu+qv}$, then

$$\int_\Omega |f(x,u,v)|^\alpha\, dx \le c\int_\Omega e^{\alpha pu} \cdot e^{\alpha qv}\, dx \le \left(\int_\Omega e^{2\alpha pu}\, dx\right)^{1/2} \left(\int_\Omega e^{2\alpha qv}\, dx\right)^{1/2},$$

that is, $f(x,u,v) \in L^\alpha$ for every $\alpha > 1$. By L^p-estimates it follows from the second equation in system (1.1) that $v \in W^{2,\alpha}(\Omega)$ for all $\alpha > 1$, and hence $v \in L^\infty$. Using now the first equation in (1.1) we conclude that also $u \in L^\infty$.

Using similar arguments we come to the same conclusions if (H_1') is assumed instead of (H_1). □

Remark 2.1. As a consequence of Theorem 1.1 and standard regularity results for elliptic equations we conclude that solutions of (1.1) in the distribution sense are, in fact, classical solutions.

3. Proof of Theorem 1.2 (uniform estimates)

Let φ_1^b be an eigenfunction associated to the first eigenvalue $\lambda_1(b)$ of the eigenvalue problem

$$-\Delta\varphi_1^b = \lambda_1(b)b(x)\varphi_1^b \quad \text{in } \Omega \quad \text{and} \quad \varphi_1^b = 0 \quad \text{on } \partial\Omega$$

which is chosen in such a way that $\varphi_1^b > 0$ and $\int_\Omega |\varphi_1^b|^2 = 1$.

Lemma 3.1. *Assume* (H_1) *and* (H_2') *(or* (H_2)*). Then for any solution* (u,v) *of system* (1.1),

$$\int_\Omega g(x,u,v)\varphi_1^b\,dx \le C, \qquad \int_\Omega f(x,u,v)\varphi_1^b\,dx \le C, \tag{3.1}$$

where the constant C *depends only on* f, g *and* Ω.

Proof. From assumption (H_2'), it follows that one can find an $\varepsilon > 0$ such that

$$a(x) \ge (\lambda_1(b)^2 + \varepsilon)b(x), \quad x \in \overline{\Omega}. \tag{3.2}$$

Multiplying the first equation in (1.1) by φ_1^b, integrating by parts and using (H_2') one obtains

$$\begin{aligned} \lambda_1(b)\int_\Omega ub(x)\varphi_1^b\,dx &= \int_\Omega g(x,u,v)\varphi_1^b\,dx \\ &\ge \int_\Omega b(x)v\varphi_1^b\,dx - \int_\Omega b_1(x)\varphi_1^b\,dx. \end{aligned} \tag{3.3}$$

Multiplying the second equation in (1.1) by φ_1^b, integrating by parts and using again (H_2') one obtains

$$\begin{aligned} \lambda_1(b)\int_\Omega vb(x)\varphi_1^b\,dx &= \int_\Omega f(x,u,v)\varphi_1^b\,dx \\ &\ge \int_\Omega a(x)u\varphi_1^b\,dx - \int_\Omega a_1(x)\varphi_1^b\,dx. \end{aligned} \tag{3.4}$$

Using (3.4) and (3.2), we get

$$\lambda_1(b)\int_\Omega vb(x)\varphi_1^b\,dx \ge (\lambda_1(b)^2+\varepsilon)\int_\Omega b(x)u\varphi_1^b\,dx - \int_\Omega a_1(x)\varphi_1^b\,dx. \tag{3.5}$$

Next using (3.5) and (3.3) we obtain

$$\lambda_1(b)\int_\Omega vb(x)\varphi_1^b\,dx \ge \frac{\lambda_1(b)^2+\varepsilon}{\lambda_1(b)}\int_\Omega vb(x)\varphi_1^b\,dx - C. \tag{3.6}$$

So it follows that $\int_\Omega vb(x)\varphi_1^b \le C$, and hence by (3.4), $\int_\Omega f(x,u,v)\varphi_1^b \le c$, and also $\int_\Omega a(x)u\varphi_1^b \le c$, from which by (3.3) also $\int_\Omega g(x,u,v)\varphi_1^b\,dx \le c$. □

Lemma 3.2. *Assume conditions* (H_1) *(or* (H_1')*),* (H_3)*,* (H_4) *and* Ω *convex (or* (H_5)*). Then there exist* $r,\delta > 0$ *such that*

$$\nabla u(x)\cdot\theta \le 0 \quad \text{and} \quad \nabla v(x)\cdot\theta \le 0 \qquad \text{for all } x \in \Omega_r,\ |\theta - \nu(x)| < \delta,$$

for each solution (u,v) *of* (1.1)*, where* Ω_r *is defined in* (H_3)*;* θ,ν *are as in* (H_4).

Proof. We can assume, without loss of generality, that

$$\Omega \subset \mathbb{R}^2_+ := \{(x,y) \in \mathbb{R}^2 : x > 0\} \quad \text{and} \quad (0,0) \in \partial\Omega.$$

Now, we consider

$$\begin{aligned} T_\lambda &:= \{(x,y) : x = \lambda\} && \text{(the hyperplane)},\\ \Sigma_\lambda &:= \{(x,y) \in \Omega : x < \lambda\} && \text{(the cap)},\\ \Sigma'_\lambda &:= \{(x_\lambda, y) : (x,y) \in \Sigma_\lambda\} && \text{(the reflected cap)}, \end{aligned}$$

where

$$x_\lambda := 2\lambda - x.$$

It follows that there exists $\overline{\lambda}$ such that $\Sigma_\lambda \cup \Sigma'_\lambda \subset \Omega_r$ for each $0 < \lambda < \overline{\lambda}$. In fact, this $\overline{\lambda}$ depends only on r and not on the particular point on the boundary.

For $0 < \lambda < \overline{\lambda}$, define in Σ_λ the auxiliary functions

$$w_\lambda(x,y) := u(x_\lambda, y) - u(x,y), \qquad z_\lambda(x,y) := v(x_\lambda, y) - v(x,y).$$

Then we have

$$\begin{aligned} \Delta w_\lambda &= \Delta u(x_\lambda, y) - \Delta u(x,y)\\ &= -g((x_\lambda,y), u(x_\lambda,y), v(x_\lambda,y)) + g((x,y), u(x,y), v(x,y))\\ &= -g((x_\lambda,y), u(x_\lambda,y), v(x_\lambda,y)) + g((x_\lambda,y), u(x,y), v(x_\lambda,y))\\ &\quad -g((x_\lambda,y), u(x,y), v(x_\lambda,y)) + g((x_\lambda,y), u(x,y), v(x,y))\\ &\quad -g((x_\lambda,y), u(x,y), v(x,y)) + g((x,y), u(x,y), v(x,y))\\ &=: J_1 + J_2 + J_3. \end{aligned}$$

It follows from (H_4) that $J_3 < 0$. The other two expressions are evaluated using the mean value theorem, namely

$$J_1 = -c_\lambda(x)[u(x_\lambda,y) - u(x,y)], \qquad J_2 = -d_\lambda(x)[v(x_\lambda,y) - v(x,y)],$$

where

$$c_\lambda(x) = \frac{\partial g}{\partial t}((x_\lambda,y), \zeta, v(x_\lambda,y)), \qquad d_\lambda(x) = \frac{\partial g}{\partial s}((x_\lambda,y), u(x,y), \xi),$$

where ζ is some value between $u(x_\lambda, y)$ and $u(x,y)$. Similarly for ξ.

Observe that $c_\lambda(x)$ and $d_\lambda(x)$ are non-negative in view of (H_3). With a similar computation for z_λ, we come to the system

$$-\Delta w_\lambda \ge c_\lambda w_\lambda + d_\lambda z_\lambda, \qquad -\Delta z_\lambda \ge c'_\lambda w_\lambda + d'_\lambda z_\lambda,$$

in a neighborhood of the boundary, where the coefficients are all non-negative in that neighborhood. By the maximum principle for cooperative systems (see [15], [13]) w_λ, z_λ are positive, and so there exists $\epsilon > 0$ such that u and v are increasing in Ω_ϵ. At this point, the proof follows the same steps as in the proof of Theorem 1.1 in [9]. □

Proof of Theorem 1.2. From Lemma 3.2 and an argument similar to the one in the proof of Theorem 1.1 in [9] we conclude that there exist $\varepsilon > 0$ and $C > 0$ such that

$$\|u\|_{L^\infty(\Omega_\varepsilon)} \leq C \quad \text{and} \quad \|v\|_{L^\infty(\Omega_\varepsilon)} \leq C$$

for all solutions (u, v) of system (1.1).

Finally, let $a := \inf\{\varphi_1^b(x) : x \in \overline{\Omega} \setminus \Omega_\varepsilon\}$, and estimate

$$\begin{aligned}\int_\Omega f(x,u,v)\,dx &= \int_{\Omega_\varepsilon} f(x,u,v)\,dx + \int_{\Omega\setminus\Omega_\varepsilon} f(x,u,v)\,dx \\ &\leq C + \frac{1}{a}\int_{\Omega\setminus\Omega_\varepsilon} f(x,u,v)\varphi_1^b(x)\,dx \\ &\leq C + \frac{1}{a}\int_\Omega f(x,u,v)\varphi_1^b(x)\,dx \leq C,\end{aligned}$$

by Lemma 3.1, and similarly for $\int_\Omega g(x,u,v)\,dx$. This completes the proof of Theorem 1.2. □

4. Proof of Theorem 1.3

In this section we rely on two inequalities, some kind of Young inequalities. The first one was introduced in [10] to treat simpler elliptic systems in dimension two. The second one generalizes the first one to the case when there are three terms.

Proposition 4.1. *For any $a, s, t \geq 1$ we have*

$$st \leq e^{t^a} - 1 + s(\log s)^{1/a}.$$

Proposition 4.2. *For any $a, b, r, s, t \geq 1$ we have*

$$rst \leq e^{t^a} + e^{r^b} + s(\log s)^{1/a+1/b}.$$

Proof. We are going to prove that for any $a, b, s \geq 1$ fixed we have

$$\sup_{r,t>1}\{rst - e^{t^a} - e^{r^b}\} \leq s(\log s)^{1/a+1/b}.$$

Let $h(r,t) := rst - e^{t^a} - e^{r^b}$. From

$$\frac{\partial}{\partial r}h(r,t) = t_s s - b r_s^{b-1} e^{r_s^b} = 0, \qquad \frac{\partial}{\partial t}h(r,t) = r_s s - a t_s^{a-1} e^{t_s^a} = 0$$

we obtain

$$\begin{aligned}\log t_s + \log s &= \log(b r_s^{b-1} e^{r_s^b}) = r_s^b + \log(b r_s^{b-1}) \geq r_s^b, \\ \log r_s + \log s &= \log(a t_s^{a-1} e^{t_s^a}) = t_s^a + \log(a t_s^{a-1}) \geq t_s^a,\end{aligned}$$

since $a t_s^{a-1} \geq 1$ and $b t_s^{b-1} \geq 1$. Thus

$$r_s \leq (\log t_s + \log s)^{1/b}, \qquad t_s \leq (\log r_s + \log s)^{1/a}.$$

Therefore,

$$h(r,t) \le s r_s t_s \le s(\log t_s + \log s)^{1/b}(\log r_s + \log s)^{1/a} \\ \le s\,(\log s)^{1/a+1/b},$$

and the proof is complete. □

Proof of Theorem 1.3. First we recall that by Theorem 1.1 each solution (u,v) of (1.1) belongs to $(L^\infty(\overline{\Omega}))^2$, and hence to $(W^{1,2}(\Omega))^2$. Let η, η' be positive parameters to be determined later. Using the first equation of system (1.1) we obtain

$$\int_\Omega |\nabla u|^2\,dx = \int_\Omega u g(x,u,v)\,dx \\ \le c + \int_{\Omega_1} \frac{g(x,u,v)}{u^\eta + v^{\eta'}}(u^\eta + v^{\eta'})u = c + I_1 + I_2,$$

where

$$I_1 := \int_{\Omega_1} \frac{g(x,u,v)}{u^\eta + v^{\eta'}} u^{\eta+1} \quad \text{and} \quad I_2 := \int_{\Omega_1} \frac{g(x,u,v)}{u^\eta + v^{\eta'}} v^{\eta'} u$$

with

$$\Omega_1 := \{x \in \Omega : u(x) \ge 1 \text{ or } v(x) \ge 1\}.$$

First we estimate I_1. We have

$$I_1 \le \|u\|^{\eta+1} \int_{\Omega_1} \frac{g(x,u,v)}{u^\eta + v^{\eta'}} \left(\frac{u}{\|u\|}\right)^{\eta+1},$$

and using Proposition 4.1 with $a = \gamma$, we obtain

$$I_1 \le \|u\|^{\eta+1} \left\{ \int_{\Omega_1} e^{(u/\|u\|)^{(\eta+1)\gamma}} + \int_{\Omega_1} \frac{g(x,u,v)}{u^\eta + v^{\eta'}} \left(\log \frac{g(x,u,v)}{u^\eta + v^{\eta'}}\right)^{1/\gamma} \right\},$$

The first term above is bounded by the Trudinger–Moser inequality [14] if we take

$$(\eta+1)\gamma \le 2. \tag{4.1}$$

To estimate the second term we notice that by assumption (H_6),

$$\int_{\Omega_1} \frac{g(x,u,v)}{u^\eta + v^{\eta'}} \left(\log \frac{g(x,u,v)}{u^\eta + v^{\eta'}}\right)^{1/\gamma} \le \int_{\Omega_1} \frac{g(x,u,v)}{u^\eta + v^{\eta'}} (\log(e^{u^\alpha + v^\beta}))^{1/\gamma} \\ \le C \int_{\Omega_1} \frac{g(x,u,v)}{u^\eta + v^{\eta'}} (u^{\alpha/\gamma} + v^{\beta/\gamma}) \le C \int_{\Omega_1} g(x,u,v) \le C$$

if we take

$$\eta \ge \alpha/\gamma \quad \text{and} \quad \eta' \ge \beta/\gamma, \tag{4.2}$$

and where we have used the elementary inequality

$$(u^\alpha + v^\beta)^{1/\gamma} \le C(u^{\alpha/\gamma} + v^{\beta/\gamma}).$$

Thus we have

$$I_1 \le C(\|u\|^{\eta+1} + 1).$$

Next we estimate I_2. By Proposition 4.2 we get

$$I_2 \le C\|u\|\,\|v\|^{\eta'}\int_{\Omega_1}\frac{g(x,u,v)}{u^\eta+v^{\eta'}}\left(\frac{v}{\|v\|}\right)^{\eta'}\left(\frac{u}{\|u\|}\right)$$
$$\le C\|u\|\,\|v\|^{\eta'}\left\{\int_{\Omega_1}e^{(v/\|v\|)^{\eta' a}}+\int_{\Omega_1}e^{(u/\|u\|)^b}+J_1\right\},$$

where

$$J_1 := \int_{\Omega_1}\frac{g(x,u,v)}{u^\eta+v^{\eta'}}\left(\log\frac{g(x,u,v)}{u^\eta+v^{\eta'}}\right)^{a^{-1}+b^{-1}}.$$

Using the Trudinger–Moser inequality [14] and taking

$$\eta' a \le 2 \quad \text{and} \quad b \le 2 \tag{4.3}$$

we obtain

$$I_2 \le\ C\|u\|\,\|v\|^{\eta'}\{C+J_1\}.$$

Finally, the integral J_1 is estimated as above:

$$J_1 \le \int_{\Omega_1}\frac{g(x,u,v)}{u^\eta+v^{\eta'}}(u^{\alpha(a^{-1}+b^{-1})}+v^{\beta(a^{-1}+b^{-1})}) \le C\int_{\Omega_1}g(x,u,v)\,dx \le C,$$

by choosing

$$\alpha\left(\frac{1}{a}+\frac{1}{b}\right)\le\eta \quad \text{and} \quad \beta\left(\frac{1}{a}+\frac{1}{b}\right)\le\eta'. \tag{4.4}$$

Thus, we have obtained

$$I_2 \le C\|v\|^{\eta'}\|u\|.$$

Hence, altogether with the above choices we have

$$\|u\|^2 = \int_\Omega |\nabla u|^2 \le C(\|u\|^{\eta+1}+\|v\|^{\eta'}\|u\|+1),$$

which implies

$$\|u\| \le C(\|u\|^\eta+\|v\|^{\eta'}+1). \tag{4.5}$$

We now analyze the inequalities (4.1)–(4.2), with a view to obtaining the possible values of the parameters η and η'. First, we eliminate γ, a and b: by using (4.1)–(4.2) we obtain

(I)
$$\eta \ge \frac{\alpha}{2-\alpha},$$

(II)
$$\eta' \ge \frac{\beta(\eta+1)}{2}.$$

By using (4.3) and (4.4) we have

$$\frac{1}{a}+\frac{1}{b}\ge\frac{\eta'+1}{2},$$

and

$$(III) \qquad \eta \geq \frac{\alpha(\eta'+1)}{2},$$

$$(IV) \qquad \eta' \geq \frac{\beta}{2-\beta}.$$

We now choose *equality* in (III) and (IV) to obtain η':

$$\eta' = \frac{\beta}{2-\beta}, \tag{4.6}$$

$$\eta = \frac{\alpha}{2-\beta}. \tag{4.7}$$

Checking the compatibility of this choice with inequality (I) yields

$$\frac{\alpha}{2-\beta} = \eta \geq \frac{\alpha}{2-\alpha},$$

which agrees with the assumption (H_6)(a), that is, $\alpha \leq \beta$, which is also compatible with (II).

We now return to inequality (4.5). To eliminate the term $\|u\|^\eta$ on the right, we need to require

$$\eta < 1 \ \Leftrightarrow \ \alpha + \beta < 2, \qquad \text{that is, hypothesis } (H_6)\text{(b)};$$

thus, inequality (4.5) reduces to

$$\|u\| \leq C(\|v\|^{\eta'} + 1). \tag{4.8}$$

Note that $2 > \alpha + \beta \geq 2\alpha$ also implies that $1 > \alpha$.

Next, we proceed similarly with the second equation in system (1.1):

$$\int_\Omega |\nabla v|^2\, dx = \int_\Omega f(x,u,v)v \leq c + \int_{\Omega_1} \frac{f(x,u,v)}{u^{\bar\eta'} + v^{\bar\eta}}(u^{\bar\eta'} + v^{\bar\eta})v$$
$$= c + J_1 + J_2,$$

where

$$J_1 := \int_{\Omega_1} \frac{f(x,u,v)}{u^{\bar\eta'} + v^{\bar\eta}} v^{\bar\eta+1}.$$

We estimate as before, using Proposition 4.1, the Trudinger–Moser inequality and Theorem 1.2:

$$J_1 \leq C\|v\|^{\bar\eta+1} \int_{\Omega_1} \frac{f(x,u,v)}{u^{\bar\eta'} + v^{\bar\eta}} \left(\frac{v}{\|v\|}\right)^{\bar\eta+1}$$
$$\leq C\|v\|^{\bar\eta+1}\left\{\int_{\Omega_1} e^{(v/\|v\|)^{(\bar\eta+1)\gamma'}} + \int_{\Omega_1} \frac{f(x,u,v)}{u^{\bar\eta'} + v^{\bar\eta}}\left(\log \frac{f(x,u,v)}{u^{\bar\eta'} + v^{\bar\eta}}\right)^{1/\gamma'}\right\}$$
$$\leq C\|v\|^{\bar\eta+1}$$

if we choose

$$(\bar\eta+1)\gamma' \leq 2, \quad \bar\eta' \geq \frac{\beta'}{\gamma'}, \quad \bar\eta \geq \frac{\alpha'}{\gamma'}. \tag{4.9}$$

The integral J_2 is estimated using Proposition 4.2, the Trudinger–Moser inequality and Theorem 1.2:

$$J_2 \leq c \int_{\Omega_1} \frac{f(x,u,v)}{u^{\bar\eta'}+v^{\bar\eta}} u^{\bar\eta'} v = c\|u\|^{\bar\eta'}\|v\| \int_{\Omega_1} \frac{f(x,u,v)}{u^{\bar\eta'}+v^{\bar\eta}} \left(\frac{u}{\|u\|}\right)^{\bar\eta'} \left(\frac{v}{\|v\|}\right)$$

$$\leq c\left\{ \int_{\Omega_1} e^{(u/\|u\|)^{\bar\eta' a'}}\,dx + \int_{\Omega_1} e^{(v/\|v\|)^{b'}}dx + \int_{\Omega_1} \frac{f(x,u,v)}{u^{\bar\eta'}+v^{\bar\eta}} \left(\log \frac{f(x,u,v)}{u^{\bar\eta'}+v^{\bar\eta}}\right)^{1/a'+1/b'} \right\}$$

$$\leq C\|u\|^{\bar\eta'}\|v\|$$

if we choose

$$\bar\eta' a' \leq 2, \qquad b' \leq 2, \qquad \bar\eta' \geq \beta'\left(\frac{1}{a'}+\frac{1}{b'}\right), \qquad \bar\eta \geq \alpha'\left(\frac{1}{a'}+\frac{1}{b'}\right). \tag{4.10}$$

Combining the estimates for J_1 and J_2 we thus obtain

$$\|v\|^2 \leq c\|v\|^{\bar\eta+1} + c\|u\|^{\bar\eta'}\|v\| + c,$$

that is,

$$\|v\| \leq c\|v\|^{\bar\eta} + c\|u\|^{\bar\eta'} + c. \tag{4.11}$$

From (4.9) and (4.10) we now get, by choosing equality in the appropriate relations, the analogues of (4.6) and (4.7):

$$\bar\eta' = \frac{\beta'}{2-\beta'}, \qquad \bar\eta = \frac{\alpha'}{2-\beta'}.$$

By condition (H_6)(b) we conclude that $\bar\eta < 1$, and then we deduce from (4.11) that

$$\|v\| \leq C(\|u\|^{\bar\eta'}+1). \tag{4.12}$$

Finally, combining inequalities (4.8) and (4.12) we get

$$\|u\| \leq C(\|v\|^{\eta'}+1) \leq C(\|u\|^{\eta'\bar\eta'}+1).$$

Condition (H_6)(c) yields $\eta'\,\bar\eta' < 1$, since

$$\begin{aligned} \eta'\,\bar\eta' < 1 &\Leftrightarrow \frac{\beta}{2-\beta}\,\frac{\beta'}{2-\beta'} < 1 \\ &\Leftrightarrow \beta\beta' < 4 - 2\beta - 2\beta' + \beta\beta' \\ &\Leftrightarrow \beta + \beta' < 2, \end{aligned}$$

and hence the solutions u are uniformly bounded in $H_0^1(\Omega)$. Combining inequalities (4.8) and (4.12) in the other order we also find that the solutions v are uniformly bounded. Finally, since $e^t \leq c_\varepsilon e^{\varepsilon t^2}$ for any $\varepsilon > 0$, we conclude by the Trudinger–Moser inequality that the right-hand sides in system (1.1) are uniformly bounded in L^p, for any $p > 1$, and hence we conclude that all solutions (u,v) are uniformly bounded in L^∞. □

5. Proof of Theorem 1.4

Assume that (u_n, v_n) is a sequence of solutions of system (1.1). Theorem 1.2 says that the sequences $(f(x, u_n, v_n))$ and $(g(x, u_n, v_n))$ are bounded in $L^1(\Omega)$. So it follows, by passing to subsequences if necessary (see [2]), that there is a non-negative Radon measure μ such that

$$f(x, u_n, v_n) + g(x, u_n, v_n) \to \mu. \tag{5.1}$$

We also observe that, as a consequence of Theorem 1.2 and assumption (H_2) (or (H_2')), the solutions (u_n, v_n) of (1.1) are bounded in $L^1(\Omega)$:

$$\|u_n\|_{L^1}, \|v_n\|_{L^1} \leq C \quad \text{for all } n. \tag{5.2}$$

Definition 5.1. We say that $x_0 \in \Omega$ is a *regular point* of the measure μ if there is a function $\psi \in C_c(\Omega)$ with $0 \leq \psi \leq 1$ and $\psi \equiv 1$ in some neighborhood of x_0 and such that

$$\int_\Omega \psi \, d\mu < 4\pi.$$

We denote by Σ_μ the set of *non-regular* points in Ω for the measure μ.

Remark 5.1. For a bounded non-negative measure μ, Σ_μ is a finite set. Indeed, if $x_0 \in \Sigma_\mu$, we have

$$\int_\Omega \psi \, d\mu \geq 4\pi$$

for any $\psi \in C_c(\Omega)$ with $0 \leq \psi \leq 1$ and $\psi \equiv 1$ in some neighborhood of x_0. Thus, $\mu(\{x_0\}) \geq 4\pi$. Finally, since $\int_\Omega d\mu \leq C$, it follows that Σ_μ is a finite set.

Let S_u be the *blow-up set* for the sequence (u_n), that is,

$$S_u := \{x \in \Omega : \exists (x_n) \subset \Omega \text{ such that } x_n \to x \text{ and } u_n(x_n) \to +\infty\}.$$

The assertions of Theorem 1.4 will be proved if we show that $S_u = S_v = \emptyset$. This will be achieved in the next lemmas.

Lemma 5.1. *Assume that x_0 is a regular point for the measure μ. Then there exist constants $\rho > 0$ and C, independent of n, such that*

$$\|u_n\|_{L^\infty(B_\rho(x_0))} \leq C \quad \text{and} \quad \|v_n\|_{L^\infty(B_\rho(x_0))} \leq C.$$

Proof. Since x_0 is a regular point of the measure μ we have a function $\psi \in C_c(\Omega)$ with $0 \leq \psi \leq 1$ and $\psi \equiv 1$ in some neighborhood V_{x_0} of x_0, such that $\int \psi \, d\mu < 4\pi$. Thus, $\int_{V_{x_0}} d\mu < 4\pi$, which implies that there exist $R, \delta > 0$ and n_0 such that for all $n \geq n_0$,

$$\int_{B_R(x_0)} [f(x, u_n, v_n) + g(x, u_n, v_n)] \, dx \leq 4\pi - \delta. \tag{5.3}$$

Let us write $u_n := u_{1,n} + u_{2,n}$ and $v_n := v_{1,n} + v_{2,n}$, where

$$-\Delta u_{1,n} = g(x, u_n, v_n) \text{ in } B_R(x_0) \quad \text{and} \quad u_{1,n} = 0 \text{ for } |x - x_0| = R,$$
$$-\Delta v_{1,n} = f(x, u_n, v_n) \text{ in } B_R(x_0) \quad \text{and} \quad v_{1,n} = 0 \text{ for } |x - x_0| = R.$$

Notice that $-\Delta u_{2,n} = 0$ and $-\Delta v_{2,n} = 0$ in $B_R(x_0)$.

Using Proposition 2.1 and (5.3), we obtain

$$c \geq \int_{B_R(x_0)} e^{(4\pi-\delta/2)\frac{u_{1,n}+v_{1,n}}{\int(f+g)}} \geq \int_{B_R(x_0)} e^{(4\pi-\delta/2)\frac{u_{1,n}+v_{1,n}}{4\pi-\delta}} = \int_{B_R(x_0)} e^{p(u_{1,n}+v_{1,n})}, \tag{5.4}$$

where $p > 1$ is a constant depending only on δ. Using the fact that $t \leq e^t$ we get

$$\|u_{1,n}\|_{L^1(B_R(x_0))} \leq C \quad \text{and} \quad \|v_{1,n}\|_{L^1(B_R(x_0))} \leq C. \tag{5.5}$$

Furthermore, as in the regularity Theorem 1.1 we get

$$\|u_{1,n}\|_{L^\infty(B_R(x_0))} \leq C \quad \text{and} \quad \|v_{1,n}\|_{L^\infty(B_R(x_0))} \leq C. \tag{5.6}$$

Since the functions $u_{2,n}$, $v_{2,n}$ are harmonic in $B_R(x_0)$, it follows from the mean value theorem for harmonic functions that

$$\|u_{2,n}\|_{L^\infty(B_{R/2}(x_0))} \leq C\|u_{2,n}\|_{L^1(B_R(x_0))},$$
$$\|v_{2,n}\|_{L^\infty(B_{R/2}(x_0))} \leq C\|v_{2,n}\|_{L^1(B_R(x_0))}.$$

On the other hand, using (5.2) and (5.5) we obtain

$$\|u_{2,n}\|_{L^1(B_R)} \leq \|u_n\|_{L^1(B_R(x_0))} + \|u_{1,n}\|_{L^1(B_R(x_0))} \leq C,$$
$$\|v_{2,n}\|_{L^1(B_R)} \leq \|v_n\|_{L^1(B_R(x_0))} + \|v_{1,n}\|_{L^1(B_R(x_0))} \leq C.$$

Thus

$$\|u_{2,n}\|_{L^\infty(B_{R/2}(x_0))} \leq C \quad \text{and} \quad \|v_{2,n}\|_{L^\infty(B_{R/2}(x_0))} \leq C. \tag{5.7}$$

From (5.6) and (5.7) we obtain

$$\|u_n\|_{L^\infty(B_{R/2}(x_0))} \leq C \quad \text{and} \quad \|v_n\|_{L^\infty(B_{R/2}(x_0))} \leq C,$$

and the proof is complete. □

Lemma 5.2. $S_u \cup S_v \subset \Sigma_\mu$.

Proof. Suppose that $x_0 \notin \Sigma_\mu$, i.e. x_0 is a regular point for μ. Then it follows from Lemma 5.1 that $\|u_n\|_{L^\infty(B_\rho(x_0))} \leq C$ and $\|v_n\|_{L^\infty(B_\rho(x_0))} \leq C$ for some $\rho > 0$, and hence $x_0 \notin S_u \cup S_v$ by the definition of the sets S_u and S_v. □

Lemma 5.3. $\Sigma_\mu \subset S_u \cup S_v$.

Proof. Let $x_0 \in \Sigma_\mu$. We claim that for each $R > 0$ we have

$$\lim_{n\to+\infty} \|u_n\|_{L^\infty(B_R(x_0))} = +\infty \quad \text{or} \quad \lim_{n\to+\infty} \|v_n\|_{L^\infty(B_R(x_0))} = +\infty. \tag{5.8}$$

Suppose by contradiction that there exists $R_0 > 0$ and a subsequence, which we denote also by (u_n, v_n), such that

$$\|u_n\|_{L^\infty(B_{R_0}(x_0))} \leq C \quad \text{and} \quad \|v_n\|_{L^\infty(B_{R_0}(x_0))} \leq C.$$

So,

$$\|e^{u_n+v_n}\|_{L^\infty(B_{R_0}(x_0))} \leq C.$$

Now using the hypothesis

$$f(x,u,v) \leq de^{t+s}, \qquad g(x,u,v) \leq de^{u+v}$$

it follows that

$$\|f(x,u_n,v_n)\|_{L^\infty(B_{R_0}(x_0))} \le C \quad \text{and} \quad \|g(x,u_n,v_n)\|_{L^\infty(B_{R_0}(x_0))} \le C,$$

which implies that for $R < R_0$ we have

$$\int_{B_R(x_0)} [f(x,u_n,v_n) + g(x,u_n,v_n)] \le CR^2.$$

Thus, there exists $R_1 > 0$ such that

$$\int_{B_{R_1}(x_0)} [f(x,u_n,v_n) + g(x,u_n,v_n)] < 4\pi.$$

This implies that x_0 is a regular point of μ, which is a contradiction. □

As a consequence of Lemmas 5.2 and 5.3 we conclude that

$$\Sigma_\mu = S_u \cup S_v.$$

Finally, we prove that this set is indeed empty, and this completes the proof of Theorem 1.4.

Lemma 5.4. *$S_u \cup S_v = \emptyset$.*

Proof. Suppose, by contradiction, that there exists $x_0 \in S_u \cup S_v$. Since x_0 is isolated, we can take $R > 0$ such that $\overline{B_R(x_0)} \cap ((S_u \cup S_v) \setminus \{x_0\}) = \emptyset$.

Next, we consider the Dirichlet problem in $B_R(x_0)$,

$$-\Delta z_n = f(x,u_n,v_n) + g(x,u_n,v_n) \text{ in } B_R(x_0) \quad \text{and} \quad z_n = 0 \text{ for } |x - x_0| = R.$$

We know that the function $u_n + v_n$ satisfies

$$-\Delta(u_n + v_n) = f(x,u_n,v_n) + g(x,u_n,v_n) \quad \text{in } B_R(x_0),$$
$$u_n + v_n \ge 0 \quad \text{for } |x - x_0| = R.$$

Thus, by the maximum principle we have

$$0 \le z_n \le u_n + v_n \quad \text{in } \overline{B_R(x_0)}.$$

Taking the limit we have $z_n \to z$, where z is a solution of the problem

$$-\Delta z = \mu \text{ in } B_R(x_0) \quad \text{and} \quad z = 0 \text{ when } |x - x_0| = R.$$

On the other hand, the problem

$$-\Delta w = 4\pi\delta_0 \text{ in } B_R(x_0) \quad \text{and} \quad w = 0 \text{ when } |x - x_0| = R$$

has the solution

$$w(x) = 2\log\frac{R}{|x - x_0|}.$$

Since x_0 is not a regular point of μ it follows that $\mu \ge 4\pi\delta_0$. So

$$z(x) \ge 2\log|x - x_0|^{-1} + o(1), \qquad x \to x_0.$$

Now, by hypothesis (H_7),

$$f(x,t,s) + g(x,t,s) \ge g(x,t,s) \ge ce^{t+s},$$

we have

$$\lim_{n\to\infty}\int_{B_R(x_0)}[f(x,u_n,v_n)+g(x,u_n,v_n)]\,dx \geq \lim_{n\to\infty} c\int_{B_R(x_0)} e^{z_n}$$
$$\geq c\int_{B_R(x_0)} e^{w}=\infty,$$

which is impossible. □

6. Proof of Theorem 1.5

We rely on the following well-known theorem by Krasnosel'skiĭ (we state it in the form given in [7, Theorem 3.1]).

Theorem 6.1. *Let $C\subset E$ denote a cone in a Banach space, and let $T: C\to C$ be a compact mapping with $T(0)=0$. Furthermore, assume that there are real numbers $0<r<R$ such that*

(1) *$x\neq tTx$ for $0\leq t\leq 1$, $\|x\|=r$, $x\in C$,*
(2) *there exists $H: C\times[0,\infty)\to C$ such that*
   (a) *$H(x,0)=Tx$, $x\in C$,*
   (b) *$H(x,t)\neq x$, $\|x\|=R$, $t\geq 0$,*
   (c) *$H(x,t)=x$ has no solution $x\in \bar B_R$, $t\geq t_0$.*

Then (1) *implies $i_C(T,B_r)=1$, and* (2) *implies $i_C(T,B_R)=0$.*

Here $i_C(T,\mathcal{O})$ is the fixed point index of T in C with respect to the (relatively) open set $\mathcal{O}$.

Note that it then follows by the excision property that T has a fixed point in $B_R\setminus \bar B_r$.

Proof of Theorem 1.5. Let $X=C^0(\bar\Omega)\times C^0(\bar\Omega)$ with norm $\|(u,v)\|=\|u\|_\infty+\|v\|_\infty$, and set

$$C=\{(u,v)\in X: u\geq 0,\ v\geq 0\}.$$

Furthermore, define the compact mapping $T: C\to C$ by

$$T\begin{pmatrix}u\\ v\end{pmatrix}=\begin{pmatrix}(-\Delta)^{-1}g(x,u,v)\\ (-\Delta)^{-1}f(x,u,v)\end{pmatrix}. \tag{5.9}$$

We now check that T satisfies hypotheses (1) and (2) of Theorem 6.1:

Hypothesis (1): We have to show that there is no solution (u,v) of

$$\begin{cases}-\Delta u = tg(x,u,v)\\ -\Delta v = tf(x,u,v)\end{cases} \quad \text{for } 0\leq t\leq 1,\ \|(u,v)\|=r,\ r<\varepsilon, \tag{5.10}$$

where ε is given in (H_8).

Indeed, suppose by contradiction that there is a solution $(u,v)\in C$ with $\|(u,v)\|=r$. Then by (H_8) we can estimate the right-hand sides of (5.10):

$$\|tg(x,u,v)\|_\infty\leq c_1\|(u,v)\|=c_1r, \qquad \|tf(x,u,v)\|_\infty\leq c_2r,$$

and obtain, by (5.10) and (1.7),

$$\|u\|_\infty \le Mc_1 r, \quad \|v\|_\infty \le Mc_2 r.$$

This implies that

$$r = \|(u,v)\| = \|u\|_\infty + \|v\|_\infty \le M(c_1+c_2)r,$$

which is a contradiction, since $M(c_1+c_2) < 1$.

Hypothesis (2): Define the homotopy $H : C \times [0,1] \to C$ by

$$H\left(\begin{pmatrix} u \\ v \end{pmatrix}, t\right) := \begin{pmatrix} (-\Delta)^{-1} g_t(x,u,v) \\ (-\Delta)^{-1} f_t(x,u,v) \end{pmatrix},$$

where

$$g_t(x,u,v) := (1-t)g(x,u,v) + t\mu(v^+ + 1),$$
$$f_t(x,u,v) := (1-t)f(x,u,v) + t\mu(u^+ + 1),$$

with $\mu > \max\{\lambda_1, \|a\|_\infty, \|b\|_\infty\}$, where a, b are as in assumption (H_2), resp. (H_2').

Note that with this choice g_t and f_t satisfy assumption (H_2) uniformly with respect to t; indeed, for $u, v \ge 0$,

$$\begin{aligned} g_t(x,u,v) &\ge (1-t)(b(x)v - b_1(x)) + t\mu(v+1) \\ &= ((1-t)b(x) + t\mu)v - (1-t)b_1(x) + t\mu \\ &\ge b(x)v - b_1(x) + tb_1(x) + t\mu \\ &\ge b(x)v - b_1(x), \end{aligned}$$

and similarly for $f_t(x,u,v)$.

Observe that the hypotheses on f and g transfer to f_t and g_t with the same parameters. So (2)(b) holds for R sufficiently large by the a priori bounds for solutions in Theorems 1.3 and 1.4.

Finally, we show (2)(c), that is, there exists no solution for $t = 1$: otherwise we would have

$$\begin{aligned} -\Delta u &= \mu(v^+ + 1), \\ -\Delta v &= \mu(u^+ + 1), \end{aligned} \quad \text{so} \quad \begin{aligned} -\Delta u &= \mu v + \mu, \\ -\Delta v &= \mu u + \mu, \end{aligned}$$

Multiplying both equations by the first eigenfunction φ_1 and integrating gives

$$\lambda_1 \int_\Omega u\varphi_1 = \mu \int_\Omega v\varphi_1 + \mu \int_\Omega \varphi_1,$$
$$\lambda_1 \int_\Omega v\varphi_1 = \mu \int_\Omega u\varphi_1 + \mu \int_\Omega \varphi_1.$$

Inserting the second equation into the first we obtain

$$\lambda_1 \int_\Omega u\varphi_1 = \mu\left\{\frac{\mu}{\lambda_1}\int_\Omega u\varphi_1 + \frac{\mu}{\lambda_1}\int_\Omega \varphi_1\right\} + \mu\int_\Omega \varphi_1,$$

and hence

$$\lambda_1^2 \int_\Omega u\varphi_1 = \mu^2 \int_\Omega u\varphi_1 + \mu^2 \int_\Omega \varphi_1 + \mu\lambda_1 \int_\Omega \varphi_1;$$

this is a contradiction, since $\mu > \lambda_1$.

Hence, the map T given in (5.9) has a non-trivial fixed point (u, v) in C, that is, (u, v) is a positive solution of system (1.1). □

References

[1] H. Berestycki and L. Nirenberg, *On the method of moving planes and the sliding method.* Bol. Soc. Brasil. Mat. (N.S.) **22** (1991), 1–37.

[2] H. Brezis, *Analyse fonctionnelle. Théorie et applications.* Masson, Paris, 1983.

[3] H. Brezis and F. Merle, *Uniform estimates and blow-up behavior for solutions of* $-\Delta u = V(x)e^u$ *in two dimensions.* Comm. Partial Differential Equations **16** (1991), 1223–1253.

[4] H. Brezis and R. E. L. Turner, *On a class of superlinear elliptic problems.* Comm. Partial Differential Equations **2** (1977), 601–614.

[5] W. X. Chen and C. M. Li, *A priori estimates for solutions to nonlinear elliptic equations*, Arch. Ration. Mech. Anal. **122** (1993), 145–157.

[6] D. G. de Figueiredo, *Semilinear elliptic systems: existence, multiplicity, symmetry of solutions.* In: Handbook of Differential Equations, Vol. 5: Stationary Partial Differential Equations, M. Chipot (ed.), Elsevier, 2008, 1–48.

[7] D. G. de Figueiredo, *Positive Solutions of Semilinear Ellpitic Equations.* Lecture Notes in Math. 957, Springer, 1982.

[8] D. G. de Figueiredo, *Monotonicity and symmetry of solutions of elliptic systems in general domains.* NoDEA Nonlinear Differential Equations Appl. **1** (1994), 119–123.

[9] D. G. de Figueiredo, P.-L. Lions and R. D. Nussbaum, *A priori estimates and existence of positive solutions of semilinear elliptic equations.* J. Math. Pures Appl. **61** (1982), 41–63.

[10] D. G. de Figueiredo, J. M. do Ó and B. Ruf, *Critical and subcritical elliptic systems in dimension two.* Indiana Univ. Math. J. **53** (2004), 1037–1054.

[11] D. G. de Figueiredo, J. M. do Ó and B. Ruf, *An Orlicz-space approach to superlinear elliptic systems.* J. Funct. Anal. **224** (2005), 471–496.

[12] D. G. de Figueiredo, J. M. do Ó and B. Ruf, *Semilinear elliptic systems with exponential nonlinearities in two dimensions.* Adv. Nonlinear Stud. **6** (2006), 199–213.

[13] D. G. de Figueiredo and E. Mitidieri, *Maximum principles for linear elliptic systems.* Rend. Ist. Mat. Univ. Trieste **22** (1990), 36-66.

[14] J. Moser, *A sharp form of an inequality by N. Trudinger.* Indiana Univ. Math. J. **20** (1971), 1077–1092.

[15] M. Protter and H. Weinberger, *Maximum Principle in Differential Equations.* Prentice-Hall, 1967.

[16] B. Ruf, *Lorentz spaces and nonlinear elliptic systems.* In: Contributions to Nonlinear Analysis, Progr. Nonlinear Differential Equations Appl. 66, Birkhäuser, Basel, 2006, 471–489.

[17] B. Ruf, *Superlinear elliptic equations and systems.* In: Handbook of Differential Equations, Vol. 5: Stationary Partial Differential Equations, M. Chipot (ed.), Elsevier, 2008, 211–276.

[18] P. Souplet, *Optimal regularity conditions for elliptic problems via L^p_δ spaces*, Duke Math. J. **127** (2005), 175–192.

[19] W. Troy, *Symmetry properties in systems of semilinear elliptic equations.* J. Differential Equations **42** (1981), 400–413.

Djairo G. de Figueiredo
IMECC – Departamento de Matemática
Universidade Estadual de Campinas
13083-970 Campinas, S.P., Brazil
e-mail: djairo@ime.unicamp.br

João Marcos do Ó
CCEN – Departamento de Matemática
Universidade Federal da Paraíba
58059-900 João Pessoa, Pb, Brazil
e-mail: jmbo@mat.ufpb.br

Bernhard Ruf
Dipartimento di Matematica
Università degli Studi
Via Saldini 50
20133 Milano, Italy
e-mail: ruf@mat.unimi.it

To access this journal online:
www.birkhauser.ch/jfpta

Annali di Matematica (2008) 187:531–545
DOI 10.1007/s10231-007-0054-1

The critical hyperbola for a Hamiltonian elliptic system with weights

Djairo G. de Figueiredo · Ireneo Peral · Julio D. Rossi

Received: 17 July 2006 / Revised: 8 February 2007 / Published online: 30 October 2007

Abstract In this paper we look for existence results for nontrivial solutions to the system,

$$\begin{cases} -\Delta u = \dfrac{v^p}{|x|^\alpha} & \text{in } \Omega, \\ -\Delta v = \dfrac{u^q}{|x|^\beta} & \text{in } \Omega, \end{cases}$$

with Dirichlet boundary conditions, $u = v = 0$ on $\partial\Omega$ and $\alpha, \beta < N$. We find the existence of a critical hyperbola in the (p, q) plane (depending on α, β and N) below which there exists nontrivial solutions. For the proof we use a variational argument (a linking theorem).

Keywords Elliptic systems · Nonlinear boundary conditions · Variational problems

Mathematics Subject Classification (2000) 35J65 · 35J20 · 35J55

I. Peral was partially supported by project MTM2004-02223 of MEC, Spain. J. D. Rossi partially supported by Universidad de Buenos Aires (grant TX066), ANPCyT and Fundacion Antorchas.

D. G. de Figueiredo
IMECC, Universidade Estadual de Campinas,
13081-970 Campinas, SP, Brazil
e-mail: djairo@ime.unicamp.br

I. Peral (✉)
Departamento de Matemáticas, U. Autonoma de Madrid,
28049 Madrid, Spain
e-mail: ireneo.peral@uam.es

J. D. Rossi
Departamento de Matemática, FCEyN, UBA (1428),
Buenos Aires, Argentina
e-mail: jrossi@dm.uba.ar

Originally published in: Ann. Mat. Pura Appl 187 (2008), 531-545,
© Springer-verlag 2007
Reprint by Springer-Verlag Berlin Heidelberg 2013,
DOI: 10.1007/978-3-319-02856-9_42

1 Introduction

In this paper we study the existence of nontrivial solutions of the following elliptic system,

$$\begin{cases} -\Delta u = \dfrac{v^p}{|x|^\alpha} & \text{in } \Omega, \\ -\Delta v = \dfrac{u^q}{|x|^\beta} & \text{in } \Omega, \end{cases} \tag{1.1}$$

with Dirichlet boundary conditions

$$u = v = 0 \quad \text{on } \partial\Omega.$$

Here Ω is a bounded smooth domain in $\mathbb{R}^N$ with $0 \in \Omega$ and $s^p := \operatorname{sgn}(s)|s|^p$. We will assume that the exponents p, q are positive and that $\alpha, \beta < N$.

Our main concern in this paper is to look at the role played by the two weights when dealing with existence of solutions. We find the existence of a critical hyperbola, given by,

$$\frac{N-\alpha}{p+1} + \frac{N-\beta}{q+1} = N-2. \tag{1.2}$$

Below this hyperbola we find existence of nontrivial solutions. Remark that when the two weights are not present, that is, for $\alpha = \beta = 0$, we recover the critical hyperbola for elliptic systems without weights that was found independently in [15] and [31] (see also [17] and [20]). Also remark that the hyperbola (1.2) is monotone with respect to α and β. The stronger the weights the smaller the hyperbola.

The main result in this paper is the following theorem.

Theorem 1.1 *Let us assume that p, q, α, β verify*

$$\frac{N-\alpha}{p+1} + \frac{N-\beta}{q+1} > N-2, \tag{1.3}$$

$$\frac{1}{p+1} + \frac{1}{q+1} < 1 \tag{1.4}$$

and

$$q+1 < \frac{2(N-\beta)}{N-4} \quad \text{and} \quad p+1 < \frac{2(N-\alpha)}{N-4} \quad \text{if } N \geq 5, \tag{1.5}$$

then there exist infinitely many strong solutions and at least one positive strong solution of (1.1).

In the case $\alpha = \beta = 0$ nonexistence of positive solutions on and above the critical hyperbola on starshaped domains could be obtained by some Pohozaev type identities, see for instance [35]. Notice that if $\alpha = \beta = 2$ and $p = q = 1$ the existence result depends on a real parameter λ. More precisely consider the linear problem

$$\begin{cases} -\Delta u = \lambda \dfrac{v}{|x|^2} & \text{in } \Omega, \\ -\Delta v = \lambda \dfrac{u}{|x|^2} & \text{in } \Omega. \end{cases} \tag{1.6}$$

By adding both equations we have that if $\lambda > (N-2)^2/4$ there is no positive distributional solution. In this sense we could conjecture that also our hyperbola is optimal.

Moreover, our ideas have no sign restriction on α and β, i.e., we are able to solve systems in which one equation (or both) are of Hénon type. In a general domain Ω, then we obtain the same hyperbola as before. However in the radial case, we can improve the result using ideas from [30], see Remark 3.1.

Notice that, if we take $p = q$, $\alpha = \beta < 0$, Theorem 1.1 gives the pioneering existence result in [30] above the critical Sobolev exponent in the presence of a weight. The new Sobolev exponent *with weight* is given by the corresponding radial Caffarelli–Kohn–Nirenberg estimate in [11]. For scalar Hénon equations and properties we refer to [6,30,33,34] and the references therein.

Note that the system (1.1) has a variational structure. In fact, it can be seen as a Hamiltonian system, since, if we consider

$$H(x,u,v) = \frac{v^{p+1}}{(p+1)|x|^{\alpha}} + \frac{u^{q+1}}{(q+1)|x|^{\beta}},$$

then we have

$$H_v(x,u,v) = \frac{v^p}{|x|^{\alpha}} \quad \text{and} \quad H_u(x,u,v) = \frac{u^q}{|x|^{\beta}}.$$

The crucial point of our arguments is to find the proper functional setting for (1.1) that allows us to treat our problem variationally. We accomplish this by considering fractional powers of the self adjoint operator $-\Delta$ with Dirichlet boundary conditions. The main ideas are taken from [17,20,23] and [28]. We also use a linking theorem in a version that can be found in [23]. See also the survey [18].

We observe that the same techniques used here can be applied to deal with more general Hamiltonian systems (with adequate hypotheses on H)

$$\begin{cases} -\Delta u = H_v(x,u,v) & \text{in } \Omega, \\ -\Delta v = H_u(x,u,v), \end{cases}$$

with Dirichlet boundary conditions. For example, we can consider, for two different points, $x_1 \neq x_2 \in \Omega$,

$$\begin{cases} -\Delta u = \dfrac{v^p}{|x-x_1|^{\alpha}} & \text{in } \Omega, \\ -\Delta v = \dfrac{u^q}{|x-x_2|^{\beta}} & \text{in } \Omega. \end{cases}$$

To clarify the exposition we will state and prove our results for (1.1) and leave the details of the general case to the reader.

We end the introduction with some bibliographical discussion. Existence results for nonlinear elliptic systems have obtained a great deal of interest in recent years. For this type of results see, among others, [8,12,14,17,19,20,22,24] and the survey [16]. Results of existence, non-existence and multiplicity for elliptic equations involving weights could be found, among other papers, in [1–5,9,10,25] and [27].

The rest of the paper is organized as follows, in Sect. 2 we establish the functional setting in which the problem will be posed and in Sect. 3 we prove our main result, Theorem 1.1.

2 The functional setting

In this section we describe the functional setting that allows us to treat (1.1) variationally. The natural functional associated to (1.1) is given by

$$J(u,v)=\int_{\Omega}\nabla u\nabla v-\int_{\Omega}\left(\frac{v^{p+1}}{(p+1)|x|^{\alpha}}+\frac{u^{q+1}}{(q+1)|x|^{\beta}}\right), \qquad (2.1)$$

in the natural space $H_0^1(\Omega)\times H_0^1(\Omega)$. However, in order to have a C^1 functional one has to impose conditions on p,q,α,β that are too restrictive. Therefore we use an idea from [17,20] and [28] and use suitable fractional powers of an operator on fractional Sobolev spaces to define the functional.

Let us consider the Hilbert space $L^2(\Omega)$ and the operator

$$-\Delta : H^2(\Omega)\cap H_0^1(\Omega)\subset L^2(\Omega)\mapsto L^2(\Omega).$$

It is well known that there exists a sequence of eigenvalues of $-\Delta$, $(\lambda_n)\subset\mathbb{R}$ with eigenfunctions $\phi_n\in H^2(\Omega)\cap H_0^1(\Omega)$ such that $0<\lambda_1\le\lambda_2\le\cdots\le\lambda_n\le\cdots\nearrow+\infty$.

Let us consider the fractional powers of $-\Delta$, namely for $0<s<1$, let

$$A^s=(-\Delta)^s,$$

that is,

$$A^s : D(A^s)\subset L^2(\Omega)\mapsto L^2(\Omega)$$

is given by

$$A^s u=\sum_{n=1}^{\infty}\lambda_n^s a_n\phi_n,$$

when u has the expansion $u=\sum a_n\phi_n$. We call

$$E^s-D(A^s)=\left\{u\in L^2(\Omega)\ :\ \sum_{n=1}^{\infty}\lambda_n^{2s}a_n^2<\infty\right\},$$

which is a Hilbert space with inner product, that we denote by $(\cdot,\cdot)_{E^s}$, given by

$$(u,v)_{E^s}=\langle A^s u, A^s v\rangle.$$

These spaces E^s are fractional Sobolev spaces, see [29]. In fact, we have

$$E^s\subset H^{2s}(\Omega),\quad 0<s\le 1.$$

By the Sobolev embedding theorem we have that

$$E^s\hookrightarrow L^r(\Omega)\quad\text{if } \frac{1}{r}\ge\frac{1}{2}-\frac{2s}{N}$$

and the embedding is compact if we have a strict inequality. Now, we use Holder's inequality to obtain

$$\int_\Omega \frac{u^{q+1}}{|x|^\beta} \leq \left(\int_\Omega u^r\right)^{(q+1)/r} \left(\int_\Omega |x|^{-\beta r/(r-(q+1))}\right)^{(r-(q+1))/r}$$
$$\leq C\left(\int_\Omega u^r\right)^{(q+1)/r}$$

if

$$\frac{\beta r}{r-(q+1)} < N.$$

That is

$$N(q+1) < (N-\beta)r. \tag{2.2}$$

Therefore, if

$$q+1 < \frac{2(N-\beta)}{N-4s}$$

we can choose r such that

$$\frac{1}{r} > \frac{1}{2} - \frac{2s}{N}$$

and (2.2) hold. Hence, we have the following inclusions

$$E^s \subset H^{2s}(\Omega) \hookrightarrow L^r(\Omega) \subset L^{q+1}(\Omega, |x|^{-\beta}).$$

More precisely, we have proved the following proposition,

Proposition 2.1 *Given $q > 1$, $\beta > 0$ and $s > 0$ so that*

$$q+1 < \frac{2(N-\beta)}{N-4s}$$

the inclusion map $i : E^s \to L^{q+1}(\Omega, |x|^{-\beta})$ is well defined and compact.

Let us set $E = E^s \times E^t$ where $s + t = 1$, with the norm

$$\|(u,v)\|_E^2 = \|u\|_{E^s}^2 + \|v\|_{E^t}^2.$$

Let the linear operator $L : E \to E$ be given by

$$L(u,v) = (A^{-s}A^t v, A^{-t}A^s u).$$

Next, we consider the eigenvalue problem $Lz = \lambda z$. We can rewrite this as

$$A^{-s}A^t v = \lambda u, \quad A^{-t}A^s u = \lambda v,$$

where $z = (u,v)$. As A^s and A^t are isomorphisms, it follows that $\lambda = 1$ or $\lambda = -1$. The associated eigenvectors are

$$\text{for } \lambda = 1,\ (u, A^{-t}A^s u)\ \forall u \in E^s,$$
$$\text{for } \lambda = -1,\ (u, -A^{-t}A^s u)\ \forall u \in E^s.$$

We can define the eigenspaces

$$E^+ = \{(u, A^{-t}A^s u) \;/\; u \in E^s\},$$
$$E^- = \{(u, -A^{-t}A^s u) \;/\; u \in E^s\},$$

which give a natural splitting

$$E = E^+ \oplus E^-.$$

We can define the functional, $\mathrm{H} : E \to \mathbb{R}$ as

$$\mathrm{H}(u, v) = \int_\Omega H(x, u, v).$$

Proposition 2.2 *The functional* H *defined above is of class* C^1 *and its derivative is given by*

$$\mathrm{H}'(u, v)(\phi, \psi) = \int_\Omega H_u(x, u, v)\phi + \int_\Omega H_v(x, u, v)\psi.$$

Moreover, H′ *is compact.*

Proof We have

$$\int_\Omega \left| \frac{\partial H}{\partial u}(x, u, v)\phi \right| = \int_\Omega \left(\frac{|u|^q}{|x|^\beta} \right) |\phi|.$$

By Holder's inequality and Proposition 2.1 we have

$$\int_\Omega \left| \frac{\partial H}{\partial u}(x, u, v)\phi \right| \le C \left(\|u\|^q_{E^s} \right) \|\phi\|_{E^s}.$$

In a similar way we obtain the analogous inequality for H_v.

Thus H′ is well defined and bounded in E. Next, a standard argument gives that H is Fréchet differentiable with H′ continuous. The fact that H′ is compact comes from Proposition 2.1 (see [32] for the details). □

We consider the functional

$$I : E \mapsto \mathbb{R}$$

given by

$$\begin{aligned} I(u, v) &= \int_\Omega A^s u A^t v - \int_\Omega \left(\frac{v^{p+1}}{(p+1)|x|^\alpha} + \frac{u^{q+1}}{(q+1)|x|^\beta} \right) \\ &= \int_\Omega A^s u A^t v - \int_\Omega H(x, u, v). \end{aligned} \tag{2.3}$$

Where we have selected $s, t > 0$ such that

$$s + t = 1, \quad q + 1 < \frac{2(N - \beta)}{N - 4s} \quad \text{and} \quad p + 1 < \frac{2(N - \alpha)}{N - 4t}.$$

Remark 2.1 Note that this selection of s, t is possible since we are below the critical hyperbola (1.2). Moreover, there exists a positive selection of s and t thanks to our last assumption on the exponents in Theorem 1.1, (1.5).

Let us now give the definition of weak solution of (1.1).

Definition 2.1 We say that $z = (u, v) \in E = E^s \times E^t$ is an (s, t)-weak solution of (1.1) if z is a critical point of I. In other words, for every $(\phi, \psi) \in E$ we have

$$\int_\Omega A^s u A^t \psi + \int_\Omega A^s \phi A^t v - \int_\Omega \frac{v^p}{|x|^\alpha}\psi - \int_\Omega \frac{u^q}{|x|^\beta}\phi = 0. \tag{2.4}$$

Now, we prove a theorem that provides regularity of (s, t)-weak solutions. In fact, every (s, t)-weak solution has two derivatives in some $L^r(\Omega)$ and hence it is a strong solution.

Theorem 2.1 *If* $(u, v) \in E^s \times E^t$ *is an* (s, t)*-weak solution of* (1.1) *then* $u \in W^{2,a}(\Omega)$, $v \in W^{2,b}(\Omega)$ *for every*

$$1 < a < \frac{2N}{p(N-4t)+2\alpha} \quad \text{and} \quad 1 < b < \frac{2N}{q(N-4s)+2\beta}.$$

Hence (u, v) *is in fact a strong solution of* (1.1).

Proof Let us first consider $\psi = 0$ in (2.4), then

$$\int_\Omega A^s \phi A^t v - \int_\Omega \frac{u^q}{|x|^\beta}\phi = 0, \tag{2.5}$$

for all $\phi \in E^s$. If we take $\phi \in H^2(\Omega) \cap H_0^1(\Omega)$, we have

$$\int_\Omega A^s \phi A^t v = -\int_\Omega \Delta\phi\, v. \tag{2.6}$$

On the other hand, since $u \in E^s$, by our previous calculations, we have

$$\frac{u^q}{|x|^\beta} \in L^b(\Omega) \quad \text{if} \quad b < \frac{2N}{q(N-4s)+2\beta}.$$

Then, from basic elliptic theory (see [26]), there exists a function $w \in W^{2,b}(\Omega)$ such that

$$\begin{cases} -\Delta w = \dfrac{u^q}{|x|^\beta} & \text{in } \Omega, \\ w = 0 & \text{on } \partial\Omega. \end{cases}$$

Now, integration by parts gives

$$0 = -\int_\Omega \Delta w \phi - \int_\Omega \frac{u^q}{|x|^\beta}\phi = -\int_\Omega w \Delta\phi - \int_\Omega \frac{u^q}{|x|^\beta}\phi. \tag{2.7}$$

Combining (2.5),(2.6) and (2.7), we obtain

$$\int_\Omega (v - w)\Delta\phi = 0,$$

with $v = w = 0$ on $\partial\Omega$. From this it follows that $v = w$ and hence v belongs to $W^{2,b}(\Omega)$. We argue similarly for u. □

Remark 2.2 Since we have

$$q+1<\frac{2(N-\beta)}{N-4s} \quad \text{and} \quad p+1<\frac{2(N-\alpha)}{N-4t}$$

we get

$$1<\frac{2N}{p(N-4t)+2\alpha} \quad \text{and} \quad 1<\frac{2N}{q(N-4s)+2\beta}.$$

3 Proof of Theorem 1.1

First, we prove that there exist infinitely many solutions to (1.1). To this end, we present an abstract theorem from critical point theory from [23] (see also [7]) that provides us with infinitely many critical points of a functional. Next, we prove that this abstract result can be applied to our functional setting stated in the previous section.

Let E be a Hilbert space with inner product $(\cdot,\cdot)_E$. Assume that E has a splitting $E = X \oplus Y$ where X and Y are both infinite dimensional subspaces. Assume there exists a sequence of finite dimensional subspaces $X_n \subset X$, $Y_n \subset Y$, $E_n = X_n \oplus Y_n$ such that $\overline{\cup_{n=1}^{\infty} E_n} = E$. Let $T : E \to E$ be a linear bounded invertible operator.

Let $I \in C^1(E,\mathbb{R})$. Instead of the usual Palais–Smale condition we will require that the functional I satisfies the so-called $(PS)^*$ conditions with respect to E_n, i.e., any sequence $z_k \in E_{n_k}$ with $n_k \to \infty$ as $k \to \infty$, satisfying $I|'_{E_{n_k}}(z_k) \to 0$ and $I(z_k) \to c$ has a subsequence that converges in E.

Then we define the basic sets over which the linking process will take place. For $\rho > 0$ we define

$$S = S_\rho = \{y \in Y |\ \|y\|_E = \rho\}$$

and for some fixed $y_1 \in Y$ with $\|y_1\|_E = 1$ and subspaces $\mathcal{X}_1$ and $\mathcal{X}_2$, we consider

$$X \oplus \operatorname{span}\{y_1\} = \mathcal{X}_1 \oplus \mathcal{X}_2.$$

Without loss of generality we may assume that $y_1 \in \mathcal{X}_2$. Next, we define for $M, \sigma > 0$

$$D = D_{M,\sigma} = \{x_1 + x_2 \in \mathcal{X}_1 \oplus \mathcal{X}_2 |\ \|x_1\|_E \le M,\ \|x_2\|_E \le \sigma\}.$$

Now we can state our abstract critical point result whose proof can be found in [23].

Theorem 3.1 *Let $I \in C^1(E,\mathbb{R})$ be an even functional satisfying the $(PS)^*$ condition with respect to E_n. Assume that $T : E_n \to E_n$, for n large. Let $\rho > 0$ and $\sigma > 0$ be such that $\sigma\|Ty_1\|_E > \rho$. Assume that there are constants $\alpha \le \beta$ such that*

$$\inf_{S\cap E_n} I \ge \alpha, \qquad \sup_{T(\partial D\cap E_n)} I < \alpha \quad \text{and} \quad \sup_{T(D\cap E_n)} I \le \beta$$

for all n large. Then I has a critical value $c \in [\alpha,\beta]$.

Next, we show how the functional setting introduced in Sect. 2 can be used to apply Theorem 3.1.

Let ϕ_n be the eigenfunctions of $-\Delta$. Let

$$E_n = \operatorname{span}\{\phi_1,\ldots,\phi_n\} \times \operatorname{span}\{\phi_1,\ldots,\phi_n\}.$$

It is easy to see that $\overline{\cup_{n=1}^{\infty} E_n} = E$. Next, we prove that I satisfies the $(PS)^*$ condition with respect to the family E_n.

Lemma 3.1 *The functional I satisfies the $(PS)^*$ condition with respect to E_n.*

Proof Let $(z_k)_{k\geq 1} = (u_k, v_k)_{k\geq 1} \subset E_{n_k}$ be a sequence such that

$$I(z_k) \to c \quad \text{and} \quad I'|_{E_{n_k}}(z_k) \to 0. \tag{3.1}$$

Let us first prove that (3.1) implies that (z_k) is bounded in E. From (3.1) it follows that there exists a sequence $\varepsilon_k \to 0$ such that

$$|I'(z_k)w| \leq \varepsilon_k \|w\|_E, \tag{3.2}$$

for all $w \in E_{n_k}$. Let us take

$$w_k = ((w_k)_1, (w_k)_2) = \frac{(q+1)(p+1)}{p+q+2}\left(\frac{1}{q+1}u_k, \frac{1}{p+1}v_k\right).$$

Now, using (3.1) and (3.2), for k large

$$\begin{aligned}
c + 1 + \varepsilon_k\|w_k\|_E &\geq I(z_k) - I'(z_k)w_k \\
&= \int_\Omega A^s u_k A^t v_k - \int_\Omega H(x, u_k, v_k) - \int_\Omega A^s u_k A^t (w_k)_2 \\
&\quad - \int_\Omega A^s (w_k)_1 A^t v_k + \int_\Omega H_u(x, u_k, v_k)(w_k)_1 + \int_\Omega H_v(x, u_k, v_k)(w_k)_2 \\
&= -\frac{(1-pq)}{p+q+2}\int_\Omega H(x, u_k, v_k).
\end{aligned}$$

Now, by (1.4) we get $pq > 1$ and hence we obtain

$$C(1 + \|z_k\|_E) \geq \int_\Omega H(x, u_k, v_k).$$

Therefore,

$$\int_\Omega \frac{|u_k|^{q+1}}{|x|^\beta} + \frac{|v_k|^{p+1}}{|x|^\alpha} \leq C(1 + \|u_k\|_{E^s} + \|v_k\|_{E^t}). \tag{3.3}$$

Next we consider $w = (\phi, 0)$ with $\phi \in E^s_{n_k}$. From (3.2) we have

$$\int_\Omega A^s \phi A^t v_k \leq \int_\Omega \frac{|u_k|^q}{|x|^\beta}|\phi| + \varepsilon_k\|\phi\|_{E^s}.$$

Now, using Holder's inequality,

$$\int_\Omega \frac{|u_k|^q}{|x|^\beta}|\phi| \leq \|u_k\|^q_{L^{q+1}(\Omega,|x|^{-\beta})}\|\phi\|_{L^{q+1}(\Omega,|x|^{-\beta})}.$$

So, by Proposition 2.1, we get that

$$|\langle A^s\phi, A^t v_k\rangle| \leq C\|\phi\|_{E^s}\left(\|u_k\|^q_{L^{q+1}(\Omega,|x|^{-\beta})} + 1\right).$$

By duality (A^s is an isometry between E^s and L^2) we get

$$\|v_k\|_{E^t} \leq C\left(\|u_k\|^q_{L^{q+1}(\Omega,|x|^{-\beta})} + 1\right). \tag{3.4}$$

Analogously, we obtain

$$\|u_k\|_{E^s} \leq C\left(\|v_k\|^p_{L^{p+1}(\Omega,|x|^{-\alpha})} + 1\right). \quad (3.5)$$

Now combining (3.3), (3.4) and (3.5), we obtain

$$\|u_k\|_{E^s} + \|v_k\|_{E^t} \leq C\left(\|u_k\|^{q/(q+1)}_{E^s} + \|v_k\|^{p/(p+1)}_{E^t} + 1\right).$$

Since all the involved exponents are less than one, we conclude that z_k in bounded.

Now, by compactness and the invertibility of L we can extract a subsequence of z_k that converges in E. Indeed, we can take a subsequence z_{k_j} that converges weakly in E, as H′ is compact, it follows that $H'(z_{k_j})$ converges strongly in E. Hence, using the fact that $I'(z_{k_j}) \to 0$ strongly and the invertibility of L, the result follows. □

Now we define the splitting of E_n. Fix $k \in \mathbb{N}$ and for $n \geq k$ let

$$X_n = \left(E_1^- \oplus \cdots \oplus E_n^-\right) \oplus \left(E_1^+ \oplus \cdots \oplus E_{k-1}^+\right) \quad \text{and} \quad Y_n = \left(E_k^+ \oplus \cdots \oplus E_n^+\right), \quad (3.6)$$

where $E_j^+ = \text{span}\{(\phi_j, A^{-t}A^s\phi_j)\}$ and $E_j^- = \text{span}\{(\phi_j, -A^{-t}A^s\phi_j)\}$. We have $E_n = X_n \oplus Y_n$.

Lemma 3.2 *There exist $\alpha_k > 0$ and $\rho_k > 0$ independent of n such that for all $n \geq k$*

$$\inf_{z \in S_{\rho_k} \cap Y_n} I(z) \geq \alpha_k$$

where $S_{\rho_k} = \{y \in E^+ \mid \|y\| = \rho_k\}$. Moreover, $\alpha_k \to \infty$ as $k \to \infty$.

Proof We first recall that by Proposition 2.1, E^s is embedded in $L^\gamma(\Omega, |x|^{-\varrho})$ for any γ such that

$$\gamma \leq \frac{2(N-\varrho)}{N-4s}.$$

Hence, there exists $a = a(\gamma)$ such that

$$\|u\|_{L^\gamma(\Omega,|x|^{-\varrho})} \leq a\|u\|_{E^s} \quad \text{for all } u \in E^s.$$

Also for $z \in E_k^+ \oplus \cdots \oplus E_j^+ \oplus \cdots$ we have

$$\|z\|_E \geq \lambda_k^{\min\{s,t\}}\|z\|_{L^2(\Omega)}$$

with $\lambda_k \to \infty$ as $k \to \infty$.

Now consider $z = (u, v) \in Y_n$. For a constant a independent of n, we observe that there exists $\kappa > 0$ such that

$$\|u\|^{q+1}_{L^{q+1}(\Omega,|x|^{-\beta})} \leq \|u\|^{2/\kappa}_{L^2(\Omega)}\|u\|^{q+1-2/\kappa}_{L^\gamma(\Omega,|x|^{-\varrho})} \leq \frac{a}{\lambda_k^{\min\{s,t\}(2/\kappa)}}\|u\|^{q+1}_E.$$

Analogously, we obtain

$$\|v\|^{p+1}_{L^{p+1}(\Omega,|x|^{-\alpha})} \leq \frac{a}{\lambda_k^{\min\{s,t\}(2/\theta)}}\|v\|^{p+1}_E$$

for some $\theta > 0$. Then for $z = (u, v)$ we have

$$I(z) \geq \|z\|^2_E - C\left(\frac{a}{\lambda_k^{\min\{s,t\}\min\{2/\kappa,\, 2/\theta\}}}\max\left\{\|z\|^{q+1}_E,\ \|z\|^{p+1}_E\right\} + 1\right).$$

Then we choose

$$\rho_k^{\max\{p+1,q+1\}} = \lambda_k^{\min\{s,t\}\min\{2/\kappa,\,2/\theta\}}$$

and observe that $\rho_k \to \infty$ as $k \to \infty$.

Therefore, for $z \in S_{\rho_k} \cap Y_n$ we find that

$$I(z) \geq \rho_k^2 - C. \tag{3.7}$$

Defining α_k as the right hand side of (3.7) and noting that both ρ_k and α_k are independent of $n \geq k$ we complete the proof of the Lemma. □

Next we define, for $z = (u, v) \in E$

$$T_\sigma(z) = (\sigma^{\mu-1}u, \sigma^{\nu-1}v) \tag{3.8}$$

where μ and ν are such that

$$\mu + \nu < \min\{\mu(p+1), \nu(q+1)\}.$$

In particular, since $pq > 1$ by (1.4), we take $\mu = q+1$ and $\nu = q+1$ for which,

$$(p+1) + (q+1) < (p+1)(q+1).$$

Lemma 3.3 *There exist $B_k > 0$, σ_k and $M_k > 0$ independent of n such that for all $n \geq k$ they satisfy $\sigma_k > \rho_k$,*

$$\sup_{T_{\sigma_k}(\partial D \cap E_n)} I \leq 0 \quad \textit{and} \quad \sup_{T_{\sigma_k}(D \cap E_n)} I \leq B_k$$

where

$$D = \{z \in E^- \oplus E_1^+ \oplus \cdots \oplus E_k^+ \mid \|z^-\| \leq M_k,\ \|z^+\| \leq \sigma_k\}.$$

Proof Let us consider $z = T_\sigma(u, v)$ with $(u, v) \in D$. Then we can write

$$z = (\sigma^{\mu-1}u^+, \sigma^{\nu-1}v^+) + (\sigma^{\mu-1}u^-, \sigma^{\nu-1}v^-).$$

Using the definition of the spaces E^+ and E^- we have

$$\int_\Omega A^s u, A^t v = \sigma^{\mu+\nu-2}(\|z^+\|^2 - \|z^-\|^2).$$

On the other hand, we have

$$\int_\Omega H(x, z) = \int_\Omega \sigma^{(q+1)(\mu-1)} \frac{|u^+ + u^-|^{q+1}}{|x|^\beta} + \sigma^{(p+1)(\nu-1)} \frac{|v^+ + v^-|^{p+1}}{|x|^\alpha}.$$

The functions u^+ and u^- can be written as

$$u^+ = \sum_{i=1}^k \theta_i \phi_i \quad \text{and} \quad u^- = \sum_{i=1}^k \gamma_i \phi_i + \tilde{u}^-,$$

where $\tilde{u}^-$ is orthogonal to ϕ_i, $i = 1, \ldots, k$ in $L^2(\Omega)$. Using Holder's inequality we get

$$\sum_{i=1}^{k} \lambda_i^{s-t}(\theta_i^2 + \theta_i \gamma_i) = \langle u^+ + u^-, A^{s-t}u^+\rangle$$

$$\leq \|u^+ + u^-\|_{L^{q+1}(\Omega,|x|^{-\beta})}\|A^{s-t}u^+\|_{L^{\frac{q+1}{q}}(\Omega,|x|^{\frac{\beta}{q}})}$$

$$\leq C(\Omega, \beta)(\|u^+ + u^-\|_{L^{q+1}(\Omega,|x|^{-\beta})}\|A^{s-t}u^+\|_{L^2(\Omega)}.$$

Then by the definition of A^{s-t}, there exists a constant C_k such that

$$\sum_{i=1}^{k} \lambda_i^{s-t}(\theta_i^2 + \theta_i \gamma_i) \leq C_k\|u^+ + u^-\|_{L^{q+1}(\Omega,|x|^{-\beta})}\|u^+\|_{L^2(\Omega)}. \tag{3.9}$$

In a similar way, using that $v^+ = A^{s-t}u^+$ and $v^- = -A^{s-t}u^-$ we have that there exists a constant C_k such that

$$\sum_{i=1}^{k} \lambda_i^{s-t}(\theta_i^2 - \theta_i \gamma_i) \leq C_k\|v^+ + v^-\|_{L^{p+1}(\Omega,|x|^{-\alpha})}\|v^+\|_{L^2(\Omega)}. \tag{3.10}$$

Depending on the sign of $\sum_{i=1}^{k} \lambda_i^{s-t}\theta_i \gamma_i$ we use (3.9) or (3.10) to conclude that

$$\|u^+\|_{L^2(\Omega)} \leq C_k\|u^+ + u^-\|_{L^{q+1}(\Omega,|x|^{-\beta})}$$

or

$$\|u^+\|_{L^2(\Omega)} \leq C_k\|v^+ + v^-\|_{L^{p+1}(\Omega,|x|^{-\alpha})}.$$

Hence,

$$I(z) \leq \sigma^{\mu+\nu-2}(\|z^+\|^2 - \|z^-\|^2) - C_k\sigma^{(q+1)(\mu-1)}\|u^+\|_{L^2(\Omega)}^{q+1}$$

or

$$I(z) \leq \sigma^{\mu+\nu-2}(\|z^+\|^2 - \|z^-\|^2) - C_k\sigma^{(p+1)(\nu-1)}\|u^+\|_{L^2(\Omega)}^{p+1}.$$

Thus we may choose $\|z^+\|_E = \sigma_k$ large enough in order to obtain $\sigma_k > \rho_k$ and, by the condition on μ and ν, $I(z) \leq 0$.

Taking $\|z^+\| \leq \sigma_k$ and $\|z^-\| = M_k$, we get

$$I(z) \leq \sigma_k^{\mu+\nu-2}\left(\sigma_k^2 - M_k^2\right)$$

and then choosing M_k large enough we find that

$$I(z) \leq 0.$$

In this way we have finished with the proof of the first part of Lemma 3.3.

Next we choose B_k large so that the second inequality holds. □

Proof of Theorem 1.1 Existence of infinitely many solutions.

For $k \geq 1$, Lemmas 3.2 and 3.3 allow us to use Theorem 3.1. As a consequence the functional I has a critical value $c_k \in [\alpha_k, B_k]$. Since $\alpha_k \to \infty$ we get infinitely many critical values of I. Therefore we have infinitely many solutions of (1.1). □

Now we turn our attention to the existence of a positive solution to (1.1). We use ideas from [17] under the functional setting of Sect. 2. We start by redefining the Hamiltonian. Let us define $\tilde{H} : \partial\Omega \times \mathbb{R} \times \mathbb{R} \to \mathbb{R}$ by

$$\tilde{H}(x,u,v) = \begin{cases} H(x,u,v) & \text{if } u, v \geq 0, \\ H(x,0,v) & \text{if } u \leq 0,\ v \geq 0, \\ H(x,u,0) & \text{if } u \geq 0,\ v \leq 0, \\ 0 & \text{if } u, v \leq 0. \end{cases} \tag{3.11}$$

We observe that if (u, v) is a nontrivial strong solution of

$$\begin{cases} -\Delta u = \tilde{H}_v(x,u,v) & \text{in } \Omega, \\ -\Delta v = \tilde{H}_u(x,u,v), \end{cases} \tag{3.12}$$

with Dirichlet boundary conditions, then by the maximum principle we have that u and v are strictly positive in $\overline{\Omega}$. Hence (u, v) is a positive strong solution of (1.1).

To find a nontrivial solution of (3.12) we want to apply the results of Sect. 3. By our assumptions, the new Hamiltonian $\tilde{H}$ is regular.

We have to adapt the proof of Theorem 1.1 to the functional I with the Hamiltonian H replaced by $\tilde{H}$. We observe that the proof of the Palais–Smale condition and the geometric conditions follows as before with some minor modifications, see [17] for the details.

Proof of Theorem 1.1 Existence of a positive strong solution.

As a consequence of the previous results the modified functional I (with the modified Hamiltonian $\tilde{H}$ instead of H) has a critical value $c \neq 0$. Hence, by the maximum principle, we obtain a positive strong solution of (1.1). □

Finally, we comment on the radial case for α and β positives.

Remark 3.1 From now on we consider $\Omega = B(0,1)$. From the results in [30], using interpolation between

$$H^2_{\text{rad}}(B(0,1)) \to L^{2(N+\alpha)/(N-4)}(B(0,1), |x|^{-\alpha}),$$

and

$$L^2(B(0,1)) \to L^2(B(0,1), |x|^{-\alpha}).$$

we have the following Proposition.

Proposition 3.1 *Let $\alpha > 0$, if*

$$p + 1 < \frac{2(N+\alpha)}{N - 4s + \alpha(1-s)},$$

then

$$E^s \hookrightarrow L^{q+1}(B(0,1), |x|^{-\alpha})$$

is compact.

Using this proposition and the same ideas as before we conclude that in the radial case with $\alpha, \beta > 0$ the critical hyperbola is given by

$$\frac{(N+|\alpha|)(|\beta|+4)}{p+1} + \frac{(N+|\beta|)(|\alpha|+4)}{q+1} = \left[4(N-2) + N\frac{|\alpha|+|\beta|}{2} + \frac{|\alpha\beta|}{2}\right].$$

Note added in proof After the revision of the paper the authors have received a preprint by F. Liu and S. Yeng with related results.

References

1. Abdellaoui, B., Peral, I.: Some results for semilinear elliptic equations with critical potential. Proc. R. Soc. Edinb. **132A**, 1–24 (2002)
2. Abdellaoui, B., Peral, I.: Existence and nonexistence results for quasilinear elliptic equations involving the p-Laplacian with a critical potential. Ann. Mat. Pura Appl. **182**, 247–270 (2003)
3. Abdellaoui, B., Peral, I.: Nonexistence results for quasilinear elliptic equations related to Caffarelli–Kohn–Nirenberg inequalities. Commun. Pure Appl. Anal. **2**, 539–566 (2003)
4. Abdellaoui, B., Colorado, E., Peral, I.: Some remarks on elliptic equations with singular potentials and mixed boundary conditions. Adv. Nonlinear Stud. **4**, 501–531 (2004)
5. Abdellaoui, B., Colorado, E., Peral, I.: Effect of the boundary conditions in the behavior of the optimal constant of some Caffarelli–Kohn–Nirenberg inequalities: applications to critical nonlinear elliptic problems. Adv. Diff. Equ. **11**, 667–720 (2006)
6. Badiale, M., Serra, E.: Multiplicity results for the supercritical Hénon equation. Adv. Nonlinear Stud. **4**, 453–467 (2004)
7. Bartsch, T., de Figueiredo, D.G.: Infinitely many solutions of Nonlinear Elliptic Systems. Prog. Nonlinear Diff. Eq. and Their Appl. vol. **35**, pp. 51–67. Birkhäuser Verlag Basel/Switzerland (1999)
8. Boccardo, L., de Figueiredo, D.G.: Some remarks on a system of quasilinear elliptic equations. NoDEA Nonlinear Diff. Equ. Appl. **9**(3), 309–323 (2002)
9. Brezis, H., Cabré, X.: Some simple nonlinear PDE's without solution. Boll. Unione. Mat. Ital. Sez. B **8**, 223–262 (1998)
10. Brezis, H., Dupaigne, L., Tesei, A.: On a semilinear equation with inverse-square potential. Selecta Math. **11**, 1–7 (2005)
11. Caffarelli, L., Kohn, R., Nirenberg, L.: First order interpolation inequalities with weights. Comp. Math. **53**, 259–275 (1984)
12. Clément, Ph., de Figueiredo, D.G., Mitidieri, E.: Positive solutions of semilinear elliptic systems. Comm. Part. Diff. Equ. **17**, 923–940 (1992)
13. Colorado, E., Peral, I.: Eigenvalues and bifurcation for elliptic equations related to Caffarelli–Kohn–Nirenberg inequalities, with mixed Dirichlet–Neumann boundary conditions. Topol. Methods Nonlinear Anal. **23**, 239–273 (2004)
14. de Figueiredo, D.G.: Positive solutions of semilinear elliptic equations. Springer Lecture Notes in Mathematics, vol. 957, pp 34–87 (1982)
15. Clément, Ph., de Figueiredo, D.G., Mitidieri, E.: Positive solutions of semilinear elliptic systems. Comm. Part. Diff. Equ. **17**, 923–940 (1992)
16. de Figueiredo, D.G.: Semilinear elliptic systems: a survey of superlinear problems. Resenhas IME-USP **2**, 373–391 (1996)
17. de Figueiredo, D.G., Felmer, P.L.: On superquadratic elliptic systems. Trans. Am. Math. Soc. **343**, 99–116 (1994)
18. de Figueiredo, D.G.: Semilinear Elliptic Systems in Nonlinear Functional Analysis and Applications to Differential Equations. In: Ambrosetti, A., Chang, K.C. Ekeland I. (eds). pp. 122–152, World Scientific (1998)
19. de Figueiredo, D.G., Magalhães, C.A.: On nonquadratic hamiltonian elliptic systems. Adv. Diff. Equ. **1**(5), 881–898 (1996)
20. de Figueiredo, D.G., Ruf, B.: Elliptic systems with nonlinearities of arbitrary growth. Mediterr. J. Math. **1**(4), 417–431 (2004)
21. Felmer, P.: Periodic solutions of 'superquadratic' Hamiltonian systems. J. Diff. Equ. **102**, 188–207 (1993)
22. Felmer, P., Manásevich, R.F., de Thélin, F.: Existence and uniqueness of positive solutions for certain quasilinear elliptic systems. Comm. Par. Diff. Equ. **17**, 2013–2029 (1992)
23. Felmer, P., Wang, Z-Q.: Multiplicity for symmetric indefinite functionals: application to Hamiltonian and elliptic systems. Topol. Methods Nonlinear Anal. **12**(2), 207–226 (1998)
24. Fernández Bonder, J., Rossi, J.D.: Existence for an elliptic system with nonlinear boundary conditions via fixed point methods. Adv. Diff. Equ. **6**(1), 1–20 (2001)
25. Garcia Azorero, J., Peral, I.: Hardy inequalities and some critical elliptic and parabolic problems. J. Diff. Equ. **144**, 441–476 (1998)
26. Gilbarg, D., Trudinger, N.S.: Elliptic Partial Differential Equations of Second Order. Springer, New York (1983)

27. Ghoussoub, N., Yuan, C.: Multiple solutions for quasi-linear involving the critical Sobolev and Hardy exponents. Trans. Am. Math. Soc. **352**, 5703–5743 (2000)
28. Hulshof, J., van der Vorst, R.C.A.M.: Differential systems with strongly indefinite variational structure. J. Funct. Anal. **114**, 32–58 (1993)
29. Lions, J.L., Magenes, E.: Non-homogenoeus Boundary Value Problems and Applications, Vol. I and II. Springer, Berlin (1972)
30. Ni, W.M.: A nonlinear Dirichlet problem on the unit ball and its applications. Indiana Univ. Math. J. **31**, 801–807 (1982)
31. Peletier, L.A., van der Vorst, R.C.A.M.: Existence and non-existence of positive solutions of non-linear elliptic systems and the Biharmonic equation. Diff. Int. Equ. **5**, 747–767 (1992)
32. Rabinowitz, P.: Minimax Methods in Critical Point Theory with Applications to Differential Equations. CBMS Regional Conf. Ser. in Math., no. 65, Am. Math. Soc., Providence, R.I (1986)
33. Serra, E.: Non radial positive solutions for the Hénon equation with critical growth. Calc. Var. **23**, 301–326 (2005)
34. Smets, D., Willem, M.: Partial symmetry and asymptotic behavior for some elliptic variational problems. Calc. Var. **18**, 57–75 (2003)
35. van der Vorst, R.: Variational identities and applications to differential systems. Arch. Rat. Mech. Anal. **116**, 375–398 (1991)

Available online at www.sciencedirect.com
ScienceDirect
Journal of Functional Analysis 257 (2009) 721–752
JOURNAL OF Functional Analysis
www.elsevier.com/locate/jfa

Local "superlinearity" and "sublinearity" for the p-Laplacian

Djairo G. de Figueiredo [a], Jean-Pierre Gossez [b,*], Pedro Ubilla [c]

[a] IMECC-UNICAMP, Caixa Postal 6065, 13081-970 Campinas, SP, Brazil
[b] Département de Mathématique, C.P. 214, Université Libre de Bruxelles, 1050 Bruxelles, Belgium
[c] Universidad de Santiago de Chile, Casilla 307, Correo 2, Santiago, Chile

Received 23 November 2008; accepted 1 April 2009
Available online 29 April 2009
Communicated by J. Coron

Abstract

We study the existence, nonexistence and multiplicity of positive solutions for a family of problems $-\Delta_p u = f_\lambda(x,u)$, $u \in W_0^{1,p}(\Omega)$, where Ω is a bounded domain in $\mathbb{R}^N$, $N > p$, and $\lambda > 0$ is a parameter. The family we consider includes the well-known nonlinearities of Ambrosetti–Brezis–Cerami type in a more general form, namely $\lambda a(x)u^q + b(x)u^r$, where $0 \leqslant q < p-1 < r \leqslant p^* - 1$. Here the coefficient $a(x)$ is assumed to be nonnegative but $b(x)$ is allowed to change sign, even in the critical case. Preliminary results of independent interest include the extension to the p-Laplacian context of the Brezis–Nirenberg result on local minimization in $W_0^{1,p}$ and C_0^1, a $C^{1,\alpha}$ estimate for equations of the form $-\Delta_p u = h(x,u)$ with h of critical growth, a strong comparison result for the p-Laplacian, and a variational approach to the method of upper–lower solutions for the p-Laplacian.

Keywords: p-Laplacian; Concave convex nonlinearities; Critical exponent; C_0^1 versus $W_0^{1,p}$ local minimization; Strong comparison principle; $C^{1,\alpha}$ estimate; Upper–lower solutions

* Corresponding author.
E-mail addresses: djairo@ime.unicamp.br (D.G. de Figueiredo), gossez@ulb.ac.be (J.-P. Gossez), pubilla@fermat.usach.cl (P. Ubilla).

Originally published in: J. Funct. Anal. 257 (2009), 721-752,

DOI: 10.1007/978-3-319-02856-9_43

1. Introduction

This paper is concerned with the existence, nonexistence and multiplicity of solutions for the family of problems

$$\begin{cases} -\Delta_p u = f_\lambda(x,u) & \text{in } \Omega, \\ u > 0 & \text{in } \Omega, \\ u = 0 & \text{on } \partial\Omega \end{cases} \tag{1.1}$$

where $\Delta_p u := \operatorname{div}(|\nabla u|^{p-2}\nabla u)$ is the usual p-Laplacian, Ω is a bounded domain in $\mathbb{R}^N$, and $\lambda > 0$ is a real parameter. A basic feature of the family considered here is its monotone dependence on λ, i.e. $f_\lambda(x,s) \leqslant f_{\lambda'}(x,s)$ if $\lambda < \lambda'$. In the context of (1.1) the conditions of local "sublinearity" at 0 and of local "superlinearity" at ∞ mean, roughly speaking, that for x in a subdomain Ω_1 of Ω, one has

$$\lim_{\substack{s\to 0\\ s>0}} f_\lambda(x,s)/s^{p-1} = +\infty,$$

while for x in another subdomain Ω_2 of Ω, one has

$$\lim_{s\to+\infty} f_\lambda(x,s)/s^{p-1} = +\infty$$

(see (H_{Ω_1}) and (H_{Ω_2}) in Section 2 for the precise statements).

There are several motivations to our study of (1.1). The main one comes from the following example:

$$\begin{cases} -\Delta_p u = \lambda a(x)u^q + b(x)u^r & \text{in } \Omega, \\ u > 0 & \text{in } \Omega, \\ u = 0 & \text{on } \partial\Omega \end{cases} \tag{1.2}$$

where $0 < q < p-1 < r$. Problem (1.2) was originally considered in [2] when $p = 2$ and $a(x) \equiv 1$, $b(x) \equiv 1$. It was in particular shown there that if $r \leqslant 2^* - 1$, then there exists $0 < \Lambda < \infty$ such that (1.2) has at least two solutions for $\lambda < \Lambda$, at least one solution for $\lambda = \Lambda$, and no solution for $\lambda > \Lambda$. This result of "global multiplicity" was extended in [15] to the case $p = 2$ and variable coefficients $a(x)$, $b(x)$. It was also extended in [19] to the case $p \neq 2$ and $a(x) \equiv 1$, $b(x) \equiv 1$, although here under some restrictions on the exponents p, q in the critical case $r = p^* - 1$. In this paper we consider the general case: $p \neq 2$ and variable coefficients $a(x)$, $b(x)$. As in [15] $a(x)$ is restricted to be $\geqslant 0$ and $b(x)$ is allowed to change sign, even in the critical case $r = p^* - 1$. For the existence of a second solution, we will however need here a stronger restriction on $a(x)$, namely $a(x) > 0$. This difference with respect to the semilinear case is connected with the use of a strong comparison principle for the p-Laplacian (see the comments after hypothesis (M) in Section 2). In the critical case $r = p^* - 1$ we will meet similar restrictions as in [19] on the exponents p, q (see Remark 2.7).

As observed on p. 454 of [7], critical problems become more delicate in the presence of variable coefficients. In this respect our basic assumption on $b(x)$ in (1.2) for $r = p^* - 1$ is of the same nature as that introduced in [15] when $p = 2$: $b(x)$ should be sufficiently close to $\|b\|_{L^\infty(\Omega)}$ on a small ball (cf. condition (b) in Theorems 2.5 and 2.6).

Our general results relative to (1.1) can also be applied to various situations rather different from (1.2). For instance the Δ_p version of example (1.3) from [15] can be handled, as we shall see in Section 6.

There are several preliminary results in our study which have an independent interest. We mention in particular the extension to the p-Laplacian context of the well-known result of Brezis and Nirenberg [8] on local minimization in C_0^1 and $W_0^{1,p}$ (cf. Proposition 3.9). Several works have been devoted to this problem, e.g. [19,23,21,16]. Our approach differs from that in these papers and is more in the line of that introduced recently in [9] in the subcritical case. It avoids in particular the consideration of equations involving two p-Laplacians. An important step in the proof of this minimization result is Proposition 3.7 which provides a $C^{1,\alpha}$ estimate in the critical case. Another result of independent interest concerns the strong comparison principle for the p-Laplacian. This is known to be a delicate question, which has not received yet a complete answer (see [29] for a recent survey). The version we present here (cf. Proposition 3.4) is based on ideas from [4] and [17]. There is finally a variational approach to the method of upper–lower solutions for the p-Laplacian (cf. Proposition 3.1). It is adapted from [30] and could also prove useful in other situations.

Our method to obtain multiple solutions to (1.1) follows the classical way of obtaining a first solution via upper–lower solutions and a second solution via the mountain pass theorem. To handle the (PS) condition in the critical case we use some of the techniques initially developed for $p = 2$ in [7,2] and later extended for $p \neq 2$ in [18].

Our results relative to (1.1) are stated in detail in Section 2 and their proofs given in Sections 4 and 5. Section 3 is devoted to various preliminaries, including those mentioned above. Section 6 is devoted to some applications of the results of Section 2, in particular to (1.2).

2. Statement of results

In this section we state our results relative to (1.1). We successively consider $f_\lambda(x,u)$ of arbitrary growth, of subcritical growth, and finally of critical growth.

Let Ω be a smooth bounded domain in R^N. We assume $1 < p < N$. Our results, however, can be easily adapted to the case $p \geqslant N$, by replacing the subcritical or critical growth by an arbitrary power growth.

Our general assumptions on the family $f_\lambda(x,s)$ are the following:

(H) For each $\lambda > 0$, $f_\lambda : \Omega \times [0,\infty[\to \mathbb{R}$ is a Carathéodory function with the property that for any $s_0 > 0$, there exists a constant A (depending on λ and s_0), such that $|f_\lambda(x,s)| \leqslant A$ for a.e. $x \in \Omega$ and all $s \in [0,s_0]$. Moreover if $\lambda < \lambda'$, then

$$f_\lambda(x,s) \leqslant f_{\lambda'}(x,s)$$

for a.e. $x \in \mathbb{R}$ and all $s \geqslant 0$.

(H_0) For each $\lambda > 0$ and any $s_0 > 0$, there exists $B \geqslant 0$ (depending on λ and s_0), such that

$$f_\lambda(x,s) \geqslant -Bs^{p-1}$$

for a.e. $x \in \Omega$ and all $s \in [0,s_0]$.

Assumption (H_0) concerns the behavior of $f_\lambda(x,s)$ near $s=0$ and is used to apply the maximum principle; it implies $f_\lambda(x,0)\geqslant 0$. From now on we will always understand that $f_\lambda(x,s)$ has been extended for $s<0$ by putting $f_\lambda(x,s)=f_\lambda(x,0)$.

At this stage, if $u\in W_0^{1,p}(\Omega)\cap L^\infty(\Omega)$ satisfies $-\Delta_p u=f_\lambda(x,u)$ in the weak sense, then (H) and the regularity theory for the p-Laplacian (cf. e.g. [26]) imply $u\in C^{1,\alpha}(\overline{\Omega})$ for some $\alpha=\alpha(N,p)\in\,]0,1[$. Moreover $u\geqslant 0$ (take $-u^-$ as testing function and use $f_\lambda(x,0)\geqslant 0$). In addition, by the strong maximum principle of [31], if $u\not\equiv 0$, then $u>0$ in Ω and $\partial u/\partial\nu<0$ on $\partial\Omega$, where ν denotes the exterior normal. We thus have a solution of (1.1). Observe also that the associated functional

$$I_\lambda(u):=\frac{1}{p}\int_\Omega |\nabla u|^p-\int_\Omega F_\lambda(x,u) \tag{2.1}$$

where $F_\lambda(x,s):=\int_0^s f_\lambda(x,t)\,dt$, is well defined for $u\in W_0^{1,p}(\Omega)\cap L^\infty(\Omega)$.

The following two assumptions will also be used throughout the paper:

(H_e) There exist $\lambda>0$ and a nondecreasing function g with $\inf\{g(s)/s^{p-1}\colon s>0\}<1/\|e\|_\infty^{p-1}$ such that

$$f_\lambda(x,s)\leqslant g(s)$$

for a.e. $x\in\Omega$ and all $s\geqslant 0$; here $e\in W_0^{1,p}(\Omega)\cap C^{1,\alpha}(\overline{\Omega})$ is the solution of $-\Delta_p e=1$ and $\|\ \|_\infty$ denotes the $L^\infty(\Omega)$ norm.

(H_{Ω_1}) For any $\lambda>0$ there exists a smooth subdomain Ω_1, $s_1>0$ and $\theta_1>\lambda_1(\Omega_1)$ such that

$$f_\lambda(x,s)\geqslant\theta_1 s^{p-1}$$

for a.e. $x\in\Omega_1$ and all $s\in[0,s_1]$; here $\lambda_1(\Omega)$ denotes the principal eigenvalue of $-\Delta_p$ on $W_0^{1,p}(\Omega)$.

Assumption (H_e) is used to guarantee the existence of an upper solution for that specific value of λ. More comments about (H_e) can be found for instance in [14]. Assumption (H_{Ω_1}) is a local (i.e. on Ω_1) "sublinearity" condition at 0, which is satisfied for instance if

$$\lim_{\substack{s\to 0\\ s>0}} f_\lambda(x,s)/s^{p-1}=+\infty$$

uniformly for $x\in\Omega_1$. (H_{Ω_1}) is used to construct a lower solution.

Theorem 2.1 *(Existence of one solution without growth condition). Under the assumptions* (H), (H_0), (H_e) *and* (H_{Ω_1}), *there exists* $0<\Lambda\leqslant+\infty$ *such that problem* (1.1) *has at least one solution* $u\in W_0^{1,p}(\Omega)\cap L^\infty(\Omega)$ *(with* $I_\lambda(u)<0$*) for* $0<\lambda<\Lambda$ *and no solution for* $\lambda>\Lambda$.

As observed in [15, p. 272], in the present generality, Λ can be $+\infty$.

Theorem 2.2 *(Nonexistence for λ large). In addition to the hypotheses of Theorem* 2.1 *assume:*

$(H_{\tilde\Omega})$ *There exist $\lambda > 0$, a smooth subdomain $\tilde\Omega$, $\tilde m \in L^\infty(\tilde\Omega)$ with $\tilde m \geqslant 0$, $\not\equiv 0$, $\mu > \lambda_1(\tilde\Omega, \tilde m)$ such that*

$$f_\lambda(x,s) \geqslant \mu \tilde m(x) s^{p-1}$$

for a.e. $x \in \tilde\Omega$ and all $s \geqslant 0$; here $\lambda_1(\tilde\Omega, \tilde m)$ denotes the principal eigenvalue of $-\Delta_p$ on $W_0^{1,p}(\tilde\Omega)$ for the weight $\tilde m$.

Then problem (1.1) *for the value of λ provided by $(H_{\tilde\Omega})$ has no solution (and consequently $\Lambda < +\infty$).*

Assumption $(H_{\tilde\Omega})$ can be seen as a localized version of the trivial sufficient condition of nonexistence for the semilinear problem $-\Delta u = l(u)$ in Ω, $u > 0$ in Ω and $u = 0$ on $\partial\Omega$, namely $\inf\{l(s)/s\colon\ s > 0\} > \lambda_1(\Omega)$, where $\lambda_1(\Omega)$ denotes here the first eigenvalue of $-\Delta$ on $H_0^1(\Omega)$. Assumption $(H_{\tilde\Omega})$ is satisfied in particular if $f_\lambda(x,s) \geqslant h(\lambda)\tilde m(x)s^{p-1}$ for some function h with $h(\lambda) \to +\infty$ as $\lambda \to \infty$.

Due to the absence of growth condition, we have up to now included in the definition of a solution u the requirement that u belongs to $L^\infty(\Omega)$. We will now assume the following growth condition on $f_\lambda(x,u)$, where $p^* := Np/(N-p)$:

(G) For any $\lambda > 0$, there exist d_1, d_2 and $\sigma \leqslant p^* - 1$ such that

$$\left|f_\lambda(x,s)\right| \leqslant d_1 + d_2 s^\sigma$$

for a.e. $x \in \Omega$ and all $s \geqslant 0$.

Condition (G) implies that any $u \in W_0^{1,p}(\Omega)$ which solves $-\Delta_p u = f_\lambda(x,u)$ in the weak sense belongs to $L^\infty(\Omega)$ (cf. e.g. [3] in the subcritical case, [22] in the critical case), and consequently, as before, belongs to $C^{1,\alpha}(\overline{\Omega})$ for some $\alpha = \alpha(p,N)$. Moreover, when $\sigma < p^* - 1$, the norm of u in $C^{1,\alpha}(\overline{\Omega})$ can be estimated in terms of the constants from (G) and the norm of u in $W_0^{1,p}(\Omega)$ (by using successively [3] and [26]). Such an estimate does not hold anymore when $\sigma = p^* - 1$, as can be seen by considering for $p = 2$ the family of instantons

$$u_\varepsilon(x) = \left(\frac{\varepsilon}{\varepsilon^2 + |x|^2}\right)^{\frac{N-2}{2}} - \left(\frac{\varepsilon}{\varepsilon^2 + 1}\right)^{\frac{N-2}{2}}$$

on $\Omega = B(0,1)$. Note that Proposition 3.7 from Section 3 will provide a substitute to this estimate in the critical case. Condition (G) also implies that the functional $I_\lambda(u)$ is now well defined for all u in $W_0^{1,p}(\Omega)$.

The following Ambrosetti–Rabinowitz type condition, introduced in [14] to handle indefinite nonlinearities, will play a role in our subsequent results:

$(AR)_d$ For any $\lambda > 0$, there exist $\theta > p$, $\rho < p$, $s_0 \geqslant 0$ and $d \geqslant 0$ such that

$$\theta F_\lambda(x,s) \leqslant s f_\lambda(x,s) + d s^\rho$$

for a.e. $x \in \Omega$ and all $s \geqslant s_0$.

Theorem 2.3 *(Existence of one solution for $\lambda = \Lambda$ in the subcritical case). In addition to the hypotheses of Theorem* 2.2 *assume the continuity of $f_\lambda(x,s)$ with respect to λ (for a.e. x and uniformly for s bounded). Assume also that* (G) *with $\sigma < p^* - 1$ and $(AR)_d$ hold uniformly on each interval $[r, R] \subset \{\lambda > 0\}$ (i.e. the various constants appearing in* (G) *and* $(AR)_d$) *can be chosen independently of λ for $\lambda \in [r, R]$). Then problem* (1.1) *has at least one solution $u \in W_0^{1,p}(\Omega) \cap L^\infty(\Omega)$ (with $I_\lambda(u) \leqslant 0$) for $\lambda = \Lambda$.*

Next we state our results on the existence of at least two solutions for $0 < \lambda < \Lambda$. We will first deal with the subcritical case $\sigma < p^* - 1$. Some of the hypothesis of Theorem 2.1 have to be strengthened. Condition (H_0) is replaced by:

$(H_0)'$ For any $\lambda > 0$ and any $s_0 > 0$, there exists $B \geqslant 0$ such that for a.e. $x \in \Omega$, the function

$$s \to f_\lambda(x,s) + B s^{p-1}$$

is nondecreasing on $[0, s_0]$; moreover $f_\lambda(x, 0) \geqslant 0$ for all $\lambda \geqslant 0$ and a.e. $x \in \Omega$.

The monotonicity of the family f_λ is also strengthened in the following way, where we write $h(x) \prec l(x)$ to mean that for any compact $K \subset \Omega$ there exists $\varepsilon > 0$ such that $h(x) + \varepsilon \leqslant l(x)$ for a.e. $x \in K$:

(M) For any $\lambda < \lambda'$ and any $u \in C_0^1(\overline{\Omega})$ with $u > 0$ in Ω, one has

$$f_\lambda\big(x, u(x)\big) \prec f_{\lambda'}\big(x, u(x)\big).$$

Note that (M) is significantly stronger than the corresponding requirement in the semilinear case: $f_\lambda(z, u(x)) \leqslant\not\equiv f_{\lambda'}(x, u(x))$ (cf. [15, p. 273]). This difference is related to the use of a strong comparison principle for the p-Laplacian (cf. Proposition 3.4 below). Our last additional assumption is:

(H_{Ω_2}) For any $\lambda > 0$, these exist a subdomain Ω_2, $s_2 > 0$ and $\theta_2 > 0$ such that

$$F_\lambda(x,s) \geqslant \theta_2 s^p$$

for a.e. $x \in \Omega_2$ and all $s \geqslant s_2$.

Assumption (H_{Ω_2}) is a local (i.e. on Ω_2) "superlinearity" condition at ∞, which is satisfied for instance if

$$\lim_{s \to +\infty} f_\lambda(x,s)/s^{p-1} = +\infty$$

uniformly for $x \in \Omega_2$. It is used in conjunction with $(AR)_d$ to derive the geometry of the mountain pass.

Theorem 2.4 *(Second solution in the subcritical case). In addition to the hypotheses of Theorem* 2.1, *assume* (G) *with* $\sigma < p^* - 1$ *as well as* $(AR)_d$, $(H_0)'$, (M) *and* (H_{Ω_2}). *Then problem* (1.1) *has at least two solutions* u, v *for* $0 < \lambda < \Lambda$, *with* $u \leqslant\not\equiv v$ *in* Ω *and* $I_\lambda(u) < 0$.

We finally consider multiplicity in the case where f_λ has critical growth. Here we write f_λ as

$$f_\lambda(x, s) = h_\lambda(x, s) + b(x)s^{p^*-1}, \tag{2.2}$$

where $h_\lambda : \Omega \times [0, \infty[\to \mathbb{R}$ is a Carathéodory function and $b \in L^\infty(\Omega)$. We will distinguish two cases depending on the growth of the subcritical part $h_\lambda(x, s)$: either (i) h_λ satisfies (G) with $\sigma < p - 1$, $b(x)$ may change sign, or (ii) h_λ satisfies (G) with $\sigma < p^* - 1$, $b(x) \geqslant 0$ in Ω. In each case $b(x)$ will be assumed to satisfy the following condition:

(b) For some $x_0 \in \Omega$, some ball $B_1 \subset \Omega$ around x_0, some constant M and some γ_b with $\gamma_b > N(N-p)/(p^2 + N - p)$, one has

$$0 \leqslant \|b\|_\infty - b(x) \leqslant M|x - x_0|^{\gamma_b}$$

for a.e. $x \in B_1$. (Recall that $\| \ \|_\infty$ denotes the $L^\infty(\Omega)$ norm.)

Moreover, when $p \geqslant 3$, the following condition on h_λ will also be assumed:

(H_h) For any $\lambda > 0$ and any $s_0 > 0$, there exist $c_0 > 0$, $\delta > 0$, and q with $p^* - \frac{2}{p-1} < q + 1 < p$ such that

$$H_\lambda(x, s + u) - H_\lambda(x, s) \geqslant h_\lambda(x, s)u + c_0 u^{q+1}$$

for all $u \geqslant 0$, $s \in [0, s_0]$ and a.e. $x \in B(x_0, \delta)$, where $H_\lambda(x, s) := \int_0^s h_\lambda(x, t)\,dt$, and x_0 is the point involved in assumption (b) above.

Assumption (b) implies some control on the negative part of b: $\|b^-\|_\infty \leqslant \|b^+\|_\infty$, with in addition some limitation on the way $b(x)$ approaches $\|b\|_\infty$. It trivially holds if $b(x) = \|b\|_\infty$ on a small ball. Assumption (H_h) provides some control on the way $H_\lambda(x, s)$ is increasing. It is used to handle the case $p \geqslant 3$ in Lemma 5.3. A simple calculation based on Lemma A4, part (4), from [18] shows that (H_h) holds for instance if $h_\lambda(x, s) = \lambda a(x)(s^q + g(s))$ with g nondecreasing and $a(x) \geqslant \varepsilon > 0$ near x_0. Note that the condition $p^* - \frac{2}{p-1} < q + 1 < p$ in (H_h) imposes a rather strong restriction on the dimension: $N > p(1 + \frac{p(p-1)}{2})$. See in this respect Remarks 2.7 and 6.4.

We first deal with the critical case (i).

Theorem 2.5 *(Second solution in the critical case with* $\sigma < p - 1$ *for* h_λ*). In addition to the assumptions of Theorem* 2.1, *assume* $f_\lambda(x, s)$ *satisfies* $(H_0)'$ *and* (M). *Suppose also that* $f_\lambda(x, s)$ *can be written as in* (2.2) *with* $h_\lambda(x, s)$ *nondecreasing with respect to* s *and satisfying* (G) *with*

$\sigma < p - 1$. *Suppose also that* $b(x)$ *in* (2.2) *is* $\not\equiv 0$, *and satisfies* (b). *Then the conclusion of Theorem* 2.4 *holds provided that in addition either* (i) $2N/(N+1) < p < 3$, *or* (ii) $p \geqslant 3$ *and* (H_h) *is satisfied.*

We now deal with the critical case (ii).

Theorem 2.6 *(Second solution in the critical case with* $\sigma < p^* - 1$ *for* h_λ*). In addition to the assumptions of Theorem* 2.1, *assume* $f_\lambda(x,s)$ *satisfies* $(H_0)'$ *and* (M). *Suppose also that* $f_\lambda(x,s)$ *can be written as in* (2.2) *with* $h_\lambda(x,s)$ *nondecreasing with respect to* s *and satisfying* (G) *with* $\sigma < p^* - 1$, *and* $h_\lambda(x,s)$ *satisfying* $(AR)_d$. *Suppose that* b *in* (2.2) *is* $\not\equiv 0$, $\geqslant 0$ *in* Ω *and satisfies* (b). *Then the conclusion of Theorem* 2.4 *holds provided that in addition either* (i) $2N/(N+1) \leqslant p < 3$, *or* (ii) $p \geqslant 3$ *and* (H_h) *is satisfied.*

In Theorem 2.6, $h_\lambda(x,s)$ is allowed any subcritical growth, at the expense of assuming $(AR)_d$ for $h_\lambda(x,s)$ and $b(x) \geqslant 0$.

Remark 2.7. The condition $p > \frac{2N}{N+1}$ in Theorems 2.5 and 2.6 is slightly more restrictive than the condition $p > \frac{2N}{N+2}$ considered in [18,19]. The condition $p^* - \frac{2}{p-1} < q+1 < p$ from (H_h) already appears in [18,19].

3. Some preliminaries

3.1. Upper–lower solutions

We start by recalling the version of the method of upper–lower solutions which we will use repeatedly. Let $g(x,s)$ be a Carathéodory function on $\Omega \times \mathbb{R}$ with the property that for any $s_0 > 0$, there exists a constant A such that $|g(x,s)| \leqslant A$ for a.e. $x \in \Omega$ and all $s \in [-s_0, s_0]$. A function $u \in W^{1,p}(\Omega) \cap L^\infty(\Omega)$ is called a (weak) lower solution of the problem

$$\begin{cases} -\Delta_p u = g(x,u) & \text{in } \Omega, \\ u = 0 & \text{on } \partial\Omega, \end{cases} \tag{3.1}$$

if $u \leqslant 0$ on $\partial\Omega$ and

$$\int_\Omega |\nabla u|^{p-2} \nabla u \nabla \varphi \leqslant \int_\Omega g(x,u)\varphi$$

for all $\varphi \in C_c^\infty(\Omega)$, $\varphi \geqslant 0$. An upper solution is defined by reversing the inequality signs.

Proposition 3.1. *Assume that* $\underline{u}$ *and* $\bar{u}$ *are respectively lower and upper solutions for* (3.1), *with* $\underline{u} \leqslant \bar{u}$ *a.e. in* Ω. *Consider the associated functional*

$$\Phi(u) := \frac{1}{p}\int_\Omega |\nabla u|^p - \int_\Omega G(x,u)$$

where $G(x,s) := \int_0^s g(x,t)\,dt$, *and the interval*

$$M := \{u \in W_0^{1,p}(\Omega):\ \underline{u} \leqslant u \leqslant \overline{u} \text{ a.e. in } \Omega\}.$$

Then the infimum of Φ *on* M *is achieved at some* u, *and such a* u *is a solution of* (3.1).

Proof. It is adapted from [30] which deals with the semilinear case. By coercivity and weak lower semicontinuity, one easily sees that the infimum of Φ on M is achieved at some u. Let $\varphi \in C_c^\infty(\Omega)$, $\varepsilon > 0$, and define

$$v_\varepsilon := \min\{\overline{u}, \max\{\underline{u}, u + \varepsilon\varphi\}\} = u + \varepsilon\varphi - \varphi^\varepsilon + \varphi_\varepsilon$$

where $\varphi^\varepsilon := \max\{0, u + \varepsilon\varphi - \overline{u}\}$ and $\varphi_\varepsilon := -\min\{0, u + \varepsilon\varphi - \underline{u}\}$. Since u minimizes Φ on M, one has $\langle \Phi'(u), v_\varepsilon - u\rangle \geqslant 0$, which gives

$$\langle \Phi'(u), \varphi\rangle \geqslant \left(\langle \Phi'(u), \varphi^\varepsilon\rangle - \langle \Phi'(u), \varphi_\varepsilon\rangle\right)/\varepsilon. \tag{3.2}$$

Since $\overline{u}$ is an upper solution and $-\Delta_p$ is monotone, one also has

$$\begin{aligned}\langle \Phi'(u), \varphi^\varepsilon\rangle &\geqslant \langle \Phi'(u) - \Phi'(\overline{u}), \varphi^\varepsilon\rangle \\ &\geqslant \varepsilon \int_{\Omega_\varepsilon} \left(|\nabla u|^{p-2}\nabla u - |\nabla \overline{u}|^{p-2}\nabla\overline{u}\right)\nabla\varphi - \varepsilon \int_{\Omega_\varepsilon} \left|g(x,u) - g(x,\overline{u})\right||\varphi|\end{aligned}$$

where $\Omega_\varepsilon := \{x \in \Omega:\ u(x) + \varepsilon\varphi(x) \geqslant \overline{u}(x) > u(x)\}$. As $|\Omega_\varepsilon| \to 0$ as $\varepsilon \to 0$, this latter relation implies $\langle \Phi'(u), \varphi^\varepsilon\rangle \geqslant o(\varepsilon)$ as $\varepsilon \to 0$. Similarly $\langle \Phi'(u), \varphi_\varepsilon\rangle \leqslant o(\varepsilon)$, and so by (3.2), $\langle \Phi'(u), \varphi\rangle \geqslant 0$. Replacing φ by $-\varphi$, one concludes that u solves (3.1). □

Other results asserting the existence of a solution between a lower solution and an upper solution for a p-Laplacian equation can be found e.g. in [12,19]. These results however do not have the level of generality we need (our local "sublinearity" assumption will lead us to deal with weak lower solutions) or do not give enough information on the solution (the minimization property will be crucial for our multiplicity results).

3.2. Integration by parts formula

We now turn to an integration by parts formula which will play a role in the use of our local "sublinearity" condition (H_{Ω_1}).

Proposition 3.2. *Let* $u \in C^1(\overline{\Omega})$ *with* $\Delta_p u \in L^1(\Omega)$ *in the distribution sense. Let* $\varphi \in C^\infty(\overline{\Omega})$. *Then*

$$\int_\Omega |\nabla u|^{p-2}\nabla u\nabla\varphi = \int_{\partial\Omega} |\nabla u|^{p-2}\nabla u\nu\varphi - \int_\Omega (\Delta_p u)\varphi \tag{3.3}$$

where ν *denotes the exterior normal vector.*

Formula (3.3) is standard when $u \in W^{2,p}(\Omega)$. Its proof in the present situation is based on the following

Lemma 3.3. *(Cf. [11].) Let $a \in C(\overline{\Omega})^N$ be a vector field such that* $\operatorname{div} a \in L^1(\Omega)$ *in the distribution sense. Then $\int_\Omega \operatorname{div} a = \int_{\partial\Omega} a\nu$.*

Proof of Proposition 3.2. Take $a := |\nabla u|^{p-2}\nabla u\varphi$. By the well-known formula for the derivation of the product of a distribution by a C^∞ function, one has

$$\operatorname{div} a = (\Delta_p u)\varphi + |\nabla u|^{p-2}\nabla u \nabla\varphi.$$

Lemma 3.3 thus applies and yields (3.3). □

3.3. Strong comparison principle

The following strong comparison principle, which is obtained by combining arguments from [4] and [17], will be of importance in our study of multiplicity. We recall that the notation $f \prec g$ was introduced before the statement of assumption (M) in Section 2; moreover, for $C_0^1(\overline{\Omega})$ functions u and v, we will write $u \ll v$ to mean $u(x) < v(x)$ in Ω and $\frac{\partial u}{\partial \nu}(x) > \frac{\partial v}{\partial \nu}(x)$ on $\partial\Omega$.

Proposition 3.4. *Let $f, g \in L^\infty(\Omega)$, and let u, v be solutions of*

$$-\Delta_p u = \mu|u|^{p-2}u + f \quad \text{in } \Omega, \qquad u = 0 \quad \text{on } \partial\Omega, \tag{3.4}$$

$$-\Delta_p v = \mu|v|^{p-2}v + g \quad \text{in } \Omega, \qquad v = 0 \quad \text{on } \partial\Omega \tag{3.5}$$

where $\mu \leqslant 0$. If $0 \leqslant f \prec g$, then $u \ll v$.

Several works have been devoted in the last years to the strong comparison principle for the p-Laplacian. In particular the strict inequality $u < v$ in Proposition 3.4 was derived recently in [4]. Note that this conclusion $u < v$ does not hold if the hypothesis $0 \leqslant f \prec g$ in Proposition 3.4 is weakened into $0 \leqslant f \leqslant g$, $f \not\equiv g$ (cf. [11]). The stronger conclusion $u \ll v$ however holds under this weakened hypothesis on f, g if $0 \leqslant \mu < \lambda_1(\Omega)$ and $\partial\Omega$ is connected (cf. [10]). This conclusion $u \ll v$ also holds if $\mu = 0$ and $0 \leqslant f \leqslant g$ with $f \not\equiv g$ on any open subset of Ω (cf. [22]). Note that the assumption $0 \leqslant f \prec g$ was also considered recently in a slightly different setting in [21].

Proof of Proposition 3.4. The assumptions imply $g \geqslant 0, \not\equiv 0$, and so, by standard arguments, $v \gg 0$ (take $-v^-$ as testing function in (3.5) to get $v \geqslant 0$ and apply the strong maximum principle of [31]). Once this is observed, the proof of Proposition 2.6 in [4] can be followed without any change to reach $u(x) < v(x)$ in Ω. Now to derive the strict inequality of the normal derivatives on $\partial\Omega$, it suffices to apply near any point of $\partial\Omega$ a local strong comparison result from [17], which we recall below. □

Lemma 3.5. *(Cf. [17].) Let $f, g \in L^\infty(\Omega)$ and let u, v be solutions of* (3.4), (3.5) *where $\mu < \lambda_1(\Omega)$. Assume $0 \leqslant f \leqslant g$ and call $\Omega_\delta := \{x \in \Omega\colon \operatorname{dist}(x, \partial\Omega) < \delta\}$. Then for every $\delta > 0$*

sufficiently small and for every component Σ *of* Ω_δ, *one has either* (i) $u \equiv v$ *in* Σ, *or* (ii) $u < v$ *in* Σ *and* $\partial u/\partial \nu > \partial v/\partial \nu$ *on* $\partial\Omega \cap \overline{\Sigma}$.

3.4. Brezis–Lieb lemma

The following version of the Brezis–Lieb lemma (cf. [6]) for vector-valued functions will also be needed.

Lemma 3.6. *Let* f_k *be a bounded sequence in* $L^p(\Omega, \mathbb{R}^n)$, *where* $1 \leqslant p < \infty$. *Assume* $f_k \to f$ *a.e. in* Ω. *Then* $\|f_k\|_{L^p}^p - \|f - f_k\|_{L^p}^p \to \|f\|_{L^p}^p$, *where* $\|g\|_{L^p}$ *denotes* $(\int_\Omega |g(x)|^p)^{1/p}$ *with* $|g(x)|$ *the Euclidean norm of* $g(x) \in \mathbb{R}^n$.

Proof. It is easily adapted from the proof given for instance in [24] in the scalar case. The only difference is the verification of

$$\big||a+b|^p - |a|^p - |b|^p\big| \leqslant \varepsilon|a|^p + C_\varepsilon|b|^p$$

for $a, b \in \mathbb{R}^N$. This latter relation follows from the observation that for $a, b \in \mathbb{R}^N$, $(|a+b|^p - |a|^p - |b|^p)/|a|^p \to 0$ as $|a| \to +\infty$, uniformly for $|b| \leqslant 1$. □

3.5. $C_0^{1,\alpha}$ estimates of weak solutions in the critical case

For the proof in the next subsection on the local minimization in C_0^1 and $W_0^{1,p}$, we will need the following estimate.

Proposition 3.7. *Let a sequence* $u_k \in W_0^{1,p}(\Omega)$ *satisfy*

$$-\Delta_p u_k = h_k(x, u_k) \tag{3.6}$$

where the Carathéodory functions h_k *verify the uniform growth condition*

$$\big|h_k(x,s)\big| \leqslant C_1 + C_2|s|^{p^*-1}. \tag{3.7}$$

Assume that u_k *remains bounded in* $W_0^{1,p}(\Omega)$. *Moreover assume* $\int_E |u_k|^{p^*} \to 0$ *as* $|E| \to 0$, *uniformly in* k. *Then* u_k *remains bounded in* $C_0^{1,\alpha}(\bar\Omega)$ *for some* $0 < \alpha < 1$.

As observed in the comments after hypothesis (G) in Section 2, the fact that each u_k above belongs to $C_0^{1,\alpha}(\bar\Omega)$ follows from [22]. We are proving here that u_k remains bounded in $C_0^{1,\alpha}(\bar\Omega)$ provided, in addition to being bounded in $W_0^{1,p}(\Omega)$, u_k is uniformly equi-integrable in $L^{p^*}(\Omega)$. The necessity of such an additional requirement is clear from the comments after hypothesis (G). Our proof below combines arguments and results from [18,25,26].

Proof of Proposition 3.7. We break the proof in three steps. (i) There exists $q > p^*$ such that the sequence u_k remains bounded in $L^q(\Omega)$. (ii) The sequence u_k remains bounded in $L^\infty(\Omega)$. (iii) There exists $0 < \alpha < 1$ such that the sequence u_k remains bounded in $C_0^{1,\alpha}(\bar\Omega)$.

Proof of step (i). This is essentially a consequence of the calculations on pp. 951–952 from [18]. Indeed a careful reading of [18], using our assumption of uniform equi-integrability, shows the existence of $R > 0$ such that for any nonnegative $\eta \in C_c^\infty(\mathbb{R}^N)$ with support of diameter $\leqslant R$, on has

$$\left(\int_\Omega \eta^{p^*}\left(u_k^+\right)^{\beta p^*}\right)^{p/p^*} \leqslant C(\eta)\int_\Omega \left(u_k^+\right)^{\beta p} + C|\Omega| \tag{3.8}$$

with constants independent of k. Here β is fixed with $1 < \beta < p^*/p$. This clearly implies that u_k^+ remains bounded in $L^{\beta p^*}(\Omega)$. And a similar argument applied to u_k^- yields the conclusion of step (i).

Proof of step (ii). Theorem 7.1 from [25] can be applied to (3.6) to derive that u_k remains bounded in $L^\infty(\Omega)$. In fact a particular case of this result from [25] suffices here, which is recalled below as Lemma 3.8. Here are some details on the application of this Lemma 3.8 to (3.6): (3.10) clearly holds with $\varphi_1 \equiv 0$; taking $q > p^*$ as given by step (i), one can verify (3.11) with $\alpha_2 = p^* - 1$ and φ_2 a suitable constant by picking r_2 sufficiently large.

Proof of step (iii). Once the L^∞ estimate of step (ii) is obtained, the global regularity result of [26] can be applied to (3.6) and gives that for some $0 < \alpha < 1$, u_k remains bounded in $C_0^{1,\alpha}(\bar{\Omega})$. This concludes the proof of Proposition 3.7. □

Lemma 3.8. *(Cf. [25].) Let $u \in W_0^{1,p}(\Omega) \cap L^q(\Omega)$ with $q \geqslant p^*$ satisfy*

$$\int_\Omega a(x,u,\nabla u)\nabla v = \int_\Omega b(x,u)v \tag{3.9}$$

for v of the form $(u-c)^+$ or $(u+c)^-$, c any positive constant. Here the functions $a(x,s,\eta)$ and $b(x,s)$ are assumed to verify for $x \in \Omega$, $s \in \mathbb{R}$ and $\eta \in \mathbb{R}^N$,

$$\langle a(x,s,\eta),\eta\rangle \geqslant \nu|\eta|^p - \left(1+|s|^{\alpha_1}\right)\varphi_1(x), \tag{3.10}$$

$$(\operatorname{sign} s)b(x,s) \leqslant \left(1+|s|^{\alpha_2}\right)\varphi_2(x), \tag{3.11}$$

with ν a positive constant, $0 \leqslant \varphi_i \in L^{r_i}(\Omega)$, $r_i > N/p$, $0 \leqslant \alpha_1 < p\frac{N+q}{N} - \frac{q}{r_1}$ and $0 \leqslant \alpha_2 < p\frac{N+q}{N} - 1 - \frac{q}{r_2}$. Then $u \in L^\infty(\Omega)$ and $\|u\|_{L^\infty}$ can be estimated in terms of $\|u\|_{L^q}$, ν, α_i, $\|\varphi_i\|_{L^{r_i}}$ and Ω.

3.6. *Local minimization in C_0^1 and $W_0^{1,p}$*

In our study of multiplicity we will use the following extension to L^p of a well-known result of Brezis and Nirenberg [8].

Proposition 3.9. *Let $\Phi(u)$ be a functional of the form $\Phi(u) := \frac{1}{p}\int_\Omega |\nabla u|^p - \int_\Omega G(x,u)$, where $G(x,s) := \int_0^s g(x,t)\,dt$ and g satisfies the growth condition*

$$\left|g(x,s)\right| \leqslant d_1 + d_2|s|^\sigma \tag{3.12}$$

for some $\sigma \leqslant p^* - 1$ and some constants d_1, d_2. Let $u_0 \in W_0^{1,p}(\Omega)$ be a local minimizer of Φ for the $C_0^1(\overline{\Omega})$ topology, i.e. $\Phi(u_0) \leqslant \Phi(u_0 + w)$ for some $\varepsilon_0 > 0$ and all $w \in C_0^1(\bar{\Omega})$ with $\|w\|_{C_0^1} \leqslant \varepsilon_0$. Then $u_0 \in C_0^{1,\alpha}(\bar{\Omega})$ for some $0 < \alpha < 1$ and u_0 is a local minimizer of Φ for the $W_0^{1,p}(\Omega)$ topology, i.e. $\Phi(u_0) \leqslant \Phi(u_0 + w)$ for some $\varepsilon_1 > 0$ and all $w \in W_0^{1,p}(\Omega)$ with $\|w\|_{W_0^{1,p}} \leqslant \varepsilon_1$.

Proposition 3.9 was proved in [8] when $p = 2$ and later extended to the case $p \neq 2$ with $\sigma < p^* - 1$ in [19,23] (see also [21,16,9] for recent related works). Its validity in the critical case $\sigma = p^* - 1$ is also suggested in [19], although not explicitly. For that matter, we feel necessary to give here a complete proof. Our proof below borrows some ideas from [8,19] but follows a different approach, which turns out to be simpler and to yield a slightly stronger result (cf. Remark 3.10). In fact, as in [9], we avoid the consideration of equations involving two p-Laplacians, equations to which the global $C^{1,\alpha}$ estimates from [26] do not seem to apply.

Proof of Proposition 3.9. Since (3.12) with $\sigma < p^* - 1$ implies a similar condition with $\sigma = p^* - 1$, it suffices to consider the latter case. Moreover since $u_0 \in W_0^{1,p}(\Omega)$ satisfies

$$-\Delta_p u_0 = g(x, u_0) \quad \text{in } \Omega, \tag{3.13}$$

it follows from Corollary 1.1 in [22] that u_0 belongs to some $C_0^{1,\alpha}(\bar{\Omega})$. Assume by contradiction that u_0 is not a local minimizer of Φ for the $W_0^{1,p}(\Omega)$ topology. This means that for any $\varepsilon > 0$ there exists $v_\varepsilon \in W_0^{1,p}(\Omega)$ with $\|v_\varepsilon - u_0\|_{W_0^{1,p}} \leqslant \varepsilon$ and $\Phi(v_\varepsilon) < \Phi(u_0)$. For later use of Lagrange multiplier rule, it will be convenient to use only, as in [9], a consequence of that, namely $\|v_\varepsilon - u_0\|_{L^{p^*}} \leqslant \varepsilon$ and $\Phi(v_\varepsilon) < \Phi(u_0)$.

We now consider as in [8] the truncated functional

$$\Phi_j(u) := \frac{1}{p} \int_\Omega |\nabla u|^p - \int_\Omega G_j(x, u)$$

for $j = 1, 2, \ldots,$ where $g_j(x, s) := g(x, T_j(s))$, $G_j(x, s) := \int_0^s g_j(x, t)\, dt$ and $T_j(s) = -j$ if $s \leqslant -j$, s if $-j \leqslant s \leqslant j$ and $+j$ if $s \geqslant j$. Note that (3.12) still holds for each g_j, with constants and exponent independent of j. This easily implies, by dominated convergence, that for each $v \in W_0^{1,p}(\Omega)$, $\Phi_j(v) \to \Phi(v)$ as $j \to \infty$. It follows that for each $\varepsilon > 0$, there is some j_ε such that $\Phi_{j_\varepsilon}(v_\varepsilon) < \Phi(u_0)$. On the other hand, since g_{j_ε} has subcritical growth and since for some constants D_1, D_2 independent of ε,

$$\Phi_{j_\varepsilon}(v) \geqslant \frac{1}{p} \int_\Omega |\nabla v|^p - \int_\Omega \left(D_1 + D_2 |v|^{p^*}\right) \tag{3.14}$$

for $v \in W_0^{1,p}(\Omega)$, one deduces that Φ_{j_ε} achieves its infimum on $\{v \in W_0^{1,p}(\Omega): \|v - u_0\|_{L^{p^*}} \leqslant \varepsilon\}$ at some u_ε; this easily follows by taking a minimizing sequence and using (3.14). One thus has

$$\Phi_{j_\varepsilon}(u_\varepsilon) \leqslant \Phi_{j_\varepsilon}(v_\varepsilon) < \Phi(u_0). \tag{3.15}$$

By construction $u_\varepsilon \to u_0$ in $L^{p^*}(\Omega)$ as $\varepsilon \to 0$, and it follows from (3.14), (3.15) that u_ε remains bounded in $W_0^{1,p}(\Omega)$.

Claim. *There exists* $0 < \alpha < 1$ *such that* u_ε *remains bounded in* $C_0^{1,\alpha}(\overline{\Omega})$ *as* $\varepsilon \to 0$.

Accepting this claim, one deduces from Ascoli–Arzela theorem that $u_\varepsilon \to u_0$ in $C_0^1(\overline{\Omega})$. It follows that for $\varepsilon > 0$ sufficiently small,

$$\Phi(u_\varepsilon) = \Phi_{j_\varepsilon}(u_\varepsilon) < \Phi(u_0),$$

which contradicts the fact that u_0 is a local minimizer of Φ for the $C_0^1(\bar{\Omega})$ topology, and the proof of Proposition 3.9 will be complete.

It remains to prove the claim. For this purpose we write the Euler equation satisfied by u_ε:

$$-\Delta_p u_\varepsilon = g_{j_\varepsilon}(x, u_\varepsilon) + \mu_\varepsilon |u_\varepsilon - u_0|^{p^*-2}(u_\varepsilon - u_0) \tag{3.16}$$

where μ_ε is a Lagrange multiplier associated to the constraint $\|u_\varepsilon - u_0\|_{L^{p^*}} \leqslant \varepsilon$. Taking $u_0 - u_\varepsilon$ as testing function in (3.16) and using the minimizing property of u_ε, one gets $\mu_\varepsilon \leqslant 0$.

We now distinguish two cases according to the behavior of μ_ε as $\varepsilon \to 0$: either (i) μ_ε remains bounded, or (ii) for a subsequence $\mu_\varepsilon \to -\infty$. In case (ii) below, for simplicity of notation, we will keep writing $\varepsilon \to 0$ instead of considering a subsequence.

Case (i). In this case the conclusion of the claim is a direct consequence of Proposition 3.7.

Case (ii). In this case there exists $\varepsilon_0 > 0$ and a constant M such that for $0 < \varepsilon < \varepsilon_0$,

$$g_{j_\varepsilon}(x, s) + \mu_\varepsilon |s - u_0(x)|^{p^*-2}(s - u_0(x)) \begin{cases} < 0 & \text{for } s > M, \\ > 0 & \text{for } s < -M. \end{cases} \tag{3.17}$$

Indeed, by (3.12) and the fact that u_0 is bounded, $|g_{j_\varepsilon}(x, s)| \leqslant \tilde{d}_1 + \tilde{d}_2|s - u_0(x)|^{p^*-1}$ for some constant $\tilde{d}_1, \tilde{d}_2$; one then picks ε_0 such that $\mu_\varepsilon \leqslant -\tilde{d}_2 - 1$ for $0 < \varepsilon < \varepsilon_0$, and observe that the left-hand side of (3.17) for $s > \|u_0\|_\infty$ (resp. $s < -\|u_0\|_\infty$) and $0 < \varepsilon < \varepsilon_0$ is $\leqslant \tilde{d}_1 - |s - u_0|^{p^*-1}$ (resp. $\geqslant \tilde{d}_1 + |s - u_0|^{p^*-1}$); consequently (3.17) follows. Taking now $(u_\varepsilon - M)^+$ and $(u_\varepsilon - M)^-$ as testing functions in (3.16), one concludes that $|u_\varepsilon(x)| \leqslant M$ for $x \in \Omega$ and $0 < \varepsilon < \varepsilon_0$. So u_ε remains bounded in $L^\infty(\Omega)$ as $\varepsilon \to 0$.

We now take $|u_\varepsilon - u_0|^{\beta-1}(u_\varepsilon - u_0)$ with $\beta \geqslant 1$ as testing function in (3.13), (3.16), and use the monotonicity of $-\Delta_p$ to get

$$\begin{aligned} 0 &\leqslant \int_\Omega \big(|\nabla u_\varepsilon|^{p-2}\nabla u_\varepsilon - |\nabla u_0|^{p-2}\nabla u_0\big)\nabla\big[|u_\varepsilon - u_0|^{\beta-1}(u_\varepsilon - u_0)\big] \\ &= \int_\Omega \big(g_{j_\varepsilon}(x, u_\varepsilon) - g(x, u_0)\big)|u_\varepsilon - u_0|^{\beta-1}(u_\varepsilon - u_0) + \mu_\varepsilon \int_\Omega |u_\varepsilon - u_0|^{p^*+\beta-1}. \end{aligned}$$

Since u_ε remains bounded in $L^\infty(\Omega)$, using Hölder inequality in the integral involving g, we obtain

$$-\mu_\varepsilon \|u_\varepsilon - u_0\|_{L^{p^*+\beta-1}}^{p^*-1} \leqslant c$$

where the constant c does not depend on β and ε. Letting $\beta \to \infty$ yields

$$-\mu_\varepsilon \|u_\varepsilon - u_0\|_{L^\infty}^{p^*-1} \leqslant c.$$

So the right-hand side of (3.16) remains bounded in $L^\infty(\Omega)$, and it follows from [26] that for some $0 < \alpha = \alpha(N, p) < 1$, u_ε remains bounded in $C_0^{1,\alpha}(\overline{\Omega})$. This concludes the proof of the claim in case (ii). □

Remark 3.10. The above proof of Proposition 3.9 in the critical case $\sigma = p^* - 1$ shows that u_0 is a local minimizer of Φ on $W_0^{1,p}(\Omega)$ for the $L^{p^*}(\Omega)$ topology, i.e. $\Phi(u_0) \leqslant \Phi(u_0 + w)$ for some $\varepsilon_1 > 0$ and all $w \in W_0^{1,p}(\Omega)$ with $\|w\|_{L^{p^*}} \leqslant \varepsilon_1$. And in the subcritical case $\sigma < p^* - 1$, one would conclude that u_0 is a local minimizer of Φ on $W_0^{1,p}(\Omega)$ for the $L^{\sigma+1}(\Omega)$ topology.

Remark 3.11. The above proof of Proposition 3.9 greatly simplifies in the subcritical case $\sigma < p^* - 1$ (cf. [9]): the fact that u_0 belongs to some $C_0^{1,\alpha}(\bar{\Omega})$ now follows by using successively [3] and [26], no truncation of Φ is needed, and Proposition 3.7 can be replaced by another direct application of [3] and [26].

4. Proofs of Theorems 2.1, 2.2 and 2.3

This section is devoted to the proof of the first theorems stated in Section 2. The general strategy in this section as well as in the following one is rather similar to that in the semilinear case [15] and we will mainly concentrate on the differences with respect to [15]. It will be convenient from now on to denote (1.1) by $(1.1)_\lambda$.

Proof of Theorem 2.1. One starts by proving the existence of an upper solution of $(1.1)_\lambda$ for the value λ provided by (H_e). With g and e as in (H_e), there exists $M > 0$ such that

$$1/\|e\|_\infty^{p-1} \geqslant g\big(M\|e\|_\infty\big)/\big(M\|e\|_\infty\big)^{p-1}$$

and so one has, for any $\varphi \in C_c^\infty(\Omega)$ with $\varphi \geqslant 0$,

$$\begin{aligned}\int_\Omega \big|\nabla(Me)\big|^{p-2}\nabla(Me)\nabla\varphi &= \int_\Omega M^{p-1}\varphi \geqslant \int_\Omega g\big(M\|e\|_\infty\big)\varphi \\ &\geqslant \int_\Omega g(Me)\varphi \geqslant \int_\Omega f_\lambda(x, Me)\varphi.\end{aligned}$$

This shows that Me is an upper solution.

We now construct a lower solution of $(1.1)_\lambda$ by using the local "sublinearity" assumption (H_{Ω_1}) at λ. Denote by φ_1 a positive principal eigenfunction of $-\Delta_p$ on $W_0^{1,p}(\Omega_1)$; one has $\varphi_1 \in C^{1,\alpha}(\overline{\Omega_1})$ and $\partial\varphi_1/\partial\nu < 0$ on $\partial\Omega_1$, where ν denotes here the exterior normal on $\partial\Omega_1$. Extending φ_1 by 0 on $\Omega \setminus \Omega_1$, the extended function, still denoted by φ_1, belongs to $W_0^{1,p}(\Omega) \cap L^\infty(\Omega)$. Call $u_\varepsilon := \varepsilon\varphi_1$ for $\varepsilon > 0$, and let $\varphi \in C_c^\infty(\Omega)$, $\varphi \geqslant 0$. One has, for ε sufficiently small so that $\|u_\varepsilon\|_\infty \leqslant s_1$ (where s_1 comes from (H_{Ω_1})),

$$\int_{\Omega} |\nabla u_\varepsilon|^{p-2}\nabla u_\varepsilon \nabla\varphi = \int_{\partial\Omega_1} |\nabla u_\varepsilon|^{p-2}\nabla u_\varepsilon \nu_\varphi - \int_{\Omega_1} (\Delta_p u_\varepsilon)\varphi$$

$$\leqslant \lambda_1(\Omega_1)\int_{\Omega_1} |u_\varepsilon|^{p-2}u_\varepsilon\varphi \leqslant \int_{\Omega_1} f_\lambda(x,u_\varepsilon)\varphi \leqslant \int_{\Omega} f_\lambda(x,u_\varepsilon)$$

where we have used successively Proposition 3.2, (H_{Ω_1}) and $f_\lambda(x,0)\geqslant 0$. This shows that u_ε is a (weak) lower solution. Moreover, taking $\varepsilon>0$ smaller if necessary, one has $u_\varepsilon \leqslant Me$ in Ω.

Proposition 3.1 can thus be applied and yields a solution $u\in W_0^{1,p}(\Omega)\cap L^\infty(\Omega)$ of $(1.1)_\lambda$ for the value of λ provided by (H_e). So at this stage, we have proved that

$$\Lambda := \sup\{\lambda>0\colon\ (1.1)_\lambda \text{ has a solution}\} > 0.$$

It remains to show that for each $0<\lambda<\Lambda$, $(1.1)_\lambda$ has a solution u with $I_\lambda(u)<0$. Let $0<\lambda<\Lambda$ and take $\bar\lambda$ with $\lambda<\bar\lambda<\Lambda$ such that $(1.1)_{\bar\lambda}$ has a solution $\bar u$. One has, by the monotonicity of the family f_λ (cf. (H)),

$$\int_{\Omega} |\nabla\bar u|^{p-2}\nabla\bar u\nabla\varphi = \int_{\Omega} f_{\bar\lambda}(x,\bar u)\varphi \geqslant \int_{\Omega} f_\lambda(x,\bar u)\varphi$$

for all $\varphi\in C_c^\infty(\Omega)$ with $\varphi\geqslant 0$. This shows that $\bar u$ is an upper solution for $(1.1)_\lambda$. The rest of the argument is then easily adapted from p. 275 in [15]: one first constructs a (weak) lower solution as above in the form $\varepsilon\varphi_1$ by using (H_{Ω_1}) at λ, and Proposition 3.1 applies again to yield a solution u_0 to $(1.1)_\lambda$; moreover the minimization property provided by Proposition 3.1 leads to $I_\lambda(u_0)\leqslant I_\lambda(\varepsilon\varphi_1)$, and by (H_{Ω_1}), $I_\lambda(\varepsilon\varphi_1)<0$ for ε sufficiently small. □

Proof of Theorem 2.2. Let λ be given by $(H_{\tilde\Omega})$ and suppose by contradiction that $(1.1)_\lambda$ admits a solution $u\in W_0^{1,p}(\Omega)\cap L^\infty(\Omega)$. Consider the eigenvalue problem with weight

$$\begin{cases} -\Delta_p v = \mu\tilde m(x)|v|^{p-2}v & \text{in } \tilde\Omega,\\ v=0 & \text{on } \partial\tilde\Omega.\end{cases} \tag{4.1}$$

Since by $(H_{\tilde\Omega})$,

$$\int_{\tilde\Omega} |\nabla u|^{p-2}\nabla u\nabla\varphi = \int_{\tilde\Omega} f_\lambda(x,u)\varphi \geqslant \mu\int_{\tilde\Omega} \tilde m(x)u^{p-1}\varphi$$

for $\varphi\in C_c^\infty(\Omega)$, $\varphi\geqslant 0$, we see that u (restricted to $\tilde\Omega$) is an upper solution for (4.1). On the other hand let φ_1 be a positive eigenfunction associated to $\lambda_1(\tilde\Omega,\tilde m)$ and call $u_\varepsilon := \varepsilon\varphi_1$. Since

$$\int_{\tilde\Omega} |\nabla u_\varepsilon|^{p-2}\nabla u_\varepsilon\nabla\varphi = \lambda_1(\tilde\Omega,\tilde m)\int_{\tilde\Omega} \tilde m(x)u_\varepsilon^{p-1}\varphi < \mu\int_{\tilde\Omega} \tilde m(x)u_\varepsilon^{p-1}\varphi$$

for $\varphi\in C_c^\infty(\Omega)$, $\varphi\geqslant 0$, we see that u_ε is a lower solution for (4.1). Clearly, for $\varepsilon>0$ sufficiently small, $u_\varepsilon\leqslant u$ on $\tilde\Omega$ (since for $x_0\in\partial\tilde\Omega$, either $u(x_0)>0$ or $u(x_0)=0$ and $\partial u/\partial\nu(x_0)<0$, where

ν is here the exterior normal on $\partial\tilde{\Omega}$). Proposition 3.1 then guarantees the existence of a solution v of (4.1) with $u_\varepsilon \leqslant v \leqslant u$ in $\tilde{\Omega}$. In particular $v \geqslant 0$, $v \not\equiv 0$, which shows that μ is a principal eigenvalue of $-\Delta_p$ on $\tilde{\Omega}$ for the weight $\tilde{m}$. This is a contradiction since $\mu > \lambda_1(\tilde{\Omega}, \tilde{m})$. □

Remark 4.1. Here is another proof of Theorem 2.2 based on Picone's identity (cf. [1]). Suppose again that $(1.1)_\lambda$ admits a solution u and let $\varphi \in C_c^\infty(\tilde{\Omega})$ with $\varphi \geqslant 0$ and $\int_{\tilde{\Omega}} \tilde{m}\varphi^p > 0$. By Picone's identity,

$$\int_\Omega |\nabla\varphi|^p - \int_\Omega \nabla\left(\varphi^p/u^{p-1}\right)|\nabla u|^{p-2}\nabla u \geqslant 0$$

and consequently, from the equation satisfied by u and $(H_{\tilde{\Omega}})$,

$$\int_\Omega |\nabla\varphi|^p \geqslant \int_\Omega f_\lambda(x,u)\varphi^p/u^{p-1} \geqslant \mu \int_{\tilde{\Omega}} \tilde{m}(x)\varphi^p.$$

Taking the infimum with respect to φ yields $\lambda_1(\tilde{\Omega}, \tilde{m}) \geqslant \mu$, a contradiction.

Proof of Theorem 2.3. Let $\mu_k \to \Lambda$ with $0 < \mu_k < \Lambda$, μ_k increasing, and let u_k be a solution of $(1.1)_{\mu_k}$ with $I_{\mu_k}(u_k) < 0$.

We first show that u_k remains bounded in $W_0^{1,p}(\Omega)$. This is carried out as on [15, p. 276]: using $I_{\mu_k}(u_k) < 0$ and $(AR)_d$, and denoting by $\|v\|$ the $W_0^{1,p}(\Omega)$ norm $(\int_\Omega |\nabla v|^p)^{\frac{1}{p}}$, one obtains

$$\frac{\theta}{p}\|u_k\|^p \leqslant \int_\Omega f_{\mu_k}(x,u_k)u_k + d\int_\Omega u_k^p + c_1$$

for some constant c_1. One then deduces, using $(1.1)_{\mu_k}$,

$$\left(\frac{\theta}{p} - 1\right)\|u_k\|^p \leqslant c_2\|\mu_k\|^p + c_1$$

for some constant c_2, which gives the required bound.

Since $\sigma < p^* - 1$, we have, for a subsequence, $u_k \to u$ in $C^1(\overline{\Omega})$. Clearly u solves $-\Delta_p u = f_\Lambda(x,u)$ in Ω, $u \geqslant 0$ in Ω, $u = 0$ on $\partial\Omega$, and one has $I_\Lambda(u) \leqslant 0$. It remains to see that $u \not\equiv 0$.

Assume by contradiction $u \equiv 0$. We will use (H_{Ω_1}) for $\lambda = \mu_1$. Let as before Ω_1 be the corresponding subdomain and φ_1 a positive eigenfunction associated to the principal eigenvalue $\lambda_1(\Omega_1)$ of $-\Delta_p$ on $W_0^{1,p}(\Omega_1)$. We have, for $\varphi \in C_c^\infty(\Omega)$, $\varphi \geqslant 0$, using the monotonicity of the family f_λ,

$$\int_{\Omega_1} |\nabla u_k|^{p-2}\nabla u_k \nabla\varphi \geqslant \int_{\Omega_1} f_{\mu_1}(x,u_k)\varphi \geqslant \theta_1 \int_{\Omega_1} u_k^{p-1}\varphi$$

for k sufficiently large (so that $0 \leqslant u_k(x) \leqslant s_1$ on Ω_1, which is possible since $u_k \to 0$ uniformly). The above relation shows that u_k (restricted to Ω_1) is an upper solution for the problem

$$\begin{cases} -\Delta_p v = \theta_1 |v|^{p-2} v & \text{in } \Omega_1, \\ v = 0 & \text{on } \partial\Omega_1. \end{cases} \tag{4.2}$$

Since $\lambda_1(\Omega_1) < \theta_1$, one also has that for $\varepsilon > 0$ sufficiently small, $\varepsilon\varphi_1$ is a lower solution of (4.2) which satisfies $\varepsilon\varphi_1 \leqslant u_k$ on Ω_1. Proposition 3.1 then implies the existence of a solution v of (4.2) with $\varepsilon\varphi_1 \leqslant v \leqslant u_k$. This is a contradiction since $\theta_1 > \lambda_1(\Omega_1)$. □

Remark 4.2. Here is another proof of Theorem 2.3 based on Picone's identity. One starts by constructing u as above. Assume again by contradiction $u \equiv 0$. With $\Omega_1, \lambda_1(\Omega_1), \varphi_1$ and k sufficiently large as above, one has

$$\lambda_1(\Omega_1) \int_{\Omega_1} \varphi_1^p = \int_{\Omega_1} |\nabla\varphi_1|^p \geqslant \int_{\Omega_1} \nabla\left(\varphi_1^p / u_k^{p-1}\right) |\nabla u_k|^{p-2} \nabla u_k$$
$$= \int_{\Omega} f_{\mu_k}(x, u_k) \varphi_1^p / u_k^{p-1} \geqslant \int_{\Omega_1} f_{\mu_k}(x, u_k) \varphi_1^p / u_k^{p-1} \geqslant \theta_1 \int_{\Omega_1} \varphi_1^p,$$

which is a contradiction since $\theta_1 > \lambda_1(\Omega_1)$. Here one has used successively the definition of $\lambda_1(\Omega_1)$, Picone's identity, the equation satisfied by u_k, the monotonicity of the family f_λ, and (H_{Ω_1}).

5. Proofs of Theorems 2.4, 2.5 and 2.6

We now turn to the proof of the multiplicity theorems from Section 2. We start with the following result which concerns the first solution.

Theorem 5.1. *In addition to the assumptions of Theorem* 2.1, *assume that* $f_\lambda(x, s)$ *satisfies* (H_0'), (M) *and the growth condition* (G) *with* $\sigma \leqslant p^* - 1$. *Let* $0 < \lambda < \Lambda$. *Then there exists a solution* $u_0 \in C^{1,\alpha}(\bar{\Omega})$ *of problem* $(1.1)_\lambda$ *which is a local minimum of* I_λ *in the* $W_0^{1,p}$ *topology.*

Proof. Pick λ_1, λ_2 with $0 < \lambda_1 < \lambda < \lambda_2 < \Lambda$. We first observe that $(1.1)_{\lambda_1}$ and $(1.1)_{\lambda_2}$ have solutions u_1 and u_2, respectively, which satisfy $u_1 \leqslant u_2$. This can be seen as follows. One starts with a solution u_2 of $(1.1)_{\lambda_2}$ and considers $(1.1)_{\lambda_1}$. By the monotonicity of the family f_λ, u_2 is an upper solution of $(1.1)_{\lambda_1}$; moreover, using (H_{Ω_1}) at λ_1 as in the proof of Theorem 2.1, one constructs a lower solution of $(1.1)_{\lambda_1}$ which is smaller than u_2. Proposition 3.1 thus applies and yields a solution u_1 of $(1.1)_{\lambda_1}$ with $u_1 \leqslant u_2$ in Ω and $I_{\lambda_1}(u_1) < 0$.

We now use u_1 and u_2 as lower and upper solutions for $(1.1)_\lambda$ and apply as before Proposition 3.1 to obtain a solution u_0 of $(1.1)_\lambda$ which minimizes I_λ on $\{u \in W^{1,p}(\Omega)\colon u_1 \leqslant u \leqslant u_2\}$ and satisfies $I_\lambda(u_0) < 0$ (the latter inequality follows from the minimization property, using $I_\lambda(u_1) \leqslant I_{\lambda_1}(u_1) < 0$). We claim that

$$\begin{cases} u_1 < u_0 < u_2 & \text{in } \Omega, \\ \partial u_1/\partial\nu > \partial u_0/\partial\nu > \partial u_2/\partial\nu & \text{on } \partial\Omega. \end{cases} \tag{5.1}$$

Let us prove the relations involving u_1 and u_0 (same argument for u_0 and u_2). We first use (M) and $(H_0)'$ at λ to derive that for a suitable $B \geqslant 0$,

$$
\begin{aligned}
-\Delta_p u_1 = f_{\lambda_1}(x,u_1) \prec f_\lambda(x,u_1) &= f_\lambda(x,u_1) + Bu_1^{p-1} - Bu_1^{p-1} \\
&\leqslant f_\lambda(x,u_0) + Bu_0^{p-1} - Bu_1^{p-1} = -\Delta_p u_0 + Bu_0^{p-1} - Bu_1^{p-1}.
\end{aligned}
$$

So

$$-\Delta_p u_1 + Bu_1^{p-1} \prec -\Delta_p u_0 + Bu_0^{p-1}.$$

Moreover, taking B larger if necessary and applying $(H_0)'$ at λ_1, one also has

$$-\Delta_p u_1 + Bu_1^{p-1} = f_{\lambda_1}(x,u_1) + Bu_1^{p-1} \geqslant f_{\lambda_1}(x,0) \geqslant 0.$$

We are thus in a position to apply the strong comparison principle of Proposition 3.4, which yields $u_1 \ll u_0$, i.e. the assertion (5.1) relative to u_1 and u_0.

It follows from (5.1) that $\{u \in W_0^{1,p}(\Omega)\colon u_1 \leqslant u \leqslant u_2\}$ contains a $C_0^1(\overline{\Omega})$ neighborhood of u_0 and consequently, u_0 is a local minimizer of I_λ on $C_0^1(\overline{\Omega})$. We now apply Proposition 3.9 to get that u_0 is a local minimizer of I_λ on $W_0^{1,p}(\Omega)$. □

The proofs of Theorems 2.4, 2.5 and 2.6 have a common part, that we will first consider. We keep assuming here the hypothesis of Theorem 5.1. In each theorem the second solution of $(1.1)_\lambda$ will be constructed in the form $u_0 + w$ where u_0 is the first solution provided by Theorem 5.1 and w is a nonzero solution of

$$
\begin{cases}
-\Delta_p(u_0 + w) = f_\lambda(x, u_0 + w^+) & \text{in } \Omega, \\
w = 0 & \text{on } \partial\Omega.
\end{cases}
\tag{5.2}
$$

And the construction of a nonzero solution of (5.2) will be carried out using the mountain pass theorem.

Observe that if $w \in W_0^{1,p}(\Omega)$ solves (5.2), then $w \geqslant 0$. Indeed, by (G) and the regularity theory, $w \in L^\infty(\Omega)$; moreover, using $(H_0)'$, one has, for a suitable $B \geqslant 0$,

$$
\begin{aligned}
-\Delta_p u_0 &= f_\lambda(x,u_0) + Bu_0^{p-1} - Bu_0^{p-1} \\
&\leqslant f_\lambda(x, u_0 + w^+) + B(u_0 + w^+)^{p-1} - Bu_0^{p-1} \\
&= -\Delta_p(u_0 + w) + B(u_0 + w^+)^{p-1} - Bu_0^{p-1}
\end{aligned}
$$

and consequently, since $(B(u_0 + w^+)^{p-1} - Bu_0^{p-1})w^- \equiv 0$, we have

$$\int_\Omega \big[|\nabla u_0|^{p-2}\nabla u_0 - |\nabla(u_0 + w)|^{p-2}\nabla(u_0 + w)\big]\nabla w^- \leqslant 0.$$

Splitting the preceding integral as an integral on $\{w > 0\}$ and an integral on $\{w \leqslant 0\}$, one obtains, by strict monotonicity, $w^- \equiv 0$, i.e. $w \geqslant 0$.

It follows that if w is a nonzero solution of (5.2), then $u_0 + w$ will be a second solution of $(1.1)_\lambda$ satisfying all the requirements in Theorem 2.4. To derive the existence of a nonzero solution of (5.2), we write the associated functional

$$J_\lambda(w) := \frac{1}{p}\int_\Omega |\nabla(u_0+w)|^p - \int_\Omega G_\lambda(x,w) \tag{5.3}$$

where $G_\lambda(x,w) := F_\lambda(x, u_0 + w^+) - F_\lambda(x, , u_0) - f_\lambda(x, u_0)w^-$. We are thus lead to look for a nonzero critical point of J_λ on $W_0^{1,p}(\Omega)$.

We now prove that 0 is a local minimizer of J_λ on $W_0^{1,p}(\Omega)$. Indeed, using the fact that u_0 is a local minimizer of I_λ on $W_0^{1,p}(\Omega)$, one obtains

$$J_\lambda(w) \geqslant \frac{1}{p}\int_\Omega |\nabla(u_0+w)|^p - \frac{1}{p}\int_\Omega \left|\nabla\left(u_0+w^+\right)\right|^p + \frac{1}{p}\int_\Omega |\nabla u_0|^p + \int_\Omega f_\lambda(x,u_0)w^- \tag{5.4}$$

for $\|w^+\|$ sufficiently small. Recall that when $p \geqslant 2$,

$$|\xi_2|^p \geqslant |\xi_1|^p + p|\xi_1|^{p-2}\langle \xi_1, \xi_2 - \xi_1\rangle + c(p)|\xi_2 - \xi_1|^p/(2^p-1) \tag{5.5}$$

for some positive constant $c(p)$ and all $\xi_1, \xi_2 \in R^N$, and that a similar relation holds when $p < 2$, with the last term of (5.5) replaced by $c(p)|\xi_1 - \xi_2|^2/(|\xi_2| + |\xi_1|)^{2-p}$ (cf. [27,28]). Using (5.5) and the fact that u_0 solves $(1.1)_\lambda$, one derives from (5.4) that when $p \geqslant 2$,

$$J_\lambda(w) \geqslant \frac{1}{p}\int_\Omega |\nabla u_0|^p + c(p)\int_\Omega |\nabla w^-|^p/(2^p-1) \geqslant J_\lambda(0),$$

i.e. the conclusion that 0 is a local minimizer of J_λ. A similar inequality can be obtained when $p < 2$.

We will now distinguish between the subcritical situation of Theorem 2.4 and the critical situation of Theorems 2.5 and 2.6.

Proof of Theorem 2.4. Assumption (G) with $\sigma < p^* - 1$ and $(AR)_d$ imply that J_λ satisfies the (PS) condition on $W_0^{1,p}(\Omega)$. Indeed, if w_k is a (PS) sequence, then, for θ as in $(AR)_d$ and for some $\varepsilon_k \to 0$ and some constant c, we have

$$\theta J_\lambda(w_k) - J'_\lambda(w_k)(u_0 + w_k) \leqslant c + \varepsilon_k\|u_0 + w_k\|.$$

Next, after some computations, using $(AR)_d$, one deduces, for another constant c',

$$\left(\frac{\theta}{p} - 1\right)\|u_0 + w_k\|^p + (\theta - 1)\int_\Omega f_\lambda(x,u_0)w_k^- \leqslant c' + d\int_\Omega \left(u_0 + w_k^+\right)^\rho + \varepsilon_k\|u_0 + w_k\|.$$

Since $\rho < p$, this implies that the sequence (w_k) remains bounded in $W_0^{1,p}(\Omega)$. Passing to a subsequence, let w_0 be the weak limit of (w_k). So it follows that $J'_\lambda(w_k)(w_k - w_0) \to 0$. Using the fact that $\sigma < p^* - 1$ and the $(S)_+$ property of $-\Delta_p$, one derives, in a rather standard way, that for a further subsequence, w_k converges in $W_0^{1,p}(\Omega)$.

From the above discussion, we know that 0 is a local minimizer of J_λ on $W_0^{1,p}(\Omega)$. One now faces the following alternative, as in [15, p. 278]: either J_λ admits near 0 another local minimizer (and we are finished), or for any $r > 0$ sufficiently small,

$$J_\lambda(0) < \inf\{J_\lambda(w)\colon w \in W_0^{1,p}(\Omega) \text{ and } \|w\| = r\} \tag{5.6}$$

(by using Theorem 5.10 from [13]). We aim in the latter case at applying the mountain pass theorem. This will be possible if for some $\varphi \in W_0^{1,p}(\Omega)$, $J_\lambda(t\varphi) \to -\infty$ as $t \to +\infty$. Assumption (H_{Ω_2}) will be used to construct such a φ. One first adapts the calculation from p. 462 in [14] to derive from (H_{Ω_2}) and $(AR)_d$ that for some s_3 and some $c > 0$,

$$F_\lambda(x, s) \geqslant cs^\theta \tag{5.7}$$

for a.e. $x \in \Omega_2$ and all $s \geqslant s_3$, where θ comes from $(AR)_d$. Take $\varphi \in C_c^\infty(\Omega_2)$ with $\varphi \geqslant 0$, $\varphi \not\equiv 0$. Since

$$J_\lambda(t\varphi) = \frac{1}{p}\int_\Omega |\nabla(u_0 + t\varphi)|^p - \int_\Omega F_\lambda(x, u_0 + t\varphi) + \int_\Omega F_\lambda(x, u_0),$$

one easily derives from (5.7) that $J_\lambda(t\varphi) \to -\infty$ as $t \to \infty$. This concludes the proof of Theorem 2.4. □

Now we study the (PS) condition for the functional J_λ under the hypotheses of either Theorem 2.5 or 2.6.

Lemma 5.2. *Assume that* 0 *is the only critical point of* J_λ. *Then* J_λ *satisfies the* $(\mathrm{PS})_c$ *condition for all levels* c *with*

$$c < c_0 := \frac{\|u_0\|^p}{p} + \frac{S^{N/p}}{N\|b\|_\infty^{(N-p)/p}} \tag{5.8}$$

where $S := \inf\{\int_\Omega |\nabla u|^p / (\int_\Omega |u|^{p^*})^{p/p^*}\colon u \in W_0^{1,p}(\Omega),\ u \not\equiv 0\}$ *is the best Sobolev constant.*

Proof. Let w_k be a $(\mathrm{PS})_c$ sequence with $c < c_0$, i.e.

$$\frac{1}{p}\|u_0 + w_k\|^p - \int_\Omega G_\lambda(x, w_k) \to c, \tag{5.9}$$

$$\left|\int_\Omega |\nabla(u_0 + w_k)|^{p-2}\nabla(u_0 + w_k)\nabla\varphi - \int_\Omega g_\lambda(x, w_k)\varphi\right| \leqslant \varepsilon_k\|\varphi\|, \quad \forall\varphi \in W_0^{1,p}(\Omega) \tag{5.10}$$

where $\varepsilon_k \to 0$. Recall that in the present situation $g_\lambda(x,s) := h_\lambda(x, u_0 + s^+) + b(x)(u_0 + s^+)^{p^*-1}$ and $G_\lambda(x,s) := H_\lambda(x, u_0 + s^+) + b(x)(u_0 + s^+)^{p^*}/p^* - H_\lambda(x, u_0) - b(x)u_0^{p^*}/p^* - h_\lambda(x, u_0)s^- - b(x)u_0^{p^*-1}s^-$, with $H_\lambda(x,s) := \int_0^s h_\lambda(x,t)\,dt$.

We first observe that w_k remains bounded in $W_0^{1,p}(\Omega)$. This follows by multiplying (5.10) with $\varphi = u_0 + w_k$ by $1/p^*$ and subtracting from (5.9): the terms of power p^* cancel, and the remaining dominating term is $\|w_k\|^p$, which easily yields the desired bound. In this argument we have used condition (G) with $\sigma < p - 1$ in the case of Theorem 2.5, and condition $(AR)_d$ as well as $b(x) \geqslant 0$ in the case of Theorem 2.6 (the argument in this latter case is a little more involved but can be easily adapted from that on p. 283 in [15]). So, for a subsequence, $w_k \to w_0$ weakly in $W_0^{1,p}(\Omega)$, strongly in $L^r(\Omega)$ for any $r < p^*$, and a.e. in Ω. Writing (5.10) as

$$-\Delta_p(u_0 + w_k) = g_\lambda(x, w_k) + f_k \tag{5.11}$$

where $f_k \in W^{-1,p'}(\Omega)$ with $\|f_k\|_{-1,p'} \leqslant \varepsilon_k$, one can apply Theorem 2.1 from [5] to go to the limit in the weak form of (5.11), and so w_0 solves (5.2). By the assumption of Lemma 5.2, one then concludes $w_0 = 0$.

We now claim that

$$\|u_0 + w_k\|^p/N + \|u_0\|^p/p^* \to c. \tag{5.12}$$

Indeed, multiplying again (5.10) with $\varphi = u_0 + w_k$ by $1/p^*$ and subtracting from (5.9), one obtains

$$\frac{1}{N}\|u_0 + w_k\|^p + \frac{1}{p^*}\int_\Omega \left(h_\lambda(x, u_0)u_0 + b(x)u_0^{p^*}\right) \to c,$$

and (5.12) follows by using the equation for u_0.

Relation (5.12) and the weak lower semi-continuity of the norm imply $c \geqslant \|u_0\|^p/p$. We distinguish two cases: either (i) $c = \|u_0\|^p/p$ or (ii) $c > \|u_0\|^p/p$.

In case (i), one deduces from (5.12) that $\|u_0 + w_k\| \to \|u_0\|$, and consequently $w_k \to 0$ in $W_0^{1,p}(\Omega)$, which shows that $(\mathrm{PS})_c$ holds.

We will now prove that case (ii) leads to $c \geqslant c_0$, which contradicts assumption (5.8). For that purpose we start from (5.10) with $\varphi = u_0 + w_k$ and use Eq. (1.1) for u_0 to obtain

$$\lim \|u_0 + w_k\|^p = \|u_0\|^p + \lim \int_\Omega b(x)\left[(u_0 + w_k^+)^{p^*} - u_0^{p^*}\right].$$

Applying (5.12) to the left-hand side and the mean-value theorem to $(u_0 + w_k^+)^{p^*} - w_k^{+p^*}$ in the right-hand side, one obtains

$$N\left(c - \|u_0\|^p/p\right) = \lim \int_\Omega b(x)\left(w_k^+\right)^{p^*},$$

which easily leads to

$$\left(c - \|u_0\|^p/p\right)^{p/p^*} \leqslant \left(\|b\|_\infty/N\right)^{p/p^*} S^{-1} \lim \|w_k\|^p. \tag{5.13}$$

On the other hand, by Lemma 3.6, $\|u_0 + w_k\|^p - \|w_k\|^p \to \|u_0\|^p$. Replacing in (5.13) and using again (5.12) together with the fact that we are in case (ii), we come to the inequality $c \geqslant c_0$, which is a contradiction. This completes the proof of Lemma 5.2. □

Proof of Theorems 2.5 and 2.6. Proceeding exactly as before, we know that 0 is a local minimizer of J_λ on $W_0^{1,p}(\Omega)$, and we look for a nonzero critical point of J_λ.

Assume by contradiction that 0 is the only critical point of J_λ. Then, for some ball $B(0,r)$ in $W_0^{1,p}(\Omega)$, we have

$$J_\lambda(0) < J_\lambda(w) \tag{5.14}$$

for all $w \in B(0,r)$ with $w \neq 0$. Using Lemma 5.2 above and Theorem 5.10 from [13] (which only requires the $(PS)_c$ condition to hold at the level of the strict local minimum, here the level $J_\lambda(0) = 0 < c_0$), one deduces from (5.14) that (5.6) holds for all $r > 0$ sufficiently small, i.e. one has mountain ranges surrounding 0. We aim again at applying the mountain pass theorem. For this purpose it remains to show the existence of $\bar{u} \in W_0^{1,p}(\Omega)$ such that $J_\lambda(\bar{u}) < 0$ and the inf max value of J_λ over the family of all continuous paths from 0 to $\bar{u}$ is $< c_0$. Once this is done, the mountain pass theorem yields the existence of a nonzero critical point of J_λ, and we reach a contradiction with the fact that 0 was supposed to be the only critical point of J_λ.

The construction of the desired $\bar{u}$ is made as follows. One considers as in [2,18] functions of the form $u_\varepsilon(x) := \rho(x)U_\varepsilon(x)$, where

$$U_\varepsilon(x) := \frac{\varepsilon^{(N-p)/p(p-1)}}{(\varepsilon^{p/(p-1)} + |x - x_0|^{p/(p-1)})^{(N-p)/p}}$$

and $\rho(x)$ is a cutoff function near x_0 (from assumption (b)). More precisely ρ is smooth, nonnegative, with $\rho \equiv 1$ near x_0 and support in a ball B_2 around x_0, where B_2 is chosen such that $\overline{B_2} \subset B_1 \cap B(x_0, \delta)$ (from assumptions (b) and (H_h)) and $b(x) \geqslant$ some $\eta > 0$ a.e. on B_2. Using Lemma 5.3 below, one easily sees that the function $\bar{u} = tu_\varepsilon$ satisfies the desired properties if one first selects $\varepsilon > 0$ sufficiently small and then t sufficiently large. □

Lemma 5.3.

(i) *For any* $\varepsilon > 0$, $J_\lambda(tu_\varepsilon) \to -\infty$ *as* $t \to +\infty$.
(ii) *One has*

$$\sup_{t \geqslant 0} J_\lambda(tu_\varepsilon) < c_0$$

for $\varepsilon > 0$ *sufficiently small.*

Proof. Part (i) easily follows from the "superlinearity" of the problem near x_0. The proof of part (ii) is separated in three cases. We use similar arguments as in [18], in particular their Lemmas A4 and A5.

Case 1: $2 \leqslant p < 3$. Let

$$J_\lambda^\flat(tu_\varepsilon) := J_\lambda(tu_\varepsilon) - \frac{1}{p}\|u_0\|^p$$

where u_0 is the first solution from Theorem 5.1. We thus have

$$J_\lambda^\flat(tu_\varepsilon) = \frac{1}{p}\|u_0 + tu_\varepsilon\|^p - \frac{1}{p}\|u_0\|^p - \frac{1}{p^*}\int_\Omega b(x)(u_0 + tu_\varepsilon)^{p^*} + \frac{1}{p^*}\int_\Omega b(x)u_0^{p^*}$$
$$- \int_\Omega H_\lambda(x, u_0 + tu_\varepsilon) + \int_\Omega H_\lambda(x, u_0).$$

Since $h_\lambda(x,\cdot)$ is nondecreasing, using Lemma A4, parts (1) and (4), with $r = p^*$, as well as the gradient estimate on p. 946 in [18], we obtain

$$J_\lambda^\flat(tu_\varepsilon) \leqslant \frac{t^p}{p}\int_\Omega |\nabla u_\varepsilon|^p + t\int_\Omega |\nabla u_0|^{p-2}\nabla u_0 \nabla u_\varepsilon + Ct^{\gamma_0}\varepsilon^\beta$$
$$- \frac{t^{p^*}}{p^*}\int_\Omega b(x)u_\varepsilon^{p^*} - t\int_\Omega b(x)u_0^{p^*-1}u_\varepsilon - t^{p^*-1}\int_\Omega b(x)u_\varepsilon^{p^*-1}u_0$$
$$- Ct^\gamma \int_\Omega u_0^{p^*-\gamma}u_\varepsilon^\gamma - t\int_\Omega h_\lambda(x, u_0)u_\varepsilon$$

for some $\beta > (N-p)/p$ and all γ, γ_0 with $1 < \gamma < p^* - 1$, $p - 1 < \gamma_0 < \frac{N(p-1)}{N-1}$, where C denotes, here and below, various positive constants possibly depending on γ and γ_0 but independent on t and ε. Moreover, as on p. 947 in [18],

$$\int_\Omega |\nabla u_\varepsilon|^p\,dx = \int_{\mathbb{R}^N} |\nabla U_1|^p\,dx + O\big(\varepsilon^{\frac{N-p}{p-1}}\big),$$

$$\int_\Omega |u_\varepsilon|^{p^*}\,dx = \int_{\mathbb{R}^N} |U_1|^{p^*}\,dx + O\big(\varepsilon^{\frac{N}{p-1}}\big)$$

when $\varepsilon \to 0$. It follows, using Lemma A5, part (1), with $\alpha = p^* - 1$, and the fact that u_0 solves (5.2), that

$$J_\lambda^\flat(tu_\varepsilon) \leqslant \frac{t^p}{p}\bigg(\int_{\mathbb{R}^N} |\nabla U_1|^p + O\big(\varepsilon^{\frac{N-p}{p-1}}\big)\bigg) - \frac{t^{p^*}}{p^*}\|b\|_\infty\bigg(\int_\Omega |U_1|^{p^*} + O\big(\varepsilon^{\frac{N}{p-1}}\big)\bigg)$$
$$+ Ct^{\gamma_0}\varepsilon^\beta - Ct^{p^*-1}\varepsilon^{\frac{N-p}{p}} - \frac{t^{p^*}}{p^*}\int_\Omega \big(b(x) - \|b\|_\infty\big)u_\varepsilon^{p^*}.$$

Note that

$$\int_{\Omega} (b(x) - \|b\|_\infty) u_\varepsilon^{p^*} = \int_{B_{\varepsilon^\delta}} (b(x) - \|b\|_\infty) u_\varepsilon^{p^*} + \int_{\Omega \setminus B_{\varepsilon^\delta}} (b(x) - \|b\|_\infty) u_\varepsilon^{p^*}$$

where B_{ε^δ} is the ball centered in x_0 and radius ε^δ. Using assumption (b), we have that $\int_{B_{\varepsilon^\delta}} (b(x) - \|b\|_\infty) u_\varepsilon^{p^*}$ is of order $\varepsilon^{\delta \gamma_b}$; we also have, for M large enough,

$$\int_{\Omega \setminus B_{\varepsilon^\delta}} (b(x) - \|b\|_\infty) u_\varepsilon^{p^*} \leqslant \int_{\varepsilon^\delta}^{M} \frac{\varepsilon^{\frac{N}{p-1}}}{(\varepsilon^{\frac{p}{p-1}} + r^{\frac{p}{p-1}})^N} r^{N-1}\, dr$$
$$= O\big(\varepsilon^{\frac{N(1-\delta)}{p-1}}\big).$$

Now from the assumption on γ_b, one can choose $\delta > 0$ such that

$$\frac{N-p}{p} < \frac{N(1-\delta)}{p-1},$$

and

$$\frac{N-p}{p} < \delta \gamma_b,$$

and so we have

$$\int_{\Omega} (\|b\|_\infty - b(x)) u_\varepsilon^{p^*} \leqslant o\big(\varepsilon^{\frac{N-p}{p}}\big).$$

Therefore

$$J_\lambda^b(tu_\varepsilon) \leqslant \frac{t^p}{p}\left(\int_{\mathbb{R}^N} |\nabla U_1|^p + O\big(\varepsilon^{\frac{N-p}{p-1}}\big)\right) - \frac{t^{p^*}}{p^*}\left(\|b\|_\infty \int_{\mathbb{R}^N} |U_1|^{p^*} + o\big(\varepsilon^{\frac{N-p}{p}}\big)\right)$$
$$+ Ct^{\gamma_0}\varepsilon^\beta - Ct^{p^*-1}\varepsilon^{\frac{N-p}{p}}.$$

Denoting

$$A_\varepsilon = \int_{\mathbb{R}^n} |\nabla U_1|^p + O\big(\varepsilon^{\frac{N-p}{p-1}}\big), \tag{5.15}$$

$$B_\varepsilon = \|b\|_\infty \int_{\mathbb{R}^n} |U_1|^{p^*} + o\big(\varepsilon^{\frac{N-p}{p}}\big),$$

we thus have $J_\lambda^{b}(tu_\varepsilon) \leqslant f(t)$, where f is defined by

$$f(t) := \frac{t^p}{p} A_\varepsilon - \frac{t^{p^*}}{p^*} B_\varepsilon + C t^{\gamma_0} \varepsilon^\beta - C t^{p^*-1} \varepsilon^{\frac{N-p}{p}}.$$

It is clear that this function f is bounded from above and reach its maximum at some $t_\varepsilon > 0$. Note that $f'(t_\varepsilon) = 0$ if and only if

$$C\gamma_0 t_\varepsilon^{\gamma_0 - p} \varepsilon^\beta + A_\varepsilon - t_\varepsilon^{p^*-p} B_\varepsilon = C\left(p^* - 1\right) t_\varepsilon^{p^*-(p+1)} \varepsilon^{\frac{N-p}{p}}. \tag{5.16}$$

Since $p^* > p$, $-1 \leqslant \gamma_0 - p < \frac{-N+p}{N-1} \leqslant 0$,

$$\lim_{\varepsilon \to 0} A_\varepsilon = \int\limits_{\mathbb{R}^N} |\nabla U_1|^p$$

and

$$\lim_{\varepsilon \to 0} B_\varepsilon = \|b\|_\infty \int\limits_{\mathbb{R}^N} |U_1|^{p^*},$$

we deduce from (5.16) that t_ε remains bounded as $\varepsilon \to 0$. If $\liminf_{\varepsilon \to 0} t_\varepsilon = 0$, then part (ii) of Lemma 5.3 is proved. Otherwise, (5.16) implies $\lim_{\varepsilon \to 0} t_\varepsilon > 0$. And so

$$f(t_\varepsilon) = \frac{t_\varepsilon{}^p}{p} \int\limits_{\mathbb{R}^N} \|\nabla U_1\|^p - \frac{t_\varepsilon^{p^*}}{p^*} \|b\|_\infty \int\limits_{\mathbb{R}^N} |U_1|^{p^*} + o\left(\varepsilon^{\frac{N-p}{p}}\right) + C\varepsilon^\beta - C\varepsilon^{\frac{N-p}{p}}. \tag{5.17}$$

But

$$\sup_{t \geqslant 0} \left\{ \frac{t^p}{p} \int\limits_{\mathbb{R}^N} |\nabla U_1|^p - \frac{t^{p^*}}{p^*} \|b\|_\infty \int\limits_{\mathbb{R}^n} |U_1|^{p^*} \right\} = \frac{1}{N} \frac{1}{\|b\|_\infty^{\frac{N-p}{p}}} S^{\frac{N}{p}}.$$

Thus, taking $\varepsilon > 0$ small enough, we have

$$I(tu_\varepsilon) < \frac{1}{N} \frac{1}{\|b\|_\infty^{\frac{N-p}{p}}} S^{\frac{N}{p}}$$

for $t \geqslant 0$. Part (ii) of Lemma 5.3 is thus proved in Case 1.

Case 2: $\frac{2N}{N+1} < p < 2$. Using Lemma A4, parts (3) and (4), with $r = p^*$ (note that $p > \frac{2N}{N+1}$ implies $p > \frac{2N}{N+2}$, i.e. $p^* > 2$), we obtain

$$J_\lambda^\flat(tu_\varepsilon) \leqslant \frac{t^p}{p}\int_\Omega |\nabla u_\varepsilon|^p + t\int_\Omega |\nabla u_0|^{p-2}\nabla u_0 \nabla u_\varepsilon + t^{\gamma_0}\int_\Omega |\nabla u_\varepsilon|^{\gamma_0}$$
$$- \frac{t^{p^*}}{p^*}\int_\Omega b(x)u_\varepsilon^{p^*} - t\int_\Omega b(x)u_0^{p^*-1}u_\varepsilon - t^{p^*-1}\int_\Omega b(x)u_\varepsilon^{p^*-1}$$
$$- Ct^\gamma \int_\Omega u_0^{p^*-\gamma}u_\varepsilon^\gamma - t\int_\Omega h_\lambda(x,u_0)u_\varepsilon \tag{5.18}$$

for all $1<\gamma<p^*-1$ and $1<\gamma_0<p$. Since $p>\frac{2N}{N+1}$, one can select γ_0 such that $\frac{N(p-1)}{N-1}<\gamma_0$ and

$$\beta := N - \frac{N}{p}\gamma_0 > \frac{N-p}{p}.$$

Then, by Lemma A5, part (2), we have

$$\int_\Omega |\nabla u_\varepsilon|^{\gamma_0} \leqslant C\varepsilon^\beta.$$

We can now proceed as in Case 1 above to conclude the proof of part (ii) of Lemma 5.3 in Case 2.

Case 3: $p \geqslant 3$. Since $p\geqslant 3$, Lemma A4, part (2), gives

$$\frac{1}{p}|\nabla u_0 + t\nabla u_\varepsilon|^p - \frac{1}{p}|\nabla u_0|^p$$
$$\leqslant \frac{t^p}{p}|\nabla u_\varepsilon|^p + t|\nabla u_0|^{p-2}\langle \nabla u_0, \nabla u_\varepsilon\rangle + Ct^2|\nabla u_\varepsilon|^2 + Ct^{p-1}|\nabla u_\varepsilon|^{p-1}.$$

Thus, using (H_h) and proceeding as in Case 1 above, using Lemma A4, part (4), we obtain

$$J_\lambda^\flat(tu_\varepsilon) \leqslant \frac{t^p}{p}\int_\Omega |\nabla u_\varepsilon|^p + t\int_\Omega |\nabla u_0|^{p-2}\nabla u_0\nabla u_\varepsilon + Ct^2\int_\Omega |\nabla u_\varepsilon|^2$$
$$+ Ct^{p-1}\int_\Omega |\nabla u_\varepsilon|^{p-1} - \frac{t^{p^*}}{p^*}\int_\Omega b(x)u_\varepsilon^{p^*} - t\int_\Omega b(x)u_0^{p^*-1}u_\varepsilon$$
$$- t^{p^*-1}\int_\Omega b(x)u_\varepsilon^{p^*-1} - Ct^\gamma\int_\Omega u_0^{p^*-\gamma}u_\varepsilon^\gamma - t\int_\Omega h_\lambda(x,u_0)u_\varepsilon - c_0t^{q+1}\int_\Omega u_\varepsilon^{q+1}$$

for all $1<\gamma<p^*-1$. And so

$$J_\lambda^\flat(tu_\varepsilon) \leqslant \frac{t^p}{p}\int_\Omega |\nabla u_\varepsilon|^p + Ct^2\int_\Omega |\nabla u_\varepsilon|^2 + Ct^{p-1}\int_\Omega |\nabla u_\varepsilon|^{p-1}$$
$$- \frac{t^*}{p^*}\int_\Omega b(x)u_\varepsilon^{p^*} - c_0t^{q+1}\int_\Omega u_\varepsilon^{q+1}$$

$$= \frac{t^p}{p}\int_\Omega |\nabla u_\varepsilon|^p - \frac{t^{p^*}}{p^*}\|b\|_\infty \int_\Omega u_\varepsilon^{p^*} + Ct^2 \int_\Omega |\nabla u_\varepsilon|^2$$

$$+ Ct^{p-1}\int_\Omega |\nabla u_\varepsilon|^{p-1} + \frac{t^{p^*}}{p^*}\int_\Omega \big(\|b\|_\infty - b(x)\big)u_\varepsilon^{p^*} - c_0 t^{q+1}\int_\Omega u_\varepsilon^{q+1}.$$

Since $p \geqslant 3$, we have $2 < \frac{N(p-1)}{N-1}$. Thus Lemma A5, part (2), implies

$$\int_\Omega |\nabla u_\varepsilon|^2 \leqslant C\varepsilon^{\frac{2(N-p)}{p(p-1)}}$$

and since $(p-1) < \frac{N(p-1)}{N-1}$, we have

$$\int_\Omega |\nabla u_\varepsilon|^{p-1} \leqslant C\varepsilon^{\frac{N-p}{p}}.$$

In addition Lemma A5, part (1), implies

$$\int_\Omega u_\varepsilon^{q+1} \geqslant C\varepsilon^{N-\frac{(N-p)}{p}(q+1)}$$

for ε small enough (note that by (H_h), $q+1 > p^* - \frac{2}{p-1} > \frac{N(p-1)}{N-p}$). Hence

$$J_\lambda^\flat(tu_\varepsilon) \leqslant \frac{t^q}{p}A_\varepsilon - \frac{t^{p^*}}{p^*}B_\varepsilon + C\big(t^2 + t^{p-1}\big)\varepsilon^{\frac{2(N-p)}{p(p-1)}} - Ct^{q+1}\varepsilon^{N-\frac{(N-p)}{p}(q+1)} =: f(t).$$

Using again the hypothesis $q+1 > p^* - \frac{2}{p-1}$, we have

$$N - \frac{(N-p)}{p}(q+1) < \frac{2(N-p)}{p(p-1)}.$$

And by the same argument as in Case 1 above, we can finish the proof of Lemma 5.3. □

6. Applications

In this section we will first show how the previous theorems apply to problem (1.2). The verification of the corresponding hypotheses are either easy or can be easily adapted from the arguments in [15]; in Theorem 6.3, the verification of (H_h) uses Lemma A4, part (4), from [18]. The functional $I_\lambda(u)$ here reads

$$I_\lambda(u) := \frac{1}{p}\int_\Omega |\nabla u|^p - \frac{\lambda}{q+1}\int_\Omega a(x)\big(u^+\big)^{q+1} - \frac{1}{r+1}\int_\Omega b(x)\big(u^+\big)^{r+1}.$$

As an application of Theorems 2.1, 2.2 and 2.3, we have

Theorem 6.1. *Let $0 \leqslant q < p-1 < r$ and assume $a, b \in L^{\infty}(\Omega)$ with*

(i) $a(x) \geqslant 0$ *a.e. in* Ω,
(ii) $a(x) \geqslant \varepsilon > 0$ *on some ball* B_1.

Then there exists $0 < \Lambda \leqslant \infty$ such that problem (1.2) *has at least one solution u (with $I_\lambda(u) < 0$) for $0 < \lambda < \Lambda$ and no solution for $\lambda > \Lambda$. If in addition*

(iii) $b(x) \geqslant 0$ *on some ball* B_2, *with* $a(x)b(x) \not\equiv 0$ *on* B_2,

then $\Lambda < \infty$. Moreover, if $r < p^* - 1$, then problem (1.2) *has at least one solution u (with $I_\lambda(u) \leqslant 0$) for $\lambda = \Lambda$.*

As an application of Theorem 2.4, we have

Theorem 6.2. *Let $0 \leqslant q < p-1 < r < p^* - 1$. Assume that $a, b \in L^{\infty}(\Omega)$ satisfy*

(iv) $a \succ 0$, *i.e.* $a(x) \geqslant \varepsilon_K > 0$ *on any compact* $K \subset \Omega$,
(v) $b(x) \geqslant \varepsilon > 0$ *on some ball* B_2.

Then problem (1.2) *has at least two solutions u, v for $0 < \lambda < \Lambda$, with $u \lneqq v$ and $I_\lambda(u) < 0$.*

Note that (iv), (v) imply assumptions (i), (ii) and (iii) of Theorem 6.1.
As an application of Theorem 2.5, we finally have

Theorem 6.3. *Let $0 \leqslant q < p-1$ and $r = p^* - 1$. Assume that $a, b \in L^{\infty}(\Omega)$ satisfy respectively condition* (iv) *of Theorem* 6.2 *and condition* (b) *of Theorem* 2.5. *Assume further that either $2N/(N+1) < p < 3$, or $p \geqslant 3$ and $p^* - \frac{2}{p-1} < q + 1$. Then the conclusion of Theorem* 6.2 *holds.*

Note that condition (b) implies assumption (v) of Theorem 6.2.

Remark 6.4. Here are some questions which remain unsolved in the context of problem (1.2). (i) In Theorem 6.1, the question of existence of at least one solution for $\lambda = \Lambda$ when $r = p^* - 1$. (ii) In Theorems 6.2 and 6.3, the question whether the two solutions u and v satisfy $u < v$ in Ω and $\partial u/\partial \nu > \partial v/\partial \nu$ on $\partial\Omega$. (iii) In Theorem 6.3, the question of the necessity of the restrictions on the exponents p, q. Note that these questions are also unsolved in the constant coefficients case.

The second application concerns the problem

$$\begin{cases} -\Delta_p u = \lambda c(x)(u+1)^r & \text{in } \Omega, \\ u > 0 & \text{in } \Omega, \\ u = 0 & \text{on } \partial\Omega \end{cases} \tag{6.1}$$

where $p-1<r$. This problem was studied in [7,20] when $c(x)\equiv 1$ and $p=2$ and more recently in [15] when $p=2$ and $c(x)$ is variable. The functional $I_\lambda(u)$ here reads

$$I_\lambda(u):=\frac{1}{p}\int_\Omega |\nabla u|^p-\frac{\lambda}{r+1}\int_\Omega c(x)\big(u^+ +1\big)^{r+1}.$$

As a consequence of Theorems 2.1, 2.2 and 2.3, we have

Theorem 6.5. *Assume that $c\in L^\infty(\Omega)$ satisfies*

(i) $c(x)\geqslant 0$ *a.e. in* Ω,
(ii) $c(x)\geqslant\varepsilon>0$ *on some ball* B.

Then there exists $0<\Lambda<+\infty$ such that problem (6.1) *has at least one solution u (with $I_\lambda(u)<0$) for $0<\lambda<\Lambda$ and no solution for $\lambda>\Lambda$. Moreover, if $r<p^*-1$, then problem* (6.1) *has at least one solution u (with $I_\lambda(u)\leqslant 0$) for $\lambda=\Lambda$.*

As an application of Theorem 2.4, we have

Theorem 6.6. *Assume $r<p^*-1$, and that $c\in L^\infty(\Omega)$ satisfies*

(iii) $c\succ 0$.

Then problem (6.1) *has at least two solutions u,v for $0<\lambda<\Lambda$, with $u\leqslant\not\equiv v$ and $I_\lambda(u)<0$.*

Finally, as an application of Theorem 2.6, we have

Theorem 6.7. *Assume $r=p^*-1$ and that $c(x)$ satisfies* (iii) *above as well as condition* (b) *of Theorem* 2.6. *Assume further that $2N/(N+1)<p<3$. Then problem* (6.1) *has at least two solutions u,v for $0<\lambda<\Lambda$, with $u\leqslant\not\equiv v$ and $I_\lambda(u)<0$.*

The critical case $r=p^*-1$ in Theorem 6.7 requires more care because the right-hand side of (6.1) is not written in the form (2.2). However, u solves (6.1) if and only if $v=\lambda^g u$ solves

$$\begin{cases} -\Delta_p v=c(x)(v+\mu)^{p^*-1} & \text{in } \Omega,\\ v>0 & \text{in } \Omega,\\ v=0 & \text{on } \partial\Omega \end{cases}\tag{6.2}$$

for $\mu=\lambda^g$ with $g=\frac{1}{p^*-p}$. This implies in particular that (6.2) has at least one solution for $\mu<\Lambda^g$ and no solutions for $\mu>\Lambda^g$. In order to apply Theorem 2.6 to (6.2), we write the nonlinearity in the following way

$$c(x)(v+\mu)^r=h_\mu(x,u)+c(x)u^r$$

where $h_\mu(x,s)=c(x)((s+\mu)^{p^*-1}-s^{p^*-1})$. And now is not difficult to verify the hypotheses of Theorem 2.6. Note that the critical case with $p\geqslant 3$ in (6.2) remains unsolved; the difficulty lies in the verification of (H_h).

Acknowledgments

Most of this work was done with the support of CNPq, FNRS, PRONEX, FAPESP and FONDECYT 1080430 at Unicamp, ULB and USACH. We wish to thank H. Brezis for several comments relative to Proposition 3.7 and the referee for some remarks on condition (H_h) in Theorems 2.5 and 2.6.

References

[1] W. Allegretto, Yin Xi Huang, A Picone's identity for the p-Laplacian and applications, Nonlinear Anal. 32 (7) (1998) 819–830.

[2] A. Ambrosetti, H. Brezis, G. Cerami, Combined effects of concave and convex nonlinearities in some elliptic problems, J. Funct. Anal. 122 (2) (1994) 519–543.

[3] A. Anane, Etudes des valeurs propres et de la résonance pour l'opérateur p-Laplacien, Thèse de doctorat, Université Libre de Bruxelles, 1988.

[4] D. Arcoya, D. Ruiz, The Ambrosetti–Prodi problem for the p-Laplacian operator, Comm. Partial Differential Equations 31 (4–6) (2006) 849–865.

[5] L. Boccardo, F. Murat, Almost everywhere convergence of the gradients of solutions to elliptic and parabolic equations, Nonlinear Anal. 19 (6) (1992) 581–597.

[6] H. Brezis, E. Lieb, A relation between pointwise convergence of functions and convergence of functionals, Proc. Amer. Math. Soc. 88 (3) (1983) 486–490.

[7] H. Brezis, L. Nirenberg, Positive solutions of nonlinear elliptic equations involving critical Sobolev exponents, Comm. Pure Appl. Math. 36 (4) (1983) 437–477.

[8] H. Brezis, L. Nirenberg, H^1 versus C^1 local minimizers, C. R. Acad. Sci. Paris Sér. I Math. 317 (5) (1993) 465–472.

[9] F. Brock, L. Iturriaga, P. Ubilla, A multiplicity result for the p-Laplacian involving a parameter, Ann. Henri Poincaré 9 (7) (2008) 1371–1386.

[10] M. Cuesta, P. Takáč, A strong comparison principle for the Dirichlet p-Laplacian, in: Reaction Diffusion Systems, Trieste, 1995, in: Lect. Notes Pure Appl. Math., vol. 194, Dekker, New York, 1998, pp. 79–87.

[11] M. Cuesta, P. Takáč, A strong comparison principle for positive solutions of degenerate elliptic equations, Differential Integral Equations 13 (4–6) (2000) 721–746.

[12] C. De Coster, M. Henrard, Existence and localization of solution for second order elliptic BVP in presence of lower and upper solutions without any order, J. Differential Equations 145 (2) (1998) 420–452.

[13] D.G. de Figueiredo, Lectures on the Ekeland Variational Principle with Applications and Detours, Tata Inst. Fund. Res. Lect. Math. Phys., vol. 81, Springer-Verlag, 1989.

[14] D.G. de Figueiredo, J.-P. Gossez, P. Ubilla, Local superlinearity and sublinearity for indefinite semilinear elliptic problems, J. Funct. Anal. 199 (2) (2003) 452–467.

[15] D.G. de Figueiredo, J.-P. Gossez, P. Ubilla, Multiplicity results for a family of semilinear elliptic problems under local superlinearity and sublinearity, J. Eur. Math. Soc. 8 (2) (2006) 269–286.

[16] X. Fan, On the sub–supersolution method for $p(x)$-Laplacian equations, J. Math. Anal. Appl. 330 (1) (2007) 665–682.

[17] J. Fleckinger-Pellé, P. Takáč, Uniqueness of positive solutions for nonlinear cooperative systems with the p-Laplacian, Indiana Univ. Math. J. 43 (4) (1994) 1227–1253.

[18] J. García, I. Peral, Some results about the existence of a second positive solution in a quasilinear critical problem, Indiana Univ. Math. J. 43 (3) (1994) 941–957.

[19] J. García, I. Peral, J. Manfredi, Sobolev versus Hölder local minimizers and global multiplicity for some quasilinear elliptic equations, Commun. Contemp. Math. 2 (3) (2000) 385–404.

[20] F. Gazzola, A. Malchiodi, Some remarks on the equation $-\Delta u = (u + 1)^p$ for varying domains, Comm. Partial Differential Equations 27 (2002) 809–845.

[21] J. Giacomoni, I. Schindler, P. Takáč, Sobolev versus Hölder local minimizers and existence of multiple solutions for a singular quasilinear equation, Ann. Sc. Norm. Super. Pisa Cl. Sci. (5) 6 (1) (2007) 117–158.

[22] M. Guedda, L. Véron, Quasilinear elliptic equations involving critical Sobolev exponents, Nonlinear Anal. 13 (8) (1989) 879–902.

[23] Zongming Guo, Zhitao Zhang, $W^{1,p}$ versus C^1 local minimizers and multiplicity results for quasilinear elliptic equations, J. Math. Anal. Appl. 286 (1) (2003) 32–50.

[24] O. Kavian, Introduction à la théorie des points critiques et applications aux problèmes elliptiques, Math. Appl., vol. 13, Springer-Verlag, 1993.

[25] O.A. Ladyženskaja, N.N. Ural'ceva, Équations aux dérivées partielles de type elliptique, Monographies Universitaires de Mathématiques, No. 31, Dunod, 1968.

[26] G.M. Lieberman, Boundary regularity for solutions of degenerate elliptic equations, Nonlinear Anal. 12 (11) (1988) 1203–1219.

[27] P. Lindqvist, On the equation $\operatorname{div}(|\nabla u|^{p-2}\nabla u) + \lambda|u|^{p-2}u = 0$, Proc. Amer. Math. Soc. 109 (1) (1990) 157–164.

[28] I. Peral, Some results on quasilinear elliptic equations: Growth versus shape, in: Nonlinear Functional Analysis and Applications to Differential Equations, Trieste, 1997, World Sci. Publ., 1998, pp. 153–202.

[29] P. Pucci, J. Serrin, The Maximum Principle, Progr. Nonlinear Differential Equations Appl., vol. 73, Birkhäuser, 2007.

[30] M. Struwe, Variational Methods. Applications to Nonlinear Partial Differential Equations and Hamiltonian Systems, third ed., Ergeb. Math. Grenzgeb., vol. 34, Springer-Verlag, 2000.

[31] J.L. Vázquez, A strong maximum principle for some quasilinear elliptic equations, Appl. Math. Optim. 12 (3) (1984) 191–202.

LIST OF PUBLICATIONS

of

Djairo G. de Figueiredo

(* denotes books and monographs)

1958

[1] * "Decompositions of the sphere." Notas de Matemática 14, Rio de Janeiro (1958), 40 pages.

1963

[2] "The Coerciveness Problem for Forms over Vector Functions", Comm..Pure Appl. Math. **XVI** (1963), 63-94.
[3] * "A Teoria Clássica do Potencial." Universidade de Brasília (1963), 166 pages.
[4] * "Extensões da desigualdade de Gårding." Rio de Janeiro (1963), 65 pages.

1964

[5] "A Simplified Proof of the Divergence Theorem", American Mathematical Monthly **71** (1964), 619-622.

1966

[5] "J-Monotone Nonlinear Operators in Banach Spaces" (with F.E. Browder), Nederl. Akad. Wetensch. Proc. Ser. A 69=Indag. Math. **28** (1966), 412-420.
[7] * "Fixed Point Theorems for Weakly Continuous Mappings." MRC Report no. 638 (1966).
[8] * "Espaços Vetoriais Topológicos e Teoria das Distribuições." Notas de Matemática 41, Rio de Janeiro (1966). Co-author F. Trèves, 301 pages.

1967

[9] "Fixed Point Theorems for Nonlinear Operators and Galerkin Approximations", J. Differential Equations **3** (1967), 271-281.
[10] "On the Radial Projection in Normed Spaces" (with L.A. Karlovitz), Bull. Amer. Math. Soc. **73** (1967), 364-368.
[11] "Some Remarks on Fixed Point Theorems for Nonlinear Operators in Banach Spaces", Anais Acad. Bras. Ciencias **39** (1967), 345-349.
[12] "On the Projection onto Convex Sets and the Extension of Contractions" (with L.A. Karlovitz), University of Maryland Tech. Note BN 524 (l967) and Proc. Conf. on Projections and Related Topics at Clemson University.
[13] * "Topics in Nonlinear Functional Analysis." Lecture Notes 48, University of Maryland (1967), 153 pages.

1968

[14] "On the Approximate Solution of Linear Functional Equations in Banach Spaces" (with L.A. Karlovitz), J. Math. Anal. Appl. **24** (1968), 654-664.

1970

[15] "On the Extension of Contractions on Normed Spaces" (with L.A. Karlovitz), Nonlinear Functional Analysis, 1970. (Proc. Sympos. Pure Math. Vol. **XVIII**, Part 1, Chicago, 1968), 95-104, Amer. Math. Soc..
[16] "Maximal Monotone Operators and Nonlinear Integral Equations of Hammerstein Type" (with F.E. Browder and C.P. Gupta), Bull. Amer. Math. Soc. **76** (1970), 700-705.
[17] * "Funções Reais." Pan American Union Monographs, Vol. 10 (1970), 153 pages.

1971

[18] "Borsuk Type Theorems for Nonlinear Noncompact Mappings in Banach Spaces" (with C.P. Gupta), Proc. Conf. Funct. Anal. and its Applic., Indian Institute of Technology, Kanpur (1971), 48-55.

DOI: 10.1007/978-3-319-02856-9,

[19] "An Existence Theorem for Pseudo-Monotone Operator Equations in Banach Spaces", J. Math. Anal. Applic. **34** (1971), 151-156.

1972

[20] "The Extension of Contractions and the Intersection of Balls in Banach Spaces" (with L.A. Karlovitz), J. Funct. Anal. **11** (1972), 168-178.
[21] "Nonlinear Integral Equations of Hammerstein Type involving Unbounded Monotone Linear Mappings" (with C.P. Gupta), J. Math. Anal. Applic. **39** (1972), 37-48.
[22] "Non-Linear Integral Equations of Hammerstein Type with Indefinite Linear Kernel in a Hilbert Space" (with C.P. Gupta), Indag. Math. **34** (1972), 335-344.
[23] "On a Boundary Value Problem Arising in Elastic Membrane Theory", Anais Acad. Bras. Ciências **44** (1972), 181-183.

1973

[24] "On the Variational Method for the Existence of Solutions of Nonlinear Equations of Hammerstein Type" (with C.P. Gupta), Proc. Amer. Math. Soc. **40** (1973), 470-476.
[25] "Extensions of Contractions and Mappings of Accretive Type", Proceedings of the 'Reunião de Análise', São Paulo, SBM (1973), 109-160.
[26] "A Remark on the Fixed Point Theorem for Nonexpansive Mappings", Trabalho de Matemática no. 31, UnB (1973).
[27] * "Análise na Reta." Course at the 9th Brazilian Congress of Mathematics, IMPA (1973), 340 pages.

1974

[28] "On the Range of Nonlinear Operators with Linear Asymptotes which are not Invertible", Comment. Math. Univ. Carolinae **15** (1974), 415-428.
[29] "Some Remarks on the Dirichlet Problem for Semilinear Elliptic Equations", Anais Acad. Bras. Ciências **46** (1974), 187-193.
[30] * "Números Irracionais e Transcendentes." Collection 'Monografias de Matemática', IMPA (1974), 101 pages.
[31] * "An Introduction to the Theory of Monotone Operators." Published by the Sociedade Brasileira de Matemática for a course delivered at the 'Reunião de Análise Funcional', UNICAMP (1974), 56 pages.

1975

[32] "On Coercive Elliptic Problems", Proceedings of the 2nd Brazilian Analysis Seminar, 1975.
[33] "The Dirichlet Problem for Nonlinear Elliptic Equations: a Hilbert Space Approach." Partial Differential Equations and Related Topics (Tulane Univ. 1974), 144-165. Lecture Notes in Math., Vol. 446, Springer, 1975.
[34] * "Analise I." Livros Tecnicos e Cientificos e Editora Universidade de Brasilia, Rio de Janeiro (1975), 259 pages.
[35] * "O Problema de Dirichlet." Proceedings of the 10th Brazilian Congress of Mathematics, IMPA (1975), 21-68.

1977

[36] "Semilinear Elliptic Problems with Nonlinearities near the First Eigenvalue." Dynamical Systems, Proceedings of University of Florida International Symposium, Edit. A.R. Bednarek and L.Cesari, Academic Press (1977), 59-70.
[37] "Perturbation Non Linéaire d'un Problème Elliptique Linéaire près de sa Première Valeur Propre" (with J.-P. Gossez), C. R. Acad. Sci. Paris Sér. A-B **284** (1977), 163-166.
[38] * "Análise de Fourier e Equações Diferenciais Parciais." Collection 'Projeto Euclides', CNPq, Edgar Blucher, São Paulo (1977), 274 pages, 3rd edition (1997).
[39] * "Equações Integrais Lineares." Notes for a course delivered at the 'Escola de Análise', published by the University of São Paulo (1977), 108 pages.
[40] * "Equações Elípticas não Lineares." Course at the 11th Brazilian Congress of Mathematics. IMPA (1977), 108 pages.

1978

[41] "Nonlinear Perturbations of a Linear Elliptic Problem near its First Eigenvalue" (with J.-P. Gossez), J. Differential Equations **30** (1978), 1-19.

1979

[42] "Perturbations of Second Order Linear Elliptic Problems by Nonlinearities without Landesman-Lazer Condition" (with W.-M. Ni), Nonlinear Anal. **3**.(1979), 629-634.
[43] * "Equações Diferenciais Aplicadas." Course at the 12th Brazilian Congress of Mathematics, IMPA (1979), 337 pages.

1980

[44] "Estimations a Priori pour les Solutions Positives de Problèmes Elliptiques Superlinéaires" (with P.-L. Lions and R.D. Nussbaum), C. R. Acad. Sci. Paris Sér. A-B **290** (1980), 217-220.
[45] * "Lectures on the Boundary Value Problems of the Ambrosetti-Prodi Type." Text for a course delivered at the 12th Brazilian Analysis Seminar (1980), 231-292.

1981

[46] "Semilinear Elliptic Equations at Resonance: Higher Eigenvalues and Unbounded Nonlinearities", Recent Advances in Differential Equations (Trieste, 1978), 89-99, Roberto Conti Editor, Academic Press, 1981.
[47] "Double Resonance in Semilinear Elliptic Problems" (with H. Berestycki), Comm. Partial Differential Equations **6** (1981), 91-120.

1982

[48] "A Priori Estimates and Existence of Positive Solutions of Semilinear Elliptic Equations" (with P.-L. Lions and R.D. Nussbaum), J. Math. Pures Appl. **61** (1982), 41-63.
[49] * "Positive Solutions of Semilinear Elliptic Problems." Proceedings of the First Latin American School of Differential Equations (S. Paulo, 1981), 34-87. Lecture Notes in Math. **957**, Springer, 1982.
[50] * "Alguns Aspectos do Cálculo das Variações." Text for a course delivered at the 15th Brazilian Analysis Seminar, 1982, 42 pages.

1984

[51] "On the Superlinear Ambrosetti-Prodi Problem", Nonlinear Anal. **8** (1984), 655-665.
[52] "A Variational Approach to Superlinear Elliptic Problems" (with S. Solimini), Comm. Partial Differential Equations **9** (1984), 699-717.
[53] "On the Uniqueness of Positive Solutions of the Dirichlet Problem for $-\Delta u = \lambda \sin u$", Nonlinear Partial Differential Equations and their Applications, College de France Seminar, Vol. **VII**, (Paris, 1983-1984), 80-83. Res. Notes in Math. **122**, Pittman, 1985.

1985

[54] "On Pairs of Positive Solutions for a Class of Semilinear Elliptic Problems" (with P.-L. Lions), Indiana Univ. Math. J. **34** (1985), 591-606.
[55] "Radial Solutions for a Dirichlet Problem in a Ball" (with D.G. Costa), J. Differential Equations **60** (1985), 80-89.
[56] * "O Princípio de Dirichlet." Matemática Universitária, no. 1 (1985), 63-84.

1986

[57] "A Maximum Principle for an Elliptic System and Applications to Semilinear Problems" (with E. Mitidieri), SIAM J. Math. Anal. **17** (1986), 836-849.
[58] "Positive Solutions for some Classes of Semilinear Elliptic Problems", Nonlinear Functional Analysis and its Applications", Proc. Sympos. Pure Math. **45**, Part 1, Amer. Math. Soc. (1986), 371-380.
[59] "Conditions de Non-résonance pour Certains Problèmes Elliptiques Semi-linéaires" (with J.-P. Gossez), C. R. Acad. Sci. Paris Sér. I Math. **302** (1986), 543-545.

1987

[60] "On the Uniqueness of Solution for a Class of Semilinear Elliptic Problems" (with D.G. Costa and J.V.A. Gonçalves), J. Math. Anal. Appl. **123** (1987), 170-180.
[61] "On the Existence of Multiple Ordered Solutions of Nonlinear Eigenvalue Problems", Nonlinear Anal..**11** (1987), 481-492.
[62] * "Lectures on the Ekeland Variational Principle with Applications and Detours." Campinas, October 1987, 97 pages. Published in the Tata Lectures on Mathematics and Physics, Vol. 81 (1989).

1988

[63] "On Superlinear Elliptic Problems with Nonlinearities Interacting only with Higher Eigenvalues", Rocky Mountain J. Math. **18** (1988), 287-303.

[64] "Nonresonance Below the First Eigenvalue for a Semilinear Elliptic Problem" (with J.-P. Gossez), Math. Ann. **281** (1988), 589-610.

[65] * "Métodos Variacionais em Equações Diferenciais." Matemática Universitária, no. 7 (1988), 21-48.

1989

[66] "Un Problème Elliptique Semi-linéaire sans Condition de Croissance" (with J.-P. Gossez), C. R. Acad. Sci. Paris, Serie I Mat..**308** (1989), 277-280.

[67] * "Problemas de Máximo e Mínimo na Geometria Euclidiana". Matemática Universitária, no. 9/10 (1989), 69-108.

[68] * "On Some Recent Variational Principles". Pro Mathematica, Vol. 5-6 (1989), 53-64.

1990

[69] "Maximum Principles for Cooperative Systems" (with E. Mitidieri), C. R. Acad. Sci. Paris Ser. I Math. **310** (1990), 49-52.

[70] "Maximum Principles for Linear Elliptic Systems" (with E. Mitidieri), Rend. Ist. Mat. Univ. Trieste, Vol. **XXII** (1990), no. 1-2, 36-66.

1991

[71] "Semilinear Elliptic Equations with the Primitive of the Nonlinearity away from the Spectrum" (with O.H. Miyagaki), Nonlinear Anal. **17** (1991), 1201-1219.

[72] "Semilinear Elliptic Equations with the Primitive of the Nonlinearity interacting with the First Eigenvalue" (with I. Massabò), J. Math. Anal. Appl. **156** (1991), 381-394.

[73] "On a Superlinear Sturm-Liouville Equation and a Related Bouncing Problem" (with B. Ruf), J. Reine Angew. Math. **421** (1991) 1-22.

1992

[74] "The Exact Number of Solutions for a Class of Ordinary Differential Equations through Morse Index Computation" (with D.G. Costa and P.N. Srikanth), J. Differential Equations **96** (1992), 185-199.

[75] "Strict Monotonicity of Eigenvalues and Unique Continuation" (with J.-P. Gossez), Comm. Partial Differential Equations **17** (1992), no. 1-2, 339-346.

[76] "Positive Solutions of Semilinear Elliptic Systems" (with Ph. Clément and E. Mitidieri), Comm. Partial Differential Equations **17** (1992), 923-940.

1993

[77] "On the Periodic Fucik Spectrum and a Superlinear Sturm-Liouville Equation" (with B. Ruf), Proc. Roy. Soc. Edinburgh Sect. A **123** (1993), 95-107.

[78] "Bifurcation from Infinity and Multiple Solutions for an Elliptic Systems" (with R. Chiappinelli), Differential Integral Equations **6** (1993), 757-771.

[79] "Sur la Première Courbe du Spectre de Fucik d'un Opérateur Elliptique" (with J.-P. Gossez), C. R. Acad. Sci. Paris Sér. I Math. **316** (1993), 1295-1298.

1994

[80] "On the First Curve of the Fucik Spectrum of an Elliptic Operator" (with J.-P. Gossez), Differential Integral Equations **7** (1994), 1285-1302.

[81] "On Superquadratic Elliptic Systems" (with P. Felmer), Trans. Amer. Math. Soc. **343** (1994), 99-116.

[82] "Monotonicity and Symmetry of Solutions of Elliptic Systems in General Domains", NoDEA Nonlinear Differential Equations Appl. **1** (1994), 119-123

[83] "A Liouville-Type Theorem for Elliptic Systems" (with P. Felmer), Ann. Scuola Norm. Sup. Pisa Vol. **XXI** (1994), 387-397.

1995

[84] "Elliptic Equations in R^2 with Nonlinearities in the Critical Growth Range", (with O.H. Miyagaki and B. Ruf), Calc. Var. and Partial Differential Equations **3** (1995), 139-153.
[85] "Existence and Non-Existence of Radial Solutions for Elliptic Equations with Critical Growth in R^2" (with B. Ruf), Comm. Pure Appl. Math. **XLVIII** (1995), 639-655.

1996

[86] "A Priori Estimates for Positive Solutions of Elliptic Systems via Hardy-Sobolev Inequalities" (with Ph. Clement and E. Mitidieri), Nonlinear Partial Differential Equations, Pitman Res. Notes Math. Ser. **343** (1996), 73-91.
[87] "On Nonquadratic Hamiltonian Elliptic Systems" (with C.A. Magalhães), Adv. Differential Equations **1** (1996), 881-898.
[88] "Quasilinear Elliptic Equations with Critical Exponents" (with Ph. Clément and E. Mitidieri), Topol. Methods Nonlinear Anal. **7** (1996), 133-170.
[89] * "Um Convite às Equações Diferenciais Elípticas", Lecture Notes of a course given at UFCe, 1996, 35 pages, Translation to English: Cubo Matematica Educacional, Vol. 5 nº 2, 2003, 159-184.
[90] * "Semilinear Elliptic Systems: a Survey of Superlinear Problems", Resenhas IME-USP, Vol. 2 (1996), 373-391.
[91] * "Análise I", 2nd. Edition, LTC, Rio de Janeiro, 1996, 256 pages.

1997

[92] * "Equações Diferenciais Aplicadas", co-author Aloisio F. Neves, Collection Matemática Universitária, IMPA, Rio de Janeiro, 1997.

1998

[93] "Decay, Symmetry and Existence of Solutions of Semilinear Elliptic Systems" (with J. Yang), Nonlinear Anal. **33** (1998), 211-234.
[94] "On Linear Perturbations of Superquadratic Elliptic Systems" (with M. Ramos), Reaction-Diffusion Systems, Lectures Notes in Pure and Appl. Math. **194** (1998), 121-130.
[95] "Sur le Spectre de Fucik du p-Laplacien" (with M. Cuesta and J.-P. Gossez), C. R. Acad. Sci. Paris **326** (1998), 681-684.
[96] * "Semilinear Elliptic Systems", a set of notes of a course offered at the "Second School on Nonlinear Functional Analysis and Applications to Differential Equations" in ICTP-Trieste, 1997, published in the Proceedings of the School by World Sci. Publ. (1998), 122-152.

1999

[97] "The Beginning of the Fucik Spectrum for the p-Laplacian" (with M. Cuesta and J.-P. Gossez), J. Differential Equations **159** (1999), 212-238.
[98] "Infinitely many Solutions of Nonlinear Elliptic Systems" (with T. Bartsch), Progr. Nonlinear Differential Equations Appl. **35** (The Herbert Amann Anniversary Volume) (1999), 51-67.
[99] "Critical Superlinear Ambrosetti-Prodi Problems" (with J. Yang), Topol. Methods Nonlinear Anal. **14** (1999), 59-80.

2000

[100] "On a Class of Quasilinear Elliptic Problems involving Critical Exponents" (with J.V.A. Gonçalves and O.H. Miyagaki), Commun. Contemp. Math. **2** (2000), 47-59.
[101] "A nodal domain property for the p-Laplacian", C. R. Acad. Sci. Paris, Ser. I Math. 330 (2000), 669-673.
[102] * "Nonlinear Elliptic Systems", invited paper for the Anais of the Brazilian Academy of Sciences, Vol. **72** (2000), 453-469.

2001

[103] "A Priori Bounds for Positive Solutions of a Non-Variational Elliptic System" (with J. Yang), Comm. Partial Differential Equations **26** (2001), no. 11-12, 2305-2322.

2002

[104] "Some Remarks on a System of Quasilinear Elliptic Equations" (with L. Boccardo), NoDEA Nonlinear Differential Equations Appl. **9** (2002), 309-323.

[105] "On an Inequality by N. Trudinger and J. Moser and Related Elliptic Equations" (with J.M. do O' and B. Ruf), Comm. Pure Appl. Math. **55** (2002), 135-152.
[106] "Solutions of a Nonlinear Schrodinger Equation" (with Y. Ding), Discrete Cont. Dyn. Syst. **8** (2002), 563-584.
[107] "Nonvariational Elliptic Systems" (with C.O. Alves), Discrete Cont. Dyn. Syst. **8** (2002), 289-302.

2003

[108] "On a Semilinear Elliptic Problem without (PS) Condition" (with J. Yang), Journal of Differential Equations. **187** (2003), 412-428.
[109] "Nonvariational Elliptic Systems via Galerkin Methods" (with C.O. Alves), Function Spaces, Differential Operators and Nonlinear Analysis (The Hans Triebel Anniversary Volume), D. Haroske, T. Runst, H.J. Schmeiser Editors, Birkhauser (2003), 47-57.
[110] "Local Superlinearity and Sublinearity for Indefinite Semilinear Elliptic Problems" (with J.-P. Gossez and P. Ubilla), J. Funct. Anal. **199** (2003), 452-467.
[111] "Strongly Indefinite Functionals and Multiple Solutions of Elliptic Systems" (with Y.H. Ding), Trans. Amer. Math. Soc. **355** (2003), 2973-2989.
[112] "On a Resonant Superlinear Elliptic Problem" (with M. Cuesta and P.N. Srikanth), Calc. Var. Partial Differential Equations **17** (2003), 221-233.
[113] * "Um Convite às Equações Diferenciais Elípticas". Lecture Notes of a course given at UFCe, 1996, 35 pages. Translation to English: Cubo Matematica Educacional Vol. 5, n° 2 (2003), 159-184.

2004

[114] "Critical and Subcritical Elliptic Systems in Dimension Two" (with J.M. do O' and B. Ruf), Indiana Univ. Math. J. **53** (2004), 1037-1054.
[115] "Semilinear Elliptic Equations with Dependence on the Gradient via Mountain-Pass Techniques" (with M. Girardi and M. Matzeu), Differential Integral Equations **17** (2004), 119-126.
[116] "Elliptic Systems with Nonlinearities of Arbitrary Growth" (with B. Ruf), Mediterr. J. Math. **1** (2004), 417-431.

2005

[117] "Multiplicity of Non-Radial Solutions of Critical Elliptic Problems in an Annulus" (with O.H. Miyagaki), Proc. Roy. Soc. Edinburgh Sect. A **135** (2005), 25-37.
[118] "Non-radially Symmetric Solutions for a Superlinear Ambrosetti-Prodi Type Problem in a Ball" (with P.N. Srikanth and S. Santra), Comm. Contemp. Math. **7** (2005), 849-866.
[119] "An Orlicz-Space Approach to Superlinear Elliptic Systems" (with J.M. do Ó and B. Ruf), J. Funct. Anal. **224** (2005), 471-496.
[120] "Liouville Type Theorems, Monotonicity Results and a Priori Bounds for Positive Solutions of Elliptic Systems" (com B. Sirakov), Math. Ann. **333** (2005), 231-260.

2006

[121] "Semilinear Elliptic Systems with Exponential Nonlinearities in Two Dimension" (with J.M. do O' e B. Ruf), Adv..Nonlinear Stud. **6** (2006), 199-214.
[122] "Multiplicity Results for a Family of Semilinear Elliptic Problems under Local Superlinearity and Sublinearity" (with J.-P. Gossez and P. Ubilla), J. Eur. Math. Soc. **8** (2006), 269-286.

2007

[123] "On the Ambrosetti-Prodi Problem for Non-Variational Elliptic Systems" (with B. Sirakov), J. Differential Equations **240** (2007), 357-374.

2008

[124] "The Critical Hyperbola for a Hamiltonean Eliiptic System with Weights" (with I. Peral and J. Rossi), Ann. Mat. Pura Appl. **187** (2008), 531-545.
[125] "Solitary Waves for some Nonlinear Schrodinger Systems" (with O. Lopes), Ann. Inst. H. Poincare, Anal. Non Lineaire **25** (2008), 149-161.

[126] "Superlinear Systems of Second-Order ODE's" (with P. Ubilla), Nonlinear Anal. **68** (2008), 1765-1773.
[127] "Non-Variational Elliptic Systems in Dimension 2: a Priori Bounds and Existence of Positive Solutions", (with J.M. do O' and B. Ruf), J. Fixed Point Theory Appl. **4** (2008), 77-96.

[128] "Semilinear Elliptic Systems: Existence, Multiplicity, Symmetry of solutions". Handbook of Differential Equations: Stationary Partial Differential Equations. Vol. V (2008), 1-48.

2009

[129] "Local 'Superlinearity' and 'Sublinearity' for the p-Laplacian" (with J.-P. Gossez and P. Ubilla), J.Funct. Anal. **257** (2009), 721-752.
[130] "Quasilinear Equations with Dependence on the Gradient" (with J. Sanchez and P. Ubilla), Nonlinear Anal. **71** (2009), 4862-4868.
[131] "Quasilinear Elliptic Equations of Henon-type: Existence of Non-Radial Solutions" (with P.C. Carriao and O.H. Miyagaki), Comm. Contemp. Math. **11** (2009), 783-798.

2010

[132] "On Best Constants for Limiting Embeddings of Fractional Sobolev Spaces" (with D.G. Costa and J. Yang), Adv. Nonlinear Stud. **10** (2010), 501–510.

2011

[133] "Sobolev Spaces of Symmetric Functions and Applications (with E.M. dos Santos and O.H. Miyagaki), J. Funct. Anal. **261** (2011), 3735-3770.
[134] "Elliptic Equations and Systems with Critical Trudinger-Moser Nonlinearities" (with J. M. do O' and B. Ruf), Discrete Contin. Dyn. Syst. **30** (2011), 455-476.

List of Ph.D. Students of Djairo G. de Figueiredo

1. JOSÉ VALDO DE ABREU GONCALVES (UnB, 1978), faculty member at Universidade de Brasilia (Brasilia/Brasil)
2. FRANCISCO JULIO DE ARAUJO CORREIA (UnB, 1986), faculty member at Universidade Federal do Pará (Belém/Brasil)
3. JOÃO CARLOS NASCIMENTO PÁDUA (UnB, 1986), faculty member at Universidade de Brasilia (Brasilia/Brasil)
4. JOÃO BATISTA XAVIER (UnB, 1988), faculty member at Universidade Federal do Paraná (Paraná/Brasil)
5. CELIUS MAGALHÃES (UnB, 1988), faculty member at Universidade de Brasilia (Brasilia/Brasil)
6. MARCO AURELIO SOARES SOUTO (UNICAMP, 1992), faculty member at Universidade Federal de Campina Grande (Campina Grande/Brasil)
7. PEDRO UBILLA LOPEZ (UNICAMP, 1992), faculty member at Universidade de Santiago (Santiago/Chile)
8. CARLOS CABEZAS MANRIQUEZ (UNICAMP, 1992), faculty member at Universidade de Concepcion (Concepcion/Chile)
9. DANIEL CORDEIRO DE MORAIS FILHO (UNICAMP, 1994), faculty member at Universidade Federal de Campina Grande (Campina Grande/Brasil)
10. JOÃO MARCOS BEZERRA DO Ó (UNICAMP, 1995), faculty member at Universidade Federal da Paraíba (João Pessoa/Brasil)
11. JESUS ALFONSO PEREZ SANCHEZ (UNICAMP, 1996), faculty member at Universidade de Los Andes (Merida/Venezuela)
12. MARCELO MONTENEGRO (UNICAMP, 1996), faculty member at Universidade Estadual de Campinas - UNICAMP (Campinas/Brasil)
13. MARCOS MONTENEGRO (UNICAMP, 1997), faculty member at Universidade Federal do Ceará (Fortaleza/Brasil)
14. HELDER CANDIDO RODRIGUES (UNICAMP, 1997), faculty member at Universidade Federal de Minas Gerais (Belo Horizonte/Brasil)

D. G. Costa (ed.), *Djairo G. de Figueiredo – Selected Papers*,
DOI: 10.1007/978-3-319-02856-9, © Springer International Publishing Switzerland 2013

15. EMERSON ALVES MENDONCA DE ABREU (UNICAMP, 2001), faculty member at Universidade Federal de Minas Gerais (Belo Horizonte/Brasil)

16. FRANCISCO ODAIR VIEIRA DE PAIVA (UNICAMP, 2002), faculty member at Universidade Federal de São Carlos (São Carlos/Brasil)

17. ILMA MARQUES (UNICAMP, 2003), faculty member at Universidade Federal do ABC (Santo André/Brasil)

18. MARCELO FERNANDES FURTADO (UNICAMP, 2004), faculty member at Universidade de Brasilia (Brasilia/Brasil)

19. EDSON ALEX ARRAZOLA IRIARTE (UNICAMP, 2004), faculty member at Universidade Federal do ABC (Santo André/Brasil)

20. HUGO ALEX CARNEIRO DINIZ (UNICAMP, 2005), faculty member at Universidade Federal do Pará (Santarem/Brasil)

21. FABIO RODRIGUES PEREIRA (UNICAMP, 2005), faculty member at Universidade Federal Fluminense (Niteroi/Brasil)

22. EDERSON MOREIRA DOS SANTOS (UNICAMP, 2007), faculty member at Universidade de São Paulo (São Carlos/Brasil)

23. EDCARLOS DOMINGOS DA SILVA (UNICAMP, 2009), faculty member at Universidade Federal de Goiás (Goiania/Brasil)

24. TAISA JUNGES MIOTTO (UNICAMP, 2009), faculty member at Universidade Federal de Santa Maria (Santa Maria/Brasil)

25. BRUNO HENRIQUE CARVALHO RIBEIRO (UNICAMP, 2010), faculty member at Universidade Federal da Paraíba (João Pessoa/Brasil)

26. EVANDRO MONTEIRO (UNICAMP, 2011), faculty member at Universidade Federal de Alfenas (Minas Gerais/Brasil)

GPSR Compliance

The European Union's (EU) General Product Safety Regulation (GPSR) is a set of rules that requires consumer products to be safe and our obligations to ensure this.

If you have any concerns about our products, you can contact us on ProductSafety@springernature.com

In case Publisher is established outside the EU, the EU authorized representative is:

Springer Nature Customer Service Center GmbH
Europaplatz 3
69115 Heidelberg, Germany

Batch number: 10614703

Printed by Printforce, the Netherlands